T0136551

EVOLUTION AND GEOLOGICAL SIGNIFICANCE OF LARGER BENTHIC FORAMINIFERA

EVOLUTION AND GEOLOGICAL SIGNIFICANCE OF LARGER BENTHIC FORAMINIFERA

SECOND EDITION

MARCELLE K. BOUDAGHER-FADEL

ᴬUCLPRESS

This Edition published in 2018 by
UCL Press
University College London
Gower Street
London WC1E 6BT

First edition published in 2008 by Elsevier.

Available to download free: www.ucl.ac.uk/ucl-press

A CIP catalogue record for this book is available from The British Library.

ISBN: 978-1-911576-95-2 (Hbk.)
ISBN: 978-1-911576-94-5 (Pbk.)
ISBN: 978-1-911576-93-8 (PDF)
ISBN: 978-1-911576-96-9 (epub)
ISBN: 978-1-911576-97-6 (mobi)
ISBN: 978-1-911576-98-3 (html)
DOI: https://doi.org/10.14324/111.9781911576938

Acknowledgements

Following the success of making the second edition of *Biostratigraphic and Geological Significance of Planktonic Foraminifera* freely available in open access format via UCL Press, I decided to publish a second edition of this book, in the same way. This second edition contains extensive revisions, additional figures, and a significant update to a large number of orders and families of the Larger Benthic Foraminifera.

During the course of writing this book, I have been helped by numerous friends and colleagues. I would like to thank Prof Pamela Hallock Muller, University of South Florida, for carefully editing and reviewing Chapter 1, and Prof Vladimir Davydov of Florida International University, for reviewing early drafts of Chapter 2. I am grateful to the late Prof Lucas Hottinger, and to Dr G. Wyn Hughes for carefully reviewing various sections of this book. I would like to thank Dr Ebrahim Mohammadi, Iran (Graduate University of Advanced Technology) for carefully reading the first edition and Prof Ahmad Aftab for sending me information on Pakistan localities. I am also grateful to Prof Felix Schlagintweit and Prof Ercan Özcan, Department of Geology, Maslak, for allowing me to publish the original illustrations of some of their new genera and species, Mr Nguyen Van Sang Vova for access to his Vietnam material, the South East Asia Group (SEA), Royal Holloway University London, for access to their Indonesian material, Gyongyver Jennifer Fischer and Pascal Kindler, University of Geneva for access to their Mayaguana Bank (SE Bahamas) material, and to Prof Hu Xiumian for access to his Cretaceous and Paleogene Tibetan material.

As with the first edition, in creating this edition I have been greatly aided and supported by my dear colleague and friend Prof David Price. I would also like to acknowledge the help of Chris Penfold from UCL Press who has been invaluable in the creation of this open access work.

There are many photographs and illustrations in this book. Most are original, but some are reproduced from standard sources. I have tried to contact or reference all potential copyright holders. If I have overlooked any or been inaccurate in any acknowledgement, I apologise unreservedly and I will ensure that suitable corrections are made in subsequent editions. The charts mentioned in this book, are freely available separately online at https://doi.org/10.14324/111.9781911576938.

In conclusion, I should repeat here the acknowledgement from the first edition (BouDagher-Fadel, 2008) as in the course of writing that edition, I was helped by numerous other friends and colleagues:

I would like to thank Prof Ron Blakey for his permission to use his exquisite palaeogeographic maps as illustrations in my book. Many more of these splendid maps can be found on his website http://jan.ucc.nau.edu/~rcb7/.

Prof Rudolf Röttger has been most supportive, and has given me photographs of living larger foraminifera, and corrected Chapter 1 of my book. I also relied heavily on his work in my discussions of the biology of the larger foraminifera as presented in Chapter 1.

I would like to refer readers to Prof Lukas Hottinger's outstanding web pages "Illustrated glossary of terms used in foraminiferal research" which can be found at http://paleopolis.rediris.es/cg/CG2006_M02/4_droite.htm. Some of these illustrations are reproduced in this book, courtesy of Prof Hottinger.

I would like to thank Prof McMillan for access to his South African collection of larger foraminifera, Dr Michelle Ferrandini, Université de Corse, for access to her Corsican collections and Prof K. Matsumaru for access to some of his original material.

I would also like to thank the Natural History Museum, London for giving me access to their excellent collection, which includes type species of many early workers. I would like to thank all scientists who contributed to this collection and thus to my book. My gratitude is also expressed to the Senckenberg-Forschungsinstitut und Naturmuseum, Germany for their Permian collection and UCL Geological Sciences, Micropalaeontology unit collections. I am particularly grateful for the assistance of Mr Jim Davy, UCL, and in Mr Clive Jones, NHM. Mr Jones was very helpful in locating specimens and his methods of filing and storing the NHM collection were so very useful.

Finally, I am especially grateful for the careful editing and reviewing carried out by Prof Alan Lord (of the Senckenberg-Forschungsinstitut und Naturmuseum, Germany) and Prof David Price (UCL). Prof Price' s advice throughout the book, and our useful discussions on the causes of extinctions gave me many ideas on the relationship between sensitive, small, living organisms, such as the larger foraminifera, and large scale geological processes. I also thank him for helping me to look into the wider processes involved in evolution and for his encouragement.

Marcelle BouDagher-Fadel
London
September 2017

Contents

Chapter 1

Biology and Evolutionary History of Larger Benthic Foraminifera

1.1 Biological Classification of Foraminifera

1.1.1 Introduction

Foraminifera are unicellular eukaryotes characterized by streaming granular ecto-plasm usually supported by an endoskeleton or "test" made of various materials. They are considered to fall within the phylum Retaria, which in turn is within the infrakingdom Rhizaria (Ruggiero et al., 2015). Their cellular cytoplasm is organised into a complex structure by internal membranes, and contains a nucleus (Plate 1.1, Figs. 1–2), mitochondria, chloroplasts (when present) and Golgi bodies (Plate 1.1, Figs. 3–5; Plate 1.2). In foraminifera, the cytoplasm is subdivided into the endo-plasm, in which the nucleus (or nuclei, as many foraminifera are multinucleate) and other organelles are concentrated, and ectoplasm, which contains microtubules and mitochondria (Hemleben et al., 1977; Anderson et al., 1979; Alexander, 1985). Foraminifera are characterised by specialized pseudopodia (temporary organic pro-jections) known as granuloreticulopodia (also called rhizopodia), which are thread-like, granular, branched and anastomosing filaments that emerge from the cell body (Fig. 1.1). The unique ability of the foraminiferal ectoplasm to assemble and dis-assemble microtubules allows them rapidly to extend or retract their rhizopodia (Bowser and Travis, 2002). The functions of the rhizopodia include movement, feed-ing, and construction of the test.

Both living and fossil foraminifera come in a wide variety of shapes and sizes. Academically, the study of their preserved tests is referred to as micropalaeontology, and although their typical size is sub-millimetric, they have occurred in the geologi-cal past with sizes up to ~150mm. In addition, they occur in many different environ-ments, from freshwater to the deep sea, and from near surface to the ocean floor. Their remains are extremely abundant in most marine sediments and they live in nearly all marine to brackish habitats (Fig. 1.2).

Foraminifera that dwell in freshwater do not produce tests (Pawlowski et al., 2003), however most marine foraminiferal species grow an elaborate test or endoskeleton made of a series of chambers (Fig. 1.3).

These single-celled organisms have inhabited the oceans for more than 500 million years. The complexity of their fossilised test structures (and their evolution in time) is the basis of their geological usefulness. The earliest known foraminifera, mostly forms that had an organic wall or produced a test by agglutinating particles within an organic or mineralized matrix, appeared in the Cambrian, and were common in the

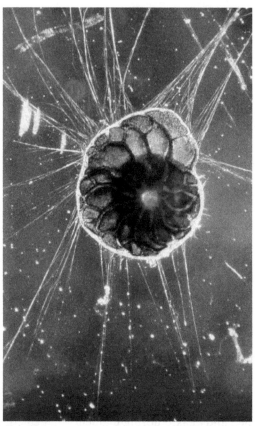

Fig. 1.1.   Larger foraminifera *Heterostegina depressa* with thread-like, granular, branched and anastomosing filaments that emerge from the cell body (courtesy of Prof Röttger).

Early Paleozoic (Platon et al., 2001). Forms with calcareous tests appeared by the Early Carboniferous, becoming diverse and abundant, with the evolutionary development of taxa with relatively large and complicated test architecture by the Late Paleozoic. Their long, diverse and well-documented evolutionary record makes Foraminifera of outstanding value in zonal stratigraphy, and in paleoenvironmental, palaeobiological and palaeoceanographic interpretation and analysis.

Fossil and living foraminifera have been known and studied for centuries. They were noted by Herodotus (in his *Histories* written in the 5th century BC) as occurring in the limestone of the Egyptian pyramids, which in fact contain fossils of the larger benthic Foraminifera *Nummulites*. The name Foraminifera derives from the apertures and the "foramen" connecting successive chambers seen in their tests. The test surfaces of many foraminiferal species are covered with microscopic holes (foramen), normally visible at about x40 magnification (Fig. 1.4). Among the earliest workers who described and drew foraminiferal tests were Anthony van Leeuwenhoek in 1600, and Robert Hooke in 1665, but an accurate description of foraminiferal architecture was not given until the 19th century (Carpenter et al., 1862).

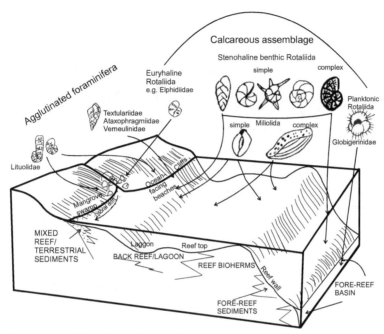

Fig. 1.2.   The ecological distribution of foraminifera.

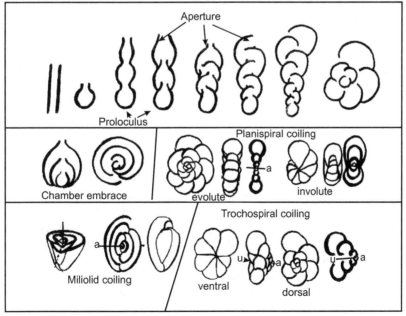

Fig. 1.3.   The different shapes of foraminiferal test; a = axis of the test; u = umbilicus.

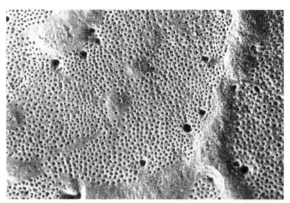

Fig. 1.4.   An enlargement of the surface of a *Heterostegina* shell showing two types of holes. 1) the many small pores which are characteristic of all foraminifera. They do not form open connections between the test lumen and the sea water, but are closed by a membrane. Only small molecules like nutrition salts may penetrate, which are important for the nutrition of the algal endosymbionts. 2) the larger openings on the lateral test surface are openings of the canal system of the chamber walls and chamberlet walls (shown in Fig. 1.17) with the outside world. In *Heterostegina depressa*, and other nummulitids, the protoplasm emerges through these openings and forms a thin veil covering the test surface in living specimens, which is also responsible for the secretion of the elastic inanimate protective sheath with radiating processes that covers the test, attaching it to the algal or rock surface. This function is described and illustrated by Röttger (1983). The apertures in the last chamber are masked in *Heterostegina*.

The first attempts to taxonomically classify Foraminifera placed them within the genus *Nautilus,* a member of the phylum Mollusca. In 1781, Spengler was among the first to note that foraminiferal chambers are in fact divided by septa. In 1826, d'Orbigny, having made the same observation, named the group Foraminifères. In 1835, Foraminifera were recognised by Dujardin as protozoa, and shortly afterwards d'Orbigny produced the first classification of foraminifera, which was based on test morphology. Modern workers normally use the structure and composition of the test wall as a basis of primary classification, and this approach will be followed in this book.

Despite the diversity and usefulness of the foraminifera, the phylogenetic relationship of Foraminifera to other eukaryotes has only recently emerged. Early genetic work on the origin of the Foraminifera postulated that the foraminiferal taxa are a divergent "alveolate" lineage, within the major eukaryotic radiation (Wray et al., 1995; Baldauf, 2003). Subsequently, many researchers have tried to determine the origin of the foraminifera, but molecular data from Foraminifera generated conflicting conclusions.

Molecular phylogenetic trees have assigned most of the characterised eukaryotes to one of eight major groups. Baldauf (2003) tried to resolve the relationships among these groups to find "the deep roots of the eukaryotes". He placed them in the "Cercozoa" group. Cercozoans are amoebae, with filose pseudopodia, often living within tests, some of which can be very elaborate. The phylum Cercozoa was originally erected by Cavalier-Smith (1998) to accommodate the euglyphid filose amoebae, along with the heterotrophic cercomonadids and thaumatomonad flagellates, which were shown to be related by Cavalier-Smith and Chao (1997).

However, the origins of both Cercozoa and Foraminifera have been evolutionary puzzles because foraminiferal ribosomal RNA gene sequences are generally divergent, and show dramatic fluctuations in evolutionary rates that conflict with fossil evidence. Ribosomal RNA gene trees have suggested that Foraminifera are closely related to slime moulds and amoebae (Pawlowski et al., 1994), or alternatively used to suggest that they are an extremely ancient eukaryotic lineage (Pawlowski et al., 1996). In 2003, Archibald et al. found that cercozoan and foraminiferal polyubiquitin genes (76 amino acid proteins) contain a shared derived character, a unique insertion, which implies that Foraminifera and Cercozoa indeed share a common ancestor. Archibald et al. (2003) proposed a cercozoan-foraminiferan supergroup to unite these two large and diverse eukaryotic groups. In recent molecular phylogenetic studies, Nikolaev et al. (2004) adopted the name "Rhizaria" (proposed first by Cavalier-Smith, (2002), which refers to the root-like filose or reticulose pseudopodia) and included the Retaria, Cercozoa and Foraminifera within this supergroup. While additional protein data, and future molecular studies on Rhizaria, Retaria, Cercozoa and Foraminifera, are necessary to provide a better insight into the evolution of the pseudopodial divisions, the placement of the Foraminifera within the Rhizaria appears to be well supported (Pawlowski and Burki, 2009; Ruggiero et al., 2015; Burki et al., 2016) (see Fig. 1.5).

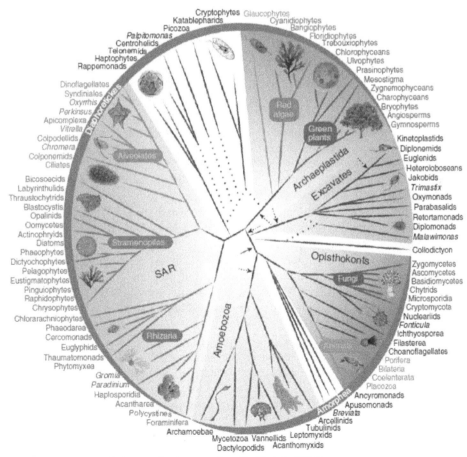

Fig. 1.5. A consensus phylogeny of eukaryotes from Burki et al., (2016).

Similarly, the higher taxonomy of the Foraminifera is still unsettled. Although proposed as the Class Foraminifères by d'Orbigny (1826), throughout most of the 20th century the group was considered as the Order Foraminiferida, and the major subdivisions were considered to be suborders. In 1992, Loeblich and Tappan recommended Lee's (1990) re-elevation of the Order Foraminiferida to Class Foraminifera, thereby elevating the suborders to orders. Sen Gupta (1999), Platon et al. (2001) adopted the class-level designation with some modifications at the order-level that have been largely supported by molecular phylogenies (Mikhalevich, 2000, 2004; Pawlowski and Burki, 2009; Groussin et al., 2011). Most recently Ruggiero et al. (2015) suggested a subphylum status for the Foraminifera.

Recognizing that the classification of Foraminifera is still in flux, in this edition (in contrast to BouDagher-Fadel (2008)) we accept the elevation of the Order Foraminiferida to Class Foraminifera, and the concomitant elevating of the previously recognized suborders to the ordinal level.

1.1.2 Larger Benthic Foraminifera

Foraminifera are separated into two groups following their life strategy, namely the planktonic and the benthic foraminifera. Fewer than 100 extant species of foraminifera are planktonic, though they occur worldwide in broad latitudinal and temperature belts. They drift in the pelagic waters of the open ocean as part of the marine zooplankton (see Fig. 1.6). Their wide distribution and rapid evolution reflect their successful colonization of the pelagic realm. When this wide geographical range, achieved through the Late Mesozoic and in the Cenozoic, is combined with a short stratigraphic time range due to their rapid evolutionary characteristic, they make excellent index fossils at family, generic and species levels (see BouDagher-Fadel, 2013, 2015).

The benthic foraminifera, however, are far more diverse, with estimates of roughly 10,000 extant species. Benthic foraminifera live, attached to a substrate or free of any attachment, at all ocean depths, and include an informal group of foraminifera with complicated internal structures known as "**larger benthic foraminifera**". It is these forms that are the principle subject of this book.

The larger benthic foraminifera are not necessarily morphologically bigger than other benthic foraminifera, although many are, but they are characterised by having internally complicated tests. While one can identify most small benthic foraminifera from their external morphology, one must study thin sections to identify many of the larger benthic foraminifera, using features of their internal test architecture (Fig. 1.7).

Larger benthic foraminifera develop characteristically complicated endoskeletons, which permit the taxa to be accurately identified, even when they are randomly thin-sectioned. The tests of dead, larger foraminifera can dominate carbonate sediments, and foraminiferal-limestones are extensively developed in the Upper Paleozoic, the Mesozoic (especially the Upper Cretaceous) and in the Cenozoic (see Fig. 1.8).

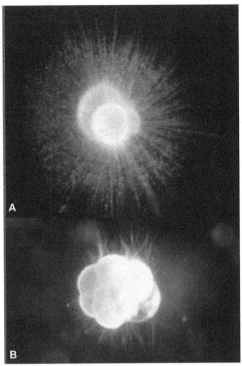

Fig. 1.6. (A) *Globigerinoides sacculifer* (Brady), a spinose species with symbionts carried out by rhizopodial streaming on the spines (courtesy of Dr Kate Darling); (B) *Neogloboquadrina dutertrei* (d'Orbigny), a non-spinose species (courtesy of Dr Kate Darling). See BouDagher-Fadel, 2015 for a detailed study of the planktonic foraminifera mode of life, classification and distribution.

Following recent taxonomic studies and reassessments of classifications, we recognise 14 different large benthic foraminiferal orders (Fig. 1.9). The orders with larger foraminiferal lineages that are discussed in detail in this volume are the:

- Parathuramminida
- Moravamminida
- Archaediscida
- Earlandiida
- Palaeotextulariida
- Tetrataxida
- Tournayellida
- Endothyrida
- Fusulinida
- Lagenida
- Involutinida
- Miliolida
- Textulariida
- Rotaliida.

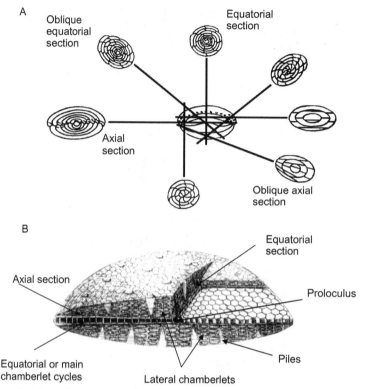

Fig. 1.7.   Examples of two dimensional sections through the three-dimensional test of a larger, three layered foraminifera. A) Sections through a milioline test (modified from Reichel, 1964). B) Three-dimensional view of *Lepidocyclina* sp., showing the distinction between equatorial or main chamberlet cycles and lateral chambers (modified from Vlerk and Umbgrove, 1927).

Throughout this book standard nomenclature is followed, so **orders** are expressed via the suffix "–ida", or generically as "–ides" (e.g. Miliolida or miliolides). The suffix of "–oidea" is used to denote **superfamilies**, rather than the older suffix "-acea" following the recommendation of the International Commission on Zoological Nomenclature (see the 'International Code of Zoological Nomenclature', 1999, p. 32, Article 29.2). **Families** are designated via the suffix "–idae". In this book, the suffix "–ids" is used to indicate a generic superfamily or family member (e.g. Fusulinoidea/ Fusulinidae or fusulinids).

The study of living larger foraminifera shows that they occur abundantly in the shelf regions of most tropical and subtropical shallow marine, especially in carbonate-rich, environments. Indeed, they seem to have done so ever since the first larger foraminifera emerged during the Carboniferous. Again, from the study of extant forms, it seems that many larger foraminifera enclose photosynthetic symbionts, which appear to be essential to the development of most of the lineages with morphologically larger forms (Hallock, 1985; BouDagher-Fadel et al., 2000; BouDagher-Fadel, 2008).

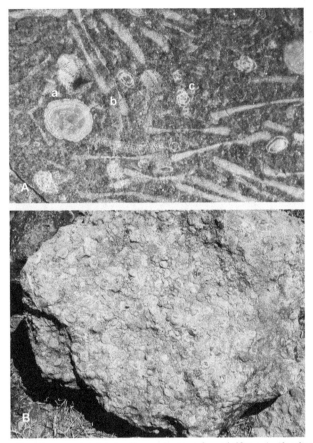

Fig. 1.8.　A. Eocene limestone containing fossil porcelaneous foraminifera; a) *Alveolina* sp., b) *Orbitolites* sp., c) *Quinqueloculina* sp., from France. B. Miocene limestone dominated by *Lepidocyclina* spp. from Indonesia, courtesy of Peter Lunt.

From their structural complexity, and because of the diversity of the shelf environments that they inhabited, fossil larger foraminifera provide unique information on palaeoenvironments and biostratigraphy of shelf limestones around the world. Generally, in such environments, calcareous nannofossils are unavailable because of the shallowness of the marine environment and because of the recrystallisation of the calcite in the limestone matrices. Furthermore, macrofossils are relatively rare in these habitats. By the late 1920s, the larger foraminifera had become the preferred fossil group for biostratigraphy in several oil-rich regions including the Indonesian area, parts of Russia, and in the United States, especially western Texas. Larger foraminifera had the advantage that they were more abundant than molluscs, and additionally a scheme was developed that utilised assemblage zones, rather than percentages of forms to be found. Using molluscs to identify and correlate sections required extensive knowledge of both living and fossil species. The larger foraminiferal assemblage zones could be identified by the presence of a few key taxa, usually with use of a hand

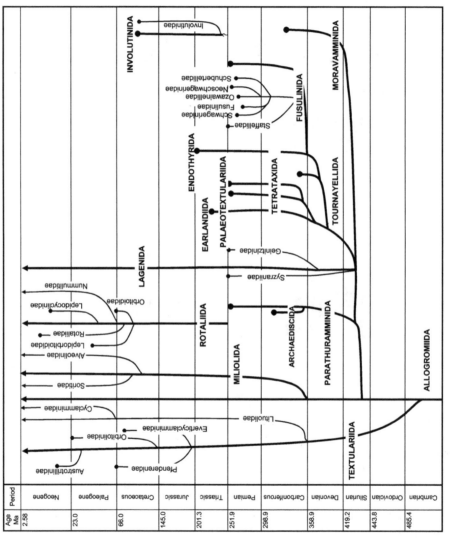

Fig. 1.9   The geological range of the larger foraminifera orders and some selected, important families.

lens in the field. Some groups of larger foraminifera provide excellent biostratigraphic markers, sometimes the only ones which can be used to date carbonate successions (e.g. the fusulinids and schwagerinids in the Upper Paleozoic (Figs. 1.10A; 1.10B), orbitoidoids in the Middle to Upper Cretaceous (Fig. 1.10C), nummulitids in the Paleogene (Fig. 1.10D), and lepidocyclinids (Fig. 1.10E) and miogypsinids in the Oligocene and Neogene (Fig. 1.10F)). Provincialism was often a problem in these groups, but this is now well understood, so that biozonal schemes applicable to certain time intervals in defined bioprovinces have recently been erected and successfully applied (BouDagher -Fadel and Price, 2010a, b, 2014; BouDagher et al., 2015).

Larger foraminifera are an ideal "group" of organisms to use in the study of general evolutionary theory. Their fossil record is so rich in individual fossils that assemblage concepts can be used, and both horizontal and vertical variation can be studied in the stratigraphic record. Their preference for certain marine environments is well understood and documented. Because representatives of most of the orders are still extant, it is also possible to infer their reproductive strategy, which as will be seen later, is quite complex.

This book does not attempt to present a comprehensive or extensive listing of all genera and species of larger foraminifera, but rather focuses on the taxonomy, phylogeny and biostratigraphic applications of the most important forms. For an almost comprehensive list, the reader can refer to Loeblich and Tappan (1988). In addition, for brevity, the complete references to genera and species are not given and again

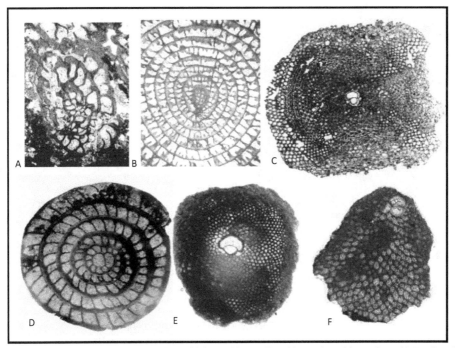

Fig. 1.10   Examples of larger foraminifera which provide excellent biostratigraphic markers, A) *Fusulina*; B) *Schwagerina*; C) *Lepidorbitoides*; D) *Nummulites*; E) *Lepidocyclina*; F) *Miogypsina*.

the reader can refer to Loeblich and Tappan (1988) and the contemporary, on-line literature. Finally, the reader can refer to Hottinger (2006) for an exhaustive set of definitions of terms used in the taxonomic description of the larger foraminifera, many of which, but inevitably not all, are also explained below.

1.1.3 Trimorphic life cycle in larger benthic foraminifera

Larger foraminifera may reproduce asexually by multiple fission, producing many hundreds of offspring, and at other times they reproduce sexually, many by broadcasting gametes. Röttger (1983) described the asexual reproduction of *Heterostegina*. He stated that the protoplasmic body leaves the test through an internally developed canal system. The spherical daughter cells are colourless and are without a calcareous test (Fig. 1.11; Plate 1.3, Fig. 3). A small part of the symbiont-containing residual protoplasm is then apportioned to each daughter cell. At this stage the test is formed and consist of two chambers, which as the juvenile grows are followed by the addition of further chambers. The growth rate of the calcareous tests of foraminifera is light-dependent (Röttger, 1983).

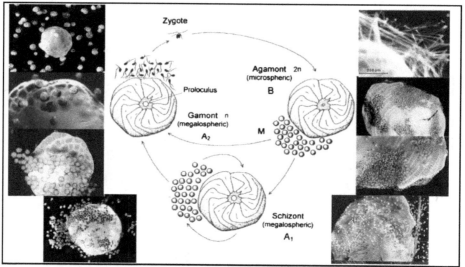

Fig. 1.11. Schematic figures (in the centre) showing a trimorphic life-cycle of the larger benthic foraminifera *Amphistegina gibbosa* from Dettmering et al. (1998). The upper part shows the dimorphic life cycle, consisting of an alternation between a haploid, megalospheric gamont with its gametes, and the microspheric diploid agamont with its offspring produced by multiple fission. The lower part represents the megalospheric generation in a trimorphic cycle reproduced by cyclic schizogony, inserted between agamont and schizont. The photographs of living *Heterostegina depressa* (courtesy of Prof R. Röttger) show an alternation of generations in which a 2-4mm sized gamont (megalospheric generation) (on the left) alternates with an 1-2cm-sized agamont (microspheric generation) (on the right). During multiple fission of the agamont, the symbionts-containing protoplasm (top right) flows out of the calcareous test and then divides into 1000 to 3000 daughter individuals, the young gamonts. In addition to gamont and agamont forms another generation, which looks like a gamont of *Heterostegina depressa*, but which reproduces asexually (Röttger, 1990).

In other taxa, the process has some small differences. For example, in the species *Amphistegina* spp., the cytoplasm exits through the aperture (see Fig. 1.12). In the soritid foraminifera, the partitions of the final chambers are dissolved to form a brood chamber in which the daughter cells form (Röttger, 1984; 1990). After asexual multiple fission, the empty parent test becomes a lifeless grain of sediment.

Larger foraminifera are dimorphic (having two forms), which is the result of the heterophasic alternation of generations between a haploid, uninucleate gamont (the sexual generation which produces gametes) and a diploid, multinucleate agamont (the asexual generation which produces daughter individuals by multiple fission) (Schaudinn, 1895; Röttger, 1990). The dimorphic forms usually exhibit different morphological characters; the two forms are called:

- The asexual **microspheric** (or B-) form, which is larger, with numerous chambers, but with a small proloculus (first chamber, see Plate 1.3 and Fig. 1.11). It is this asexual generation which produces daughter individuals by multiple fission, and
- the **megalospheric** (or A-) form, which is smaller with fewer chambers, but with a large proloculus. It is this sexual generation which usually produces gametes (see Fig. 1.11).

However, in addition to these two generations, a third generation is documented by many authors, where the agamont produces megalospheric schizonts instead of gamonts (see Fig. 1.11). This life cycle was first discovered by Rhumbler (1909), and since then has been recognised by many authors (Leutenegger, 1977; Röttger, 1990;

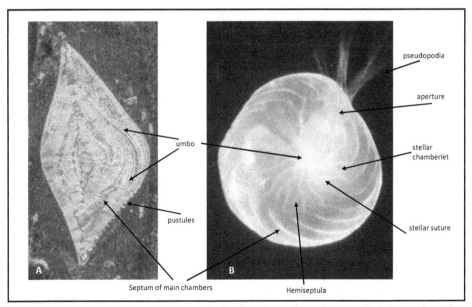

Fig. 1.12. Amphistegina: A) an axial section of a fossil specimen of *Amphistegina*; B) A live specimen showing the cytoplasm exiting through the aperture.

Dettmering et al., 1998; Harney et al., 1998). Röttger (1990) cultivated *Heterostegina depressa* (Plate 1.3, Fig. 1) in the laboratory and was able to confirm the trimorphic cycle. Dettmering et al. (1998) and Harney et al. (1998) suggested that the trimorphic cycle can account for the abundance of the megalospheric generation in many populations. The schizonts, which are produced by asexual reproduction, in contrast to zygotes, which are too small to carry symbionts, begin their ontogeny as large symbiont-bearing cells. Harney et al. (1998) also suggested that the trimorphic cycle provides tremendous colonization potential, allowing foraminifera to rapidly increase their population densities sufficient to successfully sexually reproduce by gamete broadcasting, while at the same time promoting genetic divergence by amplifying the colonizing genotypes, all of which could promote relatively rapid rates of evolution.

1.2 Morphological and Taxonomic Features Used in the Classification of Larger Foraminifera

Larger foraminifera are subdivided into six groups according to the wall structure of their tests (see Fig. 1.13):

- the **agglutinated group,** with walls composed of detrital particles held together by calcareous cement (as in the larger **Textulariida**),
- the **calcareous granular group,** with compound, microgranular walls of low-Mg calcite, in which the crystalline grains are without optical alignment (characteristic of the **Fusulinida** and related orders),
- the **porcelaneous group,** composed of three-layered calcitic, imperforate, non-lamellar walls with a high percentage of rod-like magnesium calcite that have their axes randomly oriented in the embedding organic material and with an outer layer parallel to the outer walls, as shown by the **Miliolida,**
- the **hyaline calcareous group,** a lamellar-perforate group, consisting of layers of calcite crystals, with the C-axis oriented perpendicular to the test surface (Haynes, 1981; Hallock, 1999). The magnesium ratio is low in some taxa and high in others. This wall structure is characteristic of the **Rotaliida** (e.g. Fig. 1.14). The pore canals in these perforate tests have proximal ends closed by organic membrane with micropores (Röttger, 1983). They do not, therefore, allow the passage of cytoplasm to the seawater, but they facilitate the transport of carbon dioxide, oxygen and nutrient salts in the symbiont-bearing larger benthic foraminifera.
- The **monolamellar group,** with or without secondary laminations, with radiating calcite crystals which have the crystallographic c-axis perpendicular to the surface, as shown by the **Lagenida,** and
- The **aragonitic group,** commonly they are recrystallised to give a homogeneous microgranular structure. This wall is characteristic of the **Involutinida.**

The wall structures of the larger foraminifera reflect the biological method used by their living cell to build its test. The microgranular walls developed by the fusulinides of the Paleozoic have a thin, dense outer layer forming the spirotheca (spiral outer

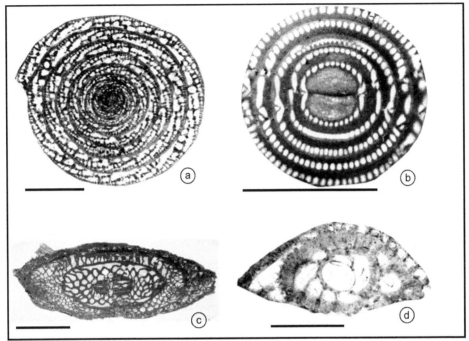

Fig. 1.13. Wall structure of the larger foraminifera. A) *Loftusia* sp. (agglutinated); B) *Alveolina* sp. (calcareous imperforate); C) *Quasifusulina* sp. (calcareous microgranular); D) *Rotalia* sp. (calcareous perforate). Scale bars (1-3) = 2mm; 4 = 0.5mm.

wall). In advanced fusulinides, the wall becomes alveolar (i.e., it develops small sacks), and has a honeycomb-like structure. The term "keriotheca" is restricted to structures with two layers of alveoles (Figs 1.13C and 1.15; see Chapter 2). The rotaliides test is made of perforate hyaline lamellar calcite, and many of the larger rotaliides are characterized by having a developed canal system (Fig. 1.14, and see Chapter 6), which gives rise to special laminations in the tests (Hottinger, 1977). The lamellar tests are formed during the process of chamber construction, where each chamber wall, consisting of secreted, Mg-calcite, covers the total test including all former chambers (see Figs. 1.14; Hohenegger et al., 2001).

The basic structural element of the test is the chamber. Larger benthic foraminifera have multichambered (plurilocular) tests, which have attained large sizes, up to ~150mm in the case of *Cycloclypeus carpenteri* (See Chapter 7). The internal space between the chamber walls is called the chamber lumen. All cavities subdividing the chambers are called chamberlets. Hottinger (2006) divided the basic architectural components of the foraminiferal test into elements that do not modify the shape of the living cell, such as the wall, and those that do modify it.

The elements that modify the shape of the living cell can be, according to Hottinger (2006), divided into three factors. The first factor is the shape of the first chamber (proloculus) and subsequently the growth of the second chamber (deuteroloculus). The chambers are separated by a wall (the septum) and connected by the intercameral

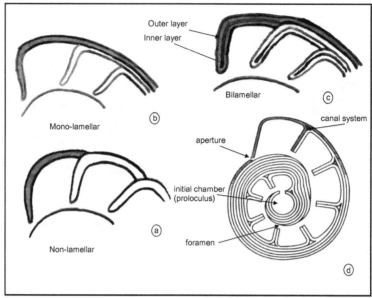

Fig. 1.14.   The structure of A) non-lamellar, B) mono-lamellar and C) bilamellar test walls, where the septum has an inner and outer primary lamella, separated by an organic layer, and is secondarily doubled distally by the "septal flap" formed from the inner lamella of the succeeding chamber, D) a rotaliine test showing that the open external spaces between juxtaposed chamber walls (intraseptal spaces), and between successive shell whorls as they become enclosed by the outer lamellae of newer chambers, thereby forming a canal system (modified after Haynes, 1981).

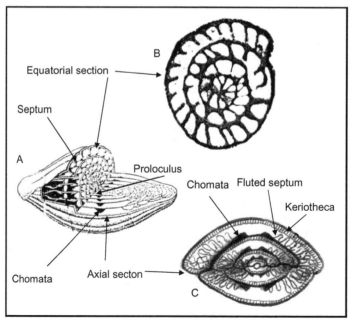

Fig. 1.15.   Views of a schematic fusulinide, A) 3-dimensional view of test; B) equatorial section of *Fusulinella* sp.; C) axial section of *Triticites* sp. illustrating the development of secondary deposits of calcite (chomata) on the chamber floor, and the development of contorted, "fluted" septa.

lumen. Tubular foramen are called stolons, and if they are wide open they are called tunnels (Fig. 1.16).

The two other factors that determine the shape of the living cells are the chamber shape and the arrangement of the chambers (Fig. 1.17). The arrangement of the chambers may form a compressed, planispiral, involute and flaring growth, e.g. *Archaias* (Fig. 1.18Aa), fusiform elongate test, e.g. *Flosculinella* (Fig. 1.18Ab), *Alveolinella* (Fig. 1.18B), annular concentric (in two dimensions), e.g., *Cyclorbiculina* (Fig. 1.18C), *Marginopora* (Fig. 1.18D), or spherical-concentric (in three dimensions) test, e.g., *Sphaerogypsina* (see Fig. 1.18F). The chambers can be developed in a serial arrangement, uniserial (chambers arranged in a single row), biserial (chambers arranged in two rows), etc., or in a spiral arrangement, such as the streptospiral arrangement, where coiling occurs in different planes, the planispiral arrangement where the spiral and umbilical

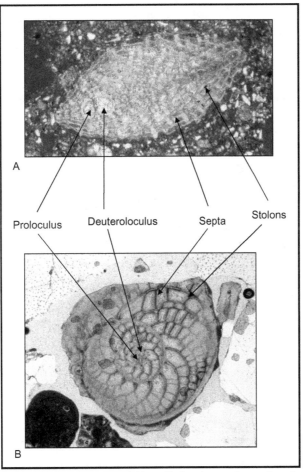

Fig. 1.16. Two different forms of larger foraminifera: A) the elongated test of a miogypsinid, where the deuteroloculus is in alignment with the proloculus and the test is elongated; B) the lenticular test of a heterostegine where the position of the deuteroloculus makes it essential for the test to enrol.

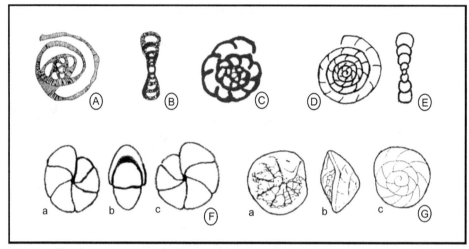

Fig. 1.17. Chamber arrangement of A) equatorial section of *Archaediscus*, with a streptospiral test, with an undivided tubular second chamber; B) an axial section of *Tournayella* showing a planispiral test in the adult; C) an equatorial section of *Endothyra* showing an initially streptospiral to planispiral test with well developed septa and characterised by the development of secondary deposits of calcite (chomata) on the chamber floor; D) an equatorial section of a planispiral evolute test; E) an axial section of an evolute test; F) a, c) side views of an involute test, b), apertural view; G) a trochospiral test, a) umbilical side, b) vertical view, c) spiral side.

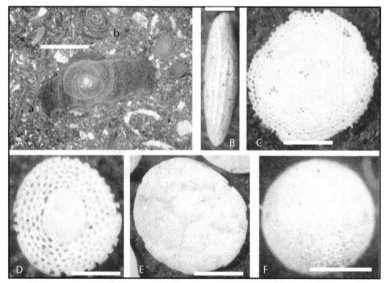

Fig. 1.18. A) Thin section photomicrograph showing a) compressed, planispiral, involute and flaring growth, *Archaias*; b) a fusiform test, *Flosculinella*; B) fusiform elongate test, *Alveolinella*; C) annular concentric growth, *Cyclorbiculina*; D) annular concentric growth, *Marginopora*; E) annular concentric growth, *Archaias*; F) spherical concentric growth in *Sphaerogypsina*. Scale bars (A-B) = 2mm; (C-E) = 1mm; (F) = 0.5mm.

sides are identical and symmetrical (Fig. 1.17F), and the trochospiral where spiral and umbilical sides are dissimilar (Fig. 1.17G). The trochospiral arrangement in the larger foraminifera exposes the umbilical region and creates a direct access to the ambient environment (Hottinger, 1978). In involute spiral forms, the lumina of the chambers in one coil cover laterally those of the preceding coil (e.g., *Nummulites,* Fig. 1.19A) and develop in some cases wing-like extensions from the lumen to the poles (alar prolongation). However, in a spirally coiled evolute form, the chamber lumina do not laterally cover those of the preceding coil (e.g., *Assilina,* Fig. 1.19C). Thus, for example, *Operculina* has a planispiral, evolute lenticular, compressed and loosely coiled test (see Fig. 1.19B), while *Heterostegina* has a planispiral, involute to evolute test with chambers divided by secondary septa to form small chamberlets (Fig. 1.19E). In *Spiroclypeus*, the heterostegine chambers increase rapidly in height and project backwards (Fig. 1.19D).

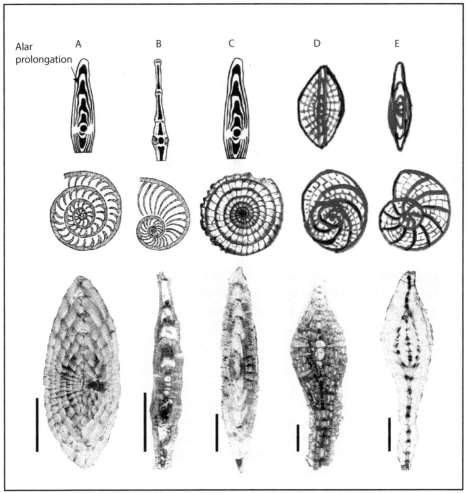

Fig. 1.19.   Differing shapes of nummulitic tests. A) Involute test, axial section of *Nummulites*; B) Evolute test, axial section of *Operculina*; C) Evolute test, axial section of *Assilina*; D) Involute test, *Spiroclypeus*; E) Test initially involute, evolute in mature stage, *Heterostegina* (see Chapter 6). Scale bars = 0.5mm.

Exoskeletal elements are developed that reflect protoplasmic flux (Hottinger, 2006). These include alveoles (honeycomb-like sacks), reticular subepidermal networks, etc., that produce multiple, blind-ended small chamberlets/tiny compartments of the chamber cavity coated by organic lining (Hottinger, 2000). Various agglutinated larger benthic foraminifera develop layers of alveoles, coating the lateral chamber wall, such as in the lituolids that have an exoskeletal layer of undivided, shallow alveoles., e.g., the early extinct representative of *Pseudocyclammina* (Fig. 1.20A) and *Everticyclammina* sp. (Fig. 1.20B) and the still living *Cyclammina* (Fig. 1.20C). The alveoles in the porcelaneous Paleogene *Alveolina* (Fig. 1.20D) are blind recesses separated by septula. Alveolinids, such as *Subalveolina* or *Bullalveolina* (See Chapter 6) have alveoles in postseptal positions over supplementary apertures in the previous septal face. The Neogene genus of *Textulariella* has branching alveoles, while in the porcelaneous foraminifera, *Austrotrillina* alveoles (Fig. 1.20E) evolve from early forms with layers of shallow, undivided alveoles (see Chapter 6) to deep and branching alveoles (*A. howchini,* Fig. 1.20E) in order to harbour symbiotic algae.

The early fusulinides had keriothecal cavities in their walls, which is described by Hottinger (2000) as an alveolar, honeycomb-like structure with a spiral wall, not filled with living chamber plasma nor coated by the organic lining. In advanced fusulinides, the keriotheca may consist of an outer and an inner "layer" produced by a split of the alveoli into narrower subunits below the tectum, while others have both alveolar structures and keriothecal wall texture (e.g. *Verbeekina,* see Chapter 2.)

Some agglutinated larger benthic foraminifera have parapores (canaliculi), straight to tortuous tubular spaces, coated and closed off internally by organic lining (e.g., *Chrysalidina,* Fig. 1.20F). Others, have combined alveolar exoskeletons with a paraporous external wall (e.g., *Dicyclina,* see Chapter 5) or with a bilamellar perforate wall (e.g., *Fabiania,* see Chapter 6).

Only agglutinated foraminifera possess exoskeletal polygonal structures called "subepidermal networks" (e.g. in *Orbitolina* and *Pseudochoffatella,* Fig. 1.21). The basal layer of the larger foraminiferal test, when thickened as in flosculinisation (Hottinger, 1960) or perforated by canalicular passages (Hottinger, 1978), exist only in the porcelaneous forms (Fig. 1.20D).

Other exoskeletal features found in larger foraminifera are the partitions of the chamber lumen, by for example beams, which are perpendicular to the septum, or rafters which are parallel to the septum (Fig. 1.22).

In many recent larger foraminifera, the exoskeletal alveoles harbour photosynthesising symbionts in internal pores. These pores are also seen in extinct Cenozoic species such as in *Miogypsina* (Fig. 1.23A). However, exoskeletal structures also exist in species, such as *Cyclammina* (see Chapter 6), that live at depths too great to have photosynthetic symbionts. Hottinger (2000) interpreted the exoskeletal structures of these deep-water larger foraminifera as providing a mechanism that permits control of gas exchange, by separating the gas diffusion from protoplasmic streaming. However, the endoskeleton of many larger foraminifera includes pillars (Fig. 1.23B), which fill the interior of the test, or continuous walls (septula), which subdivide the larger chamber lumen (see Fig. 1.23). Pillars may also be seen as providing mechanical strength to the test, so for example in discoidal forms, heavily pillared endoskeletons, as a rule, occur

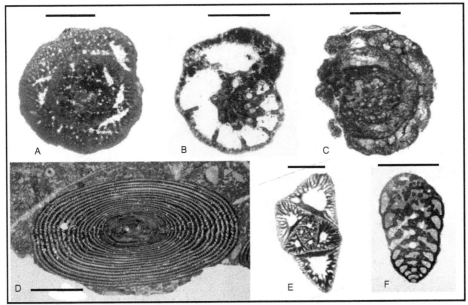

Fig. 1.20. Alveoles and parapores. A-D) Wall covered by alveoles which are subepidermal blind chamberlets/recesses coated by the organic lining: A-B) Cretaceous extinct agglutinated foraminifera, A) *Pseudocyclammina*; B) *Everticyclammina*; C) Cretaceous to Holocene *Cyclammina* sp.; D) an Eocene milioline foraminifera, *Alveolina elliptica* var. *nuttalli*, which shows also some degree of flosculinisation, (thickening of the basal layer of the early chambers). The "alveoles" are blind recesses separated by septula; E) *Austrotrillina howchini* with deep and branching alveoles which evolve from earlier forms with layers of shallow, undivided alveoles (see Chapter 7); F) Paraporous wall with tubular spaces, coated and closed off internally by the organic lining, Cretaceous *Chrysalidina*. Scale bars = 0.5mm.

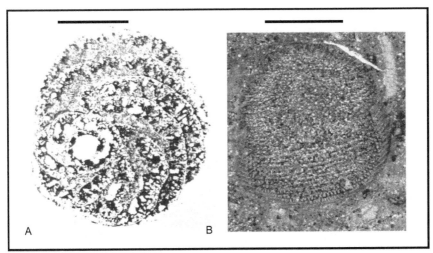

Fig. 1.21. Subepidermal networks: A) *Pseudochoffatella*; B) *Orbitolina*. Scale bars = 0.5mm.

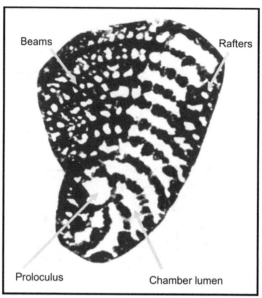

Fig. 1.22. Schematic test of *Alzonella,* showing a planispiral test with a hypodermis consisting of a coarse lattice of beams and rafters.

in forms living in very shallow, turbulent water (e.g., *Archaias*, Fig. 1.18A and E), while modestly pillared forms (e.g., *Cyclorbiculina*, Fig. 1.18C) inhabit deeper and quieter environments (Hottinger, 2000). The massive pillars in *Lepidocyclina* (see Fig. 1.23B) may also be thought to be associated with occupation of high-energy, marine environments (BouDagher-Fadel et al., 2000); however, there are many exceptions to this rule.

Larger foraminifera have many different overall adult shapes. Discoidal forms evolve progressively into flat tests, which can be generated by uniserial growth, such as in the orbitolinids (as in *Orbitolina*, Fig. 1.24A), spiral growth (as in *Choffatella,* Fig. 1.24B), and annular growth (as in *Orbitopsella*, Fig. 1.24C). An elongate form may be realized by a concentric growth pattern, such as in *Lacazina* (Fig. 1.24D), or in a planispiral-fusiform test, such as in *Fusulina* (Fig. 1.10A) and *Alveolina* (Fig. 1.13B).

As larger foraminifera increased their sizes, their internal structures became more complicated. One of the most intriguing complication occurred in the fusulinides. They subdivided the inhabited space of the test by folding their test walls, thus creating septal fluting (see Fig. 1.15). In very elongate forms the folded septa became disengaged from the chamber floor to create cunicular passages (see Chapter 2). This septal folding seems to be present in fusiform amphisteginids, such as *Boreloides* (Hottinger, 1978). In tightly coiled, elongate fusiform tests (e.g. in *Alveolina*), the function of the elongation of the fusiform test is related to motility, the test moving in the polar direction (but growing in equatorial direction) (see Chapter 6, and Hottinger, 2000).

As the tests of the foraminifera become large, the protoplasmic body must inhabit all compartments and those compartments must be interconnected. Therefore, a system of apertures or stolons is necessary to shorten the distance between the first and final chambers, and to provide a communication route between the compartments

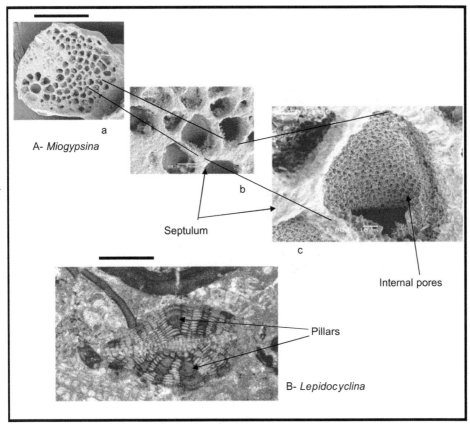

Fig. 1.23. A) SEM image of *Miogypsina*, with enlargement of the lateral chamberlets to show internal pores that harbour symbionts; B) *Lepidocyclina* sp. with pillars embedded in the lateral parts of the test. Scale bars (A) 250 μm; (B) 1mm.

(Hottinger, 1978, 2000). This can be provided by leaving a primary aperture or, in some cases, multiple apertures for extrusion of the rhizopods between the chambers during growth, and according to Hottinger, the linear nature of the rhizopodial protoplasm involved in wall building guarantees in some species that successive chambers are connected by a single foramen or aperture, the last formed chamber opening to the surrounding water via a terminal aperture. Some larger foraminifera enhanced the control of their chamberlet cycle growth by oblique, crossed-over stolon systems. Hottinger (2000) noted that in Mesozoic (see Chapters 4 and 5) non-perforate discoidal tests, the radial arrays (as in *Orbitopsella*) are more frequent than the oblique, crossed-over ones (*Ilerdorbis*), whereas, in conical forms, the latter pattern dominates without being exclusive (*Orbitolina*). During the Cenozoic, the discoidal tests with crossed-over stolon systems prevail (*Orbitolites, Marginopora*, see Chapters 6 and 7), while in uniserial-conical forms the cross-over is less well developed (*Dictyoconus, Chapmanina*, see Chapters 6 and 7). Organic lining may cover the connections between the chamberlets, creating sealed compartments (Ferrandez-Canadell, 2002). In planispiral-fusiform to elongate

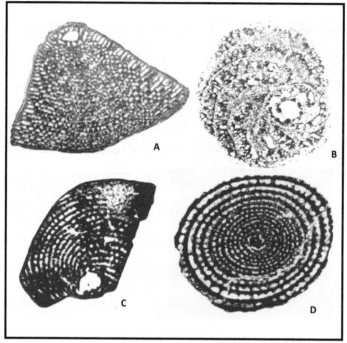

Fig. 1.24.   Examples of adult shapes A) uniserial growth as in *Orbitolina*; B) spiral growth, as in *Choffatella*; C) annular growth, as in *Orbitopsella*; D) concentric growth pattern, as in *Lacazina*.

tests (as in the fusulinides and the alveolinids) or high-trochospiral tests with a columellar structure (kurnubiids, pfenderinids, see Chapter 5), the apertures are aligned around the centre and the columella are in the polar direction. As a result, the apertural face is enlarged to admit supplementary polar apertures (Hottinger, 2006, 2007).

In the extinct fusulinides, it is believed that the rhizopods extruded from the septal pores, which replaced the main aperture, in the apertural face (Hottinger, 2001). Similarly, many species of rotaliides do not have primary apertures (Hottinger, 1997, 2000). However, the chambers communicate instead by a canal system (Fig. 1.25) that replaces the true primary and secondary apertures (Röttger et al., 1984), and feeds the different cavities by opening into the ambient seawater. In *Operculina*, the canals allow communication between the chamber cavities and the lateral surface of the walls, while *Heterostegina* has in addition a three-dimensional network of canals within the marginal cord (Murray, 1991).

Nummulites also have a three-dimensional canal system within a thickened peripheral keel (marginal cord). This canal system has multiple functions, such as locomotion, growth, excretion, reproduction and protection (Röttger, 1984). It permits the extrusion of the pseudopodia from any point of the marginal cord, provides the foraminifera with radial symmetry and enables the disposal of waste products. During sexual reproduction, it enables the release of gametes, and during asexual reproduction it allows the extrusion of the cytoplasm and symbionts to the ambient seawater.

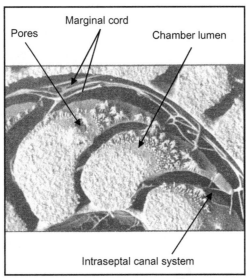

Fig. 1.25. The complicated canal system is visible within the chamber walls in an araldite cast of part of a shell of larger foraminifera (after Röttger, 1983).

1.3 Ecology of the Larger Foraminifera

Most extant larger benthic Foraminifera are marine and neritic, living largely in warm, nutrient-poor, reef and carbonate shelf environments, where they are important producers of carbonate sediments (Fig. 1.26). It is inferred that larger benthic foraminifera had similar distributions in the Mesozoic and Cenozoic. Modern taxa have geographic ranges similar to that of hermatypic corals, although some larger foraminifera certainly have a wider latitudinal distribution. Combining results from the study of the larger benthic and planktonic foraminifera provides an approximate guide to major changes in sea temperatures during the past 66 million years (McMillan, 2000). In general, the presence of larger benthic foraminifera in the fossil record indicates a warm environment, while their absence points to cooler or more nutrient-rich environments. Some extant larger foraminifera can tolerate water temperatures as low as 10-11° C, including *Amphistegina* and *Sorites* in the Mediterranean (Hallock et al., 2011), and *Amphisorus* and *Amphistegina* on the southwest Australian shelf (Li et al., 1999). Depth distributions of larger foraminifera depend upon water transparency (Hallock, 1987; Mateu-Vicence et al., 2009), with some taxa such as *Cycloclypeus* found at depths exceeding 100 m (Hohenegger, 2004). It is inferred that the larger benthic foraminifera in the Cenozoic carbonates of Tethys occupied niches analogous to those filled by modern forms.

Foraminifera typically gather food particles for extrathalamous digestion. Most larger foraminifera also have small chamberlets or cubiculae, which can act not only as a small convex lens for the focusing of sunlight, but also serve as "greenhouses" for the containment and development of symbiotic microalgae, which can provide the host foraminifera much more energy than they can consume as food (Hallock, 1981).

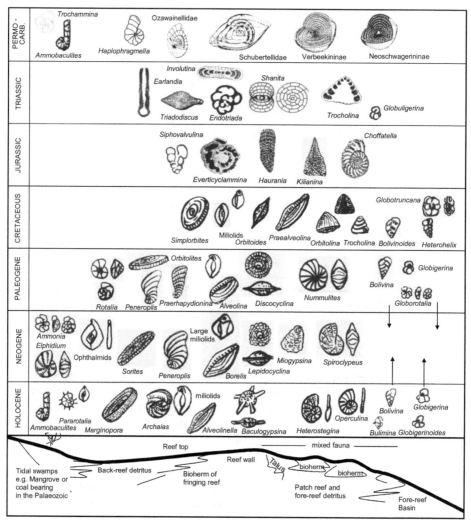

Fig. 1.26. The ecological distribution of larger and key smaller benthic and planktonic foraminifera through space and time.

Most extant larger benthic foraminifera host endosymbiotic unicellular algae, such as rhodophytes, chlorophytes, diatoms or dinoflagellates, which enhance growth and calcification in much the same way as in zooxanthellate corals (Lee and Anderson, 1991). Algal symbiosis provides the foraminifera with a reliable source of energy in water that is poor in other food sources and allows them to recycle nutrients, a necessary strategy of life in environments where nutrients, not light energy, are the limiting factor for survival (Hottinger, 2000). These symbionts also determine the colour of their host foraminifera, which tend to be brown and yellow when hosting diatoms or dinoflagellates, red to violet with rhodophytes, and green with chlorophytes (Röttger, 1983).

Larger benthic foraminifera, which are discoidal and fusiform in shape, likely have achieved their large size because of such symbiotic associations. According to ter Kuile (1991) these endosymbionts release photosynthates into their hosts, and consume CO_2 during photosynthesis, producing CO_3^{2-}, which allows high rates of $CaCO_3$ precipitation during test growth and calcification. It follows that larger benthic foraminifera are very sensitive to light levels. However, McConnaughey (1989) and McConnaughey and Whelan (1997) have proposed the reverse interpretation as a role for algal symbiosis. They suggest that lack of CO_2 limits photosynthesis in warm, shallow environments and that calcification provides protons and make CO_2 more readily available. These proposed benefits are not mutually exclusive. Because the symbionts tend to be located in the endoplasm, within the older chambers, photosynthesis does not occur in ectoplasm, where calcification primarily occurs. Thus, the primary benefit of algal symbiosis is likely the fixation of solar energy that can be used in all functions, including proton pumping and other cellular active-transport processes required for calcification (Pomar and Hallock, 2008).

Although some benthic foraminifera are "r-strategists" (well adapted to an exponential increase in population size as they have the ability to produce large numbers of offspring), larger foraminifera have largely been considered "K-strategists" (Hallock, 1985; Hottinger, 2007), where the terms r and K, come from standard ecological analyses such as the Verhulst model of population dynamics (Verhulst, 1838).

K-strategists have limited ability to rapidly increase their population densities. They have relatively long life spans, large sizes, delayed reproduction, and invest a relatively large amount of energy into each of the offspring they produce. They also exhibit low reproductive effort in maturity. This leads to morphological adaptation and increase in complexity (Gould, 1977). The long life spans of larger foraminifera is documented by authors such as Purton and Brasier (1999), who, by using oxygen and carbon isotope variation in annular cycles in the test of *Nummulites laevigatus* (see Chapter 6), were able to deduce that this species lived at least 5 years and the largest *Nummulites* could be many years older than that. As a result of the prolonged adult stage and their relatively large sizes, larger foraminifera exhibit strong hypermorphic mutations which lead to complex morphological characters (McKinney and McNamara, 1991; Lunt and Allan, 2004). Gould (1977) and McKinney and McNamara (1991) have linked Cope's (1896) rule (the increase in size of organisms during their evolutionary history) to K-strategy and hypermorphosis. The K-strategy mode of life usually occurs in relatively stable environments, as it requires delayed maturity, fewer offspring, and therefore lower reproductive potential. Gould (1977) notes that reaching sexual maturity usually marks the termination of growth and size increase of most organisms. This is certainly the case for larger foraminifera, where sexual (and asexual) reproduction usually coincides with death of the parent (Hallock, 1985). Harney et al. (1998), however, postulated that the trimorphic life cycle allows larger foraminifera to become r-strategists when subjected to ecological stress, as successive asexual generations can more rapidly build up the population density than can strict alternation of generations. Fermont (1982) documented a 50% reduction in test size in two orbitoid species that survived the Cretaceous-Paleogene extinction. However, during other periods in the evolutionary history of larger foraminifera, such as in the Triassic (see Chapter 3), only opportunistic, small, short-lived r-strategist foraminifera are believed to have survived after major, global extinctions events.

In recent years attempts have been made to understand the palaeobiology of larger benthic foraminifera using the oxygen and carbon isotopic compositions of fossil tests (Purton and Brasier, 1999), and from living assemblages (Saraswati et al., 2003). The presence of algal endosymbionts leads to disequilibrium fractionation of isotopes (Hansen and Buchardt, 1977; Saraswati et al., 2003). Carbon isotopes show this effect more than oxygen isotopes (Saraswati et al., 2003). In reef-flat environments, the size of the test of the symbiont-bearing foraminifera affects the oxygen and the carbon isotopic variations, while in deeper waters oxygen isotopic values vary little with the size of the test. These differences likely reflect differences in the variability of the environment during the life span of the foraminifera. Saraswati et al. (2003) concluded from their observations of miliolides and rotaliides that the two groups differ distinctly in their carbon isotopic fractionation; the low and medium fractionated taxa belong to Miliolida and the highly fractionated taxa belong to Rotaliida. Although microhabitat also appears to have a role in carbon isotope variation, the relative contributions of biomineralization, metabolism and microhabitat are difficult to estimate. Langer (1995) analysed five species from a lagoon in Papua New Guinea and examined isotopic composition with depth. These studies again showed a general trend in the depletion of heavier isotopes of oxygen and carbon with depth and intensity of light. Wefer and Berger (1980) recorded isotopic variation within an individual specimen of *Marginopora vertebralis* by sampling along the direction of growth (ontogeny), and they inferred that oxygen isotopes reflected seasonal variation in temperature. The carbon isotopic composition is more variable between the specimens of the same species (Saraswati, 2004). Saraswati (2004) concluded that ontogenetic oxygen isotope variation decreases progressively in deeper water species.

Studies of living larger benthic foraminifera, in controlled laboratory environments, have provided some further information regarding life strategies (e.g., by the culture of *Heterostegina depressa*; Röttger, 1984), but much has been inferred by relating test morphology to habitat (Hallock, 1981; Murray, 1991). Predators such as bristle worms, crustacea, hermit crabs, snails, gastropods, echinoderms and fish, as well as microscopic predators (including other foraminifers (Hallock and Talge, 1994), some nematodes (roundworms) and flatworms), selectively feed upon foraminifera. Such predation pressure, of course, will depress the foraminiferal populations and, in most cases, the observed thanatocoenosis (death assemblage) is not fully representative of the living population (biocoenosis).

Living distribution patterns of the symbiont-bearing larger foraminifera are confined to tropical, subtropical and warm-temperate photic marine environments, as their distribution is determined by a complex set of inter-related parameters such as temperature, nutrient availability, water transparency and light intensity (Renema, 2002). Water depth is a secondary factor related to the distribution of larger benthic foraminifera, because light intensity, temperature and hydrodynamic energy decrease with depth. Some larger foraminifera, such as *Amphistegina* (Plates 1.3 and 1.4), are known to become flatter, with thinner outer walls, with increasing water depth and decreasing light intensity (Hallock et al., 1986). Imperforate foraminifera, such as the miliolides, are generally restricted to shallower depths than perforate forms (Hallock, 1988; Hottinger, 2000). However, both perforate and imperforate larger foraminifera house symbionts, and the dependence on light for their symbionts limits their distribution to the photic zone. The depth distribution of living larger benthic forami-niferal taxa is related to the transparency of their test walls (porcelaneous versus hya-line) and the light wavelengths required by their symbionts (see Renema, 2002), e.g., *Archaias* (0–20 m, chlorophytes, red light), *Peneroplis* (0–70 m, rhodophytes, yellow light) and *Amphistegina* (0–130 m, diatoms, blue light) (Hallock, 1985). This has led many authors to use calcareous algae and larger foraminifera assemblages as proxy water-depth indicators in carbonate sediments (Banner and Simmons, 1994; Mateu-Vicens et al., 2009).

The interpretation of use of sunlight by many fossil larger benthic foraminifera, such as *Miogypsina*, *Miolepidocyclina*, is shown by the common occurrence of miogyp-sinids only in shallow water marine limestones where fossil algae also occur (Fig. 1.26). The irregular shape of many species of *Miogypsina*, often revealing apparent divi-sion of their flanges (e.g., *Miogypsina bifida*, see Chapter 7), shows they could become concavo--convex and were not growing on a flat surface. Such a shape would be devel-oped if the individual was attached to a strongly curved substrate, such as fronds of macroalgae or the stems or leaves of seagrass (the substrate would be biodegradable and seagrass is not seen preserved in thin sections). Therefore the sedentary, attached miogypsinids likely grew to accommodate the shape of the vegetable substrate to which they adhered. Only in strong ambient sunlight, which would benefit both the miogyp-sinids and their vegetable substrate, could true *Miogypsina* flourish (BouDagher-Fadel and Wilson, 2000). On the other hand, elongate forms, such as *Alveolinella*, can hide or shelter under shallow layers of coral sand, in order to regulate the illumination that they require.

The ectoplasm of foraminifera forms the template on which new chamberlets are secreted (Röttger, 1984). Larger foraminifera can modify their shape and structure to some degree depending upon environmental conditions. Moreover, morphologic trends indicate both adaptation and acclimation (Hallock et al., 1986). Wide varia-tion in test structure and morphology has resulted from such adaptation. According to Larsen and Drooger (1977) and many subsequent researchers, the diameter-thickness (D/T) ratio of many larger benthic foraminifera varies inversely with depth. The near-shore samples have a higher D/T than the more offshore species. Mateu-Vicens et al. (2009) further quantified this relationship for *Amphistegina*, which can be used to estimate paleodepth and water transparency. Thus, the mor-phology of larger benthic foraminifera has been used for depth-estimation in geo-logical facies interpretations, usually based on comparison with homeomorphs of living occurrences.

Robust and fusiform tests, as seen in *Fusulina* and *Alveolina*, and conical (e.g. orbi-tolinids) or strongly biconvex (*Amphistegina*; Plates 1.3 and 1.4) forms are adapted to a life in environments of high hydrodynamic energy, characteristic of water depths less than 10m, but usually not in mobile sands, but rather on phytal or hard sub-strates like reef rubble (Hallock, 1985). Those with very thin and flat tests, however, can only live in very calm waters that tend to have lower levels of light intensity, and some of them such as *Discospirina* are able to live in deeper and cooler waters. The larger foraminifera found attached on sediments in deeper waters tend to be flat and discoidal in shape (e.g. *Spiroclypeus*). Forms adapted for adherence to seagrass or algae tend to be flat (*Amphisorus* and *Cyclorbiculina*), and sometimes relatively small (e.g., *Peneroplis*; Plate 1.4, Fig. 6). Some foraminifera develop some kind of anchorage (e.g. the spines in calcarinids; Plate 1.4, Figs. 2–3), or an ectoplasmic sheath (*Heterostegina depressa*; Plate 1.3, Fig. 1). The numbers of *H. depressa* tend to be low in shallow waters and restricted to shaded locations (Haynes, 1981). In some taxa, such as *Amphistegina*, the morphology of the apertural face can change to increase their potential to cling to surfaces in turbulent waters (see see Chapter 7). According to Hottinger (2000), in *Alveolina* and *Fusulina* the function of the elongated fusiform test is related to motility, the test growing in equatorial direction but moving in the polar direction.

So, ever since the Carboniferous, larger foraminifera have thrived in the shallow, warm, marine environments (see Fig. 1.26). Their remarkable abundance and diversity is due to their ability to exploit a range of ecological niches by having their tests utilised as greenhouses for symbionts. However, attaining large size made some forms very specialised and vulnerable to rapid ecological changes. For this reason K-strategist, larger foraminifera show a tendency to suffer periodic major extinctions when environmental conditions change rapidly and/or substantially. This makes them valuable biostratigraphic zone fossils, and, as will be explored in the following section, also provides a valuable insight into the general process of biological evolution, including parallel and convergent evolutionary trends.

1.4 Palaeontological and Evolutionary History of the Larger Foraminifera

1.4.1 Evolution

Like all fossils, from dinoflagellates to dinosaurs, the larger benthic foraminifera are biofacies bound, and often regionally constrained. They have biotopes closely associated with carbonate environments. Large-scale changes in these biotopes occur in response to, for example, eustatic sea-level fluctuations and climate change, that cause stress to the associated fauna and flora. Palaeoecological studies have demonstrated that feeding mechanisms and reproductive strategies are key traits that affect survival rates (Twitchett, 2006). Small unspecialized and opportunistic taxa fare better than large and advanced forms during times of stress, and after a major event, the surviving primitive forms thrive in the new environment, which initially is one of low diversity and limited competition for food resources. Typically, this state is associated with the predominance of small forms (the "Lilliput effect") with very high turnover rates and low biomass. These "disaster species" can quickly take advantage of the relatively high food supplies and lack of competition (Pomar and Hallock, 2008). As environmental conditions stabilize, the survivors diversify into the new environments and eventually new larger forms evolve and colonise more specialised niches.

Larger foraminifera always preserve the juvenile stage, at least in the microspheric form, before growing into an adult stage. They show considerable similarities in evolutionary trends, and often follow parallel lines of evolution. These similarities were first recognised by van der Vlerk (1922) and Tan Sin Hok (1932), and were later summarised by Haynes (1981). Larger foraminifera, as they evolve, tend to increase the size of their test, together with increasing the size of the macrospheric embryont, especially the second embryonic chamber (the deuteroconch). The increase in size is also accompanied by a tendency to increase the number of chambers, or subdivision of the chambers. However, there is also a tendency towards the reduction in embryont size with radial growth. Hallock (1985) proposed biological interpretations for these trends that include optimizing the benefits of algal symbiosis, relatively long life span, and minimizing mortality of asexually produced juveniles by increasing the size of the embryont. Advanced forms tend to have ogival (curved diamond shape) or hexagonal equatorial chamberlets. There is also a progressive development of lateral chamberlets in the rotaliides.

During the evolutionary history of larger foraminifera many morphological features have shown convergent evolutionary trends. Identical gross morphologies, shapes or structures appear again and again within the same lineage or in parallel lineages from different stocks. An example of such convergence is the fusiform test, which appeared several times in different lineages. The elongate fusiform fusulinides made their first appearances in the shallow warm waters of the Carboniferous-Permian. They evolved from simple foraminifera with calcareous granular walls, acquiring as they did so alveolar compound walls (Chapter 2). Quite independently, and from a miliolide ancestor, fusiform elongate alveolinids made their first appearance in the Paleogene (Chapter 6), and have re-evolved many times (but showing different detailed internal features) up to the present day (Fig. 1.27).

This morphological evolution occurred as the foraminifera occupied specific environmental or ecological niches. Genes are the fundamental units of life, and determine the genotypical properties of any life form. However, environment, ontogeny and conditions during growth determine the phenotypical character of a species, determining to some degree, for example, the shape of the test. The interaction between the gene in the embryont and the selection of the external features that the foraminifera develop during growth is an example of this selection process discussed by Gould (2002) in his book *The Structure of Evolutionary Theory*. This process acts on features that emerge from complex gene interaction during ontogeny and not from individual genes. One need only look at the range of sizes and shapes of domestic dogs to see how morphological variation can be generated by selected or controlled conditions for reproduction, to realise how, despite the variety of genetic components, certain ecological conditions will favour certain morphological developments. The many iterative and convergent evolutionary trends found in larger foraminiferal lineages provide wonderful examples of the fundamental restrictions on genetic possibilities within a higher taxon, which is the basis of the concept of "bauplan" (Hintzsche, 1947) that is widely recognized in evolutionary biology and palaeontology.

Many characteristics in the evolution of foraminifera are gradual and linked to the timing of phases in growth. They could have been started by DNA mutation at one stage of foraminiferal life and amplified as a consequence of having to cope with stress and adverse environments. A good example is the gradual evolution of the lateral cubiculae/chamberlets that are created by intra-lamellar spaces and linked by pores in *Miogypsinoides*, which is accompanied with nepionic reduction that results in forms such as *Miogypsina* (see Chapters 6 and 7). Other characteristics appear suddenly, such as alveoles in the miliolide *Austrotrillina*, which appear to have evolved directly from

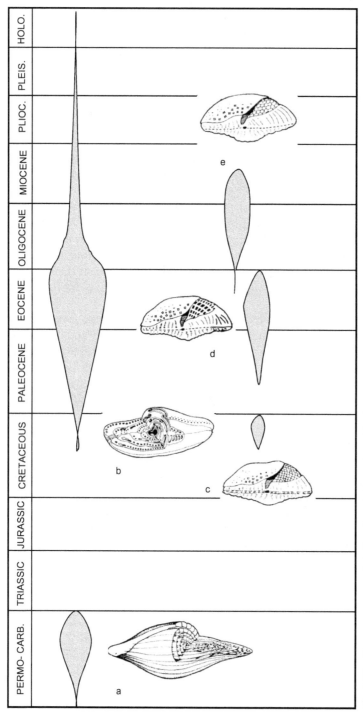

Fig. 1.27. The convergence of similar shapes of test, but with different internal structures, shown by the fusiform test in a) the calcareous granular fusulinines and b) the porcelaneous alveolinids; or the rotalides which have same shape and the same internal three-layered structures but different shapes of embryonic apparatus and shape of chamberlets, in forms such as c) the orbitoids, d) the discocyclinids, e) the lepidocyclinids.

Quinqueloculina, a small, simple benthic foraminifer. However, these characteristics had also appeared in larger miliolides earlier on in the stratigraphic column, such as the alveoles in *Alveolina*. Evolving from a quinqueloculine ancestor, *Alveolina*, did not acquire the alveoles in its ontogeny but later on in the adult stage. This suggests that the propensity to develop a mutation that allows for the development of alveoles is common to most or all orders of larger foraminifera, and that when the conditions are suitable, this feature will be selected.

Larger foraminifera, such as *Eulepidina* and *Discocyclina*, achieve large sizes and radial symmetry early in their evolutionary history. Such forms having reached the optimum size with a small nepionic stage generally do not exhibit any more changes for millions of years unless subjected to environmental change.

Some features seem to be essential to establish morphological trends in larger foraminifera. An example is the appearance of a complicated system of stolons in the lateral layers, which allow the cytoplasm to extrude from these apertures in order to build small chamberlets. The stolon system must have resulted from an adventitious gene mutation which enabled the cytoplasm to extrude and build more chamberlets. This would also have allowed the foraminiferal test to spread and have more space to house symbionts. Tan Sin Hok (1932) and Smout (1954), whose work preceded recognition that algal symbiosis was nearly ubiquitous among extant larger foraminifera, proposed that large tests are more efficient as they provide shorter lines of communication for internal processes, thus providing a competitive advantage.

The larger foraminifera, by developing alveoles, chamberlets, systems of stolons, pillars and other structures that strengthen their tests, are one of the best examples of how genetic mutation allows a wide range of stable environmental niches to be occupied. Proliferation is the reason for a species' existence, but survival is the short-term goal. However, as for all creatures, larger more specialised forms cannot as easily survive the development of adverse conditions, or adapt as readily to a different mode of life. They are therefore the first ones to become extinct. However, the nascent characteristics still exist in the genes of more primitive forms, and when a niche is re-colonized the old morphological features may reappear. Genotypical convergence gives rise to forms similar to the extinct ones, but with slight or major structural variations, such as the folded septa in the fusiform fusulinides compared with the alveoles in the fusiform alveolinids (see Fig. 1.20). However, the fusulinides and miliolides are not closely related, as the first ones have calcareous granular walls and the second ones are porcelaneous with calcitic non-perforate walls. This parallel evolution from two different stocks was nonetheless driven by comparable forces, for the same purpose and benefit for the foraminifera, as both variations likely were used to house symbionts.

Since Carboniferous times, changes in local or global conditions have caused the highly specialised forms of foraminifera to become extinct. Later, when environmental conditions that can support larger foraminifera are re-established, new taxa evolve and fill the restored niches. Based on the evolutionary records of larger foraminifera and other carbonate-producing organisms, Hallock (1987) proposed the concept of "geologically longevous" versus "geologically ephemeral" niches. Moreover, the disappearance and reappearance of larger foraminiferal taxa through geological time illustrates the converging trends highlighted so clearly by Conway Morris (2003). These convergent trends are consistent with developing similar characteristics at different times and probably under similar conditions. Larger foraminifera are, therefore, good examples of the creative powers of gene mutation and gene interaction, and can provide case studies for the important role of both genotypical and phenotypical processes.

1.4.2 Palaeontology

As a consequence of their ability to evolve rapidly and fill a range of ecological niches, larger foraminifera are very valuable guides to changes in geological environments (Jones et al., 2002; Jones et al., 2003; Stefaniuk et al., 2005; Marriner et al., 2005; Morhange et al., 2005; Lord et al., 2009; Meinhold et al., 2009; Henderson et al., 2010; BouDagher-Fadel et al., 2010c). Their study provides excellent insight into palaeoecology but also into the factors that give rise to both local and global extinction events. Eustatic sea level fluctuations and climate change have been two of the main mechanisms invoked by geologists to explain past extinction events. More recently, processes like volcanic eruption, anoxia or changes in atmospheric chemistry, and still more famously catastrophic meteorite impacts have been invoked. Hallock (1987) used a simple conceptual model to illustrate how significant perturbations to ocean circulation, whether gradual (such as climate change) or nearly instantaneous (such as a meteor impact), could eliminate high proportions of the niches occupied by taxa specialized to shallow, warm, nutrient-poor habitats within the photic zone. It is certainly the case that the ecological sensitivity, or vulnerability, of the larger foraminifera coupled with the ability of smaller primitive forms to survive and then evolve to repopulate the niches that were left vacant, means that larger foraminifera are very good tools to study global extinction processes (Fig. 1.28). The story of the evolution and periodic extinction of larger foraminifera will be told in detail in the subsequent chapters, but is summarised briefly below.

The first larger foraminifera seem to have evolved from the agglutinated foraminifera in the Carboniferous (see Chapter 2) by developing compound walls and a complicated internal structure. These lineages developed fusiform tests and gave rise to the **Fusulinida** and several related orders (such as the **Staffellida**, etc) that became very abundant rock-forming fossils. However, these Palaeozoic orders did not survive the end Permian, but some small related orders, such as the **Earlandiida** and the **Endothyrida** survived to the Early Triassic and end Triassic respectively (see Chapters 2 and 3).

The Triassic was a period of recovery, larger foraminifera re-emerged slowly. During the Early Triassic, foraminifera were simple and rare. They were initially dominated by small arenaceous forms, the "disaster forms", which are characteristic of the survival phase following a mass extinction (see Chapter 3). The recovery forms were dominated by the **miliolides** and later by the **involutinides** in the Eastern Tethys area.

The end of the Triassic saw another crisis, which again killed most of the larger foraminifera, and all of the small Permian survivors were completely wiped out (see Chapter 4). Following this mass extinction, the early Jurassic witnessed the steady evolution of the agglutinated forms from the small simple **textulariides** to those with an internally complicated form, which became abundant from the Pliensbachian onwards,

Ma	Stages	Extinction events	Possible causes
5.3	End of Messinian (Late Miocene)	Extinction of *Heterostegina* (*Vlerkina*) 	Dessication of the Mediterranean and tectonic events.
11.6	End of Serravallian (Middle Miocene)	Extinction of lepidocyclinids and miogypsinids in the Tethyan province. 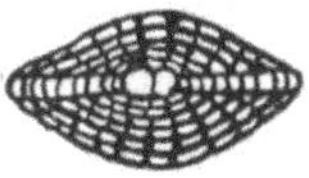	Tectonic events.
15.9	End of Burdigalian (Early Miocene)	Extinction of lepidocyclinids and miogypsinids in the American province. 	Eruption of the Columbia River basalts.
23.0	End of Oligocene	Extinction of Orbitolinoidea and Orbitoidoidea 	Plate tectonic events.
33.9	End of Eocene	Extinction of the nummulitd *Pellatispira, Biplanispira* 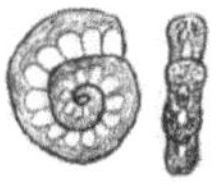	Global climate changes attributed to the expansion of the Antarctic ice cap and multiple bolide impact events.
37.8	End of Bartonian (Middle Eocene)	Extinction of *Coskinolinoidea* and the nummulitid *Assilina*  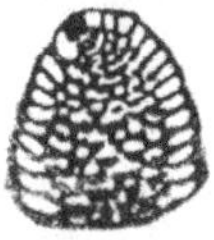	The move into an "ice house" climate may have been triggered at this time by the opening of the Tasmanian gateway.
66.0	End of Maastrichtian (Late Cretaceous)	Extinction of Cretaceous alveolinids, orbitoids, pfenderinids, large lituolids, etc. 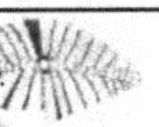  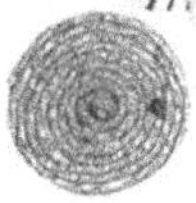	Multiple impact events, e.g.Chixculub crater impact.
93.9	End of Cenomanian (Early Cretaceous)	Extinction of the *Orbitolina* 	The Wallaby eruption and a near-peak Mesozoic eustatic sea-level highstand.
201.3	End of Triassic	Extinction of all small fusulinids and large lagenids 	Central Atlantic Magmatic Province.
251.9	End Permian	Extinction of all large fusulinids  	Siberian traps flood basalts.
259.1	End of Guadalupian (Late Permian)	Extinction of all large Schwagerinidae 	The Emersham basalts outcrop.

Fig. 1.28. Summary of the main extinction events affecting the larger benthic foraminifera through the geologic column.

thereby giving the carbonate facies of the Jurassic a characteristic that is recognizable throughout all of Tethys. By the end of the Jurassic some forms became extinct, but many robust forms from shallow, clear waters survived the Tithonian and crossed over to the Cretaceous. The Early Cretaceous biota as a whole was dominated by large agglutinated foraminifera with complicated structures such as alveoles, internal partitions and pillars (see Chapter 5). The complex and partitioned **lituolids** (of the order Textulariida, such as *Orbitolina*) dominated the biota of inner carbonate platform environments that were widespread along the western and eastern margins of the Cretaceous Tethys. They were joined by the **alveolinids** (of the order Miliolida), which showed spectacular expansion in the Middle Cretaceous, proliferating in mid-latitudes, and often becoming annular or discoid and subdivided by partitions. Many of them resembled the planispiral-fusiform fusulinides of the Permian, attaining approximately the same range of sizes, but differing fundamentally in their imperforate, porcelaneous wall structure. In the Late Cretaceous new simple **rotaliides** forms evolved into forms with complicated three-layered textures, the **orbitoids**. While the previous two groups had their main breeding ground in the Tethyan realm, the orbitoids showed provincialism and some were only found in the Caribbean.

As will be seen in Chapter 5, the Cretaceous–Tertiary crisis wiped out most of the Maastrichtian larger benthic foraminifera. The Early Paleocene was a recovery period and only by the Late Paleocene had larger **miliolides, nummulitids and orthophragminids** appeared and spread throughout Tethys. The miliolides included large fusiform alveolinids, which showed morphological convergence with the extinct Cretaceous alveolinids, and the discoid **soritids**, which became prominent throughout the Eocene. Parallel to these lines of evolution, the agglutinated foraminifera developed into forms imitating their Cretaceous ancestors by developing internal pillars (as in the textulariides), or with more complicated partitions (as in the orbitolinids). However, in the American province the rotaliides evolved into three-layered lepidocyclinids.

Towards the end of the Paleogene, the extinction of larger foraminifera was not as pronounced, however fluctuations in climate, sea-level and/or oceanic currents influenced the geographic distribution of the larger benthic foraminifera, causing biogeographical provincialism. Although most of the Miocene superfamilies are still extant, provincialism was prominent at generic and specific levels. The Late Oligocene and Early Miocene show the least provincialism, with **miogypsinids** and **lepidocyclinids** spreading from the Americas to the east (see Chapter 6). By the Middle Miocene, with the closure of Tethys, provincialism was re-established. Lepidocyclinids and miogypsinids completely disappeared from America in the latest Early Miocene and from the Mediterranean in the Serravallian (late Middle Miocene). Deep-water textulariides made their first appearance in America, while new genera of alveolinids appeared in the Indo-Pacific.

The development of the Indo-Pacific as a separate province continued in the Late Miocene with the closure of the Tethyan seaway. Neogene sedimentary sequences (Chapter 7) of this province are dominated by warm-water, shallow-marine carbonates of crucial importance as the product and record of climatic/oceanic conditions and as hydrocarbon reservoirs. Throughout the past century, larger foraminifera have been used extensively in the Far East from Japan southwards to Australia, and through the Tethyan region westward to the Caribbean and America for biostratigraphical, palae-oceanographical and palaeoclimatological analysis. In the Middle East, larger fora-minifera have been used to correlate the many biostratigraphical stages, which have been proposed for Europe. Their role as markers for biostratigraphical zonation and correlation underpins most of the drilling of marine sedimentary sequences that is cen-tral to hydrocarbon exploration.

By the Holocene, new forms of larger benthic foraminifera appeared through the tropical realm, especially in the Indo-Pacific. Present day larger foraminifera play anal-ogous roles in the ecosystem, where the tropical belt is divided into two parts, and differ-ent assemblages colonise different environments. In particular, the Indo-Pacific larger **soritids** (*Amphisorus* and *Marginopora*) are substituted in the Caribbean by *Archaias* and *Cyclorbiculina* as porcelaneous discoidal epiphytes on tropical seagrasses (Langer and Hottinger, 2000). Modern tropical western-Pacific sandy shoals and beaches can be composed of nearly pure concentrations of *Calcarina, Amphistegina,* or *Baculogypsina* tests (see Fig. 1.26), though only *Amphistegina* spp., which are directly descended from a Paleogene ancestor, are abundant across the subtropical-tropical regions worldwide.

1.5 Conclusion

From this introductory chapter, it can be seen that larger foraminifera are biologi-cally complex and highly versatile. They have repeatedly evolved from simple ances-tors since the Carboniferous, becoming highly specialised and therefore highly sensitive to environmental changes. Their study provides, as a result, considerable insight into evolutionary process as well as into the major geological mechanisms associated with extinction and recovery.

Larger foraminifera also have, and continue to occupy, the very important ecolog-ical niche of being a reef-forming group. The worldwide distribution of carbonate biota, especially reef biota, contains important information on environmental condi-tions, including oceanographic parameters, that control this most sensitive of habitats. The study of the distribution patterns of this biome, over different time slices, provides valuable information on how the climate of the Earth has evolved in the past 350 Ma.

Finally, the carbonate rich shallow marine and reef environments favoured by larger foraminifera are also those which may give rise to economically vital deposits of oil and gas. So larger foraminifera are now central to our ability to date, correlate and analyse the sedimentary basins that are currently key to the economic wellbeing of the world. A detailed understanding of the taxonomy of the larger foraminifera is essen-tial, therefore, for any applied biostratigraphic analysis.

In this book, each of the following chapters outlines the palaeobiological and the geological significance of the larger foraminifera through time. Specifically, the tax-onomy, phylogenetic evolution, palaeoecology and biogeography of the larger fora-minifera are outlined and discussed, relative to the biostratigraphical time scale (as defined by Cohen et al., 2013) of the middle and late Phanerozoic. In establishing the most suitable markers for the biozonal boundaries, new (bio)chronostratigraphic units are established. The larger benthic foraminiferal of the 'letter stages' of the Far East, as defined by BouDagher-Fadel and Banner (1999) and BouDagher-Fadel (2008), and the shallow benthic zones (SBZ) for the Paleocene–Eocene epochs as proposed by Serra-Kiel et al. (1998) are revised and correlated to the planktonic foraminiferal zonal scheme of BouDagher-Fadel (2015).

Many of the figures presented here are type figures from the Natural History Museum, London (referred to as NHM), while other types, such as those from the Permian are from the Senckenberg-Forschungsinstitut und Naturmuseum, Germany (referred to as SFN), others are deposited in the UCL Collections, while some are taken from referenced literature sources.

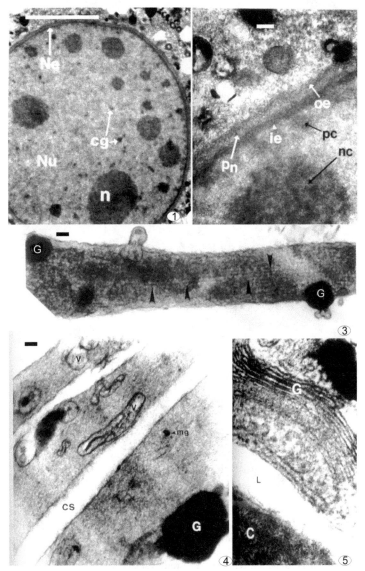

Plate 1.1   Scale bars: Fig. 1 = 0.5mm; Fig. 2 = 50µm; Fig. 3 = 0.170µm; Fig. 4 = 0.038µm. Fig. 1. A figure showing the nuclear envelope (Ne), of the nucleus which is situated in the sixth youngest chamber of *Elphidium excavatum* (Terquem), enclosing the nucleoplasm (Nu) which contain nucleoli (n) and dense chromatin granules (cg) (after Alexander, 1985). Fig. 2. Enlargement of the nucleus in Fig. 1, showing the nuclear envelope to consist of a perinuclear space (Pn) separating the outer envelope (oe) from the inner envelope. Abbreviations: pc = peripheral chromatin, nc = nucleus-associated chromatin (after Alexander, 1985). Fig. 3. Longitudinal section of fine pseudopodia of *Haynesina germanise* (Ehrenberg). Microtubules are visible in the central region. G = granules. Fig. 4. High magnification detail of a pseudopod from *Elphidium williamsoni* Haynes showing the membrane-lined canal system (cs). Fig. 5. Enlargement of the cytoplasm of *Elphidium williamsoni* Haynes to show close association of the golgi (G) with vacuoles containing chloroplasts (C), and vacuolar lumen (L).

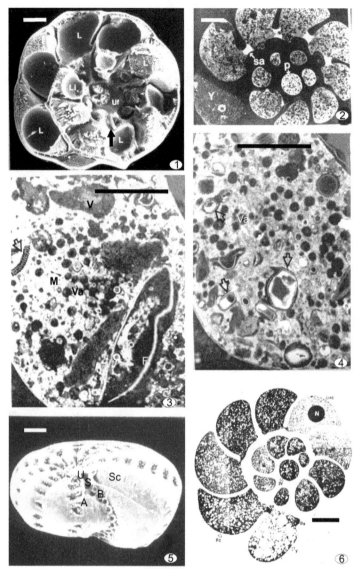

Plate 1.2 Scale bars: Figs 1, 2, 5-6 = 50µm; Figs 3-4 = 25µm. Fig. 1. *Ammonia batava* (Hofker). SEM pho-tograph of dorsal side a partial resin cast of an embedded specimen. Resin has filled the umbilical fissure (Uf) and also vertical canals (arrow) extending from the umbilical region towards the dorsal test surface. L = chamber lumina of outer whorl; LI = lumen of inner whorl chamber (after Alexander, 1985). Fig. 2. Equatorial section of *Ammonia batava* (Hofker). Y = youngest chamber; P = proloculus; sa = septal aper-ture (after Alexander, 1985). Figs 3-4. Detail of penultimate chamber in two different planes of section of *Ammonia batava* (Hofker, 1951). Large voids (V) occupy much of the cytoplasm and small spherical vacuoles (Va) are present. Diatoms (arrows), mineral-like fragments (M) and other ingested material such as filament structures (F) are also present (after Alexander, 1985). Fig. 5. Equatorial apertural view of a megalospheric form of *Elphidium williamsoni* Haynes. Last formed chamber removed. A = apertural openings located on a retral process of the previous whorl; B = apertural opening formed above a fosette (the opening to the exterior of an intraseptal interlocular space) of the previous whorl; S = "septal (rotaliine) flap"; Sc = intra-septal canal; U = umbilical lobe of chamber lumen, opening to the umbilicus (after Alexandra, 1985). Fig. 6. Equatorial section of *Elphidium excavatum* (Terquem) showing an increased density of cytoplasm within the proloculus (Pr) when compared to the sixth youngest chamber (CH6), which contains the nucleus (N), Pa = primary aperture of youngest chamber (y), Sa = septal aperture between youngest chamber and penul-timate chamber (Pc) (after Alexander, 1985).

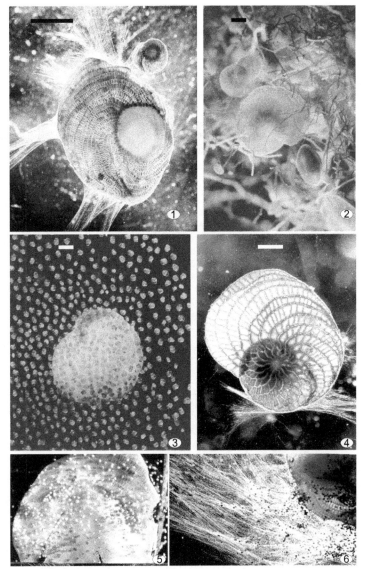

Plate 1.3 All photographs are from living specimens. Scale bars: Fig. 1 = 1mm; Figs 2-4 = 2mm. Fig. 1. *Heterostegina depressa* d'Orbigny, the large individual is an agamont (diameter 14.50 mm), the small individual a gamont (diameter 3.75 mm). In many cases, the youngest chambers of agamonts are partly broken off, and older parts of the test show signs of former damage, now healed and overgrown by later chambers. The pseudopodia protrude from the canal system of the marginal cord. The living laboratory specimens from a dredge haul from 40 m water depth off Kekaa Point, Island of Maui, Hawaii (courtesy of Prof R. Röttger). Fig. 2. Larger fora-minifera in their natural habitat, the protective algal tangle of a rock-pool at Makapuu Point, Island of Oahu, Hawaii. In the centre, *Heterostegina depressa* d'Orbigny (size 2.4 mm); top left *Amphistegina lessonii* d'Orbigny, behind it *Aphistegina lobifera* Larsen. In the lower right corner another *Heterostegina depressa* d'Orbigny can be seen. These species of larger foraminifera obtain their characteristic yellowish coloration from symbiotic diatoms which contribute to the nutrition of their hosts, (courtesy of Prof R. Röttger). Fig. 3. *Amphistegina lessonii* d'Orbigny (test size 1.57 mm), two days after the reproductive process. Specimen from the natural hab-itat, Hawaii, photographed in the laboratory. The now colourless inanimate mother cell is surrounded by its 670 daughter individuals formed by multiple fission of the parent protoplasm. Most of the offspring are in the

3-chamber stage. The majority of symbiotic diatoms transferred from the mother individual to the juveniles are located in the second and third chamber, (courtesy of Prof R. Röttger). Fig. 4. *Heterostegina depressa* d'Orbigny (test size 3.5 mm), specimen from off Pearl Harbour, Hawaii. Transmitted light reveals the planispiral test construction (courtesy of Prof R. Röttger). Figs 5, 6. *Heterostegina depressa* d'Orbigny. Microspheric agamonts after multiple fission. 5). The protoplasm of the specimen depicted in Fig. 1 has divided into ca. 1300 gamonts now in the 2-chambered embryonic stage (size 155μm). 6). Some hours later, the juveniles being transported off the inanimate mother test by streams of residual protoplasm, (courtesy of Prof R. Röttger).

Plate 1.4 All photographs are from living specimens. Scale bars: Fig. 1 = 2mm. Figs 2-6 = 1mm. Fig. 1. *Amphistegina radiata* Fichtel and Moll (test size 1.93 mm). Specimen from Kudaka Island, Ryukyu Islands, Japan (courtesy of Prof R. Röttger). Fig. 2. *Calcarina gaudichaudii* d'Orbigny, Agamont (diameter 3.5 mm). From the trochospiral test branched processes (spines) with a protoplasm-containing internal canal system radiate. On the side facing the observer the spiral of chambers is covered by a thin-walled brood chamber within which the many juveniles (the young gamonts) are formed. They are released by dissolution of the roof of this chamber. Specimen from Belau, Micronesia (courtesy of Prof R. Röttger). Fig. 3. *Calcarina gaudichaudii* d'Orbigny, Gamont (test size 2.8 mm). Gamonts are smaller than agamonts. Also the number of spines, now unbranched, is much smaller than in agamonts. Specimen from Kudaka Island, Ryukyu Islands, Japan, (courtesy of Prof R. Röttger). Fig. 4. *Baculogypsina sphaerulata* Parker and Jones (test size 2.2 mm). The subspherical test differs from other calcarinids; only the innermost part, not visible from outside, forms a spiral. All other chambers are inserted between the axes of the spines according to a different pattern of construction. The network of chambers is interrupted by transparent pillars (visible as dark spots) oriented perpendicular to the test surface. Specimen from Kudaka Island, Ryukyu Islands, Japan (courtesy of Prof R. Röttger). Fig. 5. *Calcarina gaudichaudii* d'Orbigny and one *Baculogypsina sphaerulata* Parker and Jones (lower left corner) attached to a red algal thallus from their natural habitat, the littoral algal zone of Kudaka Island (Ryukyu Islands, Japan). Attachment occurs with an organic cement secreted from the tip of the canaliferous spines. This device helps to counteract transport by wave action (size of largest specimen 3.8 mm) (courtesy of Prof R. Röttger). Fig. 6. *Peneroplis* sp., Peneroplidae (order Miliolida) (test size 1.5 mm) has formed daughter individuals by multiple fission. The test was partly dissolved during this process to release the juveniles. *Peneroplis* harbours symbiotic unicellular red algae which were transferred to the offspring (courtesy of Prof R. Röttger).

Chapter 2
The Palaeozoic Larger Benthic Foraminifera

2.1 Introduction

The first foraminifera with a hard test (i.e. one that was biomineralised and therefore had preservation potential) to appear in the fossil record are the unilocular, simple agglutinated Allogromiida. From these the calcareous agglutinated foraminifera, Textulariida, evolved during the Cambrian. The members of the Textulariida remained the dominant (but morphologically small) group in the Early Palaeozoic, but morphologically larger foraminifera with compound, microgranular walls and a complicated internal structure, exemplified by the order Fusulinida (which are generally referred to here as fusulinides), became abundant and ecologically dominant in the Late Palaeozoic. These larger fusulinides and related forms are the major fossil group in many Late Palaeozoic shallow-marine limestones.

In addition to the textulariides and the fusulinides related forms, the lagenides foraminifera evolved in the Silurian, but did not show significant evolutionary diversity until the late Gzhelian in the Carboniferous, becoming more abundant in the Permian. It appears likely that their defining morphological characteristic (having walls made of orientated calcite crystals) evolved independently several times in the Palaeozoic, and that the superfamilies Robuloidoidea and Nodosinelloidea had different phylogenetic roots. However, these forms always remained morphologically relatively small, and it was not until the Triassic that significant, larger lagenides appeared. The morphological evolution of this group is discussed in detail in Chapter 3, and here the Palaeozoic genera are only briefly discussed to enable the continuity between these forms and those of the later Triassic to be understood.

The miliolides (with porcelaneous test walls) were also present in the Palaeozoic, but again (as with the lagenides) they remained morphologically small and primitive throughout the Palaeozoic. It was not before the Triassic that they became relatively large and more prominent as rock-forming fossils (see Chapter 3). Whereas, true Involutinida really first appeared in the Triassic: their only suggested occurrences in the Permian were erroneous, based on the misidentification of the porcelaneous genera *Neohemigordius* and *Pseudovidalina* (Loeblich and Tappan, 1988).

As stated, the Palaeozoic Fusulinida and related forms first evolved from the Allogromiida in the Silurian (see Fig. 2.1), with the appearance of members of the Parathuramminida, followed directly by the Moravamminida (a disputed foraminiferal order, see Vachard, 1994; Vachard and Cozar, 2010), the Archaediscida, and the Lagenida referred to above. From the Devonian onwards fusulinides and related forms became established and included, the Earlandiida, the Palaeotextulariida, the Tetrataxida, the Tournayellida, the Endothyrida, and the Fusulinida.

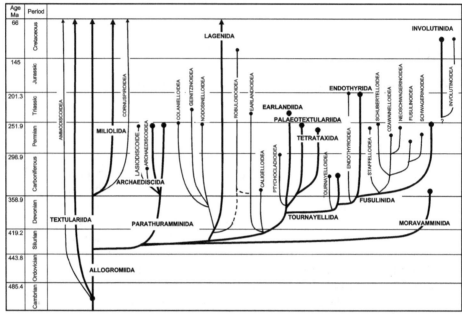

Fig. 2.1.   Evolution of the Palaeozoic larger foraminifera.

During the latter part of the Palaeozoic, the fusulinides and related forms evolved from being simple to highly-specialized, and are characterised by a history of rapid evolution that gave rise to diverse lineages. They were sensitive to their physical environment, and were the first foraminifera to develop a test with a complex internal architecture, combined with, for a single cell organism, a relatively gigantic, centimetric size. The fusulinides and related forms became cosmopolitan, colonising most shallow, warm waters of the Carboniferous and Permian. All (except members of the superfamily Fusulinoidea) occurred in temperate waters, and in shallow to deeper conditions. They are well preserved, but their preservation mode depends on post-sedimentary processes. In many areas, during the Late Carboniferous and Permian, the fusulinides far outnumber any other single marine invertebrate group.

The Palaeozoic foraminifera (excluding the lagenides) comprises more than 667 distinct genera, not including synonyms (see Loeblich and Tappan, 1988; Vdovenko et al., 1993; Rauser-Chernousova et al., 1996). In China alone, more than 3395 species belonging to 92 genera of the Fusulinida have so far been described (Jin-Zhang, 1990).

The lagenides and members of the other Palaeozoic orders were minute, measuring only 0.4 to 1.3 mm in diameter. Many of the advanced fusulinides, however, became morphologically large, reaching 1 to 2 cm in length, while some of the Permian forms reached 15 cm in length. These are amongst the largest foraminifera ever to have existed (Douglass, 1977). The larger forms are an exclusively Late Palaeozoic group that achieved their peak diversities during the Visean stage of the Carboniferous and the Cisuralian epoch of the Permian. However, at the end of the Palaeozoic, just as they began to achieve their largest sizes, with extremely complicated internal structures, these forms became extinct.

The taxonomic ranking and division of the fusulinides is controversial, and has been subjected to considerable change over time. In 1937, Wedekind first introduced the name superfamily Fusulinacea in order to represent all the widely spread unique Paleozoic forms. Fursenko (1958) subsequently proposed ranking the fusulinides as a separate order, which was later accepted in *Fundamentals of Paleontology* (Rauser-Chernousova and Fursenko, 1959; Haynes, 1981) and other works of Soviet micropaleontologists. At the same time, in the best known and widely used classification developed by Loeblich and Tappan (1964; 1980; 1987), the fusulinides were regarded as a suborder, and only in 1992 was their rank raised to that of order (Loeblich and Tappan, 1992). However, in the meantime Mikhalevich (1980) had raised the fusulinides to the rank of a superorder, including within it the orders Endothyrida and Fusulinida. The system of superorder was further developed by a team of Russian micropaleontologists headed by Rauser-Chernousova (Rauser-Chernousova et al., 1996), and additionally included the order Tournayellida. Most recently, Vachard et al. (2010) assigned the early foraminifera to the class Fusulinata, and then subdivided this class into six orders: Parathuramminida, Archaediscida and Earlandiida (forming together the subclass Afusulinana n. subcl.), and Tournayellida, Endothyrida and Fusulinida (subclass Fusulinana nom. translat.).

In this chapter, various views are referenced, and an approach is presented which attempts to provide a self-consistent and evidenced based taxonomic and phylogenetic classification. The approach presented here is intended to resolve some of the confusion and complexity surrounding the larger benthic Palaeozoic foraminiferal taxonomy. To elucidate their evolutionary trajectory and to guide their classification, previous work (e.g. Delage and Hérouard, 1896; Hohenegger and Piller, 1973; Mikhalevich, 1980; Leven, 2009; Davydov, 2011; Mikhalevich, 2013) is built upon and synthesized. The wall composition and ultrastructure, the presence or absence of the chomata, and the degree of septal folding in the test are used to define phylogenetic relationships and relative ranking (see Fig. 2.1). It is acknowledged, however, that this is an active area of research which may well develop alternative interpretations in the future as the evidence base grows.

In this the revised approach six orders are separated from the Fusulinida, namely the Parathuramminida, the Palaeotextulariida, the Archaediscida, the Tetrataxida, the Tournayellida and the Endothyrida (see below for phylogenetic description). Thus, in the next section of this chapter the main features of the taxonomy of the following orders are presented:

- Parathuramminida
- Moravamminida
- Archaediscida
- Earlandiida
- Palaeotextulariida
- Tetrataxida
- Tournayellida
- Endothyrida
- Fusulinida
- Lagenida
- Miliolida

This is followed by a discussion of the biostratigraphic significance and phylogenetic evolution of the geologically most important Palaeozoic forms, and the chapter concludes with a review of their palaeoecological significance and their palaeogeographic distribution during the Palaeozoic.

2.2 Morphology and Taxonomy of Palaeozoic Larger Benthic Foraminifera

ORDER PARATHURAMMINIDA MIKHALEVICH 1980
In this order all forms are unilocular, large globular or tubular, and occurred both as free and as attached forms. The wall is thin, calcareous microgranular, simple to bilamellar with an inner hyaline pseudofibrous layer. Apertures are terminal at the top of a hollow neck. Silurian to Permian.

Superfamily PARATHURAMMINOIDEA Bykova, 1955
In this superfamily all forms are unilocular, globular or tubular, and occurred both as free and as attached forms. The wall structure is simple calcareous microgranular or with granulo-fibrous layers (see Fig. 2.2). They range from Early Silurian to Permian.

Family Archaesphaeridae Malakhova, 1966
The representatives of this family mainly have one or more globular to elongate chamber(s), with a test with no apparent aperture. They range from Late Silurian to Early Permian.

* *Archaesphaera* Suleymanov, 1945 (Type species: *Archaesphaera minima* Suleymanov, 1945). The test is globular and smooth. Devonian to Carboniferous (Tournaisian), (Fig. 2.2).
* *Diplosphaerina* Derville, 1952 (Type species: *Diplosphaera inaequalis* Derville, 1931). The two-chambered test, has a small proloculus enveloped within a larger chamber. The wall is dark, granular, nonperforated and single layered. Middle Devonian to Carboniferous (Tournaisian), (Plate 2.1, figs 15, 16; Plate 2.2, Fig. 10; Fig. 2.3).

Family Parathuramminidae Bykova, 1955
The representatives of the Parathuramminidae include globular, or irregular forms with a simple calcareous agglutinated wall and multiple apertures at the end of a tubular neck. They range from Silurian to Carboniferous (Mississippian) and they are the ancestral group of all the fusulinids.

* *Parathurammina* Suleymanov, 1945 (Type species: *Parathurammina dagmarae* Suleymanov, 1945). The test is globular, with apertures at the ends of numerous tubular protuberances. Late Silurian to Carboniferous (Tournaisian) (Plate 2.2, fig.19).

Family Chrysothuramminidae Loeblich and Tappan, 1988
Members of this family have a globular to irregular test in outline, with the aperture at the end of tubular projections. They range from Middle Devonian to Carboniferous (Tournaisian).

- *Chrysothurammina* Neumann, Pozaryska and Vachard, 1975. (Type species: *Chrysothurammina tenuis* Neumann, Pozaryska and Vachard, 1975). The test is composed of a single subspherical chamber with an aperture on a neck-like protrusion. Middle Devonian to Carboniferous (Tournaisian) (Plate 2.3, Fig. 1B).

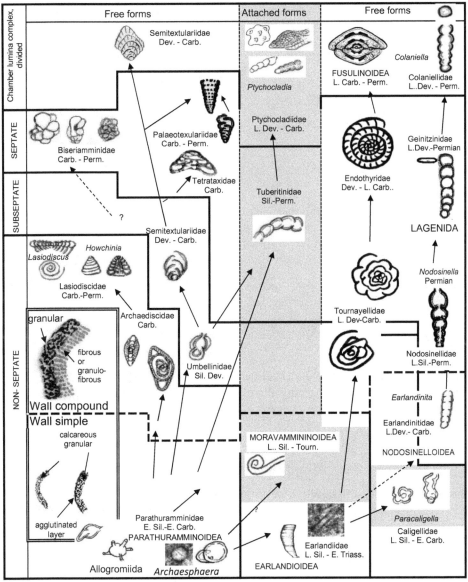

Fig. 2.2. Schematic morphological evolution of the fusulinides and lagenides from the Allogromiida. Figures are not to scale. Not all superfamilies and families discussed in the text are shown on this figure and not all superfamilies names are given (for reasons of simplicity and clarity).

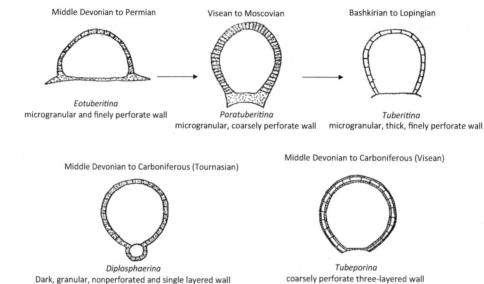

Fig. 2.3.   Schematic morphological evolution of the tuberitinids.

Family Ivanovellidae Chuvashov and Yuferev, 1984
The representatives of this family are spherical with a thick calcareous wall with two layers, a thin dark compact granular inner layer and a thick grey radially fibrous outer layer. This family ranges from Late Silurian to Early Carboniferous (Tournaisian).

• *Elenella* Pronina, 1969 (Type species: *Neoarchaesphaera* (*Elenella*) *multispinosa* Pronina, 1969). The outer surface has tiny projections. There is no aperture. Late Silurian to Carboniferous (Tournaisian) (Plate 2.3, Fig. 2B).

Family Marginaridae Loeblich and Tappan, 1986
The members of this family are globular in shape, with a wall made of three layers, the inner and outer ones are dark and the median one light grey, with canals running through it. These canals open into the chamber cavity and make conical projections outside the wall. This family ranges from the Middle to Late Devonian.

• *Marginara* Pertova, 1984 (Type species: *Parathurammina tamarae* Petrova, 1981). The test is spherical, with some projections from the surface with canals opening through the projections. Middle Devonian.

Family Uralinellidae Chuvashov, Yuferev and Zadorozhnyy, 1984
The Uralinellidae are small, unilocular forms with a three-layered wall, the median layer is thick and clear and the surrounding layers are dark. A multiple apertural neck is projected from the test. They range from Middle Devonian (Givetian) to Early Carboniferous (Visean).

• *Sogdianina* Saltovskaya, 1973 (Type species: *Sogdianina angulata* Saltovskaya, 1973). Similar to *Elenella*, but with a thicker wall and a very prominent apertural neck. Carboniferous (Visean) (Plate 2.3, Fig. 3).

- *Uralinella* Bykova, 1952 (Type species: *Uralinella bicamerata* Bykova, 1952). The test is subglobular with neck-like tubular projections. Middle Devonian to Carboniferous (Givetian to Tournaisian).

Family Auroriidae Loeblich and Tappan, 1986
The Auroriidae have an ovate test with a two-layered wall, a thin inner finely porous dark layer and a thick outer canaliculated dark layer. They range from Middle to Late Devonian.

- *Auroria* Poyarkov, 1969 (Type species: *Auroria singularis* Poyarkov, 1969). The test is irregularly globular. Middle to Late Devonian (Plate 2.3, Fig. 4B).

Family Usloniidae Miklukho-Maklay, 1963
Members of this family have a globular test with no distinct aperture. They range from Middle Devonian to Early Carboniferous.

- *Bisphaera* Birina, 1948 (Type species: *Bisphaera malevkensis* Birina, 1948). The test is subglobular, with a constriction. Middle Devonian to Carboniferous (Givetian to Tournaisian).
- *Parphia* Miklukho-Maklay, 1965 (Type species: *Cribrosphaeroides* (*Parphia*) *robusta* Miklukho-Maklay, 1965). The test is spherical to ovate with a wall that is dark and evenly porous. Late Devonian (Plate 2.3, Fig. 5).

Family Eovolutinidae Loeblich and Tappan, 1986
The representatives of this family have a globular test with a micro-granular wall and a distinct single aperture. They range from Late Silurian to Late Devonian.

- *Eovolutina* Antropov, 1950 (Type species: *Eovolutina elementa* Antropov, 1950). The test is tiny, with the proloculus completely surrounded by the second chamber. Late Silurian to Late Devonian (Plate 2.3, fig.6).

Family Tuberitinidae Miklukho-Maklay, 1958
Members of the Tuberitinidae have a simple attached test, made of one or more globular chamber with calcareous microgranular walls (see Figs 2.2, 2.3). They occur between the Silurian and the Permian.

- *Draffania* Cummings, 1957 (Type species: *Draffania biloba* Cummings, 1957). The test is flask-shaped with an aperture at the end of an elongate neck. Carboniferous (Tournaisian to Visean) (Plate 2.1, figs 6, 10, 13, 17, 18; Plate 2.2, Fig. 16; Plate 2.3, Fig. 5B).
- *Eotuberitina* Miklukho-Maklay, 1958 (Type species: *Eotuberitina reitlingerae* Miklukho-Maklay, 1958). The test is hemispherical with a basal disc, and with a wall that is microgranular and finely perforate. Middle Devonian to Permian (Lopingian) (Plate 2.2, figs 6, 8, 9; Fig. 2.3).
- *Paratuberitina* Miklukho-Maklay, 1957 (Type species: *Tuberitina collosa* Reytlinger, 1950). The test is highly hemispherical with a basal disk, with a microgranular, coarsely perforate wall. Carboniferous (Visean to Moscovian) (Plate 2.2, Fig. 17; Fig. 2.3).

- *Tubeporina* Pronina, 1960 (Type species: *Tubeporina gloriosa* Pronina, 1960). The test is hemispherical with a basal disc, and with a coarsely perforate three-layered wall, a hyaline layer between two microgranular layers. Middle Devonian to Carboniferous (Visean) (Plate 2.2, fig.7; Fig. 2.3).
- *Tuberitina* Galloway and Harlton, 1928 (Type species: *Tuberitina bulbacea* Galloway and Harlton, 1928). Shows bulbous chambers in an arcuate or straight series. The wall is thick and finely perforate. Late Carboniferous to Late Permian (Bashkirian to Lopingian) (Plate 2.2, Fig. 11; Plate 2.3, Fig. 8).

ORDER MORAVAMMINIDA POKORNY, 1951
Members of this order have an attached tubular test, irregularly septate. They may also have closer affinity to algae than foraminifera (Vachard, 1994; Vachard and Cozar, 2010). Late Silurian to Early Carboniferous (Tournaisian).

Superfamily MORAVAMMINOIDEA Pokorny, 1951
This superfamily (Fig. 2.2) had simple free or attached tests, consisting of a proloculus and a rectilinear second chamber that was subseptate (i.e. with partial septa). They had simple walls and evolved from the Parathuramminoidea in the Silurian, but died out in the Early Carboniferous (Tournaisian).

Family Moravamminidae Pokorny, 1951
In this possible foraminiferal family, the test is attached and irregularly septate with an enrolled rectilinear part. They occurred in the Middle to Late Devonian.

ORDER ARCHAEDISCIDA POJARKOV AND SKVORTSOV 1979
The members of this order have free discoidal, lenticular to conical involute or rarely evolute test. It is composed of a proloculus followed directly by a planispiral to tro-chospirally, or streptospirally tubular second chamber. The wall is calcareous, bilay-ered formed of an inner dark microgranular layer and a hyaline and radially fibrous outer layer. They range from the Early Carboniferous to the Late Permian (Visean to Lopingian).

Superfamily ARCHAEDISCOIDEA Cushman, 1928
This superfamily is characterized by having a small, free test, consisting of a proloculus followed by an enrolled chamber. The wall is formed of one or more layers. They range from the Early Carboniferous to the Late Carboniferous (Visean to Moscovian), or is unilayered and pseudofibrous.

Family Archaediscidae Cushman, 1928
The Archaediscidae (Fig. 2.2) have a discoidal test, composed of a proloculus followed by a streptospirally enrolled second chamber. The wall is formed of a dark inner gran-ular layer and a clear radially fibrous or granulo-fibrous outer layer. They range from the Early Carboniferous (Visean) to the Late Carboniferous (Moscovian).

- *Archaediscus* Brady, 1873 (Type species: *Archaediscus karreri* Brady, 1873). The test is free, streptospiral, with an undivided tubular second chamber and thickened wall.

Carboniferous (Visean to Moscovian) (Plate 2.3, fig. 6; Plate 2.4, fig. 17C; Plate 2.5, fig. 18B; Plate 2.6, figs 3-16; Plate 2.7, fig. 6; fig 2.5).

• *Asteroachaediscus* Miklukho-Maklay, 1956 (Type species: *Archaediscus baschkiricus* Krestovnikov and Teodorovitch, 1936). The test is small, lenticular with sigmoidal coiling, stellate and with occluded chamber cavity. Carboniferous (Visean to Moscovian) (Plate 2.7, Fig. 3; Plate 2.8, Fig. 12A).

• *Glomodiscus* Malakhova, 1973 (Type species: *Glomodiscus biarmicus* Malakhova, 1973). The test is discoidal, with a globular proloculus followed by an involutely coiled, undivided tubular second chamber. Carboniferous (Visean).

• *Hemiarchaediscus* Miklukho-Maklay, 1957 (Type species: *Hemiarchaediscus planus* Miklukho-Maklay, 1957). The test is small, discoidal with a glomospiral early whorl, later becoming planispiral; chamber lumen are always open. Carboniferous (late Visean to early Serpukhovian) (Plate 2.7, figs 1, 2).

• *Permodiscus* Dutkevich, 1948 (Type species: *Permodiscus vetustus* Dutkevich, in Chernysheva, 1948). The test is planispirally enrolled with a tubular undivided second chamber, later whorls have small nodosities partially filling the chamber lumen. Carboniferous (Visean to Bashkirian).

• *Planoarchaediscus* Miklukho-Maklay, 1957 (Type species: *Archaediscus spirillinoides* Rauzer-Chernousova, 1948). The test is small, discoidal with a glomospiral early stage, followed by a planispiral adult. Chamber lumen remain open. Carboniferous (Visean to Serpukhovian) (Plate 2.7, Fig. 5).

• *Propermodiscus* Miklukho-Maklay, 1953 (Type species: *Hemigordius ulmeri* Mikhaylov, 1939). The test is small, lenticular with glomospiral early whorls and a planispiral, involute adult. Carboniferous (Visean) (Plate 2.7, Fig. 4).

Superfamily LASIODISCOIDEA Reitlinger in Vdovenko et al., 1993

The members of this superfamily have a bilayered wall with finely granular dark inner layer and a radially fibrous outer layer. Tubercles of pseudofibrous fillings or pillars fill the umbilical region. The aperture is simple and terminal, but additional supplementary apertures may occur along the spiral sutures of the successive whorls. The range is between the Early Carboniferous (Visean) and the Permian.

Family Lasiodiscidae Reytlinger, 1956

In this family, the test is discoidal to conical, composed of a proloculus and an undivided enrolled tubular second chamber. The radial fibrous outer layer is mainly concentrated in the umbilical region, where it may form pillars, or a series of tubercles on the surface of one side of the test. The range is between the Early Carboniferous (Visean) and the Permian.

• *Eolasiodiscus* Reytlinger, 1956 (Type species: *Eolasiodiscus donbassicus* Reytlinger, 1956). The test is discoidal and concavo-convex, with the umbilical filling on the concave side. Middle to Late Carboniferous.

• *Glomotrocholina* Nikitina, 1977 (Type species: *Glomotrocholina pojarkovi* Nikitina, 1977). The test is conical. Second chamber is initially streptospirally enrolled, later forming an irregular trochospiral filled with shell material. Late Permian (Lopingian).

- *Howchinia* Cushman, 1927 (Type species: *Patellina bradyana* Howchin, 1888). The test is conical with septal bridges. The umbilical region is filled with fibrous calcite forming pillars (Fig. 2.2). Carboniferous (Visean to Moscovian) (Fig. 2.2; Fig. 2.22; Plate 2.1, Fig. 11A; Plate 2.5, Fig. 1; Plate 2.10, figs 1-7, 13-14).
- *Lasiodiscus* Reichel, 1946 (Type species: *Lasiodiscus granifer* Reichel, 1946). The test is planispiral with tubular extensions. Middle Carboniferous to Late Permian (Fig. 2.2).
- *Lasiotrochus* Reichel, 1946 (Type species: *Lasiotrochus tatoiensis* Reichel, 1946). The test is conical with tubular chamberlets curved towards the proloculus, with distinct pillars filling the centre of the test. Permian (Cisuralian to Lopingian).
- *Monotaxinoides* Brazhnikova and Yartseva, 1956 (Type species: *Monotaxinoides transitorius* Brazhnikova and Yartseva, 1956). The test is low conical to nearly planispiral, with a spherical proloculus followed by a semi-cylindrical second chamber. Carboniferous.

ORDER EARLANDIIDA SABIROV IN VDOVENKO ET AL., 1993 EMEND. VACHARD ET AL., 2010
This order had simple free or attached tests, consisting of a proloculus and a rectilinear second chamber. Aperture terminal, simple. Late Silurian to Early Triassic.

Superfamily EARLANDIOIDEA Cummings, 1955
This superfamily is characterised by having a free, non-septate test with a globular first chamber and a straight tubular second chamber. Members range from the Late Silurian to Early Triassic.

Family Earlandiidae Cummings, 1955
The Earlandiidae have a single free chamber (Fig. 2.2) and range from the Late Silurian to Early Triassic.

- *Aeolisaccus* Elliott, 1958 (Type species: *Aeolisaccus dunningtoni* Elliott, 1958). Free test, elongate and slightly tapering. Late Permian (Lopingian).
- *Earlandia* Plummer, 1930 (Type species: *Earlandia perparva* Plummer, 1930). Free, elongate test, composed of a globular proloculus followed by an undivided straight tubular chamber. The wall is calcareous microgranular. Late Silurian to Early Triassic (Plate 2.1, figs 1-5; Plate 2.3, Fig. 1A; Plate 3.2, fig.3).
- *Gigasbia* Strank, 1983 (Type species: *Gigasbia gigas* Strank, 1983). The test is free and elongate, consisting of a globular proloculus and an undivided tubular chamber. Early Carboniferous (Tournaisian).

Family Endotebidae Vachard, Martini, Rettori and Zaninetti, 1994
The test is free, planispiral, in early stages, but later uniserial to biserial. The walls are grey, thick, calcareous agglutinated, and the apertures are simple. Late Permian to Triassic.

- *Endoteba* Vachard and Razgallah, 1988 emend. Vachard et al. 1994 (Type species: *Endoteba controversa* Vachard et al., 1994). Axial view compressed. Late Permian to Late Triassic (Late Kungurian to Rhaetian) (Fig. 3.5).

Family Pseudoammodiscidae Conil and Lys, 1970
The Pseudoammodiscidae are small, with a simple aperture and globular proloculus followed by a planispiral, trochospiral or streptospirally coiled, undivided second chamber. They occur from the Devonian to the Permian.

- *Brunsia* Mikhaylov, 1935 (Type species: *Spirillina irregularis* von Möller, 1879). The proloculus is followed by an early streptospiral coil, and later a planispirally enrolled tubular chamber. Devonian to Carboniferous (Tournaisian) (Plate 2.3, Fig. 9).
- *Brunsiella* Reytlinger, 1950 (Type species: *Glomospira ammodiscoidea* Rauzer-Chernousova, 1938). The globular proloculus is followed by an undivided second chamber. Early whorls are streptospiral, later ones planispiral and evolute. Carboniferous (Visean) to Early Permian (Plate 2.2, Fig. 15B; Plate 2.3, figs 9,11,12; Plate 2.11, figs 7-9).

Family Pseudolituotubidae Conil and Longerstaey, 1980
The Pseudolituotubidae have attached tests, consisting of a single enrolled chamber with a simple terminal aperture and a compound calcareous microgranular wall. They range from the Early Carboniferous (Visean) to the Late Carboniferous (Moscovian).

- *Pseudolituotuba* Vdovenko, 1971 (Type species: *Lituotuba*? *gravata* Conil and Lys, 1965). The proloculus is followed by an undivided streptospirally coiled tube. Carboniferous (Visean to Moscovian).

Superfamily CALIGELLOIDEA Reytlinger 1959
This superfamily (Fig. 2.2) had simple free or attached tests, consisting of a proloculus and a rectilinear second chamber that was subseptate (i.e. with partial septa). They had simple walls and evolved from the Parathuramminoidea in the Late Silurian, but died out in the Early Carboniferous (Tournaisian).

Family Caligellidae Reytlinger, 1959
In this Family, the test is attached and subseptate. They ranged from the Late Silurian to the Early Carboniferous (Tournaisian).

- *Paracaligella* Lipina, 1955 (Type species: *Paracaligella antropovi* Lipina, 1955). The test is composed of a sub-spherical proloculus followed by an irregular tubular chamber partially divided by incipient septa. Devonian to Carboniferous (Late Tournaisian) (see Fig. 2.2).

Family Paratikhinellidae Loeblich and Tappan, 1984
This family was free with a elongate, sub-septate test and with a simple microgranular wall. They ranged from Middle Devonian to Carboniferous (Tournaisian).

- *Paratikhinella* Reytlinger, 1954 (Type species: *Tikhinella cannula* Bykova, 1952). The test is elongate and composed of a subglobular proloculus followed by a cylindrical tubular chamber with incipient septation. Late Devonian to Early Carboniferous (Plate 2.4, Fig. 5).

- *Saccaminopsis* Sollas, 1921 (Type species: *Saccammina carteri* Brady, 1871). The test is free, uniserial with ovate chambers, a thin wall and a terminal aperture. Late Devonian to Carboniferous (Tournaisian). (Plate 2.1, Fig. 11B; Plate 2.2, figs 12-15).

Superfamily PTYCHOCLADIOIDEA Elias, 1950
These were attached forms with no distinct apertures. Their walls were microgranular, banded with transverse tubuli. They range from Late Devonian to Late Carboniferous.

Family Ptychocladiidae Elias, 1950
Here the test is uniserial and attached. They ranged from the Late Devonian to Late Carboniferous (Pennsylvanian).

- *Ptychocladia* Ulrich and Bassler, 1904 (Type species: *Ptychocladia agellus* Ulrich and Bassler, 1904). The wall is calcareous microgranular, with two layers with no distinct aperture. Carboniferous (Late Pennsylvanian) (see Fig. 2.2).

ORDER PALAEOTEXTULARIIDA HOHENEGGER AND PILLER, 1975
Members of this order have biserial or uniserial tests, with a microgranular calcareous wall, commonly with an inner radial fibrous layer and a finely granular outer layer. The aperture is generally single but may be multiple in later stages. Their range is from Late Devonian to Permian.

Superfamily PALAEOTEXTULARIOIDEA Galloway, 1933
This superfamily has a test that is biserial or uniserial, with a single or two-layered microgranular calcareous wall. The aperture is generally single but may be multiple in later stages. Their range is from Late Devonian to Permian.

Family Semitextulariidae Pokorny, 1956
Here the test is biserial and flattened, but often becoming monoserial with broad chambers, and is fully septate (see Fig. 2.2). They range from the Devonian to the Carboniferous (Pennsylvanian).

- *Koskinobigenerina* Eickhoff, 1968 (Type species: *Koskinobigenerina breviseptata* Eickhoff, 1968). The test is elongate, biserial in the early stage, later becoming uniserial. Early Carboniferous to Late Carboniferous (Visean to Gzhelian).
- *Koskinotextularia* Eickhoff, 1968 (Type species: *Koskinotextularia cribriformis* Eickhoff, 1968). The test is elongate, biserial throughout with a single layered wall. The aperture is single in the early stage but later becoming cribrate. Early Carboniferous to Late Carboniferous (Visean to Gzhelian) (Plate 2.4, Fig. 6).

Family Palaeotextulariidae Galloway, 1933
This family (see Figs 2.2 and 2.3) has biserial to uniserial genera that closely resembles the Textulariidae, but they have a dark granular calcareous outer wall and an inner clear to yellowish "fibrous" layer with stacks of granules perpendicular to the surface. Cummings (1956) demonstrated that these forms evolved from simple agglutinated forms in the Devonian (see Fig. 2.4). Early Carboniferous (Tournaisian) to Permian (Lopingian).

- *Climacammina* Brady, 1873 (Type species: *Textularia antiqua* Brady, in Young and Armstrong, 1871). The test has a biserial early stage followed by a uniserial stage. The aperture in the early stage is at the base of the last chamber, later becoming areal, multiple and cribrate. Carboniferous to Permian (Visean to Early Lopingian) (Plate 2.4, Fig. 7; Plate 2.5, figs 7, 9-10).

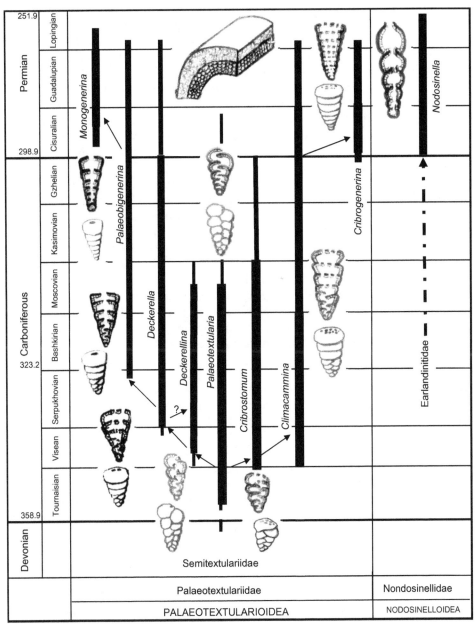

Fig. 2.4.   Evolution of the Palaeotextulariida (modified from Cummings 1956).

- *Cribrogenerina* Schubert 1908 (Type species: *Bigenerina sumatrana* Volz, 1904). The test has a short early biserial stage followed by a uniserial stage. The aperture is a slit in the early stage, later becoming areal, multiple and cribrate in the uniserial stage. Carboniferous to Permian (latest Gzhelian to Lopingian) (see Fig. 2.4).
- *Cribrostomum* von Möller, 1879 (Type species: *Cribrostomum textulariforme* von MöIler, 1879). The test is biserial throughout. The aperture in the early part is basal, but in the last two whorls becomes areal, multiple and cribrate. Carboniferous (late Tournaisian to Gzhelian) (Plate 2.4, figs 1, 5, 13, 15; see Figs 2.2, 2.4).
- *Deckerella* Cushman and Waters, 1928 (Type species: *Deckerella clavata* Cushman and Waters, 1928). The test is biserial in the early stage, later becoming uniserial and rectilinear. The aperture is a low slit in the early stage, later becoming terminal with two parallel slits separated by a partition. Carboniferous to Permian (late Visean to early Lopingian) (Plate 2.4, Fig. 8; see Fig. 2.4).
- *Deckerellina* Reytlinger, 1950 (Type species: *Deckerellina istiensis* Reytlinger, 1950). The test is biserial throughout. The aperture, in the early stage, is a low slit, but the adult aperture is two parallel slits. Carboniferous (Visean to Moscovian) (Plate 2.4, Fig. 11, 16; see Fig. 2.4).
- *Monogenerina* Spandel, 1901 (Type species: *Type species*: *Monogenerina atava* Spandel, 1901). The test is biserial in the early stage, later becoming uniserial. The aperture extends throughout the test. Permian (Cisuralian to Lopingian) (see Fig. 2.4).
- *Palaeobigenerina* Galloway, 1933 (Type species: *Bigenerina geyeri* Schellwien, 1898). The test is biserial in the early stage, but later uniserial, with a slit opening in the biserial stage, followed by a single rounded aperture. Carboniferous to Permian (late Serpukhovian to early Lopingian) (see Fig. 2.4).
- *Palaeotextularia* Schubert, 1921 (Type species: *Palaeotextularia schellwieni* Galloway and Ryniker, 1930). The test is biserial throughout. The wall is thick, and the aperture is a low slit opening at the base of the final chamber. Carboniferous to Permian (Tournaisian to Cisuralian) (Fig. 2.4; Plate 2.4, figs 2-4, 9-10, 12, 14, 17-18; Plate 2.5, Fig. 6).

Family Biseriamminidae Chernysheva, 1941

In this family (Fig. 2.2), the test is biserial becoming planar in later stages, with a microgranular wall with one or more layers. They range from the Carboniferous (Tournaisian) to Permian (Lopingian).

- *Biseriammina* Cherhysheva, 1941 (Type species: *Biseriammina uralica* Cherhysheva, 1941). The test is planispirally enrolled. Early Carboniferous (Tournaisian).
- *Biseriella* Mamet, 1974 (Type species: *Globivalvulina parva* Chernysheva, 1948). The early stage is tightly coiled, while later it has an open helicoid spire. Late Carboniferous (Bashkirian to Moscovian) (Plate 2.3, Fig. 10).
- *Dagmarita* Reytlinger, 1965 (Type species: *Dagmarita chanakchiensis* Reyltinger, 1965). The test is flattened with spine-like projections at the outer corners of the angular chambers. Late Permian (Lopingian).
- *Globispiroplectammina* Vachard, 1977 (Type species: *Globispiroplectammina mameti* Vachard, 1977). The test is biserial with the early stage enrolled, later becoming uncoiled. Carboniferous (Visean).

- *Globivalvulina* Schubert, 1921 (Type species: *Valvulina bulloides* Brady, 1876). The test is planispirally coiled. Middle Carboniferous (Bashkirian) to Late Permian (Wuchiapingian).
- *Lipinella* Malakhova, 1975 (Type species: *Lipinella notata* Malakhova, 1975). The test is planispirally enrolled following a globular proloculus. Middle Carboniferous to Late Permian (Bashkirian to Lopingian).
- *Louisettita* Altiner and Brönnimann, 1980 (Type species: *Louisettita elegantissima* Altiner and Brönnimann, 1980). The test is trochospirally enrolled in the early stage. In the biserial stage, chambers are subdivided by vertical partitions perpendicular to the septa and have spine-like projections. Late Permian (Wuchiapingian).
- *Paraglobivalvulina* Reytlinger, 1965 (Type species: *Paraglobivalvulina mira* Reytlinger, 1965). The test is spherical with the biserial chambers enrolled trochospirally. Late Permian (Wuchiapingian).
- *Paradagmarita* Lys, 1978 (Type species: *Paradagmarita monodi* Lys, 1978). The test is small with the early stage completely enrolled, but the later stage is uncoiled. Late Permian (Wuchiapingian).
- *Paraglobivalvulinoides* Zaninetti and Jenny-Deshusses, 1985 (Type species: *Paraglobivalvulina? septulifera* Zaninetti and Jenny-Deshusses, 1981). The test is globular with biserially, enrolled, strongly enveloping chambers. The wall is calcareous, microgranular, a single layer. The aperture is rimmed with a well-developed tongue that bends inward. Late Permian (Wuchiapingian).

ORDER TETRATAXIDA MIKHALEVICH 1981
Members of this order have a conical, trochospiral test with an evolute spiral side and involute umbilical side. The wall is microgranular calcareous with one or two distinct layers. Early Carboniferous to Late Permian (Tournaisian to Wuchiapingian).

Superfamily TETRATAXOIDEA Galloway, 1933
The Tetrataxoidea have a conical to spreading test with secondary internal partitions. The walls are microgranular with one or two layers. They range from the Early Carboniferous to Late Permian (Tournaisian to Wuchiapingian).

Family Tetrataxidae Galloway, 1933
The test of the Tetrataxidae is conical leaving an open central umbilicus where the chambers are partially overlapping (Fig. 2.2). The wall is calcareous, microgranular, and in two layers. They occur from the Early Carboniferous to the Late Carboniferous (Tournaisian to Moscovian).

- *Globotetrataxis* Brazhnikova, 1983 (Type species: *Tetrataxis* (*Globotetrataxis*) *elegantula* Brazhnikova, in Brazhnikova and Vdovenko, 1983). The test enlarges gradually with a last large hemispherical umbilical chamber that forms a convex base. Early Carboniferous (late Visean).
- *Polytaxis* Cushman and Waters, 1928 (Type species: *Polytaxis laheei* Cushman and Waters, 1928). The test is low conical. Later chambers form many whorls and result in a spreading test. Late Carboniferous (Moscovian).

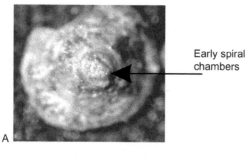

Early spiral
chambers

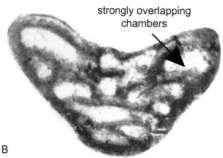

strongly overlapping
chambers

Fig. 2.5. Main morphological features of *Tetrataxis*. A) The early spires of a solid specimen, B) An oblique thin section through the axis of the test.

- *Tetrataxis* Ehrenberg, 1854 (Type species: *Tetrataxis conica* Ehrenberg, 1854). The test is circular in plan with chambers strongly overlapping on the umbilical side. The outer layer is dark and microgranular and the inner layer is light and fibrous. Carboniferous (late Tournaisian to Visean) (Fig. 2.5; Plate 2.2, figs 1, 2, 5; Plate 2.4, Fig. 17B; Plate 2.7, figs 8-18).

Family Pseudotaxidae Mamet, 1974
In this family, the wall is calcareous, microgranular and single layered. They range from Early Carboniferous to Late Carboniferous (Late Tournaisian to Early Bashkirian).

- *Pseudotaxis* Mamet, 1974 (Type species: *Tetrataxis eominima* Rauzer-Chernousova, 1948). The test is conical with the proloculus followed by irregular trochospiral coil. Early Carboniferous to Late Carboniferous (late Tournaisian to Early Bashkirian).
- *Vissariotaxis* Mamet, 1970 (Type species: *Monotaxis exilis* Vissarionova, 1948). The test is conical with trochospirally coiled chambers and a wide aperture. Carboniferous (early Bashkirian) (Plate 2.10, figs 8-12).

Family Valvulinellidae Loeblich and Tappan, 1984
Here the test is conical with subdivided chambers and a single layer microgranular wall. The range is from the Early to Late Carboniferous (Visean to early Bashkirian).

- *Valvulinella* Schubert, 1908 (Type species: *Valvulina youngi* Brady, 1876). Chambers are subdivided by numerous vertical pillars and one or two horizontal ones. Carboniferous (Visean to early Bashkirian). (Plate 2.2, figs 3-4; Plate 2.9, fig. 4; Fig. 2.22C).

Family Abadehellidae Loeblich and Tappan, 1984

Representatives of this family differ from the Valvulinellidae in having two double-layered walls instead of a single layered wall. They occur in the Late Permian (Wuchiapingian).

- *Abadehella* Okimura and Ishii, 1975 (Type species: *Abadehella tarazi* Okimura and Ishi, in Okimura et al., 1975). The test is conical, with numerous whorls. Chambers are subdivided by regularly spaced pillars. Late Permian (Wuchiapingian) (Plate 2.3, Fig. 12).

ORDER TOURNAYELLIDA HOHENEGGER AND PILLER 1973

The representatives of this order have walls that range from homogeneously calcareous microgranular to those that are differentiated into two or more layers in the most advanced forms. Late Devonian to Carboniferous.

Superfamily TOURNAYELLOIDEA Dain, 1953

In this superfamily, the proloculus is followed by a planispiral or streptospiral non-septate to fully septate test. The wall is made of microgranular calcite evolving into forms with a thick outer layer, called a tectum, and an inner granulo-fibrous layer. Late Devonian to Carboniferous (Frasnian to Early Bashkirian).

Family Tournayellidae Glaessner, 1945

The Tournayellidae are small, with early streptospiral or trochospiral coiling, tending to become planispiral in the adult. The aperture is simple and open at the end of a tubular chamber. The Tournayellidae evolved in the Devonian from simple planispiral, non-septate forms (e.g. *Eotournayella;* see Fig. 2.6) with an initial proloculus and an aperture at the end of an uncoiled tube, to forms with a compound wall and rudimentary septa in the Late Tournaisian (e.g. *Tournayella* Dain, 1953, see Fig. 2.6; *Carbonella* Dain, 1953, see Fig. 2.6; Plate 2.9, Fig. 14) to almost complete septa in the late Visean (*Mstinia* Dain, Plate 2.9, figs 8, 10). The streptospiral tubular forms (*Glomospiranella* Lipina, 1951, see Fig. 2.6) gradually also developed rudimentary septa in the Late Tournaisian (e.g. *Brunsiina* Lipina, 1953, see Fig. 2.6). The Tournayellidae range from the Devonian to the Carboniferous.

- *Bogushella* Conil and Lys, 1977 (Type species: *Mistinia ziganensis* Grozdilova and Lebedeva, 1960). The test is streptospirally enrolled in the early stage, later becoming rectilinear with no true septa. The aperture is cribrate. Early Carboniferous (Visean).
- *Brunsiina* Lipina, 1953 (Type species: *Brunsiina uralica* Lipina, 1953). The test is streptospirally enrolled in early stage, later becoming planispiral, without rudimentary constrictions. Late Devonian to Early Carboniferous (Fammenian to Visean) (Fig. 2.6).

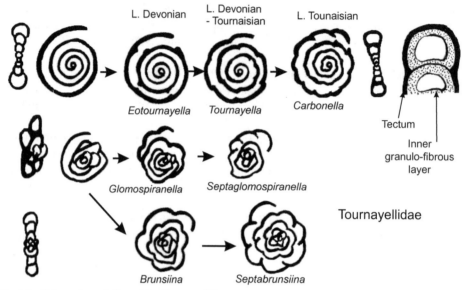

Fig. 2.6.   Schematic evolution of the Tournayellidae.

- *Carbonella* Dain, 1953 (Type species: *Carbonella spectabilis* Dain, 1953). The test is planispiral with septa that are well developed only in the final whorl. Early Carboniferous (Late Tournaisian to Early Visean) (Fig. 2.6; Plate 2.9, Fig. 14).
- *Conilites* Vdovenko, 1970 (Type species: *Ammobaculites? Dinantii* Conil and Lys, 1964). The test is planispirally enrolled and undivided in the early undivided part, but the later part is uncoiled rectilinear with distinct septa. The aperture is multiple and cribrate. Late Carboniferous (Late Tournaisian to Early Visean).
- *Costayella* Conil and Lys, 1977 (Type species: *Tournayella costata* Lipina, 1955). The test has an early part that is streptospiral, but is later planispiral, with later whorls having slight constrictions opposite basal supplementary deposits. Early Carboniferous (Tournaisian to Visean).
- *Chernobaculites* Conil and Lys, 1977 (Type species: *Ammobaculites sarbaicus* Malakhova subsp. *beschevensis* Brazhnikova, in Brazhnikova et al., 1967). The test is streptospirally enrolled in the early part, later becoming rectilinear with straight and horizontal septa. Late Carboniferous (Bashkirian).
- *Chernyshinella* Lipina, 1955 (Type species: *Endothyra glomiformis* Lipina, 1948). The test is streptospirally enrolled throughout with a tubular undivided early portion, and a final whorl with strongly asymmetrical and teardrop-shaped chambers. Late Devonian to Early Carboniferous (Late Famennian to Visean).
- *Chernyshinellina* Reytlinger, 1959 (Type species: *Ammobaculites? pygmaeus* Malakhova, 1954). The test is streptospirally enrolled in the early stage and has teardrop-like chambers in the later whorl, which are separated by distinct horizontal septa. Carboniferous (Tournaisian).
- *Condrustella* Conil and Longerstaey, 1977 (Type species: *"Mstinia" modavensis* Conil and Lys, 1967). The test is streptospirally enrolled in the early part, later becoming planispiral with teardrop-shaped chambers. Early Carboniferous (Late Tournaisian to Visean).

- *Elbanaia* Conil and Marchant, 1977 (Type species: *Plectogyra michoti* Conil and Lys, 1964). The test is streptospirally enrolled in the early part, but later evolute and septate. Early Carboniferous (Late Tournaisian to Early Visean).
- *Endochernella* Conil and Lys, 1980 (Type species: *Plectogyra* (*Latiendothyra*) *quaesita* Ganelina, 1966). The test is streptospirally enrolled in the early part, but later with radial septa and numerous chambers in the whorl. Early Carboniferous (Early Tournaisian to Visean).
- *Eoforschia* Mamet, 1970 (Type species: *Tournayella moelleri* Malakhova, in Dain and Grozdilova, 1953). The test is planispirally enrolled lacking definite septa. Early Carboniferous (Tournaisian to Visean).
- *Eotournayella* Lipina and Pronina, 1964 (Type species *Tournayella* (*Eotournayella*) *jubra* Lipina and Pronina, 1964). The test is streptospiral to planispiral in later stages, with slight growth constrictions. Late Devonian (Fig. 2.6).
- *Forschia* Mikhaylov, 1935 (Type species: *Spirillina subangulata* von Möller, 1879). The test is evolutely coiled without true septa. The aperture is cribrate. Early Carboniferous (Visean).
- *Forschiella* Mikhaylov, 1935 (Type species: *Forschiella prisca* Mikhaylov, 1935). The test is evolutely coiled without true septa in the early stage, later it is uncoiled. The aperture is cribrate. Early Carboniferous (Visean).
- *Glomospiranella* Lipina, 1951 (Type species: *Glomospiranella Asiatica* Lipina, 1951). The test is streptospiral in the early stage, later becoming planispiral, without distinct septation, but with slight constrictions. Late Devonian to Early Carboniferous (Visean) (Fig. 2.6).
- *Glomospiroides* Reytlinger, 1950 (Type species: *Glomospiroides fursenkoi* Reytlinger, 1950). The test is streptospirally enrolled, being undivided in the early stage, but becoming rectilinear and irregularly divided with thin pseudosepta. Middle to Late Carboniferous.
- *Laxoseptabrunsiina* Vachard, 1977 (Type species: *Laxoseptabrunsiina valuzierensis* Vachard,1977). The test is streptospirally enrolled and undivided, later becoming planispiral with true septa. Early Carboniferous (Visean).
- *Lituotubella* Rauzer-Chernousova, 1948 (Type species: *Lituotubella glomospiroides* Rauzer-Chernousova, 1948). The test is streptospirally coiled in the early stage, later becoming rectilinear with no true septa. Early Carboniferous (Visean).
- *Mstinia* Dain, 1953 (Type species: *Mstinia bulloides* Dain, in Dain and Grozdilova, 1953). The test is involute, with teardrop-like chambers. The aperture is cribrate. Early Carboniferous (Late Visean).
- *Mstiniella* Conil and Lys, 1977 (Type species: *Mstinia fursenkoi* Dain, in Dain and Grozdilova, 1953). The test is streptospirally, undivided in the early part, but later planispiral with septa. The aperture is cribrate. Early Carboniferous (Visean) (Plate 2.9, figs 8, 10).
- *Neobrunsiina* Lipina, 1965 (Type species: *Glomospiranella finitima* Grozdilova and Lebedeva, 1954). The test is streptospirally enrolled in the early stage, but later planispiral with constrictions in the wall. Early Carboniferous (Tournaisian to Visean).
- *Nevillea* Conil and Lys, 1980 (Type species: *Georgella dytica* Conil and Lys, 1977). The test is streptospirally enrolled in the early stage, but later rectilinear, divided by arched septa. The aperture is multiple with a cribrate aperture occupying the entire terminal chamber surface. Early Carboniferous (Visean).

- *Nodochernyshinella* Conil and Lys, 1977 (Type species: *Chernyshinella tumulosa* Lipina, 1955). The test is streptospirally enrolled in the early part, but later planispiral with teardrop-like chambers. Early Carboniferous (Tournaisian).
- *Pohlia* Conil and Lys, 1977 (type species: *Septatournayella henbesti* Skipp, in Skipp et al., 1966). The test is planispirally enrolled with distinct septa in the final whorl. Early Carboniferous (Visean).
- *Pseudolituotubella* Vdovenko, 1967 (Type species: *Pseudolituotubella multicamerata* Vdovenko, 1967). The test is streptospirally enrolled in the early stage, but later is planispiral becoming rectilinear with true septa. The aperture is cribrate. Early Carboniferous (Late Tournaisian to Early Visean).
- *Rectoseptatournayella* Brazhnikova and Rostovceva, 1963 (Type species: *Rectoseptatournayella stylaensis* Brazhnikova and Rostovceva, 1963). The test is planispirally coiled in the early stage, but uncoiled in the final stage with septa and distinct chambers. Early Carboniferous (Tournaisian).
- *Rectotournayellina* Lipina, 1965 (Type species: *Tournayellina* (*Rectotournayellina*) Lipina, 1965). The test is streptospirally enrolled in the early part, but later rectilinear. Late Devonian to Early Carboniferous (Famennian to Tournaisian).
- *Septabrunsiina* Lipina, 1955 (Type species: *Endothyra*? *Krainica* Lipina, 1948). The test is streptospirally enrolled in the early nonseptate stage, but later it is planispiral with distinct septa. Early Carboniferous (Tournaisian) (Fig. 2.6).
- *Septaforschia* Conil and Lys, 1977 (Type species: *Tournayella questita* Malakhova, in Dain and Grozdilova, 1953). The test is enrolled, with later stages having true septa. Early Carboniferous (Tournaisian).
- *Septaglomospiranella* Lipina, 1955 (Type species: *Endothyra primaeva* Rauzer-Chernousova, 1948) Early Carboniferous (Tournaisian) (Fig. 2.6). Test with a non-septate early stage later becoming planispiral and septate.
- *Septatournayella* Lipina, 1955 (Type species: *Tournayella segmentata* Dain, in Dain and Grozdilova, 1953). The test is a planispirally enrolled tube, with slight constrictions in the early stage, but later with complete septa forming distinct chambers. Late Devonian (Famennian) to Early Carboniferous (Visean).
- *Spinobrunsiina* Conil and Longerstaey, 1980 (Type species: *Septabrunsiina* (*Spinobrunsiina*) *ramsbottomi* Conil and Longerstaey, 1980). The test is streptospirally enrolled in the early stage, but later becoming planispiral and septate. Early Carboniferous (Tournaisian to Visean).
- *Spinolaxina* Conil and Naum, 1977 (Type species: *Plectogyra pauli* Conil and Lys, 1964). The test is streptospirally coiled in the early stage, later becoming nearly planispiral. Early Carboniferous (Visean).
- *Spinotournayella* Mamet, 1970 (Type species: *Plectogyra tumula* Zeller, 1957). The test is involute with early parts being streptospiral, but later becoming planispiral with endothyroid chambers. Early Carboniferous (Late Tournaisian).
- *Tournayella* Dain, 1953 (Type species: *Tournayella discoidea* Dain, 1953). The test is enrolled without definite septa and supplementary deposits. Early Carboniferous (Tournaisian to Visean) (Fig. 2.5).
- *Tournayellina* Lipina, 1955 (Type species: *Tournayellina vulgaris* Lipina, 1955). The test is enrolled with few chambers enlarging rapidly. Late Devonian to Early Carboniferous (Famennian to Tournaisian).

- *Uviella* Ganelina, 1966 (Type species: *Uviella aborigena* Ganelina, 1966). The test is streptospirally enrolled in the early undivided stage, but later is planispiral with constrictions in the early part followed in the final whorl by true septa. Early Carboniferous (Tournaisian to Visean).
- *Viseina* Conil and Lys, 1977 (Type species: *Septatournayella? Conspecta* Conil and Lys, 1977). The test is enrolled with later chambers divided by true septa. The aperture is cribrate in the final stage. Early Carboniferous (Visean).

Family Palaeospiroplectamminidae Loeblich and Tappan, 1984
The members of this family represent a biserial development from the Tournayellidae, and may be ancestral to the Palaeotextularioidea. They are initially streptospiral with teardrop-like or endothyroid-like chambers, then become planispiral and in later stages the test become biserial. The range is Late Devonian to Early Carboniferous.

- *Endospiroplectammina* Lippina, 1970 (Type species: *Spiroplectammina venusta* Vdovenko, 1954). The test is streptospirally enrolled, with distinctly endothyroid chambers, later becoming planispiral, then uncoiled with a later biserial stage with a single aperture. Early Carboniferous (Late Tournaisian to Middle Visean).
- *Eotextularia* Mamet, 1970 (Type species: *Palaeotextularia diversa* Chernysheva, 1948). The test is irregularly coiled in the early part, later becoming uncoiled with a few pairs of biserial chambers. Early Carboniferous (Late Tournaisian to Middle Visean).
- *Halenia* Conil, 1980 (Type species: *Halenia legrandi* Conil, 1980). The test has a biserial elongate stage, followed by a broad uniserial stage with horizontal septa. The aperture is cribrate in the adult. Early Carboniferous (Visean).
- *Palaeospiroplectammina* Lipina, 1965 (Type species: *Spiroplectammina tchernyshinensis* Lipina, 1948). The test has a biserial uncoiled stage with curved septa, but may become uniserial with last chambers. Late Devonian to Carboniferous (Early Visean) (Plate 2.3, Fig. 11).
- *Rectochernyshinella* Lipina, 1960 (Type species: *Spiroplectammina mirabilis* Lipina, 1948). The test is streptospirally enrolled in a large early stage, the later stage uncoiled with few biserially arranged chambers. Late Devonian to Early Carboniferous (Late Famennian to Early Tournaisian).

ORDER ENDOTHYRIDA FURSENKO 1958
Members of this order have a lenticular test, planispirally coiled. The wall is dark microgranular, but sometimes can be bilayered or multilayered. Aperture simple, basal or cribrate. Late Devonian to Triassic.

Superfamily ENDOTHYROIDEA Brady, 1884 nom. translat. Fursenko, 1958
Members of this superfamily have typical endothyroid coiling with streptospiral to planispiral tests and constant deviation of the axis of coiling, with many chambers followed by a rectilinear stage in some forms (see Fig. 2.9). The walls are microgranular and calcareous, but some forms evolved two to three distinct layers, others may develop an inner perforate or keriothecal layer. Notably, they ranged from the Late Devonian to the Triassic (Famennian to Rhaetian).

Family Endotebidae Vachard, Martini, Rettori and Zaninetti, 1994
Test is free planispiral, in early stages, later uniserial to biserial. Wall is grey, thick, calcareous agglutinated. Aperture: simple. Late Permian to Triassic.

- *Endoteba* Vachard and Razgallah, 1988 emend. Vachard, Martini, Rettori and Zaninetti, 1994 (Type species: *Endoteba controversa* Vachard and Razgallah, 1988 emend. Vachard, Martini, Rettori and Zaninetti, 1994). The axial view is compressed. Late Permian to Late Triassic (Late Kungurian to Rhaetian) (Fig. 3.4).

Family Endothyridae Brady, 1884
Here the test is small, multilocular, enrolled, with evolute planispiral coiling and well developed septa in early whorls (e.g. *Loeblichia* Cummings 1955; in the Visean), or initially streptospiral to planispiral characterised by the development of secondary deposits of calcite (chomata) on the chamber floor (e.g. *Endothyra* Phillips, 1846, Early Carboniferous to Permian). They exhibit a low equatorial aperture. Advanced forms may become uniserial (e.g. *Haplophragmella* Rauzer-Chernousova, 1936, Visean, see Fig. 2.7) with a cribrate terminal aperture. Their range is Late Devonian to Permian.

- *Andrejella* Malakhova, 1975 (Type species: *Andrejella laxiformis* Malakhova, 1975). The test is streptospirally coiled in the early part, with later chambers rapidly increasing in height. Early Carboniferous (Visean).
- *Banffella* Mamet, 1970 (Type species: *Endothyra? banffensis* McKay and Green, 1963). The test is streptospirally coiled in early whorl, later it is nearly planispiral and evolute, with straight and slightly oblique septa well developed only in early whorls. Early Carboniferous (Visean).
- *Corrigotubella* Ganelina, 1966 (Type species: *Corrigotubella posneri* Ganelina, 1966). The test is streptospirally enrolled in the early stage, later it is planispiral and rectilinear, with a few short and broad chambers. Septa in the enrolled stage are short, while those of the rectilinear portion are horizontal. Early Carboniferous (Late Tournaisian to Early Visean)
- *Cribranopsis* Conil and Longerstaey, 1980 (Type species: *Cribranopsis fossa* Conil and Longerstaey, in Conil et al., 1980). The test is enrolled, with a streptospiral early stage and a planispiral late coil, expanding rapidly. The septa are short and thick. Early Carboniferous (Visean).
- *Cribrospira* von Möller, 1878 (Type species: *Cribrospira panderi* von Möller, 1878). The test is streptospirally enrolled in the early stage, but later coiling is planispiral with numerous chambers per whorl. Septa are nearly radial, short and thick. Early Carboniferous (Visean).
- *Elergella* Conil, 1984 (Type species: *Elergella simakoyi* Conil, 1984). The test is streptospirally enrolled in the early stage, later becoming planispiral and evolute with short, slightly oblique septa. Early Carboniferous (Tournaisian)
- *Endostaffella* Rozovskaya, 1961 (Type species: *Endothyra parya* von Möller, 1879). The test is streptospirally enrolled in the early stage, later becoming planispiral and evolute. Early Carboniferous (Late Tournaisian to Visean).
- *Endothyra* Phillips, 1846 (Type species: *Endothyra bowmani* Phillips, 1846). The test is partially involute, planispiral with an early part streptospirally enrolled. It is formed

by few whorls, with strong development of secondary deposits, such as nodes, ridges, or hooks on the chamber floor. The wall is calcareous, microgranular, with two or three layers, a thin dark outer layer or tectum and a thicker, fibrous to alveolar inner layer or diaphanotheca. The aperture is simple, basal. Early Carboniferous to Permian (Plate 2.5, fig. 20; Plate 2.6, fig. 1-2; Plate 2.12, figs 8; Plate 2.13, figs 1-3; see Figs 2.7, 2.8).

- *Endothyranella* Galloway and Harlton, 1930 (Type species: *Ammobaculites powersi* Harlton, 1927). Streptospiral test becoming planispiral and then rectilinear. The aperture is simple and rounded. Late Carboniferous (Moscovian to Ladinian) (Plate 2.9, Fig. 1; Plate 3.5, Fig. 10).
- *Endothyranopsis* Cummings, 1955 (Type species: *Involutina crassa* Brady, in Moore, 1870). The test is subglobular involute, being almost planispiral with not more than 3 and a half whorls. They have a thick wall with no secondary deposits. Early coiling

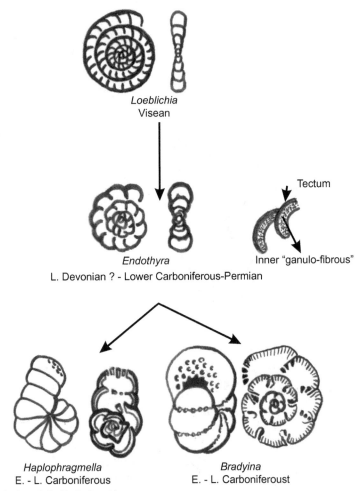

Fig. 2.7 Evolution of the Endothyroidea.

plectogyroid with a large proloculus. Sutures are perpendicular to the spirotheca. The aperture is a low basal equatorial arch. Carboniferous (Visean) (Plate 2.9, figs 3, 6, 7; Plate 2.12, figs 7, 10, 11; Plate 2.14, figs 1, 4, 5).

• *Eoendothyra* Miklukho-Maklay, 1960 (Type species: *Endothyra communis* Rauzer-Chernousova, 1948). The test is involute and streptospirally enrolled in the early stage, later becoming nearly planispiral, with numerous chambers. Late Devonian to Early Carboniferous (Famennian to Visean).

• *Eoendothyranopsis* Reytlinger and Rostovzeva, 1966 (Type species: *Parastaffella pressa* Grozdilova, in Lebedeva, 1954). The test is planispirally enrolled, involute, with oblique septa and secondary deposits at the base of the chambers that appear as a forward projecting hook or spine in the final chamber. Early Carboniferous (Early Visean).

• *Eoquasiendothyra* Durkina, 1963 (Type species: *Endothyra bella* Chernysheva, 1952). The test is streptospirally enrolled in the early stage, but later becoming planispiral, with inflated chambers and septa slightly oblique to the outer wall. Late Devonian to Early Carboniferous (Tournaisian).

• *Euxinita* Conil and Dil, 1980 (Type species: *Dainella? efremoyi* Vdovenko and Rostovtseva, in Brazhnikova et al., 1967). The test is streptospirally enrolled in the

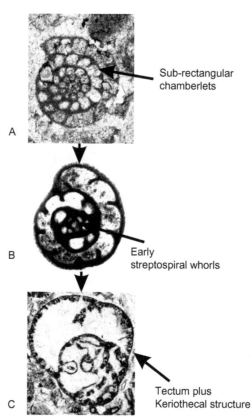

Fig. 2.8   Main morphological features of A) *Loeblichia,* B) *Endothyra* and C) *Bradyina.*

early stage with numerous chambers per whorl, later it becomes nearly planispiral. Carboniferous (Visean to Bashkirian).

- *Globochernella* Hance, 1983 (Type species: *Globochernella braibanti* Hance, 1983). The test is enrolled, with chambers enlarging rapidly as added. The septa are thick at the base and tapering rapidly to a very thin at the inner edge. Early Carboniferous (Visean).

- *Globoendothyra* Bogush and Yuferev, 1962 (Type species: *Globoendothyra pseudo-globulus* Bogush and Yuferev, 1962). The test is streptospirally enrolled in the early stage, but later is nearly planispiral with oblique septa. Secondary deposits on the chamber floors. Carboniferous (Visean to Moscovian).

- *Granuliferella* Zeller, 1957 (Type species: *Granuliferella granulosa* Zeller, 1957). The test is involute, streptospirally enrolled in the early stage, but later nearly planispiral, with relatively few chambers per whorl. The septa are short and slightly oblique. Late Devonian (Famennian) to Early Carboniferous (Visean).

- *Granuliferelloides* McKay and Green, 1963 (Type species: *Granuliferelloides jasperensis* McKay and Green, 1963). The test is streptospirally enrolled in the early stage, but later are nearly planispiral, with a few slightly inflated chambers and oblique septa, becoming uncoiled, with short cylindrical chambers and nearly horizontal septa. Early Carboniferous (Late Tournaisian).

- *Haplophragmella* Rauzer-Chernousova and Reytlinger, 1936 (Type species: *Endothyra panderi* von Möller,1879). The test is streptospirally enrolled in the early stage, with few chambers per whorl, later chambers become uncoiled and rectilinear. The aperture is simple in the early stage, becoming multiple and cribrate in the later stages. Early Carboniferous (Visean) (Fig. 2.7).

- *Haplophragmina* Reytlinger, 1950 (Type species: *Haplophragmina kashkirica* Reytlinger, 1950). The test is involute, planispirally enrolled in the early stage, becoming uncoiled in the late stage. Late Carboniferous (Moscovian).

- *Holkeria* Strank, 1982 (Type species: *Rhodesina avonensis* Conil and Longerstaey, in Conil et al., 1980). The test is streptospirally enrolled with the plane of coiling oscillating in the early stage, but later it is nearly planispiral and evolute with short septa short, which follow the curvature of the chambers. Early Carboniferous (Middle Visean).

- *Klubonibelia* Conil, 1980 (Type species: *Klubonibelia immanis* Conil, 1980). The test is streptospirally enrolled in the early stage, later becoming planispiral, with numerous sub-quadrate chambers per whorl. The final stage is uncoiled and rectilinear. Early Carboniferous (Middle to Late Visean).

- *Latiendothyra* Lipina, 1963 (Type species: *Endothyra latispiralis* Lipina, in Grozdilova and Lebedeva, 1954). The test is inflated, with a broadly rounded periphery and flattened sides. The early stage is streptospiral, later becoming planispiral, with rapidly enlarging whorls, with a moderate number of chambers per whorl (about six to eight in the final whorl). The septa are short, thick, and slightly oblique, projecting toward the aperture. The wall is calcareous, thick, microgranular, dark, single layered, with secondary deposits that result in septal thickening. The aperture is simple at the base of the apertural face. Early Carboniferous (Late Tournaisian to Early Visean).

- *Latiendothyranopsis* Lipina, 1977 (Type species: *Endothyra latispiralis* Lipina, in Grozdilova and Lebedeva, 1954). The test is streptospiral and enrolled in the early stage, later becoming planispiral, with rapidly enlarging whorl and short, thick, slightly oblique septa. Early Carboniferous (Late Tournaisian to Early Visean).
- *Laxoendothyra* Brazhnikova and Vdovenko, 1972 (Type species: *Endothyra parakosyensis* Lipina, 1955). The test is streptospirally enrolled in the early stage, later is planispiral, with elongate chambers in rapidly enlarging whorls and short septa. Early Carboniferous (Tournaisian to Visean).
- *Mediopsis* Bogush, 1984 (Type species: *Planoendothyra*? *kharaulakhensis* Bogush and Yuferev, 1966). The test is involute, streptospirally enrolled in the first half to one and a half whorls, later being planispiral. Early Carboniferous (Late Tournaisian to Visean).
- *Melatolla* Strank, 1983 (Type species: *Melatolla whitfieldensis* Strank, 1983). The test is streptospirally enrolled in the early stage, later becoming planispiral, and finally tending to uncoil, with massive and robust chomatal deposits in the last chambers. The aperture is areal and cribrate in the uncoiled part. Early Carboniferous (Visean).
- *Mikhailovella* Ganelina, 1956 (Type species: *Endothyrina*? *gracilis* Rauzer-Chernousova, 1948). The test is streptospiral to rectilinear with a short rectilinear part. The aperture is a low basal slit in the early enrolled part, but cribrate in the uncoiled stage. Carboniferous (Middle Visean) (Plate 2.3, Fig. 19).
- *Mirifica* Shlykova, 1969 (Type species: *Endothyra mirifica* Rauzer-Chernousova, 1948). The test is involute, streptospirally enrolled in the early stage, later nearly planispiral with numerous chambers per whorl and short, hooked sutures. The aperture is cribrate in the final whorl. Early Carboniferous (Late Visean).
- *Omphalotis* Shlykova, 1969 (Type species: *Endothyra omphalota* Rauzer-Chernousova and Reytlinger, in Rauzer-Chernousova and Fursenko, 1937). The test is involute, with early streptospiral coiling, later becoming planispiral. The sutures are radial to slightly oblique. Carboniferous (Middle Visean to Early Bashkirian).
- *Paradainella* Brazhnikova, 1971 (Type species: *Paradainella dainelliformis* Brazhnikova and Vdovenko, 1971). The test is streptospirally enrolled in the early stage, but the final whorls are nearly planispiral with secondary deposits forming massive chomata that may cover the inner surface of the whorls. The aperture is simple, basal. Early Carboniferous (Late Tournaisian).
- *Paraendothyra* Chernysheva, 1940 (Type species: *Paraendothyra nalivkini* Chernysheva, 1940). The test is enrolled and biumbilicate. The early stage is slightly streptospiral, later becoming nearly completely planispiral. The sutures are radial with a slight serrate appearance, and the final few septa are hook-like in section. The aperture is in a median septal concavity. Early Carboniferous (Middle Tournaisian).
- *Paraplectogyra* Okimura, 1958 (Type species: *Paraplectogyra masanae* Okimura, 1958). The test is streptospirally enrolled in the early stage, involute, later becoming planispiral with whorls expanding rapidly with straight, radial septa. Early Carboniferous (Late Tournaisian to Early Visean).
- *Planoendothyra* Reytlinger, 1959 (Type species: *Endothyra aljutovica* Reytlinger, 1950). The test is compressed, streptospiral in the early stages, later becoming

planispiral and evolute, and slightly asymmetrical with supplementary deposits. Carboniferous (Visean to Bashkirian) (Plate 2.9, Fig. 18; Plate 2.12, Fig. 12).

- *Plectogyra* Zeller, 1950 (Type species: *Plectogyra plectogyra* Zeller, 1950). A streptospiral test with a two-layered wall. There are 4 to 12 chambers in the final whorl. Late Devonian/Early Carboniferous to Permian (Plate 2.3, Fig. 2C; Plate 2.12, figs 1, 3-5; Plate 2.13, Fig. 10B; Plate 2.14, figs 3, 6-9, 11, 12).
- *Plectogyranopsis* Vachard, 1977 (Type species: *Endothyra convexa* Rauzer-Chernousova, 1948). The test is planispiral, biumbilicate with short, straight, thick septa. Early Carboniferous (Early Visean) to Late Carboniferous (Early Bashkirian).
- *Pojarkovella* Simonova and Zub, 1975 (Type species: *Pojarkovella honesta* Simonova and Zub, 1975). The test is streptospirally enrolled in the early stage, biumbilicate, tightly coiled and involute, later becoming planispiral with radiate and straight septa and massive secondary deposits consisting of angular chomata. Early Carboniferous (Visean).
- *Priscella* Mamet, 1974 (Type species: *Endothyra prisca* Rauzer-Chernousova and Reytlinger, in Rauzer-Chernousova et al., 1936). The test is streptospirally enrolled in the early stage, later becoming nearly planispiral with long strongly oblique septa and no distinct basal secondary deposits. Carboniferous (Tournaisian to Bashkirian).
- *Quasiendothyra* Rauzer-Chernousova, 1948 (Type species: *Endothyra kobeitusana* Rauzer-Chernousova, 1948). The test is discoidal with concave sides and slightly streptospiral early whorls, with septa as thick as the outer wall. Secondary deposits are chomata-like, at each side of the median line against the previous whorl. Late Devonian (Famennian) to Early Carboniferous (Tournaisian).
- *Rectoendothyra* Brazhnikova, 1983 (Type species: *Endothyra (Rectoendothyra) donbassica* Brazhnikova, 1983). The test is streptospirally coiled in the early stage, becoming planispiral with the final whorl enlarging rapidly and may tend to uncoil. Septa are straight and radial. Chomata-like deposits occur on chamber floor. The aperture is multiple in the final chamber. Early Carboniferous (Tournaisian).
- *Rhodesinella* Conil and Longerstaey, 1980 (Type species: *Cribrospira pansa* Conil and Lys, 1965). The test is streptospiral enrolled in the early part, later becoming planispiral with a tendency to uncoil in the later stage. Septa moderately short, straight, and inclined toward the aperture. The aperture is multiple in the final chamber. Early Carboniferous (Tournaisian).
- *Semiendothyra* Reytlinger, 1980 (Type species: *Semiendothyra surenica* Reytlinger, 1980). The test is streptospirally enrolled with early whorls, involute, but later slightly evolute, planispiral with slightly arched septa. Secondary deposits are extensive, small chomata-like mounds near the aperture, spine-like in the final chamber. The aperture is simple, basal, and low. Late Carboniferous (Bashkirian).
- *Spinoendothyra* Lipina, 1963 (Type species: *Endothyra costifera* Lipina, in Grozdilova and Lebedeva, 1954). The test is closely coiled in the early stage, with later whorls planispiral. Sutures are slightly curved without any thickenings at the ends. Secondary deposits are spine-like, well developed on chamber floors, anteriorly directed. The aperture is simple and basal. Carboniferous (Late Tournaisian to Early Visean) (Plate 2.9, Fig. 15).

- *Spinothyra* Mamet, 1976 (Type species: *Endothyra pauciseptata* Rauzer-Chernousova, 1948). The test is streptospirally enrolled, involute, biumbilicate, with short thick septa, whorls enlarge rapidly and secondary chomata-like deposits occur on the floor of the chambers. Carboniferous (Middle Visean to Early Bashkirian).
- *Timanella* Reytlinger, 1981 (Type species: *Endothyra eostaffelloides* Reytlinger, 1950). The test is planispirally enrolled with rapidly enlarging whorls and straight oblique septa, with deposits covering the chamber floor and filling the lateral areas. Late Carboniferous (Moscovian).
- *Tuberendothyra* Skipp, 1969 (Type species: *Endothyra tuberculata* Lipina, 1948). The test is streptospirally enrolled in the early stage, later becoming planispiral with long, curved septa, long and secondary spine-like deposits on the floor of all chambers. Early Carboniferous (Late Tournaisian to Early Visean).
- *Urbanella* Malakhova, 1963 (Type species: *Quasiendothyra urbana* Malakhova, 1954). The test is streptospirally enrolled in the early whorls, with later whorls increasing in height, becoming planispiral and evolute, with secondary small rounded chomata at the margins of the aperture. Early Carboniferous (Tournaisian to Early Visean).
- *Zellerinella* Mamet, 1981 (Type species: *Endothyra discoidea* Girty, 1915). The test is streptospirally enrolled in early stage, later becoming planispiral and evolute with pseudochomata. Early Carboniferous (Late Visean to Early Bashkirian).

Family Bradyinidae Reytlinger, 1950
The Endothyridae evolved into forms such as *Bradyina* von Möller, 1878 (see Fig. 2.9). These forms have nautiloid planispiral coiling with few chambers. The microgranular calcareous wall is perforate, with distinct radial lamellae with supplementary septal pores opening into the septal aperture. The primary aperture is also cribrate. They occur in the Carboniferous (Visean) to the Permian (Sakmarian).

- *Bibradya* Strank, 1983 (Type species: *Bibradya inflata* Strank, 1983). The test is streptospirally enrolled with short, thick septa which bifurcate close to the outer wall. Early Carboniferous (Late Visean)
- *Bradyina* von Möller, 1878 (Type species: *Bradyina nautiliformis* Möller, 1878). The test is planispiral and involute with supplementary pores and septal partitions. The wall has a microgranular tectum and keriothecal structure. Carboniferous (Tournaisian to Gzhelian) (Figs 2.7, 2.8; Plate 2.8, Fig. 13; Plate 2.9, Fig. 20; Plate 2.12, Fig. 2; Plate 2.13, Fig. 10A).
- *Glyphostomella* Cushman and Waters, 1928 (Type species: *Ammochilostoma? triloculina* Cushman and Waters, 1927). The test is planispirally enrolled, involute, with a complex wall, with plates and partitions and large and branching pores in the wall. Late Carboniferous to Permian (Kasimovian to Changhsingian).
- *Janischewskina* Mikhaylov, 1935 (Type species: *Janischewskina typica* Mikhaylov, 1935). The test is planispirally enrolled, and involute with septal chamberlets as in *Bradyina*, but without the alveolar structure of *Bradyina*. Early Carboniferous (Late Visean).
- *Postendothyra* Lin, 1984 (Type species: *Postendothyra scabra* J. X. Lin, 1984). The test is planispirally enrolled and involute, with outer tectum and inner coarsely alveolar keriotheca. Early Permian (Asselian).

Family Loeblichiidae Cummings, 1955
Members of this family have a typical endothyroid planispiral test with numerous chambers, pseudochomata, a basal aperture, two-layered walls, an inner clear fibrous light layer (called early tectorium) and a dark granular outer layer (called tectum) (Fig. 2.8A). They range from Middle Devonian to Late Carboniferous (Moscovian).

- *Dainella* Brazhnikova, 1962 (Type species: *Endothyra? chomatica* Dain, in Brazhnikova, 1962). The test is streptospiral and involute throughout, with secondary deposits in the form of massive chomata. Early Carboniferous (Early Visean)
- *Loeblichia* Cummings, 1955 (Type species: *Endothyra ammonoides* Brady, 1873). The test is small, flattened, discoidal and is planispiral throughout, with up to 20 sub-rectangular chambers in the final whorl, but lacks secondary deposits. Early Carboniferous (Visean) (Figs 2.7, 2.8; Plate 2.7, Fig. 7; Plate 2.8, Fig. 12B; Plate 2.13, Fig. 11B).
- *Lysella* Bozorgnia, 1973 (Type species: *Lysella gadukensis* Bozorgnia, 1973). The test is streptospirally enrolled in the early stage, later becoming planispiral, involute with many chambers per whorl that appear sub-quadrate in section. Septa are slightly inclined toward the aperture. The wall has chomata-like structures. Early Carboniferous (Visean).
- *Novella* Grozdilova and Lebedeva, 1950 (Type species: *Novella evoluta* Grozdilova and Lebedeva, 1950). The test is planispirally enrolled and evolute with straight septa and well-developed chomata in later whorls. Late Carboniferous (Bashkirian to Early Moscovian).
- *Seminovella* Rauzer-Chernousova, 1951 (Type species: *Eostaffella* (*Seminovella*) *elegantula* Rauzer-Chernousova, in Rauzer-Chernousova et al., 1951). The test is planispiral, involute becoming later evolute, with weakly developed pseudochomata. Late Carboniferous (Late Bashkirian to Moscovian).

ORDER FUSULINIDA WEDEKIND, 1937
This order includes all larger benthic foraminifera with a homogeneously microgranular primary test made of low-Mg calcite, in which the crystal units have no optical alignment and various foreign particles might be incorporated (see Rauzer-Chernousova, 1948; Loeblich and Tappan 1964; Brazhnikova and Vdovenko 1973; Tappan and Loeblich 1988; Rigaud et al., 2014). Advanced forms have two or more differentiated layers in the wall. They range from the Carboniferous to the Permian.

The Fusulinida have special morphological diagnostic characters, which are unique to this order (see Haynes, 1981; Loeblich and Tappan, 1988; and Figs 2.9, 2.10). The traces of the septa on the external surface are called "**septal furrows**". These furrows extend from pole to pole of the test and marks the early part of the partitions between the chambers, called "**septa**". The apertural face of the test is the "**antetheca**" while the external wall of the test is the "**spirotheca**". Both spirotheca and antetheca are finely perforate by numerous small openings "**septal pores**" and the test lacks primary apertures. Communication between chambers occurs at the base of septa and aided by resorption of a **tunnel** in the central part of the test in many fusulinides or of several small circular tunnels, **foramina** throughout the length of the test in others. In some advanced fusulinides, the antetheca is corrugated into uniformly or irregular spaced waves, called **fluting**. Thin sections of the fusulinides reveal highly complicated internal structures which are essential for identifying and classifying the genera and species. The first chamber is spherical to sub-spherical, called **proloculus**. Dense calcite, called **chomata**, were deposited along the margins of the tunnel in many fusulinides, and ridges of dense calcite, **parachomata** were developed between adjacent foramina in those with **multiple foramina**). Deposition of dense calcite and simultaneous formations of chomata and parachomata, **axial fillings** occurred in axial regions of some fusulinides. The spirotheca consists of a thin, dense, primary layer, **tectum,** which is sometimes covered by layers of **tectoria**. In more advanced forms, the tectum is augmented by a transparent layer, the **diaphanotheca**, or a thick layer of honeycomb-like structure, **keriotheca**. Ridges, called **septula** spread down from the lower surface of the spirotheca to subdivide the chambers in the neoschwagerinoids.

Since the studies made by Loeblich and Tappan (1964, 1980) and Haynes (1981), many authors have tried to subdivide the fusulinides by considering the highly complicated internal structures, such as the wall structure, the presence or absence of the chomata, and the degree of septal folding in the test. On the basis of these main morphological features, the Fusulinida are divided here into six superfamilies (see Figs 2.1 and 2.2), the **Fusulinoidea**, the **Ozawainelloidea**, the **Staffelloidea**, the **Schubertelloidea**, the **Schwagerinoidea**, and the **Neoschwagerinoidea** (= Verbeekinoidea). These superfamilies are considered as orders by Rauzer-Chernousova et al. (1996) and Mikhalevich (2013).

The structure of the spirotheca plays an important role in differentiating the fusulinids. Five types of wall structures (see Fig. 2.10) characterise these superfamilies:

1. Forms with two layered walls, an inner clear fibrous light layer (called early tectorium) and a dark granular outer layer (called tectum) (Fig. 2.10A)

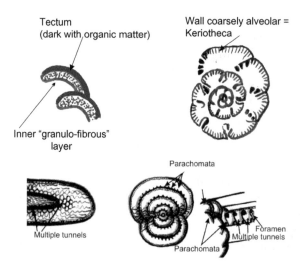

Fig. 2.9.   Schematic figures showing a microgranular wall and a *Bradyina*-type of wall with keriotheca, multiple tunnels and parachomata in more advanced forms.

Types A-D, e.g. Ozawainellidae, Staffellidae, Fusulinidae, Schubertellidae

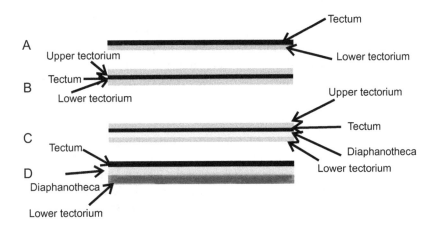

A — Tectum / Upper tectorium / Lower tectorium

B — Tectum / Lower tectorium

C — Upper tectorium / Tectum / Diaphanotheca / Lower tectorium

D — Tectum / Diaphanotheca / Lower tectorium

E   E.g. Schwagernidae, Verbeekinidae, Neoschwagerinidae

Tectum / keriotheca

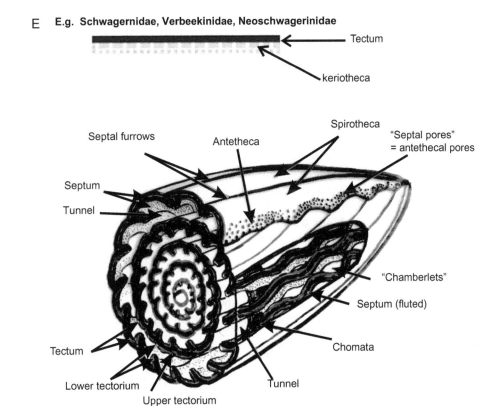

Fig. 2.10. Schematic figures showing important features of the fusulinides.

2. Forms with three layered walls, the late and early tectorium surrounding the dark layer of the tectum (Fig. 2.10B);
3. Forms with four layered walls, with the late and early tectorium surrounding the tectum and a glossy layer called the diaphanotheca (Fig. 2.10C);
4. Forms with three layered walls, where the late tectorium disappears and they are left with the tectum, diaphanotheca and the early tectorium (Fig. 2.10D);
5. Forms composed of a tectum and glossy layer with alveoles running through it (the keriotheca) (Fig. 2.10E).

The first four types of wall characterise the ozawainelloids, staffelloids, the schubertulloids and the fusulinoids and the most advanced final type of wall belong to the schwagerinoids and neoschwagerinoids (discussed below).

Over 100 fusulinides genera are recognised. Below are the main taxonomic classifications of the fusulinides, but for an even more detailed taxonomic description at the generic and specific levels see Loeblich and Tappan (1988) and Rauser-Chernousova et al. (1996).

Superfamily OZAWAINELLOIDEA Slovieva, 1978
The test is small, lenticular to rhombic to discoid and involute. The early coiling may be streptospiral. The wall is simple in the early forms, composed of tectum with upper and lower tectoria (Fig. 2.10B), but with diaphanotheca occurring between tectum and lower tectorium (Fig. 2.10C) in advanced forms. The proloculus is small and spherical. The pseudochomata develop into weak to strong chomata. The aperture is simple and basal. This superfamily ranges from the Early Carboniferous to the Late Permian (Visean to Capitanian).

Family Ozawainellidae Thompson and Foster, 1937
Here, the tests are discoidal to elongate in shape. They are involute to slightly evolute with moderate to strong secondary deposits of calcite (chomata) (Fig. 2.11). The wall is of a primitive fusulinid type, with a thin dense, primary layer (tectum) dark with organic matter, surrounded by a thicker, less dense layer of upper and lower tectoria (Fig. 2.10B). Carboniferous to the Late Permian (Late Tournaisian to Capitanian).
This family includes genera such as:

• *Chenella* Miklukho-Maklay, 1959 (Type species: *Orobias kueichihensis* S. Chen, 1934). The test is planispirally enrolled, with the final whorl increasing abruptly in height. Late Permian (Capitanian).
• *Eoparastaffella* Vdovenko, 1954 (Type species. *Parastaffella* (*Eoparastaffella*) *simplex* Vdovenko, 1954). Test is almost planispiral. The wall is of two very thin, and dark microgranular tecta surrounded by thick, pale brown-gray zone. Early Caroboniferous (Late Tournaisian to Early Visean).
• *Eostaffella* Rauser-Chernousova, 1948 (Type species: *Staffella* (*Eostaffella*) *parastruvei* Rauser-Chernousova, 1948). The test is lenticular, involute and planispiral with a sub-acute periphery. Septa are thick, curved, and often with truncated edges.

The wall is layered with tectum and upper and lower tectoria. There are discontinuous knobs on either side of the tunnel but no continuous chomata. Carboniferous to Permian (Visean to Capitanian) (Plate 2.3, Fig. 4A; Plate 2.5, Fig. 19; Plate 2.9, Fig. 5; Plate 2.12, figs 4, 9, 13, 14; Plate 2.14, figs 14, 15; Plate 2.15, Fig. 4).

- *Eostaffelloides* Miklukho-Maklay, 1959 (Type species: *Eostaffelloides orientalis* Miklukho-Maklay, 1959). The test is lenticular, planispirally enrolled with secondary deposits forming triangular chomata. Late Permian (Capitanian).

- *Millerella* Thompson, 1942 (Type species: *Millerella marblensis* Thompson, 1942; see Fig. 2.9). Minute discoidal and partially evolute test. The septa are slightly arched forward. Carboniferous (Serpukhovian to Gzhelian) (Fig. 2.13; Plate 2.3, Fig. 2A; Plate 2.8, Fig. 11; Plate 2.12, Fig. 6; Plate 2.14, Fig. 10).

- *Ozawainella* Thompson, 1935 (Type species: *Fusulinella angulata* Colani, 1924). The test is angular with massive chomata. Late Carboniferous to Permian (Serpukhovian to Capitanian).

- *Pamirina* Leven, 1970 (Type species: *Pamirina darvasica* Leven, 1970). The test is small and varies in shape from nautiloid to subspherical. The spirotheca is thin, one-layered in inner whorls and two-layered consisting of tectum and protheca in later whorls. The septa are straight and flat. The chomata vary from rudimentary to well developed. Early Permian (Sakmarian to Kungurian).

- *Pseudonovella* Kireeva, 1949 (Type species: *Pseudonovella irregularis* Kireeva, 1949). The test is lenticular with early whorls evolute and final whorl increasing rapidly in height and becoming partially evolute. The pseudochomata are weakly developed. Late Carboniferous (Late Bashkirian to Moscovian).

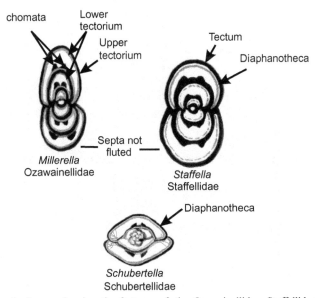

Fig. 2.11. Schematic figures showing the features of the Ozawainellidae, Staffellidae, Schubertellidae. Drawings not to scale.

- *Quasireichelina* Ueno 1992 (Type species: *Quasireichelina expansa* Ueno, 1992). The test is involute in the early part, but later is uncoiled with numerous unfluted septa. Secondary deposits form small chomata. Middle Permian (Guadalupian).
- *Reichelina* Erk, 1942 (Type species: *Reichelina cribroseptata* Erk, 1942). The test is planispirally enrolled in the early stage, but the final whorl uncoils to appear almost peneropliform. Secondary deposits consist of broad chomata, extending and thickening poleward from the tunnel or the small circular foramina. Late Permian (Capitanian).
- *Sichotenella* Tumanskaya, 1953 (Type species: *Sichotenella sutschanica* Tumanskaya, 1953). The test is lenticular, resembling *Chenella,* but the final whorl increases rapidly in height and uncoils. The uncoiled stage is relatively large. The wall is three layered, with diaphanotheca. Late Permian (Capitanian).
- *Zarodella* Sosnina, 1981 (Type species: *Zarodella zhamoidai* Sosnina, 1981). The test is small, slightly biumbilicate. The proloculus is large in size compare to the adult test which comprises three to five whorls. The septa are flat and the spirotheca is weakly differentiated. Early Permian (Sakmarian).

Family Pseudostaffellidae Putrya 1956
The Pseudostaffellidae have an involute test which may be a streptospiral in the early stage. Early walls are weakly differentiated, later ones have a tectum, thin diaphanotheca and two tectoria (Fig. 2.10D) with well-developed chomata. Carboniferous to Permian (Serpukhovian to Capitanian).

- *Hubeiella* Lin, 1977 (Type species: *Hubeiella simplex* J. X. Lin, 1977). The test is planispirally enrolled and involute in the early stage, later stages may be evolute, and the final whorl is greatly expanded in height and flares. The septa are straight, and the wall shows a tectum and fine keriotheca with small but distinct chomata present only in the inner whorls. Early Permian (Serpukhovian).
- *Kangvarella* Saurin, 1962 (Type species: *Kangvarella irregularis* Saurin, 1962). The test is involute in the early part, and later whorls increase rapidly in height. The wall has a tectum, diaphanotheca, and inner and outer tectoria. Irregular and asymmetrical parachomata may be present in the second and third whorls. Late Permian (Capitanian).
- *Mediocris* Rozovskaya, 1961 (Type species: *Eostaffella mediocris* Vissarionova, 1948). The test is streptospirally enrolled in the early part, later becoming planispiral and involute. The wall has a dark tectum and a poorly defined tectoria, with strong secondary deposits in the axial region. Carboniferous (Middle Visean to Early Moscovian).
- *Neostaffella* Miklukho-Maklay, 1959 (Type species: *Pseudostaffella sphaeroidea* Ehrenberg, in Rauzer-Chernousova et al., 1951). Chomata are very broad, extending from the edge of the narrow tunnel outward to the umbilical region. Late Carboniferous (Moscovian) (Plate 2.16, Fig. 9).
- *Ninella* Malakhova, 1975 (Type species: *Endothyra staffelliformis* Chernysheva, 1948). The test is streptospirally enrolled in the early part, later becoming nearly planispiral, with straight septa and widely spaced rounded pseudochomata in all whorls. Early Carboniferous (Visean).

- *Pseudostaffella* Thompson, 1942 (Type species: *Pseudostaffella needhami* Thompson, 1942). The test is streptospirally coiled in early whorls, later becoming planispiral with straight, long, unfluted and perpendicular to the outer wall septa, slightly curved in the polar regions. Chomata are well developed. Late Carboniferous (Bashkirian to Moscovian).
- *Rauserella* Dunbar, 1944 (Type species: *Rauserella erratica* Dunbar, 1944). The test is planispirally enrolled in the early whorls. The whorls become much wider and the test fusiform with irregular septa in outer whorls. Septal pores are present in the outer whorls. Permian (Kungurian)
- *Toriyamaia* Kanmera, 1956 (Type species: *Toriyamaia latiseptata* Kanmera, 1956). The test is fusiform, the first whorl is evolute, later whorls are sub-cylindrical, involute, increasing rapidly in height. Septa are sparse and unfluted. Early Permian (Cisuralian).

Superfamily STAFFELLOIDEA Slovieva, 1978 emend. Vachard et al., 2013
In this superfamily, the tests are small, spherical to discoidal and strong chomata are present. The wall structure of this order has been discussed by many authors, but is still ill-resolved (see Jin-Zhang, 1990). The wall is usually composed of tectum and diaphanotheca (Fig. 2.11) and may have tectoria, but it is mainly found recrystallized and altered by secondary mineralisation, suggesting that they secreted an aragonite or high-Mg calcite wall (Groves et al., 2005). Early Carboniferous to Late Permian (Visean to Wuchiapingian).

Family Staffellidae Miklukho-Maklay, 1949
Staffellidae have a primitive wall structure, consisting of a tectum followed by a transparent layer (diaphanotheca) which is formed of clear calcite (Fig. 2.11). They occur from the Early Carboniferous to Late Permian (Visean to Wuchiapingian).

- *Chenia* Sheng, 1963 (Type species: *Chenia kwangsiensis* Sheng, 1963). The test is lenticular with plane, unfluted septa. Chomata are well developed in all whorls with poorly developed and discontinuous parachomata present in the outer four or five whorls. Middle Permian (Kungurian).
- *Nankinella* Lee, 1934 (Type species: *Staffella discoides* Lee, 1931). The test is discoidal with plane septa and distinct chomata. Early to Late Permian (Sakmarian to Wuchiapingian).
- *Pisolina* Lee, 1934 (Type species: *Pisolina excessa* Lee, 1934). The test is spherical with thick, plane septa and well defined, low, asymmetrical chomata. Early Permian (Asselian to Kungurian).
- *Pseudoendothyra* Mikhaylov, 1939 (Type species: *Fusulinella struvii* von Möller, 1879). The test is lenticular, compressed, planispiral and involute, with plane septa and supplementary deposits adjacent to the aperture but not forming continuous chomata. The aperture is low and broad. Early Carboniferous to Early Permian (Visean to Sakmarian). (Plate 2.9, figs 2, 13; Plate 2.12, Fig. 15A; Plate 2.14, figs 2, 13).
- *Staffella* Ozawa, 1925 (Type species: *Fusulina sphaerica* Abich, 1859). It is the main representative of this group. It has a small lenticular test with a wall composed of a tectum and diaphanotheca. Permian (Artinskian to Kungurian) (Fig. 2.11).

Family Kahlerinidae Leven, 1963

This family is monotypic; the test is globose with minor chomata and rarely with small parachomata. Middle Permian.

- *Kahlerina* Kochansky-Devidé and Ramovš, 1955 (Type species: *Kahlerina pachytheca* Kochansky-Devidé and Ramovš, 1955). Septa are long, planar and unfluted. Chomata are only rudimentarily present in the initial whorls. Middle Permian (Guadalupian).

Superfamily SCHUBERTELLOIDEA Vachard in Vachard et al., 1993

The shell is minute, fusiform or irregular in shape. The wall is composed of tectum with lower tectorium only (Fig. 2.10A), or tectum surrounded by an upper and lower tectoria (Fig. 2.10B), or tectum with diaphanotheca and tectoria (Fig. 2.10C). Late Carboniferous to Late Permian (Bashkirian to Changhsingian).

Family Schubertellidae Skinner, 1931

Schubertellidae have fusiform tests with a streptospiral early coiling. Early septa are flat but evolve into fluted septa in advanced forms. The wall varies, as in some forms the diaphanotheca is present (Fig. 2.10B), while in others only the tectum and early tectorium are present (see Fig. 2.10A; Fig. 2.11). This family includes spirothecal structures of the post-keriotheca phase. Carboniferous to Permian (Bashkirian to Changhsingian)

Significant genera include:

- *Biwaella* Morikawa and Isomi, 1960 (Type species: *Biwaella omiensis* Morikawa and Isomi, 1960). The test is large, fusiform to sub-cylindrical, with subglobose and tightly coiled initial whorls followed by ovoid whorls that increase rapidly in size. The wall in first whorls is composed of a thin, dark tectum and a thicker but lighter tectorium. The wall in the final whorl is perforated with coarse mural pores, which do not develop into keriotheca. The septa are nearly straight and widely separated. The chomata are small and prominent in all the test, except for the final whorl. (Davydov, 2007, 2010). Late Carboniferous to Early Permian (Early Gzhelian to Kungurian).
- *Boultonia* Lee, 1927 (Type species: *Boultonia willsi* Lee, 1927). The test is fusiform, with strongly fluted septa throughout and asymmetric chomata. Early Permian (Asselian to Sakmarian).
- *Codonofusiella* Dunbar and Skinner, 1937 (Type species: *Codonofusiella paradoxica* Dunbar and Skinner, 1937). The test is irregularly coiled, becoming uncoiled in later stages, small, fusiform, with strongly fluted septa. Chomata are present but tunnels are not clearly defined. Late Permian (Lopingian).
- *Dunbarula* Ciry, 1948 (Type species: *Dunbarula mathieui* Ciry, 1948). The test is an elongate ellipsoid, irregularly coiled, with septa strongly fluted throughout. Late Permian (Capitanian).
- *Dutkevichites* Daydov, 1984 (Type species: *Dutkevichites darvasica* Davydov, 1984). The test is cylindrical. Septa are few, but strongly fluted with an increase in intensity towards the poles. Minute chomata are present in outer whorls. Late Carboniferous to Early Permian (Gzhelian to Asselian) (Fig. 2.14).

- *Eoschubertella* Thompson, 1937 (Type species: *Schubertella lata* Lee and Chen, in Lee et al., 1930). The test is elongate with early whorls close coiled and showing a broad tunnel, bordered by low chomata. The wall is unilayered. Late Carboniferous (Moscovian).
- *Fusiella* Lee and Chen, 1930 (Type species: *Fusiella typica* Lee and Chen, in Lee et al., 1930). The test is elongate, early whorls are discoidal and endothyroid, later becoming fusiform with axial fillings prominent. Late Carboniferous (Moscovian) (Plate 2.15, Fig. 6).
- *Gallowaiina* Chen, 1934 (Type species: *Gallowaiina meitiensis* Chen, 1934). The test is fusiform with very thin septa, closely fluted throughout, with folds extending over most of the chamber height. No chomata are present, fillings only found in the axial region. Late Permian (Changhsingian).
- *Grovesella* Davydov and Arefifard, 2007 (Type species: *Grovesella tabasensis* Davydov and Arefifard, 2007). Test: Discoidal, nearly planispirally coiled with a large proloculus and two-layered poorly developed wall lacking chomata. Late Carboniferous to Permian (middle Bashkirian to Wordian).
- *Kwantoella* Sakagami and Omata, 1957 (Type species: *Kwantoella fujimotoi* Sakagami and Omata, 1957). The test is fusiform with numerous, straight and unfluted septa. Chomata are poorly developed, and axial fillings are more extensive in later whorls. Early Permian (Sakmarian).
- *Lantschichites* Tumanskaya, 1952 (Type species: *Codonofusiella* (*Lantschichites*) *maslennikovi* Tumanskaya, 1953). The test is cylindrical with strongly developed and fluted septa. Late Permian (Lopingian).
- *Mesoschubertella* Kanuma and Sakagami, 1957 (Type species: *Mesoschubertella thompsoni* Sakagami, in Kanuma and Sakagami, 1957). The test is fusiform with straight to slightly fluted septa in the polar ends. The wall is thick and composed of a thin, dark tectum and a diaphanotheca between a well-developed upper tectorium and a lower tectorium. Chomata are small, but well developed and asymmetrical. Early Permian (Cisuralian to Guadalupian).
- *Minojapanella* Fujimoto and Kanuma, 1953 (Type species: *Minojapanella elongata* Fujimoto and Kanuma, 1953). The test is subcylindrical, with intensely fluted septa, but without developing cuniculi (i.e. multiple tunnels, see Fig. 2.9). Chomata are massive. Permian (Capitanian to Changhsingian) (see Fig. 2.14).
- *Nanlingella* Rui and Sheng, 1981 (Type species: *Nanlingella meridionalis* Rui and Sheng, 1981). The test has a small proloculus. Septa are strongly fluted except in the median part. The wall is thin and composed of a tectum and diaphanotheca. Chomata are present only in the first two whorls. Late Permian (see Fig. 2.14).
- *Neofusulinella* Deprat, 1912 (Type species: *Neofusulinella praecursor* Deprat, 1913). The test is ovoid in the early stage, later becoming fusiform, planispirally coiled throughout with flat, slightly fluted septa. Chomata are large and asymmetrical. Late Carboniferous (Early Moscovian).
- *Neoschubertella* Saurin, 1962 (Type species: *Neoschubertella sisophonensis* Saurin, 1962). The test is fusiform, with slightly curved but not fluted septa and continuous pseudochomata. Permian (Sakmarian to Capitanian).
- *Palaeofusulina* Deprat, 1912 (Type species: *Palaeofusulina prisca* Deprat, 1913). The test has the spirotheca composed of a tectum and an early transparent diaphanotheca.

The *Palaeofusulina*-type fusulinids became abundant after most of the *Schwagerina-Verbeekina-Neoschwagerina* types had disappeared (Jin-Zhang, 1990). Lattermost Permian (Changhsingian) (see Fig. 2.14).

- *Paradoxiella* Skinner and Wilde, 1955 (Type species: *Paradoxiella pratti* Skinner and Wilde, 1955). The test is discoidal, flaring and uncoiling with intensely fluted septa, and with well-developed cuniculi and low chomata. Late Permian (Capitanian).
- *Paradunbarula* Skinner, 1969 (Type species: *Paradunbarula dallyi* Skinner, 1969). The test is fusiform with intensely fluted septa and moderately wide, weak chomata. Late Permian (Capitanian)
- *Russiella* Miklukho-Maklay, 1957 (Type species: *Russiella pulchra* Mildukho-Maklay, 1957). The test is fusiform, and the first two whorls endothyroid. Septa are strongly fluted, with heavy axial fillings. Chomata are weakly developed except in the last two revolutions. Permian (Capitanian to Changhsingian).
- *Schubertella* Staff and Wedekind, 1910 (Type species: *Schubertella transitoria* Staff and Wedekind, 1910). The test is small becoming fusiform in later stages, with straight to weakly fluted septa at the polar ends. The wall is differentiated into three layers that are penetrated by relatively coarse pores. Chomata are prominent, low and asymmetrical. Late Carboniferous to Permian (Moscovian to Wordian) (see Fig. 2.11; Plate 2.17, Fig. 1).
- *Schubertina* Marshall, 1969 (Type species: *Schubertina circuli* Marshall, 1969). The test is subglobose to ovoid in outline, with a thin wall clearly differentiated into two layers, poorly developed secondary deposits and straight septa. Late Carboniferous to Permian (Bashkirian to Lopingian).
- *Sphaeroschwagerina* Miklukho-Maklay, 1959 (Type species: *Schwagerina sphaerica* var. *karnica* Shcherbovich, in Rauzer-Chernousova and Shcherbovich, 1949). The test is almost spherical with weakly fluted septa at the poles and small chomata. Permian (Plate 2.18, figs 6, 7, 9).
- *Yangchienia* Lee, 1934 (Type species: *Yangchienia iniqua* Lee, 1934). The test is fusiform with plane unfluted septa and massive, asymmetrical chomata that extend nearly to the poles. Permian (Asselian to Capitanian).

Superfamily FUSULINOIDEA von Möller, 1878
The test of the Fusulinoidea is large and varies from spherical to fusiform. The development of secondary deposits of calcite (chomata) is prominent in most forms, but the numerous chambers are subdivided by folds or septula (Figs 2.10, 2.12). Late Carboniferous to Permian (Bashkirian to Kungurian).

Family Fusulinidae von Möller, 1878
In the Fusulinidae, the test is spherical or elongate fusiform, and mostly planispiral. The Fusulinidae family belongs to the pre-keriotheca stage of fusulinides development. It includes genera with a wall composed of a tectum surrounded by a late and an early tectorium (Fig. 2.10B) (e.g., the type of wall typical for *Profusulinella*, see Fig. 2.12). However, it also includes more evolved genera such as the cylindrical *Fusulina* and the fusiform *Beedeina* (see Fig. 2.12) with four layered walls, with the late and early tectorium surrounding the tectum and a glossy layer called the diaphanotheca (Fig. 2.10C). Carboniferous to Permian (Bashkirian to Kungurian).

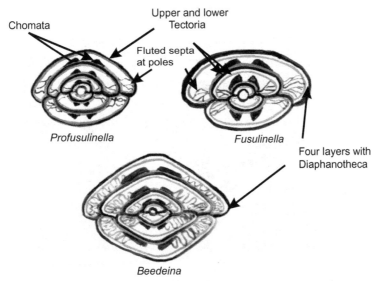

Fig. 2.12. Schematic drawings showing the features of the fusulinides (not to scale).

Significant genera include:

- *Beedeina* Galloway, 1933 (Type species: *Fusulinella girtyi* Dunbar and Condra, 1928). The test is fusiform with plane to weakly fluted septa in the early whorls, later septa are strongly fluted. Secondary fillings also coat the septa. Chomata are massive, high and broad in the early stage, but lower in the adult. Late Carboniferous (Moscovian) (Fig. 2.12).
- *Eofusulina* Rauzer-Chernousova, 1951 (Type species: *Fusulina triangula* Rauzer-Chernousova and Belyaev, in Rauzer-Chernousova et al.,1936). The test is fusiform with septa strongly fluted throughout, with high and narrow folds. Carboniferous (Early Moscovian).
- *Eowedekindellina* Ektova, 1977 (Type species: *Eowedekindellina fusiformis* Ektova, 1977). The test is fusiform with fluted septa, mainly in final whorl. Chomata are extended poleward, forming a basal layer. Late Carboniferous (Bashkirian).
- *Fusulina* Fisher de Waldheim, 1829 (Type species: *Fusulina cylindrica* Fischer de Waldheim, 1830). The test is cylindrical with a spirotheca composed of a tectum, diaphanotheca and thick late and early tectoria. The septa are fluted throughout the test but more intensely near the polar ends of the test and the chomata are strong. Late Carboniferous (Moscovian) (Plate 2.8, figs 6, 9; Plate 2.19, Fig. 2).
- *Fusulinella* von Möller, 1877 (Type species: *Fusulinella bocki* von Möller, 1878). The test is fusiform with the septa being fluted in the polar region. Chomata are prominent and asymmetrical. Late Carboniferous (Moscovian) (Figs 2.12, 2.13).
- *Hemifusulina* von Möller, 1877 (Type species: *Hemifusulina bocki* von Möller, 1878). The test has its early whorls closely coiled, later they become more loosely coiled. Septa are slightly to moderately fluted. Chomata are rounded and well developed. Middle Carboniferous (Moscovian).

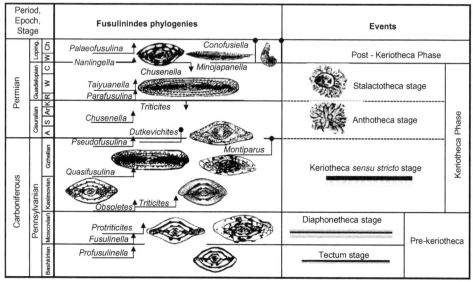

Fig. 2.13. The stratigraphic occurrence of the keriotheca, antetheca and stalactotheca stages (modified after Vachard *et al.,* 2004).

- *Obsoletes* Kireeva, 1950 (Type species: *Fusulina obsoleta* Schellwien, 1908). The test is elongate fusiform with tightly coiled early whorls becoming rapidly enlarged in the last whorls. Septa are fluted at the poles. Chomata are well developed and wide. Late Carboniferous (Kasimovian) (Fig. 2.13).

- *Paraeofusulina* Putrya, 1956 (Type species: *Eofusulina (Paraeofusulina) trianguliformis* Putrya, 1956). The test is elongate fusiform with strongly fluted septa forming strongly arched loops and poorly differentiated pseudochomata in the early whorls. Late Carboniferous (Moscovian).

- *Parafusulinella* Stewart, 1970 (Type species: *Parafusulinella propria* Stewart, 1970). The test is small, fusiform with septa undulating in the central region and slightly fluted towards the apices. Late Carboniferous (Moscovian).

- *Profusulinella* Rauzer-Chernousova and Belyaev, 1936 (Type species: *Profusulinella pararhomboides* Rauzer-Chernousova and Belyaev, in Rauzer-Chernousova et al., 1936). The test is fusiform with septa fluted in the polar regions. Chomata are prominent, asymmetrical with secondary thickening extending laterally towards the poles. Late Carboniferous (Middle Bashkirian to Early Moscovian) (Figs 2.12, 2.13; Plate 2.8, figs 1-4).

- *Protriticites* Putrya, 1948 (Type species: *Protriticites globulus* Putrya, 1948). The test is fusiform, and septa are moderately fluted in the axial region. The wall has four-layers, with massive asymmetrical chomata. Late Carboniferous (lattermost Moscovian to Kasimovian) (Fig. 2.13).

- *Quasifusulina* Chen, 1934 (Type species: *Fusulina longissima* Moeller, 1878). The test is elongate with poles bluntly rounded. Septa are intensely fluted and may form cuniculi. The wall is very thin, with a tectum and diaphanotheca and a poorly defined tectorium. Chomata are weakly developed, axial filling is heavy. Late Carboniferous (Moscovian) to Early Permian (Sakmarian) (Plate 2.15, Fig. 8; Plate 2.16, figs 3, 5, 8, 10; Plate 2.20, figs 2-3; Fig. 2.13).

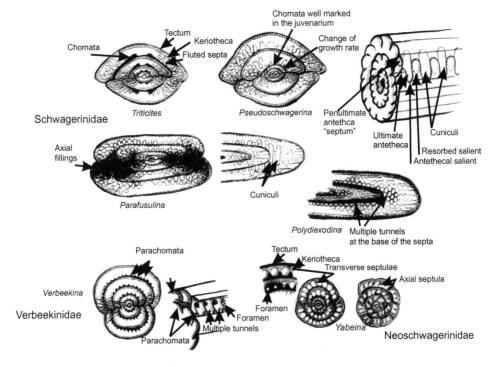

Fig. 2.14. Schematic features of the advanced forms of fusulinides: the Schwagerinidae, the Verbeekinidae and the Neoschwagerinidae.

- *Skinnerella* Coogan, 1960 emend. Skinner, 1971 (Type species: *Parafusulina schuberti* Dunbar and Skinner, 1937). The test is elongate with strong secondary deposits that may completely fill the chambers of the early whorls and tend to spread away from the axis to outer whorls. Low cuniculi are well developed. Early Permian (Kungurian).
- *Wedekindellina* Dunbar and Henbest, 1933 (Type species: *Fusulinella euthusepta* Henbest, 1928). The test is fusiform with straight septa in the central area of the test, but may be slightly fluted toward the extremities. Late Carboniferous (Moscovian).

Superfamily SCHWAGERINOIDEA Vachard, Pille and Gaillot, 2010
The schwagerinoids are large and fusiform to irregularly cylindrical. The test is planispirally enrolled, involute to irregularly coiled. The spirotheca is thick, composed of tectum and alveolar keriotheca (Fig. 2.10E). The septa are fluted in last whorls in primitive genera and completely across the test, forming multiple tunnels or cuniculi in more advanced genera (Fig. 2.14). Late Carboniferous to Permian (Kasimovian to Capitanian).

Family Schwagerinidae Dunbar and Henbest, 1930
The Schwagerinidae have tests that are large, fusiform to sub-cylindrical, planispiral and involute. They have spirothecal structures showing a tectum and a glossy layer, keriotheca (including a finely or coarsely alveolar) phase (Fig. 2.10E). In the most

advanced forms, such *Polydiexodina* Dunbar and Skinner 1931, the folded septa with a reduced single-layer wall have multiple tunnels or cuniculi throughout the test and heavy fillings along the axis (see Fig. 2.14; Plates 2.19, 2.21, 2.22). During the early phase of the keriotheca, the Schwagerinidae possess chomata that range from thin to massive (Figs 2.14–2.16). The intensity of the folded septa varies from genus to genus. Late Carboniferous to Permian (Kasimovian to Capitanian).
Significant genera include:

- *Carbonoschwagerina* Ozawa, Watanabe and Kobayashi, 1992 (Type species: *Pseudoschwagerina morikawai* Igo, 1957). The test is subspherical, septa are closely spaced, and weakly fluted. Chomata are distinct, and massive in inner whorls. This form has a smaller proloculus and much heavier chomata than in *Pseudoschwagerina*. Late Carboniferous (Gzhelian).
- *Chalaroschwagerina* Skinner and Wilde, 1965 (Type species: *Chalaroschwagerina inflata* Skinner and Wilde, 1965). The test is small fusiform, septa are strongly fluted in the early part, but merely wavy in the late part. Weak chomata occur on the proloculus. Early Permian (Artinskian to Kungurian).
- *Chusenella* Hsu, 1942 (Type species: *Chusenella ishanensis* Hsu, 1942). The test is fusiform, and later septa are tightly fluted. The wall has a tectum and keriotheca. Chomata may be lacking throughout but axial filling is prominent. Early to Middle Permian (Sakmarian to Capitanian) (Fig. 2.13).
- *Cuniculinella* Skinner and Wilde, 1965 (Type species: *Cuniculinella tumida* Skinner and Wilde, 1965). The test is fusiform, and septa are intensely fluted with folds reaching the top of the septa. Those of adjacent septa join to form chamberlets. Chomata are present only on the proloculus. Permian (Artinskian to Kungurian).
- *Daixina* Rozovskaya, 1949 (Type species: *Daixina ruzhencevi* Rozovskaya, 1949). The test is fusiform, having septa with strong and irregular fluting. Chomata are rarely present in the early whorls. Late Carboniferous to Early Permian (Gzhelian to Asselian)

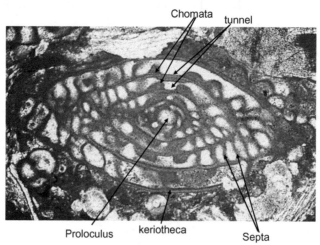

Fig. 2.15.   Morphological features of an axial section of *Triticites*. Wolfcamp Beds, Wolfcamp Hills, N.E. Marathon Texas, UCL collection.

- *Dunbarinella* Thompson, 1942 (Type species: *Dunbarinella ervinensis* Thompson, 1942). The test is fusiform with a straight axis of coiling. It differs from *Schwagerina* and *Pseudofusulina* in the presence of heavy axial fillings and distinct chomata. Late Carboniferous to Early Permian (Gzhelian to Sakmarian) (Plate 2.8, Fig. 5; Plate 2.16, Fig. 11; Plate 2.23, Fig. 8; Plate 2.24, Fig. 4).
- *Eoparafusulina* Coogan, 1960 (Type species: *Fusulina gracilis* Meek, 1864). The test is subglobular, tightly coiled and expanding uniformly. Late parts of the septa are planar, while early parts are strongly fluted throughout the test, producing chamberlets, with low tunnel bordered by pseudochomata. Rudimentary chomata are present in in first whorls. Early Permian (Late Asselian to Artinskian and?Kungurian).
- *Eopolydiexodina* Wilde, 1975 (Type species: *Polydiexodina afghanensis* Thompson, 1946). Septa are intensely fluted throughout the entire length and in all whorls of the test, with cuniculi well developed. It differs from *Polydiexodina* in lacking a median tunnel and in having only sporadic supplementary tunnels. Permian (Middle Guadalupian).
- *Monodiexodina* Sosnina, 1956 (Type species: *Schwagerina wanneri* var. *sutschanica* Dutkevich, 1934). The test is subcylindrical, with septal fluting with folds restricted to the early part, forming low cuniculi. Early whorls have chomata. Early to Middle Permian (Cisuralian to Late Guadalupian).
- *Montiparus* Rosovskaya, 1948 (Type species: *Alveolina montipara* Ehrenberg, 1854, emend. Möller, 1878). The test is fusiform. Septa are irregularly plicate. Chomata are sub-quadrate. Carboniferous (Kasimovian) (Fig. 2.13; Plate 2.8, Fig. 7).
- *Parafusulina* Dunbar and Skinner, 1931 (Type species: *Parafusulina wordensis* Dunbar and Skinner, 1931). The test is very large with intense folding but with no chomata (Fig. 2.9). Middle to Late Permian (Guadalupian to Lopingian) (Fig. 2.14; Plate 2.16, figs 1-2; Plate 2.19, figs 5, 7).
- *Paraschwagerina* Dunbar and Skinner, 1936 (Type species: *Schwagerina gigantea* White, 1932). The test is fusiform to subspherical. The septa are distinctly fluted throughout test, with high, strong, and parallel sided folds. The wall is thin, composed of thin tectum and coarsely alveolar keriotheca. The chomata are small in the first whorls, but absent in the later whorls. Early Permian (Asselian to Sakmarian) (Plate 2.17, figs 4, 6; Plate 2.18, Fig. 10).
- *Polydiexodina* Dunbar and Skinner, 1931 (Type species: *Polydiexodina capitanensis* Dunbar and Skinner, 1931). Septa are numerous, regularly and strongly fluted, with narrow septal folds, the fluting of adjacent septa producing cuniculi where in contact. Permian (Late Guadalupian) (Fig. 2.14; Plate 2.17, Figs 3, 7).
- *Pseudofusulina* Dunbar and Skinner, 1931 (Type species: *Pseudofusulina huecoensis* Dunbar and Skinner, 1931). The test is fusiform with a large proloculus and fluted septa, most strongly towards the poles where it is densely folded to produce chamberlets. Late Carboniferous to Permian (lattermost Gzhelian to Kungurian) (Fig. 2.13; Plate 2.18, Fig. 13).
- *Pseudoschwagerina* Dunbar and Skinner, 1936 (Type species: *Schwagerina uddeni* Beede and Kniker, 1924). The test has inflated outer whorls and strongly folded septa throughout (Fig. 2.9). Chomata are well marked at juvenile stage only. They

are rudimentary or lacking in the final stage. Early to Middle Permian (Asselian to Kungurian) (Fig. 2.14; Plate 2.19, Fig. 5B; Plate 2.18, figs 3, 4, 8, 11; Pate 2.22, figs 3, 5).

- *Schwagerina* von Mölle, 1877 (Type species: *Borelis princeps* Ehrenberg, 1842). Septa are regularly and intensely fluted, and folds of septa divide the early part of the chambers into small chamberlets. Chomata are weak or absent, with narrow tunnels. Early Permian (Sakmarian to Artinskian) (Plate 2.15, Fig. 5; Plate 2.16, figs 4,6; Plate 2.17, Fig. 5A; Plate 2.20, figs 4, 5; Plate 2.23, figs 3-4, 6; Plate 2.24, Fig. 6; Plate 2.25, figs 1, 3-4).
- *Skinnerina* Ross, 1964 (Type species: *Skinnerina typicalis* Ross, 1964). Septa are numerous and intensely fluted from pole to pole, resulting in high and well developed cuniculi. Permian (Early Guadalupian).
- *Taiyuanella* Zhuang, 1989 (Type species: *Taiyuanella subsphaerica* Zhuang, 1989). The test is fusiform, with spirotheca composed of a tectum and stalactotheca. Septa are strongly fluted throughout the length of the test. Chomata are weakly developed. Middle Permian (Wordian) (Fig. 2.13).
- *Triticites* Girty, 1904 (Type species: *Miliolites secalicus Say,* in James, 1823). Septa are weakly folded over the equator, with massive chomata (Fig. 2.16). Late Carboniferous

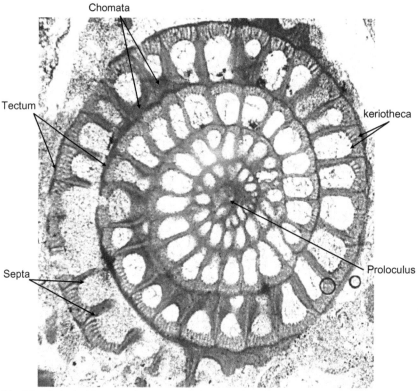

Fig. 2.16. Morphological features of an equatorial section of *Triticites.*

to Early Permian (Kasimovian to Sakmarian) (Figs 2.14–2.16; Plate 2.8, Fig. 10; Plate 2.15, figs 1-3, 7; Plate 2.20, Fig. 6; Plate 2.11, figs 1-2; Plate 2.15, figs 1- 3, 7; Plate 2.20, Fig. 6; Plate 2.25, figs 5-6).

- *Zellia* Kahler and Kahler, 1937 (Type species: *Pseudoschwagerina (Zellia) heritschi* Kahler and Kahler, 1937). The test is globular with a thick wall composed of a tectum and coarsely alveolar keriotheca, with low, widely spread chomata. Permian (Sakmarian) (Plate 2.25, Fig. 2).

Superfamily NEOSCHWAGERINOIDEA Solovieva, 1978
Found in the Late Permian showing the last stage of the keriotheca phase, these Neoschwagerinoids are different from the fusulinoids in having a wall made of a tectum and a glossy layer with alveoles running through it (the keriotheca) (Fig. 2.10E), and they differ from the schwagerinoids in having secondary layers and multiple chomata or parachomata, with multiple tunnels. Permian (Asselian to Wordian).

Family Verbeekinidae Staff and Wedekind, 1910
The test is globose with discontinuous parachomata. Permian (Asselian to Early Lopingian).

- *Verbeekina* Staff, 1909 (Type species: *Fusulina verbeeki* Geinitz, in Geinitz and von der Marck, 1876). Spherical with a tiny proloculus, parachomata are absent or rudimentary (Fig. 2.9). Permian (Roadian to Capitanian) (Figs 2.9, 2.14; Plate 2.15, Fig. 9).
- *Brevaxina* Schenck and Thompson, 1940 (Type species: *Doliolina compressa* Deprat, 1915). The test is subspherical with a small proloculus, followed by an endothyroid early stage. The wall is thick, with tectum, high and broad parachomata and numerous foramina. Permian (Kungurian to early Wordian).

Family Misellinidae Miklukho-Maklay, 1958
The test is fusiform with well-developed parachomata. Permian (Sakmarian to Early Lopingian).

- *Misellina* Schenck and Thompson, 1940 (Type species: *Doliolina ovalis* Deprat, 1915). The test is ellipsoidal with numerous chambers, and numerous plane, unfluted septa. The keriotheca is thick and the parachomata are "saddle-shaped". Permian (Kungurian to early Wordian).

Family Pseudodoliolinidae Leven, 1963
Ellipsoidal test with planar septa and narrow, high parachomata. The parachomata supplement the primary pair of ridges in the equatorial zone of the shell laterally and polewards, regularly intercalated between supplementary tunnels until the polar end of the chamber, Permian (Artinskian to Capitanian).

- *Pseudodoliolina* Yabe and Hanzawa, 1932 (Type species: *Pseudodoliolina ozawai* Yabe and Hanzawa, 1932). The test is bluntly circular with thin wall and reduced keriotheca. Septa are planar and unfluted. The parachomata are high, continuous and may reach the top of the chambers adjacent to the septa, appearing as septula in axial section. Permian (Artinskian to Capitanian).

Family Neoschwagerinidae Dunbara and Condra, 1927

In the Neoschwagerinidae (see Fig. 2.11) the finely alveolar keriotheca join the para-chomata to form transverse septula which form complete partitions in the chamber. They occur in the Permian (Asselian to Early Lopingian).

- *Cancellina* Hayden, 1909 (Type species: *Fusulina (Neoschwagerina) primigena* Hayden, 1909). The test is fusiform and broadly inflated in the centre. The septa are arcuate and broad transverse septula are formed by the extension of the lower part of the kerio-theca. Broad parachomata are present. Early to Late Permian (Roadian to Capitanian).
- *Neoschwagerina* Yabe, 1903 (Type species: *Schwagerina craticulifera* Schwager, 1883). The test is subspherical. Chambers are divided into chamberlets by transverse septula of first order which connect with the poorly developed parachomata. The wall con-sists of a tectum and a very thick keriotheca that may be differentiated into an upper finely alveolar layer and a lower layer with fewer and coarser alveoli and with upper and lower tectoria. Early to Middle Permian (Wordian to Capitanian) (Plate 2.20, Fig. 1; Plate 2.23, figs 1-2; Plate 2.26, figs 5-7).
- *Sumatrina* Volz, 1904 (Type species: *Sumatrina annae* Volz, 1904). The test is elon-gate fusiform with a large proloculus and long, thin septa. Up to four short sec-ondary septula develop between primary septula. The wall is very thin with a thin keriotheca. Massive parachomata and numerous foramina occur throughout the length of the test. Middle Permian (Capitanian).
- *Yabeina* Deprat, 1914 (Type species: *Neoschwagerina (Yabeina) inouyei* Deprat, 1914). The test is large and fusiform with well-developed parachomata connected to the early ends of the septula. Up to three secondary spiral septula in outer whorls and nine axial septula per chamber. Middle Permian (Guadalupian) (Fig. 2.14; Plate 2.24, figs 1-2).

ORDER LAGENIDA DELAGE AND HÉROUARD, 1896

As an order the Lagenida are characterised by having walls made of orientated calcite crystals. It appears that this characteristically evolved independently, in parallel several times in the Palaeozoic, and that the superfamilies Robuloidoidea and Nodosinelloidea have different phylogenetic roots (see Fig. 2.1).

Members of this order (Late Silurian to Holocene) were referred to the Nodosariida Calkins 1926 by Vachard et al. (2010) and Mikhalevich (2013). They have alveo-les. Primitive taxa are without secondary lamination, more advanced forms have a secondary lamination. The wall structures in selected lagenides have been restudied recently (Groves et al., 2004). "Monolamellar" refers to the primary single-layered septal wall. In many lagenides, secondary lamellar develop, where extensions of the primary wall of a given chamber overlap some of the previous chambers. Reiss (1963) proposed an extensive reclassification of lamellar forms, and was among the first to note that radial walls are secreted on an organic substrate. Grønland and Hansen (1976) by examining Holocene lagenides developed a new terminology for lamellarity:

- "Atelo-monolamellar" refers to forms that lack secondary lamellarity and they rep-resent the earliest lagenides, e.g. the simple *Nodosaria* (Plate 2.5, Fig. 2.11);

- "plesio-monolamellar" refers to forms in which secondary lamellarity envelops some but not all previous chambers;
- "ortho-monolamellar" refers to forms in which the primary wall of each chamber secondarily envelops all previous chambers; and
- "poly-monolamellar" refers to forms in which yet another lamella secondarily covers the entire test.

The plesio-monolamellar forms are thought to have originated in Early Jurassic, followed by ortho-monolamellar and poly-monolamellar types in the Cretaceous and Cenozoic (Groves et al., 2004).

Superfamily ROBULOIDOIDEA Reiss, 1963
The tests of members of this superfamily are those of a typical Lagenida, but without secondarily lamellar or with slight lamination in younger taxa. The aperture is primitive and cylindrical. Late Silurian to Early Cretaceous.

Family Syzraniidae Vachard, 1981
The test has two chambers and is elongate, composed of a proloculus and an undivided or subseptate tubular chamber partitioned to varying degrees by internal thickenings of the wall, which is calcareous and may have two layers. Late Silurian to Late Permian (Pridolian to Lochkovian; Tournaisian to Changhsingian).

- *Amphoratheka* Mamet and Pinard, 1992 (Type species: *Amphoratheka iniqua* Mamet and Pinard, 1992). The septa are slightly more pronounced than in *Tezaquina*. Late Carboniferous to Early Permian (Gzhelian to Sakmarian).
- *Rectostipulina* Jenny-Deshusses, 1985 (Type species: *Rectostipulina quadrata* Jenny-Deshusses, 1985). The test consists of a long narrow tapering tube with a polygonal transverse section. Septa are short and nearly perpendicular to the test wall. The wall is atelo-monolamellar. The aperture is a simple opening in the center of the apertural face. Late Permian (Wuchiapingian).
- *Syzrania* Reytlinger, 1950 (Type species: *Syzrania bella* Reytlinger, 1950). The test has a spherical proloculus followed by a non-septate tubular chamber. The wall has a "fibrous" layer and an inner microgranular layer. The aperture is a simple terminal opening. This genus may have evolved from another simple robuloidoid or most likely from the earlandiides *Earlandia* in Middle Pennsylvanian. Carboniferous to Late Triassic (Moscovian to Rhaetian) (Fig. 3.3).
- *Tezaquina* Vachard, 1980 (Type species: *Tezaquina clivuli* Vachard, 1980). Groves in Groves and Boardman (1999) emended *Tezaquina* to include nonseptate syzraniid forms whose tubular second chamber is subdivided into pseudo-chambers by weak internal thickenings of the wall, which is thick hyaline fibrous and finely perforate. The test consists of a spherical proloculus followed by a second chamber that is partitioned into elongate, subcylindrical pseudo-chambers by weak internal thickenings of the wall. The aperture is a simple terminal opening, and the wall is atelo-monolamellar. Carboniferous to Late Triassic (late Moscovian to Rhaetian) (Fig. 3.3).

- *Tuborecta* Pronina, 1980 (Type species: *Tuborecta vagranica* Pronina, in Petrova and Pronina, 1980). The test is small, forming a long narrow tube. The wall is calcareous, of two layers, a thin inner finely granular, dark layer and a thick radial, light outer layer. The aperture is terminal, at the end of the tube. Late Silurian to Early Devonian (Pridolian to Lochkovian).

Family Protonodosariidae Mamet and Pinard 1992
This family includes the uniserial Lagenida with atelo-monolamellar or plesio-monolamellar wall structures consisting of radial-fibrous forms. Late Carboniferous to Late Permian (Kasimovian to Changhsingian).

- *Protonodosaria* Gerke, 1959 (Type species: *Nodosaria proceraformis* Gerke, 1952). The test is elongate, circular in thin section. Sutures are horizontal and straight. The aperture is round. Late Carboniferous to Late Permian to Late Triassic (Kasimovian to Rhaetian).

Family Ichtyolariidae Loeblich and Tappan, 1986
The test is elongate uniserial with a single layered wall, and may show some secondary lamination, with a simple terminal aperture. Early Permian to Early Cretaceous (Artinskian to Albian).

- *Pseudotristix* Miklukho-Maklay, 1960 (Type species: *Tristix (Pseudotrislix) tcherdvnzevi* Miklukho-Maklay, 1960). The test consists of a spherical proloculus followed by a uniserial arrangement of triangular chambers. The sutures are nearly horizontal, the aperture is a simple opening, and the wall is atelo-monolamellar. Late Permian (Lopingian).

Family Robuloididae Reiss, 1963
The test is uniserial, enrolled with an atelo-monolamellar wall and a terminal aperture. Middle Permian to Late Jurassic (Guadalupian to Kimmeridgian).

- *Calvezina* Sellier de Civrieux and Dessauvagie, 1965 (Type species: *Calvezina ottoman* Sellier de Civrieux and Dessauvagie, 1965). The test consists of a proloculus followed by five or more uniserial chambers that are weakly trochospirally enrolled in the early stage, later becoming rectilinear. The chambers are slightly compressed laterally and their size increases rapidly from early hemispherical ones to later lobate or irregular ones. The aperture is a simple terminal opening accompanied by a slight thickening of the wall. Late Permian (Lopingian).
- *Eocristellaria* Miklukho-Maklay, 1954 (Type species: *Eocristellaria permica* Miklukho-Maklay, 1954). The test is planispirally enrolled with chambers increasing rapidly in breadth, but slowly in height, so the apertural face extend back towards the proloculus. The calcareous wall is composed of an outer hyaline radiate layer and an inner dark microgrannular layer. Late Permian (Lopingian).
- *Robuloides* Reichel, 1946 (Type species: *Robuloides lens* Reichel, 1946). The test is planispiral with oblique septa, arching back towards the periphery. Permian to Middle Triassic (Artinskian to Ladinian).

Superfamily NODOSINELLOIDEA Rhumbler, 1895

The Nodosinelloidea evolved from the Earlandiida which in turn evolved from the Parathuramminoidea (see Fig. 2.2). Their tests were free, with more than one distinct chamber, partially septate or fully septate, with a simple microgranular wall or evolving into two layers with an extra fibrous inner layer. The Nodosinelloidea range from the Late Silurian to the Permian.

Family Earlandinitidae Loeblich and Tappan, 1988

The Earlandinitidae (Fig. 2.2) evolved from an earlandioid, Caligellidae, by becoming elongate, uniserial, semi-septate, with a simple microgranular wall, and range from Late Devonian to Late Carboniferous (Bashkirian).

- *Darjella* Malakhova, 1964 (Type species: *Darjella monilis* Malakhova, 1964). The test is uniserial, with globular chambers and strongly constricted sutures. Early Carboniferous (Early Visean).
- *Earlandinita* Cummings, 1955 (Type species: *Nodosinella perelegans* Plummer, 1930). The test is subseptate with a terminal round aperture. Carboniferous (See Fig. 2.2).
- *Lugtonia* Cummings, 1955 (Type species: *Nodosinella concinna* Brady, 1876). The test is rectilinear with depressed chambers overhanging the sutures. Carboniferous.

Family Nodosinellidae Rhumbler, 1985

This family (Fig. 2.2) evolved from the Earlandioidea, and is characterised by having a uniserial test which develops in the Late Carboniferous to having a double layered wall, with an inner fibrous layer and an outer microgranular layer. They range from the Late Silurian to Permian.

- *Biparietata* Zolotova, 1980 (Type species: *Biparietata ampula* Zolotova, 1980). The test is rectilinear with elongate chambers and sutures not apparent externally. Middle Permian (Guadalupian).
- *Nodosinella* Brady, 1876 (Type species: *Nodosinella digitata* Brady, 1876). The test is free and fully septate. Permian (see Figs 2.2, 2.4).

Superfamily COLANIELLOIDEA Fursenko, 1959

In this superfamily, the tests are uniserial and fully septate, and chambers are strongly overlapping with partitions. The walls occur with an outer vitreous layer and an inner granular layer. They range over the Late Devonian to Late Permian.

Family Colaniellidae Fursenko, 1959

The family evolved from the Nodosinellidae by developing strong internal partitions or septa (Fig. 2.2). They range from the Late Devonian to Late Permian.

- *Colaniella* Likharev, 1939 (Type species: *Pyramis parva* Colani, 1924). Elongate and uniserial, having a calcareous wall with a fibrous structure with strongly overlapping sutures. The aperture is terminal radiate. Late Permian (see Fig. 2.2).

- *Cylindrocolaniella* Loeblich and Tappan, 1985 (Type species: *Wanganella ussuriensis* Sosnina, in Kiparisova et al., 1956). The test is slightly arcuate, narrow and cylindrical in form with numerous short septa. Late Permian (Lopingian).
- *Pseudowanganella* Sosnina, 1983 (Type species: *Pseudowanganella tenuitheca* Sosnina, 1983). The test is narrow, flattened and elliptical in cross section. Chambers are curved centrally and overlapping near the periphery. Septa are more thickened near the aperture. Late Permian (Lopingian).

Superfamily GEINITZINOIDEA Bozorgnia, 1973
The tests of this superfamily are uniserial and appear similar to the Nodosinelloidea, but the microgranular layer is the inner dark layer while the radially fibrous layer is an outer layer. Advanced forms have secondary lamellarity. Late Devonian to Middle Triassic.

Family Geinitzinidae Bozorgnia, 1973
The tests are free and uniserial. They evolve from species that are rounded in cross-section to species that are compressed with a terminal aperture. The range is Late Devonian to Late Permian.

- *Geinitzina* Spandel, 1901 (Type species: *Geinitzella (Lunucammina) permiana* Spandel, 1898). The test is uniserial elongate, laterally compressed, exhibiting in transverse view two planes of bilateral symmetry and an oval aperture in the centre of the apertural face. Permian (Cisuralian to Lopingian) (Figs 3.3 and 3.4).
- *Howchinella* Palmieri, 1985 (Type species: *Frondicularia woodwardi* Howchin, 1895). The test is laterally compressed, with arched or chevron sutures. The wall is calcareous granular with a dark thin inner organic layer and an outer hyaline layer of optically radial calcite. The aperture is radiate with a tooth. Early Permian (Sakmarian).
- *Nodosinelloides* Mamet and Pinard, 1992 (Type species: *Nodosinelloides potievskayae* Mamet and Pinard, 1996 (for *Nodosaria gracilis* Potievskaya, 1962 preoccupied)). The test consists of a spherical proloculus followed by as many as 10 or more uniserial circular chambers with pronounced septal inflection and a simple aperture. The wall lacks secondary lamellarity (atelo-monolamellar) and the aperture is a simple terminal opening. Carboniferous to Triassic (Kasimovian to Rhaetian) (Fig. 3.3).

Family Pachyphloiidae Loeblich and Tappan, 1984
This family has a free uniserial compressed test with a simple calcareous microgranular wall, but with secondary thickenings on both sides of the test. The range is from Early to Late Permian.

- *Pachyphloia* Lange, 1925 (Type species: *Pachyphloia ovata* Lange, 1925). The test consists of a globular proloculus followed by a uniserial succession of laterally compressed, strongly overlapping chambers. The wall is plesio- or ortho-monolamellar with thin dark inner layer and thicker hyaline outer layer. The aperture is terminal, oval, radiate (toothed). Early Permian to Late Permian (Sakmarian to Changhsingian) (Plate 2.5, figs 2-4; Fig. 3.3).

ORDER MILIOLIDA DELAGE AND HÉROUARD, 1896

The miliolides have tests that are porcellaneous and imperforate, which are made of high Mg-calcite with fine randomly oriented crystals. They range from the Carboniferous to the Holocene.

Superfamily CORNUSPIROIDEA Schultze, 1854

The test is free or attached, and composed of a globular proloculus followed by a tubular enrolled chamber. The coiling is planispiral or trochospiral, evolute or involute and may become irregular. There is a simple aperture at end if the tube. Early Carboniferous to Holocene.

Family Cornuspiridae Schultze, 1854

The tests are free or attached, composed of a proloculus followed by an undivided planispiral to streptospiral, involute or evolute second tubular chamber. Early Carboniferous (Visean) to Holocene.

- *Cornuspira* Schultze, 1854 (Type species: *Orbis foliaceus* Philippi, 1844). The test is discoidal with a globular proloculus and undivided planispirally enrolled and tubular second chamber. Carboniferous (Moscovian) to Holocene.
- *Rectocornuspira* Warthin, 1930 (Type species: *Rectocornuspira lituiformis* Warthin, 1930). Test elongate, with a globular proloculus and planispirally enrolled to later uncoiling and rectilinear undivided tubular second chamber. Late Carboniferous to Early Triassic (Moscovian to Induan).

Family Hemigordiidae Reitlinger, 1993

Characterized by a test with at least an early trochospiral stage, but that may later become streptospiral. Carboniferous.

- *Hemigordius* Schubert, 1908 (Type species: *Cornuspira schlumbergeri* Howchin, 1895). The test has early whorls that are streptospiral, later becoming planispiral and evolute. Early Carboniferous to Late Carboniferous (Late Visean to Gzhelian) (Fig. 3.5).
- *Neohemigordius* Wang, 1973 (Type species: *Neohemigordius maopingensis* Wang and Sun, 1973). The test is lenticular, with a lamellar thickening over the umbilical area. Early Permian (Asselian to Sakmarian) (Fig. 3.5).
- *Pseudovidalina* Sosnina, 1978 (Type species: *Pseudovidalina* Sosnina, 1978). The test is discoid with a globular proloculus followed by planispirally, evolute, enrolled tubular undivided second chamber with secondary thickenings in the umbilical region. Middle Permian (Kungurian).

Family Hemigordiopsidae Nikitina, 1969

Forms in this family have a test with at least an early streptospiral stage, that later may be planispiral. Early Carboniferous (Visean) to Holocene.

- *Hemigordiopsis* Reichel, 1945 (Type species: *Hemigordiopsis renzi* Reichel, 1945). The test is globular with a proloculus followed by an undivided tube, which is strongly involute. Early to Middle Permian (Guadalupian) (Fig. 3.3).

• *Shanita* Brönnimann, Whittaker and Zaninetti, 1978 (Type species: *Shanita amosi* Brönnimann, Whittaker and Zaninetti, 1978). The proloculus followed by an undivided enrolled second chamber, with a streptospiral in early coiling, becoming planispiral and involute. Later whorls with alternating vertical pillars. Late Permian (Wuchiapingian to Changhsingian) (Fig. 3.6).

2.3 Biostratigraphy and Phylogenetic Evolution

2.3.1 The fusulinides and related orders

The organic-walled allogromiides gave rise to the most primitive Palaeozoic larger benthic foraminifera, the **Parathuramminida,** in the Silurian (Figs 2.1, 2.2). The simple form parathuramminides gave rise rapidly to the pseudoseptate **Pseudoammodiscidae** belonging to the **earlandiides** in the Devonian, and eventually to the pseudoseptate **tournayellides** in the Late Devonian (Fig. 2.6). The latter gave rise to the completely septate **endothyrides**. The endothyrides include coiled forms and advanced uniserial genera with uniserial adults (e.g. *Haplophragmella*). They also exhibit species with a strong development of secondary deposits of calcite (chomata) on the chamber floor (e.g. *Endothyra, Quasiendothyra*). The chomata enfold the wall to give septal chamberlets in the **Bradyinidae,** which communicate with the exterior via septal pits (e.g. *Bradyina,* Figs 2.9, 2.10). The endothyrides reached their acme and largest sizes in the Early Carboniferous (Visean), with trends towards evolute planispiral coiling and well developed septa in early whorls in the **Loeblichiidae** (e.g. *Loeblichia*). The latter finally gave rise to the higher **Fusilinida** (see Fig 2.17).

The **Loeblichiidae** developed walls that were differentiated into two or more layers in the most advanced forms, leading to the first appearance of the **Ozawainelloidea** lineage of the fusulinides in the Early Carboniferous (Visean) (Còzar and Vachard, 2001). The fusulinides started as simple tiny organisms in the Early Carboniferous, but had evolved into large and complicated forms by the end of the Carboniferous and Permian. Their very large sizes in the Early Permian, some up to 8cm length, are related to the microspheric generations, as dimorphism becomes evident in larger benthic foraminifera. The rapid proliferation of genera, the increasingly strongly folded septa, the insertion of septula, the development of chomata and axial deposits, the replacement of the central tunnel by numerous foramina, the development of axial infilling in the very large species, and the development of parachomata in the most advanced forms, led to the evolution of the different superfamilies of the fusulinides, the **Staffelloidea,** the **Schubertelloidea,** the **Fusulinoidea** and the more advanced **Schwagerinoidea,** and **Neoschwagerinoidea** (Fig. 2.18). Because of their complexity, size and stratigraphical usefulness, the evolution and range of the Fusulinida superfamilies deserve more detailed discussion, and are the focus of the remainder of this section.

The Fusulinida and related orders are the only significant forms of foraminifera without living representatives (Groves et al., 2005). The evolutionary relationships between many members of the Fusulinida and other Palaeozoic groups are summarized in Figs 2.1 and 2.2, while their biostratigraphic ranges are shown later in Figs. 2.17 to 2.19.

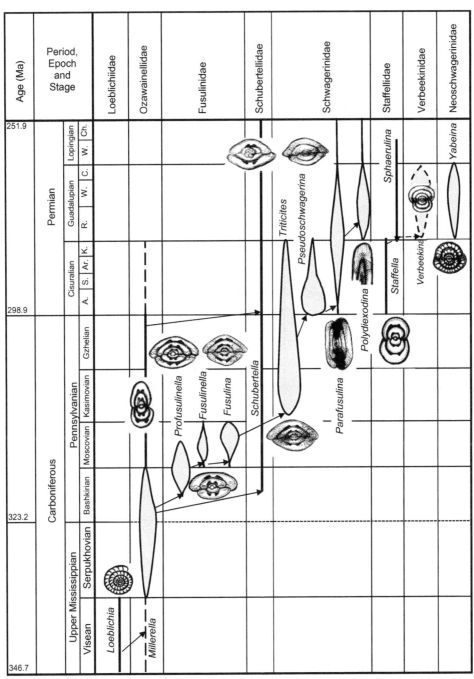

Fig. 2.17. Evolution the fusulinides. Abbreviations: A. = Asselian, S. = Sakmarian, Ar. = Artinskian, K. = Kungurian, R. = Roadian, W. = Wordian, C. = Capitanian, W. = Wuchiapingian, Ch. = Changhsingian.

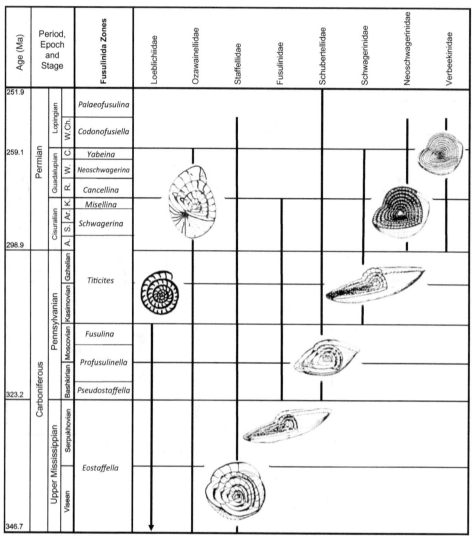

Fig. 2.18. Evolution and biozonations of the fusulinides (drawings of foraminifera from Jin-Zhang (1990) and are not here to scale).

One of the most important morphological features of the fusulinides and the related orders is their mode of coiling. Planispiral coiling is dominant (70%; Haynes, 1981) especially in the Early Carboniferous. On the other hand, a fusiform chamber arrangement is common in the most advanced forms of fusulinides from Late Carboniferous to Late Permian. Many of the structural features of the large fusulinides tests can be interpreted from external observation. The traces of the septa on the external surface are called septal furrows. The apertural face of the last chamber is the antetheca and the external wall is the spirotheca (see Figs 2.15 and 2.16). The antetheca lacks true apertures, but the internal septa show single foramen or multiple foramina that are formed by resorption of the wall (Haynes, 1981; Leppig et al., 2005). In the **Fusulinidae** (except for few genera such as *Paraeofusulina*) and the **Schwagerinidae** (except for few genera such as *Polydiexodina* and *Eopolydiexodina*) a single large foramen is formed, while in the neoschwagerinids **Verbeekinidae** and **Neoschwagerinidae** multiple foramina are characteristic. However, like all other larger foraminifera, they cannot be identified solely on external features, and thin sections are essential for the identification of fusulinids, neoschwagerinids and staffellids. Two sections cut through the test reveal all the complicated structures needed for classification (see Figs 1.7, 2.15, 2.16). One of these is the cut through the axis of coiling (the axial section), and the other is the cut normal to the axis of coiling (the equatorial or sagittal section).

One of the diagnostic features to be seen in either section is the morphological development shown by the wall structures of the Late Carboniferous and Permian fusulinides. These wall structures range from simple calcareous granulated, in the primitive fusulinides, to compound microgranular, and then very complicated in the large fusulinids and schwagerinids of the Permian. As described above, the fusulinides can exhibit one of five types of wall structure (see Figs 2.10), which can be recognised as major evolutionary events (see Fig. 2.13):

• a Pre-keriotheca Phase, with a thin spirotheca, composed of a tectum only, the tectum stage, such as in the fusulinid *Profusulinella* and an early thicker, less dense layer, the diaphanotheca. This diaphanotheca stage includes forms such as *Fusulinella, Ozawainella* and *Protriticites* (see Fig. 2.10, A-D; Fig. 2.13). This phase corresponds to the Middle Carboniferous.

• a second phase (the Keriotheca Phase), in which the spirotheca is composed of a tectum and a coarsely alveolar keriotheca (Fig. 2.10E), is divided according to Vachard et al (2004) into three stages, the keriotheca s.s., the anthotheca and the stalactotheca (see Fig. 2.13). These stages evolve respectively from the Late Carboniferous to the late Middle Permian. During the first two stages in the Late Carboniferous and Early Permian, the family **Schwagerinidae** dominated. On the chamber floors of these advanced fusulinides secondary deposits, or chomata, were laid down and are related to the form of the chamber communications (see Figs 2.13, 2.15). During the stalactotheca stage, in the Middle Permian, the **Verbeekinidae** and **Neoschwagerinidae** were the dominant fusulinides. They had multiple foramina and multiple small chomata, thus developing parachomata (see Figs 2.9, 2.14). In the Neoschwagerinidae, the parachomata became linked with the primary spiral septula to form complete partitions.

• a Post-keriotheca Phase, exemplified by *Nanlingella* and *Palaeofusulina,* which became abundant after most of the *Schwagerina-Verbeekina-Neoschwagerina* types had disappeared (see Fig. 2.19).

The oldest and most primitive keriothecal wall structure is seen in *Triticites* (Hageman and Kaesler, 1998). The wall composition is microgranular and the alveoles are filled by a structureless microspar (Vachard et al., 2004). In other forms, such as *Sakmarella* (*Pseudofusulina*-like form)*, Praeskinerella* (*Parafusulina*-like form), and *Taiyuanella* (*Chusenella*-like), the alveoles are occupied by radial calcite prisms, which

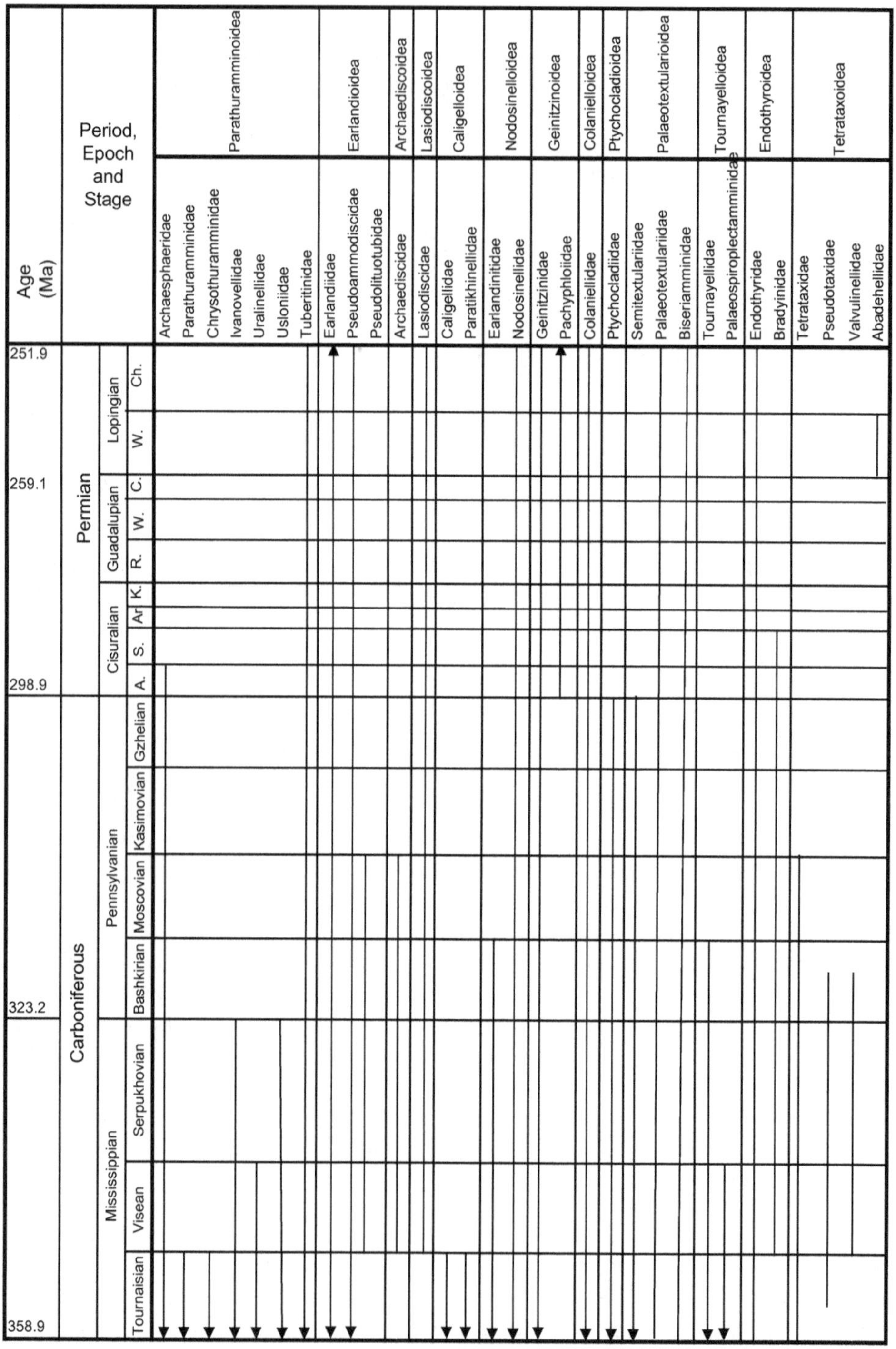

Fig. 2.19. Ranges of the Parathuramminoidea and Robuloidoidea in the Carboniferous and Permian.

are arranged like narrow-centred petals (see Fig. 2.13), called anthotheca (Vachard et al., 2004). This structure is thought to represent the calcification of some bacteria or cyanobacteria (Riding, 1977; Feldmann and McKenzie, 1998). Finally, another texture appearing like large-centred petals is known as the stalactotheca (Zhuang, 1989). In this structure, the alveoles are filled by a blocky granular calcite, but radiating prisms of calcite also exist at the periphery (Fig. 2.13). Some more advanced forms have well developed perforations (Haynes, 1981). Algal symbionts, common in modern larger foraminifera, had evolved by the Carboniferous (Tappan, 1971). Larger perforations were developed in taxa with an anthotheca, so cyanobacteria could calcify within the pores of the wall. The diaphanotheca (internal clear layer in the wall) was probably partially filled in the living organism by cyanobacterial endosymbionts and microgranules that were possibly aragonitic (Vachard et al., 2004). This hypothesis of a mixture of organic and mineral components has also been suggested by Bender and Hemleben (1988). Vachard et al. (2004) also suggested a phylogenetic lineage from the keriotheca to anthotheca, and stalactotheca stages (see Fig. 2.13). In the Neoschwagerinidae, the alveolar keriotheca seem to appear independently from those of the fusulinids. The inner structures, the arrangement of internal foramina, and the straight septa which divide the chambers completely into chamberlets seen in the neoschwagerinids and verbeekinids are morphologically similar to those seen in the later alveolinids (Mikhalevich, 2004; 2009), and provide an example of parallel evolution over time.

In advanced forms of the Fusulinoidea the septa become strongly folded. The folding first appeared near the poles, then migrated up to the centre of the test near the sutures and subsequently fill the test. In the **Schwagerinoidea** the septa became very strongly folded and produced tunnel-like cuniculi (see Fig. 2.10) running in the direction of the coiling (e.g. in the Carboniferous *Paraeofusulina,* and also in the Permian *Parafusulina*) (see Plate 2.16, figs 1-2).

Due to the problem of recrystallisation, tracing possible evolutionary relationships between the different genera within the fusulinides has been difficult, because detailed ultra-structure is over printed by diagenetic processes. Their evolution, however, has been discussed by many authors (Haynes, 1981; Rauzer-Chernosova, 1963; Loeblich and Tappan, 1980). Jin-Zjang (1990) considered the small lenticular endothyrides as the ancestral form to the Chinese fusulinids. They evolved into the **Ozawainelloidea** in the Late Mississippian, which are the earliest primitive ancestors of the superfamily **Fusulinoidea**. The Ozawainellidae family are characterised by small discoidal shells with unfluted septa. The most common genera of this family are minute, spherical forms belonging to *Ozawainella* and *Millerella* (Fig. 2.11), and are mostly common in the mid-Carboniferous (Serpukhovian and Bashkirian). The wall structure of these genera is of the *Ozawainella*-type. The number of the genera of the ozawainellids decreased in the Late Carboniferous, but the family persisted to the Late Permian.

The **Fusulinidae** family (as well as the less significant **Staffellidae** family) arose from the Ozawainellidae in the Middle Carboniferous (Moscovian), where there was a rapid increase in the number of genera. The septa became increasingly folded and the family persisted into the Middle Permian, but with a marked decrease in the number of genera. Most of the genera of the Middle Carboniferous had a wall structure of the

Profusulinella-type, including genera with a wall composed of a tectum, late and early tectorium, or the *Fusulina*-type, where the spirotheca is composed of a tectum, diaphanotheca and thick late and early tectoria. This family persists into the Middle Permian but with a decreasing diversity of genera (Fig. 2.12).

The **Schubertellidae** evolved from the Ozawainellidae in the middle Bashkirian (Sinitsyna and Sinitsyn 1987; Nikolaev 2005). The earliest primitive schubertellid, *Grovesella,* may have evolved from the ozawainellid *Eostaffella* by developing loosely coiled whorls, a two−layered wall (as opposed to the undifferentiated wall in *Eostaffella*), planispiral coiling and an absence of chomata or pseudochomata that are always present in *Eostaffella* (Davydov, 2011). The affinity between *Grovesella* and ozawainellids or staffellids or schubertellids is still disputable. *Grovesella* has been considered by Leven (2009) as belonging to the *Zarodella* Sosnina lineage of the ozawainellids, however Davydov (2011) demonstrated that this taxon belongs to the schubertellids. *Grovesella* evolved rapidly into the larger *Schubertina* in the late Bashkirian (Sinitsyna and Sinitsyn 1987; Nikolaev 2005). *Schubertina,* with its wall differentiated into two layers, survived up to the Wordian (Davydov, 2011). The fusiform *Schubertella* first appeared in the Moscovian (Rauser−Chernousova et al. 1951). *Schubertella,* with a wall differentiated into three layers, lived from the Moscovian through to the Lopingian with several acme zones in the Moscovian–Kasimovian, late Asselian–early Sakmarian and late Artinskian times (Davydov, 2011). In the early Gzhelian, *Schubertella* evolved into the relatively large schubertellid *Biwaella,* which survived through Artinskian–Kungurian time. In the latest Gzhelian, the latter developed fluted septa in *Dutkevitchites,* which in turn developed into a highly-specialized form in the Permian, *Sphaeroschwagerina* (Davydov 1984). Another advanced schubertellid, *Mesoschubertella,* is documented throughout the Permian, but Davydov (2012) speculated that its origination could have been in the Sakmarian–Asselian or even in the late Gzhelian.

The **Schwagerinidae** arose from *Fusulina* in the Late Carboniferous (Fig. 2.17). The Schwagerinidae of the Late Carboniferous are dominated by *Triticites,* a moderately folded form with strong chomata and a spirotheca with an alveolar keriotheca (Figs 2.15 and 2.16). There was a progressive decrease in the number of wall layers, accompanied with a progressive increase in the opacity of the diaphanotheca, from *Fusulinella* to *Prototriticites, Montiparus* and *Triticites* in the Late Carboniferous (van Ginkel and Villa, 1999) (see Fig. 2.13). Most of the Late Carboniferous genera became extinct at the top of the Gzhelian, except members of the Schubertellidae, which did not become extinct until the end of the Permian. *Schubertella* itself is a Permian form (see Fig. 2.17).

In the Upper Carboniferous strata, the marker species of the middle Kasimovian is *Montiparus montiparus* (Fig. 2.13), and is widely distributed in the East European Basin and Tethyan province. In the Gzhelian, *Montiparus* evolves into *Carbonoschwagerina* in the peri-Gondwana realm (Ozawa et al., 1990). The precise position of the Carboniferous-Permian boundary has long been the subject of debate (Leven and Gorgij, 2006; Ozawa et al., 1990). However, many studies agree that the Carboniferous-Permian boundary in the East European Basin is defined on the first appearance of *Pseudoschwagerina* (see Fig. 2.19). The latter evolved from *Triticites* at the base of the Permian. In the Tethyan realm and the peri-Gondwana part of Tethys the base of the Permian coincides with the appearance of *Sphaeroschwagerina fusiformis* (Ozawa et al., 1992).

Schwagerinids evolved slowly during the Late Carboniferous. However, during the Asselian and Sakmarian part of the Early Permian a dramatic evolution of the Schwagerinidae produced 50 genera (Leven, 2003). These forms became intensely folded (Fig. 2.14) as well as becoming increasingly abundant and diverse. While *Triticites* decreased in abundance and importance (until it disappeared completely at the top of the Early Permian) about half of the schwagerinids became extinct before the Kungurian. From Early Permian to Middle Permian, Schwagerinids exhibited evolutionary lineages with chamber inflation, such as in genera *Pseudoschwagerina*, *Sphaeroschwagerina* and *Zellia*. *Sphaeroschwagerina* is considered to have evolved in the early Asselian from a species of *Dutkevitchites*, a member of the Schubertellidae (Ozawa et al., 1992).

Cuniculi also appeared in two separate lineages. The earliest lineage contains the North American *Eoparafusulina* in the Early Permian and the cosmopolitan *Monodiexodina* in the Early to Middle Permian and, appearing slightly later, the second lineage contains the cosmopolitan *Parafusulina* (Guadalupian-Lopingian), and the North American *Skinnerina* and *Polydiexodina* (late Kungurian to late Guadalupian). The wall structure of the Schwagerinidae became more complicated and of the *Schwagerina* type. Most of these forms became extinct in the Middle Permian. The Schwagerinidae evolved into even more complicated forms in the Late Permian, such as the *Polydiexodina* and its related genera (Fig. 2.14). These forms are characterised by a single-layer wall but a very large test with intensely folded septa.

A rapid evolution of the **Neoschwagerinoidea** followed the first appearance of the **Misellinidae** in the Kungurian; *Misellina* evolved directly from the ozawainellid *Pamirina* in the Kungurian by developing "saddle-shaped" parachomata (Leven, 2009). The *Sumatrina* group of the **Neoschwagerinidae** sprung directly from *Misellina* at the base of the Guadalupian. This horizon also marks the first occurrence of the **Verbeekinidae**, which formed one of the most distinctive lineages in the Tethyan faunal realm. The earliest verbeekinid, *Brevaxina,* which first appeared during the earliest part of the Guadalupian, witnessed a rapid evolution to lineages that culminated with *Verbeekina* and *Yabeina*. It descended either indirectly from *Staffella* via the intermediate form *Sphaerulina* (Ross, 1967), or from the ozawainellids *Pamirina* (Kanmera et al., 1976) and rapidly evolved into a succession of specialised genera (see Fig. 2.17). These fusulinides tests have parachomata and evolved into forms with complete systems of foramina in each chamber (see Fig. 2.14). The parachomata are discontinuous in *Verbeekina*, but continuous in the neoschwagerinid *Yabeina*.

The end Guadalupian saw the extinction of all large and morphologically complex forms assigned to the **Schwagerinidae** and **Neoschwagerinidae** (Leven and Korchagin, 2001), and only 15 genera in the families **Schubertellidae** and **Staffellidae** persisted into the Lopingian (Groves and Altiner, 2005). The diversity of schubertellids reached its maximum at the end of the Guadalupian, and it underwent a minor burst of evolutionary diversity during the Lopingian (see Fig. 2.20). The new genus, *Codonofusiella* has irregularly coiled shells with strongly fluted septa, becoming uncoiled at maturity (see Fig. 2.13), while others developed thick fusiform shells having thickened walls, such as *Palaeofusulina*. However, this late diversification was followed by the extinction

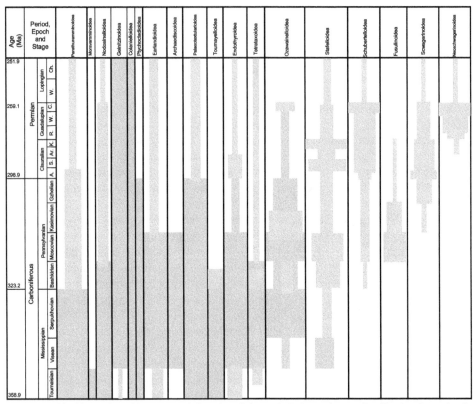

Fig. 2.20. The biostratigraphic range and diversity of the main Palaeozoic superfamilies (as shown by the horizontal scale of the spindles) found in the Carboniferous-Permian.

of the remaining schwagerinids, verbeekinids, and neoschwagerinids at the end of the Permian.

2.3.2 Other Palaeozoic Larger Foraminifera

The **Archaediscida** evolved from the parathuramminides in the Carboniferous. They include the bilocular **Archaediscoidea** (Fig. 2.21) and **Lasiodiscoidea** (Fig. 2.22A). The archaediscids have a streptospirally enrolled second chamber, while the lasiodiscids are conical with some genera (e.g. *Howchinia*) developing a high helicoidal spire. The archaediscids remain undivided and differ from the fusulinides in that the outer layer of their wall has a radial hyaline structure. They reach their maximum abundance in the Visean.

In addition to the fusulinides described above, the **earlandiides** gave rise to many other diversified descendants, such as the plurilocular **Semitextulariidae** of the **Palaeotextulariida** in the Devonian. The **Semitextulariidae** include biserial, flattened forms, often becoming monoserial with broad, fully septate chambers. Members of the **Palaeospiroplectamminidae** represent a biserial development from the **Tournayellidae**, and may be ancestral to the **Palaeotextulariidae.** *Palaeospiroplectammina* have an initial coil and a two-layered wall indicating a possible origin from a planispiral endothyrid (Haynes, 1981). It gave rise to the first **Palaeotextulariidae,** *Palaeotextularia* in the Early Carboniferous (Tournaisian), which in turn evolved into *Palaeobigenerina* and related genera in the Visean. The Palaeotextulariidae (see Figs 2.2 and 2.4) have biserial to uni-serial genera that closely resembles the Textulariidae, but have a dark granular calcar-eous outer wall and an inner clear to yellowish "fibrous" layer with stacks of granules perpendicular to the surface. Cummings (1956) demonstrated that these forms evolved from simple agglutinated forms in the Devonian where, in some cases, a partially aggluti-nated outer layer is still present (e. g. *Cribrogenerina*). The aperture is basal in the biserial forms, becoming terminal and cribrate in the uniserial part (e.g. *Climacammina*). Like the archaediscids, the paleotextulariids reach their maximum abundance in the Visean.

The **Tetrataxida** (Fig. 2.5) are highly conical, multilocular forms with trochospiral arrangements and an open central umbilicus at the base of the cone. The wall structure is two-layered and very similar to the paleotextulariids from which they are inferred to have evolved. They are divided into the single-layered **pseudotaxids** and **valvulinellids** and the two layered **tetrataxids**. The valvulinellids differ from the pseudotaxids and the tetrataxids in having subdivided chambers with vertical and horizontal partitions.

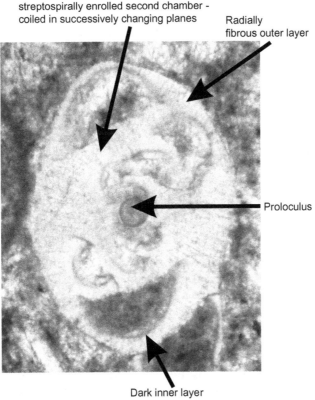

streptospirally enrolled second chamber -
coiled in successively changing planes

Radially
fibrous outer layer

Proloculus

Dark inner layer

Fig. 2.21. Main morphological features of *Archaediscus* test. Scale bars = 0.5mm.

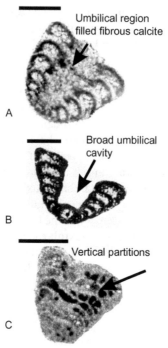

Fig. 2.22. Main morphological features of A) *Howchinia bradyana (Howchin)*; B) *Vissariotaxis cummings Hallet;* C) *Valvulinella youngi* (Brady). Scale bars = 0.3mm.

The two-layered walls of the tetrataxids suggests a phylogenetic relationship with the endothyrides but this has not been proven yet.

The partially to fully spetate **nodosinelloids** of the Palaeozoic **Lagenida** evolved from the parathuramminides in the Silurian via the subseptate **caligellids** of the **earlandiides**. The nodosinelloids include the earlandinitids with the single-layered wall and the nodosinellids with the double-layered wall.

The **Syzraniidae** evolved within the **Robuloidoidea** in the Upper Silurian of North Urals in Russia with the first appearance of *Tuborecta*. The latter has a two-layered wall structure resembling the other Syzraniidae, but with a simpler morphology. *Tuborecta* died out soon after the beginning of the Devonian and it is only in the Late Carboniferous, the Moscovian-Kasimovian interval, that the Syzraniidae reappeared. The evolutionary radiation of this family, with its oldest known Carboniferous genus *Syzrania*, might have evolved from the another simple robuloidoids or most likely from the earlandiides *Earlandia* in Middle Pennsylvanian time (see Chapter 3), through the addition of a hyaline-radial layer external to the ancestral microgranular wall (Groves et al., 2004). Palaeozoic lagenides are poorly studied in comparison with most other foraminiferal groups. They did not show significant evolutionary radiation until the Moscovian (Middle Pennsylvanian), but early lagenides were minor members of foraminiferal faunas for most of the remaining Palaeozoic Era, while fusulinides were spectacularly diverse (Groves et al., 2003). Lagenides became conspicuous only after fusulinides suffered steep declines in both diversity and abundance at the end of the Guadalupian epoch (Leven and Korchagin, 2001) and their almost complete disappearance at the end of the Permian (Groves, 2005).

It is noteworthy that the Palaeozoic lagenides survived the end Permian extinction, and gave rise to the Mesozoic-Cenozoic lineages (Groves et al., 2003; Hallam and Wignall, 1997). Groves (2005), in studying the assemblage of lagenides in lattermost Permian rocks of the central Taurides in Turkey, stated that out of the 22 species in 16 genera, only two identifiable species in the primitive long-ranging "*Nodosaria*" (such as *N. radicula* (Linné), see Plate 2.5, Fig. 11) and indeterminate *Syzrania* (see Chapter 3) survived the end Permian mass extinction. The last occurrences of most taxa fall within the last half-meter of the Permian strata, a pattern consistent with abrupt extinction when tested for the Signor-Lipps effect. Generic diversity within the lagenides exceeded that of all other calcareous foraminiferal groups throughout most of the Mesozoic Era prior to the Late Cretaceous rapid diversification of rotaliines. The suborder includes approximately 120 extant genera and ranks behind only the Miliolida and Rotaliida as the most diverse group of living calcareous foraminifers (Tappan and Loeblich, 1988). They will be discussed further in the following chapter.

The **miliolides** of the Triassic made their first appearance in the Carboniferous and Permian. During this time, they were small foraminifera, and not until the Triassic, after the extinction of most of the larger fusulinides, did they evolve into different larger forms, filling the empty niches left by the extinction of the Permian larger benthic foraminifera. The evolutionary relationships within all of these groups are also explored further in Chapter 3.

2.4 Palaeoecology

The dominant earlier Palaeozoic foraminifera were mainly characterised by having undivided tubular chambers with diverse types of coiling. The uncoiled planispiral, biserial to uniserial, or uniserial genera represented the infaunal assemblage of the Carboniferous. They were mainly infaunas living an endobenthic mode of life within the sediment or at the seawater/sediment interface (Vachard et al., 2010). The unilocular parathuramminides, and the plurilocular semitextulariids of the palaeotextulariids were mainly distributed along the Late Silurian and Devonian reef, restricted to back-reefs (lagoons) and fore-reef (middle ramps), or off-reef environments (Krebs, 1972; Vachard, 1974, 1994; Préat and Kasimi, 1995), while the moravamminoides typically lived on the Devonian reefs or inner ramps (Vachard et al., 2010). The palaeotextulariides (e.g. *Palaeotextularia*) are believed to have thrive mostly in low energy environments, where turbidity, current and wave action were minimal, with the most favorable conditions being in water deeper than 20m (Stevens, 1971). Also, palaeotextulariides could live in shallower environments (5–10m depth), where they probably found sheltered niches among algal thalli (Gallagher, 1998). In contrast, the distribution of large palaeotextulariides (e.g., *Climacammina, Cribrogenerina*) shows a preference for wave-agitated environments (Porta et al., 2005).

In the Carboniferous, the first trochospirally coiled, conical Tetrataxida (e.g., *Tetrataxis, Pseudotaxis* and *Abadehella*) occur in a wider range of environments than most fusulinides. They have a greater tolerance to decreased light. They are abundant in the inner platform facies and are common in shallow reefal facies (Toomey and Winland, 1973). Attached forms are rare, and forms such as *Tetrataxis* might have lived attached during a part of its life cycle (Kochansky-Devidé, 1970; Cossey and Mundy, 1990; Gallagher, 1998; Vachard and Krainer, 2001; Pille, 2008; Vachard et al., 2010). The late Tournaisian-Visean *Tetrataxis* could also tolerate decreased light conditions at water depth of about 200m (Lees, 1997; Gallagher, 1998). Similarly, archaediscoids could thrive in turbulent environments (Haynes, 1965; Brenckle et al., 1987; Gallagher, 1998), and in the relatively lower energy and deeper (below wave base) environments where they frequently occur with lasiodiscoids and pseudoammodiscids (Porta et al., 2005). It was suggested by Gallagher (1998) that the lenticular shape of the archaediscoid test contributed to the stability in turbulent environment and the calcareous prismatic wall enabled algae symbiosis.

In the late Tournaisian and Visean, the Endothyrida and Tournayellida became diversified and spread on the entire inner ramp (Vachard et al., 2010). They appear to have been mostly endobenthic (Pille, 2008). The earliest forms *Eotournayella* and *Septatournayella* were found in deep water limestones (Vachard, 1973, 1974; Gallagher, 1998; Cózar and Rodríguez, 2003; Mohtat-Aghai et al., 2009). Compressed small endothyrids and tournayellids are most abundant in near-shore and shelf limestones (Ross, 1973), flourishing in warm shallow, moderately high energy environments (Skipp, 1969). In comparing them with modern benthic foraminifera, such as *Elphidium,* Haynes (1981) suggested that the endothyrids lived on the sea bed "clinging by their pseudopods to various plant and animal substrates, crinoids, bryozoans, etc."

During the Late Carboniferous (Bashkirian to Gzhelian), the larger benthic foraminifera extended their habitats from the more confined, shallower areas of the shelf to inner ramp. The endothyrids, staffellids and bradynids may have made their habitats in or near high-energy environments, patch-reef facies (Dingle et al., 1993; Porta et al., 2005). The small lenticular/subspherical test facilitate their transport (Rich, 1969). The endothyrids were common in low- and high-energy settings. The occurrences of secondary deposits in the endothyrids could have stabilized their tests in turbulent environments (Haynes, 1981). The exclusion of the endothyrids from lagoonal environments with restricted circulation and variable salinity suggests that these forms preferred open-marine environments. (Porta et al., 2005). The lenticular staffellids, e.g., *Pseudoendothyra,* are found to be abundant in higher energy, patch-reef facies, whereas the subspherical forms, e.g. (*Staffella*) were common in the quieter, back reef facies (Dingle et al., 1993), in the shallowest setting and in paleoenvironments characterized by abnormally high temperatures and salinity. From the Bashkirian to the Middle Permian, the epiphytic staffellids increasingly adopted the ecological setting and taphonomic behaviour of its modern equivalents, the Recent miliolid genus *Peneroplis,* as it progressively occupied the semi-restricted platform environments (Vachard et al., 2010).

The Bradyinidae were probably epiphytes and *Bradyina* is interpreted as a shallow-water taxon adapted to life in current-swept environments (Haynes, 1981; Gallagher, 1998; Gallagher and Somerville, 2003). This contrasts with its occurrence in the Pennsylvanian Minturn Formation (Colorado, U.S.A.), where *Bradyina* seemed to have lived in deeper, open shelf environments at a depth of at least 15m, and relatively deeper than most fusulinids (Stevens, 1971). Malakhova (1961) suggested that the presence of

Bradyina in deeper settings indicates the possibility of an adaptation to a planktonic life. However, this is contradicted by its highest abundance in shallow environments. Ponta et al. (2005) suggested that the lateral "chamberlets" of *Bradyina* were part of a channel system separated from the life chambers and acted as a hydrostatic function which would allow these globose forms to be adapted to a planktonic way of life.

The Carboniferous witnessed also the first appearance of the large groups of the fusulinides which gradually occupied a more extensive environmental spectrum than their ancestors. As the fusulinides are an extinct group, their palaeoecology has to be inferred by comparing their shape and their faunal and floral associations with morphologically similar groups from the Holocene. They seem to be very well adapted to several distinct ecological and environmental conditions. Their fossil shells occur in light grey, shallow-water limestones or calcareous shales. They are associated with phylloid algae, corals, chaetetids, crinoids, bryozoans, and brachiopods (Thompson, 1964; Ross, 1965, 1967, 1969; Stevens, 1971; Wilson, 1975; Haynes, 1981; Gallagher, 1998; Wahlman, 2002; Wahlman and Konovalova, 2002; Almazán et al., 2007, Davydov, 2011). This type of benthic carbonate production prevails in shallow, oligotrophic, warm and sunlit environments, with minimal silicilastic input, and with production rates highest in very shallow waters above a maximum depth of 25-30m (Weidlich, 2007). However, at some localities they are also abundant in sandstones (Loeblich and Tappan, 1988).

The wide variety of coiling modes, from lenticular to subcylindrical or globose, indicates many life strategies in shallow water carbonate sub-environments were adopted by the fusulinides (Ross, 1969). Ross found in his study of the palaeoecology of the fusulinids that shape was an important factor in their distribution. Elongate, sub-cylindrical forms (e.g. *Parafusulina*) occurred in very shallow sub-tidal environments, lagoons and bays, whereas, small forms with inflated chambers occurred in a wide variety of sediments suggesting deeper water to wave base depths (see Fig. 2.23). However, only rare fusulinids occurred in sediments interpreted as forming in deep water environments.

The large proloculus in the fusulinides is comparable to those of extant forms such as *Marginopora* (Ross, 1969). This suggests that the fusulinidean proloculus was formed within the parent shell and released. According to Ross this restricted fusulinidean dispersal to nearby shallow shelf areas as the released proloculus could only have been carried a short distance by local currents. During the Late Carboniferous, fusulinides forms, such as *Profusulinella*, acquired heavy chomata. These heavy secondary deposits may completely infill large parts of early chambers so that the protoplasm is concentrated in the outer whorls. The benefit of these deposits might be the redistribution of protoplasm or, perhaps, the additional weight of the shell may have kept the individual from being easily dislodged by waves or currents (Ross, 1969). Along the upper Bashkirian platform, the abundance of *Profusulinella, Pseudostaffella* and *Ozawainella* increases from wave swept areas to the lower-energy facies, whereas, *Eostaffella* decreases in abundance in decreasing environmental energy (Porta et al, 2005).

The marine biota of the fusulinides limestones frequently contains fragments of algae, corals and bryozoa (Plate 2.27). Although in some cases the calcareous algae were abundant, they were mostly facies sensitive and too long-ranging to permit the establishment of a precise stratigraphic zonation (Mamet and Zhu, 2005), but their occurrences led to the interpretation that larger foraminifera adopted a symbiotic mode of life, with the fusulinides behaving similarly to Holocene larger foraminifera (Vachard et al., 2010).

The fusulinides tests are certainly constructed so that they could have hosted photosynthetic symbionts. Many modern larger benthic foraminifera are apparently limited in their distribution by the water temperature, sunlight intensity, and the physical and chemical requirements of their symbiont (Murray, 2007). In fact, many of the fusulinides have a thin, two-layered wall in the last whorl that lacks the tectorial layers of the early ones. Others, such as many staffellids and boultonids, have translucent walls, and most of the schwagerinids and neoschwagerinids have keriothecal walls with alveoli that could have housed symbionts analogous to extant dinoflagellates (Ross, 1969). Housing symbionts suggests that the occurrence of these faunas is limited to shallow tropical to subtropical waters on shelves, reefs and platforms in the photic zone. Their distribution would be limited by cold oceans and deep oceanic barriers (Haynes, 1981). Vachard et al. (2004) suggested that the cyanobacteria, reported in Holocene *Marginopora*, are the probable symbionts in the keriothecal walls of the fusulinides, which offered them more light than previously received within the chambers of the early forms with only a tectum.

The distribution of fusulinids, staffellids and schubertellids shows varied trends through the Moscovian platform. *Fusiella* and *Schubertella* were the only schubertellids observed in restricted peri-tidal lithofacies (Baranova and Kabanov, 2003), whereas *Fusulinella* and *Fusulina* were dominant in shallow, normal-marine environments. *Hemifusulina* occurred in subtidal zones affected by storms (Porta et al., 2005). *Profusulinella* show a preference for shallower settings regardless of environmental energy and salinity (Porta et al., 2005). *Schubertella* was most common in shallow, higher energy settings (Baranova and Kabanov, 2003), but also tolerated restricted marine conditions. Schubertellids are also common in cooler/deeper water environments (Teodorovich 1949; Rauser−Chernousova 1951; Baranova and Kabanov 2003; Davydov, 2011). The presence of *Schubertella*, in the deeper and shallowest facies indicate its tolerance of all energy levels and restricted marine conditions. The dominant planispiral fusulinides of the Late Carboniferous to Early Permian exhibits the typical fusiform shape shown by some Holocene larger foraminifera, such as the miliolides (e.g. *Alveolinella*) that today appear to be confined to normal shallow marine (down to depths of 80m), well-oxygenated, nutrient-rich, tropical and subtropical waters. It is thus inferred that the fusulinides required normal marine salinity and thrived in shallow warm well-oxygenated, nutrient-rich waters. Their variations in shape, from elongate fusiform to sub-spherical globular may be directly related to specific adaptation to varying conditions in the shallow water environment. The Carboniferous foraminifera lived on the substrata and were primarily sensitive to nutrients availability, physical and chemical ecological changes, such as temperature, water currents, and wave intensity. Large forms of *Triticites* are associated with shallow water algal meadows and banks of crinoidal fragments, which suggest an environment similar to the modern Gulf of Mexico. Some elongate forms of *Triticites* are closely associated with sediment of impure silty limestone and fine to medium sandstone that indicates shallow bays, lagoons and wave-built bars and terraces. The occurrence of thick-shelled subglobose *Triticites* suggests the thickened shell wall was an adaptation to slightly more energetic environments, where resistance to abrasion, crushing, and breaking would be a positive selection factor. Large subglobose species, that have high chambers and only gently folded septa, are common in clay deposits formed in considerably less vigorously agitated environments. Small fusiform *Triticites* are mostly common in poorly sorted limestones that most

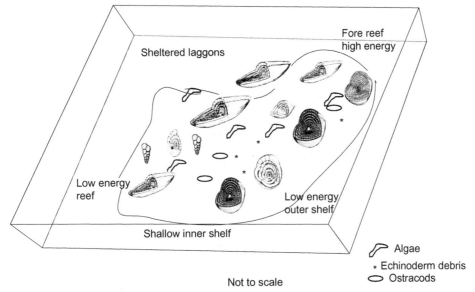

Fig. 2.23. Schematic figure showing the palaeoecological distribution of the fusulinides.

probably were deposited in deeper, shelf waters (Ross, 1969), located between 3–8m deep (Ross, 1971). Another example of shape variation driven by adaptation to different environments are the Early Permian *Eoparafusulina*, an elongate fusiform form, found living on sand bars in very shallow tidal environments with relatively strong current and wave conditions. The Early Permian *Pseudoschwagerina*, with inflated chambers with a large protoplasm volume, may even have been pelagic as it is found in a wide variety of sediment types (Ross, 1969). During Late Carboniferous to Permian (Kasimovian to Capitanian), the Schwagerinidae reached the outer limit of the inner or middle platform as they have been commonly reworked in calciturbidites (Vachard et al., 2010).

During the Middle Permian, the Neoschwagerinoidea spread rapidly, replacing the Schwagerinoidea in their habitats. The epiphytic verbeekinids (e.g. *Verbeekina*) and the neoschwagerinids (e.g. *Yabeina*) have similar shapes and arrangements of their internal foramina, as in the alveolinids (see Chapter 6) which facilitate the direct movement of both equatorial and radial flows of the cytoplasm (Davydov, 2011). They are also very sensitive to temperature fluctuations and survived only in warm environments where surface water exceeds a yearly temperature of about 22° C (Davydov and Arefifard, 2013).

Very few taxa survived the hypersaline environments of the Late Permian (Vachard et al., 2010). Only some bradynids (e.g. *Glyphostomella*), pachyphloiids (e.g. *Pachyphloia*), schubertellids (e.g. *Russiella*) and pseudodoliolinids (e.g. *Pseudodoliolina*) persisted in the very shallow, evaporitic environments of the Lopingian.

2.5 Palaeogeographic Distribution of the Fusulinides and Related Forms

The fusulinides and their related forms are found in the Late Palaeozoic basins and adjacent marine shelves of Eurasia and the Western Hemisphere (Ross, 1967). Geographic

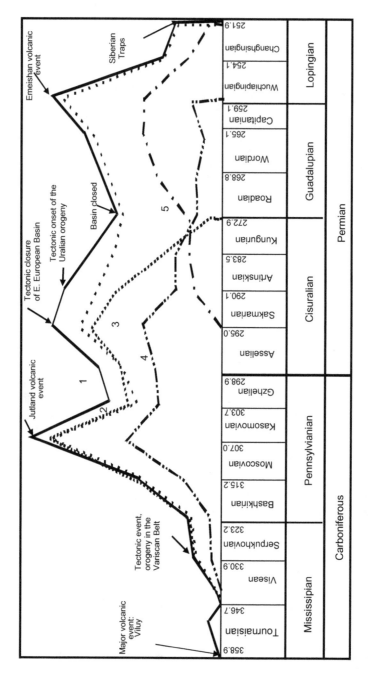

Fig. 2.24. The number of fusulinides genera as a function of time, after Leven (2003). 1) all data, 2) Tethys, 3) East European basin, 4) N America, 5) peri-Gondwana part of Tethys (S. Afghanistan, S. Pamirs, Karakorum, S. Tibet).

distribution of these genera forms the basis for recognizing faunal associations, which can be grouped into four phases (see Figs 2.24 and 2.25):

- the Tethyan realm (Middle Carboniferous (Visean) to Late Permian), reaching peaks in the Late Carboniferous and Middle Permian,
- the East European Basin realm (Middle Carboniferous (Visean) to Late Permian, reaching its peaks in the Late Carboniferous and Early Permian;
- the N American realm (Middle Carboniferous (Visean) to Middle Permian, (Capitanian)), reaching its peaks in the Late Carboniferous and Early Permian;
- the peri-Gondwana realm (Early Permian (Cisuralian) to Late Permian), reaching its peak in the Middle to Late Permian.

The Middle Silurian witnessed an important event in the history of the Palaeozoic foraminifera, namely the appearance of the microgranular wall as a building component of the test. These foraminifera evolved slowly during the Silurian and Devonian and it was not until after the Devonian-Carboniferous boundary that they evolved into many distinctive lineages. Their evolution in the Early Carboniferous was delayed by approximately 12.2 Ma as the survivors of the Devonian-Carboniferous boundary extinction event adapted to their new environments. This boundary event is referred to as the Late Devonian Hangenberg extinction (358.9 Ma ago) that occurred at the end of the Famennian and is globally associated with black shale facies (Streel, 1986; McLaren,

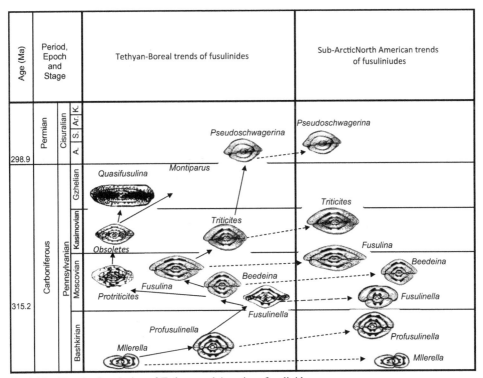

Fig. 2.25. Trends of evolution of Tethyan and American fusulinides.

1990). It is marked by the extinction of diverse marine groups, anoxia and rapid sea level fluctuations, and is named from the Hangenberg Shale. Some major groups, such as the ammonoids, stromaporoids and trilobites, suffered an entire extinction (Bambach 2006) and the event saw the collapse of the reef ecosystems (Copper, 2002; Bambach, 2006). The timing of the Hangenberg extinction coincides with the last phase of the Devonian Southern Hemisphere glaciation (Sandberg et al., 2002; Chen and Tucker, 2003; Haq, 2005). Several causes for this extinction have been suggested, including the widespread development of anoxia and a mini-glaciation (Caplan and Bustin, 1999; Kalvoda, 1989, 2002; Kaiser et al., 2006), possibly triggered or amplified by the development of the volcanic Viluy Traps in Siberia (Courtillot and Renne, 2003), or an impact event (McLaren and Goodfellow, 1990).

The Hangenberg event, however, did not seem to affect the sustainability of the Parathuramminida, Tournayellida or the Endothyrida. As in every extinction event, the small resilient foraminifera ("disaster forms") survived the end Devonian, but it took the whole of the Tournaisian for recovery, and it was not until the Visean (346.7 Ma) that the shallow reefal benthic foraminifera began to recover and evolve into different lineages. The Parathuramminida declined in number and virtually disappeared within the Early Carboniferous, with only the Tuberitinidae surviving into the Permian. The Tournayellida survived to the Bashkirian, while the endothyrid *Loeblichia* gave rise to the Fusulinida at the Tournaisian-Visean boundary.

The Visean lasted for about 15.8 Ma. During this time, major changes in ocean circulation, biogeographic differentiation and high bio-provincialism contributed to the diversification of new groups of fauna, such as in the ammonoids, fresh water pelecypods, gastropods (Davydov et al., 2004), and of course the fusulinides. During that time animal life, both vertebrate and invertebrate, consolidated its position on land, as plant life had during the Devonian. Euramerica and western Gondwana drifted northwards and moved closer together (Fig. 2.26). This movement eventually gave rise to collision, leading to the Variscan-Hercynian orogeny.

The fusulinides thrived from the Visean to the Permian, and gradually filled the reefal niches at the expense of other smaller foraminifera. At the onset of the Visean, they were represented by few genera, but soon they became global in their geographic distribution. They have been found on all continents except Australia, India and Antarctica. This can be explained by the fact that Australia and India were connected to Antarctica throughout the Late Carboniferous and Permian, and were at southern latitudes at which the equatorial fusulinides could not thrive.

Based on the reference work of Rauser-Chernousova et al. (1996), Leven generated a plot showing the generic diversity of fusulinides over the period from their first appearance in the Visean to their complete extinction in the Late Permian. Leven's data is recast in Fig. 2.27 to show the number of new fusulinides genera throughout the Carboniferous and Permian. It is clear that in Tethys, East Europe and North America a large increase in the number of genera occurred in the Moscovian. Two more peaks of generic increase in the Tethys occur in the Asselian and Capitanian. On the other hand, fusulinides were absent from East Europe and North America in the Late Permian. Fig. 2.28 shows that a high number of extinctions of genera occurred at the Moscovian-Kasimovian boundary, but these were instantly replaced by new

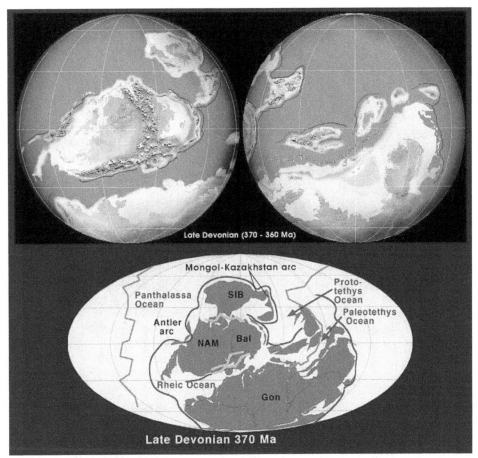

Fig. 2.26. Palaeogeographic and tectonic reconstruction of the Late Devonian (by R. Blakey, http://jan.ucc. nau.edu/~rcb7/paleogeographic.html).

genera. On the other hand, the highest number of extinctions, combined with a low number of new genera, is seen at the Capitanian and Wuchiapingian boundary.

The fusulinides were almost cosmopolitan, but the American mid-continental province showed less diversity than those of Tethys and East European Basin (Fig. 2.24). Fusulinides are only found rarely in North Africa, and as the North American margins were largely connected to Palaeotethys (Figs 2.26, 2.29), the fusulinides of the Early Visean showed less provinciality. However, the diversity of fusulinides genera in the North American Basin decreases as the connection with the Tethyan and East European Basin became limited or intermittent from the Late Visean to Permian time. This explains the similarities between the American trend in fusulinides numbers and that of the Tethyan-East European trend plotted by Leven (Fig. 2.24 and 2.27).

Tethys was the fusulinides main breeding ground, with more than 900 species being described from this realm (Haynes, 1981). The geographically specific curves of Leven (Fig. 2.24) show that the distribution of the fusulinides greatly depends on the evolutionary history of Tethys, and predominantly the equatorial part of that ocean. The

Permo-Carboniferous Total Genera

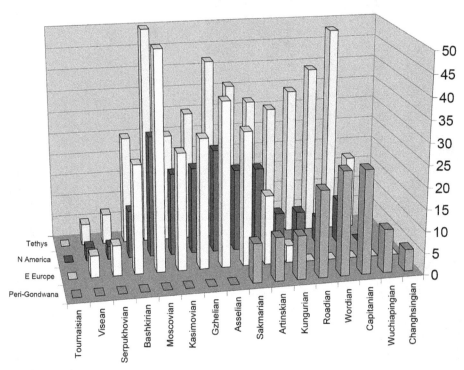

Fig. 2.27. Number of fusulinides genera through the Carboniferous and Permian.

fact that the northern hemispheric continental mass was closer to the equator in the Carboniferous (Fig. 2.29) than the southern hemispheric mass explains the wider distribution of the fusulinides in the former hemisphere.

    The Tournaisian-Visean boundary coincided with the first occurrence of *Eoparastaffella simplex* Vdovenko, 1954 (Davydov et al., 2004; Devuyst and Kaldova, 2007). The Visean foraminifera were dominated by the endothyrids, which are particularly common in algal limestones (Haynes, 1981). The endothyrids increased steadily during the Visean, however, their increase stopped abruptly at the end of the Serpukhovian. According to Walliser (1995) the faunal overturn was not an abrupt change in all affected fossil groups, but rather just a rapid transition that he correlated with the final orogeny in the Variscan Belt (Figs 2.29 and 2.30). This orogeny was not a catastrophic event and did not cause major changes in the faunal distribution, but rather a slight reduction in the diversity of the faunas. It was associated with cooling associated with changes in ocean circulation patterns and the closing of the equatorial seaway (Daydov et al., 2004). This palaeotectonic event (equatorial seaway closure) resulted in the partial, thermal isolation of the Paleo-Tethys, which became a great semi-tropical bay. The climate, which had been fairly equable throughout most of the Mississippian, became more strongly zonal. These dramatic climate changes in the late Serpukhovian occurred at just about the same time as the sudden spread of

Permo-Carboniferous Speciations and Extinctions

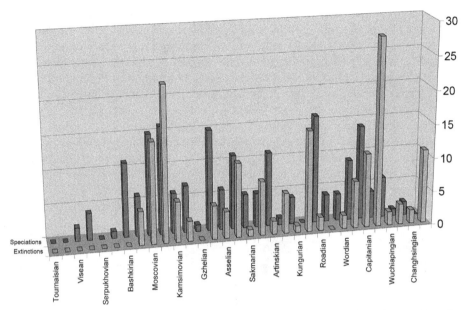

Fig. 2.28. The number of fusulinides generic extinctions compared to the speciation of genera throughout the Carboniferous and Permian.

the fusulinides and the onset of their provincialism. During the latter part of the Early Carboniferous and much of the Middle Carboniferous, seas gradually expanded and flooded low lying portions of continental areas to increase greatly their aerial extent (Ross, 1967). During the Early Carboniferous, genera and species of Ozawainellidae and Staffellidae were common on shallow carbonate shelves and in basins in nearly all parts of the world.

At the Serpukhovian-Bashkirian boundary, the diversity of the fusulinides and their related forms increased, with three different provinces becoming recognisable. The East European Basin, characterised by the abundance of *Eostaffella* and *Bradyina*; the Tethyan realm, where the palaeotextulariides were abundant; and the North American realm, where *Bradyina* did not appear before the Bashkirian. The basal beds of the Bashkirian are characterized by the appearance of the foraminiferal species *Pseudostaffella antiqua*.

During the Pennsylvanian, diversity increased steadily in the three provinces of Eastern Europe, Tethys and, to a lesser degree, the Americas (Figs. 2.24, 2.25 and 2.27). New fusulinides, with complicated internal structures, evolved rapidly and new forms such as *Fusulina* and *Fusulinella* appeared globally. The increase in size of the fusulinids during the Carboniferous coincides with a large increase of atmospheric oxygen levels (Fig. 3.10). This is also confirmed by the presence of giant insects, *Meganeura* in the Carboniferous (Chapelle and Peck, 1999). According to Moore and Thompson (1949) and Groves et al. (1999), the base of the Moscovian approximates with the first appearance of the genus *Profusulinella* in the sub-Arctic region of the North American province. However, this genus had occurred earlier, roughly 4–5 Ma, on the Russian platform of the East European basin. The genus originated in the latter area in the late Early Bashkirian and then underwent significant diversification, so that by the early Moscovian a range of shell morphologies existed (Groves et al., 2007). Although previous work suggested that North American *Profusulinella* spp. may have been derived from a local ancestor such as *Eoschubertella*, Groves et al. (2007) interpret the first sub-Arctic North American species of the genus as immigrants from Eurasia, with their migration through the Franklinian corridor having been facilitated by generally east-to-west currents during a glacio-eustatic flooding event (Fig. 2.31). This analysis has recently been greatly expanded by Davydov (2014), who points out that the first occurrences of Tethyan fusulinides in North America are associated with paleoclimatic warming events, and that the time of the delay of the first occurrences (see Fig. 2.25) depended on the scale and intensity of the warming episodes during those periods.

Fusulinides were by now common in shallow water carbonate banks, which transgressed far on to the edges of the land masses along the Tethyan seaway, found today

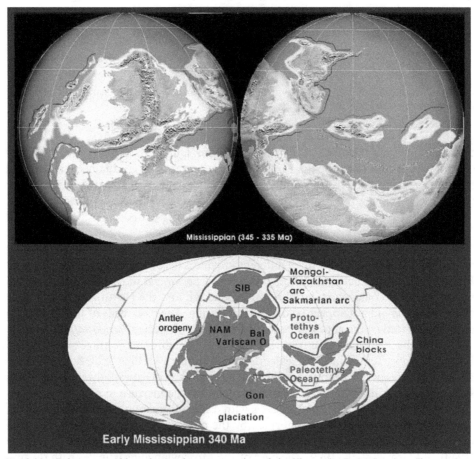

Fig. 2.29. Palaeogeographic and tectonic reconstruction of the Visean (by R. Blakey, http://jan.ucc.nau. edu/~rcb7/paleogeographic.html).

in northern Spain, northern China, Manchuria, and Korea, as well as in North and South America. Towards the end of the Moscovian, a sharp decline in the diversity of the fusulinids in all the three provinces is clearly seen in Figs. 2.27 and 2.28. Most of the fusiform species (e.g. *Fusulina, Fusulinella*) disappear at the top of the Moscovian and only the small staffellids survived. This extinction may have been related to significant climate change, which is also thought to have driven the Carboniferous Rainforest Collapse, dated around 305Ma, due to a trend toward increased aridity and changes in global glacial cover (e.g. Groves and Lee, 2008).

In the Late Carboniferous, the Tethyan realm contained relatively few fusulinides genera in comparison with earlier (Moscovian) or later (Asselian and Sakmarian) assemblages. The general distribution of the fusulinides in the East European Basin followed the same trend as those in Tethys from the Visean towards the Sakmarian (Fig. 2.24). This can be explained by the fact that the pre-Sakmarian East European basin was closely connected to Tethys (Fig. 2.29).

After the partial extinction of the fusulinides, at the end of the Moscovian, new forms appeared such as *Triticites*. These forms dominate the fusulinides assemblages of the Late Pennsylvanian of North America (Loeblich and Tappan, 1988). Wide diversification within *Triticites* first became pronounced only near the end of the Carboniferous as new, more specialized branches appeared. However, the presence of a few species, at scattered localities in the Western Hemisphere, that are similar to those abundant in eastern Europe (such as *Daixina*), suggests that the dispersal processes for these groups were limited at this time. Also, the rare occurrence at that time of a species of *Triticites* in Eastern Europe that is similar to one abundant in the North American realm, suggests that dispersal was also limited in the other direction.

Unusually, the Carboniferous-Permian boundary (298.8 Ma) is not marked by any major foraminiferal extinction event. However, it does closely correlate with a large scale volcanism (Fig. 2.32), the Jutland basalt event, which affected Europe and North Africa at this time (Smythe et al., 1995). This event was associated with the development of the Oslo graben and, amongst others, the Whin Sill in Britain.

In the Early Permian (Fig. 2.33) the fusulinides again became diverse and cosmopolitan, but a few genera were not widely distributed. Some are endemic to North American, e.g. *Chalaroschwagerina* and *Cuniculinella,* and others are only found in the Tethyan realm, e.g. *Sphaeroschwagerina* (Plate 2.18, figs 6,7,9) and *Zellia* (Plate 2.25, Fig. 2). In the Asselian of the Tethyan province, *Pseudoschwagerina* and *Sphaeroschwagerina* occur together, while only *Pseudoschwagerina* occur in both the Tethyan and North American realms.

At the top of the Asselian the tectonic closure of the East European basins isolated the fusulinides and their numbers dwindled until, at the end of the Kungurian, the tectonic closure became complete, and the East European province fusulinids disappeared completely (Fig. 2.24). During the latter part of the Early Permian, the Schubertelloidea evolved new genera of which *Russiella* and *Minojapanella* were endemic to Tethys, while *Boultonia* was restricted to the Tethyan faunal realm in the early part of the Late Permian, but later it was briefly cosmopolitan before becoming extinct towards the end of the Permian.

During this same time, the peri-Gondwanan parts of Tethys (now found in southern Afghanistan, southern Pamirs, eastern Hindu Kush, Karakorum, southern Tibet,

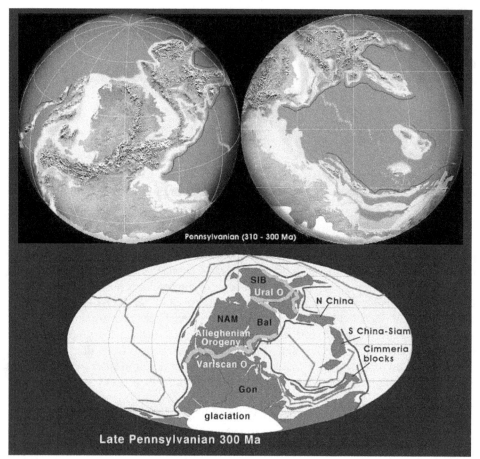

Fig. 2.30. Palaeogeographic and tectonic reconstruction of the Late Pennsylvanian world (by R. Blakey, http://jan.ucc.nau.edu/~rcb7/paleogeographic.html).

Himalayas and the Salt Range) moved away from the near glacial conditions in Gondwana towards the equator (Fig. 2.33), thus creating a warm breeding ground for the fusulinides. Their diversity increased in the Sakmarian, before it decreased again at the end of the stage. Many species then disappeared gradually towards the end of the Early Permian, such as the cosmopolitan *Pseudoschwagerina* (Plate 2.18, figs 3,4,8,11), which appearance defines the base of Permian boundary and disappearance the top of the Kungurian (see Fig. 2.17).

   Newly recorded occurrences of fusulinides assemblages from the Sakmarian of Central Oman (Angiolini et al., 2006) have extended the distribution of the *Pseudofusulina* (Plate 2.18, Fig. 13) species to the South Tethyan seas. This assemblage compares with those described by Leven from the peri-Gondwanan parts of Tethys (Leven, 1993, 1997). Their presence in Oman expands the area of distribution of the warm water fusulinides during the Sakmarian, indicating the onset of the beginning of the "greenhouse" climate which resulted in the Gondwanan deglaciation.

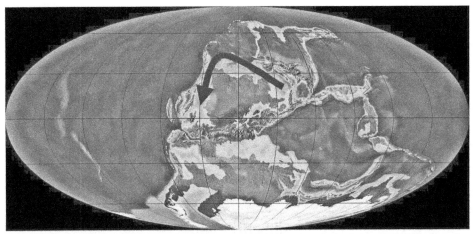

Fig. 2.31. Migration route of *Profusulinella* spp. from Eurasia to the sub-Arctic North American province through the Franklinian corridor (modified from R. Blakey http://jan.ucc.nau.edu/~rcb7/paleogeographic. html).

The decrease in the diversity at the end of the Sakmarian probably coincides with the Uralian orogeny and the end of the Permo-Carboniferous glaciation (Erwin, 1996). It affected the diversity in the East European basins, and the development of the Eastern Tethyan foraminifera. These new foraminiferal assemblages, which filled in the empty niches after the Early Permian (Kungurian), were dominated by the verbeekinids. These latter reached their peak in the Wordian (Leven, 2003). The verbeekinid association forms the main part of the Permian Tethyan fusulinides faunal realm and is found from modern day Tunis to Greece, Yugoslavia, Sicily, Afghanistan and Timor (and New Zealand) and in eastern Asia and the Japanese islands. However, many species such as *Verbeekina* reach as far south as south California in the North American realm.

The Capitanian saw a slight increase in the diversity of the fusulinides during a short-term transgression in that stage (Leven, 2003). Immediately after the transgression, the North American basin became isolated and underwent a rapid salinization that caused the complete extinction of the fusulinides in North America. The Late Guadalupian saw an extinction that was one of the largest in the Palaeozoic. All large and morphologically complex forms assigned to the Schwagerinidae and Neoschwagerinidae were eliminated. It has been suggested that this extinction could have been triggered to a flood basal event which occurred in the Late Permian in southwestern China. The Emeishan basalts extend (Fig. 2.34) over an area in excess of half a million square kilometres (Courtillot and Renne, 2003). The Late Guadalupian crisis affected the fusulinides more than the other major extinction event in the Late Moscovian. According to Vachard et al. (2010), two causes can explain the total disappearance of the keriothecal forms, Schwagerinoidea and Neoschwagerinoidea, the disappearance of the endosymbionts of the giant foraminifers and the oceanic cooling event, the high-productivity "Kamura" event. This event is estimated to have lasted over 3–4 Ma (Isozaki, 2007).

Fig. 2.32.   The Jutland Flood Basalt event at the end of the Carboniferous.

The remaining fusulinides biota were under stress during the Lopingian and diversity was in decline. The end Lopingian is, however, marked by the end Permian extinction: the most severe of the entire Phanerozoic. All of the inshore taxa (98%), including the large fusulinides, became extinct. Some small endothyrids survived, only to die out subsequently in the Early Triassic. Globally, 90 to 96% of all marine invertebrate species went extinct (Sepkoski, 1986) as did all but one of 90 genera of reptiles (McLaren and Goodfellow, 1990), most corals, brachiopods and large terrestrial species (see Benton (2005) for an extensive review).

In a comprehensive study of the end Permian mass extinction horizon recorded in the Meishan section, South China, Kaiho et al. (2001) recorded that the extinction event was characterized by the abrupt and catastrophic disappearance of major benthos, and according to these authors, the extinction horizon coincides with an abrupt decrease in the $^{34}S/^{32}S$ ratios of seawater sulphate, $^{87}Sr/^{86}Sr$ ratio, and an increase in Fe – Ni grains. There was also a pronounced negative excursion in the carbon recorded in P–Tr boundary carbonate rocks and organic matter (e.g. Berner, 2002).

The cause of the end Permian mass extinction remains debatable, and numerous theories have been formulated to explain the events of the extinction. Historically, geologists invoked in the past theories such as climate change, global warming, marine

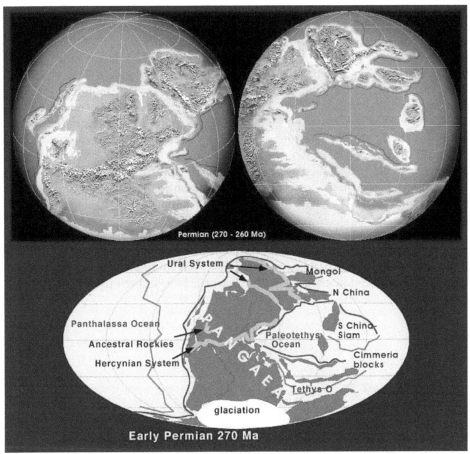

Fig. 2.33 Palaeogeographic and tectonic reconstruction of the Early Permian world (by R. Blakey, http://jan.ucc.nau.edu/~rcb7/paleogeographic.html).

anoxia or tectonic processes as the cause of such global mass extinctions. But these are difficult to reconcile with the relatively abrupt nature of such extinction events. Furthermore, cause and effect are often difficult to differentiate, and indeed, the need to make a distinction between kill and trigger mechanism was stressed by Knoll et al. (2007), who referred to the first as the physiologically disruptive process that causes death, and to the second by the critical disturbance that brings one or more kill mechanisms into play. So, if gradualist processes cannot provide an explanation for the rapidity of the observed event, then catastrophic causes of the P-Tr mass extinction must be invoked. These have included either an asteroid impact or flood basalt volcanism. However, even though many of the ecological features of the P-Tr event compare with the results of the asteroid impact that famously is thought to have concluded the Mesozoic, the end Palaeozoic strata have not yielded an unambiguous signature of a bolide impact (Wardlaw et al., 2004). This leads many researchers to consider that a major volcanic event was the more probable cause of the end Permian extinction

(Benton and Twitchett, 2003). The P-Tr mass extinction coincides with the eruption of the Siberian traps flood basalts (Fig. 2.26), the largest known Phanerozoic continental igneous province (Reichow et al., 2002; Courtillot et al., 2003). The eruption of this vast volume of basalt, in a short time (< 1 Ma), would have released aerosols and greenhouse gases, which could have triggered a rapid climate change that would have caused a mass extinction of both marine and continental biota (Erwin et al., 1994; Benton and Twitchett, 2003; Chen and Benton, 2012). A sudden release of huge volumes of carbon dioxide might have poisoned all marine and terrestrial life (Ward and Brownlee, 2000). The event may have also triggered widespread marine anoxia, and the destabilization of seafloor methane clathrates, which could account for the carbon isotope excursion noted by Berner (2002). Siberian Trap volcanism was a major trigger of these extinctions, especially if it was combined with the profound sea-level low stand, unprecedented global high temperatures and marine deep-water stagnation and anoxia (Macleod, 2013).

Despite the globally devastating effect of the end Permian event on foraminiferal life, some creatures survived. For the forms related to the fusulinides, the survival of the endothyrides can perhaps be explained by their small size and by their not needing symbionts or much oxygen for existence. This allowed them to survive the

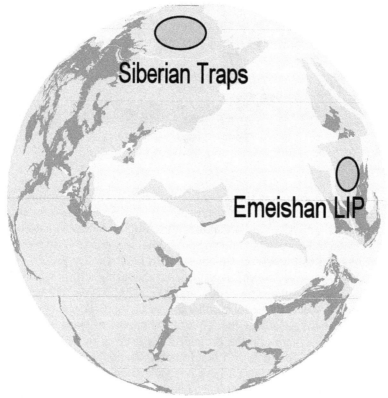

Fig. 2.34   The Emeishan Large Igneous Province and the Siberians Traps.

adverse conditions of the end of the Permian, and to occupy the empty niches and the new ecosystems found in the Early Triassic (see Chapter 3). According to Knoll et al. (2007), hypercapnia (being able to cope with the physiological effects of elevated P_{CO_2}) is a feature that seemed to be common to the survivors of the P-Tr event. This may provide an explanation of the significant loss of most calcareous foraminifera, but the survival of the small agglutinated foraminifera. It also seems that other reefal groups were affected in the same way, so that the Late Permian corals disappeared but their unskeletonized relatives, the sea anemones (which would give rise to scleractinian corals in the Triassic) survived (Knoll et al., 2007). Similarly, the skeletonized dasyclad green algae disappeared, but multiple unskeletonized sister groups survived (Aguirre and Riding, 2005).

Whatever the catastrophic cause and the physiological needs of the survivors, the large fusulinides never recovered after the end Permian event. The occurrence of two consecutive extinction events in a relatively short time (the end Guadalupian crisis and the end Permian crisis were only ~10 Ma apart) exhausted this group and completely destroyed their ability to continue to fill their shallow marine niche. This left a major opening for new fauna and ecosystems to develop. As will be seen in Chapter 3, the Triassic was a very different world (Chen and Benton, 2012).

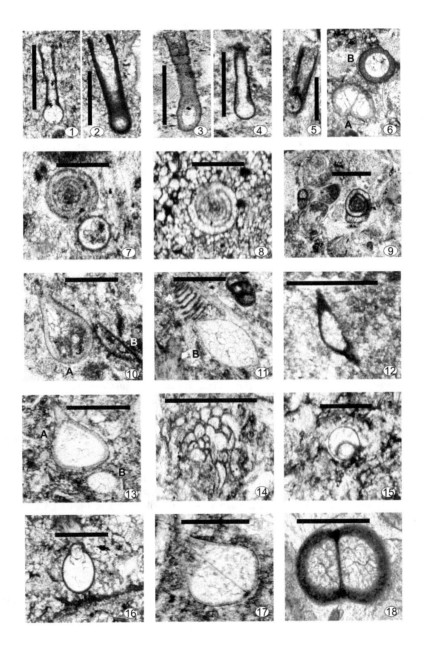

Plate 2.1 Scale bars = 0.4mm Fig. 1-3, 5. *Earlandia elegans* Rauser-Chernousova and Reitlinger, Vertical sections, Simonstone Limestone, Whitfield Gill, Yorkshire, England, 1) UCL coll., DH 109; 2) UCL coll., DH46; 3) UCL coll., D150 A; 5) UCL coll., DH688. Fig. 4. *Earlandia vulgaris* (Rauser-Chernousova and Reitlinger), Gayle Limestone, Duerley Beck, Gayle, Yorkshire, England, UCL coll., DH81. Fig. 6. A) *Draffania biloba* Cummings, B) *Earlandia* sp., Transverse section, Gayle Limestone, Duerley Beck, Gayle, Yorkshire, England, UCL coll., DH49. Figs 7-9 *Brunsiella* sp., 7) UCL coll., DH 109; 8) UCL coll., DH46; 9) UCL coll., DH82. Fig. 10. A) *Draffania biloba* Cummings B) *Palaeonubecularia* cf. *uniserialis* Reitlinger, Hawes Limestone, Gayle Beck, Yorkshire, England, UCL coll., DH 33a. Fig. 11. A) *Howchinia* sp., B) *Saccaminopsis fusulinaformis* (M'Coy), Hawes Limestone, Gayle Beck, Yorkshire, England, UCL coll., DH 33a. Fig. 12. *Palaeonubecularia* cf. *uniserialis* Reitlinger, Gayle Limestone, Duerley Beck, Gayle, Yorkshire, England, UCL coll., DH48. Figs 13, 17, 18. *Draffania biloba* Cummings, Gayle Limestone, Duerley Beck, Gayle, Yorkshire, England, 13A) axial section, 13B) transverse section, UCL coll., DH 81; 17) UCL coll., DH 272; 18) transverse section, Gayle Limestone, Duerley Beck, Gayle, Yorkshire, England, UCL coll., DH 811. Fig. 14. Alga incerta, Gayle Limestone, Duerley Beck, Gayle, Yorkshire, England, UCL coll., DH 81. Fig. 15. *Diplosphaerina sphaerica* (Derville), Underset Limestone, Howgate Head, Sleddale, UCL coll., DH327. Fig. 16. *Diplosphaerina inaequalis* (Derville), Simonstone Limestone, Whitfield Gill, Yorkshire, England, UCL coll., DH117.

Plate 2.2 Scale bars: Figs 1-2, 4, 6-11, 17-19 = 0.25mm; Figs 3, 5, 13 = 0.4mm; Figs 12, 15 = 1mm. Figs 1, 5. *Tetrataxis conica* Ehrenberg. 1) figured by Al-Habeeb from Mumbles, South Wales, M14b/B; 52) Gayle limestone, Duerley Beck, Gayle, Yorkshire England. Fig. 2. *Tetrataxis pusillus* Conyl and Lys, Norton Quarry, South Wales, N6A/A, UCL coll. Fig. 3. *Valvulinella lata* Grozdilova and Lebedeva, Gayle limestone, Duerley Beck, Gayle, UCL coll., DH 46. Fig. 4. *Valvulinella tchotchiai* Grozdilova and Lebedeva, Gayle limestone, Duerley Beck, Gayle, Yorkshire England, UCL coll. Fig. 6. *Eotuberitina cornuta* Hallet, Moss Kennels, Northumberland, England, UCL coll., DH349. Fig. 7. *Tubeporina magnifica* Hallet, Simonstone Limestone, Whitfield Gill, Yorkshire, England, UCL coll., DH117. Fig. 8. *Eotuberitina* sp., figured by Hallet (1966) from Trowbarrow Quarry, Silverdale, UCL coll., DH531. Fig. 9. *Eotuberitina reitlingerae* Miklucho-Maclay, Simonstone Limestone, Whitfield Gill, Yorkshire, England, UCL coll., JER 867. Fig. 10. *Diplosphaerina inaequalis* (Derville), figured by Hallet (1966) from Arngill Beck, Askrigg, Yorkshire, England, UCL coll., DH586. Fig. 11. *Tuberitina* sp., Simonstone Limestone, Whitfield Gill, Yorkshire, England, UCL coll., JER 867. Figs 12-15. *Saccaminopsis fusulinaformis* (M'Coy), 12-14) figured by Hallet (1966) from 12) Buckden Beck, UCL coll., DH4; 13-14) Hawes Limestone, Duerley Beck, UCL coll., DH35; 15) Gayle Beck, Hawes Limestone, Yorkshire, England, UCL coll., DH47. Fig. 16. *Draffania biloba* Cummings, figured by Hallet (1966) from the Simonstone Limestone, Whitfield Gill, Yorkshire, England, UCL coll., DH613. Fig. 17. *Paratuberitina* sp., Hawes Limestone, Duerley Beck, UCL coll., DH35. Fig. 18. *Eotuberitina fornicata* Hallett, Gayle limestone, Duerley Beck, Gayle, Yorkshire, England, UCL coll., DH10. Fig. 19. *Parathurammina aperturata* Pronina. Middle Devonian, hypotype, from Zadorozhnyy and Yuferev (1984), Tomsk District, USSR.

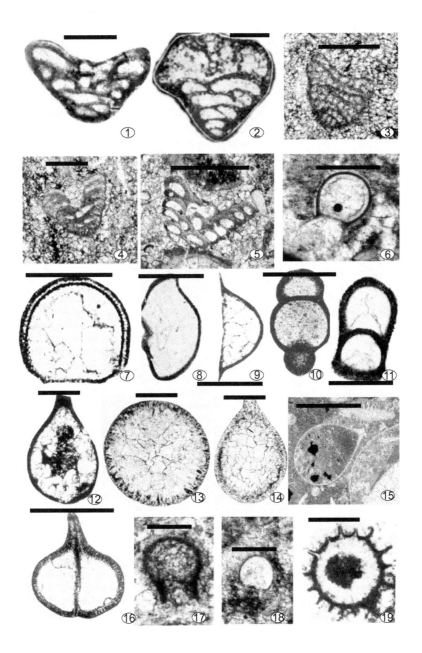

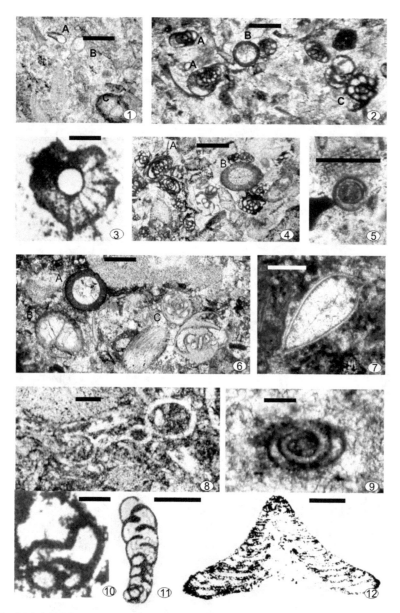

Plate 2.3 Scale bars: Figs 1 - 12 = 0.25mm. Fig. 1. Thin section photomicrographs of A) *Earlandia vulgaris* (Rauser-Chernousova), B) *Chrysothurammina* sp., C) *Draffania biloba* Cummings, Gayle Beck, Hawes Limestone, Yorkshire, England, UCL coll., DH50. Fig. 2. A) *Millerella designata* Zeller, B) *Elenella* Pronina, C) *Plectogyra irregularis* Zeller, Hardraw Scar Limestone, Muker, Yorkshire Limestone, England, UCL coll., DH 708. Fig. 3. *Sogdianina angulata* Saltovskaya, Early Carboniferous (Visean), Tadzhikistan, USSR, centred section, from Petrova (1981). Fig. 4. A) *Eostaffella* spp., B) *Auroria*, Gayle Beck, Hawes Limestone, Yorkshire, England, UCL coll., DH50. Fig. 5. *Eovolutina* sp., Antropov, Hardraw Scar Limestone, Muker, Yorkshire Limestone, England, UCL coll., DH 708. Fig. 6. *Parphia* Miklukho-Maklay, *Draffania biloba* Cummings, *Archaediscus* sp., Hardraw Scar Limestone, Muker, Yorkshire Limestone, England, UCL coll.,

DH 708. Fig. 7. Ostracod sp., Simonstone Limestone, Whitfield Gill, Yorkshire, England, UCL coll., DH115. Fig. 8 *Tuberitina* Galloway and Harlton, a specimen showing bulbous chambers in straight series, Gayle Beck, Hawes Limestone, Yorkshire, England, UCL coll., DH152. Fig. 9. *Brunsia* Mikhaylov, Gayle Beck, Hawes Limestone, Yorkshire, England, UCL coll., DH33. Fig. 10. *Biseriella parva,* late Visean, S. Urals, USSR from Chernysheva (1948). Fig. 11. *Palaeospiroplectammina* Lipina, Tournaisian section showing early coil, Russian Platform, USSR, from Lipina (1965). Fig. 12. *Abadehella tarazi* Okimura and Ishi, axial section of holotype from Okimura *et al.*, (1975), Abadeh Formation, central Iran.

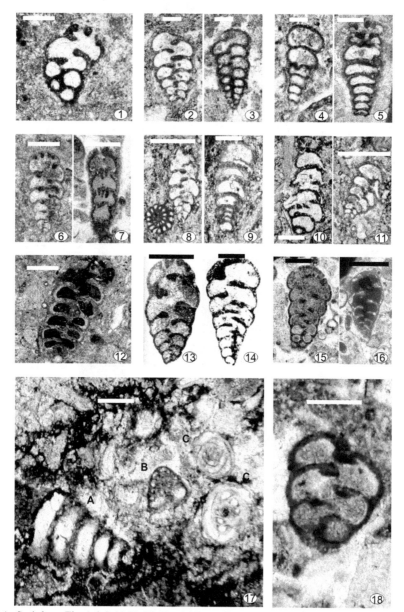

Plate 2.4   Scale bars: Figs 1-6, 9, 12-15, 17, 18 = 0.25mm; Figs 7, 8, 10, 11, 16 = 1mm. Fig. 1. *Cribrostomum inflatum* Cummings, Gayle Limestone, Duerley Beck, Gayle, Yorkshire, England, UCL coll., DH48. Figs 2-4, 9, 14. *Palaeotextularia longiseptata* Lipina, 2, 4, 9) Gayle Limestone, Duerley Beck, Gayle, UCL coll., DH44; 3) Hawes Limestone, UCL coll., DH22; 14) figured by Al-Habeeb (1977) from Port Eynon, South Wales. Figs 5, 13, 15. *Cribrostomum* sp., 5) Gayle Limestone, Duerley Beck, Gayle, UCL coll., DH11; 13) figured by Al-Habeeb (1977) from Norton Quarry, South Wales; 15) Carboniferous Limestone, West Sahara, B188. Fig. 6. *Koskinotextularia* sp., Simonstone Limestone, Whitfield Gill, UCL coll., DH120. Fig. 7. *Climacammina* sp., Carboniferous Limestone, West Sahara, B1 288. Fig. 8. *Deckerella quadrata* Cummings, Simonstone Limestone, Whitfield Gill, Yorkshire, England, UCL coll., DH110. Fig. 10. *Palaeotextularia* sp., Simonstone Limestone, Whitfield Gill, UCL coll., DH33. Figs 11, 16. *Deckerellina* sp., 11) Simonstone Limestone, Whitfield Gill, UCL coll., DH3. 16) *Deckerellina* sp., Carboniferous Limestone, West Sahara, B1 426, UCL coll.. Figs 12. *Palaeotextularia* sp., Schlykovn, Limestone, Duerley Beck, Gayle, UCL coll., DH 50. Fig. 17. A) *Palaeotextularia longiseptata* Lipina, B) *Tetrataxis conica* Ehrenberg, C) *Archaediscus inflatus* Schlykovn, Limestone, Duerley Beck, Gayle, UCL coll., DH 50. Fig. 18. *Palaeotextularia angulata* Cummings, Cowey Sike, Grindon Hills, Northumberland, England, UCL coll., JER 1086.

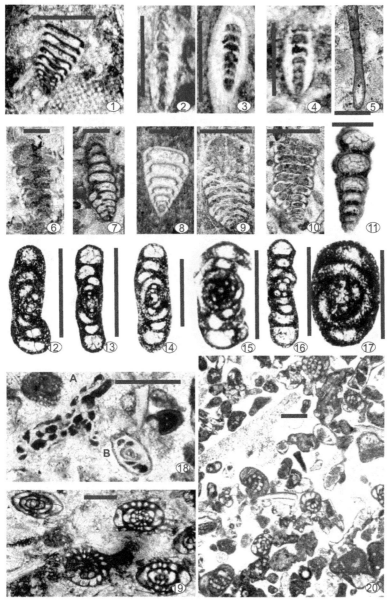

Plate 2.5 Scale bars: Figs 1-20 = 0.25mm. Fig. 1. *Howchinia bradyana* (Howchin). Gayle Beck, Hawes Limestone, Yorkshire, England, UCL coll. Figs 2. *Pachyphloia asymmetrica* (Miklukho-Maklay), Permian, Thai Peninsula, Thailand, AM 35b, NHM OGS coll. 231. Fig. 3. *Pachyphloia depressa* (Miklukho-Maklay), Permian, Thai Peninsula, Thailand, AM 35b, NHM OGS coll. 231. Fig. 4. *Pachyphloia magna* (Miklukho-Maklay), Permian, Thai Peninsula, Thailand, AM 35b, NHM OGS coll. 231. Fig. 5. *Paratikhinella cylindrica* (Brady), Main Limestone, Fossdale Gill, UCL coll., DH150. Fig. 6. *Palaeotextularia* sp., TP18, Permian, Khlong Pha Saeng, Thailand, Map. 4737 I. 808879, AHG Mitchell OGS coll. 231. Fig. 7. *Climacammina* sp., a twisted vertical section, Gayle Beck, Hawes Limestone, Yorkshire, England, UCL coll. Fig. 8. *Lucammina jonesi* (Brady), Hydraulic Limestone, Staffordshire, UK, UCL coll., F6. Figs 9-10. *Climacammina antiqua*

(Brady), Permian, Thitsipin Limestone Formation., Shan States Burma, NHM coll., 1970/24. Fig. 11. *Nodosaria radicula* (Linné) showing atelo-monolamellar wall structure. Permian, Brady, NHM coll., figured in Carboniferous and Permian Monograph, 1876. Figs 12-17. *Glomospirella paseudopulchra* Lipina, 12-13, 15) Gayle Limestone, Duerley Beck, Gayle, UCL coll., DH49; 14) Hardraw Limestone, Clough River, JER868; 16-17) Hawes Limestone, Duerley Beck, UCL coll., DH38. Fig. 18. A) *Calcifolium punctatum* Maslov, a codiacean alga; B) *Archaediscus karreri* Bradi, Carboniferous, UK, UCL coll. Fig. 19. A packstone of *Eostaffella mosquensis* Vissarionova, Hawes Limestone, Duerley Beck, Yorkshire, England, UCL coll., DH802. Fig. 20. A packstone of *Endothyra* spp., Carboniferous Limestone, Staffordshire, UCL coll..

Plate 2.6   Scale bars; Figs 1-16 = 0.4mm. Figs 1-2. *Endothyra* sp., SEM photographs figured by Al-Habeeb (1977) from Kittle, South Wales K85, 1) of etched polished surface of a specimen in limestone; 2) Part of the outer wall. Figs 3, 7-9. *Archaediscus complanatus* Conil, Gayle Limestone, Duerley Beck, Hawes, Yorkshire, England, UCL coll., DH 50; 9) figured by Al-Habeeb (1977) from Oxwich, South Wales. Fig. 4. *Archaediscus gigas* Rauser-Chernousova, Gayle Limestone, Duerley Beck, Hawes, Yorkshire, England, UCL coll., DH32. Fig. 5. *Archaediscus electus* Ganelina, Simonstone Limestone, Whitfield Gill, Yorkshire, England, UCL coll., DH115. Fig. 6. *Archaediscus* sp., an SEM enlargement of the outer wall figured by Al-Habeeb (1977) from Wrexham, South Wales, W3. Figs 10, 12, 14, 15, 16. *Archaediscus karreri* Brady, 10) figured by Al-Habeeb (1977) from Mumbles, South Wales, M5B/A; Gayle Limestone, Duerley Beck, Hawes, Yorkshire, England, UCL coll., DH50. Fig. 11. *Archaediscus inflatus* Schlykova, Gayle Limestone, Duerley Beck, Hawes, Yorkshire, England, UCL coll., DH50. Fig. 13. *Archaediscus* sp., Carboniferous Limestone, West Sahara, UCL coll., B1-496.

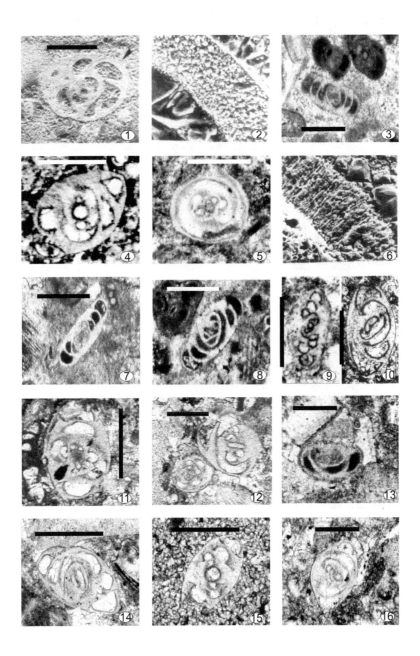

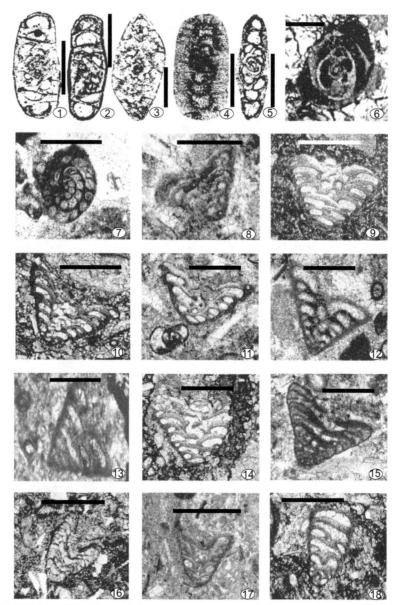

Plate 2.7 Scale bars: Figs 1-6 = 0.15; Figs 7-18 = 0.4mm Fig. 1. *Hemiarchaediscus angulatus* (Soanina), figured by Al-Habeeb (1977) from Three Yard Limestone, Gray Gill, Yorkshire, England, UCL coll., DH227. Fig. 2. *Hemiarchaediscus compressus* Al-Habeeb, 1977. Paratype, Underset Limestone, Cowgill Beck, Widdale Fell, UCL coll., DH322. Fig. 3. *Asteroachaediscus pressulus* (Grozdilova and Lebedeva), figured by Al-Habeeb (1977) from Main Limestone, Gunnerside Beck, Swaledale, UCL coll., DH714. Fig. 4. *Propermodiscus* sp., Simonstone Limestone, Whitfield Gill, Yorkshire, England, UCL coll., DH 109. Fig. 5. *Planoarchaediscus emphaticus* Al-Habeeb, holotype, Limestone IVA, River Cover, UCL coll., DH567. Fig. 6. *Archaediscus* sp., Gayle Beck, Hawes Limestone, Yorkshire, England, UCL coll., DH 36a. Fig.7. *Loeblichia* sp., Carboniferous Limestone, Sahara, UCL coll., B1693. Figs 8, 12, 13. *Tetrataxis Bradyi* Hallett, 1966, 8) AM 82 Permian Thitsipin Limestone Formation, 469491, southern Shan States Burma. NHM coll., OGS 197/11; 12, 13) Gayle Limestone, Duerley Beck, Gayle, UCL coll., DH52. Figs 9, 10, 11, 14. *Tetrataxis conica* Ehrenberg, Hawes Limestone, Duerley Beck, Hawes, Yorkshire, England, 9) UCL coll., DH137A; 11) UCL coll., DH349; 14) UCL coll., DH50. Figs 16, 17. *Tetrataxis* sp., Yorkshire, England, 16) East Stone Gill, Coverdale, UCL coll., DH272; 17) Hardraw Scar Limestone, Hardraw Force, UCL coll., DH77. Fig. 18. *Tetrataxis subcylindricus* Conil and Lys, Gayle Beck, Hawes Limestone, Yorkshire, England, UCL coll., DH 50.

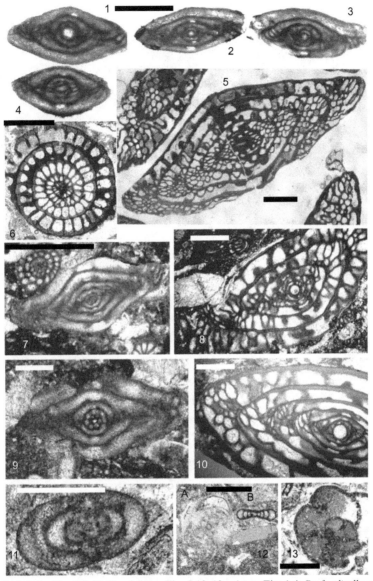

Plate 2.8   Scale bars: Figs 1-4, 11,12 = 0.3mm; Figs 5-10, 13 = 1mm. Figs 1-4. *Profusulinella* sp., fusulinid Beds Upper Coal Measures, Southern Iowa, UCL coll. Figs 5. ***Dunbarinella ervinensis*** Thompson, Camp Creek foraminifera, on Saddle Creek, 9 m south of Rockwood, Texas, UCL coll. Fig. 6. *Fusulina cylindrica* Fischer de Walheim, Limestone, Elmdale Formation Upper Gzhelian, Oklahoma, USA, UCL coll. Fig. 7. *Montiparus montiparus* (Rozovskaya), Medvedka River, USSR, UCL coll. Figs 8. *Pseudofusulinella occidentalis* (Thompson and Wheeler), Early Permian (Sakmarian), McCloud Limestone, California, USA, UCL coll. Fig. 9. *Fusulina* sp., Wolfcamp Beds, Wolfcamp Hills, 15m N. E. Marathon, Texas, UCL coll. Fig. 10. *Triticites ventricosus* (Meek and Hayden), equatorial and axial sections, Early Permian, Wolfcamp Fm., Wolfcamp Bed, 15m N.E. Marathon, Texas, UCL coll. Fig. 11. *Millerella tortula* Zeller, Carboniferous Limestone, West Sahara, UCL coll., Bl 288. Fig. 12. A) *Asteroachaediscus karreri* Brady, B) *Loeblichia* sp., Carboniferous Limestone, West Sahara, UCL coll., Bl 1426. Fig. 13. *Bradyina* sp., Carboniferous Limestone, West Sahara, UCL coll., Bl 1647a,

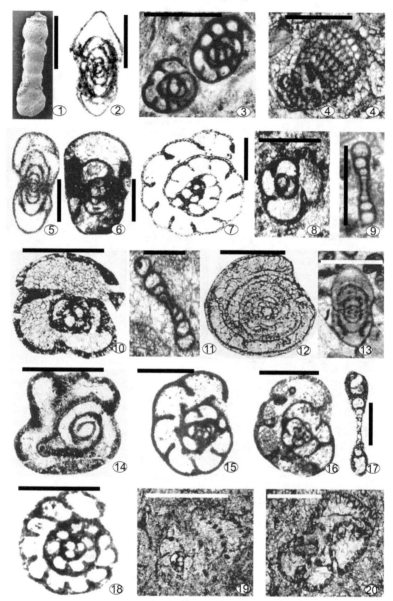

Plate 2.9 Scale bars Figs 1, 12 = 0.25mm; Figs 2-11; 13-20 = 0.4mm. Fig. 1. *Endothyranella* sp., solid specimen, Carboniferous SEM, UCL coll. Fig. 2. *Pseudoendothyra luminosa* Ganelina, figured by Al-Habeeb (1977) from Port Eynon, South Wales, P35/B. Fig. 3. *Endothyranopsis* sp., Carboniferous Limestone, West Sahara, UCL coll., B1 288g. Fig. 4. *Valvulinella youngi* (Brady). Hawes Limestone, Duerley Beck, Hawes, Yorkshire, England, UCL coll., DH37. Fig. 5. *Eostaffella arcuata* (Durkina, 1959) figured by Al-Habeeb (1977) from Pant Mawr, South Wales, TR6/A. Fig. 6, 7. *Endothyranopsis crassa* (Brady), figured by Al-Habeeb (1977) from Port Eynon, South Wales, 6) P28/A; 7) K30/B. Fig. 8, 10. *Mstinia* cf. *bulloides* Mikhailov, 8) Gayle Beck, Hawes Limestone, Yorkshire, England, UCL coll., DH 36a; 10) figured by Hallet (1966) from

Gayle Limestone, Duerley Beck, Gayle, Yorkshire, England, UCL coll., DH4905. Fig. 9, 11, 12. *Brunsiella buskensis* (Brazhnikova), Hardraw Scar Limestone, Muker, Swaledale, Yorkshire, England, UCL coll., DH 708, 9) oblique axial section; 12) equatorial section; 11) axial section, Gayle Limestone, Duerley Beck, Gayle, Yorkshire, England, UCL coll., DH 46. Fig. 13. *Pseudoendothyra composite* (Dutkevitch), Hardraw Scar Limestone, Muker, Swaledale, Yorkshire, England, UCL coll., DH 708. Fig. 14. *Carbonella* sp., figured by Hallett (1966) from Gayle Limestone, Duerley Beck, Gayle, Yorkshire, England, UCL coll., DH4905. Fig. 15. *Spinoendothyra phrissa* (Zeller), figured by Al-Habeeb (1977) from Pwll Du, South Wales, DP3/D. Fig. 16. *Chernyshinella exelikta* (Conil and Lys), figured by Al-Habeeb (1977) from Kittle section, South Wales, K75/E. Fig. 17. *Forschia* cf. *subangulata* Moller, figured by Al-Habeeb (1977) from Kittle section, South Wales, K85/A. Fig. 18. *Planoendothyra* cf. *aljutovica* Reitlinger, figured by Al-Habeeb (1977) from Port Eynon, South Wales, 35/6. Fig. 19. *Mikhailovella* sp., Simonstone Limestone, Whitfield Gill, Yorkshire, England, UCL coll., DH 109. Fig. 20. *Bradyina rotula* (Eichwald), Simonstone Limestone, Whitfield Gill, Yorkshire, England, UCL coll., DH 110.

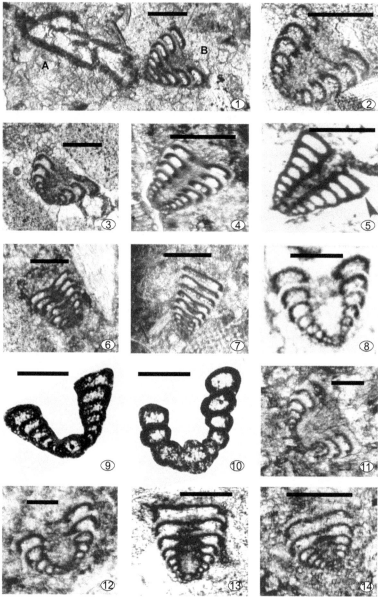

Plate 2.10 Scale bars: Figs 1-14 = 0.3mm Figs 1-7. *Howchinia bradyana* (Howchin), 1) Hawes Limestone, Duerley Beck, Hawes, Yorkshire, England, UCL coll., DH47; 2-5) Simonstone Limestone, Whitfield Gill, Yorkshire, England, UCL coll., DH117. Figs 8-12. *Vissariotaxis cummingsi* Hallet, Yorkshire, England, UCL coll., 8) Hawes Limestone; 9) Simonstone Limestone; 10) Gayle Limestone, Duerley Beck, Gayle, Yorkshire, England; 11-12) Moss Kennels, Northumberland, UCL coll., DH 349. Figs 13-14. *Howchinia nuda* Hallet, Moss Kennels, Northumberland, Yorkshire, England, UCL coll.

Plate 2.11 Scale bars: Figs 1-2 = 2mm. Fig. 1. Assemblages of *Triticites* sp., Carboniferous Gap Tank strata (bed 10), 17m S.E. of *Gap Tank*, 23m N.N.E. of Marathon, Texas, UCL coll. Fig. 2. Assemblages of *Triticites patulus* Dunbar and Newell, Wolfcamp Beds, Wolfcamp Hills, 15 m. N. E. Marathon Texas, UCL coll.

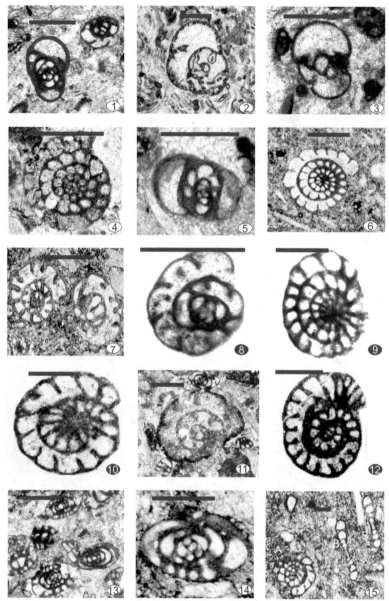

Plate 2.12 Scale bars: 1-15 = 1mm. Figs 1, 3. *Plectogyra* sp., Carboniferous Limestone, West Sahara, UCL coll., 1) B1 1630 (b); 3) B1 1647 (a). Fig. 2. *Bradyina rotula* (Eichwald), Carboniferous Limestone, West Sahara, UCL coll., B1 1647a. Fig. 4. *Eostaffella radiata* (Brady), Hardraw Scar Limestone, Hardraw Force, Yorkshire, England, UCL coll. Fig. 5. *Plectogyra* cf. *geniculata* (Ganelina), Gayle Beck, Hawes Limestone, Yorkshire, England, UCL coll., DH37. Fig. 6. *Millerella tortula* Zeller, Hunter's Stone Bank, Coverdale, Yorkshire, England, UCL coll., DH562. Fig. 7. *Endothyranopsis crassa* (Brady), Main Limestone, Fossdale Gill, Yorkshire, England, UCL coll., DH151. Fig. 8. *Endothyra bowmani* Brown, Indiana, USA, UCL coll. Fig. 9. *Eostaffella ornata* (Brady), figured by Al- Habeeb (1977) from Oxwich, South Wales, K28/A,. Fig. 10. *Endothyranopsis* aff. *E. macrus* (Zeller), figured by Al -Habeeb (1977) from Kittle, South Wales. Fig. 11. *Endothyranopsis pechorica* (Rauser-Chernousova), Hardraw Scar Limestone, Hardraw Force, Yorkshire, England, UCL coll., DH77. Fig. 12. *Planoendothyra mameti* Al Habeeb, holotype, figured by Al- Habeeb (1977), Kittle, South Wales K38/A. Figs 13, 14. *Eostaffella* spp. 13) Hardraw Limestone, Hardraw Force, Yorkshire England, UCL coll., DH77; 14) Three Yard Limestone, Walden Beck, Yorkshire Limestone, England, UCL coll., DH699. Fig. 15. A) *Pseudoendothyra struvii* (von Möller), B) *Brunsiella buskensis* (Brazhnikova), Three Yard Limestone, Walden Beck, Yorkshire Limestone, England, UCL coll., DH699.

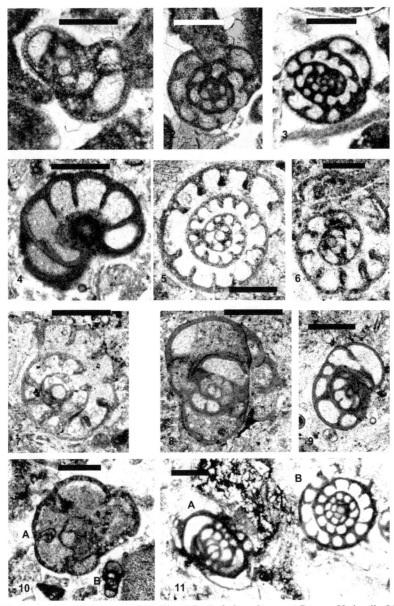

Plate 2.13 Scale bars: Figs 1-11 = 0.25 Figs 1-3. *Endothyra bowmani* Brown, Hydraulic Limestone, Waterhouses, Leek, Staffordshire, UCL coll. Fig. 4-9. *Endothyranopsis crassa* (Brady), 4) Carboniferous Limestone, West Sahara UCL coll., B1-288e; 5) Moss Kennels, Northumberland, England, UCL coll., DH349; 6-9) Carboniferous Limestone, West Sahara, UCL coll., B1-1630. Fig. 10. A) *Bradyina* sp., B) *Plectogyra* sp., Carboniferous Limestone, West Sahara, UCL coll., B1647. Fig. 11. A) *Endothyranopsis crassa* (Brady), B) *Loeblichia* sp., Carboniferous Limestone, West Sahara, UCL coll., B1-1630.

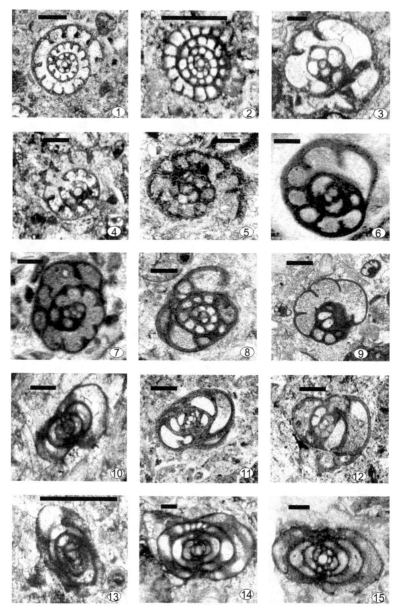

Plate 2.14   Scale bar: Figs 1-15 = 0.4mm. Figs 1, 4, 5. *Endothyranopsis crassa* (Brady), Main Limestone, Fossdale Gill, Yorkshire, England, UCL coll. Figs 2, 13. *Pseudoendothyra struvii* (von Möller), Yorkshire, England, UCL coll., 2) Main Limestone, Fossdale Gill, UCL coll., DH79; 13) Hardraw Limestone, UCL coll., DH78. Fig. 3. *Plectogyra* cf. *pandorae* Zeller, Main Limestone, Fossdale Gill, Yorkshire England, UCL coll., DH 146 A2. Fig. 6. *Plectogyra bradyi* (Mikhailov), Carboniferous Limestone, West Sahara, UCL coll., B1 288 (g). Figs 7, 11. *Plectogyra irregularis* Zeller, Carboniferous Limestone, West Sahara UCL coll., 7) B1 1661; 11) B1 288(a). Fig. 8. *Plectogyra excellens* Zeller, Gayle Beck, Hawes Limestone, Yorkshire, England, UCL coll., DH33 3a. Fig. 9. *Plectogyra phrissa* Zeller, Gayle Beck, Hawes Limestone, Yorkshire, England, UCL coll., DH33a Fig. 10. *Millerella designata* Zeller, Hardraw Scar Limestone, Muker, Yorkshire, England, UCL coll., DH 708. Fig. 12. *Plectogyra* sp., Carboniferous Limestone, West Sahara UCL coll., B1 288(d). Figs 14, 15. *Eostaffella mosquensis* Vissarionova, Hardraw Scar Limestone, Hardraw Force, Yorkshire, England, UCL coll., DH80.

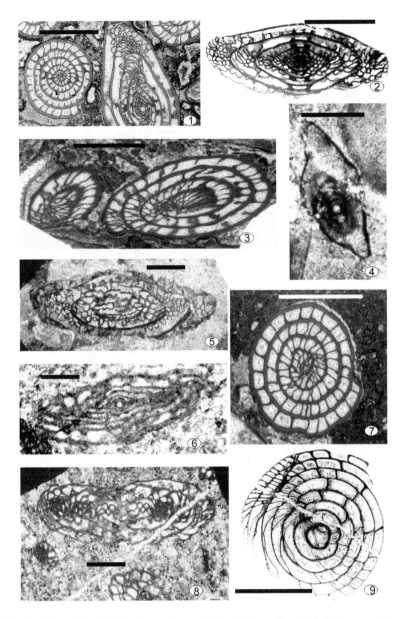

Plate 2.15 Scale bars: Figs 1-3, 6, 8 = 2mm; Figs 4, 5, 7, 9 = 0.25mm. Fig. 1-3, 7. *Triticites ventricosus* (Meek and Hayden), equatorial and axial sections, Early Permian, Wolfcamp Formation, Wolfcamp Bed, 15m N.E. Marathon, Texas, UCL coll. Fig. 4. *Eostaffella* sp., Macal Shale Group, Rio Trio, British Honduras, axial section, NHM P44725 (where wrongly identified as *Ozawainella* spp.). Fig. 5. *Schwagerina* sp., oblique axial section, Macal Shale Group, Rio Trio, British Honduras, NHM 99147. Fig.6. *Fusiella* sp., Am 87, Permian, Thitsipin Limestone Formation, 477 484, South Shan States Burma, NHM OGS coll. (where wrongly identified as *Wedekindellina* sp.), 1970/12. Fig. 8. *Quasifusulina* sp., Am 87, Permian, Thitsipin Limestone Formation, 477 484, South Shan States Burma. NHM OGS coll. (where it is identified as *Fusulina prima* Thompson), 1970/12. Fig. 9. *Verbeekina verbeeki* (Geinitz), Permian, Sumatra H.B., Brady, NHM coll.

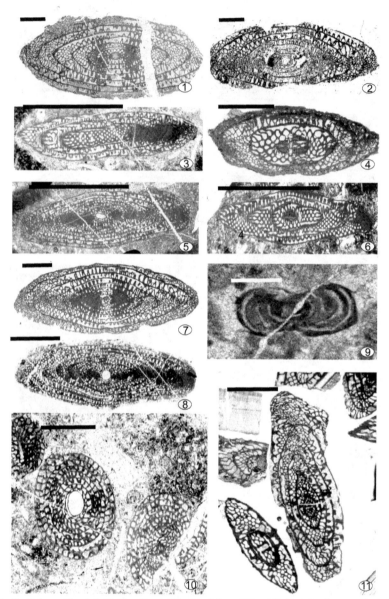

Plate 2.16 Scale bars: Figs 1-8, 10-11 = 2mm; Fig. 9 = 0.5mm. Figs 1-2. *Parafusulina* sp., Early Permian, N.E. Marathon, Texas, UCL coll. Figs 3, 5, 8, 10. *Quasifusulina* sp., Permian, Thitsipin Limestone Formation, South Shan State Burma, NHM OGS coll., 1970/24. Figs 4, 6. *Schwagerina* sp., 4, 6) Permian, Thitsipin Limestone Formation, South Shan State Burma, NHM OGS coll. (where wrongly identified as *Quasifulina*), 4) 1970/24; 6) 1970/29. Fig. 7. *Praeskinerella* sp., Early Permian, USSR, UCL coll. Fig. 9. *Neostaffella* sp., Moscovian, Khlong Pha Saeng, Thai Peninsula, Thailand. Map. 4737 I. 808879, AHG Mitchell NHM OGS coll., 231. Fig. 11. *Dunbarinella ervinensis* Thompson, Camp Creek foraminifera, on Saddle Creek, 9m south of Rockwood Texas, UCL coll.

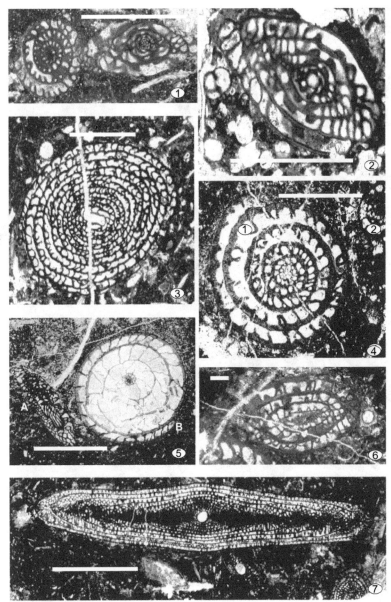

Plate 2.17 Scale bars: Figs 1-7 = 2mm. Fig. 1 *Schubertella* sp., Late Carboniferous, Yorkshire, England, UCL coll. Fig. 2. *Beedeina* sp., Late Carboniferous, Yorkshire, England, UCL coll. Fig. 3. *Polydiexodina praecursor* Lloyd, holotype, Geli Khana Section, North Kurdistan, Permian, NHM, P44395. Figs 4, 6. *Paraschwagerina* sp., Am 88b, Permian, Thitsipin Limestone Formation, 477 484, South Shan States Burma, NHM OGS coll., 1970/24. Fig. 5. A) *Schwagerina adamsi* Ross, Early Permian, oblique axial sections, B) *Pseudoschwagerina* cf. *P. fusiformis* (Krotow), Early Permian, an equatorial section of a hypotype from Geli Khana Section, Zinnar Formation, North Kurdistan, Iraq NHM, P44410. Fig.7. *Polydiexodina praecursor* Lloyd, Permian (Wordian), axial section of paratype, Kaista Section, Zinnar Formation, North Kurdistan, NHM P44386.

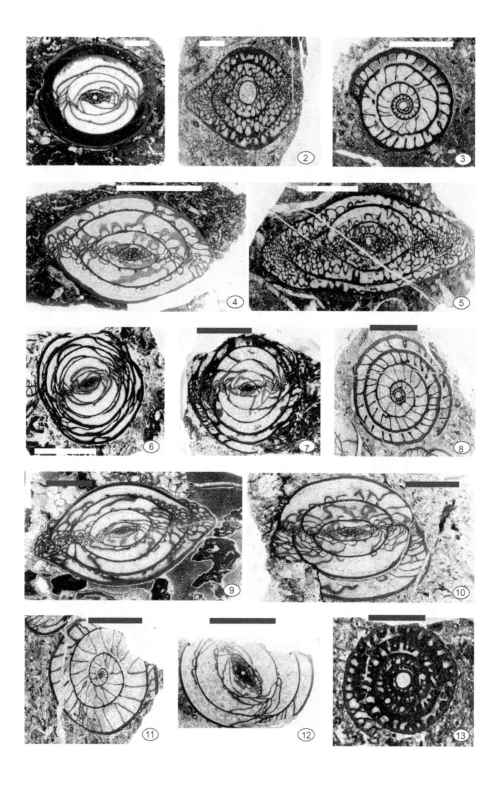

Plate 2.18 Scale bars: Figs 1-13 = 2mm. Fig. 1. *Robustoschwagerina geyeri* (Kahler and Kahler), holotype, axial section, Carnic Alps, SFN coll. Fig. 2. *Chalaroschwagerina stachei* (Kahler and Kahler), identified as *Paraschwagerina stachei* Kahler and Kahler, Paratype, oblique axial section, Carnic Alps, SFN coll. Fig. 3. *Pseudoschwagerina nitida* Kahler and Kahler, holotype, equatorial section, Carnic Alps, SFN coll. Fig. 4. *Pseudoschwagerina aequalis* Kahler and Kahler, holotype, Carnic Alps, equatorial section, SFN coll. Fig. 5. *Occidentoschwagerina alpina* Kahler and Kahler, paratype, early part of the UPL, ZK, Carnic Alps, SFN coll. Fig. 6. *Sphaeroschwagerina pulchra* (Kahler and Kahler), identified wrongly as *Pseudoschwagerina pulchra* Kahler and Kahler, holotype, Carnic Alps, SFN coll. Fig. 7. *Sphaeroschwagerina carniolica* (Kahler and Kahler), paratype, axial section, Carnic Alps, SFN coll. Fig. 8. *Pseudoschwagerina elegans* Kahler and Khaler, paratype, equatorial section, Carnic Alps, SFN coll. Fig. 9. *Sphaeroschwagerina citriformis* (Kahler and Kahler), paratype, axial section, Carnic Alps, SFN coll. Fig. 10. *Paraschwagerina lata* Kahler and Kahler, paratype, Col Mezzodi Formation, (Forni Avoltri), SFN coll. Fig. 11 .*Pseudoschwagerina lata* Kahler and Kahler, holotype, equatorial section, Carnic Alps, SFN coll. Fig. 12. *Robustoschwagerina tumida* (Likharev), wrongly identified as *Pseudoschwagerina schellwieni* Hanzawa, Kahler and Kahler, axial section, Carnic Alps, SFN coll. Fig. 13. *Pseudofusulina* sp. Kahler and Kahler, equatorial section, Carnic Alps, SFN coll.

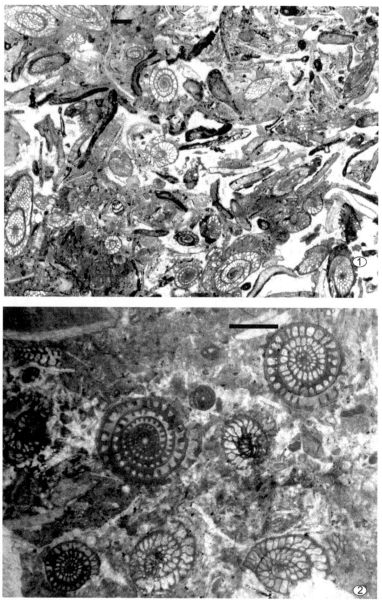

Plate 2.19 Scale bars: Figs 1-2 = 1mm. Fig. 1. Assemblages of schwagerinids sp., Wolfcamp Beds, Wolfcamp Hills, 15 m. N. E. Marathon, Texas, UCL coll. Fig. 2. Assemblages of *Fusulina cylindrica* Fischer de Walheim, Limestone Elmdale Formation, Upper Gzhelian, Oklahoma, USA, UCL coll.

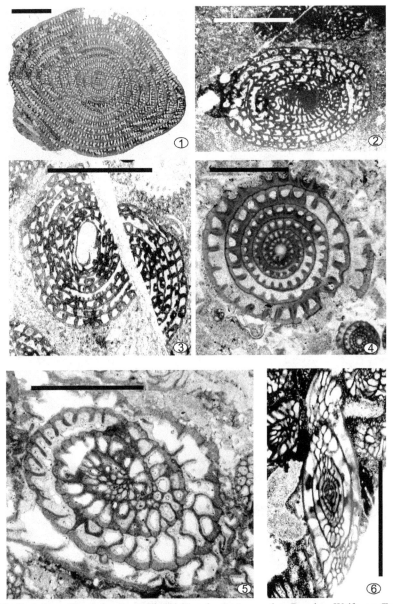

Plate 2.20   Scale bars: Figs 1-6 = 2mm. Fig. 1. *Neoschwagerina* sp., late Permian, Wolfcamp Formation, Wolfcamp Bed, 15m N.E. Marathon, Texas, UCL coll. Figs 2-3. *Quasifusulina* sp., Permian, Thitsipin Limestone Formation, South Shan State, Burma, NHM OGS coll., 2) 1970/24; 3) 1970/29. Fig. 4. *Schwagerina* sp., equatorial section, NHM coll. (where it is identified wrongly as *Neoschwagerina* sp.). Fig. 5. *Schwagerina* sp., Late Carboniferous Limestone, USSR, oblique equatorial section, UCL coll. Fig. 6. *Triticites* sp., Guadalupian, USSR, oblique axial section, UCL coll.

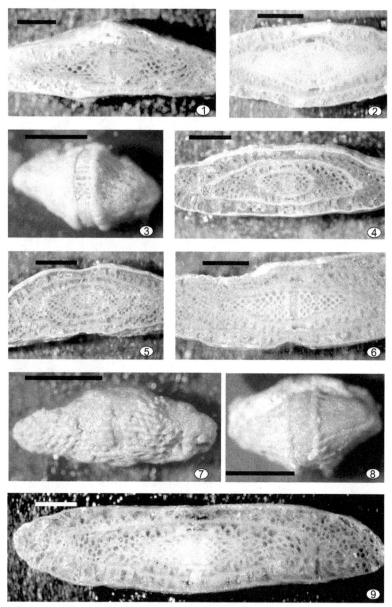

Plate 2.21   Scale bars: Figs 1-9 = 2mm. Figs 1-2, 4-6, 9. Enlargement of tangential sections of solid specimens of fusulinids displaying septal fluting, Italy, UCL coll. Figs 3, 8. Fusulinid specimens showing deposits of levee-like chomata around tunnel, USSR, UCL coll. Fig. 7. Enlargement of a fusulinid test showing the development of the tunnel and futures giving fold, UCL coll.

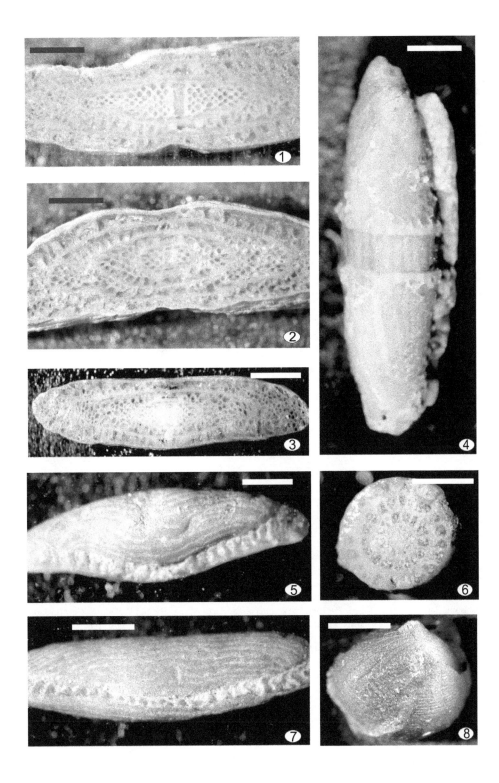

Plate 2.22 Scale bars: Figs 1-8 = 2mm. Figs 1-3. Enlargement of tangential sections of solid specimens of fusulinids displaying septal fluting, Italy, UCL coll. Fig. 4. Fusulinid specimen showing deposits of levee-like chomata around tunnel, USSR, UCL coll. Figs 5, 6. Fusulinid specimens showing antetheca and septal plications, Italy, UCL coll. Fig. 7. Fusulinid specimen showing the equatorial section with septa and spiral theca, UCL coll. Fig. 8. Solid specimen of *Schwagerina* sp. UCL coll.

Plate 2.23 Scale bars: Figs 1-8 = 2mm. Fig. 1. *Neoschwagerina* aff. *craticulifera* (Schwager), Permian, Tebaga, S Tunisia, NHM P43989. Fig. 2. *Neoschwagerina* sp., Limestone from Bukit Kepayang Quarry, Pahang, NHM P42182. Figs 3-4. *Schwagerina adamsi* Ross, Early Permian, 3) holotype, NHM, P42647, 4) paratype, NHM P42648, oblique axial sections showing the thick secondary deposits on the septa, Macusani, Peru, Permian. Fig. 5. *Parafusulina* sp., Permian, Kuzu Machi Tochigi Prefecture, Central Japan, UCL coll.. Fig. 6. *Schwagerina* sp., Permian, UCL coll. Fig. 7. *Parafusulina kaerimizensis* (Ozawa), Permian, Kuzu area, Central Japan, NHM coll. Fig. 8 *Dunbarinella ervinensis* Thompson, Camp Creek, on Saddle Creek, 9 m south of Rockwood, Texas, UCL coll.

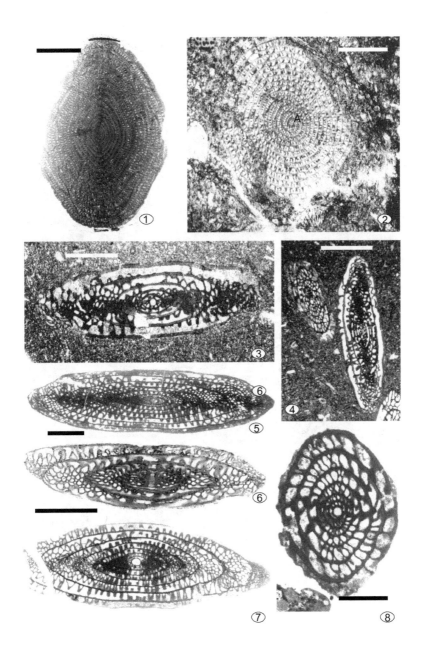

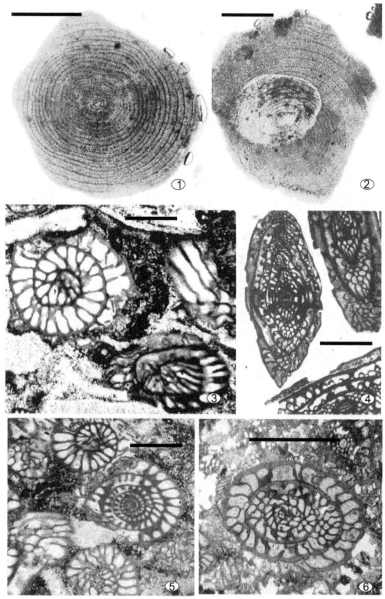

Plate 2.24   Scale bars: Figs 1, 2 = 1mm; Figs 3, 4-6 = 2mm. Figs 1-2. *Yabeina globosa* (Yabe), Georgia, Grimsdale coll. Figs 3, 5. *Pseudoschwagerina* sp., Guadalupian, USSR, oblique axial section, NHM coll. Fig. 4. *Dunbarinella ervinensis* Thompson, Camp Creek, on Saddle Creek, 9m south of Rockwood, Texas, UCL coll. Fig. 6. *Schwagerina* sp., Late Carboniferous Limestone, USSR, oblique equatorial section, UCL coll.

Plate 2.25 Scale bars: Figs 1-6 = 2mm. Fig. 1, 3-4. *Schwagerina* sp., Permian, Pontafel Austria, Stürtz, identified wrongly as *Fusulina* sp., NHM coll., P5111. Fig. 2. *Zellia heritschi mira* Kahler and Kahler, holotype, Early Permian, Carnic Alps, SFN coll. Fig. 5-6. *Triticites patulus* Dunbar and Newell, Permian, Copacabana Group (bed at top of hill), Nacusani, Peru, NHM, P412656-P42659.

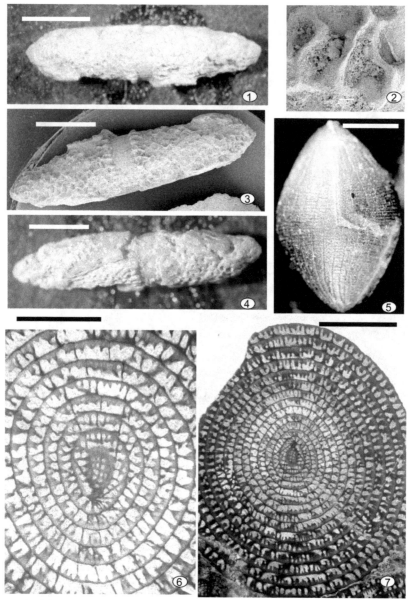

Plate 2.26 Scale bars: Figs 1-7 = 2mm. Figs 1-2. SEM photographs of a fusulinid test, 1) showing antetheca and septal placations, 2) enlargement of the septal fluting. Figs 3-4. Fusulinid specimens showing antetheca and septal placations, Italy, UCL coll. Fig. 5-7. 5) solid specimen of *Neoschwagerina* sp., 6-7) tangential thin sections of the same specimens.

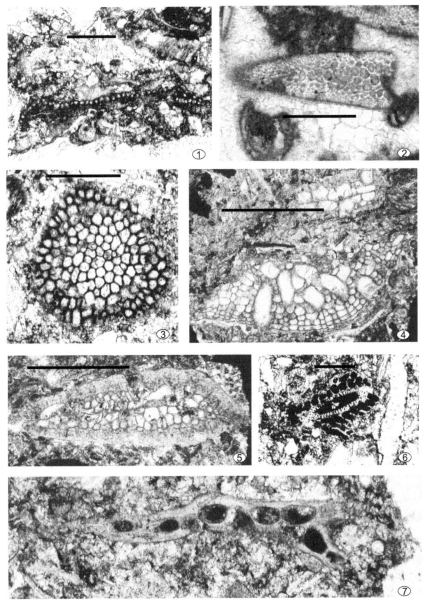

Plate 2.27 (Algae, Corals, Bryozoa) Scale bars: Figs 1-7 = 0.5mm. Fig. 1. *Calcifolium okense* Shvetzov and Birina, Northumberland, UCL coll., JER 110. Fig. 2. *Oligoporella* sp., Hardraw Limestone, Yorkshire, England, UCL coll. Fig. 3. *Chaetetes depressus* (Fleming), corals, Hardraw Limestone, Yorkshire, England, UCL coll. Figs 4-5. Bryozoa sp., Gayle Limestone, Yorkshire, UCL coll. Fig. 6. *Nanopora anglica* Wood, Fossdale Gill, Yorkshire, UCL coll., DH146. Fig. 7. *Calcifolium* sp., Swaledale, Yorkshire, UCL coll., DH714.

Chapter 3

The Mesozoic Larger Benthic Foraminifera: The Triassic

3.1 Introduction

As seen in the previous chapter, the end of the Palaeozoic saw one of the most significant events in the history of life on Earth, with two mass extinctions occurring within a period of 10 Ma of each other (Kaiho et al., 2001; Chen and Benton, 2012). As a result of these events, about 90% of calcareous foraminiferal genera became extinct. The most affected were the large Fusulinida, which were wiped out. The only survivors of the Palaeozoic were a single member of each of the Endothyroidea and Earlandioidea superfamilies, which at that time were morphologically minute. In contrast, the extinction event had less impact on, for example, the simple Textulariida, which lost only 30% of its genera (Loeblich and Tappan, 1988). Forms from the Allogromiida, Miliolida and Lagenida also survived the end Permian extinction, albeit with significantly reduced diversity and as morphologically small forms. The Involutinida with aragonitic tests made their first appearance in the Triassic (Olenekian), persisting to the Early Cretaceous (Cenomanian).

In comparison with Permian larger benthic foraminifera, the Triassic larger foraminifera have not been systematically studied on a global scale. A revision of the taxonomy of the Early and Middle Triassic taxa was presented by Rettori (1995), Rigaud et al. (2015), and a stratigraphic summary of larger benthic foraminifera of the Tethyan realm was presented by Pybernes (in De Gracianski et al., 1998). The relationship between the microgranular Palaeozoic and agglutinated textulariides was explored by Rigaud et al. (2015).

In this chapter, the taxonomy of the main genera of the Triassic larger foraminifera is presented, and a number of revisions suggested. Although most of the Triassic foraminifera are relatively morphologically small they have complex internal structures that are distinguishable in thin section, and so for the purposes of this study they are considered as "larger foraminifera". Most of the superfamilies and families found in the Mesozoic are long ranging, but this chapter is only concerned with the genera characteristic of the Triassic.

3.2 Morphology and Taxonomy of Triassic Larger Benthic Foraminifera

The Triassic larger forms are found developed in six orders:

- Textulariida
- Lagenida
- Earlandiida

- Endothyrida
- Miliolida
- Involutinida

The development and evolution of the superfamilies of these orders is schematically shown in Fig. 3.1. Below, are presented the morphological characteristics and taxonomic relationships of the major Triassic forms, while in the next section their biostratigraphic significance and their phylogenetic relations are discussed.

ORDER TEXTULARIIDA Delage and Hérouard, 1896
The tests of these agglutinated foraminifera are made of foreign particles bound by organic cement. They range from Early Cambrian to Holocene.

Superfamily AMMODISCOIDEA Reuss, 1862
The test is sub-spherical or tubular, with an aperture at the end of a tube. Cambrian to Holocene

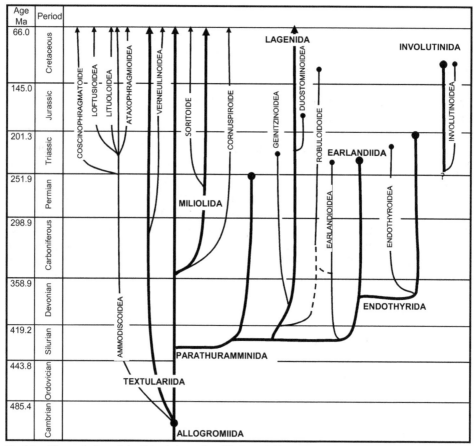

Fig. 3.1.   The evolution of the Triassic larger benthic foraminifer orders and superfamilies.

Family Ammodiscidae Reuss, 1862
Members of this family have a proloculus that is followed by an uncoiled non-septate tubular second chamber. Early Cambrian to Holocene.

- *Gandinella* Ciarapica and Zaninetti, 1985 (Type species: *Gandinella apenninica* Ciarapica and Zaninetti, 1985). The test is partly streptospiral. Early Triassic to Late Triassic (Olenekian to Rhaetian) (Plate 3.1, fig. 8).
- *Glomospira* Rzehak, 1885 (Type species: *Trochammina squamata* Jones and Parker var. *gordialis* Jones and Parker, 1860). The first chamber is followed by a streptospirally irregularly coiled chamber. The aperture is terminal. Early Carboniferous (Visean) to Holocene (Plate 3.2, figs 1-3, 5-9).
- *Glomospirella* Plummer, 1945 (Type species: *Glomospira umbilicata* Cushman and Waters, 1927). Similar to *Glomospira* but later becoming planispirally coiled. Late Carboniferous (Bashkirian) to Miocene (Plate 3.1, fig. 1; Plate 3.4, fig. 7; Plate 3.5, fig. 15; Plate 3.6, figs 6-7).
- *Paulbronnimannia* Rettori and Zaninetti, 1993 (Type species: *Agathammina judicariensis* Premoli Silva, 1971). The test is fusiform and compressed, composed of a globular proloculus followed by a long non-septate tubular chamber. Middle Triassic (Anisian) (Plate 3.5, fig. 12).
- *Pilammina* Pantic, 1965 (Type species: *Pilammina densa*. Pantić, 1965). The spherical proloculus is followed by a narrow non-septate, elongate second chamber. Early Triassic to Middle Triassic (Induan to Anisian) (Plate 3.5, fig. 9).
- *Pilamminella* Salaj, 1978 (Type species: *Pilammina grandis* Salaj, in Salaj et al., 1967). The initial coiling is the same as *Pilammina,* then the second chamber changes 90° to the plane of coiling, followed by two or three oscillating coils. Middle to Late Triassic (Anisian to Carnian).

Superfamily ATAXOPHRAGMIOIDEA Schwager, 1877
Members of this superfamily have a multilocular, trochospiral test becoming biserial or uniserial in later stages. Middle Triassic to Holocene.

Family Ataxophragmiidae Schwager, 1877
Members of this family have three or more chambers per whorl in the early stages. The aperture is high and terminal. Late Triassic to Palaeocene.

- *Palaeolituonella* Bérczi-Makk, 1981 (Type species: *Palaeolituonella majzoni* Berczi-Makk, 1981). The elongate conical test, having initial whorls with four to five chambers, is followed by a biserial stage and uniserial stages with internal rudimentary radiations. Middle to Late Triassic (Anisian to Carnian).

Superfamily LOFTUSIOIDEA Bradey, 1884
The test is planispiral, may uncoil in later stage. The wall is agglutinated with differentiated outer layer and inner alveolar layer. Late Triassic (Carnian) to Holocene.

Family Mesoendothyridae Voloshinova in Bykova et al., 1958
The test is strepto- or planispirally coiled, has involute initial chambers, and later is uncoiled. Adult chambers are cylindrical or flattened, falciform to cyclical. They are

simple, with radial partitions or with pillars. Walls may have alveoles or a hypodermic network. Late Triassic (?Carnian to early Norian).

- *Wernlina* Rigaud et al., 2014 (Type species: *Wernlina reidae* Rigaud et al., 2014). The test is symmetrical and planispirally coiled. The wall is dark, thick and microgranular, formed by an inner alveolar layer sealed by an outer imperforate layer. The septa are thick and non-alveolar. The aperture is single and basal. Rigaud et al (2014) included in *Wernlina*, the species *Everticyclammina praevirgulina* described and illustrated from the Sinemurian–early Pliensbachian of Spain by BouDagher-Fadel and Bosence (2007, pl. 3, fig. 6). However, the latter differs from typical *Wernlina* in having an initial streptospiral coiling, fewer and proportionally thicker chamber walls and septa (see Chapter 4). Late Triassic (?late Carnian to early Norian).

Superfamily LITUOLOIDEA de Blainville, 1825

Members of this superfamily have a conical, multilocular, rectilinear and uniserial test. The early stage has plani- (strepto-) or trochospiral coiling. The periphery of the chambers has radial partitions; but centrally they are with or without scattered, separated pillars. The septa are arched into hummocks (almost solid masses) between the apertures, with bases of the arches that can fuse to the hummocks of the previous septum, with the apertures then opening at the suture. The alignment of the apertures and thickening of the hummock walls produces the appearance of a series of "gutters". No true pillars are formed. The walls are solid, non-alveolar, non-canaliculate. The aperture is simple, with no internal tooth plates, areal or multiple, cribrate. Late Triassic (Carnian) to Holocene.

Family Lituolidae de Blainville, 1827

The early stages of the tests are enrolled, but later they may become rectilinear. Walls are formed from agglutinated foreign particles. There are few chambers (less than 10) per whorl. Carboniferous to Holocene.

Subfamily Ammomarginulininae Podobina, 1978

The early stage of the test is coiled, but it becomes uncoiled in later stages. Apertures are single. Carboniferous (Early Mississippian) to Holocene.

- *Ammobaculites* Cushman, 1910 (Type species: *Spirolina agglutinans* d'Orbigny, 1846). The test is simple, not compressed and uncoils in the adult. Apertures are single, areal. Carboniferous (Mississippian) to Holocene (Plate 5.6, fig. 18).

Subfamily Lituolinae de Blainville, 1827

Members differs from Ammomarginulininae in having multiple apertures. Late Triassic to Holocene.

- *Lituola* Lamarck, 1804 (Type species: *Lituolites nautiloidea* Lamarck, 1804). These forms have no internal partitions and a multiple cribrate aperture. Late Triassic to Holocene (Plate 5.5, fig. 14; Plate 5.6, fig. 9).

Superfamily COSCINOPHRAGMATOIDEA Thalmann, 1951
Members of this superfamily are attached and may be coiled in their early stages, but later are uncoiled or branched. Triassic to Holocene.

Family Coscinophragmatidae Thalmann, 1951
The wall is canaliculated, and perforated with alveoles. Late Triassic to Holocene.

- *Alpinophragmium* Flügel, 1967 (Type species: *Alpinophragmium perforatum* Flügel, 1967). The chambers are numerous. The test is uniserial with a terminal cribrate aperture. Late Triassic (Carnian to Rhaetian) (Plate 3.6, figs 1-5)

Superfamily VERNEUILINOIDEA Cushman, 1911
Representatives of this superfamily have a trochospiral test throughout, or only in the early stage. They may be triserial, biserial or uniserial. Some forms have a streptospiral initial part. The aperture is single or multiple. Late Carboniferous to Holocene.

Family Piallinidae Rettori, Zaninetti in Rettori et al., 1993
The test is multilocular and elongated, with a proloculus followed by a trochospiral development, or with a small streptospiral in the early stage followed by a high trochospire in the later stage. Late Triassic (Carnian).

- *Piallina* Rettori, Zaninetti in Rettori et al., 1993 (Type species: *Piallina tethydis* Rettori and Zaninetti, 1993). The elongated test, with a globular proloculus, is followed by a trochospiral or initially streptospiral part. Late Triassic (Carnian) (Plate 3.5, fig. 8).

ORDER LAGENIDA Delage and Hérouard, 1896
This order is characterised by having monolamellar walls, composed of low-Mg calcite in which the optical c-axes of the crystal units are perpendicular to the outer surface of the test. Primitive taxa are without secondary lamination, but more advanced forms are found with secondary lamination and a thin microgranular inner layer (see special terminology defined in Chapter 2). They range from Late Silurian to Holocene.

Superfamily DUOSTOMINOIDEA Brotzen, 1963
The test is enrolled, planispiral to trochospiral. The aperture is single or double, and interiomarginal. Triassic (Anisian) to Early Jurassic (Hettangian).

Family Duostominidae Brotzen, 1963
Characterised by two interiomarginal apertures in the final chamber. Triassic (Anisian to Rhaetian).

- *Duostomina* Kristan-Tollmann, 1960 (Type species: *Duostomina biconvexa* Kristan-Tollmann, 1960). The test is lenticular, and trochospirally coiled. Triassic (Anisian to Rhaetian) (Plate 3.2, fig. 4).

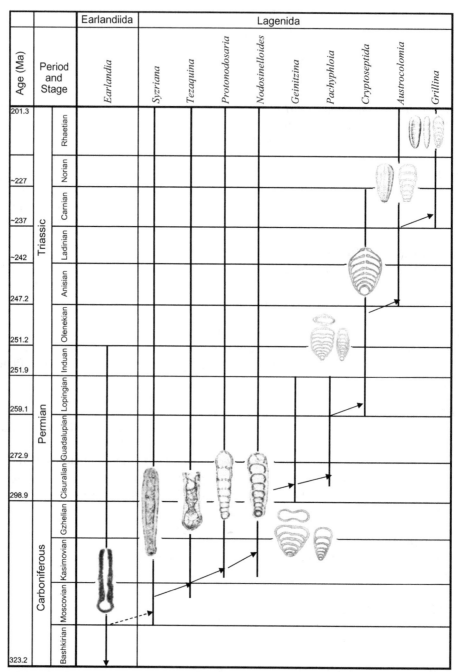

Fig. 3.2. The phylogenetic development of the lagenides through the Palaeozoic and Triassic (sections of some lagenides are modified from Sellier de Civrieux and Dessauvagie, 1965).

Family Variostomatidae Kristan-Tollmann, 1963, emend. Rigaud, Martini and Vachard, 2015

- *Cassianopapillaria* di Bari and Rettori, 1998 (nom. subst. for *Papillaria* di Bari and Rettori, 1996, preoccupied (Type species: *Papillaria laghii*, di Bari and Rettori, 1996). The test is trochospirally coiled, with pillars filling the umbilical region. Triassic (Carnian).

Superfamily GEINITZINOIDEA Bozorgnia, 1973
The tests of this superfamily are uniserial, and similar to the Palaeozoic form Nodosinelloidea, but the microgranular layer is the inner dark layer, while the radially fibrous layer is the outer layer. Late Devonian to Middle Triassic.

Family Abriolinidae Zaninetti and Rettori in Zaninetti et al., 1992
Morphologically small foraminifera, being trochospiral, with slightly curved sutures and subglobular chambers. The aperture is not observed. Middle Triassic (Anisian to Ladinian).

- *Abriolina* Luperto, 1963 (Type species: *Abriolina mediterranea* Luperto, 1963). The test has a large sub-spherical proloculus. Middle Triassic (Anisian to Ladinian) (Plate 3.5, fig. 10).

Superfamily ROBULOIDOIDEA Reiss, 1963
The tests of members of this superfamily are those of a typical Lagenida, but without secondary lamellar or with slight lamination in younger taxa. The aperture is primitive and cylindrical. Late Silurian to Early Cretaceous.

Family Ichtyolariidae Loeblich and Tappan, 1986.
The test is elongate and uniserial with a single layered wall, and may show some secondary lamination. The aperture is simple and terminal. Early Permian to Early Cretaceous.

- *Austrocolomia* Oberhauser, 1960 (Type species: *Austrocolomia marschalli* Oberhauser, 1960). The chambers are cylindrical, gradually enlarging and divided by straight sutures, perpendicular to the long axis of test. Triassic (Anisian to Rhaetian) (Figs 3.2; 3.3).
- *Cryptoseptida* de Civrieux and Dessauvagie, 1965 (Type species: *Cryptoseptida anatoliensis* de Civrieux and Dessauvagie, 1965). Loeblich and Tappan (1986) included *Pachyphloides* de Civrieux and Dessauvagie, 1965, with a question mark in the synonyms of *Cryptoseptida*, as the off-centred illustrated specimen of *Pachyphloides* might have strong lamination as in *Cryptoseptida*. Permian to Triassic (middle Sakmarian to Carnian) (Figs 3.2; 3.3).
- *Grillina* Kristan-Tollmann, 1964 = *Geinitzinita* de Civrieux and Dessauvagie, 1965 (Type species: *Geinitzinita oberhauseri* Sellier de Civrieux and Dessauvagie 1965). The tricarinate, flattened or biconcave cross-sections are without internal protective lamellae. Triassic (Carnian to Rhaetian) (Figs 3.2; 3.3; Plate 3.5, fig. 6).

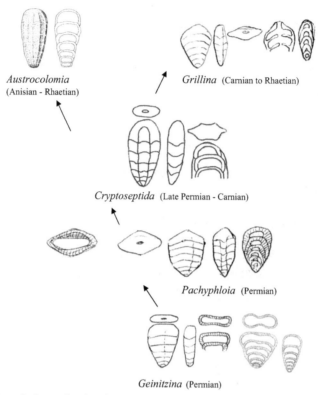

Austrocolomia
(Anisian - Rhaetian)

Grillina (Carnian to Rhaetian)

Cryptoseptida (Late Permian - Carnian)

Pachyphloia (Permian)

Geinitzina (Permian)

Fig. 3.3. A schematic figure showing the convergence of lagenides features in the Late Permian and the Late Triassic forms.

Family Robuloididae Reiss, 1963
The test is uniserial, enrolled with an atelo-monolamellar wall construction. The aperture is terminal. Middle Permian to Late Jurassic.

• *Robuloides* Reichel, 1946 (Type species: *Robuloides lens* Reichel, 1946). The test is nautiloid to lenticular with an acute periphery. It is involute and planispirally, or nearly planispirally, coiled. A primary monolamellar wall covers previous chambers (weakly to markedly plesio-monolamellar), so that the umbilical region of the test appears as a nearly solid mass of calcite and the peripheral septa are oblique. The aperture is simple. Middle Permian to Middle Triassic (Ladinian) (Plate 3.5, fig. 11).

ORDER EARLANDIIDA SABIROV IN VDOVENKO ET AL., 1993 EMEND. VACHARD ET AL., 2010
This order has simple, free or attached tests, consisting of a proloculus and a rectilinear second chamber. The aperture is terminal and simple. Late Silurian to Early Cretaceous.

Superfamily EARLANDIOIDEA Cummings, 1955

This superfamily is characterised by having a free, non-septate test with a globular first chamber followed by a straight tubular one. It ranges from the Late Silurian to Early Triassic.

Family Earlandiidae Cummings, 1955

The Earlandiidae have a single free chamber and range from the Late Silurian to Early Triassic.

- *Earlandia* Plummer, 1930 (Type species: *Earlandia perparva* Plummer, 1930). The test is free, elongate, and composed of a globular proloculus followed by an undivided straight tubular chamber. The wall is calcareous and microgranular. Late Silurian to Early Triassic (Induan) (Plate 3.3, fig. 2, Fig. 3.2).

ORDER ENDOTHYRIDA FURSENKO 1958

Members of this order have a lenticular, planispirally coiled test. The wall is dark and microgranular, but sometimes can be bilayered or multilayered. Apertures are simple, basal or cribrate. Late Devonian to Triassic.

Superfamily ENDOTHYROIDEA Brady, 1884 nom. translat. Fursenko, 1958

Members of this superfamily have streptospiral to planispiral tests with many chambers, followed by a rectilinear stage, which is biserial or uniserial in some forms. The wall is microgranular, calcareous, but some forms show two to three distinct layers; others may develop an inner perforate or keriothecal layer. Late Devonian to Triassic.

Family Endotebidae Vachard, Martini, Rettori and Zaninetti, 1994

The test is free, and planispiral in early stages, but later uniserial to biserial. Walls are calcareous, grey, thick, and calcareous agglutinated. Apertures are simple. Late Permian to Triassic.

- *Endoteba* Vachard and Razgallah, 1988 emend. Vachard, Martini, Rettori and Zaninetti, 1994 (Type species: *Endoteba controversa* Vachard and Razgallah, 1988 emend. Vachard, Martini, Rettori and Zaninetti, 1994). The axial view is compressed. Late Permian to Late Triassic (Late Kungurian to Rhaetian) (Fig. 3.4).
- *Endotebanella* Vachard, Martini, Rettori and Zaninetti, 1994 (Type species: *Endothyranella lwcaeliensis* Dager, 1978). Similar to *Endoteba,* but the final stage is uniserial. Middle Triassic (latest Olenekian to Ladinian) (Fig. 3.4).

Family Endotriadidae Vachard, Martini, Rettori and Zaninetti, 1994

Tests are trochospiral, almost planispiral, and may be uniserial in later stages with a microgranular wall. Apertures are basal, simple. Middle to Late Triassic.

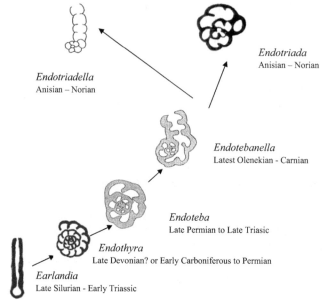

Fig. 3.4. A schematic evolution of the Endotebidae (drawings taken from Vachard *et al.* 1994) from the Endothyridae (*Earlandia*).

- *Endotriada* Vachard, Martini, Rettori and Zaninetti, 1994 (Type species: *Endotriada tyrrhenica* Vachard, Martini, Rettori and Zaninetti, 1994). The chambers are hemispherical. Triassic (Anisian to Norian) (Fig. 3.4).
- *Endotriadella* Vachard, Martini, Rettori and Zaninetti, 1994 (Type species: *Ammobaculites wirzi* Koehn-Zaninetti, 1968). The test has uniserial later stages. Triassic (Anisian to Norian) (Fig. 3.4).

ORDER MILIOLIDA Delage and Hérouard, 1896
The miliolides have tests that are porcelaneous and imperforate, made of high Mg-calcite with fine, randomly oriented crystals. They range from the Carboniferous to the Holocene.

Superfamily CORNUSPIROIDEA Schultze, 1854
The test is free or attached, and composed of a globular proloculus followed by a tubular enrolled chamber. The coiling is planispiral or trochospiral, evolute or involute, and may become irregular. Apertures are simple, at end of the tube. Early Carboniferous to Holocene.

Family Arenovidalinidae Zaninetti, Rettori in Zaninetti, Rettori, He and Martini, 1991
Tests are lenticular with a globular proloculus and a second undivided chamber. Early to Middle Triassic.

- *Arenovidalina* He, 1959 (Type species: *Arenovidalina chialingehian-gensis* He, 1959). The test is lenticular with a globular proloculus and a second tubular undivided chamber. The coiling is planispiral throughout and involute, with a lamellar umbonal region on each side of the test. Early to Middle Triassic (Induan to Anisian, ?early Ladinian) (Fig. 3.5).

• *Paratriasina* He, 1980 (Type species: *Paratriasina jiangyouensis* He, 1980*)*. The test is planispiral, involute with a possibly irregular central area. Internal pillars may be present. The wall is porcelaneous. Early whorls show short zigzag bends as in *Meandrospira* Loeblich and Tappan, followed by later planispiral developments. Triassic (lattermost Anisian to early Ladinian) (Fig. 3.5).

Family Ophthalmidiidae Wiesner, 1920
Tests are free, composed of a proloculus followed by an undivided coiled second chamber and chambers that commonly are one-half coil in length. The wall is porcelaneous and the aperture is terminal and simple. Triassic (Anisian) to Holocene.

• *Eoophthalmidium* Langer, 1968 (Type species: *Praeophthalmidium (Eoophthalmidium) tricki* Langer, 1968). The coiling is involute. Middle Triassic (Anisian) (Plate 3.1, figs 3,5).

• *Karaburunia* Langer, 1968 (Type species: *Karaburunia rendeli* Langer, 1968). The proloculus is followed by sigmoidal coiling of two chambers per whorl, the final pair of chambers are approximately 180° apart. Middle Triassic (late Anisian) (Plate 3.1 figs 2,4).

• *Ophthalmidium* Kübler and Zwingli, 1870 (Type species: *Oculina liasica* Kübler and Zwingli, 1870). The coiling is planispiral and chambers have distinct floors and lateral extensions. Late Triassic (Carnian) to Late Jurassic (Kimmeridgian) (Fig. 3.5).

Family Meandrospiridae Sadova, 1961 emend. Zaninetti et al., 1987
Tests are composed of a proloculus and an undivided second chamber with zigzag coiling. Apertures are simple and terminal. Permian to Holocene.

• *Meandrospira* Loeblich and Tappan, 1964 (Type species: *Meandrospira washitensis* Loeblich and Tappan, 1946). The undivided second chamber forms an involute planispiral zigzag of bends, with only those of the final whorl visible from the exterior. Early Permian (Artinskian) to Holocene (Plate 3.5, figs 1,2).

• *Meandrospiranella* Salaj, 1969 (Type species: *Meandrospiranella samueli* Salaj, in Salaj et al., 1967). The undivided tubular chamber coils streptospirally in short zigzag bends of about 5 whorls, later becoming somewhat irregular and uncoiling. Triassic (Anisian) to Late Cretaceous (Cenomanian) (Plate 3.5, fig. 3).

• *Turriglomina* Zaninetti in Limongi, Panzanelli-Fratoni, Ciarapica, Cirilli, Martini, Salvini-Bonnard and Zaninetti, 1987 (Type species: *Turritellella mesotriasica* Koehn-Zaninetti, 1968). The test is elongated and composed of a globular proloculus and a second undivided tubular chamber with a first stage that is *Meandrospira*-like, followed by a long helicoidally compressed stage. Triassic (Anisian to Rhaetian) (Plate 3.4, fig. 9).

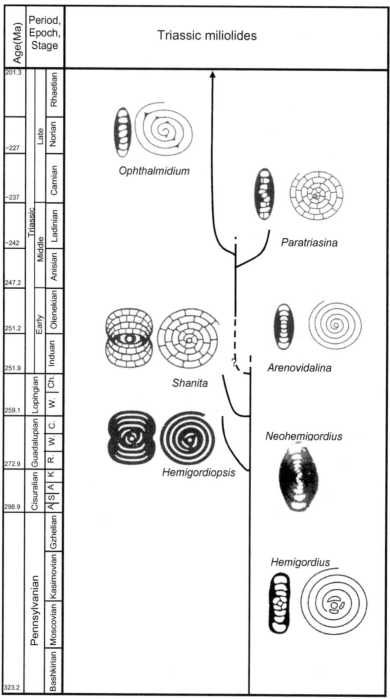

Fig. 3.5. The evolution of the Triassic miliolides from a Permian ancestor (modified from Zaninetti *et al.*, 1991; and Rettori, 1995).

Family Cornuspiridae Schultze, 1854

The tests are free or attached, composed of a proloculus followed by an undivided planispiral to streptospiral, involute or evolute second tubular chamber. Early Carboniferous (Visean) to Holocene.

• *Planiinvoluta* Leischner, 1961 (Type species: *Planiinvoluta carinata* Leischner, 1961). The test is flat, attached, with tubular chambers, nonseptate, planispirally enrolled with an evolute coiling on the flat attached side and an involute coiling on the opposite side. Late Triassic (Rhaetian) (Plate 3.8, fig. 7).

Family Hemigordiopsidae Nikitina, 1969

These forms are found with a test that has at least an early streptospiral stage, that later may be planispiral. Early Carboniferous (Visean) to Holocene.

• *Agathammina* Neumayr, 1887 (Type species: *Serpula pusilla* Geinitz, in Geinitz and Gutbier, 1848). The proloculus is followed by an undivided, non-septate tubular chamber, coiling in five planes. Middle Carboniferous (Namurian) to Triassic (Plate 3.4, fig. 3).
• *Multidiscus*. Miklukho-Maklay, 1953 (Type species: *Nummulostegina padangensis* Lange, 1925). The involute early test is followed by an evolute last whorl. Carboniferous to Permian (Kasimovian-Asselian).

Family Hoynellidae Rettori, 1994

A free test, with a globular proloculus followed by an early miliolid stage and an undivided tubular chamber arranged on several vertical planes. Triassic to ?Early Jurassic.

• *Hoynella* Rettori, 1994 (Type species: *Hoynella* s*inensis* Rettori, 1994). The test is ovoid, the tubular second chamber is followed by a planispiral evolute stage. Triassic to ?Early Jurassic (Plate 3.5, fig. 7).

Superfamily SORITOIDEA Ehrenberg, 1839

These forms have chambers that are planispiral, uncoiling, flabelliform or cyclical, and may be subdivided by partitions or pillars. Late Permian to Holocene.

Family Milioliporidae Brönnimann and Zaninetti, 1971

The tests are free or attached with a proloculus and tubular chambers arranged in various planes of coiling, which may be irregular, oscillating, or sigmoidal. Late Permian to Late Triassic (Rhaetian).

• *Galeanella* Kristan, 1958 (Type species: *Galea tollmanni* Kristan, 1957). The test is ovate in outline, planoconvex, planispiral to involute, and with thick coarsely perforate walls. Late Triassic (Norian to Rhaetian) (Plate 3.4, fig. 11).
• *Kamurana* Altiner and Zaninetti, 1977 (Type species: *Kamurana bronnimanni* Altiner and Zaninetti, 1977). The test is spherical in outline, with a globular proloculus followed by an undivided enrolled tubular second chamber. The aperture is terminal, and simple. Late Permian to Early Triassic (Wuchiapingian to Olenekian).

ORDER INVOLUTINIDA HOHENEGGER AND PILLER, 1977

This order includes all forms with an enrolled second chamber. They have walls that are aragonitic, but commonly they are recrystallised to give a homogeneous micro-granular structure. They have an umbilical region with pillar-like structures on one or both sides of the test. They range from Triassic to Cretaceous.

Superfamily INVOLUTINOIDEA Bütschi, 1880

These forms consist of a first chamber, followed by a planispiral to trochospiral enrolled tubular second chamber. Originally thought to date from Early Permian (Loeblich and Tappan, 1988; BouDagher-Fadel, 2008), but here defined from Triassic to Late Cretaceous (Cenomanian).

Family Triadodiscidae Zaninetti, 1984

The umbilical area is covered with additional lamellae added with each new whorl. Early Triassic to Late Triassic (Olenekian to Carnian).

- *Lamelliconus* Piller, 1978 (Type species: *Trocholina (Trocholina) biconvex* Oberhauser, 1957). The test is lenticular with a globular proloculus, followed by a trochospiral undivided second tubular chamber. Triassic (Ladinian to Carnian) (Plate 3.5, fig. 13).
- *Triadodiscus* Piller, 1983 (Type species: *Trocholina (Paratrocholina) eomesozoicus* Oberhauser, 1957). The test is lenticular with a globular proloculus, followed by a planispiral to slightly trochospiral and involute second undivided tubular chamber that is planispiral. The whorls are formed by single lamella. Triassic (Olenekian to Norian/?Rhaetian) (Plate 3.7, fig. 2B; Plate 3.8, fig. 6).

Family Aulotortidae Zaninetti, 1984

These forms have a lenticular to conical test with calcareous imperforate walls, and consist of a globular proloculus followed by a tubular, enrolled and undivided second chamber, lying against the previous whorl to form planispiral, oscillating or trocho-spiral coiling. Each half whorl is followed by the lamellar deposition of one or two layers. The aperture is at the open end of the tubular chamber. Triassic (Anisian) to Middle Jurassic.

- *Angulodiscus* Kristan, 1957 (Type species: *Angulodiscus communis* Kristan, 1957). The second chamber is involutedly coiled. In axial section, the chamber lumen is curved and slightly tapering towards the umbilical area. Triassic (Norian to Rhaetian) (Fig. 3.6).
- *Auloconus* Piller, 1978 (Type species: *Trocholina permodiscoides* Oberhauser, 1964). The second chamber is trochospirally enrolled and tubular. Each half whorl forms a lumina that covers the umbilicus, resulting in the build-up of a thick and solid umbil-ical filling. Triassic (Norian to Rhaetian) (Plate 3.7, figs 1,6).
- *Aulotortus* Weynschenk, 1956 (Type species: *Trocholina (Paratrocholina) oscillens* Oberhauser. 1957). The enrolled second chamber is planispiral to slightly strepto-spiral, oscillating around the proloculus. *Aulotortus* has less regular planispiral coil-ing than *Angulodiscus*. Triassic (Anisian) to Middle Jurassic (Fig. 3.6; Plate 3.3, figs 3,5; Plate 3.5, figs 4,5; Plate 3.7, figs 2A, 9B, 11, 12B; Plate 3.8, figs 2, 3).

Family Triasinidae Loeblich and Tappan, 1986
The tests have a proloculus, followed by a broad tubular chamber. The interior is filled with cylindrical pillars. Triassic (Ladinian to Rhaetian).

- *Triasina* Majzon, 1954 (Type species: *Triasina hantkeni* Majzon, 1954). Secondary thickening may occur in the umbilical area. Triassic (Norian to Rhaetian) (Plate 3.4, fig. 12).

Family Involutinidae Bütschli, 1880
The globular proloculus is followed by a trochospiral coiled tubular second chamber. They show secondary lamellar thickenings on one or both umbilical regions. The aperture is at the open end of the tube. Members of this family may have shell material deposited on the umbilical area at one or both sides of the test. Triassic to Late Cretaceous (Olenekian to Cenomanian).

- *Aulosina* Rigaud, Martini and Rettori, 2013 (Type species: *Triasina oberhauseri* Koehn−Zaninetti and Brönnimann, 1968). The test is lenticular, with no internal pillars, but with an evolute tubular chamber. The umbilical mass is wide. Late Triassic (Norian to Rhaetian).
- *Involutina* Terquem, 1862 (Type species: *Involutina jonesi* Terquem and Piette, in Terquem, 1862). The test is planispiral and evolute. Both umbilical regions are filled with lamellar deposits. Triassic to Late Cretaceous (Norian to Cenomanian) (Fig. 3.8; Plate 3.7, figs 3-5, 7, 8B, 9A, 10, 12A).
- *Parvalamella* Rigaud, Martini and Rettori, 2012 (Type species: *Glomospirella friedli* Kristan-Tollmann, 1962). The test is lenticular to globular, lacking any internal structures. The small proloculus is directly followed by a ball-like, enrolled, undivided tubular chamber. The simple aperture is terminal. Triassic (late Ladinian to Rhaetian).
- *Semiinvoluta* Kristan, 1957 (Type species: *Semiinvoluta clari* Kristan, 1957). Only the flat umbilical side is filled with pillar-like deposits. Triassic (Norian to Rhaetian) (Fig. 3.6).
- *Trocholina* Paalzow, 1922 (Type species: *Involutina conica* Schlumberger, 1898). A conical test consisting of a globular proloculus followed by a trochospirally enrolled tubular second chamber with pillars filling the umbilical area. The aperture is at the end of a tubular chamber. *Trocholina* closely resembles the simple small benthic *Conicospirillina* Cushman, but differs from the latter in having a solid plug instead of an open umbilical cavity. *Trocholina* is also less involute than *Conicospirillina*. Triassic to Cretaceous (Norian to Cenomanian) (Fig. 3.6; Plate 3.1, figs 6-7).

3.3 Biostratigraphy and Phylogenetic Evolution

3.3.1 General Biostratigraphy

The Permian-Triassic transition represents a critical period in foraminiferal evolutionary history. Without question, the end Permian extinction resulted in the greatest reduction of biodiversity in Earth history, claiming roughly 85 percent of all genera then in existence (Erwin et al., 2002; Chen and Benton, 2012). The dominant Palaeozoic order, Fusulinida, was eliminated, while diversity within the Lagenida, Miliolida and

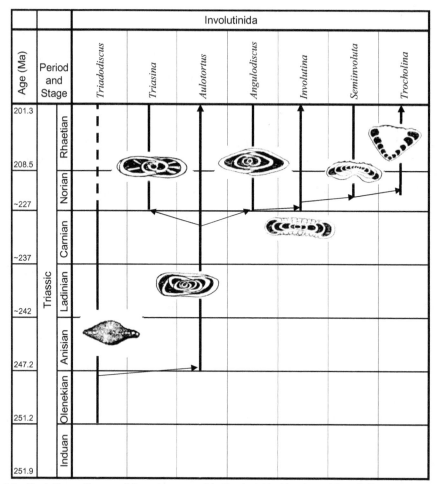

Fig. 3.6.   Evolutionary lineages of the main genera of the Involutinida. (drawings of foraminifera are not to scale).

Textulariida declined by 30, 50, and 80 percent respectively. As discussed in Chap. 2, the true Involutinida appeared as new forms in the Triassic.

The end Permian mass extinction, and the following survival and recovery intervals, resulted in a dramatic turnover in the taxonomic composition of calcareous benthic assemblages (Groves and Altiner, 2004). During the Early Triassic, shallow marine foraminifera were simple and rare, with the first recorded forms appearing in the eastern Palaeo-Tethyan realm (Sweet et al. 1992; Márquez, 2005), resulting elsewhere in a taphonomically induced gap in the understanding of the transition of the Permian survivors (Groves, 2000). They were initially dominated by small arenaceous forms (Tong and Shi, 2000). These are the "disaster forms" which are, as defined by Fischer and Arthur (1977), 'opportunistic species in the sense of MacArthur (1955) and Levinton (1970), and characteristic of the survival phase following a mass extinction (Hallam and Wignall, 1997). The subsequent disappearance of such forms, along with the proliferation of the survivors and the reappearance of 'Lazarus taxa', marks the beginning

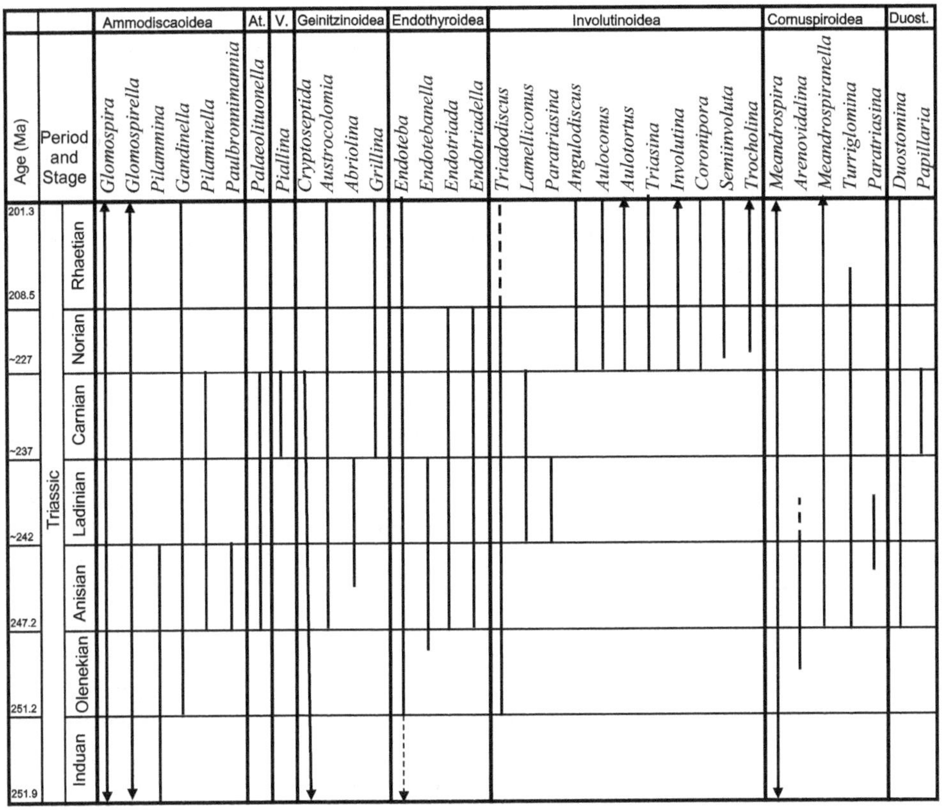

Fig. 3.7. Biostratigraphic ranges of the main Triassic genera.

of the recovery phase in the Ladinian (Groves and Altiner, 2004). The recovered forms seem to be dominated by the miliolides and later by the involutides in the Eastern Tethys area.

The larger benthic foraminifera in the Early and Middle Triassic evolved slowly, and it was not till the Ladinian stage that highly diversified communities, such as the Ammodiscoidea and the Endothyroidea thrived. Among the most evolved foraminifera along with these communities are the Involutinida, such as *Lamelliconus* which became diverse and common in the Norian to the Rhaetian. Towards the end of the Triassic a gradual decline of the larger benthic foraminifera groups occurred, which climaxed with an end Triassic mass extinction event (Tanner et al., 2004), where most remaining larger benthic foraminifera disappeared. The biostratigraphic range of the main Triassic forms are summarized in Fig. 3.7.

3.3.2 The Palaeozoic-Triassic Lagenides

One of the most important groups which survived the Permian extinction was the lagenides. Early Triassic lagenides are direct descendants of Permian forms and originated with the family Syzraniidae in Moscovian time. The wall in the earliest species of *Syzrania* (a form of Robuloidoidea) is dominantly microgranular, with only an incipiently developed hyaline-radial layer. This led authors such as Altiner and Savini (1997) to retain the Syzraniidae within the Fusulinida. Similarly, in certain reference books (e.g. Loeblich and Tappan, 1998) and in previous analyses of foraminiferal diversity across the P-T boundary, a number of lagenides genera were also regarded as either belonging to the Fusulinida or Rotaliida. In our modification of the phylogenetic of the Carboniferous forms, we would suggest that they may have indeed derived from the Earlandiida. However, more recent work (Groves et al., 2003, 2004) suggests that the order Lagenida is monophyletic. Authors, such as Palmieri (1983), Groves and Wahlman (1997), Pinard and Mamet (1998), Groves (2000) and Groves et al. (2004), suggested there were very close similarities in wall structures between Palaeozoic and Mesozoic-Cenozoic lagenides, with no change in wall structure at the Permian-Triassic boundary. If so, the Triassic Robuloidoidea would have had its origins in the Parathuramminida (see Fig 3.1). This is still an unresolved issue.

Studies by Grove et al. (2004) and Grove (2005) divided the lagenides, at the highest level, into nonseptate and septate taxa. Within the nonseptate group there are forms whose tubular second chamber is unpartitioned (e.g. *Syzrania*, see Chapter 2), and forms whose second chamber is partitioned, to varying degrees, by constrictions or thickenings of the wall (e.g. *Tezaquina*, (Fig. 3.2); see also Chapter 2). Within the septate group there are uniserial (e.g. *Nodosinelloides*, see Fig. 3.2) and coiled forms (*Calvezina*, see Chapter 2). The uniserial forms are further differentiated on the basis of chamber symmetry when seen in transverse section: i.e., radial circular (*Nodosinelloides*); radial bilateral flattened (*Geinitzina*); triangular (*Pseudotristix*); and polygonal (*Rectostipulina*). The phylogenetic evolution of the lagenides, as shown in Fig. 3.2, follows a morphological trend varying from uniserial tubular, non-septate forms to uniserial septate forms with rounded transverse section which become flattened and flaring in advanced forms.

The Moscovian-Kasimovian interval of the Carboniferous saw the initial evolutionary radiation of the early complex lagenides with the oldest known genus *Syzrania*, which as indicated above might have evolved from the Earlandioidea (*Earlandia*) in the Middle Pennsylvanian (see Figs 3.1 and 3.2), through the addition of a hyaline-radial layer external to the ancestral microgranular wall (Groves et al., 2004). This is shown in the earliest species of *Syzrania* where the wall was still dominantly microgranular with only an incipiently developed hyaline-radial layer (Fig. 3.2). The Syzranidae evolved in the Late Pennsylvanian and Permian into sub-septate and fully septate uniserial forms. In the late Ghzelian, the ancestor to all laterally compressed early lagenides, *Geinitzina*, made its first appearance, but it was not until the late Sakmarian that the *Pachyphloia – Howchinella* group finally developed.

The Late Permian (Lopingian) lagenides were diverse and abundant throughout the Tethyan region and northern higher paleolatitudes, filling the niches that became vacant after the larger fusulinides became extinct. Among these lagenides were the compressed forms *Geinitzina*, *Pachyphloia*, *Nodosinelloides*, *Protonodosaria* and *Robuloides*.

However, Early Triassic (Induan and Olenekian) rocks contain very few lagenides. These Early Triassic lagenides are morphologically simple forms, so-called "disaster forms". It was not until the late Anisian that true recovery really began, with the reappearance of more evolved Permian-like features, such as the laterally compressed test and the subdivided chambers, with radiate or modified apertures and secondary lamellarity of the wall (e.g. *Grillina*, see Fig. 3.3). The reappearance of these Permian features in the later Triassic and younger taxa is attributed to evolutionary morphological convergence (Groves et al., 2004). Indeed, this convergence almost certainly accounts for the many examples of so called "Elvis taxa" (Erwin and Droser, 1993) among Palaeozoic and Mesozoic-Cenozoic forms.

3.3.3 The Triassic Earlandiida and Endothyrida

As discussed in Chapter 2, the initial decline of the fusulinides coincided with the end Guadalupian in the Late Permian. This crisis event eliminated all large and morphologically complex forms of fusulinides, and by the end of the Permian all fusulinides were extinct. Indeed, only two genera (*Earlandia, Endoteba*), from the earlandiides and the endothyrides respectively, survived the end Permian mass extinction (Fig. 3.4), but they never experienced significant taxonomic recovery (Groves and Altiner, 2004). The Endotebidae have their origin in the Permian where they evolved from the true Endothyridae. The type species *Endoteba controversa* Vachard and Razgallah, 1988 is known from Late Permian rocks in Tunisia, Yugoslavia, Italy, Afghanistan and Japan. However, according to Vachard et al. (1993) the Triassic occurrence of this genus is known mainly from the Anisian to Ladinian. They are very rare in the Carnian and Rhaetian throughout the Tethyan realm, with the only documented Early Triassic (Olenekian) occurrence being that of *E. ex gr. bithynica* Vachard et al., 1993. *Endoteba* is regarded as a Lazarus taxon, as no known Induan occurrences have been documented to date. A slight increase in the number of genera characterized the Middle Triassic, there being five genera but with no more than ten species in all. In Tethys, *Endoteba* evolved into three genera or species-groups, *Endotebanella* (uppermost Olenekian through Carnian), *Endotriada* (Middle Triassic), and *Endotriadella* (Middle Triassic). The endothyrides became entirely extinct in the Late Triassic, while the earlandiides died out within the Early Triassic (see Fig. 3.4). *Earlandia* have been suggested as having survived into the Early Cretaceous, but this has not been confirmed (Arnaud-Vanneau, 1980; Vachard et al., 2010).

3.3.4 The Triassic Miliolides

Foraminifera in the order Miliolida first made their appearance at the beginning of the Carboniferous (Ross and Ross, 1991). Analysis by Flakowski et al. (2005) of SSU (small sub-unit) rRNA genes showed that the miliolides cluster with the calcareous Spirillinidae and agglutinated Ammodiscidae in an independent, fast evolving lineage (Pawlowski et al., 2003). This suggests that miliolides evolved directly from some allogromiide-like ancestor, which is consistent with earlier morphological observations (Arnold, 1978). This was subsequently questioned by Gaillot and Vachard (2007) and Vachard et al. (2010). The latter argued that the early attached miliolides derived probably from the attached Tournayellida and not from the Ammodiscidae, which are homeomorph of the Tournayellida. The question is still open, but in Fig 2.1 and 3.1 they are shown as having an allogromiide-like ancestor.

Although the miliolides proliferated during Middle and Late Permian, the end Permian extinction nearly eliminated the order, as almost all species disappeared. However, a few minute disaster forms survived the extinction, and forms such as *Cornuspira* and *Rectocornuspira* (see Chapter 2) are found in the Induan, apparently because of their tolerance of environmental stress (Grove and Altiner, 2004). Indeed, the middle Induan is zoned on the occurrence of *R. kahlori* by many authors, such as de Graciansky et al. (1998).

Like other forms, the miliolides were rare in the Early Triassic. The appearance of *Meandrospira* in the Olenekian represented a probable continuation of the Permian form assigned to *Streblospira* (which is according to Loeblich and Tappan is a synonym to *Meandrospira*), and is considered by Groves and Altiner (2004) as a Lazarus taxon. The late Olenekian of Tethys is zoned by *M. pusilla* (de Graciansky et al., 1998), and it was not before the middle Anisian that *Meandrospira* expanded, and subsequently evolved into forms such as *Meandrospiranella* and *Turriglomina* in the late Anisian. The appearance of the latter coincided with the disappearance of the *Meandrospira* species group from Tethys at the end of the Anisian.

A parallel evolution to that of *Meandrospira* (= *Streblospira*) also took place (according to Zaninetti et al., 1991, and Rettori, 1995) in the *Neohemigordius – Arenovidalina* lineage. *Neohemigordius* in the Early Permian loses its prominent umbilical thickening and acquires a streptospiral early coil as is in *Hemigordius* and its species group of the Permian. The *Hemigordius* group disappears in the Late Permian. However, similar forms to *Hemigordius,* but with planispiral tests and thickening in the umbilical region, as in *Neohemigordius,* and referred to as *Arenovidalina,* reappear in the Olenekian. *Arenovidalina* could be regarded as a Lazarus taxon that gave rise, just before disappearing in the early Ladinian, but giving rise to the Mesozoic group ophthalmidiids, by losing the lamellar umbonal region on each side of the test and developing a streptospiral early coil with much irregular later coiling. The tubular undivided planispiral second chamber of *Arenovidalina* becomes subdivided into distinct chambers in *Ophthalmidium.* Another short lineage also to develop from *Arenovidalina* in the late Anisian is that of *Paratriasina,* with a streptospiral early coil that later becomes planispiral. This phylogenetic evolution is shown schematically in Fig. 3.5.

3.3.5 The Triassic Involutinides

The Early Triassic witnessed the first appearance of the aragonitic involutinides (older texts such as Loeblich and Tappan (1988) identify them as originating in the Permian, but this is based on the erroneous classification of the porcelaneous genera *Neohemigordius* (Fig. 3.5) and *Pseudovidalina* as involutinides rather than miliolides, see Chapter 2). The order Involutinida includes all forms with an enrolled second chamber and pillar-like structures in the umbilical region on one or both sides of the test (see Fig. 3.6). The oldest Triassic representative of the group is *Triadodiscus,* which first appeared abruptly in the Olenekian with no known occurrences in the Induan. The ancestor of *Triadodiscus* is still unknown.

One of the common hypothesis is that involutinides had an Archaediscoidean origin (Groves and Altiner, 2004). However, Archaediscoidean forms were confined to the Palaeozoic and were never recorded in the Triassic. Gargouri and Vachard (1988) proposed the ammodiscoides *Glomospirella* as being a possible ancestor, but other authors favoured a hemigordiopsid *Multidiscus* ancestor (Altiner et al., 2005). However, *Multidiscus* is not definitely known from the earliest Triassic. Furthermore, rDNA analyses suggest a significant genetic and evolutionary separation between miliolids and other testate foraminifera (Pawloski, 2000), which would mitigate against this proposal.

In the Anisian, the planispirally coiled *Triadodiscus* evolved into streptospirally coiled *Aulotortus* (see Fig. 3.6). The late Ladinian of Tethys is recognised (de Graciansky et al., 1998) by the appearance of *A. praegaschei* (Koehn-Zaninetti). However, it was not before the Norian that significant diversification occurred, with changes in the shape of the coiled tests and the shape and occurrences of pillars. Several evolutionary trends in the involutinides occurred from Middle to Late Triassic, which are biostratigraphically useful. The Norian and Rhaetian are zoned using *Triasina oberhauseri* and *T. hantkeni* (Plate 3.4, fig. 12) respectively.

The development of *Involutina* (see Figs 3.6; 3.8) has been debated by many authors. Kristan-Tollman (1963) proposed a direct development from the trochospiral *Lamelliconus*. However, this theory was refuted by Rigaud et al. (2013) who demonstrated that trochospirally coiled involutinides are a separate group. On the other hand, *Involutina* is planispiral-evolute, non-septate, perforate, and possess at least two laminae per whorl, that are, as in *Aulotortus,* interfingered in the median part of the umbilical region (Rigaud et al., 2015). It was also postulated that *Involutina* evolved from *Triadodiscus* (Piller, 1978; Gaździcki, 1983; di Bari and Laghi, 1994), however, the lamina depositions of *Triadodiscus* are more discontinuous than those of *Involutina*. In the latter the interfingering laminar arrangement, the type of coiling with possible irregularities and tubular chamber morphology are more closely phylogenetically linked to *Aulotortus* (Rigaud et al. 2015) (see Figs 3.6; 3.8). Most involutinides disappeared at the end of the Triassic and only three species continued into the Mesozoic.

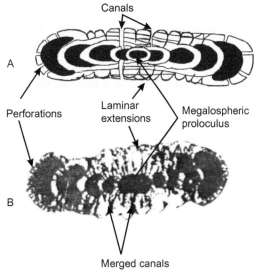

Fig. 3.8. Sections of *Involutina* sp., A) Schematic axial section: B) Photomicrograph of an Axial section.

3.4 Palaeoecology of the Triassic Larger Foraminifera

The Triassic saw the single vast super-continent of Pangea straddling the equator, with the Tethys Ocean intruding into Pangea from the east, leading to the division between Laurasia in the North and Gondwana in the South. Land animals were free to migrate from Pole to Pole (Fig. 3.9). The interior of Pangea was hot and for the most part dry and very arid. Warm temperate climates extended to the poles. Sea levels were low and the seas had little or no dissolved oxygen and were possibly alkaline (Woods, 2005). Rapid global warming at the very end of the Permian expanded the deserts and created a "super-hothouse" world, that may have contributed to or exacerbated the great Permian-Triassic extinction. The carbon dioxide content of the atmosphere was very high and the level of atmospheric oxygen was unusually low (Berner et al., 2003), see Fig. 3.10

The end Permian extinction led to a profound decline in reef production and caused the near-total elimination of metazoan reef systems (Hallam and Wignall, 1997), giving rise to a so-called "reef gap" (Fagerstrom, 1987; Flügel, 1994) during the Induan and early Olenekian. Recovery was very slow for foraminifera, and ecological recovery was delayed until the middle Olenekian. The start of the Triassic was characterised by extremely impoverished taxa, dominated by opportunists and generalists (Rodland and Bottjer, 2001). Many genera which seemed to have disappeared at the Palaeozoic-Mesozoic boundary remained hidden, possibly in geographically restricted environments, before returning as "Lazarus" taxa in the Olenekian or Middle to Late Triassic. Loeblich and Tappan (1988) suggested that the morphologies of the surviving faunas are characteristic of infaunal detrital feeders, which would have been therefore less susceptible to primary productivity crashes, in contrast to the larger benthic foraminifera which lived symbiotically in reefal environments. On the other hand, Chen and Benton

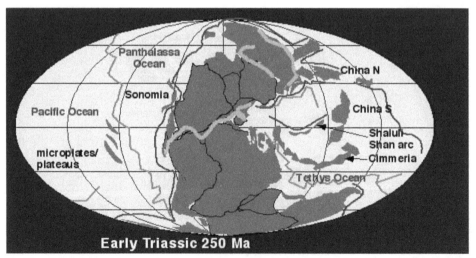

Fig. 3.9. Palaeogeographic reconstruction of the Early Triassic (by R. Blakey, http://jan.ucc.nau.edu/~rcb7/paleogeographic.html).

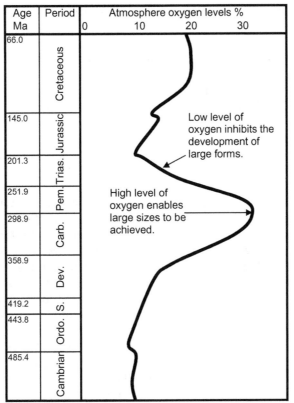

Fig. 3.10. Oxygen variation over time, highlighting the low level of oxygen during Triassic time (modified after Berner *et al.*, 2003).

(2012) postulated that recovery of complex life in the Triassic could have been delayed either by a complex multispecies interaction combined with physical perturbations.

It was not before the Anisian that larger benthic foraminifera became more widespread and cosmopolitan. During this time, a tropical sea formed between the supercontinents of Gondwana (in the south) and Laurasia (in the north). This sea, called the Tethys (Fig 3.9), became home to large diverse assemblages of foraminifera. However, the larger Triassic benthic foraminifera never recovered the relatively gigantic sizes seen in the Permian, but rather remained small and indistinctive. This was caused most probably by the drop in atmospheric and marine oxygen levels at the Permian-Triassic boundary (Berner et al., 2003), see Fig. 3.10. Increasing atmospheric oxygen levels have, for example, been implicated in the rise of late pre-Cambrian and Cambrian marine faunas (Cloud, 1976; Runnegar, 1982; McMenamin and McMenamin, 1990) and are consistent with the presence of Permian-Carboniferous giant insects. Similarly, anoxia has been regularly correlated with the dramatic Permian-Triassic extinction events in the ocean and the small size of the larger foraminifera during the Triassic (Erwin, 1993; Knoll et al., 1996; Wignall and Twitchett, 1996). In the oceans today, many benthic invertebrates compensate for hypoxic conditions via a broad spectrum of behavioural and morphological adaptations (Rogers, 2000). Similarly, infaunal penetration by the foraminifera may be a response to deplete sediment oxygenation in the Triassic (Berner et al., 2003).

During the Anisian, carbonate platform development reoccurred with the important contribution of reefal debris, which as noted above are almost totally absent in Early Triassic carbonate platforms. This was accompanied by the appearance of scleractinian corals and the increasing importance of dasycladoidean green algae (e.g. *Griphoporella* sp., Plate 3.1, fig. 9), the increasing frequency of crinoids and molluscs, and the scarcity of ooids (Kiessling et al., 2003). During the Ladinian there was a clear stabilization of marine ecosystems, accompanied by the development of wide carbonate platforms bearing highly diversified communities (Márquez and Trifonova, 2000; Márquez, 2005).

Several foraminiferal genera and species that emerged and/or developed during the Anisian and Ladinian gave rise to characteristic faunal associations that have enabled some authors to establish biozones (Salaj et al.,1983; Trifonova, 1984; Salaj et al., 1988; Márquez, 2005). Among the Anisian forms are *Meandrospira dinarica* Kochansky-Devide and Pantic (Pl. 3.5, fig. 2) in the Carpathian-Balkan zone and in the Alps (Rettori, 1995), and *Pilammina densa* Pantic (Pl. 3.5, fig.9) in the Pelsonian of the Alps (Zaninetti et al., 1991). Among the most frequently found forms in the Ladinian of western Tethys are numerous representatives of the Involutinida, such as *Lamelliconus* gr. *biconvexus-ventroplanus* (Oberhauser), *L. multispirus* (Oberhauser), *L. procerus* (Liebus), *L. cordevolicus* (Oberhauser), *Aulotortus praegaschei* (Koehn-Zaninetti), *A. sinuosus* (Eynschenk) (Pl. 3.7, figs 2A, 9B, 11) and *A. pragsoides* (Oberhauser) (Pl. 3.8, fig. 2; Trifonova, 1992, 1993, 1994; Salaj et al., 1983; Oravecz-Scheffer, 1987; Márquez, 2005). The species *Pilamminella gemerica* Salaj appears in abundance in the Ladinian characterizing the *P. gemerica* zone, and is characteristic of reef facies and dasycladoidean platform carbonate sediments (Salaj et al., 1988; Márquez, 2005).

The climate in the Carnian was unusually arid and hot and the seaway itself appears to have experienced a "salinity crisis" (general evaporation) and a breakdown of the reef system. In the Carnian, foraminiferal assemblages were dominated by endothyrides and miliolides . The Late Carnian-Norian interval saw major reef extinctions in the Tethys province, with the loss of older coral species, but diversity was maintained with reciprocal replacement by new taxa (Stanley, 2001). These in turn became extinct, possibly a result of a sudden global cooling. This is mirrored by the disappearance of many foraminiferal genera in western Tethys, which were subsequently replaced by new forms of Involutinida in the Norian. The early Norian Tethyan reef systems were replaced by much larger reef developments during the Late Triassic (mid Norian to Rhaetian) coinciding with a major rise in global sea-level. At this time the new foraminiferal taxa of Involutinida became dominant and corals diversified, constituting a new reef building consortium. There was a period of world-wide expansion of carbonate platforms and maximum reef diversity (Stanley, 2003). Associations of Late Triassic benthic foraminifera dominated by Ammodiscoidea, Involutinoidea, and Duostominoidea are mainly found in shallow marine facies, often restricted and ecologically highly unstable (Márquez, 2005). They are generally typical of low energy, bay or lagoon-type, protected settings, with a salinity sometimes higher than normal, on shallow carbonate ramps. The different reef facies include forms common to the Tethys such as *Triadodiscus mesotriasica* (Koehn-Zaninetti), *Palaeolituonella meridionalis* (Luperto), *Endoteba wirzi* (Koehn-Zaninetti), *Duostomina alta* Kristan-Tollmann and *D. astrofimbriata* Fuchs, to mention a few. In these reef environments, the foraminiferal fauna are more numerous and diverse than in the restricted carbonate ramps of the Triassic basins (Márquez, 2005). Many of the Late Triassic species have morphological characteristics, such as the irregularly distributed papillose lamellae and tube infolding of *Involutina,* which have been interpreted as being rudimentary features associated with photosymbiosis (Rigaud et al., 2015), typically developed in shallow, high-energy tropical carbonate platforms (Martini et al., 2009).

3.5 Palaeogeographic Distribution of the Triassic Larger Foraminifera

The Early Triassic was characterised by small and low-diversity taxa, and is viewed as a time of delayed biotic recovery that persisted until the Middle Triassic (Hallam, 1991; Galfetti et al., 2008; Chen and Benton, 2012). Explanations for this vary from invoking the sheer scale of the Permian extinction to the prolonged stresses of the environment post end-Permian extinction (Hallam, 1991; Payne et al., 2004; Chen and Benton, 2012), or the effects of further extinction crises in the Early Triassic (Orchard, 2007; Stanley, 2009; Song et al., 2011).

The Early Triassic larger benthic recovery shows roughly the same trajectory everywhere. The foraminifera were mainly small, and they followed the model of the "Lilliput effect" described by Urbanek (1993), where he refers to the pattern of size change through extinction events, specifically the temporary appearance of dwarfish organisms (Twitchett, 2006). There was a significant decrease in maximum and mean size among species and genera across the Permian/Triassic boundary. This was followed by gradual size increase through the late Induan and into Olenekian. The recovery of reef-building larger benthic foraminifera began 4 Ma later, in the Anisian, but they never recovered the pre-extinction sizes of the Permian (Payne et al., 2011; Song et al., 2011; Rego et al., 2012). The anoxia and global warming that prevailed through the Permian–Triassic period may have contributed to this "Lilliput effect" on larger benthic foraminifera in the aftermath of the end-Permian crisis.

The distribution of the Triassic genera is plotted in charts (see Figs 3.11 to 3.13) in order to understand their palaeogeographic distribution. The genera are divided into three provinces: Europe and Russia, China and East Asia, and North America. There is a general and progressive increase in diversity from the Induan to the Anisian, with the end Olenekian being marked by the increase in the number of orders that exhibit larger morphologies. A possible minor extinction event occurred at the end of the Anisian affecting the Ammodiscoidea in Europe. The ancestor of the Involutinida appeared in all the three provinces, and the order increased in diversity in China and East Asia during the Ladinian, but it was not before the Norian that they became widespread in Europe. The Endothyroidea were common during the Anisian and Ladinian in China, East Asia and Europe, but they decreased gradually toward the end of the Norian. However, they did not disappear completely from these provinces until the end of the Rhaetian.

Triassic Total Genera

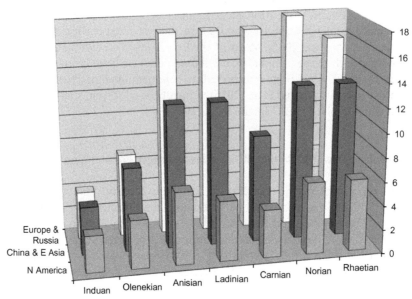

Fig. 3.11. The number of larger benthic foraminifera genera found throughout the Triassic in the three main palaeogeographic provinces.

Triassic Speciation

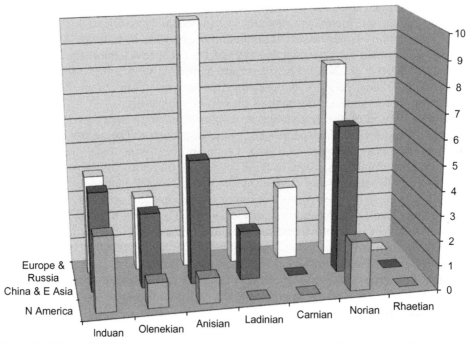

Fig. 3.12. The number of new larger benthic foraminifera genera occurring throughout the Triassic.

Triassic Extinctions

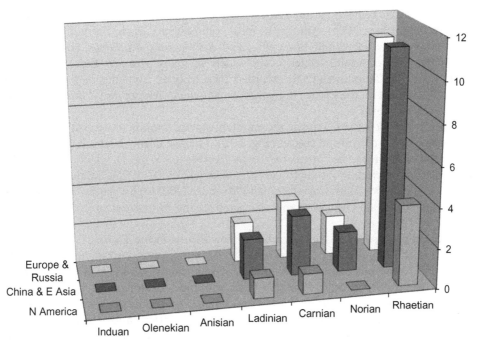

Fig. 3.13. The number of larger benthic foraminifera genera extinctions occurring throughout the Triassic.

From Fig. 3.11, it can be seen that throughout the Triassic, the diversity in North America was constantly low, but there was a distinct increase in the number of genera throughout Europe, and to a lesser extent in China and East Asia, in the Anisian and Norian. From Fig. 3.12, we see that there was a sharp decline in speciation in the Ladinian, and that no new forms appeared at all in the Rhaetian. However, during the Anisian and the Norian the larger benthic foraminifera speciation was at its most productive. North American foraminifera speciation mirrored generic diversity but remained poor throughout the Triassic. There was a gradual increase in the number of extinctions from the Ladinian to the Carnian in all three provinces, that was followed by a large extinction event at the end of the Rhaetian (Fig. 3.13).

The trends described above in all three provinces, can be understood to some extent from an analysis of the palaeogeography of the Triassic. After the great extinction of the end Permian, foraminifera were struggling to survive during the Induan interval. The recovery process was very slow in the Induan, but new forms were appearing, 80 % of the Induan foraminifera were new Triassic forms with the first records of foraminifera appearing in the eastern Tethys area (China) and Eastern Europe. They are dominated by opportunist, euryfacial forms, giving rise to associations that are sometimes numerous but of low species diversity, since they are mostly comprised of arenaceous genera (Tong and Shi, 2000) from the Ammodiscoidea (see Fig. 3.7). During the Early Triassic, the diversity of the foraminifera increases gradually in all three provinces,

North America having the least diversity. This is probably due to its relative isolation during the Early Triassic (see Fig. 3.9).

In the eastern Tethys area, the recovery process seems to be dominated by forms of Miliolida (Márquez, 2005). Among the most common species are *Cornuspira mahajeri* (Brönnimann, Zaninetti and Bozorgnia) and *Rectocornuspira kalhori* (Brönnimann, Zaninetti and Bozorgnia), which occur widely in many areas of Europe and Asia (Rettori, 1995). Trifonova (1993) recorded these species from the Early Triassic of Bulgaria, along with the species *Kamurana bronnimanni* (Altiner and Zaninetti) and *K. chatalovi* (Trifonova).

During the Olenekian, Triassic foraminifera started rapidly to diversify, leading to the large number of Anisian species (Fig. 3.12). The late Olenekian is characterised by the presence of the species *Meandrospira pusilla* (Plate 3.5, fig. 1), widely described by many authors including Zaninetti (1976), Gaździcki et al. (1975), Salaj et al. (1983), Trifonova (1978, 1993), Rettori (1995) and Márquez (2005). This species is particularly abundant in the late Olenekian and characterizes the *M. pusilla* Zone of Salaj et al. (1988).

The Anisian saw the highest number of new genera in the Triassic (see Fig. 3.12). However, at the end of the Anisian a small extinction occurred, mostly among European Ammodiscoidea (affecting 35% of the genera), and within the Ladinian, 60% of the Cornuspiroidea from all three provinces disappeared. This may have been associated with environmental stress caused by the large volcanic event (ca. 235 Ma) along the northwest margin of North America, and which formed the Wrangellia Terrane. After this event, the foraminifera in North America and Europe did not recover completely until the end of the Carnian. The Far East seems not to have been affected by this event. The Wrangellia volcanic sequences were oceanic events, and so may not have had the same globally devastating environmental and climatic effects inferred to be associated with sub-aerial volcanic eruptions (e.g. the Siberian Traps).

At the end of the Ladinian, there was a general decline in speciation after the blooming of the Early Triassic recovery, except after the end Carnian event which saw the expansion in the Norian of the superfamily Involutinida in European Tethys. The ancestral form of this family appeared in the Anisian, but only attained importance in the Norian and Rhaetian. In addition, the first forms of *Lamelliconus* emerged in the Ladinian, a genus that was to develop substantially during the Norian. The miliolide genus *Turriglomina,* which developed during the Anisian, gave rise to many species of stratigraphic significance, such as *T. conica* (He), *T. mesotriasica* (Koehn-Zaninetti) and *T. scandonei* (Zaninetti, Ciarapica, Martini, Salvini-Bonard and Rettori). The species *Palaeolituonella meridionalis* (Luperto), widely cited for the Middle Triassic from several Tethyan localities, is especially common in the Ladinian (Trifonova, 1992; Márquez, 2005).

Early Carnian Triassic foraminifera had become cosmopolitan, as a small seaway connected the Tethys Ocean with North America (see Fig. 3.14). Foraminifera thrived until the end of the Carnian. The high number of extinctions at the end of the Carnian coincide with the occurrence of set of multiple of impacts (Spray et al., 1998). Five terrestrial impact structures have been found to possess comparable ages

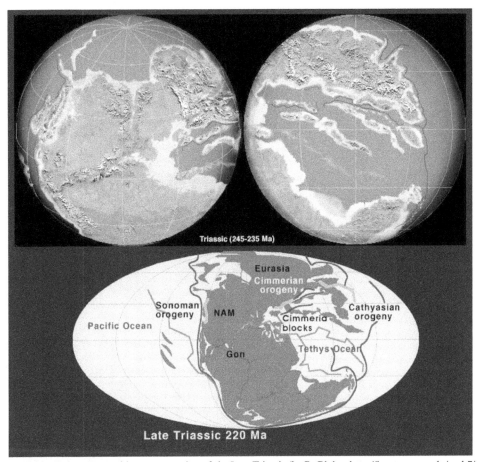

Fig. 3.14. Palaeogeographic reconstruction of the Late Triassic (by R. Blakey http://jan.ucc.nau.edu/~rcb7/ paleogeographic.html).

(214+/-2 Ma), close to Carnian-Norian boundary (see Fig. 3.15). These craters are Rochechouart (France), Manicouagan and Saint Martin (Canada), Obolon (Ukraine) and Red Wing (USA). These five large, chronologically-close impact events might have triggered adverse ecological changes that in turn might have led to the high extinction rates observed.

However, foraminifera were soon again to recover (see Fig. 3.12), with new forms filling the empty niches. The Norian saw the expansion of the Involutinida in European Tethys. The Late Triassic also witnessed the appearance in the West Pangean reefs of reportedly Permian Lazarus taxa (Stanley, 2001). Presumably these forms had remained rare and geographically isolated during most of Triassic time. Panthalassan volcanic islands might have served as refuges during time of crisis when the Tethys was affected (Stanley, 2003), and probably middle Norian conditions were similar to those of the Permian.

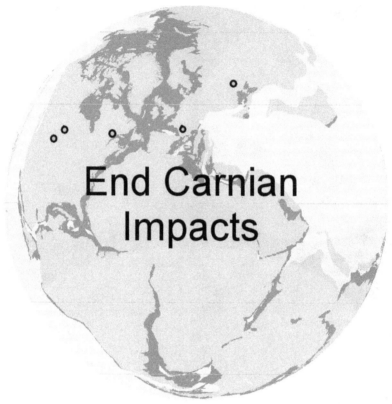

Fig. 3.15. A map showing the position of the major, Carnian-Norian impact craters.

During the Rhaetian, foraminiferal diversity in all provinces dwindled gradually to near full extinctions in the three provinces (Tanner et al., 2004; Scherreiks et al., 2010). These extinctions mirror the abrupt extinction seen for all other groups in the marine realm at the end of the Triassic. This Triassic-Jurassic extinction is generally recognized as being one of the five largest in the Phaenorozoic (Hallam, 1990). Only 16% of the Triassic foraminifera survived the end Triassic extinction. There has been much speculation as to the cause of this extinction event, and debate as to whether there was a catastrophic event at the end of the Triassic, or if it was preceded by a gradual decline. Certainly, a gradual decline of the foraminifera (see Fig. 3.12) was apparent. According to Tanner et al. (2004) the Late Triassic extinction appears protracted rather than catastrophic. Marine regression reduced the available shallow marine habitats for the Triassic foraminifera, and consequent competition may have been the forcing mechanism for extinction (Newell, 1967). Tucker and Benton (1982) specifically cited climate-induced changes occurring during the Late Triassic as a possible cause for mass extinction. Evidence of anoxia has also been documented in Triassic-Jurassic sections and has been suggested to be a significant factor in this event (Hallam and Wignall, 1997; 2000).

However, the identification of clear evidence of iridium anomalies at the Triassic-Jurassic boundary (Tanner and Kyte, 2005) again raises the possibility that the mass extinction resulted from an extraterrestrial impact, or was triggered by flood basalt volcanism. The Triassic-Jurassic boundary coincides with large scale eruptions that created a flood basalt province that covered at least 5×10^5 km^2 of northeastern North America. Deckart et al. (1997) proposed a large igneous province related to Pangaean rifting, based on correlation of outcrops in French Guyana, Guinea and Surinam. Marzoli et al. (1999, 2004) extended the range of the province, which they named Central Atlantic Magmatic Province (CAMP). The CAMP therefore includes eastern North America, northern South America, western Africa and southwestern Iberia (Tanner et al., 2004, see also Fig. 3.16). The synchronicity between the CAMP volcanism and of marine faunal extinction at the Triassic-Jurassic boundary was confirmed by Schaltegger et al. (2007), who carried out biostratigraphic correlation between zonal ammonites and zircon ages. CAMP volcanism would have created anoxic and dysoxic environments caused by rising CO_2 levels and related greenhouse effects during the rapid eruption of the flood basalt. The eruption of the CAMP volcanics amplified the rapid global sea fluctuations and is associated with global cooling and glaciation and is

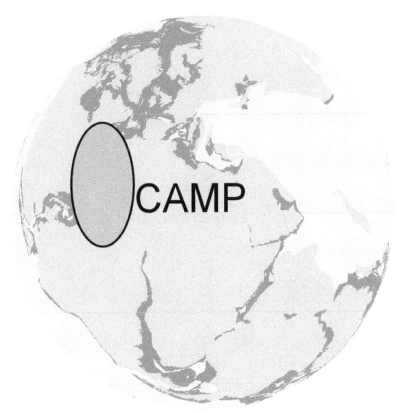

Fig. 3.16. A map showing the location of the end Triassic Central Atlantic Magmatic Province (CAMP).

closely associated with the end-Triassic extinction (Bonis et al., 2009; Whiteside et al., 2010; Schoene et al. 2010; Ruhl et al., 2011; Macleod, 2013; Blackburn et al., 2013; Percival et al., 2017), and specifically Deenen et al., (2010) claim that the extinction of the larger benthic foraminiferal correlate well with the detail timing of the CAMP eruptions.

Thus, it would seem that a globally significant volcanic event at the end of the Triassic saw another catastrophic decline in larger foraminiferal assemblages and their associated corals. This, as will be seen in Chapter 4, was followed by the now familiar trend of a "reef gap" in the Early Jurassic.

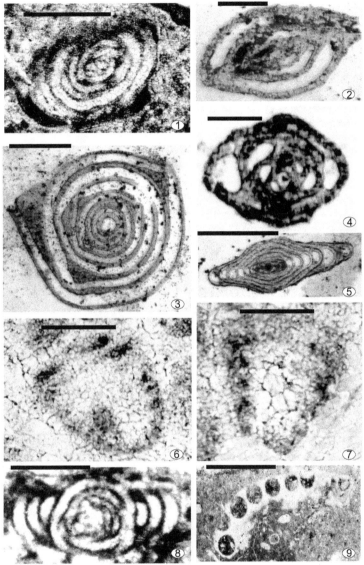

Plate 3.1　Scale bars: Figs 1-9 = 0.2mm. Fig. 1. *Glomospirella irregularis* Moeller, Sample YA 840.1/2,?Middle Triassic, Nwabangyi Dolomite Formation, Shan States, Burma, NHM P4874. Figs 2, 4. *Karaburunia rendeli* Langer, paratypes figured by Langer (1968) from the Middle Triassic of Turkey, SMF XXVII 5825-6 5826. Figs 3, 5. *Eoophthalmidium tricki* Langer, paratypes figured by Langer (1968) from the Middle Triassic of Turkey, SMF XXVII 5821, 23. Figs 6-7. *Trocholina multispira* Oberhauser, Late Triassic, Idd-el-Shargi-1 core, 10540 ft., Gulailah, UCL coll. Fig. 8. *Gandinella falsofriedli* (Salaj, Borza and Samuel), figured by Zamparelli, Iannace, Rettori (1995) from the Triassic of Italy. Fig. 9. *Griphoporella* sp. A fragment of dasy-clad algae, Late Triassic (?Norian), Burma, UCL col.

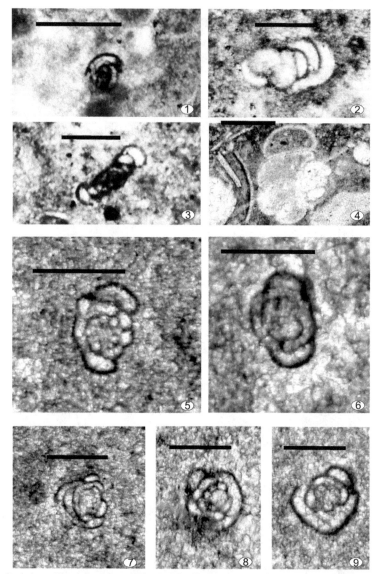

Plate 3.2 Scale bars: Figs 1-9 = 0.2mm. Figs 1, 5-9. *Glomospira meandrospiroides* Zaninetti and Whittaker, 1) Late Triassic, Lebanon; 5-9) Anisian, Nwabangyi Dolomite Formation, Yadanatheingi Area, East. Burma, Sample Am 43 (2), 5-6) NHM P50818; 7-9) NHM P50820. Fig. 2. *Glomospira* cf. *tenuifistula* Ho, Triassic, Anisian, Sample AM 37(7) Nwabangyi Dolomite Formation, East. Burma, B.J. Amos collection, NHM P50816. Fig. 3. *Glomospira sp.*, Late Triassic, Lebanon, UCL coll. Fig. 4. *Duostomina* sp. Brönnimann, Whittaker and Zaninetti, Late Triassic (?Norian), Burma, NHM P49916-49923.

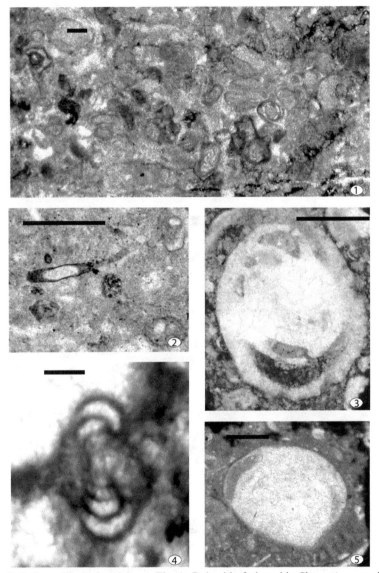

Plate 3.3 Scale bars: Figs 1-5 = 0.2mm. Fig. 1. Dolomitic facies with *Glomospira meandrospiroides* Zaninetti and Whittaker, Triassic, Nwabangyi Dolomite Formation 105 943, YA 840-1, N. Shan State Burma, NHM 1973/2. Fig. 2. *Earlandia tintinniformis* (Misik), Whittaker and Zaninetti, Late Triassic (?Norian), Nwabangyi Dolomite Formation, Burma, Sample YA196, NHM P48777. Figs 3, 5. *Aulotortus sinuosus* Weynschenk, Sample YA 672, Late Triassic (?Norian), Northern Shan States, Burma, NHM 49876-49877. Fig. 4. *Arenovidalina chialingchiangensis* Ho, Triassic (Anisian), East Burma, Amos collection, NHM P50814.

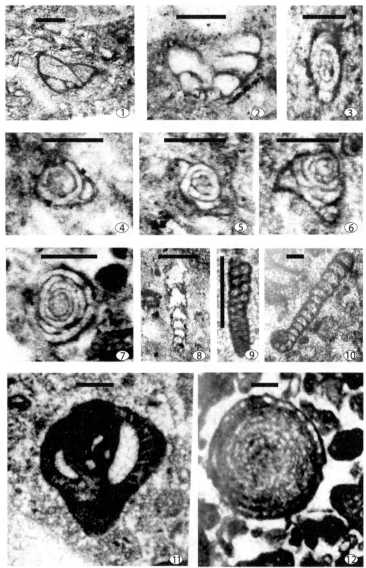

Plate 3.4 Scale bars: Figs 1-12 = 0.2mm. Figs 1-2. *Duotaxis birmanica* Brönnimann, Whittaker and Zaninetti, Sample Y673, Late Triassic (Norian), Northern Shan States, Burma, NHM 48765. Fig. 3. *Agathammina?austroalpina* Kristan-Tollmann and Tollmann, Late Triassic (?Norian), Nwabangyi Dolomite Formation, Burma, Sample YA196, NHM P48779. Figs 4-6 *Glomospira* spp., Late Triassic (Norian), Nwabangyi Dolomite Formation, Burma, Mitchell collection NHM AM 37. Fig. 7. *Glomospirella irregularis* Moeller, Sample YA 840.1/2,?Middle Triassic, Nwabangyi Dolomite Formation, Shan States, Burma, NHM P48745. Fig. 8. *Ammobaculites* sp. Late Triassic (?Norian), Nwabangyi Dolomite Formation, Burma, Sample YA196, NHM P48780. Fig. 9. *Turriglomina mesotriasica* (Koehn-Zaninetti), figured by Sartorio and Venturini (1988) from the Ladinian of Abriola, Basilicata, Italy. Fig. 10. *Endothyranella* sp., figured by Sartorio and Venrurini (1988) from the Ladinian of the Adriatic Sea Fig. 11. *Galeanella* sp., figured by Sartorio and Venrurini (1988) from the Rhaetian of Germany. Fig. 12. *Triasina hantkeni* Mason, figured by Sartorio and Venrurini (1988) from the Rhaetian of Italy.

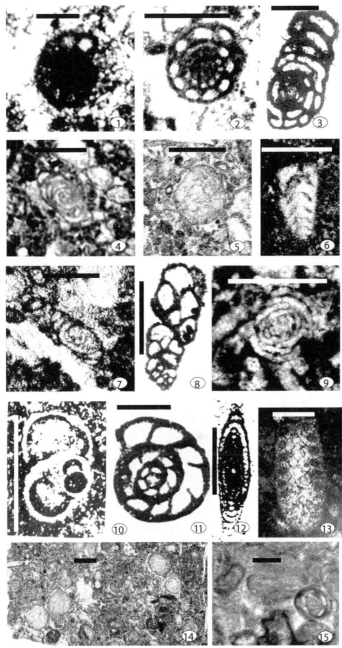

Plate 3.5 Scale bars Figs 1-7, 14-15 = 0.2mm; Fig. 8 = 0.1mm; Figs 9-13 = 0.3mm. Fig. 1. *Meandrospira pusilla* (Ho), figured by Sartorio and Venturini (1988) from Late Olenekian, Trentino, Italy. Fig. 2. *Meandrospira dinarica* Kochansky-Devidé and Pantić, figured by Sartorio and Venrurini (1988) from the Anisian of Yugoslavia. Fig. 3. *Meandrospiranella samueli* Salaj, Anisian, Czechoslovakia, holotype figured by Salaj (1983). Figs 4, 5. *Aulotortus friedli* (Kristan-Tollmann), CQ1, Rhaetian, Chalkis Quarry, Greece, UCL coll. Fig. 6. *Grillina* sp., CQ22, Rhaetian, Chalkis Quarry, Greece, UCL coll. Fig. 7. *Hoynella* gr. *sinensis* (Ho),

CQ27, Rhaetian, Chalkis Quarry, Greece, UCL coll. Fig. 8. *Piallina bronnimanni* Martini *et al.*, holotype figured by Martini et al. (1995), Carnian, Carpatho-Balkanides Belt. Fig. 9. *Pilammina densa* Pantic, YA 840-1 Triassic, Nwabangyi Dolomite Formation, 105 943, North Shan State Burma, NHM collection, 1973/2. Fig. 10. *Abriolina mediterranea* Luperto, 1963, figured by Rettori (1995) from the Middle Triassic of Central Apennines, Italy. Fig. 11. *Robuloides reicheli* (Reytlinger), figured by Reytlinger (1965) from the Triassic of Precaucasus. Fig. 12. *Paulbronnimannia judicariensis* (Premoli Silva), figured by Zaninetti et al. (1994) from the Recoaro Limestone of Italy. Fig. 13. *Lamelliconus multispirus* (Oberhauser), figured by Sartorio and Venrurini (1988) from the Carnian of Pietratagliata, Friuli, Italy. Fig. 14. Thin section photomicrograph of *Aulotortus* sp., Rhaetian, Chalkis Quarry, Greece, UCL coll. Fig. 15. Thin section photomicrograph of *Glomospirella* sp., Triassic, Nwabangyi Dolomite Formation, 105 943, YA 840-1, N. Shan State Burma, NHM coll. 1973/2.

Plate 3.6   Scale bars Figs 1-5 = 0.5mm; Figs 6-7 = 0.1mm. Figs 1-5. *Alpinophragmium perforatum* Flügel, paratypes from the Late Triassic reef limestones of the Northern Alps, SMF XXV11 4981-5. Figs 6-7. *Glomospirella irregularis* Moeller, Sample YA 840.1/2, ?Middle Triassic, Nwabangyi Dolomite Formation, Shan States, Burma, NHM P4874.

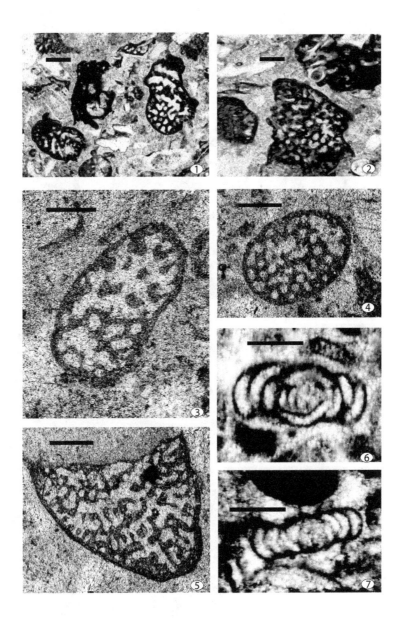

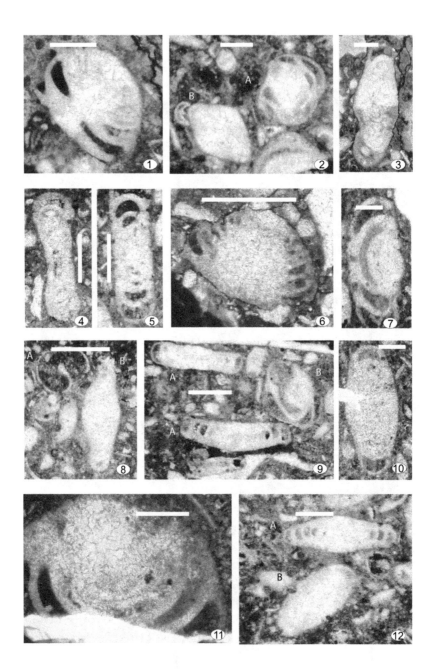

Plate 3.7 Scale bars: Figs 1-12 = 0.2mm. Figs 1, 6. *Auloconus permodiscoides* (Oberhauser), Sample YA 672, Late Triassic (?Norian), Northern Shan States, Burma, NHM 49879. Figs 2. A) *Aulotortus sinuosus* Weynschenk, B) *Triadodiscus eomesozoicus* (Oberhauser), Sample YA 672, Late Triassic (?Norian), Northern Shan States, Burma, NHM 49880. Figs 3, 7. *Involutina communis* (Kristan), Sample YA 672, Late Triassic (?Norian), Northern Shan States, Burma, NHM 49880. Figs 4, 5. *Involutina tenuis* (Kristan), Sample YA 672, Late Triassic (?Norian), Northern Shan States, Burma, NHM 49887. Fig. 8. A) Gastropod sp., B) *Involutina communis* (Kristan), Sample YA 672, Late Triassic (?Norian), Northern Shan States, Burma, NHM P49895. Fig. 9. A) *Involutina tenuis* (Kristan), B) *Aulotortus sinuosus* Weynschenk, Sample YA 672, Late Triassic (?Norian), Northern Shan States, Burma, NHM P49896. Fig. 10. A) *Involutina tumida* (Kristan-Tollmann), B) *Involutina communis* (Kristan), Sample YA 672, Late Triassic (?Norian), Northern Shan States, Burma, NHM 49915. Fig. 11. *Aulotortus sinuosus* Weynschenk, Sample YA 672, Late Triassic (?Norian), Northern Shan States, Burma, NHM 49878. Fig. 12. A) *Involutina tenuis* (Kristan), B) *Aulotortus sinuosus* Weynschenk, Sample YA 672, Late Triassic (?Norian), Northern Shan States, Burma, NHM 49879.

Plate 3.8 Scale bars: 1-8 = 0.2mm. Figs 1,4 . *Involutina communis* (Kristan), Sample YA 672, Late Triassic (?Norian), Northern Shan States, Burma, NHM 49881. Fig. 2. *Aulotortus pragsoides* (Oberhauser), Sample YA 672, Late Triassic (?Norian), Northern Shan States, Burma, NHM 49882. Fig. 3. *Aulotortus sinuosa* (Weynschenk), Sample YA 672, Late Triassic (?Norian), Northern Shan States, Burma, NHM 49883. Fig. 5. *Duotaxis birmanica* Brönnimann, Whittaker and Zaninetti, holotype, Sample Y673, Late Triassic (?Norian), Northern Shan States, Burma, NHM 48765. Fig. 6. *Triadodiscus eomesozoicus* (Oberhauser), Late Triassic (?Norian), Northern Shan States, Burma, NHM 49871. Fig. 7. *Planiinvoluta?mesotriasica* Baud, Zaninetti and Brönnimann, Late Triassic (?Norian), Northern Shan States, Burma, NHM 49872. Fig. 8. *Trochammina* sp., Late Triassic (?Norian), Northern Shan States, Burma, NHM 49870.

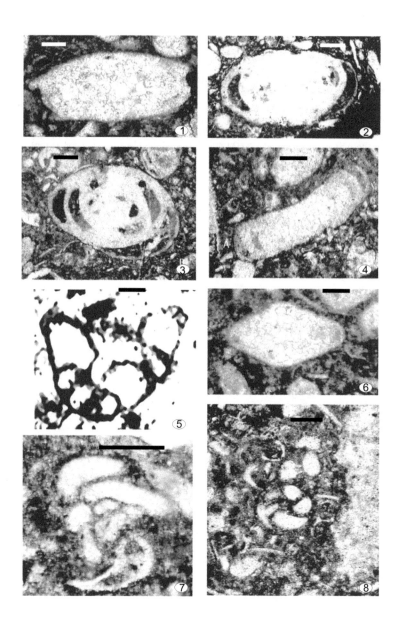

Chapter 4

The Mesozoic Larger Benthic Foraminifera: The Jurassic

4.1 Introduction

The mass extinction in the marine realm at the end of the Triassic affected all groups of the larger benthic foraminifera that had previously survived the end Palaeozoic. Most notable was the total extinction of the Endothyrida. Having survived the much greater extinction event at the end of the Permian, small Triassic endothyrids persisted but with a steadily declining diversity through the Triassic, until their final demise, in the European realm, at the end of the Triassic. This fate was also shared by most other benthic foraminifera, except for a few small forms from the Textulariida, Involutinida and Miliolida orders. Of the different orders of foraminifera with large representatives present in the Jurassic, however, only the agglutinated textulariides exhibited important evolutionary developments; becoming large, complicated and forming many lineages. The aragonitic/calcareous microgranular walls of their Palaeozoic ancestors were by now replaced by calcareous walls bonded by organic cement. The early Jurassic witnessed the steady evolution of the agglutinated forms from being small and simple to being internally complicated, which became abundant from the Pliensbachian onwards, thereby giving the carbonate facies of the Jurassic a characteristic appearance that is recognizable throughout Tethys (Fig. 4.1).

In contrast to the Triassic benthic foraminifera, the Jurassic larger benthic foraminifera have been systematically studied on a regional scale by a number of authors. Early Jurassic Hettangian-Sinemurian foraminifera from the present-day Mediterranean region were studied by Septfontaine (1981), BouDagher-Fadel (2000), Fugagnoli (2000), BouDagher-Fadel et al. (2001), Noujaim Clark and BouDagher-Fadel (2001, 2004, 2005), BouDagher-Fadel et al. (2001), BouDagher-Fadel and Lord (2002), Scherreiks et al. (2006, 2010, 2016), BouDagher-Fadel and Bosence (2007), and by BouDagher-Fadel (2016). These authors have proposed standard Jurassic biozonations for the Mesozoic realm on the basis of foraminiferal generic ranges and assemblages. Benthic forms in comparable facies have also been illustrated and described from the Early Jurassic of north Italy (Sartoni and Crescenti, 1962; Bosellini and Broglio Loriga, 1971; Castellarin, 1972), and south and central Italy (e.g. Chiocchini et al., 1994; Barattolo and Bigozzi, 1996). Jurassic larger benthic foraminifera have also been studied from Saudi Arabia (e.g. Redmond, 1964, 1965; Banner and Whittaker, 1991; Wyn and Hughes, 2004), Morocco (Hottinger, 1967; Septfontaine, 1984), the southern Tethyan realm (Sartorio and Venturini, 1988), from the 'vast carbonate platform' (e.g. García-Hernández et al., 1978; Vera, 1988; Rey, 1997) that included the external zones of the Betic Cordillera of southern Iberia (e.g. González-Donoso et al., 1974; Braga et al., 1981), Gibraltar (BouDagher-Fadel et al., 2001), and the Western Mediterranean (BouDagher-Fadel and Bosence, 2007).

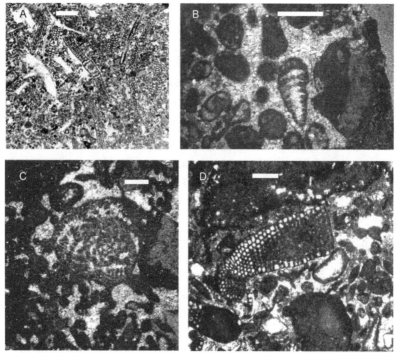

Fig. 4.1.   Scale bars = 0.5mm. A) Thin section of a clayey limestone containing microspheric forms of *Orbitopsella*, Sinemurian, Morocco; B) *Andersenolina elongata* (Leupold), figured by Vincent et al. (2014) from the Tithonian of North-East Turkey; C) *Alveosepta jaccardi* (Schrodt), figured by Vincent et al. (2014) from the Kimmeridgian of North-East Turkey; D) *Meyendorffina* sp., Bathonian, France.

In this chapter, a review and revision of the taxonomy of the main genera of the Jurassic larger foraminifera is presented, and their evolutionary lineages and phylogenetic relationships are discussed. Finally, revised and updated biostratigraphic ages, palaeogeographic and palaeoenvironmental interpretations are presented.

4.2 Morphology and Taxonomy of Jurassic Larger Benthic Foraminifera

The dominant larger foraminifera of the Jurassic were the agglutinated Textulariida. The Involutinida persisted throughout, while the Miliolida were present but were mainly composed of morphologically small genera, and it was not before the Cenomanian that larger miliolides played an important role in the benthic assemblages of carbonate platforms. The Lagenida were small and simple in the Jurassic and will not be discussed in this chapter, which will instead focus on the three orders:

- Involutinida
- Textulariida
- Miliolida

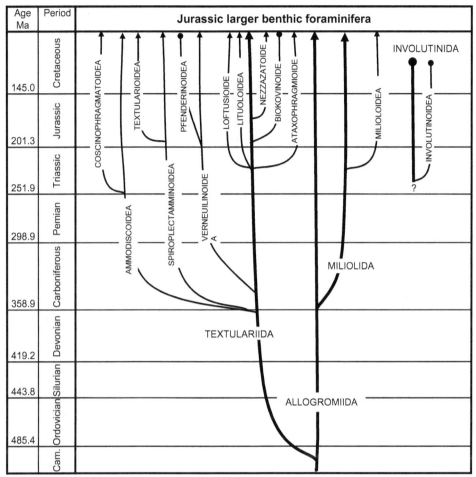

Fig. 4.2. The evolution of the Jurassic orders (thick lines) and superfamilies (thin lines) of larger foraminifera.

The development and evolution of the major superfamilies of these orders is schematically shown in Fig. 4.2.

Order Involutinida Hohenegger and Piller, 1977

An order including all forms with an enrolled second chamber. They have walls that are aragonitic, but commonly they are recrystallised to give a homogenous microgranular structure. They show an umbilical region with pillar-like structures on one or both sides of the test. They range from Triassic to Late Cretaceous (Cenomanian).

Superfamily Involutinoidea Bütschli, 1880

Forms consisting of a first chamber followed by a planispiral to trochospiral enrolled tubular second chamber. Triassic to Late Cretaceous (Cenomanian).

Family Involutinidae Bütschli, 1880

The globular proloculus is followed by a trochospiral, coiled tubular second chamber. Secondary lamellar thickenings on one or both umbilical regions. The aperture is at the open end of the tube. Late Triassic (Norian) to Late Cretaceous (Cenomanian).

- *Andersenolina* Neagu, 1994 (Type species: *Andersenolina perconigi* Neagu, 1994). The test maybe lenticular, and conical in shape. The spherical proloculus is followed by a tubular, trochospiral second chamber. The umbilical side is covered by perforated lamellae added with each whorl and surrounded by small rounded "collarette" margins. A primary aperture is absent. It differs from *Trocholina* (Fig. 4.3) and *Auloconus* (see Chapter 3) by the presence of a perforated umbilical plate relative to the former, and from the latter by the substitution of the primary aperture by pores and the absence of external lamellae. Middle Jurassic to Early Cretaceous (Bathonian to Aptian) (Plate 4.1, figs 4-5; Fig. 4.8).
- *Involutina* Terquem, 1862 (Type species: *Involutina jonesi* Terquem and Piette, in Terquem, 1862). Both umbilical regions are filled with lamellar deposits (Figs 4.4; 4.5). Late Triassic to Late Cretaceous (Norian to Cenomanian) (see Chapter 3; Plate 4.1, fig. 1; Fig. 4.8).
- *Neotrocholina* Reichel, 1956 (Type species: *Involutina conica* Schlumberger, 1898). Loeblich and Tappan (1988, p. 300) considered *Neotrocholina* Reichel as a synonym of *Trocholina* Paalzow. However, the two forms are different and should be separated. The outer wall of *Neotrocholina* is thick (approximately as thick as the spiral septum), perforate and the umbilicus is deeply fissured. On the other hand, the outer wall of *Trocholina* (= *Coscinoconus alpinus* Leupold) is thin (usually much thinner than the spiral septum, and often eroded), imperforate, and the test lacks the deeply fissured, canaliculated umbilical structure of *Neotrocholina* (Fig. 4.6). Late Middle Jurassic to Cretaceous (Bathonian to Cenomanian) (Plate 4.1, figs 2-3; Fig. 4.8).

Primary Characteristic		Specific Characteristic		Species
Biconvex test		Apical angle 130 - 150 degrees; pustulate / weakly pillared on both sides		*Neotrocholina lenticularis* (Cenomanian)
Conical tests	Base convex	large, robust; apical angle ca. 80 degrees; base pustular	Perforated umbilical plate	*Andersenolina alpina* (Bathonian - Barremian)
	Base flat	large, robust; apical angle ca. 80 degrees; base pustular	 base strongly pillared and deeply fissured	*T. arabica* (Cenomanian)
		Smaller more delicate; apical angle obtuse		*T. conica* (Bajocian - Oxfordian) (Plate 4.1, fig. 6)
	Small, delicate; spiral chamber <0.1mm high			*T. multispira* (Late Triassic) (Plate 3.1, figs 6-7)
High cones tending to be parallel sided	Large robust; spiral chamber	Base flat or weakly convex; pustules or small pillars	Perforated umbilical plate	*A. elongata* (Bathonian - Barremian) (Plate 4.1, fig. 4)
	>0.1 mm high	Base strongly convex, pitted and/or granular		*T. altispira* (Cenomanian)

Fig. 4.3. Key to main species of *Trocholina* (*T.*) and *Andersenolina* (*A.*).

Primary Characteristics	Specific Characteristics	Species
Streptospiral coiling (chamber - plane oscillates through ca. 45 degrees		*Aulotortus* Wenschenk (Middle Triassic to Middle Jurassic)
Planispiral coiling	umbilici pillared incised	*Involutina* Terquem (Upper Triassic to Cenomanian)
	umbilici umbonate, smooth	*Vidalina* Schlumberger (Cenomanian to Santonian)
Trochospiral coiling	Outer wall thick, perforate; umbilicus deeply fissured	*Neotrocholina* (Jurassic - Cretaceous)
	Outer wall thin, imperforate	*Trocholina* (Upper Triassic to Cenomanian)

Fig. 4.4. Key to main involutinides genera.

- *Septatrocholina* BouDagher-Fadel 2016 (Type species: *Septatrocholina banneri* BouDagher-Fadel and Banner 2008, type species designated in BouDagher-Fadel, 2016). The test is conical, consisting of a globular proloculus followed by a trocho-spirally enrolled, divided tubular second chamber with rudimentary septa around a solid core of pillars, filling the umbilical area. Secondary lamellar thickenings are on one or both umbilical regions. The aperture is terminal at the open end of the tube. This species is distinguished from other involutinides by the solid core of pillars filling the umbilicus and the rudimentary septa. Rigaud et al. 2013 included the genus *Septatrocholina* within the synonyms of *Coscinoconus* Leupold in Leupold and Bigler 1936, on the basis that *Coscinoconus* has "possibly slightly constricted endoskeletal structures or wall thickenings" similar to the rudimentary septa of *Septatrocholina*. However, the septa in the latter are almost non-existent, while those of *Coscinoconus* are complete (see Rigaud et al. 2013, Figs 7–11, new illustrations of syntypes of *Coscinoconus chouberti*). In addition, *Septatrocholina* lacks the complex canal system which form polygonal nodes at the umbilical surface of *Coscinoconus*. Jurassic (Callovian to Oxfordian) (Plate 4.2, figs 1-4; Plate 4.3, figs 1-6; Fig. 4.8).
- *Trocholina* Paalzow, 1922 (Type species: *Involutina conica* Schlumberger, 1898). *Trocholina* may be conical, plano-convex or lenticular in shape. It consists of a globular proloculus followed by a trochospirally enrolled undivided tubular second chamber around a solid core of pillars, filling the umbilical area. The aperture is at the end of a tubular chamber. *Trocholina* is distinguished from *Neotrocholina* by having a thinner, imperforate wall, and in lacking the deeply fissured, canaliculated umbilical structure of the latter (Fig. 4.6). Late Triassic to Late Cretaceous (Norian to Cenomanian) (Plate 3.1, figs 6-7; Plate 4.1, figs 6-9; Plate 4.4, fig. 1).

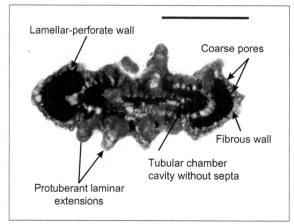

Fig. 4.5.   *Involutina liassica* (Jones), Early Jurassic, Italy, axial section, scale bar = 0.5mm.

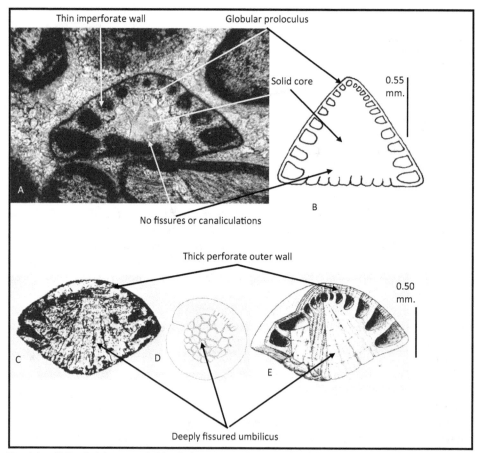

Fig. 4.6.   A) *Trocholina conica* (Schlumberger) Umm Shaif, core 3883 ft, Bajocian; B) *Trocholina* sp. sketch; C-E) *Neotrocholina friburgensis* Guillaume and Reichel, type figures, late Barremian – early Aptian, west-central Switzerland.

Family Ventrolaminidae Weynschenck, 1950

Members of this superfamily have a lenticular test, which is planispiral to low trocho-spiral with numerous chambers in a rapidly enlarging whorl. The wall is calcareous with two layers, an inner microgranular one and an outer hyaline radial layer. Middle Jurassic to Early Cretaceous (late Bajocian to Berriasian).

- *Protopeneroplis* Weynschenck, 1950 (Type species: *Peneroplis senoniensis* Hofker, 1949). The test is enrolled in about two rapidly enlarging and loosely coiled whorls. The final whorl has twelve to sixteen chambers, and is involute with a rounded to angular periphery. The aperture is areal, in the early part of the test and slightly pro-truding. Middle Jurassic to Early Cretaceous (Berriasian) (Plate 4.5, fig. 7; Fig. 4.8).

Order Textulariida Delage and Hérouard, 1896

The tests of these agglutinated foraminifera are made of foreign particles bound by organic cement. They range from Early Cambrian to Holocene.

Superfamily Verneuilinoidea Cushman, 1911

Members of this family have a trochospiral early stage that is triserial or biserial, which later may be uniserial. Walls are non-canaliculate. Late Carboniferous to Holocene.

Family Verneuilinidae Cushman, 1911

Members of this family have biserial tests, at least in early stages. Chambers are glob-ular with a terminal aperture.

- *Duotaxis* Kristan, 1957 (Type species: *Duotaxis metula* Kristan, 1957). Originally regarded as a member of the Tetrataxidae, *Duotaxis* differs in having an aggluti-nated rather than a two-layered microgranular calcareous wall, so it was reassigned to the Family Verneuilinidae by Loeblich and Tappan (1988). The genus had earlier (Loeblich and Tappan 1964) been considered a synonym of *Valvulina*, but differs by not having an early triangular stage and in lacking a truly valvular tooth (Loeblich and Tappan 1988). Triassic to Early Jurassic (?Ladinian-Rhaetian to Pliensbachian) (Plate 3.4, figs 1-2; Plate 3.8, fig. 5; Plate 4.6, fig. 1).

Superfamily Pfenderinoidea Smout and Sugden, 1962

Members of this superfamily (see Fig. 4.7) have a trochospiral test throughout, or one that may become uncoiled. Some forms have a siphonal canal, others develop a central composite columella, with pillars between apertural plates and septa. Early Jurassic (Hettangian) to Cretaceous.

Family Pfenderinidae Smout and Sugden, 1962

Members of this family have a loose trochospiral conical test with siphonal canals that connect successive apertures in primitive forms. Some forms develop a central compos-ite columella composed of thickened innermost septal ends ("septal buttons") with or without additional pillars and a spiral canal between the columella and the thickened inner parts of the adjacent septa. A subcameral tunnel (simple or multiple) is present in advanced forms. The chamber interior of advanced taxa is subdivided by vertical or horizontal (or both) exoskeletal partitions, resulting in a reticulate subepidermal layer. In the trochospires, the spiral and intracameral septa are strongly oblique to the long (coiling) axis (resembling *Arenobulimina*), but rectilinear, uniserial developments (if present) produce peneropliform, conical ("coskinoliniform") or lituoliform tests. The aperture is always cribrate, areal. Early Jurassic to Late Cretaceous (Sinemurian to Maastrichtian).

Subfamily *Pseudopfenderininae* Septfontaine, 1988

The test is trochospirally coiled throughout with no subcameral tunnel. The middle of the test has a siphonal canal, or is filled in with a columella made by interseptal pillars and calcitic infilled material. No peripheral partitions or chamberlets are present. Early Jurassic (Hettangian) to Late Cretaceous.

- *Pseudopfenderina* Hottinger, 1967 (Type species: *Pfenderina butterlini* Brun, 1962). The test is a high trochospiral with numerous chambers. The umbilical part of the chamber interior is filled with numerous pillars that are continuous from chamber to chamber. Early Jurassic (Sinemurian to Bathonian) (Plate 4.6, figs 10-16).
- *Siphovalvulina* Septfontaine, 1988 (Type species: *Siphovalvulina variabilis* Septfontaine, 1988). The test is trochospirally coiled (high or low) in general with three chambers per whorl; the test wall is canaliculate, but rarely visible as such; the interiors of the chambers are free; a twisted siphonal canal connects successive apertures; the aperture is unique, interiomarginal, but may become cribrate in the last chambers of advanced forms. Early Jurassic (Hettangian) to Early Cretaceous to Late Cretaceous? (Plate 4.6, figs 2B, 3-8, 9A).

Subfamily *Paleopfenderininae* Septfontaine, 1988

The adult chambers of this subfamily may be uncoiled with various shapes. A simple or multiple spiral subcameral canal may be present. There are pillars between apertural plates and septa have calcitic fillings, giving the appearance of a columella. No peripheral partitions or chamberlets are present. The aperture is terminal and multiple in the apertural plate. Jurassic (Bajocian to early Oxfordian).

- *Conicopfenderina* Septfontaine, 1988 (Type species: *Lituonella mesojurassica* Maync, 1972). The genus includes conical forms, with a trochospiral early part followed by a rectilinear uniserial part with separated irregular pillars filling the centres of the chambers. The marginal zone of the chambers is not subdivided. *Conicopfenderina* differs from *Parurgonina* Cuvillier, Foury and Pignatti Morano, 1968, by the uniserial arrangement of its chambers, which in the latter is trochospiral. Apertures occur multiply. Middle Jurassic (Bajocian to Callovian) (Plate 4.7, figs 8-11).
- *Chablaisia* Septfontaine, 1978 (Type species: *Pfenderina? chablaisensis* Septfontaine, 1977). This genus differs from other pfenderinoids in having a low trochospiral test and by the presence of a spiral canal and calcitic fillings (septal knobs) in the chambers. Jurassic (late Bathonian to early Oxfordian) (Plate 4.7, figs 12-13).
- *Palaeopfenderina* Septfontaine, 1988 (Type species: *Pfenderina salernitana* Sarton and Crescenti, 1962). This genus includes forms with tight trochospiral coiling, which increase in diameter with spiral height, with a central microgranular

columella twisted along the coiling direction. The apertures are multiple and crib-rate. *Palaeopfenderina* differs from *Pfenderina* Henson, 1948 (Plate 4.6, figs 6-7) by the superficial position of the tunnel, as a groove at the surface of the columella, and by the filling of the inter-pillar spaces with calcitic material. In *Pfenderina* the filling is less developed and the pillars are clearly visible. Jurassic (Bathonian to Callovian) (Fig. 4.7).

- *Pseudoeggerella* Septfontaine, 1988 (Type species: *Pseudoeggerella elongata* Septfontaine, 1988). This genus includes forms with high trochospiral coiling and a narrow columella in the axis of the test. The stalagmitic protuberance is deeply incised against the columella. This genus lacks the subcameral tunnel of *Palaeopfenderina* and differs from *Pseudopfenderina* by the presence of calcitic protuberances in the chambers. Jurassic (Bathonian) (Plate 4.7, fig. 14).

- *Sanderella* Redmond 1964 (Type species: *Sanderella laynei* Redmond, 1964). The test has long and narrow chambers without secondary septa. The initial portion of the test is a low trochospire with a single subcameral tunnel. The later portion of the test becomes discoid, with multiple subcameral tunnels. Jurassic (Bathonian to Oxfordian) (Plate 4.7, fig. 17).

- *Satorina* Fourcade and Chorowicz, 1980 (Type species: *Satorina apuliensis* Fourcade and Chorowicz, 1980). This genus differs from *Conicopfenderina* by possessing radial pillars at the margin of the central mass formed by interseptal pillars. Jurassic (late Bathonian to early Oxfordian) (Plate 4.7, figs 15-16).

- *Steinekella* Redmond, 1964 (Type species: *Steinekella steinekei* Redmond, 1964). This form is a pfenderinine with a massive, central, continuous columella of fused or coalescent pillars, as in *Pfenderina*, not the discontinuous end skeletons of thickened septal buttons, as in *Kurnubia*. It has a high trochospiral test with very long and narrow chambers. Multiple subcameral tunnels combined with strong transverse partitions and very fine and weak subepidermal structure are present throughout the test. Jurassic (late Callovian to Oxfordian) (Plate 4.7, figs 18-20).

Subfamily *Pfenderininae* Smout and Sudgen, 1962
Members of this subfamily have a test with a single subcameral tunnel that is always buried in the columella, which is formed by pillars and calcitic deposits. No peripheral partitions or chamberlets are present. The aperture is multiple, cribrate, and on the apertural plate. Early Cretaceous (Valanginian to Barremian or?Aptian).

- *Pfenderella* Redmond, 1964 (Type species: *Pfenderella arabica* Redmond, 1964). The test is an elongate cone consisting of a trochoid spiral of relatively short chambers arranged around the axis of coiling in such a manner that each chamber overlaps approximately one-half of its predecessor. Successive chambers are indirectly connected by a tunnel. Secondary septa are absent. The test is, without a solid core. *Pfenderella* differs from *Kurnubia*, *Praekurnubia* and *Steinekella* (Plate 4.7, figs 18-20) by lacking secondary partitions in the chambers. It differs from *Pfenderina* in not having a solid central core, and its chambers are also broader in proportion to their length than are those of *Pfenderina*. Jurassic (Bajocian to Callovian) (Plate 4.7, figs 21-24).

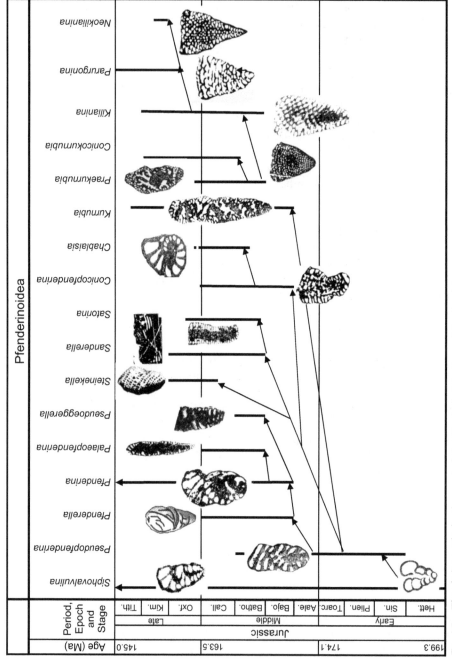

Fig. 4.7. The phylogenetic development of the Pfenderinoidea through the Jurassic.

- *Pfenderina* Henson, 1948 (Type species: *Eorupertia neocomiensis* Pfender, 1938). This genus differs from *Pfenderella* in having long narrow chambers communicating with each other, and with the exterior of the test, by means of a spiral subcameral tunnel which lies mostly beneath the surface of a solid core. Jurassic (Bajocian to Valanginian) (Plate 4.7, figs 6-7).

Subfamily *Kurnubiinae* Redmond, 1964
Members of this subfamily have a test with no subcameral tunnel or tunnels, and may or may not have a solid core. The peripheral zone is divided by radial partitions.

- *Conicokurnubia* Septfontaine, 1988 (Type species: *Conicokurnubia orbitolinifor- mis* Septfontaine, 1988). The test is conical. Vertical partitions join the centre of the chambers and coalesce with the pillars. The apertures are multiple, cribrate. It differs from *Dictyoconus* Blanckenhorn, 1900 (see Chapter 5) by having a less com- plicated marginal zone of the chambers. Jurassic (late Bathonian to Kimmeridgian) (Plate 4.8, figs 1-2).
- *Kurnubia* Henson, 1948 (Type species: *Kurnubia palastiniensis* Henson, 1948). The test is conical, cylindrical or elongate fusiform, consisting of chambers arranged in an elongate trochoid spiral. The inner surfaces of individual chambers are divided into chamberlets by several sets of inward-projecting transverse and longitudinal secondary partitions. The aperture is represented by a number of closely-set pores on the base of the test. Jurassic (Bajocian to early Tithonian) (Plate 4.8, figs 3-21; Plate 4.9, fig. 13C).
- *Praekurnubia* Redmond, 1964 (Type species: *Praekurnubia crusei* Redmond, 1964). This differs from *Kurnubia* in having transverse partitions only. Jurassic (Bathonian to early Oxfordian) (Plate 4.8, figs 22-24).

Family Valvulinidae Berthelin, 1880
The test is trochospirally coiled, and generally triserial in the early stage. The wall is microgranular, and may be alveolar. The interior of chambers is simple. The aperture is interiomarginal, with a large valvular tooth (simple or complicated). The Valvulinidae show the first known example of a crosswise-oblique stolons system (the margino- poriform structure of Hottinger and Caus (1982)) among the Pfenderinoidea. This disposition of stolons appears as a morphological convergence in different groups of lituoloids: the valvulinids (trochospirally coiled), the orbitolinids (trocho- to uniserial) and the reniform-discoidal lituoloids (planispiral to uniserial).

Subfamily *Valvulininae* Berthelin, 1880
This subfamily includes forms with a simple or complicated valvular tooth plate. Aperture: simple or cribrate. Jurassic (Pliensbachian) to Holocene

- *Kilianina* Pfender, 1933 (Type species: *Kilianina blancheti* Pfender, 1933). This form is distinguished by its conical test, where the central zone septa thicken and coalesce into an almost solid mass (hummocks). In some forms the latter half of the test is formed by thin irregular plates intergrown by finer broadly-spaced pillars. The early chambers are trochospiral. They coil along a vertical axis and occupy almost half the test, later they become rectilinear, with the outer parts of chamberlets subdivided by numerous pillars. Jurassic (Bathonian to Kimmeridgian) (Plate 4.4, fig. 3; Plate 4.6, fig. 9B; Plate 4.10, figs 12, 14).

- *Valvulina* d'Orbigny, 1826 (Type species: *Valvulina triangularis* d'Orbigny, 1826). The test is trochospiral throughout, and triserial. The valvular tooth plate is simple or perforated by supplementary apertures. The primary aperture is a basal slit. Jurassic (Pliensbachian) to Holocene (Plate 4.11, fig. 15).

Subfamily *Parurgonininae* Septfontaine, 1988
This subfamily includes genera with eight or more chambers per whorl. The valvular tooth plate is complicated by pillars. The apertures are multiple, with a crosswise-oblique stolons system. Late Jurassic.

- *Parurgonina* Cuvillier, Foury and Pignatti Morano, 1968 (Type species: *Urgonina (Parurgonina) caelinensis* Cuvillier et al., 1968). This genus has a highly conical test with numerous chamberlets, separated by curved septal extensions of the outer wall in a low trochospire. The septa break in the umbilical region to form sub-conical pillars. Jurassic (late Oxfordian to Tithonian) (Plate 4.10, figs 17-18).
- *Neokilianina* Septfontaine, 1988 (Type species: *Paravalvulina complicata* Septfontaine 1988). This genus differs from *Kilianina* in having a higher number of chambers per whorl in the adult stage (25 instead of 8), and a more pronounced development of pillars in the central part of the test. Jurassic (early Kimmeridgian) (Plate 4.10, figs 13, 15-16).

Superfamily Lituoloidea de Blainville, 1825
Members of this superfamily have conical, multilocular, rectilinear and uniserial tests. The early stage has plani- (strepto-) or trochospiral coiling. The periphery of the chambers has radial partitions; but centrally they are with or without scattered, separated pillars. The septa are arched into hummocks (almost solid mass) between the apertures. The bases of the arches can be fused to the hummocks of previous septum, with the apertures then opening at the suture. The alignment of the apertures and thickening of the hummock walls produces the appearance of a series of "gutters". No true pillars are formed. The walls are solid, non-alveolar, non-canaliculate. The aperture is simple, with no internal tooth-plates, areal or multiple, cribrate. Late Triassic (Carnian) to Holocene.

Family Hauraniidae Septfontaine, 1988
The test is uncoiled, uniserial or planispirally coiled. The wall is microgranular with a hypodermic network. The septa are simple or with complicated microstructures. The interior of the chambers is simple or with pillars. The aperture is multiple. Jurassic (late Sinemurian) to Cretaceous.

Subfamily *Hauraniinae* Septfontaine, 1988
The test is uncoiled, uniserial or planispirally coiled. The septa are simple or with complicated microstructures, and the interior of the chambers have fine pillars in the central zone. Jurassic (late Sinemurian to Bathonian).

- *Ataxella* Bassoullet and Lorenz, 1988 (*Paracoskinolina occitanica* Peybernès, 1974). The test is conical to cylindrical with an early streptospiral stage followed by a rectilinear, uniserial stage. The marginal zone is divided by alternating radial partitions. The central zones is full with pillars which are often coalescent. Middle Jurassic (Bathonian) (Plate 4.4, fig. 10).

- *Cymbriaella* Fugagnoli, 1999 (Type species: *Cymbriaella lorigae* Fugagnoli, 1999). The test is coarsely agglutinated with a coarse irregular subepidermal polygonal network, locally appearing as bifurcated alveolar cavities with a blind ending of polygonal outline below a thin epidermis. Megalospheric forms possessing a complex embryonic apparatus represented by a spherical proloculus enveloped by a subspherical deuteroconch, which is characterized by having short beams (exoskeletal partitions of the chamber lumen perpendicular to the septa) perpendicular to the wall. *Cymbriaella* is the oldest representative of the Hauraniidae that developed a complex megalospheric form. It differs from *Amijiella* by being larger, and having a more irregular, coarser subepidermal network. Jurassic (Hettangian to Pliensbachian) (Plate 4.11, fig. 14).

- *Gutnicella* Moullade, Haman and Huddleston, 1981 (Type species: *Coskinolina (Meyendorffina) minoricensis* Bourrouilh and Moullade, 1964). *Gutnicella* is a new name for *Lucasella* Gutnic and Moullade, 1967. The test is highly conical with a large spherical proloculus, enclosed in an early planispiral and involute coil, that is later briefly trochospiral, followed by being uniserial and rectilinear. Chambers are subdivided in the outer part by many short radial vertical partitions, creating a narrow peripheral zone of quadrate chamberlets; one row to each chamber. The broad central zone is filled with irregular scattered separated pillars of different sizes. Jurassic (Pliensbachian to Callovian) (Plate 4.11, figs 16-17; Plate 4.12, figs 7-9).

- *Haurania* Henson, 1948 (Type species: *Haurania deserta* Henson, 1948). The radial partitions are delicate, often bifurcating vertically to form a partial tier of peripheral chamberlets. Jurassic (late Sinemurian to Bathonian) (Plate 4.11, figs 18-20; Plate 4.12, figs 1-4).

- *Meyendorffina* Aurouze and Bizon, 1958 (Type species: *Meyendorffina bathonica* Aurouze and Bizon, 1958). The test is initially enrolled planispirally with a low, long initial trochospire, later becoming uniserial. Chambers are subdivided by vertical pillars that project a short distance inward from the outer wall, but lack horizontal rafters (exoskeletal partitions of the chamber lumen parallel to the septa and perpendicular to the beams and the lateral chamber wall, see Hottinger (2006)). Central zone filled with pillars. Jurassic (Bathonian to early Oxfordian) (Plate 4.8, fig. 23B; Plate 4.12, figs 5-6, 19).

- *Platyhaurania* Bassoullet and Boutakiout, 1996 (Type species: type species *Haurania (Platyhaurania) subcompressa* Bassoullet and Boutakiout, 1996). This form differs from *Haurania* in possessing cylindrical chambers in the uncoiled part. Jurassic (late Sinemurian to Toarcian) (Plate 4.12, fig. 12).

- *Robustoconus* Schlagintweit, Velić and Sokač 2013 (Type species: *Robustoconus tisljari* Schlagintweit et al., 2013). The test is conical with an early planispiral stage. Chambers are subdivided into a marginal zone formed of radial and intercalary beams with rafters or horizontal partitions forming a network of chamberlets. The central zone is large and complex, consisting of anastomising septal excrescences (with constrictions and swellings). Middle Jurassic (early Bajocian) (Plate 4.4, fig.12).

- *Socotraina* Banner, Whittaker, BouDagher-Fadel and Samuel, 1997 (Type species: *Socotraina serpentina* Banner et al., 1997). Tests are non-canaliculate, septate, initially coiled planispirally or in a low trochospire. Uniserial chambers are filled with near-vertical, subradial partitions with a sinuous serpentine form. In the central areas of the chambers the partitions fuse laterally. The aperture consists of many small pores between the partitions and is situated subterminally. Jurassic (Pliensbachian to Toarcian) (Plate 4.4, fig. 6; Plate 4.12, figs 10-11).

- *Timidonella* Bassoullet, Chabrier and Fourcade, 1974 (Type species: *Timidonella sarda* Bassoullet et al., 1974). Chambers have a subepidermal network of horizontal and vertical pillars connecting consecutive septa. This network is followed by a zone of quadrangular chamberlets formed by the pillars, an undivided zone termed the annular canal, and a zone of interseptal pillars. *Timidonella* differs from *Orbitopsella* in its alveolar hypodermis, which compare with the Cretaceous *Loftusia* (see Chapter 5). Jurassic (Aalenian to early Bathonian) (Plate 4.12, figs 13-16).

- *Trochamijiella* Athersuch, Banner and Simmons, 1992 (Type species: *Trochamijiella gollesstanehi* Athersuch, et al., 1992). This genus is similar to *Socotraina*, *Haurania* and *Platyhaurania*, but has an initial trochospiral coil. Late Jurassic (Bajocian to Callovian) (Plate 4.12, fig. 17).

Subfamily *Amijiellinae* Septfontaine, 1988

Members of this subfamily have an uncoiled or planispirally coiled test. The septa are simple or have a complicated microstructure. The interior of the chambers is simple, however, some genera may develop pillars. Walls may have alveoles. The apertures are multiple. Jurassic (late Sinemurian) to Cretaceous.

- *Alzonella* Banner and Neumann, 1970 (Type species: *Alzonella cuvillieri* Bernier and Neumann, 1970). A planispiral test with a hypodermis consisting of a coarse lattice of beams and rafters. Jurassic (Bathonian to Callovian) (Plate 4.13, figs 6-8).

- *Alzonorbitopsella* BouDagher-Fadel, 2008 (Type species: *Alzonorbitopsella arabia* BouDagher-Fadel, 2008, genus and species validated in BouDagher-Fadel, 2016). A planispiral, annular and discoidal test with no septulae. Any subepidermal reticulate mesh is absent in septa, which are thickened around cribrate apertures, but lacking the true pillars linking septum to septum as in *Orbitopsella* There is a delicate reticulate hypodermis of beams and rafters as in *Alzonella*, but there is a lack of continuation of this structure on to the septa. Chambers are annular, immediately following the large megalospheric proloculus, as in *Cyclorbitopsella*, but not in the megalospheric *Alzonella*. The alveolar hypodermis and septa, and the annular A-form differentiate *Alzonorbitopsella* from *Orbitopsella* with no (hypodermal) network, and from *Timidonella* with subepidermal network and partitions. The absence of medial pillars and the presence of subepidermal beams and rafters differentiate *Alzonorbitopsella* from *Cyclorbitopsella*. Jurassic (late Bathonian) (Plate 4.2, figs 5-10; Plate 4.14, figs 1-6).

- *Amijiella* Loeblich and Tappan, 1985 (Type species: *Haurania amiji* Henson, 1948). The test is straight and uniserial or planispiral, to uniserial in megalospheric generations. The chambers have no pillars in the central zone. The radial partitions are strong, tending to become thickened towards the central zone. They bifurcate vertically to form a few, scattered chamberlets. Jurassic (late Sinemurian to Bathonian or?Callovian) (Plate 4.15, figs 2-4).

- *Anchispirocyclina* Jordan and Applin, 1952 (Type species: *Anchispirocyclina henbesti* Jordan and Applin, 1952). Tests are compressed and peneropliform. The central zone is filled by a complex reticulum of densely spaced pillars. Jurassic to Early Cretaceous (Oxfordian to early Campanian) (Plate 4.15, figs 6-7, 8-9).

- *Bostia* Bassoullet, 1998 (Type species: *Bostia irregularis* Bassoullet, 1998). A dimorphic test with clearly distinct microspheric and megalospheric generations. It is characterized by a complicated embryonic apparatus and a lack of pillars in the central zone. Walls have a subepidermal network formed by irregular radial and transverse partitions that constitute a superficial extension of the chambers. It is similar to the unpillared *Amijiella* and *Alzonella*, but differs in possessing an ammobaculoid test morphology in both generations. It has a wall structure similar to *Spiraloconulus*, but differs by having an orbitoliniform test. Jurassic (Bathonian) (Plate 4.15, figs 10-11).

- *Dhrumella* Redmond, 1965 (Type species: *Dhrumella evoluta* Redmond, 1965). The initial part of the test is a low concavo-convex trochoid spiral, with chambers that are completely involute on the convex side and moderately evolute on the concave side. The later chambers become strongly evolute, becoming flattened and uniserial. Chambers are subdivided into chamberlets by prominent transverse partitions that generally alternate in position from chamber to chamber. Jurassic (Bajocian to Bathonian) (Plate 4.5, fig. 4).

- *Ijdranella* Bassoullet, Boutakiout and Echarfaoui, 1999 (Type species: *Ijdranella atlasica* Bassoullet, et al., 1999). An hauraniid with a compressed peneropliform uncoiled stage and an exoskeleton containing long radial pillars superficially united by a coarse network. *Ijdranella* differs from *Amijiella* in its compressed test and from *Trochamijiella* in its trochospiral initial coiling and canaliculated wall. Jurassic (Pliensbachian) (Plate 4.15, figs 12-13).

- *Kastamonina* Sirel, 1993 (Type species: *Kastamonina abanica* Sirel, 1993). Tests are elongated to conical, morphologically similar to *Amijiella*, but with a much reduced initial coiled portion and a more complex internal structure. The marginal zone of each chamber is subdivided by an irregular polygonal subepidermal network, consisting of vertical partitions (beams) and horizontal partitions (rafters), forming numerous irregular alveolar compartments. This genus differs from *Haurania* by the absence of a planispiral part and by lacking endoskeletal pillars in the centre of the test. It differs from *Rectocyclammina* (Plate 4.12, fig. 18) in its early chambers and cribrate aperture. Jurassic (Kimmeridgian to Tithonian) (Plate 4.10, fig. 1).

- *Pseudospirocyclina* Hottinger, 1967 (Type species: *Pseudospirocyclina maynci* Hottinger, 1967). Tests are planispiral, then rectilinear or flabelliform with an alveolar hypodermis. There are thick septa with equally thick, scattered irregular pillars. Jurassic (Bajocian to Tithonian) (Plate 4.5, figs 9-15).

- *Spiraloconulus* Allemann and Schroeder, 1980 (Type species: *Spiroconulus perconigi* Alleman and Schroeder, 1972). The test is planispiral to low trochospiral, then uncoiling to become thick and conical, with a coarsely reticulate hypodermis. Thin septa, linked by thick heavy pillars, are irregularly and widely spaced. The central zone possesses endoskeletons of pillars from septum to septum. The septa are not alveolar. *Spiraloconulus* differs from *Pseudospirocyclina* by its thick, flat initial spire and its thick, conical rectilinear stage, by its reticulate hypodermis of "*Alzonella*-like" structure, and its thin, non-alveolar septa. *Spiraloconulus* differs from *Robustoconus* by its exoskeleton, which consist of a narrow marginal zone of thin-walled chamberlets and the large central zone with septa typically agglutinating large grains (Schlagintweit et al., 2013). Jurassic (Bajocian to Callovian) (Plate 4.10, fig. 11).
- *Streptocyclammina* Hottinger, 1967 (Type species: *Pseudocyclammina (Streptocyclammina) parvula* Hottinger, 1967). A peneropliform, low streptospiral test with an empty central zone, lacking pillars. It differs from *Pseudocyclammina* only in its streptospirality. Jurassic (Pliensbachian to Kimmeridgian) (Plate 4.16, fig. 7).

Family Lituolidae de Blainville, 1827

The early stages of the tests are enrolled, but later they may become rectilinear. Walls are made from agglutinating foreign particles. Few chambers (less than 10) per whorl. Carboniferous to Holocene.

Subfamily *Ammomarginulininae* Podobina, 1978

Tests in the early stage are coiled, but later show uncoiling. Aperture are single. Carboniferous (Early Mississippian) to Holocene.

- *Ammobaculites* Cushman, 1910 (Type species: *Spirolina agglutinans* d'Orbigny, 1846). The test is simple, not compressed and uncoils in the adult. Apertures are single and areal. Carboniferous (Mississippian) to Holocene (Plate 5.6, fig. 18).

Subfamily *Lituolinae* de Blainville, 1827

Members differs from the Ammomarginulininae in having a multiple aperture. Late Triassic to Holocene.

- *Lituola* Lamarck, 1804 (Type species: *Lituolites nautiloidea* Lamarck, 1804). These forms have no internal partitions and a multiple cribrate aperture. Late Triassic to Holocene (Plate 5.9, fig. 14; Plate 5.10, fig. 9).

Superfamily Loftusioidea Bradey, 1884

The test is planispiral, but may uncoil in later stage. The wall is agglutinated with differentiated outer and inner alveolar layers. Late Triassic (Carnian) to Holocene.

Family Choffatellidae Maync, 1958

This family is characterised by having hypodermal alveoles. The test is planispiral, but the early part may be streptospiral, lacking continuously developed endoskeletal pillars. Early Jurassic (Pliensbachian) to Late Cretaceous (Coniacian).

- *Alveosepta* Hottinger, 1967 (Type species: *Cyclammina jaccardi* Schrodt, 1894). A planispiral test, that may have a streptospiral early stage. The wall is finely and complexly alveolar. The chambers are low with curved septa, perforated fine apertures and often have a clear line and/or median lamella (or pillars) only in equatorial plane. Jurassic (Oxfordian to Kimmeridgian) (Plate 4.12, fig. 20; Plate 4.13, figs 1-3).
- *Choffatella* Schlumberger, 1905 (Type species: *Choffatella decipiens* Schlumberger, 1905). A planispiral test, with a wall that is finely and complexly alveolar. The septa have many fine apertures in the median line and are as complex and thick as the hypodermis of the wall. The chambers are high with the central zone being empty with no pillars. Late Jurassic to Late Cretaceous (Oxfordian to Santonian) (Plate 4.5, figs 1-2; Plate 5.5, Figs. 8–9; Fig. 5.3).
- *Feurtillia* Maync, 1958 (Type species: *Feurtillia frequens* Maync, 1958). The test is planispiral with an involute early stage and a rectilinear later stage with no basal layer. The chamber walls are with narrow, shallow alveoles surmounted by an *Alzonella*-like (Plate 4.12, figs 6-8) reticulum of beams and rafters. The septa are weakly alveolar with single areal aperture possessing a thickened, invaginated rim. Early Jurassic to Late Cretaceous (Tithonian to Valanginian) (Plate 5.3, fig. 8)
- *Palaeocyclammina* Bassoullet, Boutakiout and Echarfaoui, 1999 (Type species: *Palaeocyclammina complanata* Bassoullet et al., 1999). The test is planispiral, compressed and involute with long low chambers. It differs from *Pseudocyclammina* in having a reticulate subepidermal skeleton comprised of an irregular superficial network, made of short radial blades perpendicular to the septa. Early Jurassic (Pliensbachian) (Plate 4.15, fig. 1).
- *Pseudocyclammina* Yabe and Hanzawa, 1926 (Type species: *Cyclammina lituus* Yokoyama, 1890). A planispiral nautiliform test, sometimes becoming uncoiled. The central zone of the test lacks the continuous endoskeletal pillars (discontinuous, columnar partitions on the inner surface of the chamber wall, see Hottinger (2006) for definition). Walls are coarsely alveolar and labyrinthic. Apertures are spread over the apertural face. Early Jurassic to Cretaceous (Pliensbachian to early Maastrichtian) (Plate 4.10, figs 3-10).
- *Redmondellina* Banner and Whittaker, 1991 (Type species: *Pseudocyclammina powersi* Redmond, 1964). This form has a compressed test and differs from *Alveosepta* in having pillar-like hypodermal extensions linking the septal hypodermis to the anterior of the epidermis of each preceding septum. The division of the hypodermal alveoles distally produces finer and finer alveoles, as they bifurcate or trifurcate on approaching the epidermis. The pillars are present only in the median, equatorial plane, and are not spread throughout the chamber space as in *Pseudospirocyclina*. Late Jurassic (late Oxfordian to Kimmeridgian) (Plate 4.9, fig. 13A; Plate 4.13, figs 4-5).
- *Torinosuella* Maync, 1959 (Type species: *Choffatella peneropliformis* Yabe and Hanzawa, 1926). The test is flabelliform, with simplified, thin septa. Late Jurassic to Early Cretaceous (Oxfordian to Barremian) (Plate 4.15, fig. 5).

Family Everticyclamminidae Septfontaine, 1988

Tests are streptospiral or planispiral in early stages, but uncoiled in the adult or uniserial throughout. Walls are microgranular with an alveolar microstructure. Interiors of chambers are simple. Apertures are unique and terminal. Early Jurassic to Early Cretaceous (Sinemurian to Aptian).

• *Everticyclammina* Redmond, 1964 (Type species: *Everticyclammina hensoni* Redmond, 1964). A planispiral test with an alveolar wall. The septal aperture is single, and the initial coil is very short and missing in most random sections, which only show the elongate uniserial part. Loeblich and Tappan (1988) assigned *Everticyclammina* to the Family Cyclamminidae, but Septfontaine (1988) assigned it to his new, monogeneric Family Everticyclamminidae. This proposal, seemingly overlooked by Banner and Highton (1990), when revising the genus to accommodate five named species other than the type, is adopted by Fugagnoli (2000) and herein. Fugagnoli's amended phylogeny and BouDagher-Fadel et al. (2001) illustrate a known origin of the genus *Everticyclammina* from *E. praevirguliana* (Plate 4.16, fig. 3) in the Early Jurassic (late Sinemurian) rather than from species of the Middle or Late Jurassic (Callovian or Oxfordian) as generally accepted by previous authors, with a supposed ancestry in near coeval *Ammobaculites*. Therefore, the earliest known *Everticyclammina* is of middle Sinemurian age, and this suggests that by this time the genus was more widely dispersed in the Tethyan realm than recorded hitherto. Jurassic to Cretaceous (mid Sinemurian to Aptian) (Plate 4.4, fig. 11; Plate 4.16, figs 3-6).
• *Buccicrenata* Loeblich and Tappan, 1949 (Type species: *Ammobaculites subgoodlandensis* Vanderpool, 1933). *Buccicrenata* was erected by Loeblich and Tappan, 1949 with *Ammobaculites subgoodlandensis* Vanderpool, 1933 as its type species. Loeblich and Tappan (1985, p.100; 1988, p.99) redefined the genus *Buccicrenata* to include an alveolar wall, which was first illustrated for *B. libyca* Gohrbandt (1966, p.67, pl.1, fig.11). However, they denied the presence of alveoles in the septum of *Buccicrenata*. Nevertheless, in examining numerous randomly thin sectioned specimens of *Buccicrenata* from different localities, some of them hypotypes of both *B. subgoodlandensis* and *B. hedbergi*, the alveoles are clearly seen to be present, and were used by Banner and Highton (1990) to distinguish this genus from *Everticyclammina*. Subsequently BouDagher-Fadel (2001) reviewed these forms and traced their evolution from a primitive form, *B. primitiva*, from the Kimmeridgian of Lebanon. It is characterized by a planispiral test with septa that are a continuation of chamber walls, but with alveoles reduced and with no basal layer. Jurassic to Cretaceous (Kimmeridgian to Cenomanian) (Plate 4.5, fig. 8; Plate 5.6, fig.16).
• *Rectocyclammina* Hottinger, 1967 (Type species: *Rectocyclammina chouberti* Hottinger, 1967). The test is elongate, the early stage is planispiral, later uncoiling and becoming rectilinear. The septa are thick, and the aperture is terminal and circular in the centre of the apertural face. Jurassic to Early Cretaceous (Pliensbachian to early Berriasian) (Plate 4.12, fig. 18).

Family Mesoendothyridae Voloshinova, 1958
Members of this family have a strepto- or planispirally coiled test, that has involute initial chambers, and later is uncoiled. Adult chambers are cylindrical or flattened, falciform to cyclical. They are simple, with radial partitions or with pillars. The wall may have alveoles or a hypodermic network. Early Jurassic (Sinemurian) to Holocene.

Subfamily *Mesoendothyrinae* Voloshinova, 1956

Tests are biumbilicate, or very compressed, strepto- or planispirally coiled, but later they are uncoiled to various degrees. Chambers increase rapidly in height with mono-lamellar septa. Adult chambers are cylindrical or flattened. Walls are microgranular to agglutinated, imperforate or alveolar, simple or with radial partitions, with or without pillars. Apertures are a single slit or made of numerous pores. Jurassic (Sinemurian) to Cretaceous.

- *Mesoendothyra* Dain, 1958 (Type species: *Mesoendothyra izjumiana* Dain, 1958). The test is involute, globose, and twisted in a tight streptospire. The wall has an outer imperforate layer and an inner alveolar layer. The aperture is a basal slit. Middle Triassic to Jurassic (Ladinian to Kimmeridgian) (Plate 4.5, figs 5-6).

Subfamily *Orbitopsellinae* Hottinger and Caus, 1982

Tests have an early planispiral coil followed by a uniserial part in the adult, becoming cylindrical, or falciform to cyclical. The wall has an alveolar microstructure. The interior of the chambers are divided by vertical radial partitions, with pillars in the central zone. Early Jurassic (late Sinemurian to Toarcian).

- *Cyclorbitopsella* Cherchi, Schroeder and Zhang, 1984 (Type species: *Cyclorbitopsella tibetica* Cherchi et al., 1984). An orbitopsellinine with adult chambers becoming annular. It has medial pillars in the central zone, but no subepidermal beams and rafters, in contrast to *Alzonorbitopsella*. Apertures are cribrate. Early Jurassic (late Sinemurian to Toarcian) (Plate 4.17, figs 8-9).
- *Lituosepta* Cati, 1959 (Type species: *Lituosepta recoarensis* Cati, 1959). This species is characterized by having a simple proloculus followed by a planispiral stage, and a well-developed uncoiled part, which becomes fan shaped particularly in the micro-spheric forms. The test possesses multiple cribrate apertures. The apertural openings between the pillars penetrate the height of the chamber space. Peripheral partitions are seen in transverse sections. The flaring flabelliform test and the canaliculated wall distinguishes it from *Labyrinthina*. Early Jurassic (late Sinemurian to Aalenian) (Plate 4.16, figs 8-9).
- *Orbitopsella* Munier-Chalmas, 1902 (Type species: *Orbitulites praecursor* Gümbel, 1872). An orbitopsellinine with a discoidal test, the first stage being planispiral followed by a flaring, flabelliform stage with 35 to 40 annular chambers. The wall has simple endoskeletal and exoskeletal pillars. Early Jurassic (late Sinemurian to Toarcian) (Plate 4.16, fig.10; Plate 4.17, figs 1-6).

Subfamily *Planiseptinae* Septfontaine, 1988

Tests are planispiral, compressed laterally and involute. Walls are microgranular with an alveolar microstructure. Interior of chambers are simple, or with vertical partitions and pillars in the median layer. Apertures are multiple. Early Jurassic (late Sinemurian to Pliensbachian).

- *Planisepta* Septfontaine, 1988 (Type species: *Lituola compressa* Hottinger, 1967). Tests have vertical radial partitions and pillars. Early Jurassic (Pliensbachian) (Plate 4.9, figs 1-3).

- *Palaeomayncina* Septfontaine, 1988 (Type species: *Mayncina termieri* Hottinger, 1967). Differs from *Planisepta* by the absence of the pillars and partitions. Septfontaine (1988) named *Mayncina ternieri* Hottinger, 1967 as the type species and proposed that this new genus is closely related to *Lituolipora* (Plate 4.11, fig. 7). Early Jurassic (Late Sinemurian-Pliensbachian) (Plate 4.10, fig. 2).

Subfamily *Labyrinthininae* Septfontaine, 1988
Tests are planispirally coiled in early stages, but uncoiled in the adult. Walls are thick and may have alveoles. Interior of chambers are subdivided by vertical, radial partitions. Pillars may be present in the central zone. Late Jurassic.

- *Labyrinthina* Wenschenk, 1951 (Type species: *Labyrinthina mirabilis* Wenschenk, 1951). A labyrinthine with multiple apertures, with internal radial partitions and pillars. Late Jurassic (Plate 4.9, figs 4-5).
- *Orbitammina* Berthelin (Type species: *Orbicula elliptica* d'Archiac, 1843). The test is compressed, with a planispiral early stage. The early chambers are narrow and elongate, increasing rapidly in size to produce a reniform test. The aperture is cribrate and radially aligned. The peripheral margin lack the thickness of *Orbitopsella,* while the internal structure is similar as in *Orbitopsella* but with finer pillars. Hottinger (1967) and Septfontaine (1981) believe *O. elliptica* to be the microspheric form, and the senior synonym of *Meyendorffina bathonica s.s.* but no comparable B-form is known to match *M. bathonica* (Plate 4.11, figs 5-6). Also *O. elliptica* is known in late Bajocian, while *M. bathonica* is widely regarded as being Bathonian only. Middle Jurassic (late Bajocian to Bathonian) (Plate 4.12, figs 5-6).

Subfamily *Levantinellinae* Fourcade, Mouti and Teherani, 1997
Characterised by having a single layer of cyclical chambers, undivided by pillars. Jurassic.

- *Levantinella* Fourcade, Mouti and Teherani, 1997 (Type species: *Mangashtia? egyptiensis* Fourcade, Arafa and Sigal, 1984). Peneropliform and compressed axially. A simple proloculus followed by a planispiral evolute stage and a later uniserial stage. The first chambers are simple, and then subdivided by zigzag pillars with perforations. Late Jurassic (Oxfordian to Kimmeridgian) (Plate 4.4, fig. 2).
- *Syriana* Fourcade and Mouti, 1995 (Type species: *Syriana khouryi* Fourcade and Mouti, 1995). This form is characterised by a compressed flaring test with bilateral symmetry, with multiple apertures successively disposed in rows. The numerous chambers are subdivided by many vertical radial subepidermal partitions arranged in a well-developed uniserial stage. Middle Jurassic (late Callovian) (Plate 4.9, fig. 6B).

Superfamily Nezzazatoidea Hamaoui and Saint-Marc, 1970
The test is trochospiral or planispiral, later occasionally uncoiled. The wall is simple nonlamellar, microgranular, and may possess internal plates or simple partitions. The aperture is simple or multiple with tooth plates. Middle Jurassic (Bajocian) to Holocene.

Family Mayncinidae Loeblich and Tappan, 1985

The test is biumbilicate or very compressed, planispirally enrolled, and rarely uncoiling. Chambers increase rapidly in height. The septa are monolamellar. The wall is simple, solid and microgranular. Middle Jurassic to Late Cretaceous (Bajocian to Santonian).

- *Daxia* Cuvillier and Szakall, 1949 (Type species: *Daxia orbignyi* Cuvillier and Szakall, 1949). The test is planispiral, completely involute, lenticular, bilaterally symmetrical with a moderately sharp margin. Chambers are very narrow. There is a single aperture, that is basal, rounded and central. Late Jurassic to Late Cretaceous (Kimmeridgian to Cenomanian) (Plate 5.3, fig. 11; Plate 5.4, fig. 7).
- *Freixialina* Ramalho, 1969 (Type species: *Freixialina planispiralis* Ramalho, 1969). The test is biumbilicate, planispiral, with sutures that are slightly curved. The wall is finely agglutinated. The aperture is an areal slit in the middle of apertural face. Jurassic (Bajocian to Tithonian) (Plate 4.5, fig. 3).

Superfamily Biokovinoidea Gušič, 1977

Members of this superfamily have a free test with a trochospiral or planispiral early stage, that later may be uncoiled. The septa are homogenous and massive. The walls have an imperforate outer layer and a canaliculate inner layer. The aperture is basal to areal, single to multiple. Early Jurassic (Sinemurian) to Late Cretaceous (Maastrichtian).

Family Biokovinidae Gušič, 1977

The test is planispirally coiled, later it may be uncoiled. Endoskeletal pillars may be present. Walls are canaliculate, with alveoles that open both to the exterior or interior. Early Jurassic to Cretaceous.

- *Bosniella* Gušič, 1977 (Type species: *Bosniella oenensis* Gušič, 1977). This genus is characterised by having a well-developed uncoiled later stage with thick, widely spaced and gently curved septa. The aperture is single in the coiled stage becoming cribrate in the uncoiled part. The presence or absence of alveoles in the wall is still debatable. Septfontaine (1988, p. 242) put the genus *Bosniella* in synonymy with *Mesoendothyra* Dain, 1958, as "the presence or absence of keriotheca which is not always visible due to diagenesis, is not a reliable criterion for the distinction between the taxa *Mesoendothyra* and *Bosniella*". The type species of *Mesoendothyra* had been originally described as having alveolar within the wall structure. However, in my specimens the wall structure is solid and I still consider the two taxa as being separate. Early Jurassic (late Sinemurian to early Pliensbachian) (Plate 4.16, figs 1-2).

Family Charentiidae Loeblich and Tappan, 1985

Early stages are planispiral or streptospiral. Walls are finely canaliculated. Apertures are single or multiple. Middle Jurassic to Late Cretaceous (Callovian to Maastrichtian).

- *Karaisella* Kurbatov, 1971 (Type species: *Karaisella uzbekistanica* Kurbatov, 1971). Tests have streptospiral early coiling but later are planispiral. The base of septum against the previous whorl is thickened and chomata-like as in *Charentia* (see Chapter 5). Late Jurassic (Oxfordian) (Plate 4.16, fig. 11).

Family Lituoliporidae Gŭsić and Velić, 1978.
Tests are planispirally coiled but later may be uncoiled. Walls have coarse alveoles. Early Jurassic.

• *Lituolipora* Gŭsić and Velić, 1978 (Type species: *Lituolipora polymorpha* Gŭsić and Velić, 1978). Tests have a later stage that may be irregularly coiled, uncoiled and rectilinear. Walls are microgranular with coarse alveoles. Early Jurassic (Sinemurian to Pliensbachian) (Plate 4.17, fig. 7).

Superfamily Spiroplectamminoidea Cushman, 1927
Tests are planispirally coiled or biserial in early stages, but later biserial. Walls are agglutinated, non-canaliculate. Carboniferous to Holocene.

Family Textulariopsidae Loeblich and Tappan, 1982
Members of this family have a biserial early stage, later they may be loosely biserial or uniserial. Walls are agglutinated. Early Jurassic (Sinemurian) to Late Cretaceous (Maastrichtian).

• *Textulariopsis* Banner and Pereira, 1981 (Type species: *Textulariopsis portsdownensis* Banner and Pereira, 1981). On the basis of a single (type) species, Banner and Pereira (1981, p. 98) defined their new genus as "Wall: agglutinated with calcareous cement, solid, imperforate, lacking canaliculi or pseudopores; proloculus succeeded by a rectilinear series of chambers, all biserially arranged; aperture: anteriomarginal, a simple basal, narrow slit. Differs from other small benthic *Spirorutilus* Hofker in its lack of a planispiral initial stage, from *Textularia* Defrance in its lack of canaliculate, pseudoporous walls, and from *Pseudobolivina* Wiesner by its low aperture and structurally insignificant chitinous endoskeleton". Two new Cretaceous species were assigned to the genus by Loeblich and Tappan (1982), plus three other Cretaceous species from the North America, and *T. areoplecta* from the Early Jurassic (Pliensbachian to Toarcian) of northern Alaska, which had previously been assigned to *Textularia*. BouDagher-Fadel et al. 2001 confirmed that this seemingly largely Cretaceous genus had originated by the Sinemurian. Early Jurassic to Cretaceous (Sinemurian to Maastrichtian) (Plate 4.8, fig. 25).

Superfamily Textularioidea Ehrenberg, 1838
Tests are trochospiral, biserial or triserial in early stages but later may be uniserial or biserial. Walls are agglutinated and canaliculated. Early Jurassic (Sinemurian) to Holocene.

Family Chrysalidinidae Neagu, 1968
Tests are high trochospiral, with quinqueserial, quadriserial, triserial or biserial coiling modes, or with certain consecutive pairs of these. The aperture is central along the axis of coiling. In quadriserial or quinqueserial forms, an umbilicus is present and the aperture is covered with a broad umbilical flap, which may be penetrated by multiple accessory apertures. Internal pillars may develop between successive intraumbilical

flaps. Banner et al. (1991) divided this family into two subfamilies; the mainly Cretaceous Chrysalidininae (see Chapter 5), which evolved as a single lineage developing pillars in the centre of the test, and the mainly Jurassic Paravalvulininae (see below) which survived into the Valanginian and possibly the Hauterivian. This classification is followed here and in the Chapter 5.

Subfamily *Paravalvulininae* Banner, Simmons and Whittaker, 1991
This subfamily was created by Banner et al. (1991) to include all initially quadriserial or quinqueserial forms, becoming quadriserial or quinqueserial in neanic growth (a postnepionic growth stage with the architecture of an adult shell, see Hottinger, 2006), and then quinqueserial, quadriserial or triserial in the adult. Jurassic (Sinemurian to Kimmeridgian).

- *Paravalvulina* Septfontaine, 1988 (Type species: *Paravalvulina complicata* Septfontaine, 1988). Internal pillars between successive umbilical flaps. The early test is quadriserial, later becoming uniserial. Jurassic to?Cretaceous (Bathonian to?Hauterivian) (Plate 4.11, fig. 1).
- *Pseudomarssonella* Redmond, 1965 (Type species: *Pseudomarssonella maxima* Redmond, 1965). Forms with cribrate apertures and an umbilicus that is concave with no internal umbilical pillars. Umbilical apertural flaps of successive whorls are broad and are axially separated by a narrow space. Adult tests are quadriserial or quinqueserial. Jurassic (Bathonian to Callovian) (Plate 4.8, fig. 26; Plate 4.11, figs 2-4).
- *Redmondoides* Banner, Simmons and Whittaker, 1991 (Type species: *Pseudomarssonella media* Redmond, 1965). The test is quadriserial in the adult form. The umbilicus is concave with no internal umbilical pillars. Apertural flaps of successive whorls are well separated. Septa are flattened with narrow umbilicus. Jurassic (Bajocian to Kimmeridgian) (Plate 4.11, figs 5-7).
- *Riyadhella* Redmond, 1965 (Type species: *Riyadhella regularis* Redmond, 1965). Forms with septa and terminal faces that are highly convex. Redmond's species of *Riyadhella* were known to him only as solid specimens extracted from their matrix. Banner et al. (1991) revised the genus and its four assigned species partly on the basis of thin sections of type material, providing amended descriptions of the Redmond (1965) species *R. arabica*, *R. elongata*, *R. inflata* and *R. regularis* and placing his species *R. hemeri*, *R. intermedia* and *R. nana* in synonymy with *R. regularis*. They reassigned the genus to the Family Chrysalidinidae rather than to the Family Prolixoplectidae (as by Loeblich and Tappan 1988). BouDagher-Fadel et al. (2001) recorded a new, more primitive species *R. praeregularis* in the Sinemurian of the western Mediterranean that reveals canaliculi within the test wall and extended the stratigraphic range of the genus down into the Early Jurassic, and its geographic range significantly westwards within Tethys. Jurassic (Sinemurian to Kimmeridgian) (Plate 4.11, figs 10-13).
- *Riyadhoides* Banner, Simmons and Whittaker, 1991 (Type species: *Pseudomarssonella mcclurei* Redmond, 1965). Tests are quadriserial in the adult with flattened septa, or with septa and terminal face concave. Jurassic (late Bajocian – Tithonian) (Plate 4.11, figs 8-9).

Order Miliolida Delage and Hérouard, 1896

The miliolines have tests that are porcelaneous and imperforate, made of high Mg-calcite with fine randomly oriented crystals. They range from the Carboniferous to the Holocene.

Superfamily Milioloidea Ehrenberg, 1839

Tests are coiled commonly with two or more chambers arranged in varying planes about the longitudinal axis, later they may become involute. Advanced forms may have secondary partitions within the chambers. Late Triassic (Norian) to Holocene.

Family Nautiloculinidae Loeblich and Tappan, 1985

Tests are free, lenticular, planispiral, and involute with secondary thickening in the umbilical region. Apertures are equatorial. Middle Jurassic to Late Cretaceous.

- *Nautiloculina* Mohler, 1938 (Type species: *Nautiloculina oolithica* Mohler, 1938). It has been variously placed in previous systematics. It was originally described by Mohler (1938, p. 18) as calcareous and imperforate, compared to porcelaneous *Planispirina.* Loeblich and Tappan (1964b, p. 443) placed the genus into the miliolines. They subsequently (1985, p. 92; 1988, p. 71) reallocated this genus into the lituoloids. I do not agree with this classification as *Nautiloculina* does not possess a microgranular agglutinated wall. Yet, it is not a member of the calcareous rotaliides with the hyaline calcitic perforate wall. The multi-layered imperforate wall of *Nautiloculina* is very difficult to separate from simple fusulinines such as the Carboniferous *Millerella* Thompson, 1942. Nonetheless, the considerable time gap between the Palaeozoic Loeblichiidae and the Jurassic-Cretaceous Nautiloculinidae would make it difficult to explain a relationship between them. Presently, *Nautiloculina* has been placed systematically in the Milioloidea (see Noujaim Clark and BouDagher-Fadel, 2001) because some of the miliolides (such as *Austrotrillina, Quinqueloculina,* etc.) display double layered walls like *Nautiloculina,* although the thin dark layer is internal to the thick transparent layer of the wall of the normal miliolides, but it is external to the wall of *Nautiloculina.* Consequently, the thin dark layer may be basal to the wall of the normal miliolids, while the later part of the wall grow inwards toward the chamber lumen; in *Nautiloculina* the thin dark layer of the wall could equally be basal to the development of the thicker translucent part of the wall which also would have grown inwards towards the chamber lumen. Tests are biumbonate, nautiliform with chambers increasing slowly in height. Late Jurassic (Bajocian) to Early Cretaceous (Aptian) (Plate 4.9, figs 6A, 7-12; see Chapter 5, Fig. 5.23).

4.3 Biostratigraphy and Phylogenetic Evolution

4.3.1 General Biostratigraphy

Following the mass extinction in the marine realm at the end of the Triassic, foraminifera did not make a full recovery until the middle of the Sinemurian. Hettangian foraminifera were rare and simple, and were predominately small Involutinoidea and Pfenderinoidea (see Figs 4.7, 4.8). *Siphovalvulina* and *Textularia* dominated the early Sinemurian (BouDagher-Fadel, 2000; BouDagher-Fadel et al., 2001; Bosence et al., 2009; Scherreiks et al., 2010). However, from the Sinemurian onwards, the Early Jurassic witnessed the steady development of these textulariides from small simple forms to internally complicated forms, which became abundant from the Pliensbachian onwards, yielding a high biostratigraphic resolution for the Tethyan carbonate facies. The Pfenderinoidea and Textularioidea dominated assemblages in the Bathonian and Callovian, but the Lituoloidea/Loftusioidea seem to have taken over during the Late Jurassic (see Fig. 4.9). The abundance of the agglutinated textulariides (see Chart 4.1) in the Jurassic and their short ranges makes them an invaluable biostratigraphical resource (e.g. Noujaim Clark and BouDagher-Fadel, 2004; BouDagher-Fadel and Bosence, 2007).

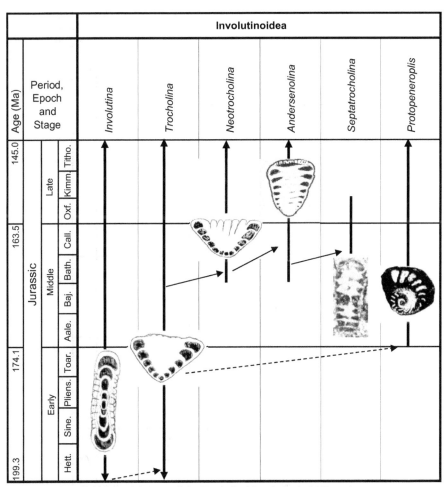

Fig. 4.8. The phylogenetic development of the Involutinoidea through the Jurassic.

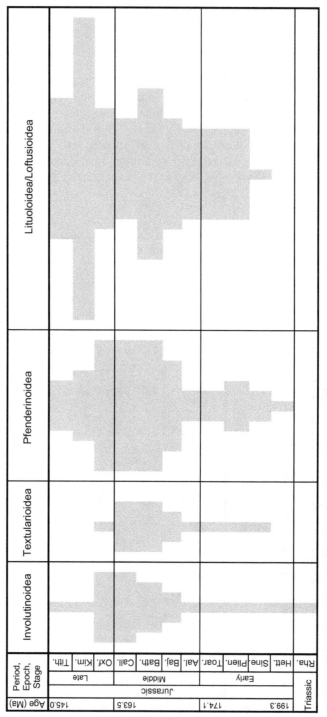

Fig. 4.9. Biostratigraphic ranges and diversity of the main Jurassic superfamilies of Tethys. Details of the genera and their ranges are shown in Chart 4.1 online.

Importantly in this epoch, dasyclad algae became abundant and flourished in a reefal environment. As foraminifera became large, they developed blind chamberlets (alveoles) in their test walls, which housed these symbiotic algae (see Fig. 4.10). From the middle to late Sinemurian, forms with internal pillars, such as *Pseudopfenderina* (Plate 4.6, figs 10-16), and fine alveoles in the walls, as in *Everticyclammina* (Fig. 4.10), began to appear in the Western Mediterranean. Such features developed in later forms, with larger and more consistent alveoles, in the Pliensbachian to the Late Jurassic and Cretaceous, becoming very important components of Mesozoic rocks.

Septfontaine (1981) had proposed a standard Jurassic biozonation on the basis of 56 foraminiferal generic ranges, which took account of work by earlier authors. His scheme began with an *Orbitopsella* range zone approximately equivalent to the Pliensbachian: none of his genera were confidently recorded from the underlying Sinemurian. Later, Septfontaine (1984) recorded '*Siphovalvulina*' from the Sinemurian-Pliensbachian (Domerian) of Morocco and distinguished a biozone of '*Siphovalvulina*' and *Mesoendothyra,* characterized by a fauna of small, primitive lituoloids: the two index genera plus *Everticyclammina praevirguliana,* Plate 4.16, fig.3), *Glomospira* sp. and *Earlandia* sp. These taxa occurred in all six biozones that he distinguished in the Moroccan Early Jurassic (Hettangian to Pliensbachian), but the lower boundary of this 'interval zone' was defined by the first appearance of *Lituosepta recoarensis* (Plate 4.16, figs 7-8), the index fossil for the overlying zone. A similar scheme was applied to the Buffadero Member of the Gibraltar Limestone (BouDagher-Fadel et al., 2001). More recently, Fugagnoli (2004) recorded that the Hettangian-early Sinemurian interval belongs to his Dasyclades Zone, early most late Sinemurian to?*Lituosepta recoarensis* zone, while the lower part of the Sinemurian and early Pliensbachian belongs to the *Orbitopsella* zone and finally the late Pliensbachian belongs to the *Planisepta compressa* zone. On the other hand, the Toarcian is dominated by forms with internal radial partitions (e.g. *Haurania*).

Later, BouDagher-Fadel and Bosence (2007), while systematically studying the Early Jurassic foraminifera, divided the Hettangian to Pliensbachian into four biozones, that correlate with strontium isotope dating. In this book, these biozones are revised and the entire Jurassic biozones are plotted (Fig. 4.11). These Jurassic biozones are:

- the *Siphovalvulina colomi* biozone, corresponding to the early Sinemurian and includes *S. colomi* (Plate 4.1, figs 3-6), *S. gibraltarensis* (Plate 4.1, figs 7-8), *Duotaxis metula* (Plate 4.6, fig. 1), *Riyadhella praeregularis* (Plate 4.11, figs 10-11), *Involutina liassica, Pseudopfenderina butterlini* (Plate 4.6, fig. 10-16);
- the *Everticyclammina praevirguliana* biozone, which corresponds to the mid Sinemurian and coincides with the first appearance of *E. praevirguliana* (Plate 4.16, fig. 3) and includes *Siphovalvulina colomi, S. gibraltarensis, Textulariopsis sinemurensis* (Plate 4.8, Fig. 25), *Riyadhella praeregularis, Duotaxis metula.* Foraminifera of the *Everticyclammina praevirguliana* biozone occur in micritic limestones with associated algae/cyanobacteria (*Cayeuxia,* Plate 4.18, figs 1-2; *Thaumatoporella,* Plate 4.18, figs 3-4 and *Palaeodasycladus mediterraneus,* Plate 4.18, figs 5-7);

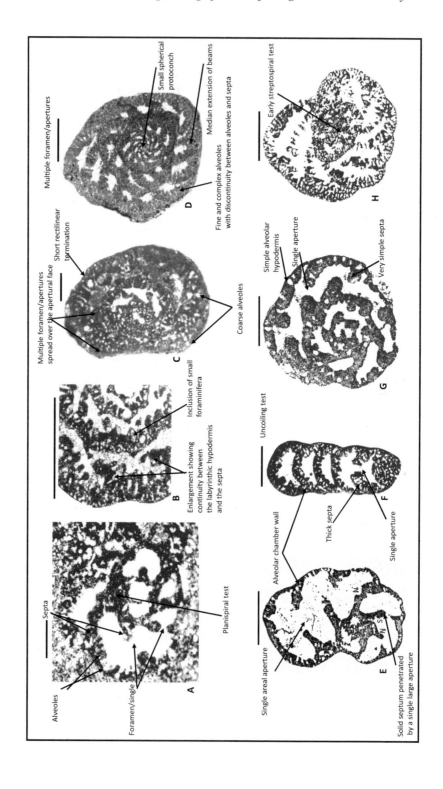

Fig. 4.10. Alveolar exoskeleton and polygonal network. Scale bars = 0.5mm. A) simple alveolar layer in *Everticyclammina virguliana* (Koechlin), Zakum-1, core 9536 ft, ARAB, Kimmeridgian; B) *Pseudocyclammina lituus* (Yokoyama), equatorial section, Hauterivian, Umm Shaif -3, cuttings 6710 ft, THAMAMA IV; C) Enlargement of *P. lituus* (Yokoyama), from the Early Cretaceous of the Persian Gulf, showing continuity between the labyrinthic hypodermis and the septa, and the wholly areal position of the septal apertures. It shows the inclusion of small exotic benthic foraminifera in the labyrinthic part of the hypodermis; D) *Alveosepta jaccardi* (Schrodt), from early Kimmeridgian of the Persian Gulf, Dukhan-51, core 7444 ft, stratotype DARB (Darb 2/3 boundary), equatorial section of a microspheric specimen showing delicately labyrinthic hypodermis and discontinuity between the alveolae of successive septa; E) *Everticyclammina virguliana* (Koechlin), early Kimmeridgian, after Hottinger (1967), showing the alveolar wall, closer spacing of narrow alveolae like these would mimic the canaliculate wall of *Charentia* (see Chap. 5). Note that the lower parts of the septa (S) often points forward (as in *Spinoendothyra*), producing very convex, almost conical septa, in which the aperture may be surrounded by a short, thick neck (N); these features are not typical of the Cretaceous *Everticyclammina greigi* (Henson) (see Chap. 5), the probable descendants of *E. virguliana;* F) *Rectocyclammina* cf. *chouberti* after Ramalho (1971), "Purbeckian", Tithonian, Portugal, much larger and with elaborately structured hypodermis than in the type *R. chouberti;* G) *Pseudocyclammina sphaeroidalis* Hottinger, type specimen from the Kimmeridgian of east Morocco, after Hottinger (1967). It is distinguished from *P. lituus* (above) by having a simpler alveolar hypodermis and spheroidal test; note the very simple septa, often showing only one aperture in any one equatorial cut; H) *Streptocyclammina liasica* Hottinger, syntypes figured from the Early Jurassic of Morocco. Only the streptospirality readily distinguishes the heavily agglutinated forms of this species from *Pseudocyclammina bukowiensis* (Plate 4.10, figs 7-10) and the exceptionally rectilinear forms of this species (see Hottinger, 1967) from *P. vasconica* (Plate 4.10, fig. 9).

- *Lituosepta recoarensis* and *Orbitopsella* spp. biozone, corresponding to the late Sinemurian. It coincides with the first appearance of *L. recoarensis* (Plate 4.16, figs 8-9) and includes *Siphovalvulina* sp., *Haurania deserta* (Plate 4.11, figs 18-20; Plate 4.12, figs 1-4), *Orbitopsella praecursor* (see Figs 4.18, 4.19; Plate 4.17, figs 2-3), *Amijiella amiji* (Plate 4.4, figs 8-9), *Pseudopfenderina sp.*, and *Bosniella oenensis* (Plate 4.16, figs 1-2);
- *Planisepta compressa* biozone which corresponds to the early Pliensbachian and coincides with the first appearance of *P. compressa* (Plate 4.9, figs 1-3). It includes *Pseudocyclammina* sp., *Haurania deserta*, *Amijiella amiji*, *Riyadhella* sp., *Siphovalvulina* sp., *Siphovalvulina colomi*, *Siphovalvulina gibraltarensis*, *Textulariopsis* sp., *Duotaxis metula*. *Everticyclammina* sp., *Pseudocyclammina* sp., *Orbitopsella* sp., *Haurania* sp., *Lituosepta recoarensis*, *Orbitopsella "circumvulvata"*, *Textularia* sp., *Siphovalvulina* sp., small miliolids, *Orbitopsella praecursor*, *Bosniella oenensis*, *Amijiella amiji*, *Haurania deserta*, *Pseudopfenderina* sp., and *Buccicrenata* sp. (see Chart 4.1);
- *Socotraina serpentina* biozone which corresponds to the Toarcian. The beginning of this biozone coincides with the last appearance of *Planisepta*. It includes forms with pillared or partitioned central zone, e.g. *Socotraina serpentina*, *Haurania deserta*, *Amijiella amiji*, and *Cyclorbitopsella tibetica*. The top of this biozone marks the disappearance of *Orbitopsella* (see Fig. 4.11);
- *Gutnicella cayeuxi* biozone which corresponds to the Aalenian. This biozone coincides with the first appearance of *Timidonella* (Plate 4.12, figs 13-16), which replaced *Orbitopsella* and *Gutnicella*. It includes *Gutnicella bizonorum*, *G. minoricensis*, *G. cayeuxi* and *Timidonella sarda*;
- *Kurnubia palastiniensis* biozone, corresponding to the Bajocian. This biozone coincides with the first appearance of *Kurnubia*, *Pfenderina* and *Conicopfenderina*. It includes *Pseudopfenderina butterlini*, *Kurnubia palastiniensis*, *K. jurassica*, *Pfenderina salernitana*, *P. trochoidea*, *Conicopfenderina mesojurassica*, *Gutnicella cayeuxi*, *Timidonella sarda*, *Amijiella slingeri*, *Rectocyclammina ammobaculitiformis*, *Pseudocyclammina maynci* and *Pseudocyclammina bukowiensis*;
- *Ataxella occitanica* biozone, corresponding to the Early Bathonian. It coincides with appearance of *Andersenolina elongata* (Plate 4.1, fig. 4) and includes *Trocholina granosa*, *T. conica*, *Redmondoides lugeoni*, *Kurnubia palastiniensis*, *K. jurassica*, *Pfenderina salernitana*, *P. trochoidea*, *Conicopfenderina mesojurassica*, *Paravalvulina complicata*, *Pseudocyclammina maynci*, *P. bukowiensis* and *Spiraloconulus perconigi*.
- *Alzonorbitopsella arabia* biozone, corresponding to the Late Bathonian. It coincides with the range of *Alzonorbitopsella arabia* (Plate 4.2, figs 5-10) and it includes *Redmondoides medius*, *R. inflatus*, *Pseudomarssonella maxima*, *P. bipartita*, *Pseudocyclammina maynci* and *P. bukowiensis*. The top of this biozone witnesses the disappearance of *Haurania deserta*;
- *Kilianina preblancheti* biozone, corresponding to the Early Callovian. It includes *Trocholina transversaria*, *Andersenolina alpina*, *Kurnubia palastiniensis*, *K. jurassica*, *Pfenderina salernitana*, *P. trochoidea*, *Pseudospirocyclina smouti* and *Kilianina blancheti*;
- *Pseudospirocyclina smouti* biozone, corresponding to the Late Callovian. It includes *Rectocyclammina ammobaculitiformis*, *Pseudocyclammina maynci*, *Pseudocyclammina bukowiensis*, *Kurnubia palastiniensis*, *K. jurassica*, *Pfenderina salernitana*, *P. trochoidea*, *Pseudospirocyclina smouti* and *Kilianina blancheti*;

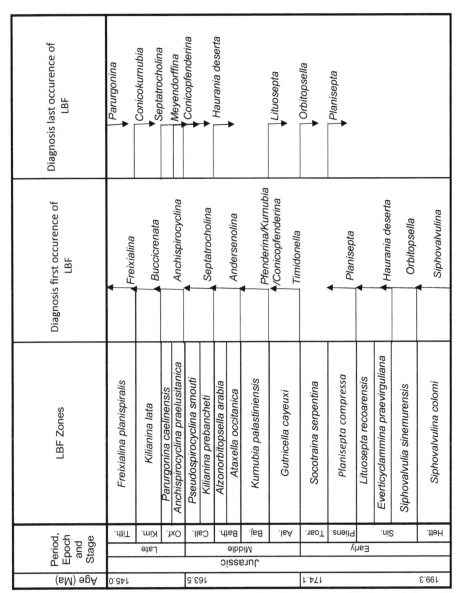

Fig. 4.11. Jurassic larger benthic foraminifera biozones defined in this study with diagnostic first and last occurrences.

- *Anchispirocyclina praelusitanica* biozone, corresponding to the Early Oxfordian. It coincides with the first appearance of *A. praelusitanica* and includes *Septatrocholina banneri, Rectocyclammina ammobaculitiformis, Pseudocyclammina maynci* and *Pseudocyclammina bukowiensis*;
- *Parurgonina caelinensis* biozone, corresponding to the Late Oxfordian. It includes *Choffatella tingitana, Torinosuella peneropliiformis, Anchispirocyclina praelusitanica* and *Alveosepta jaccardi*. The top of this biozone witnesses the disappearance of *Septatrocholina*;
- *Kilianina lata* biozone, corresponding to the Kimmeridgian. It includes *Kilianina lata, Neokilianina rahonensis, Freixialina atlasica, Buccicrenata primitiva, Pseudocyclammina ukrainica, P. sphaeroidalis, Rectocyclammina chouberti, Redmondellina powersi* and *Pseudospirocyclina mauretanica*. The top of this biozone marks the disappearance of *Conicokurnubia*;
- *Freixialina planispiralis* biozone, corresponding to the Tithonian. It includes *Anchispirocyclina neumanni, Everticyclammina virguliana, E. praekelleri, Pseudocyclammina lituus* and *Anchispirocyclina lusitanica*. The early part of this biozone witnesses the disappearance of *Kurnubia* (Plate 4.8, figs 3-21), while the top is marked by the disappearance of *Parurgonina, Pseudospirocyclina, Kastamonina* and *Labyrinthina*.

The Middle and Late Jurassic have been intensively studied from the Middle East (Henson, 1948), Eastern Mediterranean (Noujaim Clark and BouDagher-Fadel, 2002, BouDagher-Fadel and Lord, 2002; Noujaim and BouDagher-Fadel, 2004) and the Western Mediterranean (Septfontaine, 1988, BouDagher-Fadel and Bosence, 2007; Bosence et al., 2009), and phylogenetic evolutions have been traced for different groups, such as the Textularioidea (Banner et al., 1991) and the Lituoloidea (Septfontaine, 1988). The Middle Jurassic is rich with foraminifera filled with internal pillars (e.g. *Haurania*), while for assemblages in the Late Jurassic, foraminifera with narrow internal alveoles dominate (see Fig. 4.12). Chart 4.1 summarises the ranges of most important species in the Jurassic, and Chart 4.2 shows the range of the major textulariine superfamilies.

Below are discussed the evolutionary lineages and revised phylogenetic evolutions of the most important superfamilies in the Jurassic, namely the Pfenderinoidea, the Lituoloidea, the Textularioidea, and the Involutinoidea.

4.3.2 The Pfenderinoidea of the Jurassic

In the Hettangian, small simple textulariides developed a twisted siphonal canal connecting the successive apertures (Figs, 4.13, 4.14, 4.15). These forms evolved gradually from the Sinemurian to the Bathonian into *Pseudopfenderina*, a form with high, loose spires with separate pillars filling the narrow central zone (see Fig. 4.7), which in turn evolved forms, such as *Pfenderella*, with single subcameral tunnels, short chambers with secondarily deposited material but without secondary septa (Plate 4.7, figs 21-24). These forms are probably the ancestors of *Pfenderina*, which developed a solid central core in the Bajocian. The pillars in the centre of *Pfenderina* fuse and coalesce in a strong central zone (Fig. 4.13A, Plate 4.7, figs 6-7). *Pfenderina* persists into the Early Cretaceous. In the Bathonian, *Pfenderina* evolved into *Palaeopfenderina*, with a

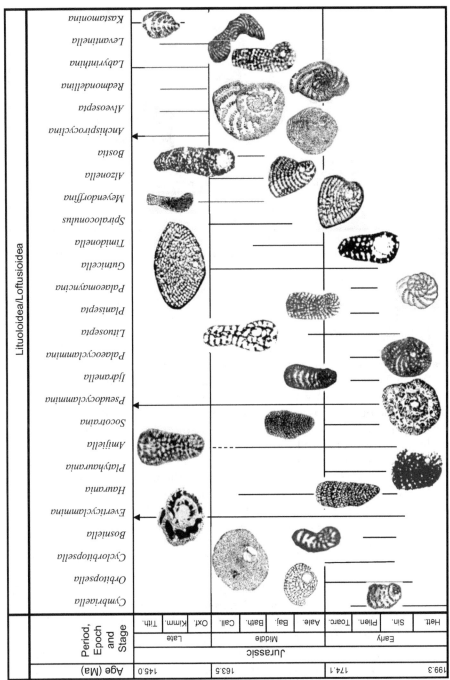

Fig. 4.12. The biostratigraphic ranges of the Lituoloidea in the Jurassic.

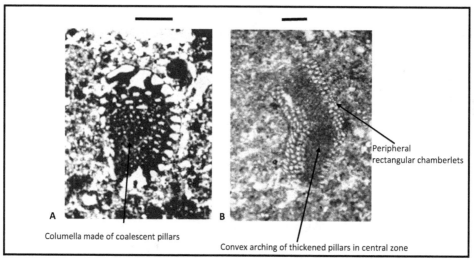

Fig. 4.13. Difference in morphological structure between: A) *Pfenderina trochoidea* Smout and Sudgen, Umm Shaif-4, core 9835 ft; B) *Meyendorffina bathonica* Aurouze and Bizon, Umm Shaif, core 9370 ft. Scale bars = 0.5mm.

superficial position of the tunnel, with a groove at the surface of the columella, and by the filling of the inter-pillar spaces with calcitic material. In the Bathonian, *Sanderella* (Plate 4.7, fig. 17), the ancestral form of the terminally uncoiled forms, evolved from *Pfenderina* by developing a flaring, flattened, peneropliform test. In *Sanderella*, the spiral canal may bifurcate in the rectilinear portion forming multiple subcameral tunnels. In the late Bathonian, forms with a subcylindrical, lituoliform test evolved from *Satorina* (Plate 4.7, figs 15-16), with the tunnel partly or wholly surrounding the rectilinear columella, and other forms appear at the same time, with conical, coskinoliniform tests, but with pillars becoming separated in a broad central zone and the tunnel of the rectilinear part being reduced and discontinuous. In the late Bathonian, *Chablaisia* (Plate 4.7, figs 12-13) may have evolved directly from *Pfenderina* by developing a spiral canal and calcitic fillings in the chambers. In the Bathonian, *Pseudoeggerella* (Plate 4.7, fig. 14) evolved from *Pfenderella* (Plate 4.7, figs 21-24) by developing a narrow columella and calcitic protuberances in the chambers. In the Oxfordian, *Steinekella* (Plate 4.7, figs 18-20) replaces *Pfenderella* by developing a massive, central, continuous columella and multiple subcameral tunnels.

Pfenderella (Plate 4.7, figs 21-24) probably is also ancestral to *Kurnubia* (Plate 4.8, figs 3-8), which developed a peripheral zone divided by radial partitions in the Bajocian (Fig. 4.14). The high, loose, slender trochospiral test with a columella made of thickened innermost septal buttons of *Kurnubia* developed transverse partitions to form *Praekurnubia* in the Bathonian to Callovian. This latter in turn developed a broad central zone of a rectilinear cone, with scattered pillars and thickened inner ends of radial partitions, to form *Conicokurnubia* (Plate 4.8, figs 1-2) in the Oxfordian to Kimmeridgian. *Pfenderina*, with secondary infillings and internal partitions, appeared in the Late Jurassic (Oxfordian) and ranged into the Early Cretaceous (Valanginian) of southern Europe, but in the Middle East, survived until the Late Cretaceous (Chart 4.2).

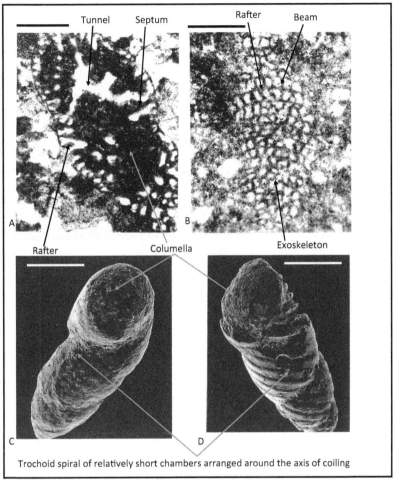

Fig. 4.14. A) *Kurnubia wellingsi* (Henson), Umm Shaif, core 8886 ft; B) *Kurnubia jurassica* (Henson), Umm Shaif 3, cutting 10.072feet; C-D) *Pfenderina salernitana* Sartonia and Crescenti, Bathonian, Uwainat, Qatar, NHM P43720-21. Scale bars: 0.4mm.

The Valvulinidae show the first known example of a crosswise-oblique stolons system (the marginoporiform structure of Hottinger and Caus, 1982) (Fig. 4.16). They evolved a trochospiral test, with a valvular tooth plate, *Valvulina*. This form gives rise to conical tests, where central septa are thickened in a broad central zone (*Kilianina*, Plate 4.1, fig.9B) in the Bathonian to Callovian. In the Oxfordian to Tithonian they are replaced with highly conical forms, possessing septa breaking in the umbilical region to form sub-conical pillars (*Parurgonina*, Plate 4.10, figs 17-18), which in turn gives rise in the early Kimmeridgian to forms with a high number of chambers per whorl and a strongly pillared centre (*Neokilianina*, Plate 4.10, Figs 13, 15-16), while compressed tests with cyclical chambers with numerous pillars evolved in the Oxfordian to Kimmeridgian (*Levantinella,* Plate 4.10, fig. 19).

Principle Characteristics		Generic characteristics and forms	Species
No pillars: a low trochospire. Wall possibly alveolar (canaliculate).		*Chablaisia* Septfontaine	*C. chablaisensis* Septfontaine (Bath. - Call. -? Oxf.)
High, loose "arenobuliminid" spires with broad chamber spaces	Pillars separated, discrete, in narrow central zone	*Pseudopfenderina* Hottinger	Loose coil, strong pillars: *P. butterlini* (Brun) (Sin. - Bath.)
High or low trochospires with reduced chamber spaces	Pillars fuse, coalescent in strong central zone	*Pfenderina* Henson	Low trochospire: *P. trochoidea* Smout and Sudgen (Baj. - Cal.)
			High trochospire, shallow sutures, smooth outlines: *P. salernitana* Sartoni and Crescenti (Baj. - Oxf.)
			Deep sutures, lobulate outline: *P. neocomensis* (Ber. Val.)
Test flaring, flattened, perneropliform	Pillars coalescent into massive columella	*Sanderella* Redmond — Tunnel may bifurcate in rectilinear portion	*S. laynei* (Cal.)
Test subcylindrical, lituoliform		*Satorina* Fourcade and Chorowicz — Tunnel partly or wholly surrounds rectilinear columella	*S. apuliensis* Fourcade and Chorowicz Bath? - Cal.
Test conical, coskinoliniform	Pillars discrete separated, in a broad central zone	*Conicopfenderina* Septfontaine, 1988 — Tunnel of rectilinear part is reduced and discontinuous	*C. mesojurassica* (Maync) (Baj. - Cal.)
High loose, slender trochospires	Columella made of thickened innermost septal ends "buttons" without pillars	*Praekurnubia* Redmond, 1964 — radial peripheral partitions only, no chamberlets in series	*P. crusei* Redmond Bath. - E. Oxf.
		Kurnubia Henson — with lateral chamberlets alternating in several tiers	6/8/ ch/whorl: *K. jurassica* (when so loosely coiled last chambers are in uniseries, called *K. palastiniensis*) *Baj.-Kim - ?Tith* ca. 20 ch/whorl: *K. wellingsi* (Baj.-Kim.-?Tith.)
Trochospire followed by conical, "coskinoline" rectilinear low chambers	Broad central zone of rectilinear cone with scattered pillars and thickened inner ends of radial partitions	*Conicokurnubia* Septfontaine — radial partitions produce one tier of peripheral chamberlets in each rectilinear chamber	*Conicokurnubia orbitoliniformis* Septfontaine, 1988 Oxf. - Kim.

Vertical classification labels (left margin):

- Trochospiral throughout / Terminally uncoiling rectilinear
- With pillars in columella between "septal buttons" of successive whorls
- **Paleopfenderininae/Pfenderininae** — No peripheral partitions or chamberlets
- **Kurnubiinae** — Peripheral zone divided by radial partitions

Fig. 4.15. Morphological characteristics of the key species of the pfenderinids.

Principle Characteristics		Generic Characteristics	Specific Characteristics	
Hummocks and arches convexly curved; no rectangular peripheral chamberlets	Initial high trochospire of 2 to 4 whorls	Walls of hummocks not thickened or coalescent: no massive central zone	Parurgonina Cuvillier et al., 1968	
	Initial spire reduced to one whorl only	Wall of hummocks are thickened in a broad central zone	Kilianina Pfender, 1933	Urgonina (Parurgonina) caelinensis Cuvillier et al., 1968 late Oxfordian to Tithonian
				Test conical; central zone arches not thickened into an almost solid mass — K. rahonensis Foury and Cincent (Early Kim.)
				Test flaring, concavo=convex; central zone still with open passage ways; chambers highly arched — K. lata Oberhauser (Kim.)
				Test conical, central zone arches thickened and coalescent into an almost solid mass — K. blancheti Pfender (Bath-Early Oxf.)

Fig. 4.16.   Morphological characteristics of key species of the valvulinids.

4.3.3 The Lituoloidea of the Jurassic

Jurassic lituoloids (see Figs 4.12, 4.17) evolved rapidly, exhibiting a marked dimorphism between microspheric and megalospheric generations. Their rapid evolution, combined with their short ranges, gives the group a very useful biostratigraphic role in the Jurassic. In the Hettangian, the ancestor form of the lituoloids evolved from a simple form, *Lituola,* with early planispirally enrolled whorls, which later became uncoiled and rectilinear with a solid wall. *Lituola* evolved into *Cymbriaella* at the beginning of the Jurassic by developing a coarse irregular subepidermal polygonal network. In the late Sinemurian, *Cymbriaella* evolved into a form with radial partitions, *Haurania*. This form evolved into *Socotraina* in the Pliensbachian by filling the uniserial chambers with vertical subradial partitions. *Haurania* in turn gives rise to *Platyhaurania* in the latest Sinemurian, which possesses cylindrical chambers in the uncoiled part. *Socotraina* gives rise, in the Middle Jurassic, to forms with open flabelliform chambers filled centrally with pillars, with subepidermal nets of alternating horizontal and vertical pillars in *Timidonella*, and a central zone with irregular pillars of different sizes in *Gutnicella. Meyendorffina* replaced *Timidonella* in the Bathonian by losing the horizontal pillars.

Early forms, such as *Haurania, Amijiella* had a comparatively coarse structure with no clear differentiation of an epiderm, just a thin outer wall covering the polygonal network (Fig. 4.20). They also have radial partitions and pillars in the central zone with no orderly differentiation of beams and rafters. This group evolved into forms such as *Orbitopsella,* which developed an alveolar microstructure in the Sinemurian (Fig. 4.18). Large, complex, internally complicated agglutinating benthic foraminifera with pillars and/or intramural alveoles, such as *Orbitopsella* and *Cyclorbitopsella,* did not appear until latest Sinemurian to Pliensbachian times (Fig. 4.19). *Orbitopsella* have a flaring second stage, that becomes a well-developed uncoiled part in *Lituosepta,* and completely annular in *Cyclorbitopsella. Lituosepta* became planispiral and laterally compressed in *Palaeomayncina,* and developed vertical partitions and pillars in *Planisepta.* In the Late Jurassic, *Labyrinthina* developed a more important spiral stage than that in *Lituosepta.*

Amijiella evolved in parallel to *Haurania* in the late Sinemurian by having strong radial partitions, but unlike *Haurania* it has no pillars in the central zone (Fig. 4.20).

Other planispiral forms became uncoiled and developed pillars superficially united by a coarse network (*Ijdranella*), or in the central zone with thin septa and uncoiled test (*Spiraloconulus*), or with a central zone filled by complex reticulum of densely spaced pillars (e.g. *Pseudospirocyclina, Anchispirocyclina*). There appear to be a grade through *Pseudospirocyclina smouti* (Plate 4.5, figs 9-10), and/or *P. mauretanica* (Plate 4.5, figs 11-12), and/or *P. maynci* (Plate 4.5, figs 13, 15), into *Anchispirocyclina lusitanica* (Plate 4.15, figs 6-9) in Early Kimmeridgian. Other uncoiled forms are found without pillars, but develop a subepidermal reticulate mesh, which continues in the septa (*Alzonella*), or have irregular transverse radial partitions (*Bostia*). *Kastamonina* replaces *Amijiella* in the Kimmeridgian to Tithonian by developing a much reduced early coil and a more complex internal structure.

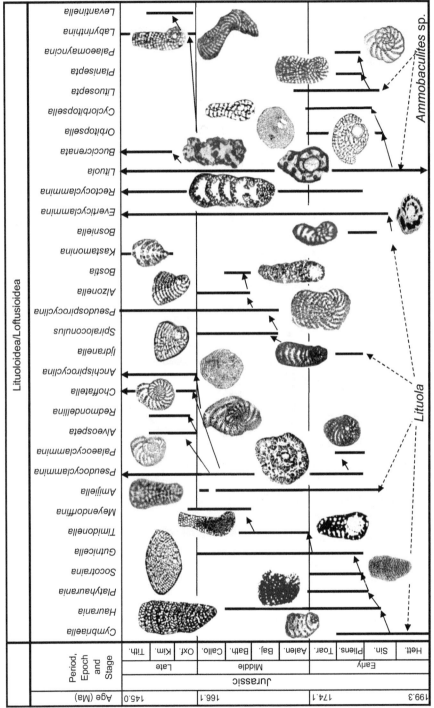

Fig. 4.17. The phylogenetic development of the Lituoloidea through the Jurassic.

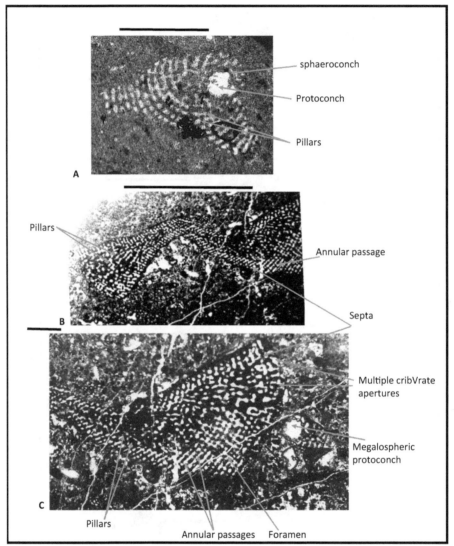

Fig. 4.18. Scale bars = 0.5mm. The structure of *Orbitopsella*, having a discoidal test formed by a sim-
ple exoskeleton and a pillared endoskeleton: A) Oblique section of a megalospheric *Orbitopsella primaeva*
(Henson) (= *Coskinolinopsis primaevus* Henson, 1948), type species figured by Henson (1948) from the Early
Jurassic of the Musandan limestone, Oman. B) Oblique section of a microspheric *Orbitopsella praecursor*
(Gümbel), Milhala, Oman; C) An enlargement of B showing (a) the alternating pattern in the disposition of
the foramen, which are the openings that allow communication between the consecutive chambers, providing
passages for functional endoplasm, and (b) oblique centered section of a megalospheric specimen showing
the spherical protoconch.

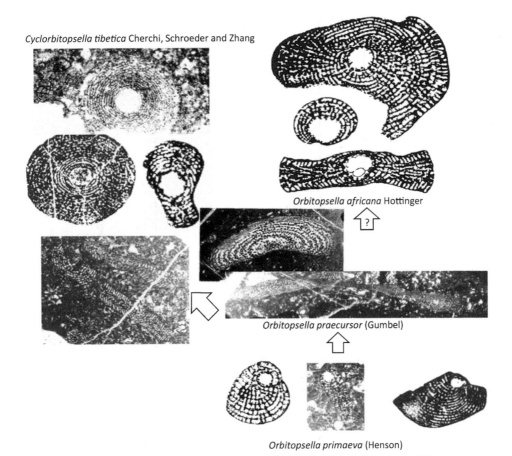

Cyclorbitopsella tibetica Cherchi, Schroeder and Zhang

Orbitopsella africana Hottinger

Orbitopsella praecursor (Gumbel)

Orbitopsella primaeva (Henson)

Fig. 4.19. Example of evolutionary lineages of some Orbitopsellinae species (Hottinger, 1967).

Microspheric forms:
   Number of spiral/annular chambers: Orbitopsella primaeva = 35; O. praecursor = +/- 12; O. africana = +/- 12; Cyclorbitopsella tibetica = 10-15 (Initial peneropline spire) + 10-12 (neanic stage of the test) + 35-45 (annular chambers). Diameter of the equatorial section of the test (mm): O. primaeva = (+/-) 5; O. praecursor: 8-10; O. africana = +/-10; C. tibetica = 6-8.

Megalospheric forms:
   Diameter of protoconch (mm): O. primaeva = 0.24-0.32; O. praecursor = 0.3-0.45; O. africana = 0.5-0.6; C. tibetica = 0.32-0.5.
   Maximum observed number of spiral/annular chambers: O. primaeva = 31; O. praecursor = 8-9; O. africana = 3-5; C. tibetica = 20-25.
   Diameter of the equatorial section of the test (mm): O. primaeva = 1.2-2.4; O. praecursor = 2-4; O. africana = 2.4-3.6; C. tibetica = 1.2-2.

4.3.4 The Loftusioidea of the Jurassic

The Jurassic planispiral forms with no pillars in the central zone, similar to *Lituola*, but with an alveolar wall, first appeared in the Sinemurian (*Everticyclammina prae-virguliana*, Plate 4.16, fig. 3). This was immediately followed in the Pliensbachian by *Pseudocyclammina,* forms with areal cribrate apertures spread over the apertural face, a labyrinthic hypodermis, and a reticulate subepidermal skeleton. The early primitive

Princple Characteristics	Generic Characteristics	Specific Characteristics	
With fine pillars in central zone	Radial partitions delicate, often bifurcating vertically to form a partial tier of peripheral chamberlets	*Haurania* Henson	*H. deserta* Henson
No pillars in central zone	Radial partitions strong, tending to become thickened in transverse section towards central zone, only rarely bifurcate vertically to form few, scattered chamberlets	*Amijiella* Loeblich and Tappan (late Sin. to Bath. or ?Cal.)	Chambers high, septa very arched, initial coil very small → *A. amiji* (Henson) Late Sin. – Bath.
			Chambers low, septa thick and weakly arched; initial coil larger → *A.* sp. = "*Iraniaca slingeri*" (Gollesstaneh) (Baj.-Bath-?Cal.)
Vertical pillars that project a short distance inward from the outer wall	Vertical partitions thickened in central zone to form a solid or vesicular central mass, arching becoming rectangular peripherally	*Meyendorffina* Arouze and Bizon, 1958 (Bath)	Septa arched convexly in central zone, but quadrately peripherally, producing rectangular peripheral chamberlets → *Myendorffina* n. sp. aff. *bathonica* Cal. /Oxf. / *Meyendorffina bathonica* Arouze and Bizon (Bath.)

Fig. 4.20. Morphological characteristics of key species of the hauraniids.

Pseudocyclammina has coarse simple alveoles, but soon several independent evolutionary modifications of the test occur, such as an increased complexity of the hypodermis which reached acme in the Santonian with *Martiguesia* (see Chapter 5), and to increased test compression and coiling rate (e.g. *Torinosuella,* Plate 4.15, fig. 5). Other lineages deriving from *Pseudocyclammina* in the Oxfordian to Kimmeridgian are forms with walls with fine and complex alveoles and septa with many apertures (*Alveosepta*), and with pillar-like hypodermal extensions (*Redmondellina*). A very few forms from the Jurassic survived into the Early Cretaceous, (e.g. *Everticyclammina, Pseudocyclammina*).

Rectocyclammina in the Kimmeridgian has the apertural characters of *Everticyclammina* and probably grades into it; *R. chouberti* (Plate 4.12, fig. 18) is very close to *E.* gr. *virguliana* (Plate 4.16, fig. 6), however, its internally thickened aperture is close to that of *Feurtillia*. The latter evolved, in latest Jurassic possibly from *Ammobaculites* by the development of an alveolar hypodermis and thick septa, but with the retention of the simple aperture with the strong apertural neck (see Chapter 5).

4.3.5 The Biokovinoidea of the Jurassic

Quite independently a group of foraminifera with canaliculate walls appear in the Sinemurian, the uncoiled Early Jurassic biokovinids and lituoliporids and the coiled planispiral to streptospiral charentiids. The charentiids that appear later in the Callovian have planispiral to streptospiral walls and are finely canaliculate.

4.3.6 The Nezzazatoidea of the Jurassic

In the Kimmeridgian, the biumbilicate nezzazatoids made their first appearance. They are distinguished by their simple nonlamellar, microgranular walls and by the simple internal partitions and their apertural tooth plates. Their oldest representative, *Freixialina* is essentially a Jurassic form, ranging from the Bajocian to the Tithonian, while *Daxia* that appear much later in the Kimmeridgian has a completely involute test and more curved sutures. It survived the Jurassic-Cretaceous boundary, only to die out in the Late Cretaceous.

4.3.7 The Spiroplectamminoidea of the Jurassic

The representative of the Jurassic spiroplectamminoids, *Textulariopsis* first appeared in the Sinemurian (BouDagher-Fadel et al, 2001). This essentially Cretaceous genus lacks the early planispiral coil and canaliculi.

4.3.8. The Textularioidea of the Jurassic

Within the Textularioidea, the chrysalidinoids evolved in the Jurassic from simple textulariide forms with triserial, quadriserial or quinqeserial tests (with simple interiors) into forms with convex septa and canaliculi in the walls (*Riyadhella*) (see Fig. 4.21).

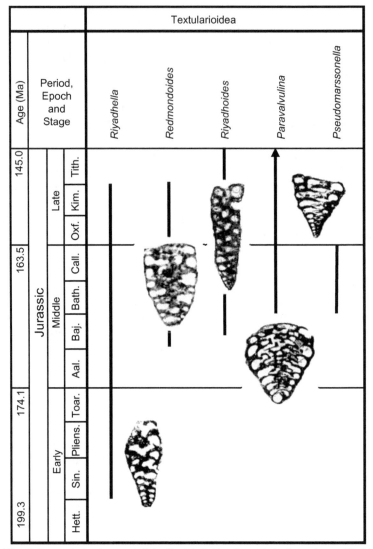

Fig. 4.21. The phylogenetic development of the Textularioidea through the Jurassic, modified from Banner *et al.* (1991).

These forms developed umbilical apertural flaps with a narrow umbilicus in the Bajocian to Kimmeridgian (*Redmondoides*), and flattened septa in the late Bajocian to Tithonian (*Riyadhoides*). Internal pillars between successive umbilical flaps appeared in the Bathonian (*Paravalvulina*), while in the Bathonian to Kimmeridgian forms with no internal pillars, but with apertural flaps separated only by a narrow space, thrived (*Pseudomarssonella*).

4.3.9. The Involutinoidea of the Jurassic

After their first appearance in the Late Triassic, *Trocholina* (Figs. 4.3, 4.6) and *Involutina* (Figs 4.4, 4.5) continue to thrive in the Jurassic. Forms, with rapidly enlarging and loosely coiled planispiral with many chambers (*Protopeneroplis*), first appeared in the Middle Jurassic, at the beginning of the Aalenian. In the Bathonian, *Trocholina* gave rise to forms with a thick outer wall and a deeply fissured umbilicus, *Neotrocholina*, and forms with a perforated plate covering the umbilicus. In the Callovian, *Septatrocholina* developed rudimentary septa and survived into the Oxfordian. Most of the involutinids survived the Jurassic-Cretaceous boundary, but this order completely disappeared at the end of the Cenomanian.

4.3.10. The Milioloidea of the Jurassic

The large miliolids are rare in the Jurassic, however, in the Late Bajocian, a lenticular, planispiral, and involute form, *Nautiloculina* with secondary thickening in the umbilical region made its first appearance. This form has been variously placed in previous systematics, but it is placed here because it displays double layered walls like the miliolides (see explanation above). *Nautiloculina* survives the Jurassic - Cretaceous boundary and is commonly found in the backreef/reefal environments of the Cretaceous.

4.4 Palaeoecology of the Jurassic Larger Foraminifera

The Jurassic period saw warm tropical greenhouse conditions world-wide. The sea level gradually rose (O'Dogherty et al., 2000) and the shallow warm waters of Tethys and the Proto-Atlantic flooded large portions of the continents and spread across Europe. The level of oxygen in the atmosphere was recovering gradually through the Jurassic (see Fig. 3.9). The Jurassic sedimentary sequences around the Mediterranean were dominated by warm-water, shallow-marine carbonates that are of crucial importance both as a record of climatic/oceanic conditions but also as hydrocarbon reservoirs. These deposits are dominantly biogenic in origin, consisting primarily of larger benthic foraminifera and algae, with hermatypic corals. Coral reefs were common in the Jurassic, just as they are today.

In the Hettangian, the diversity of Tethyan foraminifera was poor, and was made up of small agglutinated forms such as *Textularia, Siphovalvulina, Duotaxis* (Plate 4.6, fig. 1), and small microgranular and porcelaneous forms such as *Glomospira*. These small foraminifera were widespread on the platform and have been considered as very tolerant (Septfontaine, 1984; Fugagnoli, 2004). They occur in marginal environments (oligohaline, with terrestrial influx) and in deeper marine environments that reflect elevated levels of organic carbon influx (Rettori, 1995), and with high rates of micritic production (Fugagnoli, 2004). The presence of these small foraminifera, with the complete absence of larger forms, points to a stressed environment or an ecosystem that suffered severe environmental fluctuations.

In the early Sinemurian, larger benthic foraminifera were mostly textulariides. The biota as a whole is characteristic of inner carbonate platform environments that were widespread along the rifted western margins of the Early Jurassic Tethys. They are more primitive than species well-known from the later Early Jurassic (Pliensbachian). These Sinemurian assemblages still included distinctive smaller foraminifera, such as *Siphovalvulina* (with depressed chambers). Microflora are present as the probable cyano-bacterium *Cayeuxia piae* (Plate 4.18, fig 1-2), the dasyclad green alga *Palaeodasycladus mediterraneus* (Plate 18, figs 5-7), and the disputed alga *Thaumatoporella parvovesicu-lifera* (Plate 4.18, figs 3-4, 9). *Thaumatoporella* were in general widespread in Early Jurassic platform carbonate of Tethys. *Palaeodasycladus* was abundant and is well-preserved, consistent with deposition in shallow marine, inner platform conditions. These small foraminifera and dasycladacean algae are all found within limestones that show a range of shallow carbonate platform lithologies, largely packstones and grainstones that were subjected to periodic emergence, calcrete formation, and ero-sion along the margin of Tethys (BouDagher-Fadel et al., 2001; BouDagher-Fadel and Bosence, 2007).

The late Sinemurian is characterized by the presence of larger benthic foramin-ifera, with large, test-wall surface area, as in *Orbitopsella*, with many small cham-berlets, which may have been used for hosting endosymbionts (Hottinger, 1982). Such larger foraminifera are highly adapted to mesotrophic and oligotrophic (nutrient-deficient) conditions (Fugagnoli, 2004). The *Orbitopsella* assemblages are present in peloidal wackestone/packstone deposits of the inner platform, together with *Lituosepta*, *Pseudopfenderina*, *Everticyclammina* and *Haurania* in Morocco and Spain (BouDagher-Fadel and Bosence, 2007). Such an assemblage is not very tolerant of environmental change (Fugagnoli, 2004), and is encountered in shal-low water carbonate facies all along the southern Tethyan margin (Septfontaine, 1984). These complex and highly diverse faunas point to the establishment of stable ecosystems.

Orbitopsella is found throughout the Pliensbachian, together with the appearance of new forms with a so-called subepidermal network, which is described by Hottinger (1996) as a shell architecture adapted to avoid photoinhibition in the lowermost photic zone. Forms with coarsely alveolar tests, such as *Pseudocyclammina*, made their first appearance. These large hypodermal alveoles may well have had a function of har-bouring photosymbionts, which would have thrived in the shallow palaeoenvironment populated by these foraminifera (Banner and Whittaker, 1991). These forms such as *Haurania, Amijiella* (which appeared in the late Sinemurian), *Socotraina* and *Lituosepta* (which first appeared in the Pliensbachian) indicate established shallow warm waters along Tethys in that period of the Early Jurassic.

The early Toarcian transgression is marked by a brief period of global warming (Jenkyns, 2003) and the occurrence of organic carbon rich shales in large parts of western Europe and other parts of the world. There is a positive carbon isotope excur-sion of pelagic limestones in several of the Tethyan sections. The widespread occur-rence of the early Toarcian shales is explained by an Oceanic Anoxic Event (OAE) (Jenkyns, 1988; Jenkyns and Clayton, 1997). Despite these highly unfavourable living conditions for benthic foraminifera, the number of forms going extinct is low, with new forms appearing at the Pliensbachian-Toarcian boundary possessing the so-called subepidermal network and alveolar walls. Extinctions may have occurred regionally or locally, however, where relative sea level low stands resulted in enclosed stagnated basins with adverse environments.

The Aalenian saw the gradual recovery for the shallow carbonate environment. However, it was not before the Bajocian to Bathonian that an explosion of these large complicated forms became prominent in the shallow carbonate facies of Tethys, extending from the shallow carbonates of Japan to those of the Middle East, Europe, and Tanzania.

The "cold snap" at the Callovian-Oxfordian (Jenkyns, 2003) might have triggered the extinction of many elongated, internally complicated forms such as the pfenderinoids. Shallow water forms with large intramural alveoles persisted into the Tithonian and were in association with green algae (e.g. *Clypeina jurassica,* Plate 4.18, fig. 11) and stromatoporoids (e.g. *Cladocoropsis mirabilis,* Plate 4.18, fig. 12). Forms with narrow alveoles and a regularly labyrinthic hypodermis (e.g. *Alveosepta, Redmondellina, Choffatella*) occurred from Portugal through North Africa and southern Europe to the Middle East, ranging from the late Oxfordian to Kimmeridgian and occupied a deeper water palaeoenvironment than the contemporaneous forms with larger alveoles and irregularly labyrinthic hypodermis (e.g. *Everticyclammina*). Their appearance seems to have extended the distribution of these foraminifera further into outer neritic waters, as they appear to have inhabited deeper waters (outer neritic) than *Pseudocyclammina* (with large hypodermal alveoles), or to have tolerated water richer in argillaceous suspensions. According to Banner and Whittaker (1991), they seem to have thrived under reduced illumination, in which conditions codiacean and dasyclad algae were rare or absent. It is possible that the narrow hypodermal alveoles allowed ionic exchange between internal cytoplasm and surrounding seawater, through the extremely thin hypodermis. Hughes (2004) in analyzing the significance of alveoles in *Pseudocyclammina*, and in the light of its occurrence with deep water foraminifera in the Hanifa Formation, Saudi Arabia, argued that the interpreted function of broader alveoles needs further consideration. While agreeing with Banner and Whittaker (1991) that the presence of alveoles may have enabled the organism to construct a test of the required size in a muddy environment, it is also possible low oxygen availability may have been their main rationale. The alveoles gave the organism greater protection from anoxic, hostile sulphide-enriched bottom waters. Low oxygen availability has also been discussed by Preece et al. (2000), who considered that the complex wall structure and presence of alveoles are a means to increase the surface area to volume ratio for gaseous exchange under conditions of low oxygen availability. On the other hand, the increased internal surface area provided by the alveoles may have increased the efficiency of the symbiotic algae, by sheltering them from abnormal water chemistry within the photic zone. *Pseudospirocyclina* and *Anchispirocyclina* are not recorded from the Kimmeridgian (at the *Alveosepta jaccardi* horizon) of the eastern Mid East Gulf; probably both these genera were restricted to shallower, more inner shelf environmental than *Alveosepta jaccardi*.

By the end of the Kimmeridgian crisis the short lived forms with narrow alveoles had disappeared (except for *Choffatella,* Fig. 4.22), and only robust forms characteristic of shallow clear waters survived the Tithonian and crossed over to the Cretaceous.

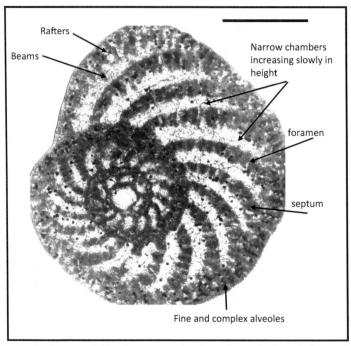

Fig. 4.22. Choffatella decipiens Schlumberger, megalospheric form, Aptian of the Persian Gulf. Scale bar = 0.3mm.

4.5 Palaeogeographic Distribution of the Jurassic Larger Foraminifera

During the Jurassic, Pangea continued to disintegrate and the extent of the oceans was far more widespread than in the Triassic. The supercontinent fragments began to drift in different directions forming rift valleys, and one of these opened to form the southern part of the North Atlantic Ocean (Fig. 4.23). Polar ice caps were still lacking throughout the Jurassic period (Hallam, 1995), and larger foraminiferal distribution indicates that warm conditions extended to much higher latitudes than today.

The Jurassic, which according to Hallam (1978) marks the end of the ancient stages of Earth evolution, was the period where newly evolved characters in the foraminiferal communities became established, thrived and went on to dominate the Cretaceous shallow marine environment. The development of this new biota in the Jurassic occurred against the background of significant events in the Earth's history: the termination of the early Cimmerian orogeny, the opening of the North Atlantic, the Triassic-Jurassic reef destruction of the Tethyan carbonate platform, the pre-Cretaceous late Cimmerian uplifts, and climatic changes that resulted in increased differences between the microfaunas in the different palaeogeographic provinces (Basov and Kuznetsova, 2000). Nonetheless, there is no clear evidence of a catastrophic extinction event during the Jurassic. On the contrary, as is evident from larger benthic foraminifera evolution, many forms experienced expansions in distribution, and forms appeared in new niches, such as the appearance of deeper-water larger benthic foraminifera (*Choffatella*) in

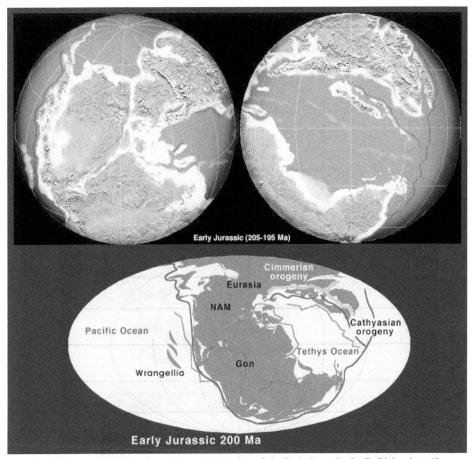

Fig. 4.23. Palaeogeographic and tectonic reconstruction of the Early Jurassic, (by R. Blakey http://jan.ucc. nau.edu/~rcb7/paleogeographic.html).

the Late Jurassic, and most notably the first occurrence (see Fig. 4.24) in the Middle Jurassic of **planktonic foraminifera** (see BouDagher-Fadel et al., 1997; 2015).

The similarity of the Mediterranean Jurassic assemblages with those recorded from Southern Turkey, Iran, Saudi Arabia, Iraq and the Gulf States, Syria and Israel is remarkable (Noujaim Clark and BouDagher-Fadel, 2001). This commonality may be surprising given the presence in the Tethys of palaeo-highs, and vertical tectonic movements affecting differentially subsided and uplifted platform blocks, but obviously oceanic circulation must have allowed a cosmopolitan distribution of larger benthic foraminifera. Despite their widespread occurrence, the larger Jurassic forms had a significant evolutionary history. The factors driving their evolution involved climatic, paleo-oceanographic, tectonic (or impact) processes, and below is outlined how all of these factors affected their test structure, phylogenetic evolution and distribution.

As discussed in the previous Chapter, the Triassic-Jurassic boundary marks one of the five largest mass extinctions in the past 500 Ma. However, there is still debate as to

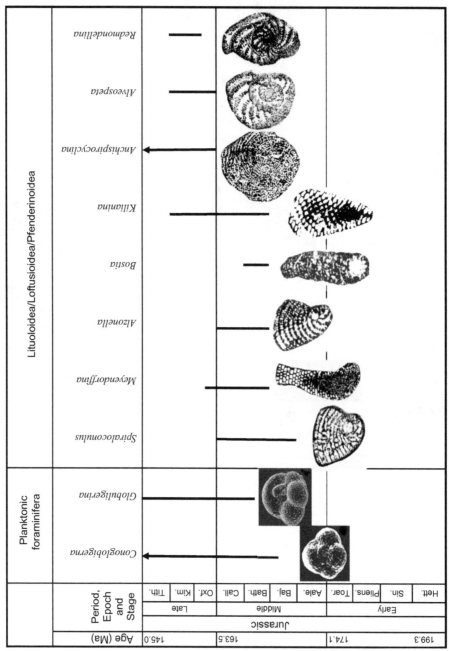

Fig. 4.24.   The parallel evolution of the early planktonic foraminifera compared with the lituoloids in the Jurassic

the cause of this extinction. This might have happened, as explained in Chapter 3, as a response to an impact at the Triassic-Jurassic boundary (e.g. Boslough et al., 1996), which in turn might have triggered the eruption of a large igneous flood basalt province (e.g. Jones et al., 2002). The Tr-J boundary, is contemporaneous with the eruption of one of the world's largest known continental igneous province, the Central Atlantic Magmatic Province (Marzoli et al., 2004). This event would have caused a climatic and biotic crisis at the Tr-J boundary, triggered by the emission volcanic gases and would have been responsible for the abrupt turnover of faunas that occurred at the Rhaetian-Hettangian boundary (Fowell and Olsen, 1993; Fowell and Traverse, 1995; McElwain et al., 1999; Olsen et al., 2002a, 2002b; Percival et al., 2017). These gases could have also caused the short-lived global warming, of possibly 2–4 °C, that has been implicated as a cause of the Tr-J mass extinction (McElwain et al., 1999; Beerling and Berner, 2002), and that has been inferred from a marked negative carbon isotope (δ^{13}C) anomaly observed in marine and terrestrial Tr-J boundary strata from Hungary, Canada, and England (Ward et al., 2001; Hesselbo et al., 2002). This crisis at the Tr-J boundary would have inhibited photosynthesis in the shallow seas of Tethys. According to Vermeij (2004) starvation and habitat loss triggered by these conditions would lead to widespread collateral extinction of species. Vermeij argues that conditions traditionally identified by palaeontologists as initiating mass extinction, such as oxygen deprivation, oversupply of nutrients and poisoning (by carbon dioxide, methane and sulphides) are manifestations of the ecological avalanche triggered by a crisis among primary producers, and are therefore considered consequences rather than primary causes of extinction.

In the aftermath of the Tr-J event, the small involutinides (see Fig. 4.8) with a planispiral to trochospiral enrolled tubular second chamber survived and continued into the Jurassic as "disaster forms". Common in pre-extinction shallow marine assemblages, *Involutina* and *Trocholina* are rare and small in the immediate post-extinction aftermath. Urbanek (1993) introduced the term "Lilliput effect", which describes a temporary, within lineage size decrease of the surviving organisms through an extinction event. This effect, which is the morphological manifestation of a post-crisis ecological stress, explains the pattern of size change which is seen through the Hettangian and early Sinemurian.

Following the Kauffman and Erwin (1995) post-extinction event recovery model, Twitchett (2006) divided the post-extinction "repopulation interval" into an initial "survival interval" and a later "recovery interval". In the Hettangian, blooms of small opportunistic genera such as *Duotaxis* dominate sparse assemblages of new, small textulariine forms, such as *Siphovalvulina,* that represent the "survival stage". By the middle Sinemurian, there was a recovery, with an increase in new genera, as new forms evolved to fill the niches which had been vacated by the end Triassic extinction event. This recovery continued through the Sinemurian to the Pliensbachian with no major extinction at the boundary, but with more evolved taxa becoming established through the Pliensbachian. The filling of the larger benthic foraminiferal niches continued through the Pliensbachian, where new forms of lituoloids appeared that went on to play an important role in the evolutionary development of this group throughout the Jurassic and Cretaceous.

The development of the carbonate facies of the western Mediterranean, from the late Pliensbachian onwards, was affected by events which led to the drowning of the Trento Platform (Zempolich, 1993). This stopped the further development of the benthic foraminifera in the Jurassic carbonates of the western Mediterranean. However, despite well documented mass extinctions of other faunas, the Pliensbachian and Toarcian show only modestly enhanced (but seemingly not catastrophic) levels of extinction amongst the larger foraminifera. Towards the end of the Pliensbachian, 33% of the larger foraminifera became extinct (see Fig. 4.25). This extinction was followed by rapid diversification in the late Toarcian. Hallam (1986) proposed that to speak of an end Pliensbachian extinction was misleading, and that it was in fact a low-level event, particularly among benthic marine invertebrates, and not focused at the Pliensbachian-Toarcian boundary, but spreading over into the early Toarcian. This was, he suggested, caused by ocean bottom-water anoxia in Western Europe, evidenced by the development of widespread units of laminated organic-rich shale (see Vermeij, 2004). In addition, Hallam (1986) argued that there is no evidence for contemporary organic-rich shale sequences (reflecting low bottom-water oxygenation) or extinctions, in South America. Thus, he concluded that the early Toarcian extinction was a regional European event only and that global explanations were irrelevant, although even in Western Europe organic-rich shale facies is not universal. However, there were

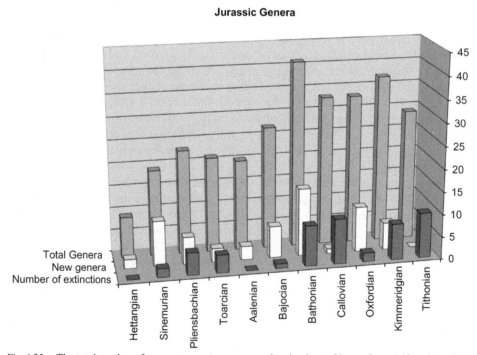

Fig. 4.25.   The total number of genera, new appearances and extinctions of larger foraminifera throughout the Jurassic. The extinctions correspond to the end of each stage and the appearances of new genera with the beginning of the following stage.

major global extinction events in the early Toarcian, for example in benthic ostracods (Boomer et al. 2008).

Little and Benton (1995) in studying the distribution of the macrofauna in Europe at the Pliensbachian-Toarcian boundary argued that there is no evidence for a single family-level mass-extinction event at the end of the Pliensbachian stage (Sepkoski, 1989, 1990). Rather, there is a five-zone phase of extinction from the late Pliensbachian to early Toarcian. The event has a global distribution because, although the majority of the family extinctions occurred within Boreal north-western Europe, there were also extinctions in the Tethyan and Austral realms.

Of the many events that happened during the Pliensbachian, one of the drivers for possible ecological stress, which may have caused the observed, but diffuse, enhanced extinction rates, is the Karoo-Ferrar flood basalt event (Palfy and Smith, 2000). The Karoo province in South Africa and the Ferrar province in Antarctica (Fig. 4.26) are disjunct parts of a once contiguous Gondwanan large igneous province. It ranks among the most voluminous flood basalt provinces of the Phanerozoic (Rampino and Stothers, 1988) and extends for 4000 km^2. The vast majority of the lava volume appears to have been extruded at 183 Ma in about 1 Ma. The anoxic event identified by Hallam (1961) and Wignall (2001) in NW European marine sections and in South America,

Fig. 4.26. The Pliensbachian world showing the position of the Karoo-Ferrar flood basalts.

may have been triggered by the eruption of volcanic CO_2 and ensuing global warming (Jenkyns, 1999; Wignall et al., 2005). Unlike the Tr-J boundary, where the CAMP volcanism seems to have triggered a major catastrophic extinction, the Pliensbachian event is more diffuse. The reduced impact of the Karoo-Ferrar large igneous province (LIP) eruption may be related to the fact that it occurred at high paleolatitudes, and so may not have globally affected the paleoclimate so rapidly.

As conditions ameliorated during the Aalenian, the biota became more diverse and gradually began to resemble pre-extinction biotas (Harries and Little, 1999). Many forms which survived the Pliensbachian-Toarcian event flourished in established niches during the Toarcian and only very few new forms appeared. The Aalenian, through to the Bathonian, shows a major expansion of larger benthic genera. This may be associated with the global recovery in O_2 levels (Fig. 3.10), or the opening up of new habitats as the incipient North Atlantic began to widen as a result of the rifting induced by the CAMP volcanism at the end of the Triassic (see Figure 4.27).

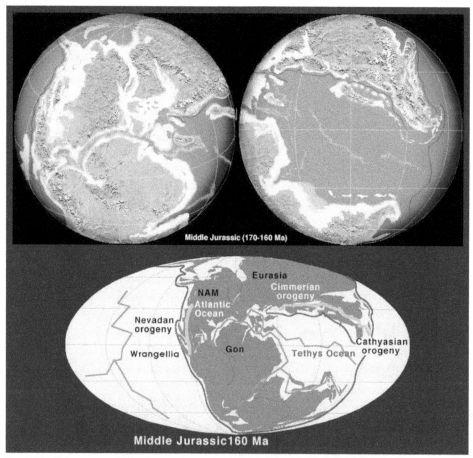

Fig. 4.27. Palaeogeographic and tectonic reconstruction of the Middle Jurassic (by R. Blakey http://jan. ucc.nau.edu/~rcb7/paleogeographic.html).

Following the Aalenian-Bajocian regional anoxic event, which is recorded in the Carpathian part of the Western Tethys, the most intense foraminiferal turnover took place at the Bajocian-Bathonian boundary, and coincides with a maximum diversity of Ammonites (O'Dogherty et al., 2000, see here Plate 4.18, fig. 8). In the Bajocian-Bathonian, the diversity of the larger benthic foraminifera is also at its maximum in the carbonate platforms and reefs of Tethyan margins. There was a major influx of new genera (~30%, see Fig. 4.25) at the beginning of the Bajocian and continued towards the Bathonian (~42% of new genera). These foraminifera were mainly agglutinated and were characterized by the noticeable development of multiple alveoles in their walls (Basov and Kuznetsova, 2000). The Bajocian saw also the most important event in the history of foraminiferal evolution, namely the appearance of planktonic foraminifera (BouDagher-Fadel et al., 1997; BouDagher-Fadel, 2015). These foraminifera, for unknown reasons, began a meroplanktonic mode of life (benthic in the early stage, becoming planktonic in the last stage). These foraminifera were represented by *Conoglobigerina* (Fig. 4.24) which had a restricted geographic occurrence; they all occurred in present-day South-Central and Eastern Europe. They did not become cosmopolitan and holoplanktonic (fully planktonic) until the late Bathonian, with the occurrence of *Globuligerina* (see BouDagher-Fadel et al., 1997; BouDagher-Fadel, 2015).

The end Bathonian and Callovian were also associated with enhanced extinction rates (see Fig. 4.25 and Chart 4.2). This might just reflect the vigorous increase in the number of genera to be found, which reached a maximum in the Bathonian, and which would have produced a more competitive evolutionary environment, and hence a higher background rate of extinctions. However, this Middle Jurassic epoch also coincides with at least two major impact events that gave rise to the 80km diameter Puchezh-Katunki crater in Russia and the 20 km diameter Obolon crater in the Ukraine. These events might also have contributed to enhanced environmental stress that could have been responsible for part of the enhanced extinction rate in these stages.

The transition from the Middle to the Late Jurassic was characterized by significant changes in oceanography and climate. These changes were accompanied by modifications in the global carbon cycle as shown in the carbon isotope record (Louis-Schmid et al., 2007). They were triggered by the opening and/or widening of the Tethys–Atlantic–Pacific seaway and a massive spread of shallow-marine carbonate production leading to higher P_{CO2}, and according to Louis-Schmid et al. (2007) this increase in P_{CO2} may have triggered changes in the biological carbon pump and in organic carbon burial in the mid Oxfordian.

The Oxfordian and Kimmeridgian show another burst of larger foraminifera speciation (Fig. 4.25) that maintained the overall number of genera at a high level, but increased extinction rates at the end of the Kimmeridgian and Tithonian saw numbers of genera decline as the Jurassic came to an end. This general decline may be related to the final opening of the proto-North Atlantic, and a consequent change in global circulation patterns. The larger foraminifera in the Oxfordian developed special features, such as narrow alveoles and a regularly labyrinthic hypodermis (e.g. *Alveosepta*), which helped them to occupy deeper waters than the contemporaneous forms with larger alveoles and irregularly labyrinthic hypodermis (e.g. *Everticyclammina*). They became cosmopolitan and can be found from Portugal through North Africa and southern Europe to the Middle East, and range from the late Oxfordian to Kimmeridgian. In parallel to their evolution, planktonic foraminifera occupying the upper waters of the oceans became more established, wholly planktonic and cosmopolitan.

The end Kimmeridgian saw only a minor extinction, and these were of forms which colonized deeper waters, while no new larger foraminifera appeared to occupy these empty niches in the Tithonian. Around 30% of the larger benthic foraminifera became extinct towards the end of the Tithonian. The end Jurassic (~145Ma) coincides with two major events: (a) a series of large terrestrial impact events (see Glikson, 2005) including Moroweng (70 km), Mjolnir (40 km) and Gosses Bluff (24 km), and (b) a major sub-marine flood basalt event that created the Shatsky Rise, which is the oldest of the great Pacific plateaus with an estimated flood basalt volume of 4.3×10^6 km^3. Mahoney et al. (2005) suggest that this feature is consistent with an impact origin. These events might have been the reasons for the disappearance of long ranging, well established Jurassic larger foraminifera, such as *Pseudospirocyclina*. However, a number of Jurassic agglutinated foraminifera continued through to the Cretaceous where they flourished and thrived, before their final extinctions within the early Cretaceous.

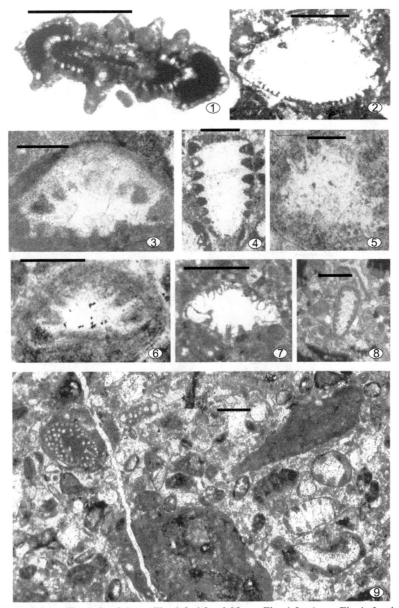

Plate 4.1 Scale bars: Figs 1, 9 = 0.5mm; Figs 2-3, 6-8 = 0.25mm; Figs 4-5 = 1mm. Fig. 1. *Involutina liassica* (Jones), axial section, Early Jurassic, Italy, UCL coll. Fig. 2. *Neotrocholina* sp., Callovian/Oxfordian, Saudi Arabia, UCL coll. Fig. 3. *Neotrocholina valdensis* Reichel, Callovian, Saudi Arabia, UCL coll. Fig. 4. *Andersenolina elongata* (Leupold), late Bathonian, Saudi Arabia, UCL coll. Fig. 5. *Andersenolina alpina* (Leupold), Kimmeridgian/Berriasian, Lebanon, UCL coll. Fig. 6. *Trocholina conica* (Schlumberger), Callovian/Oxfordian, Saudi Arabia, UCL coll. Fig. 7. *Trocholina* cf. *granosa* (Frentzen), Bajocian/Bathonian, Saudi Arabia, UCL coll. Fig. 8. *Trocholina palastiniensis* Henson, holotype, Jurassic, Kurnub, South Israel, NHM P38477. Fig. 9. *Haplophragmoides* sp., *Trocholina palastiniensis* Henson, holotype, *Kurnubia jurassica* (Henson), Jurassic, Kurnub, South Israel, NHM P38477.

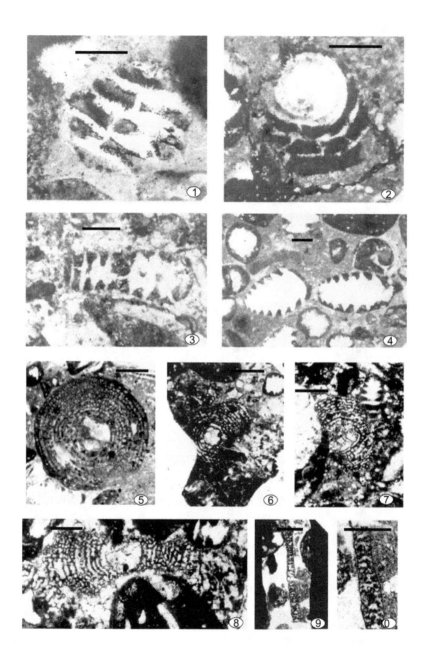

◄

Plate 4.2 Scale bars: Figs 1-4 = 0.25; Figs 5-10 = 0.5mm. *Septatrocholina banneri,* BouDagher-Fadel, first described by BouDagher-Fadel (2008) and validated in BouDagher-Fadel (2016); 1-3) NHM coll., BP 7702, from 8172 ft in Juh-1 core, Qatar; Callovian-Oxfordian, upper Araej Formation; fig. 4, NHM coll., BP 7701, from 9880 ft in Um-Shaif-4 core; Upper Bathonian, basal Uweinat Formation; Abu Dhabi. 1) Paratype, tangential section showing rudimentary Septa; 2) Holotype, equatorial section showing the globular proloculus followed by a trochospirally enroled divided tubular second chamber with rudimentary septa; 3) Paratype, tangential axial section showing the rudimentary septa; 4) Paratype, axial section in which septa are not visible. Figs 5-10. *Alzonorbitopsella arabia,* BouDagher-Fadel, first described by BouDagher-Fadel (2008) and validated in BouDagher-Fadel (2016), 5) holotype NHM BP 6626, equatorial section of the annular holotype with large megalospheric proloculus; 6, 9, 10) NHM BP 6627, from 9879 ¼ ft, 6) Paratype, equatorial section showing the annular test with no septulae, 9) Paratype, oblique axial section showing the delicate reticulate hypodermis of beams and rafters, 10) Paratype, enlargement of the axial section to show that the reticulate hypodermis of beams and rafters does not continue onto the septa; 7, 8) NHM BP 6623, from 9880 ft; Umm-Shaif-4 core; Upper Bathonian, basal Uweinat Formation, Abu Dhabi, 7) Paratype, equatorial section showing the large megalospheric proloculus, 8) Paratype, oblique equatorial section showing the annular chambers immediately following the large megalospheric proloculus.

►

Plate 4.3 Scale bars: Figs 1-6 = 0.6mm. Figs 1-6. *Septatrocholina banneri* BouDagher-Fadel, 1-3, 5-6) Callovian-Oxfordian, Upper Araej Formation, from 8172 ft in Juh-1 core, Qatar, NHM BP 7702; 4) late Bathonian, basal Uweinat Formation, from 9880 ft in Um-Shaif-4 core Abu Dhabi, BP 7701.

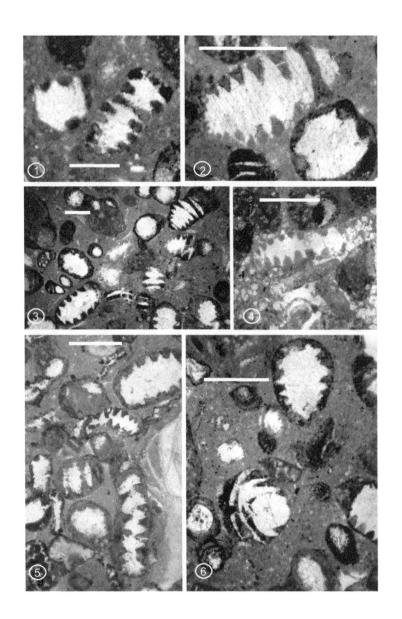

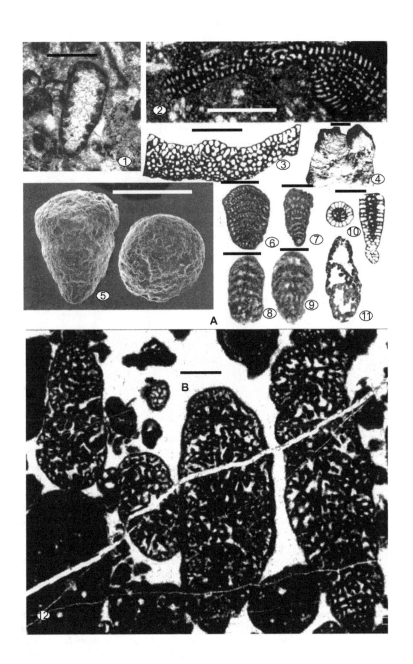

Plate 4.4   Scale bars: Fig. 1 = 0.25mm; Figs 3- 4 = 1mm; Figs 5-7, 10-12 = 0.5mm; Figs 8-9 = 0.3mm. Fig. 1. *Trocholina palastiniensis* Henson, paratype, Late Callovian, Kurnub Anticline, Israel, NHM P38477. Fig. 2. *Levantinella egyptiensis* Fourcade, Arafa and Sigal, type figured by Fourcade et al. (1997), Oxfordian, Jibal As Sahilyeh, Syria. Fig. 3. *Kilianina lata* Oberhauser, paratype figured by Oberhauser (1956), Kimmeridgian, Karadag, West Taurus, Turkey. Fig. 4. *Sanderella laynei* Redmond, holotype figured by Redmond (1964), Bathonian/Callovian (probably Early Callovian), basal Upper Dhruma, ARAMCO T-60-60A, 40-50ft. Fig. 5. *Haurania deserta* Henson, paratype, figured by Henson (1948), Bathonian, Muhaiwir Formation, West Iraq, NHM P35859. Fig. 6. *Socotraina serpentina* Banner et al., identified wrongly as *Milahaina tortuosa* Smout unpublished species and genus by Smout MS, probably Bajocian, Wadi Milaha, Oman. Fig. 7. *Amijiella* sp., identified wrongly as *Iranica slingeri* Gollestaneh MS (1965), probably Bajocian, Wadi Milaha, Oman. Gollestaneh wrongly described it as possessing an initial high trochospire, but is undoubtedly an *Amijiella*. The species name (nomen nudum) was published by Gollestaneh (1974), but the species has never been validly named. Figs 8-9. *Amijiella amiji* (Henson), paratypes, Bathonian, Muhaiwir Formation, Wadi Amij well, West Iraq, NHM M/3869-3870. Fig. 10. *Ataxella occitanica* (Peybernés), figured by Pelissié et al. (1984) as "*Paracoskinolina occitana*", late Bathonian, Pyrénées, France. Fig. 11. *Everticyclammina praekelleri* Banner and Highton, paratype, Kimmeridgian to Tithonian, Broumana, Lebanon, NHM P52255. Fig. 12. *Robustoconus tisljari* Schlagintweit, Velić and Solač, axial, oblique sections showing microspheric specimens (courtesy of Dr Schlagintweit), figured by Schlagintweit (2013), early Bajocian, Croatia.

Plate 4.5   Scale bars: Figs 1-6, 8-16 = 0.5mm; Fig. 7 = 0.3mm. Figs 1-2. *Choffatella tingitana* Hottinger, type specimens figured by Hottinger (1967), Kimmeridgian-Tithonian, Morocco, 1) holotype, equatorial B-form; 2) paratype, off-centered axial B-form. Fig. 3. *Freixialina planispiralis* Ramalho, holotype, figured by Ramalho (1969), Kimmeridgian-Tithonian,NNw of Freixial, Portugal. Fig. 4. *Dhrumella evoluta* Redmond, paratype figured by Loeblich and Tappan (1986), Bathonian, Saudi Arabia. Figs 5-6. *Mesoendothyra izumiana* Dain, type specimens figured by Dain (1958), Kimmeridgian, Russia, 5) equatorial section; 6) axial section. Fig. 7. *Protopeneroplis striata* Weynschenk, holotype, figured by Weynschenk (1950), Middle/Late Jurassic, Austria. Fig.8. *Buccicrenata primitiva* BouDagher-Fadel, holotype, equatorial section of microspheric form, NHM 66907. Figs 9-10. *Pseudospirocyclina smouti* (Banner), late Callovian-early Oxfordian, 9) holotype figured by Banner (1970), Zakum 1, Upper Araej, Umm Shaif, Persian Gulf showing irregular, sporadic extensions of the hypodermis; 10) oblique vertical section, Lebanon, UCL coll. Figs 11-12. *Pseudospirocyclina mauretanica* Hottinger, types specimens figured by Hottinger (1967), Kimmeridgian, Morocco, 11) axial section; 12) equatorial section. Fig. 14. *Pseudospirocyclina muluchensis* (Hottinger), types figured by Hottinger (1967), Kimmeridgian, Morocco. Figs 13, 15. *Pseudospirocyclina maynci* Hottinger, type specimens figured by Hottinger (1967), Kimmeridgian, Morocco, 13) paratype; 15) holotype. Fig. 16. *Pseudocyclammina sphaeroidalis* Hottinger, type specimen figured by Hottinger (1967), Kimmeridgian, East Morocco (distinguished from *P. lituus* by simpler alveolar hypodermis and sub-spheroidal test).

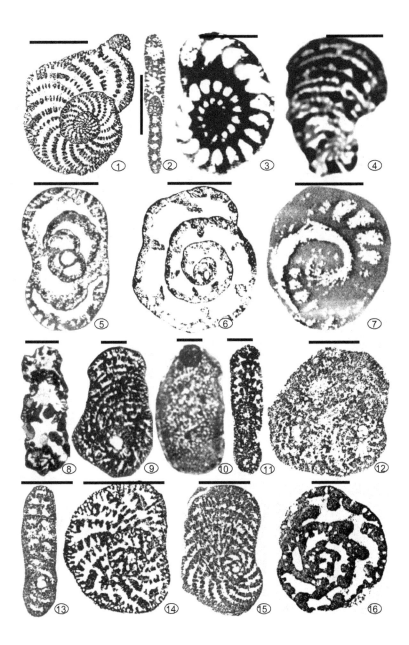

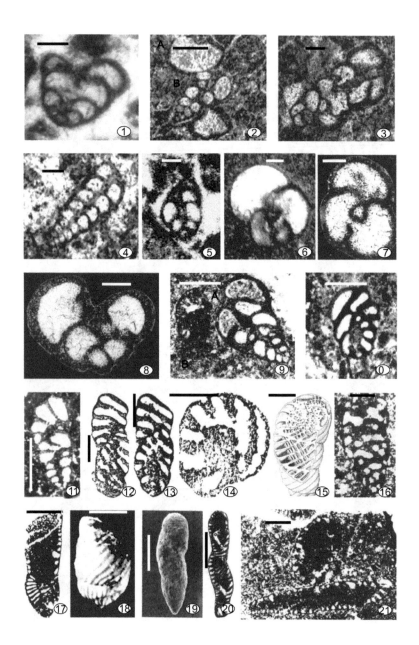

Plate 4.6 Scale bars: Figs 1-8, 16 = 0.15mm; Figs 9-15, 17-21 = 0.5mm. Fig. 1. *Duotaxis metula* Kristan, vertical section showing a conical multi-serial test, BouDagher-Fadel et al., 2001, Sinemurian, NHM P.66938, Sample L7, Gibraltar, UCL coll. Fig. 2. A). *Thaumatoporella?parvovesiculifera* (Raineri), B) *Siphovalvulina* sp., Sinemurian, showing the columellar-siphon, Sample CQ 87,UCL coll. Figs 3-6. *Siphovalvulina colomi* BouDagher-Fadel *et al.* 2001, Sinemurian, Apennines, Mt Bove, 3) megalospheric form showing the initial coiled part of the test, Sample MB 63; 4) microspheric form showing the nearly parallel sides in later growth; 5) holotype, vertical section, NHM P66910, Sample G27; 6) paratype, transverse section, Sample L6, NHM P66911b. Figs 7-8. *Siphovalvulina gibraltarensis* BouDagher-Fadel *et al.,* 2001 Sinemurian; 7) figured holotype, NHMP66912, Sample G8; 8), figured paratype, NHM P66930, Sample D20. Fig. 9. A) *Siphovalvulina beydouni* BouDagher-Fadel, holotype, B) *Kilianina blancheti* Pfender, figured by Noujaim Clark and BouDagher-Fadel (2004), Kesrouane Limestone Formation, Lebanon, UCL coll. Fig. 10. *Pseudopfenderina butterlini* (Brun), Sinemurian, Sample Ad56g, High Atlas, Jebel Rat Formation, Morocco, UCL coll. Fig. 11. A specimen supposed by Sepfontaine (1967) to be intermediate between *Siphovalvulina* and *Pseudopfenderina*, Bathonian, Cevennes, France. Figs 12-14. *Pseudopfenderina butterlini* (Brun), type species figured by Hottinger (1967), Sinemurian-Pliensbachian, Morocco. Fig. 15. A reconstruction by Hottinger (1967) of *Pseudopfenderina*. The columella reconstructed by Hottinger seems to be too elaborate for typical *P. butterlini*, and the incipient tunnels of *Pseudopfenderina* are not drawn in the reconstruction. Fig. 16. *Pseudopfenderina butterlini* (Brun), oblique vertical section figured by Noujaim Clark and BouDagher-Fadel (2004), Bathonian, Kesrouane Limestone Formation, Lebanon, UCL coll. Figs 17-21. *Palaeopfenderina salernitana* (Sartonia and Crescenti), Bathonian, Uwainat, Qatar, NHM P44639.

Plate 4.7 Scale bars: Figs 1, 2 = 0.15mm; Figs 3-5, 12-13, 20, 21 = 0.5mm; Figs 6-11, 14 = 0.3mm; Figs 18-19 = 1mm. Figs 1-2. *Palaeopfenderina salernitana* (Sartonia and Crescenti), Bathonian, Uwainat, Qatar, NHM P43712, 1) vertical section; 2) transverse section. Figs 3-5. *Palaeopfenderina trochoidea* (Smout and Sugden), 3) type figure, Bathonian, Uwainat Limestone, Qatar, showing sub-cameral tunnel and the pillared coalescent septal structure of the pfenderid columella; 4) paratype, NHM P43715; 5) paratype, NHM P42967. Figs 6-7. *Pfenderina neocomiensis* (Pfender), syntypes of *Eorupertia neocomiensis* Pfender (1938), Valanginian, near Toulon, France, 6) vertical section; 7) transverse section. Figs 8-11 . *Conicopfenderina mesojurassica* (Maync), 8) type figure, late Bathonian, Switzerland, figured by Septfontaine (1981); 9) late Bathonian of Upper Kesrouane Limestone Formation, Lebanon,UCL coll; 10 -11) Bathonian, Switzerland, 10) figured by Septfontaine (1981); 11) figured by Septfontaine (1978). Figs 12-13. *Chablaisia chablaisensis* (Septfontaine), types figured by Septfontaine (1977), Bathonian?-Callovian, French Pre-Alps. Fig. 14. *Pseudoeggerella elongata* Septfontaine, type figured by Septfontaine (1988), Bathonian, Pre-Alps, Switzerland. Figs 15-16. *Satorina apuliensis* Fourcade and Chorowicz, type figures, Bathonian-Callovian, Yugoslavia, 15) holotype. Fig. 17. *Sanderella* sp. Section showing bifurcating sub-cameral tunnel in the flaring, rectilinear growth stage, figured by Altiner and Sepfontaine (1979), Callovian, Tauras, Turkey. Figs 18-20. *Steinekella steinekei* Redmond, Oxfordian, Tuwaiq Mountain Formation, figured by Redmond (1964), 18-19) holotype; 20) superficially eroded paratype showing exoskeletal partitions and peripheral chamberlets. Figs 21-24. *Pfenderella arabica* Redmond, 21) type species from Middle Jurassic (Bathonian or Callovian), Saudia Arabia; 22-24) sketches showing: 22) high trochospiral, test with somewhat inflated chambers; 23) the aperture covered by a finely perforate hemispherical apertural plate; 24) secondary intercameral foramina connected by subcameral tunnel that spirals around the axis of coiling.

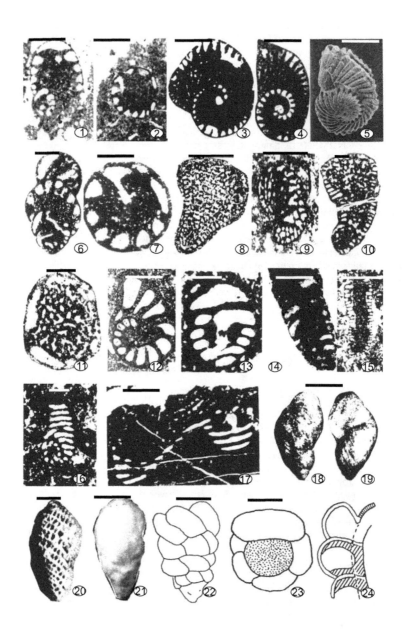

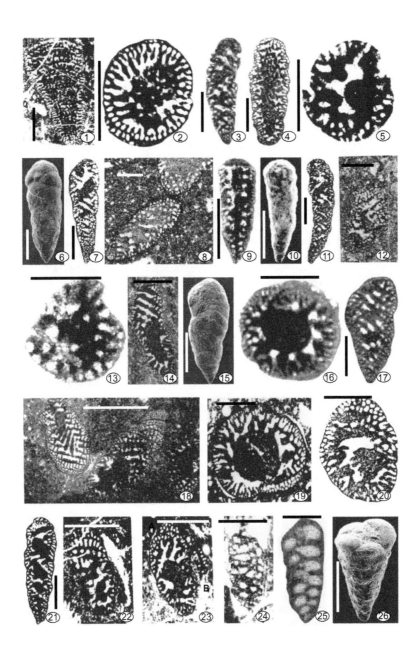

Plate 4.8   Scale bars: Figs 1-2,8-12, 14-25 = 0.5mm; Figs 3-7, 13, 26 = 0.3mm. Fig. 1. *Conicokurnubia orbitoliniformis* Septfontaine, figured by Septfontaine (1988), Oxfordian-Kimmeridgian, Pre-Alps, Switzerland. Fig. 2. *Conicokurnubia* sp., figured by Septfontaine (1981), Kimmeridgian, Turkey. Figs 3-8. *Kurnubia jurassica* (Henson), late Oxfordian, Shuqraia Beds, Kurnub, Israel, 3) NHM P39087; 4) NHM P39129; 5) NHM P39086; 6) paratype, Jurassic, Israel,NHM M38/40;7- 8) Kimmeridgian, Saudi Arabia, UCL coll. Figs 9-14. *Kurnubia palastiniensis* Henson, 9) paratype, NHM M/3836; 10) holotype, revised by Maync (1966), late Oxfordian, Kurnub, Israel, NHM P39089; 11-12) oblique vertical section of "B-form", figured by Noujaim Clark and BouDagher-Fadel (2004), Oxfordian-early Kimmeridgian, Bhannes Complex, Lebanon; 13) transverse section of a paratype; 14) figured by Noujaim Clark and BouDagher-Fadel (2004), Oxfordian, Kesrouane Limestone Formation, Lebanon. Figs 15-21. *Kurnubia wellingsi* Henson. 15-17), figured paratypes, late Oxfordian, Shuqraia Beds, Kurnub, Israel; 15) solid specimen, NHM P39083; 16) transverse section, NHM P43718; 17) vertical section, NHM P43718; 18-19) figured by Noujaim Clark and BouDagher-Fadel (2004), Oxfordian, Kesrouane Limestone Formation, Lebanon;18) oblique vertical sections; 19) transverse section; 20-21) figured by Hottinger (1967), Oxfordian, Morocco, 20) transverse section; 21) vertical section. Figs 22, 24. *Praekurnubia crusei* Redmond, figured by Noujaim Clark and BouDagher-Fadel (2004), Oxfordian, Kesrouane Limestone Formation, Lebanon. Fig. 23. A) *Praekurnubia crusei* Redmond, B) *Meyendorffina bathonica* Aurouze and Bizon, figured by Noujaim Clark and BouDagher-Fadel (2004), Oxfordian, Kesrouane Limestone Formation, Lebanon. Fig. 25. *Textulariopsis sinemurensis* BouDagher-Fadel and Bosence, vertical section, holotype, Sample G8, Gibraltar, NHM P66936. Fig. 26. *Pseudomarssonella maxima* Redmond, late Bathonian-early Callovian, paratype, solid specimen, American Museum of Natural History (AMNH) FT-1270.

Plate 4.9   Scale bars: Figs 1-6, 11-14 = 0.5mm; Figs 7-10 = 0.15mm. Figs 1-3. *Planisepta compressa* (Hottinger), 1-2) types figured by Hottinger (1967), Sinemurian, Morocco, microspheric specimens, 1) holotype, 2) paratype; 3) megalospheric specimen from Betics, Gavillan Formation, Southern Spain, UCL coll. Figs 4-5. *Labyrinthina mirabilis* Weynschenk, figured by Fourcade and Neumann (1966), Kimmeridgian, Spain, 4) vertical section; 5) oblique transverse section. Fig. 6. A) *Nautiloculina circularis* (Said and Barakat), B) *Syriana khouri* Fourcade *et al.*, figured by Noujaim Clark and BouDagher-Fadel (2004), Callovian, Lebanon. Figs 7-8. *Nautiloculina oolithca* (Mohler), Bajocian-Bathonian, Persian Gulf, NHM coll; 7) oblique axial section, Um Shaif-4, core, 9961ft, Lower ARAEJ; 8) equatorial section showing double septa, Umm Shaif-4, core, 9969ft, Lower ARAEJ. Figs 9-10. *Nautiloculina circularis* (Said and Barakat), Callovian, Persian Gulf, NHM coll, Um Shaif-4, 9) core, 9705ft; 9586ft. Figs 11,13. *Nautiloculina cretacea* Arnaud-Vanneau and Peybernés, types figures from Arnaud-Vanneau and Peybernés (1978), Berriasian-Aptian, France. Fig. 12. *Otbitammina elliptica* (d'Archaic), figured by Hottinger (1967), late Bajocian, Chaumont, France. Fig. 14. A) *Redmondellina powersi* (Redmond), B, D) *Nautiloculina oolithica* (Mohler), C) *Kurnubia wellingsi* (Henson), early Kimmeridgian, Lebanon, UCL coll.

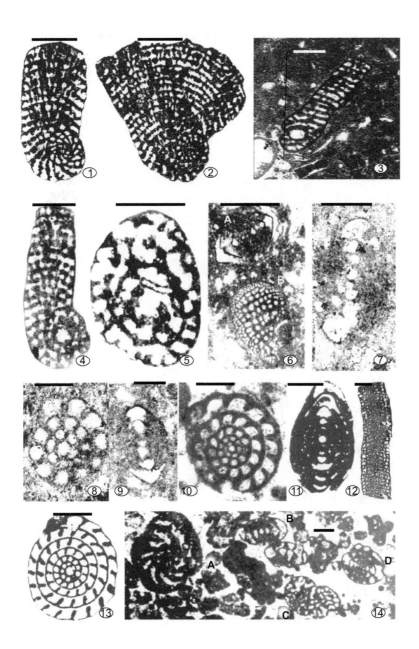

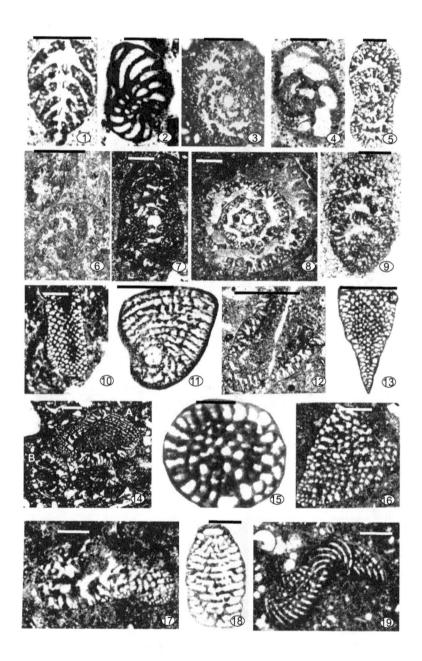

Plate 4.10   Scale bars: Figs 1-2 = 0.25mm; Figs 3, 5, 7-10, 12-13; Figs 4, 11, 14-19 = 0.5mm. Fig. 6 = 0.15mm. Fig. 1. *Kastamonina abanica* Sirrel, type specimens figured by Sirel (1993), Kimmeridgian-Tithonian, Turkey. Fig. 2. *Palaeomayncina termieri* (Hottinger), figured by Septfontaine (1988), Sinemurian-Pliensbachian, Swiss Pre-Alps. Figs 3, 5, 6. *Pseudocyclammina lituus* (Yokoyama), 3) equatorial section with the beginnings of a rectilinear terminal chamber series; this is rare in *P. lituus* but common in *P. vasconica* (see Chapter 5), Hauterivian, Saudi Arabia; 5) axial section of a topotype figured by Maync (1959), Kimmeridgian, Japan; 6) Early Cretaceous, Iran, NHM P52300. Fig. 4. *Pseudocyclammina kelleri* Henson, syntype, Awasil no.5, Ramadi, Iraq, Middle Jurassic, NHM P35968. Figs 7-8. *Pseudocyclammina bukowiensis* Cushman and Glazewski, Kimmeridgian, Saudi Arabia, 7) axial section; 8) equatorial section; 8) axial section;) late Valanginian, uppermost Zangura Formation, NHM P52301; 9) axial section; 10) equatorial section. The thick-walled, coarsely agglutinated, inner hypodermal alveolae distinguish this form from *P. lituus*. Fig. 9 *Pseudocyclammina vasconica* Maync, Kuwait, Mutriba-2, cuttings, 12,540ft, MINAGISH-D, Valanginian, rectilinear growth stages and initial coil in tangential section. Fig. 10. *Meyendorffina* n. sp. aff. *Bathonica* (see key in Fig. 4.20), showing peripheral rectangular chamberlets, Callovian/Oxfordian, Upper Araej, Umm Shaif-3, NHM coll. Note the smaller test size and less complex alveolar wall than in *P. lituus*. Note the range of *P. vasconica* in the Tethyan carbonate shelf seems to be demonstrated to be throughout the Valanginian to Aptian, although it occurs in floods as a local Valanginian index in Saudi Arabia and SE Mid East Gulf. Fig. 11. *Spiraloconulus perconigi* Alleman and Shroeder, holotype figured by Allemann and Shroeder (1972), Bathonian, Spain. Fig. 12. *Kilianina blancheti* Pfender, Oxfordian, Upper Kesrouane Limestone, Lebanon, UCL coll. Figs 13, 15-16. *Neokilianina rahonensis* Foury and Vincent, 13, 15) types figured, Kimmeridgian, Chaussin, 13) holotype, axial section; 15) transverse section; 16) vertical section, basal Kimmeridgian, Lebanon, UCL coll. Fig. 14. *Kilianina preblancheti* BouDagher-Fadel and Noujaim Clark, types figured by Noujaim Clark and BouDagher-Fadel (2004), Early Callovian;14A) holotype, axial section; 14B) oblique equatorial section. Figs 17-18. *Parurgonina coelinensis* Cuvillier, Foury and Romano, 17) type specimen figured by Cuvillier *et al.* (1968), Kimmeridgian, Italy; 18) an oblique central section showing the septa breaking in the umbilical region to form pillars, Oxfordian, Lebanon, UCL coll. Fig. 19. *Levantinella egyptiensis* (Fourcade *et al.*), oblique equatorial section, Oxfordian, Lebanon, UCL coll.

Plate 4.11   Scale bars: Figs 1-4, 6, 8-9, 11-13, 20 = 0.3mm; Figs 7, 10, 17-19 = 0.5mm; Figs 5, 15-16 = 0.15mm; Fig. 14 = 0.25mm. Fig. 1. *Paravalvulina* sp., Bathonian, Uwainat Formation, United Arab Emirates, NHM P52625. Fig. 2. *Pseudomarssonella maxima* Redmond, paratype sectioned and figured by Banner et al. (1991), late Bathonian-early Callovian, Saudi Arabia, American Museum of Natural History (AMNH) FT-1270. Fig. 3. *Pseudomarssonella plicata* Redmond, Bajocian-Bathonian, Persian Gulf, Umm Shail-3, cuttings, 10,084ft (caved from Lower ARAEJ). Fig. 4. *Pseudomarssonella bipartia* Redmond grading into *P. inflata* Redmond, Callovian, Idd-el-Shargi-1, core, 8740ft, Upper Uwainat Formation, United Arab Emirates, NHM coll. Fig. 5. *Redmondoides rotundatus* (Redmond), paratype figured and sectioned by Banner et al. (1991), mid/late Bathonian, Saudi Arabia, AMNH FT-1293A. Fig. 6. *Redmondoides medius* (Redmond), note chamber walls straight, perpendicular to the septa, Callovian-Oxfordian, Persian Gulf, Umm Shaif-4, cores, basal Upper Araej, NHM coll. Fig. 7. *Redmondoides lugeoni* (Septfontaine), metatypic topotypes, axial sectioning showing broad, plate-like apertural lips, Bathonian-Callovian, near Chablais, France, NHM P52616. Figs 8-9. *Riyadhoides mcclurei* (Redmond), late Bajocian, 8) paratype, AMNH FT-1272, 9) sectioned and figured by Banner *et al.* (1991), Saudi Arabia, Dhruma, NHM coll. Figs 10-11. *Riyadhella praeregularis* BouDagher-Fadel *et al.*, 2001, figured types, Sinemurian-Early Pliensbachian, Gibraltar, 10) holotype, axial section showing thin, convexly curved septa and canaliculated test wall, NHM P66947, Sample D20; 11) paratype, NHM P66916. Fig. 12. *Riyhadella regularis* Redmond, mid/late Bathonian, paratype of the synonymous *R. nana* Redmond, figured by Banner *et al.* (1991), Aramco well, Saudi Arabia. Fig. 13. *Riyadhella* sp., Callovian, Upper Uweinat, Persian Gulf, NHM coll. Fig. 14. *Cymbriaella lorigae* Fuganoli, figured by Fugagnoli (1999), Sinemurian-Pliensbachian, Italy. Fig. 15. A sketch showing *Valvulina* sp. and the tooth-like aperture from Banner's collection, UCL. Figs 16-17. *Gutnicella cayeuxi* (Lucas), 16) type specimen figured by Lucas (1939), Aalenian, Algeria; 17) figured by Gutnic and Moullade (1967), Aalenian, West Taurus, Turkey. Figs 18-20. *Haurania deserta* Henson, Bathonian, 18) paratype from Muhaiwir Formation, NHM P35859; 19) paratype, Jurassic, Wadi Amij well, West Iraq M/3846, NHM P35863; 20) paratype, transverse thin section, NHM P3856.

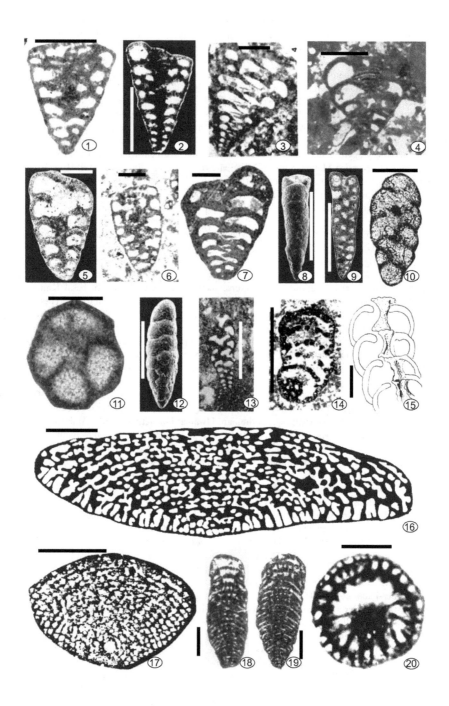

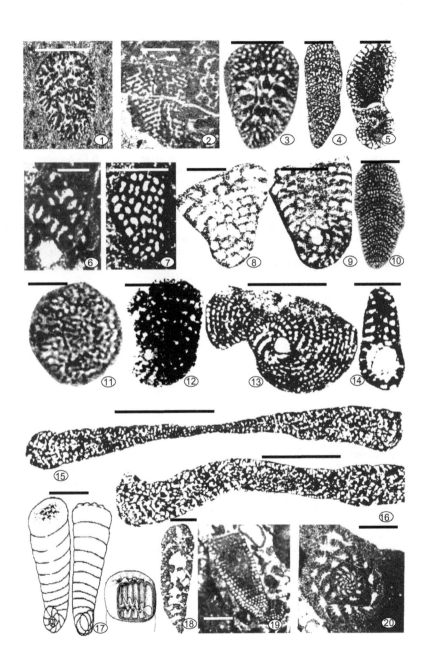

Plate 4.12 Scale bars: Figs 1-3, 5, 7, 9, 14, 17-20 = 0.25mm; Fig. 4 = 1mm; Figs 6, 8, 11 = 0.15mm; Figs 10, 12-13 = 0.3mm; Fig. 15 = 1mm; Fig. 16 = 2mm. Figs 1-4. *Haurania deserta* Henson, 1948, 1-2) figured by BouDagher-Fadel and Bosence (2007), Middle Jurassic, 1) Betics, Gavillan Formation, Southern Spain, Sample RA-01-203; 2) High Atlas, Jebel Rat Formation, Morocco; 3) longitudinal section, Toarcian, Yemen, NHM P53894; 4) vertical section figured by Hottinger (1967), Morocco. Fig. 5-6 *Meyendorffina bathonica* Arouze and Bizon, 5) figured by Furrer and Septfontaine (1977), late Bathonian, Swiss Pre-Alps; 6) figured by Noujaim Clark and BouDagher-Fadel (2004), late Bathonian, Lebanon. Fig. 7. *Gutnicella oxitanica* (Peybernes), late Bathonian-late Callovian, Upper Kessrouane Limestone, Lebanon, UCL coll. Fig. 8. *Gutnicella bizonorum* (Bourrouilf and Moullade), figured paratype, Bathonian, Minorca. Fig. 9. *Gutnicella minoricensis* (Bourrouilh and Moullade), type figured, Bathonian, Algeria. Figs 10-11. *Socotraina serpentina* Banner et al., Early Jurassic, Socotra, Yemen, 10) holotype, microspheric specimen, vertical section, NHM P53883; 11) paratype, transverse section, NHM P53892. Fig. 12. *Platyhaurania subcompressa* Bassoullet and Boutakiout, type specimen figured by Bassoullet and Boutakiout (1996), late Sinemurian, Morocco. Figs 13-16. *Timidonella sarda* Bassoulet, Chabrier and Fourcade, type figures, Bajocian-Bathonian, Sardinia, 13-14) A-forms; 15-16) B-forms. Fig. 17. *Trochamijiella gollesstanehi* Athersuch et al., type specimens figured, Bathonian, Oman. Fig. 18. *Rectocyclammina chouberti* Hottinger, type specimen figured by Hottinger (1967), early Kimmeridgian, Morocco. Fig. 19. *Meyendorffina* sp., Bathonian, France, UCL coll. Fig. 20. *Alveosepta jaccardi* (Schrodt), equatorial section of a B-form, early Kimmeridgian, Qatar, Dukhan-51, 74441/2ft, NHM coll.

Plate 4.13 Scale bars: Figs 1-8 = 0.5mm Figs 1- 3. *Alveosepta jaccardi* (Schrodt), late Oxfordian, 1) topotype figured by Hottinger (1967), Switzerland; 2-3) figured by Noujaim Clark and BouDagher-Fadel (2002) from Bhannes Complex, Lebanon, 2) axial section; 3) oblique equatorial section. Figs 4-5. *Redmondellina powersi* (Redmond), Early Kimmeridgian, figured by Noujaim Clark and BouDagher-Fadel (2002) from Bhannes Complex, Lebanon, 4) oblique equatorial section showing the "clear line" in septa, and median lamella cut lengthwise in last chamber; 5) transverse section of a microspheric form showing pillar-like hypodermal extensions, UCL coll. Figs 6-8. *Alzonella cuvillier* Bernier and Neumann, type figures, 6) showing partitions "beams" which subdivide the chamberlets, Bathonian, Alzon, France; 7) Bathonian, Mas-del-Pont Well, France; 8) Callovian, Uweinat, Umm Shaif-3, Persian Gulf, NHM coll.

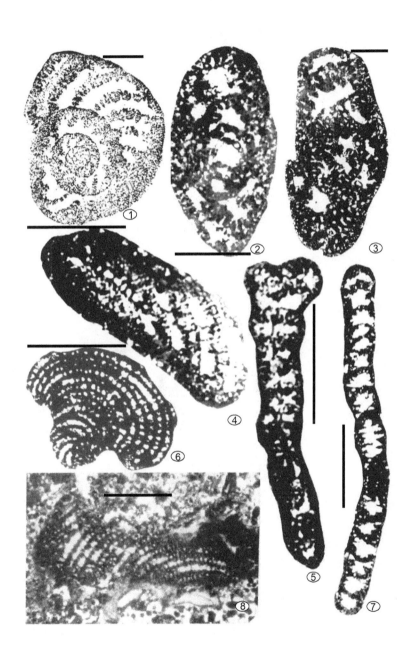

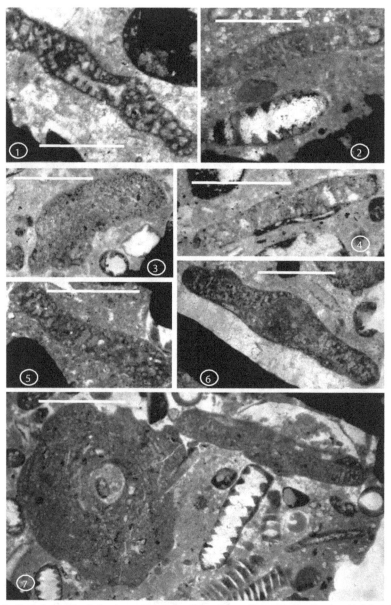

Plate 4.14 Scale bars: Figs 1-6 = 1mm. Figs 1-6. *Alzonorbitopsella arabia* BouDagher-Fadel, 5) BP 6626; 6,9,10) BP 6627, from 9879 ¼ ft; 7, 8) BP 6623, from 9880 ft; Um-Shaif-4 core; Upper Bathonian, basal Uweinat Formation, Abu Dhabi, NHM coll..

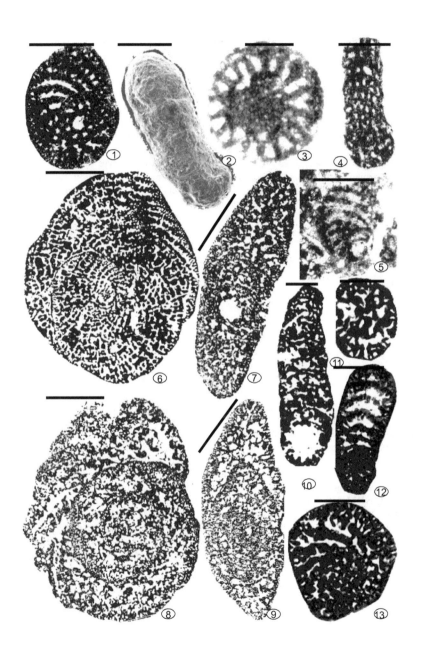

Plate 4.15 Scale bars: Figs 1, 7-13 = 0.5mm; Figs 2, 5 = 0.3mm; Figs 3-4 =0.15mm; Fig. 6 = 1mm. Fig. 1. *Palaeocyclammina complanata* Bassoullet, Boutakiout and Echarfaoui, holotype F62025, figured by Bassoullet *et al.* (1999), Pliensbachian, Morocco. Figs 2-4. *Amijiella amiji* (Henson),, type specimens, Bathonian, Muhaiwir Formation, Wadi Amij well, West Iraq, 2) solid specimen, holotype, NHM P35869; 3-4) paratypes, 3) transverse section, 4) axial section, NHM P35866. Fig. 5. *Torinosuella peneropliformis* (Yabe and Hanzawa), Thamama Formation, Oman, Umm Shaif-3, cuttings, 7255ft, NHM coll. Figs 6-7. *Anchispirocyclina lusitanica* (Egger), 6) figured by Ramalho (1971), equatorial section of B-form, Kimmeridgian, Portugal; 7) figured by Hottinger (1967), Kimmeridgian, Morocco. Figs 8-9. *Anchispirocyclina praelusitanica* (Maync), near topotypes, B-forms figured by Hottinger (1967), Oxfordian, Israel (note: Hottinger referred *praelusitanica* to *Alveosepta (=Redmondellina powersi)*, but *praelusitanica* lacks the septal "clear line" of *Redmondellina* and has pillars spread throughout the whole chamber lumen, not merely in the median plane). Figs 10-11. *Bostia irregularis* Bassoullet, type figured by Bassoullet et al. (1999), late Bathonian, France, 10) holotype, axial section; 11) paratype, axial section. Figs 12-13. *Ijdranella altasica* Bassoullet et al., type specimens figured by Bassoullet et al. (1999), Pliensbachian, Morocco, 12) holotype, equatorial section; 13) oblique vertical section.

Plate 4.16 Scale bars: Figs 1-6, 8, 10 = 0.5mm; Figs 7, 9, 11 = 0.3mm. Figs 1-2. *Bosniella oenensis* Gušiç, Sinemurian-Pliensbachian, Jebel Rat Formation, southern High Atlas, Morocco, UCL coll. Fig. 3. *Everticyclammina praevirguliana* Fugagnoli, Sinemurian-Pliensbachian, Gibraltar, UCL coll.. Fig. 4. *Everticyclammina kelleri* (Henson), paralectotype, equatorial section, Berriasian-Valanginian, Zangura Formation, Iraq, NHM P35968. Fig. 5. *Everticyclammina greigi* (Henson), paratype, basal Cretaceous, Qatar, NHM P35795. Fig. 6. *Everticyclammina virguliana* (Koechlin), Kimmeridgian, Persian Gulf, Zakum-1, core, 9536ft, NHM coll. Fig. 7. *Streptocyclammina parvula* (Hottinger), figured by Hottinger (1967), Kimmeridgian, Morocco. Figs 8-9. *Lituosepta recoarensis* (Cati), 8) microspheric form figured by Hottinger (1967), Middle Jurassic, Morocco; 9) megalospheric form, Betics, Gavillan Formation, Spain. Fig. 10. *Orbitopsella primaeva* (Henson, 1948), microspheric forms, Betics, Gavillan Formation, Spain,UCL coll.. Fig. 11. *Karaisella uzbekistanica* Kurbatov, figured from Kurbatov (1971), Oxfordian, Uzbekistan.

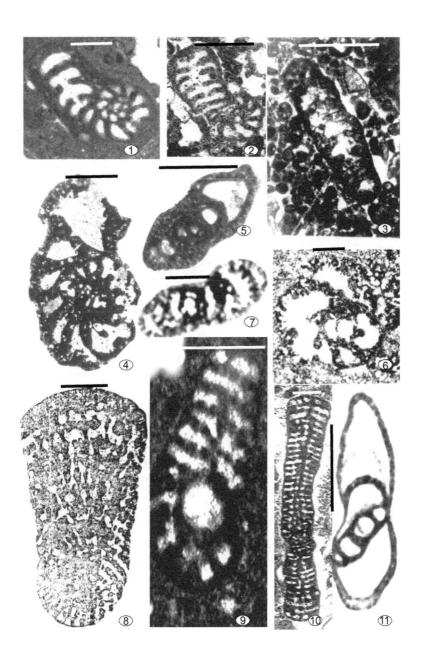

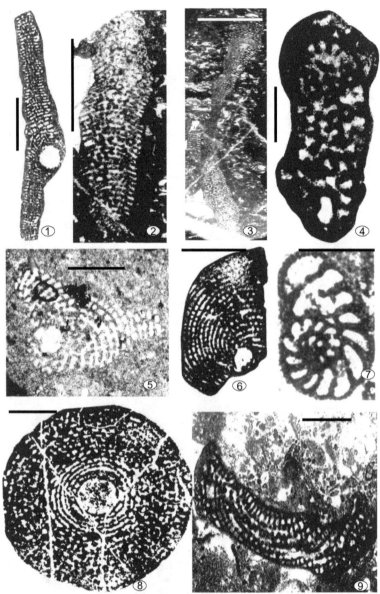

Plate 4.17   Scale bars: Fig. 1 = 1mm; Figs 2-3 = 0.3mm; Figs 4-9 = 0.5mm. Fig. 1. *Orbitopsella dubari* Hottinger, megalospheric form, figured by Hottinger (1967), Middle Jurassic, Morocco. Fig. 2-3. *Orbitopsella praecursor* (Gümbel), syntypes, Sinemurian, Jabal Milaha, Oman, NHM P35780. Figs 4-6 . *Orbitopsella primaeva*, (Henson), syntypes, Middle Jurassic, Jabal Milhaha Oman, Davies collection, 4) NHM 35791; 5) NHM P35789; 6) NHM 35788. Fig. 7. *Lituolipora polymorpha* Gǔsić and Velić, figured by Gǔsić and Velić (1978), Early Jurassic, Yugoslavia. Fig. 8-9. *Cyclorbitopsella tibetica* Cherchi, Schroeder and Zhang, Pliensbachian, 8) holotype, figured by Cherchi, Schroeder and Zhang (1984), Pupuga Formation, South Tibet; 9) Betics, Gavillan Formation, Spain, UCL coll

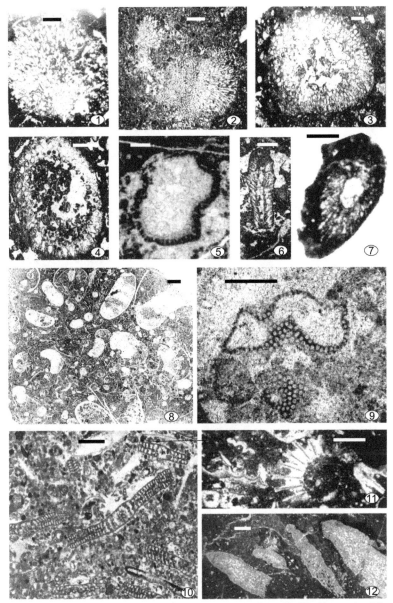

Plate 4.18 Scale bars: Figs 1-12 = 0.5mm. Figs 1-2. *Cayeuxia?piae* Frollo, Sinemurian-Pliensbachian, 1) Lebanon, UCL coll.; 2) figured by BouDagher-Fadel and Bosence (2007), High Atlas, Jebel Rat Formation, Morocco. Figs 3-4, 9. *Thaumatoporella?parvovesiculifera* (Raineri), Sinemurian, 3) Betics, Gavillan Formation, Spain, figured by BouDagher-Fadel and Bosence (2007); 4) Gibraltar Limestone Formation, NHM P66949; 9) sample MB 49, UCL coll. Figs 5-7. *Palaeodasycladus?mediterraneus* (Pia), figured by BouDagher-Fadel et al. (2001), Sinemurian, Gibraltar Limestone Formation 5) NHM P66945; 6-7) NHM P66931-32. Fig. 8. *Ammonites* spp, Pliensbachian, Italy, UCL coll. Fig. 10. *Orbitopsella* spp., Early Jurassic, Pakistan, UCL coll. Fig. 11. *Clypeina jurassica* Favre, early Kimmeridgian, Kesrouane Limestone, UCL coll. Fig. 12. *Cladocoropsis mirabilis* Felix, Kimmeridgian, Greece, UCL coll.

Chapter 5

The Mesozoic Larger Benthic Foraminifera: The Cretaceous

5.1 Introduction

Although more than a third of the larger benthic foraminifera became extinct towards the end of the Tithonian, the complex and partitioned lituolids of the order Textulariida continued to dominate the biota of the inner carbonate platform environments that were widespread along the rifted western and eastern margins of the Cretaceous Tethys. Many of the Cretaceous lituolid species have quite limited stratigraphic ranges and are valuable as index fossils. They were joined by a new group, which made its first appearance in the late Early Cretaceous, the alveolinids. These were large milioolides that showed spectacular expansion in the middle Cretaceous, proliferating in mid-latitudes, and often becoming annular or discoid and subdivided by partitions. Many of them resembled the planispiral-fusiform fusulinides (see Chapter 2) of the Permian, attaining approximately the same range of sizes, but differing fundamentally in their imperforate, porcelaneous wall structure. In the Late Cretaceous new, simple rotaliides evolved into forms with complicated three-layered structures, the orbitoids. While the previous two groups had their main breeding ground in the Tethyan realm, the orbitoids showed provincialism and some were only found in the Caribbean.

Cretaceous larger benthic foraminifera have been closely studied by many workers (for example, Maync, 1952; Brönnimann, 1954a, b, 1958; Schroeder, 1962, 1964, 1975; Banner, 1970; Bergquist, 1971; Decrouez and Moullade, 1974; van Gorsel, 1975, 1978; Caus, 1988; Hottinger and Drobne, 1989; Hottinger et al., 1989; Matsumaru, 1991; Özcan, 1995; Görög and Arnaud Vanneau, 1996; Noujaim Clark and Boudagher- Fadel, 2001; BouDagher-Fadel and Lord, 2001; Hughes, 2004; Kaminski, 2010; Schroeder et al., 2010; Albrich et al., 2014; Scherreiks et al., 2015a, 2015b; BouDagher- Fadel et al., 2015; Sun et al., 2015; Schlagintweit et al., 2016; BouDagher-Fadel., 2008, 2015; BouDagher-Fadel and Price, 2017; Schlagintweit and Rashidi, 2017). In this chapter, the taxonomy of the major Cretaceous larger benthic foraminifera is summarized, and then their biostratigraphic, paleoenvironmental and paleogeographic significance is discussed.

5.2 Morphology and Taxonomy of Cretaceous Larger Benthic Foraminifera

As in the Jurassic, the dominant larger foraminifera of the Early Cretaceous were the agglutinated Textulariida. The miliolides became abundant in the Middle Cretaceous, but it was not before the Cenomanian that larger miliolides played an important role in the benthic assemblages of carbonate platforms. The rotaliides became common in the Late Cretaceous.

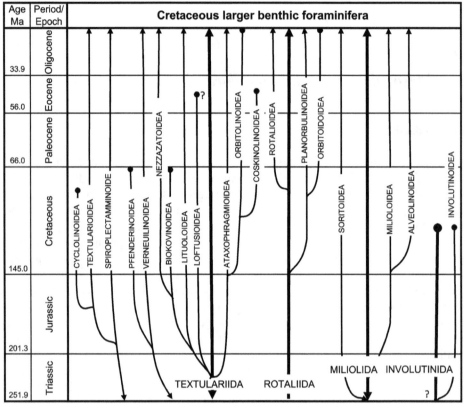

Figure 5.1. The evolution of the Cretaceous orders (thick lines) and superfamilies (thin lines) of the larger benthic foraminifera.

In this section, the following four orders will be discussed:

• Involutinida
• Textulariida
• Miliolida
• Rotaliida

The development and evolution of the superfamilies of these orders is schematically shown in Fig. 5.1. Below are given the morphological characteristics and taxonomic relationships of these major Cretaceous forms. In this chapter only the foraminifera which appeared in the Cretaceous are dealt with, those which made their first appearance in the Jurassic have been described in Chapter 4.

ORDER INVOLUTINIDA Hohenegger and Piller, 1977
This order includes all forms with an enrolled second chamber. They have walls that are aragonitic, but commonly they are recrystallised to give a homogenous microgranular structure. The umbilical region has pillar-like structures on one or both sides of the test. They range from Triassic to Late Cretaceous (Cenomanian).

Superfamily INVOLUTINOIDEA Bütschi, 1880

This superfamily consists of forms with a first chamber that is followed by a planispiral to trochospiral enrolled tubular second chamber. Triassic to Late Cretaceous (Cenomanian).

All genera to be found in the Cretaceous originated in the Jurassic or earlier, and have been described in Chapter 4.

ORDER TEXTULARIIDA Delage and Hérouard, 1896

The tests of these agglutinated foraminifera are made of foreign particles bound by organic cement. They range from Early Cambrian to Holocene.

Superfamily NEZZAZATOIDEA Hamaoui and Saint-Marc, 1970

The test is trochospiral or planispiral, later occasionally uncoiled. The wall is simple, non-lamellar, and microgranular, and may possess internal plates or simple partitions (Fig. 5.2). The aperture is simple or multiple. Middle Jurassic (Bajocian) to Holocene.

Family Nezzazatidae Hamaoui and Saint-Marc, 1970

The test is trochospiral or planispiral, later it may be uncoiled. The interior of each chamber has an internal plate and may be basally digitate. The aperture is single or multiple with an internal tooth plate. Cretaceous (Barremian to Maastrichtian).

Subfamily Nezzazatinae Hamaoui and Saint-Marc, 1970

The test is trochospiral or planispiral, later it may be uncoiled. The interior of each chamber has an internal plate and is basally digitate. The aperture is single or multiple. Cretaceous (Barremian to Turonian) (see Fig. 5.2).

- *Nezzazata* Omara, 1956 (Type species: *Nezzazata simplex* Omara, 1956). The test is trochospiral with a closed umbilicus. A narrow plate extends backwards from one septum to the previous one. The aperture is single, extending from the umbilicus to the periphery, then bending sharply with an apertural tooth. Cretaceous (Albian to Turonian) (Plate 5.1, fig. 12).
- *Nezzazatinella* Darmoian, 1976 (Type species: *Nezzazatinella adhami* Darmoian, 1976). The test is a low trochospiral, with a flattened spiral side, and the opposite side convex and involute. The chambers in the final whorl are narrow, elongate, with the final chamber flaring backwards. The aperture is a large curved slit, and may be accompanied by secondary apertures either in a curved row or scattered over the apertural face. Cretaceous (Barremian to Turonian) (Plate 5.1, figs 10-11).
- *Lupertosinnia* Farinacci, 1996 (Type species: *Lupertosinnia pallinii* Farinacci, 1996). The test is conical and trochospirally enrolled. The central portion of the test is occupied by alveolar thin layers, communicating with the chambers by perpendicular canals. The wall is simple and finely agglutinated. Cretaceous (Cenomanian).

Figure 5.2. Key morphological characteristics of the Lituoloidea, the Biokovinoidea, the Nezzazatoidea as compared to the Haplophragmioidea (representatives of the simple benthic foraminifera, see Loeblich and Tappan, 1988).

Lituoloidea	Planispiral, sometimes with last few chambers uncoiling; coiled test nautiliform to globose	Planispiral with solid, non-alveolar, non-canaliculate walls and no internal toothplates	Aperture multiple, areal; test sometimes uncoiling	*Lituola*
Biokovinoidea		Canaliculate walls	Aperture cribrate at base of apertural face	Debarina
Haplophragmioidea			Aperture a basal slit	*Haplophragmoides*
Nezzazatoidea	Planispiral, involute or nearly so; biumbilicate, compressed; many (15+) chambers per wholrl, usually rapidly increasing in height	Aperture single, interiomarginal (basal)	rounded, central basal equatorial slit	*Daxia* *Stomatostoecha*
		Aperture single, areal	in middle of apertural face	*Friexialina*
			slit at top of apertural face	*Phenacophragma*
		Aperture multiple	test thickness/diameter approximately 1/3, apertures scattered over apertural face	*Mayncina*
			test thickness/diameter approximately 1/7, apertures in vertical row in face	*Gendrotella*

Subfamily Coxitinae Hamaoui and Saint-Marc, 1970

The test is trochospiral or planispiral with an internal median plane between consecutive septa, the plates being terminally bifurcate or digitate. Apertures are multiple. Cretaceous (Cenomanian to Maastrichtian).

- *Coxites* Smout, 1956 (Type species: *Coxites zubairensis* Smout, 1956). The test is trochospiral with a bilocular embryonal stage, followed by numerous low chambers, that are oblique on the spiral side and strongly arched on the umbilical side around a central depression. There are incomplete partitions within the chambers parallel to the septa. The aperture is multiple. Cretaceous (late Cenomanian to early Turonian) (Plate 5.2, figs 6-7).
- *Rabanitina* Smout, 1956 (Type species: *Rabanitina basraenesis* Smout, 1956). The test is planoconvex, trochospiral to completely involute. A complex perforated plate within the adult chamber parallels the spiral wall, but twists and is attached to both chamber floor and roof. The aperture is interiomarginal and multiple. Cretaceous (early Cenomanian) (Plate 5.4, fig. 3).

- *Demirina* Özcan, 1994 (Type species: *Demirina meridionalis* Özcan, 1994). The test is biumbilicate, peneropliform with strongly thickened low and broad adult chambers, and has a simple inner structure with one set of partitions. Cretaceous (Cenomanian) (Plate 5.3, fig. 4).

Family Mayncinidae Loeblich and Tappan, 1985
The test is biumbilicate or very compressed, planispirally enrolled, but rarely uncoiling. The chambers increase rapidly in height. The septa are monolamellar. The wall is simple, solid and microgranular. Late Jurassic to Late Cretaceous (Kimmeridgian to Santonian).

- *Gendrotella* Maync, 1972 (Type species: *Choffatella rugoretis* Gendrot, 1968). The aperture is multiple in vertical rows. The absence of an alveolar hypodermis differentiates this form from homeomorph such as *Choffatella* or *Torinosuella*. Late Cretaceous (Santonian) (Plate 5.4, fig. 9).
- *Mayncina* Meumann 1965. (Type species: *Daxia orbignyi* Cuvillier and Szakall, 1949). The aperture is multiple and scattered. Cretaceous (Valanginian to Cenomanian) (Plate 5.4, fig. 8).
- *Phenacophragma* Applin, Loeblich and Tappan, 1950 (Type species: *Phenacophragma assurgens* Applin, Loeblich and Tappan, 1950). Similar to *Freixialina* (see Chapter 4) but with an aperture at the top of apertural face. Cretaceous Early (Albian).
- *Stomatostoecha* Loeblich and Tappan, 1950 (Type species: *Stomatostoecha plummerae* Applin et al., 1950). This form is similar to *Daxia* (see Chapter 4) but with an extensive slit-like aperture (close to the rounded, basal aperture of *Daxia*). Early Cretaceous (Albian) (Fig. 5.3; Plate 5.4, fig. 10).

Superfamily BIOKOVINOIDEA Gušiç, 1977
Members of this superfamily have a free test with a trochospiral or planispiral early stage, that later may become uncoiled. Septa are homogeneous and massive. The aperture is basal to areal, and single to multiple (Fig. 5.2). Early Jurassic (Sinemurian) to Late Cretaceous (Maastrichtian).

Family Biokovinidae Gušiç, 1977
The test is planispirally coiled, later it may uncoil. Endoskeletal pillars may be present. Walls are canaliculate with alveoles that open both to the exterior or interior. Early Jurassic to Cretaceous.

- *Zagrosella* Schlagintweit and Rashidi, 2017 (Type species: *Zagrosella rigaudii* Schlagintweit and Rashidi, 2017). The test has an oscillating coiling plane in the early stage, later it is planispiral, and finally may be rectilinear. Few and irregularly distributed endoskeletal pillars are present in the chamber interior. Cretaceous (late Maastrichtian).

Family Charentiidae Loeblich and Tappan, 1985
Early stage tests are planispiral. Walls are finely canaliculated. Apertures are single or multiple. Middle Jurassic to Late Cretaceous (Callovian to Maastrichtian).

- *Charentia* Neumann, 1965 (Type species: *Charentia cuyillieri* Neumann, 1965). Planispiral throughout with a single aperture, becoming enlarged with growth.

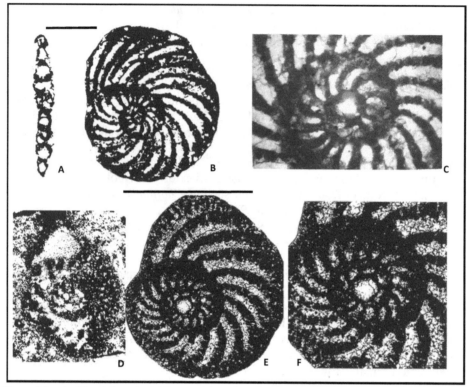

Figure 5.3.   A, D) axial section/oblique axial section; B, E) equatorial sections; C, F) part of same equatorial section (scale bar = 0.5mm). A-C are *Stomatostoecha* sp. from the Santonian of France showing gross "choffatelloid" form of tests, chambers and apertures, but lack of a labyrinthic or alveolar hypodermis of D-F *Choffatella decipiens* Schlumberger, from the Aptian of the Persian Gulf. Note: The original description of *Stomatostoecha* is misleading. It has a single, basal, slit-like aperture only and no areal apertures.

The finely canaliculate wall differs from the alveolar wall of *Hemicyclammina* and from *Everticyclammina* in that the alveoles are thicker, more separately spaced, branch and dichotomize. Early Cretaceous to earliest Late Cretaceous (Barremian to Cenomanian) (Plate 5.5, fig. 12).

- *Debarina* Fourcade, Raoult and Vila, 1972 (Type species: *Debarina hahounerensis* Fourcade et al., 1972). Planispiral in later whorls at least. Apertures are multiple, cribrate, and basal. *Debarina* differs from *Charentia* by its cribrate aperture and possible intitial streptospirality. Early Cretaceous (late Barremian to Aptian) (Plate 5.5, fig. 13).

Superfamily LITUOLOIDEA de Blainville, 1825

Members of this superfamily have multilocular, rectilinear and uniserial tests. The wall has no alveolar or reticulate hypodermis. The early stage has plani- (strepto-) or trochospiral coiling. The periphery of the chambers have radial partitions; centrally with or without scattered, separated pillars (Fig. 5.2). The aperture is simple or multiple cribrate, with no tooth plate. Early Jurassic (Sinemurian) to Holocene.

Family Hauraniidae Septfontaine, 1988
Tests may be uncoiled, uniserial or planispirally coiled. Walls are microgranular, with a hypodermic network. Septa are simple or with complicated microstructures. The interiors of chambers are simple or with pillars. Apertures are multiple. Jurassic (late Sinemurian) to Cretaceous.

Subfamily Amijiellinae Septfontaine, 1988
Members of this subfamily have an uncoiled or planispirally coiled test. The septa are simple or with a complicated microstructure. The interiors of the chambers are simple, however some genera may develop pillars. Walls may have alveoles. The aperture is multiple. Jurassic (late Sinemurian) to Cretaceous.

- *Spirocyclina* Munier-Chalmas, 1887 (Type species: *Spirocyclina choffati* Munier-Chalmas, 1887). Tests are flat coils, becoming peneropliform. There is a coarse early epidermis, with rapid development of subepidermal alveolae. Central pillars are few and scattered. Late Cretaceous (late Cenomanian to Santonian) (Plate 5.3, Fig. 4).
- *Martiguesia* Maync, 1959 (Type species: *Martiguesia cyclamminiformis* Maync, 1959). Coiled, nautiliform, with a central zone having delicate pillars bridging from septum to septum. Late Cretaceous (Santonian) (Plate 5.4, fig. 4).
- *Montsechiana* Aubert, Coustau and Gendrot, 1963 (Type species: *Montsechiana martiguae* Aubert et al., 1963). Strong central pillars span flabelliform chambers. There is a finely reticulate hypodermal mesh, with beams below. The aperture is cribrate. Early to Late Cretaceous (Valanginian to Santonian) (Plate 5.2, fig. 1).
- *Qataria* Henson, 1948 (Type species: *Qataria dukhani* Henson,1948). The early stage is planispiral, later chambers are cyclical. The outer parts of the chambers are subdivided into numerous small chamberlets. Late Cretaceous (late Cenomanian to Turonian) (Plate 5.4, figs 5-6).
- *Sornayina* Marié, 1960 (Type species: *Sornayina foissacensis* Marie, 1960). The wall has a subepidermal meshwork. Chambers are subdivided into small chamberlets by bifurcating septula perpendicular to the septa. Late Cretaceous (Coniacian) (Plate 5.5, Fig. 11).

Family Everticyclamminidae Septfontaine, 1988
Tests are streptospiral or planispiral in the early stage, uncoiled in the adult or uniserial throughout. Walls are microgranular, with an alveolar microstructure. Interiors of chambers are simple. Apertures are areal. Early Jurassic (Sinemurian) to Cretaceous (early Cenomanian).

- *Hemicyclammina* Maync, 1953 (*Hemicyclammina sigali* Maync, 1953). Tests are planispiral, with a single aperture occupying all, or nearly all, of the total height of the apertural face in equatorial section, reducing the solid septa which are clearly different in structure from spiral wall. The basal layer is deposited over the chamber floors. Cretaceous (late Aptian to early Cenomanian) (Fig. 5.4; Plate 5.6, figs 12-15).

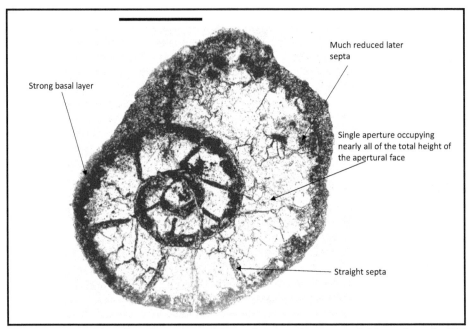

Figure 5.4 Equatorial section of an advanced form of *Hemicyclammina sigali* Maync from the Early Cenomanian of the Persian Gulf. Scale bar = 0.25mm.

- *Hottingerita* Loeblich and Tappan, 1985 (Type species: *Mesoendothyra complanata* Hottinger, 1967). Tests are evolute, biumbilicate, loosely coiled in a low, almost planar streptospire. It differs from *Alveosepta* (see Chapter 4) by having a single aperture rather than a multiple one, and differs from *Mesoendothyra* (see Chapter 4) by having complex septa and smaller and more numerous chambers, and from both genera in being evolute. Early Cretaceous (Barremian) (Plate 5.2, fig. 2).

Family Lituolidae de Blainville, 1827
The early stages of the tests are enrolled, but later they may become rectilinear. Walls are made from agglutinated foreign particles. There are few chambers (less than 10) per whorl. Carboniferous to Holocene.

Subfamily Ammomarginulininae Podobina, 1978
Test have early stages that are coiled, but later uncoiling. Apertures are single. Carboniferous (Early Mississippian) to Holocene.

- *Ammobaculites* Cushman, 1910 (Type species: *Spirolina agglutinans* d'Orbigny, 1846). The test is simple, not compressed and uncoils in the adult. Aperture is single, areal. Carboniferous (Mississippian) to Holocene (Plate 5.6, fig. 18).

Subfamily Lituolinae de Blainville, 1827
Members differs from the Ammomarginulininae in having a multiple apertures. Late Triassic to Holocene.

- *Lituola* Lamarck, 1804 (Type species: *Lituolites nautiloidea* Lamarck, 1804). These forms have no internal partitions and a multiple cribrate aperture. Late Triassic to Holocene (Plate 5.5, fig. 14; Plate 5.6, fig. 9).

Superfamily PFENDERINOIDEA Smout and Sugden, 1962
Members of this superfamily have a trochospiral test throughout, or may become uncoiled. Some forms have a siphonal canal, others develop a central composite columella with pillars between the apertural plates and septa. Early Jurassic (Hettangian) to Cretaceous.

Family Pfenderenidae Smout & Sugden, 1962
Members of this subfamily have a loose trochospiral test with siphonal canals that connect successive apertures in primitive forms. Some forms develop a central composite columella, composed of thickened innermost septal ends ("septal buttons") with or without additional pillars and a spiral canal between the columella and the thickened septa. A subcameral tunnel (either simple or multiple) is present in advanced forms. The chamber interior of advanced taxa is subdivided by vertical or horizontal (or both) exoskeletal partitions, resulting in a reticulate subepidermal layer. The apertures are always cribrate, areal. Early Jurassic (Sinemurian) to Late Cretaceous (Maastrichtian).

Subfamily Kurnubiinae Redmond, 1964
Members of this subfamily have a test without subcameral tunnels, and may or may not have a solid core. The peripheral zone is divided by radial partitions. Late Cretaceous (Maastrichtian).

- *Gyroconulina* Schroeder and Darmoian, 1977 (Type species: *Gyroconulina columellifera* Schroeder and Darmoian, 1977). The early stage is conical with a globular proloculus, followed by a cylindrical stage. The proloculus is globular and followed by chambers subdivided by vertical beams and horizontal rafters, both perpendicular to the outer surface of the test, forming a honeycomb, subepidermal network. The central zone has irregularly distributed pillars. Late Cretaceous (late Maastrichtian) (Plate 5.2, fig. 5).

Subfamily Pfenderininae Smout and Sudgen, 1962
Members of this subfamily have a test with a single subcameral tunnel that is always buried in the columella, which is formed by pillars and calcitic deposits. No peripheral partitions or chamberlets are present. The aperture is multiple, cribrate, and on the apertural plate. Early Cretaceous (Valanginian to Barremian or?Aptian).

- *Banatia* Schlagintweit and Bucur, 2017 (Type species: *Banatia aninensis* Schlagintweit and Bucur, 2017). The test is low, asymmetric conical to almost planispiral, formed by a short low trochospiral followed by flaring chambers. The marginal part of the chambers is undivided. The axial region has pillars that are continuous between successive chambers, grading into a system of numerous fine endoskeletal horizontal plates with irregularly arranged fine vertical elements (pillars or buttresses). In transverse sections this zone displays a reticulate-labyrinthic network. Cretaceous (late Barremian).

Superfamily LOFTUSIOIDEA Brady, 1884

The test is planispiral, but may uncoil in later stages. The wall is agglutinated with a differentiated outer layer and an inner alveolar layer. Late Triassic (Carnian) to Holocene.

Family Choffatellidae Maync, 1958

This family is characterised by having hypodermal alveoles. The tests are planispiral, lacking continuously developed endoskeletal pillars. Walls are finely and complexly alveolar. The septa have many fine apertures. Early Jurassic (Pliensbachian) to Late Cretaceous (Maastrichtian).

- *Balkhania* Mamontova, 1966 (Type species: *Balkhania balkhanica* Mamontova, 1966). Annular test with simplified, thin septa. Early Cretaceous (Barremian to Aptian) (Plate 5.7, fig. 12).
- *Broeckinella* Henson, 1948 (Type species: *Broeckinella arabica* Henson, 1948). The test is flabelliform, with a large undivided globular proloculus followed by a planispiral evolute stage. There are subepidermal vertical and transverse partitions resulting in a polygonal meshwork near the interior margins of the chambers. Late Cretaceous (Santonian to Maastrichtian) (Plate 5.7, fig. 13).
- *Pseudochoffatella* Deloffre, 1961 (Type species: *Pseudochoffatella cuvillieri* Deloffre, 1961). Flabelliform test with high chambers and septa that are as complex and thick as the hypodermis of the wall. It has a more coarsely agglutinated epidermis and less delicately alveolar epidermis than in the more choffatelloid B-forms (see *Choffatella*). The septa of *Pseudochoffatella* are much thicker than their narrow "tubular" apertures, whereas in *Balkhania* the septa are thinner than the broad apertures. Early Cretaceous (Aptian to Albian) (Fig. 5.5; Plate 5.3, 13; Plate 5.5, fig. 10).

Family Loftusiidae Brady, 1884

The test is globose to fusiform with septula perpendicular to septa, and parallel to the coiling direction. An alveolar hypodermis is weakly or fully developed. A subepidermal reticulate mesh is absent in the septa. Late Cretaceous (Maastrichtian) to?Eocene.

The following genera have a globose, spherical test with interseptal horizontal lamellae, and a very weakly developed alveolar hypodermis:

- *Feurtillia* Maync, 1958 (Type species: *Feurtillia frequens* Maync, 1958). This form is characterised by thick septa and a strong apertural neck even in small, wholly coiled specimens. The wall has a distinct subepidermal network of shallow alveoles that appear polygonal in tangential section. Late Jurassic to Early Cretaceous (Tithonian to Valanginian) (Plate 5.3, fig. 8)
- *Praereticulinella* Deloffre and Hamaoui, 1970 (Type species: *Praereticulinella cuvillieri* Deloffre and Hamaoui, 1970). Septula aligned from chamber to chamber, but without a basal pre-septal canal. There is a pre-septal tunnel at the top of the septum. Early Cretaceous (Barremian) (Plate 5.4, figs 1-2).
- *Reticulinella* Cuvillier, Bonnefous, Hamaoui and Tixier, 1970 (Type species: *Reticulina reicheli* Cuvillier et al., 1969). Secondary septula are intercalated between primary septula in later chambers and pre-septal canals are present at the base of the septum. Late Cretaceous (Cenomanian to Maastrichtian) (Plate 5.4, fig. 3).

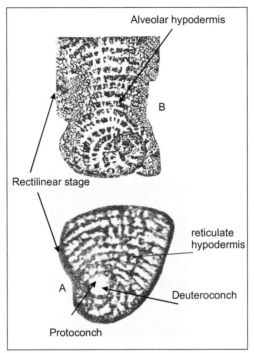

Figure 5.5. A) *Spiroconulus perconigi* Allemann and Schroeder, holotype figured by Allemann and Schroeder (1972), the Bathonian of Cadiz, South Spain; B) *Pseudochoffatella cuvillieri* Deloffre, latest Aptian to earliest Albian, North Croatia, after Gušič (1975), transverse section of a microspheric form (B-form). Note: The more coarsely agglutinated epidermis and less delicately alveolar epidermis than in other megalospheric choffatelloid (A-forms) (e.g. Plate 5.5, fig. 10)

While a fusiform test with interseptal pillars and a fully developed alveolar hypodermis (but with no pre-septal canal or tunnel) is shown by:

- *Loftusia* Brady, 1870 (T*ype* species: *Loftusia persica* Brady, in Carpenter and Brady, 1870). The test is irregularly fusiform or cylindrical in shape and may have either rounded or pointed ends. It is made of a series of chambers that are planispirally coiled around an elongate axis. The front end of each chamber slopes gently down to meet the previous whorl. The portion behind the frontal slope merges smoothly with the previous chamber at the point where its own frontal slope merges smoothly into the area of equal elevation. There is, thus, no part of the exterior wall possessing a configuration that would set it off as an apertural face in the normal sense, although there are a number of pores in the surface of the chamber adjacent to the previous whorl. The interiors of the chambers are partially filled by networks of irregular projections that cover the inner surface of the exterior walls, some of the projections reach the surface of the preceding whorl thus forming interseptal pillars or partial septa, depending on their shape. The spiral wall is made up of a thin outer calcareous layer and a thick arenaceous layer, the forward extension of the latter forms the gently-inclined frontal slope of each chamber. Late Cretaceous (Maastrichtian) to?Eocene (Fig. 5.6; Plate 5.8, figs 1-8; Plate 5.9, figs 6-7, with fig. 7

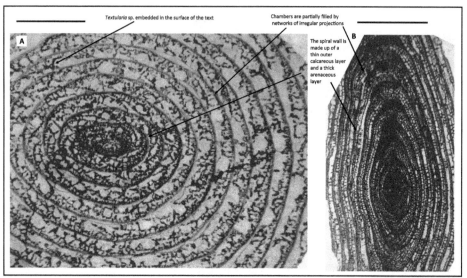

Figure 5.6. A) Oblique equatorial section of *Loftusia persica* Brady, Maastrichtian of Iran; B) *Loftusia morgani* Douville, Maastrichtian of Turkey. Scale bar = 0.5mm.

showing *Turborotalia* sp. trapped in a chamber, hence the suggested?Eocene age, see BouDagher-Fadel and Price, 2009).

Superfamily ATAXOPHRAGMIOIDEA Schwager, 1877
Members of this superfamily have multilocular, high trochospiral tests that become biserial or uniserial in later stages. Middle Triassic to Holocene.

Family Ataxophragmiidae Schwager, 1877
Tests are initially trochospirally enrolled, and often triserial or biserial with no continuous composite central columella. Apertures are high and terminal (Figs 5.7 and 5.8). Late Triassic to Palaeocene.

- *Ataxophragmium* Reuss, 1860 (Type species: *Bulimina variabilis* d'Orbigny). The chambers are formed in a low trochospiral with a simple undivided interior. Late Cretaceous (Cenomanian) to Palaeocene.
- *Opertum* Voloshina, 1972 (Type species: *Ataxophragmium (Opertum) incognitum* Voloshina, 1972). The shape of the test is similar to *Ataxophragmium* but the interior of the chambers are subdivided by numerous buttresses of internal partial partitions, that form distinct chamberlets. Cretaceous (Aptian to Maastrichtian) (Plate 5.10, figs 1-2).
- *Voloshinoides* Barnard and Banner, 1980 (Type species: *Arenobulimina labirynthica* Voloshina, 1961). The test is trochospirally enrolled with whorls enlarging rapidly. It has thick walls with the interior of the chambers being subdivided by numerous arched peripheral partitions. Often, they have intercalary buttresses attached to the outer wall that may meet to form irregular chamberlets. Cretaceous (Albian to Maastrichtian) (Plate 5.10, figs 3-4).

Figure 5.7. Key characteristics of Mesozoic genera with high comes. Trochospiral, often triserial to biserial with no continuous composite columella (the Ataxophragmiidae); with continuous composite columella (the Pfenderenidae); high trochospiral, with quinqueserial or quadriserial or triserial or biserial coiling modes, or with certain consecutive pairs of these (the Chrysalidinidae).

Coiling / chamber arrangement	Fine characteristic	Genus
Initially 3 chambers/whorl then 4 or more (tightly coiled, low spire)	No peripheral partitions	*Ataxophragmium*
	with peripheral partitions	*Opertum*
Initially 3 chambers/whorl then 4 or more (loosely coiled, high spire)	No peripheral partitions	*Arenobulimina*
	with peripheral partitions	*Voloshinoides*
3 chambers/whorl throughout	Low chambers, bo umbilical tube	*Verneuilinoides*
	High chambers, with umbilical tube	*Siphovalvulina*
4 ch/wh in adult	6/7 ch/wh to 4 ch/wh	*Orientalia*
	4 ch/wh throughout	*Verneuilinella*
4 to 3 ch/wh	4 ch/wh to 3 ch/wh	*Riyadhella*
4 ch/wh reducing to biserial	Septa convex — No peripheral partitions	*Dorothia*
	With peripheral partitions	*Cuneolina*
	Septa flat or concave — No peripheral partitions	*Marssonella*
	With peripheral partitions	*Pseudotextulariella*
adult with 3 to 4 ch/wh	Imperforate apertural plate — No pillars	*Pseudomarssonella*
	Scattered pillars on apertural plate	*Praechrysalidina*
adult with 3 ch/wh	Abundant pillars crowded centrally on broad plate	*Dukhania*
	Perforate apertural plate (cribrate apertures)	*Accordiella*

Higher-level divisions:

- Spiral suture and septa strongly oblique to the long axis
- Spiral suture almost at right angles to long axis (septa from oblique to parallel to long axis)
- Interiomarginal apertures have lip – only a thin rim
- Interiomarginal apertural lip extended across umbilicus as a broad plate
- Initially 4 or more chambers per whorl, often reducing

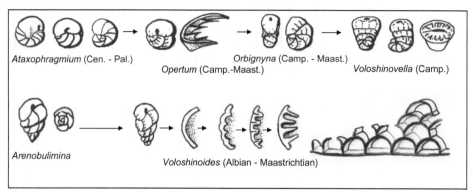

Figure 5.8.   The evolution of the internal structure of the simple Cretaceous ataxophragmids.

Family Cuneolinidae Saidova, 1981

Tests are conical to sub-flabelliform, and trochospiral in the early stage with as many as five chambers per whorl, rapidly reduced to biserial. Chambers are subdivided by radial partitions and may have horizontal partitions (see Fig. 5.9). Apertures are simple slits or series of pores along the base of the final chamber face. Middle Triassic (Anisian) to Late Cretaceous (Coniacian).

• *Cuneolina* d'Orbigny, 1839 (Type species: *Cuneolina pavonia* d'Orbigny, 1846). The test is compressed. The interiors of the chambers are divided into nearly rectangular chamberlets by both vertical and horizontal partitions perpendicular to the outer wall. Apertures form an aligned series of pores, along the base of the final chambers. Cretaceous (Valanginian to Maastrichtian) (Plate 5.10, figs 5-13, 15).

• *Pseudotextulariella* Barnard, 1953 (Type species: *Textulariella cretosa* Cushman, 1932). The initial stage is trochospiral, becoming triserial, then biserial. The chambers are subdivided by vertical and horizontal partitions, resulting in up to six tiers of chamberlets per chamber. The aperture is a single low slit at the base of the flattened apertural face. Cretaceous (Berriasian to Cenomanian) (Plate 5.10, figs 16, 17).

• *Sabaudia* Charollais and Brönnimann, 1965 (Type species: *Textulariella minuta* Hofker Jr., 1965). The test has a flattened base, a small initial trochospiral of three to four subglobular chambers, followed by later triserial and biserial final stages. Biserial chambers are subdivided by vertical and horizontal partitions resulting in a subepidermal network of chamberlets. The aperture is a single slit. *Sabaudia* differs from *Pseudotextulariella* in having a thick, hyaline calcareous layer enveloping the embryonts, which is often found isolated, and by the lack of foreign, agglutinated grains in the wall of species with tiered chamberlets. Cretaceous (Valanginian to early Cenomanian) (Plate 5.10, figs 18-19).

• *Vercorsella* Arnaud-Vanneau, 1980 (Type species: *Vercorsella arenata* Arnaud-Vanneau, 1980). The test is flaring, with an early trochospiral stage followed by a biserial stage. Chambers are subdivided by radial partitions that increase in number in successive chambers. Less well developed horizontal partitions are present in later chambers. The aperture is a single, simple slit in the basal groove. Early Cretaceous (Valanginian to Albian) (Plate 5.10, fig. 14).

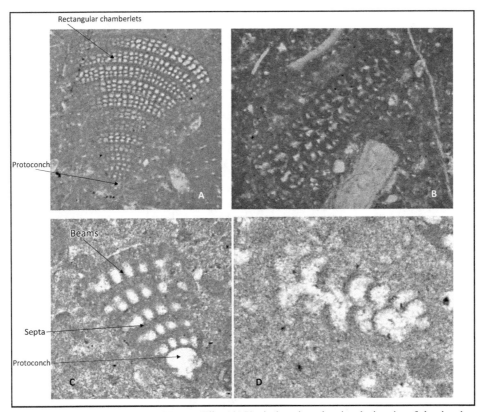

Figure 5.9. A-B) *Cuneolina parva* Henson, Tibet, A) Vertical section, showing the interior of the chambers which is divided into nearly rectangular chamberlets by both vertical and horizontal partitions perpendicular to the outer wall, B) horizontal section; C-D) *Vercorsella arenata* Arnaud-Vanneau, C) Vertical section, D) horizontal section.

Family Dicyclinidae Loeblich and Tappan, 1964
The test is discoidal with cyclical chambers subdivided by transverse and radial partitions that form numerous small chamberlets. Cretaceous (Albian to Santonian).

- *Dicyclina* Munier-Chalmas, 1887 (Type species: *Dicyclina schlumbergeri* Munier-Chalmas, 1887). Forms have a flattened initial stage, that is often formed with one proloculus, with later chambers becoming annular, and alternately added on the two sides of the test. The interiors of the chambers are divided by numerous thin radial partitions. The aperture is made of multiple pores at the periphery. Cretaceous (Albian to Santonian) (Plate 5.11, figs 1-3).

Superfamily ORBITOLINOIDEA Martin, 1890
Tests are conical with numerous chambers, partially subdivided by radial or transverse partitions or with pillars. Early Cretaceous to Oligocene.

Family Orbitolinidae Martin, 1890
Forms have an initial low trochospire, that is usually very much reduced and later is rectilinear, broad and conical (Fig. 5.10). They have low uniserial chambers, subdivided

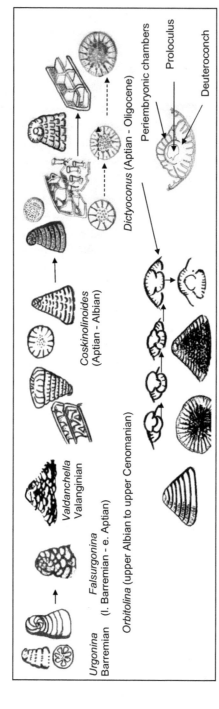

Figure 5.10.   The gradual evolution from *Coskinolinoides* to *Orbitolina*.

with pillars or peripheral vertical partitions. The aperture is cribrate. Early Cretaceous to Oligocene.

(i)   No complex central zone:
  • *Abrardia* Neumann and Damotte, 1960 (Type species: *Dictyoconus mosae* Hofker, 1955). Occasional radial partitions extending across the central region of the test. Late Cretaceous (Campanian to Maastrichtian) (Plate 5.7, fig. 9).
  • *Campanellula* De Castro, 1964 (Type species: *Campanellula capuensis* De Castro, 1964). The test is narrow, lacking thick radial partitions but having chamberlets in the uniserial part, separated by strongly undulating septa. Early Cretaceous (Valanginian to Barremian) (Plate 5.7, fig. 10).
  • *Coskinolinoides* Keijzer, 1942 (*Coskinolinoides texanus* Keijzer, 1942). Radial partitions alternate in successive chambers and extend from the periphery to the centre, with an areal aperture. Early Cretaceous (Aptian to Albian) (Plate 5.12, fig.10).
  • *Cribellopsis* Arnaud-Vanneau, 1980 (*Orbitolinopsis*? *neoelongata* Cherchi and Schroeder, 1978). With peripheral rectangular chamberlets made by alternating longer and shorter radial partitions. Early Cretaceous (late Hauterivian to early Aptian) (Plate 5.7, fig. 7).
  • *Falsurgonina* Arnaud-Vanneau and Argot, 1973 (Type species: *Falsurgonina pileola* Arnaud-Vanneau and Argot, 1973). Occurs without tiers of peripheral chamberlets and no true pillars in the central zone. Early Cretaceous (late Barremian to early Aptian) (Fig. 5.10).
  • *Pseudorbitolina* Douvillé, 1910 (Type species: *Pseudorbitolina marthae* H. Douvillé, 1910). The cyclic chambers in the median zone are subdivided into chamberlets by long radial partitions, thickening towards the lower side of the test, but leaving a large undivided zone. Late Cretaceous (Campanian to Maastrichtian) (Plate 5.7, figs 8, 14).

(ii) A complex central zone and radial partitions thickening away from the periphery, that become broken up into pillars in the central zone:
  • *Dictyoconus* Blanckenhorn, 1900 (Type species: *Dictyoconus egyptiensis* (Chapman) = *Patellina egyptiensis* (Chapman). Aptian to Oligocene (Plate 5.13, figs 11-12).
  • *Montseciella* Cherchi and Schroeder, 1999 (Type species: *Paleodictyoconus glanensis* Foury, 1968). Cretaceous (late Hauterivian-early Aptian) (Plate 5.9, figs 2-4; Plate 5.13, figs 7-10).
  • *Paracoskinolina* Moullade, 1965 (Type species: *Coskinolina sunnilandensis* Mayne, 1955). Without tiers of peripheral rectangular chamberlets. Early Cretaceous (Barremian to Albian) (Plate 5.2, fig. 3; Plate 5.12, fig. 11).
  • *Paleodictyoconus* Moullade, 1965 (Type species: *Dictyoconus cuvillieri* Foury, 1963). With peripheral tiered rectangular chamberlets in a narrow marginal zone subdivided by many radial beams. Main partitions are numerous and concentrated more towards the base of the test, those of adjacent rows alternating in position. Early Cretaceous (Valanginian to early Aptian) (Fig. 5.11; Plate 5.13, fig. 13).

- *Urgonina* Foury and Moullade, 1966 (Type species: *Urgonina protuberans* Foury and Moullade, 1966). Apex twisted and outer parts of the chambers lack partitions. Early Cretaceous (Barremian) (Fig. 5.10).

(iii) Radial partitions thicken away from periphery to anastomose centrally around the aperture and form a reticulate zone in transverse section:

- *Calveziconus* Caus and Cornella, 1982 (Type species: *Calveziconus lecalvezae* Caus and Cornella, 1982). Thick peripheral wall with numerous long radial partitions with shorter partitions intercalated among the longer ones near the periphery. Late Cretaceous (Campanian) (Plate 5.7, fig. 11).
- *Iraqia* Henson, 1948 (Type species: *Iraqia simplex* Henson, 1948). No tiers of peripheral chamberlets. With main partitions that are reticulate in the central area with transverse interconnections. Early Cretaceous (Aptian to Albian) (Plate 5.13, figs 2-3).
- *Neorbitolinopsis* Schroeder 1964 (Type species: *Orbitolina conulus* H. Douvillé, 1912). With tiers of peripheral rectangular chamberlets. Cretaceous (late Albian to early Cenomanian) (Plate 5.13, figs 5-6).
- *Orbitolinella* Henson, 1948 (Type species: *Orbitolinella depressa* Henson, 1948). The interior of the test is subdivided by numerous radial beams, with those of successive chambers alternating in position with shorter and thinner secondary beams. Some beams fuse in the central zone. Cretaceous (Late Cenomanian to Turonian) (Plate 5.13, fig. 14).
- *Orbitolinopsis* Henson, 1948 (Type species: *Orbitolinopsis kiliani* Henson, 1948). Without tiers of peripheral chamberlets. Early Cretaceous (Aptian to early Albian) (Plate 5.5, fig. 4).
- *Valdanchella* Canerot and Moullade, 1971 (Type species: *Simplorbitolina (?) miliani* Schroeder, 1968). The peripheral zones of chambers are subdivided into rectangular chamberlets by fine radial partitions. Long partitions alternate with smaller ones of constant thickness. Early Cretaceous (Valanginian) (Fig. 5.10).
- *Valserina* Schroeder, 1968 (Type species: *Valserina broennimanni* Schroeder and Conrad, 1968). Similar to *Neorbitolinopsis* except for the asymmetry of its embryon. Early Cretaceous (middle Barremian).

(iv) Radial partitions become zigzag, thickened and fused centrally, giving a stellate appearance in transverse section:

- *Simplorbitolina* Ciry and Rat, 1953 (Type species: *Simplorbitolina manasi* Ciry and Rat, 1953). Without tiers of peripheral chamberlets. Radial partitions are partly discontinuous and broken up into pillars. Tests are conical. Early Cretaceous (Aptian to Albian) (Plate 5.13, fig. 1).
- *Dictyoconella* Henson, 1948 (Type species: *Dictyoconella complanata* Henson, 1948). The test is laterally compressed, but with a central zone having pillars and tiered peripheral chamberlets. Late Cretaceous (late Cenomanian to Maastrichtian) (Plate 5.7, figs 1-3).

(v) No pillars, but with peripheral tiered, rectangular chamberlets in two or more series. Radial partitions thicken, with triangular cross-section, away from the periphery and anastomose in the central area (Fig. 5.11). The earliest formed chambers of the megalospheric generation can form a complex embryonic apparatus that can be divided into a protoconch, deuteroconch, a sub-embryonic zone and peri-embryonic chamberlets depending on the genera involved (see Schroeder, 1975, Simmons et al. 2000):

- *Conicorbitolina* Schroeder, 1973 (Type species: *Orbitolites conica* d'Archiac, 1837). The large proloculus is divided into a protoconch and deuteroconch with the marginal zone becoming extensively divided by vertical and horizontal partitions. Cretaceous (late Albian to early Cenomanian) (Plate 5.12, fig. 5).

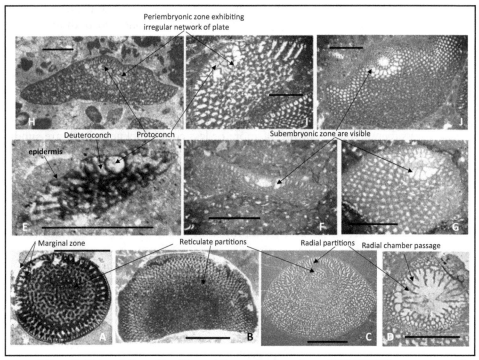

Figure 5.11. A) *Montseciella arabica* (Henson) Barremian from Qatar, NHM P35805, transverse section through the embryonic apparatus showing the reticulate zone; B) *Palorbitolina lenticularis* (Blum), sub-horizontal section, pyrite filled from the early Aptian of Venezuela, figured by Hofker (1963); C) *Orbitolina sp.*, a random tangential section through chamberlets, radial and central zones; D, I), *Mesorbitolina aperta* (Erman, 1854) from the Albian of Tibet, figured by BouDagher-Fadel et al. (2017), the upper part of the deuteroconch is subdivided by several sets of alveoli of different length and breath, whereas the basal part exhibits a more or less developed irregular network of plates; E) *Praeorbitolina cormyi* Schroeder, from the Aptian of Tibet figured by BouDagher-Fadel et al. (2017), the oldest representative of this lineage, shows a clearly developed initial spiral and a bilateral symmetrical embryonic apparatus, whose deuteroconch is still undivided; F) *Palorbitolinoides hedini* Cherchi and Schroeder from the Albian of Tibet, axial section through the large embryonic apparatus which is in a central position and consisting of alveolar protoconch and fused peri-embryonic chamberlets. The megalospheric embryonic apparatus is in a central position; G, J) *Mesorbitolina texana* (Roemer), from the Albian of Tibet, tangential section showing the embryonic apparatus with the regular partitions within the sub-embryonic zone followed by chambers triangular in shape; H) *Orbitolina qatarica* Henson, from early Cenomanian of Qatar, NHM Henson coll., showing the complex embryonic apparatus with a deuteroconch about three times thicker than the sub-embryonic zone, both are subdivided by radial beams. Scale bars = 0.5mm.

- *Eopalorbitolina* Schroeder, 1968 (Type species: *Eopalorbitolina charollaisi* Schroeder, 1968). The embryonic apparatus is present at one side of the apex, and consists of a globular proloculus and a lateral deuteroconch. Early Cretaceous (Barremian).
- *Eygalierina* Foury, 1967 (Type species: *Eygalierina turbinata* Foury, 1968). Embryonic apparatus consisting of a flattened protoconch and a larger deuteroconch, surrounded by a hemispherical zone of eight to ten periembryonal chambers. Early Cretaceous (Barremian).
- *Orbitolina* d'Orbigny 1850 (Type species: *Orbulites concava* Lamarck, 1816). There is a large apically situated embryonic apparatus. The deuteroconch is about three times thicker than the sub-embryonic zone and both are subdivided by radial beams. Zigzag radial partitions occur in the central zone and there is a series of marginal and vertical plates. Cretaceous (late Albian to Cenomanian) (Figs 5.11C, 5.11H; Fig. 5.12; Plate 5.11, figs 4-6; Plate 5.12, figs 1, 4, 6; Plate 5.14, figs 5, 14-15).
- *Mesorbitolina* Schroeder, 1962 (Type species: *Orbitulites texanus* Roemer, 1849*)*. There is an apically situated embryonic zone in which the deuteroconch and subembryonic zone are of more or less equal thickness. Both are subdivided by vertical exoskeletal beams. Early Cretaceous (early Aptian to early Cenomanian) (Plate 5.12, fig.2; Plate 5.14, figs 6-12).
- *Naupliella* Decrouez and Moullade, 1974 (Type species: *Naupliella insolita* Decrouez and Moullade, 1974). There is an embryonic apparatus near the apex, consisting of an undivided protoconch and a deuteroconch divided in the upper part by vertical partitions. Early Cretaceous (late Albian) (Plate 5.16, fig. 9).
- *Neoiraqia* Danilova, 1963 (Type species: *Neoiraqia convexa* Danilova, 1963). The embryonic apparatus is near the apex of a conical test. Later chambers are uniserial and discoidal with a narrow cellular subepidermal marginal zone

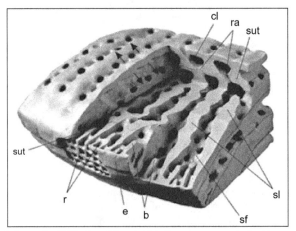

Figure 5.12. Endoskeleton in *Orbitolina*. The model shown was made in the years around 1955 by M. Reichel and was later published by Hottinger (2006). Abbreviations: **b:** beam; **cl:** chamberlet; **e:** epidermis; **r:** rafter; **ra:** ramp; **sf:** septal face; **sl:** septulum; **sut:** suture of the chambers. Double arrows in E and F: crosswise oblique foraminal axes.

formed by alternating radial beams. The central zone is wide and reticulate. Late Cretaceous (late Cenomanian) (Plate 5.2, fig. 4).

- *Palorbitolina* Schroeder, 1963 (Type species: *Madreporites lenticularis* Blumenbach, 1805). There is a large, apically sited, globular, fused protoconch and deuteroconch forming a relatively simple embryonic apparatus, which can be surrounded by a peri-embryonic ring of obliquely arranged chamberlets. Early Cretaceous (late Barremian to Aptian) (Fig. 5.28 Plate 5.14, figs 1-4, 13, 16).
- *Palorbitolinoides* Cherchi and Schroeder, 1980 (Type species: *Palorbitolinoides hedini* Cherchi and Schroeder, 1980). Similar to *Palorbitolina*, but the peri-embryonic ring of chambers is expanded and fused to form a sub-embryonic area with a few beams, which divide into irregular chamberlets. Early Cretaceous (late Aptian to early Albian) (Fig. 5.11 F; Fig. 5.29).
- *Praeorbitolina* Schroeder, 1965 (Type species: *Praeorbitolina cormyi* Schroeder, 1965). A small embryonic apparatus is eccentrically situated and not completely surrounded by the earliest post-embryonic chambers. Early Cretaceous (early Aptian to late Albian) (see BouDagher-Fadel *et al.*, 2017) (Fig. 5.11 F; Fig. 5.29).
- *Rectodictyoconus* Schroeder, 1964 (Type species: *Rectodictyoconus giganteus* Schroeder, 1964). The test is without an early coiled stage. An embryonic apparatus consists of a protoconch symmetrically overlying a deuteroconch, the latter is surrounded by a marginal zone subdivided by vertical beams. The central zone has incomplete and anastomosing pillars. Early Cretaceous (Barremian).

Superfamily COSKINOLINOIDEA Moullade, 1965
The test is conical, with an early stage trochospiral, then becoming uniserial and rectilinear. Apertures are basal, and cribrate. Late Cretaceous (Cenomanian) to Eocene (Lutetian).

Family Coskinolinidae Moullade, 1965
Forms have a rectilinear part with broad, low chambers, subdivided by pillars or irregular partitions. Late Cretaceous (Cenomanian) to Eocene (Lutetian).

- *Lituonelloides* Henson, 1948 (Type species: *Lituonelloides compressus* Henson, 1948). The test is elongate, with an early conical stage and trochospirally enrolled chambers. The central regions of the chambers in the rectilinear part have irregular pillars. Late Cretaceous (Maastrichtian) (Plate 5.5, fig. 15).
- *Pseudolituonella* Marie, 1955 (Type species: *Pseudolituonella reicheli* Marie, 1955). Initially there is a short, trochospiral stage, but the later stage shows broad and low uniserial chambers. The chamber interior has short tubular pillars. Late Cretaceous (Cenomanian to Campanian) and Eocene (Plate 5.6, fig. 17).

Superfamily TEXTULARIOIDEA Ehrenberg, 1838
In this superfamily, the test is trochospiral, biserial or triserial in the early stages and later may be uniserial or biserial. Walls are agglutinated and canaliculated. Early Jurassic (Sinemurian) to Holocene.

Family Chrysalidinidae Neagu, 1968
Tests form high trochospirals, with quinqueserial or quadriserial or triserial or biserial coiling modes, or with certain consecutive pairs of these (Figs 5.7). The aperture is central along the axis of coiling. In quadriserial or quinqueserial forms, an umbilicus is

present and the aperture is covered with a broad umbilical flap, which may be penetrated by multiple accessory apertures. Internal pillars may develop between successive intra-umbilical flaps. Banner et al. (1991) divided this family into two subfamilies; the mainly Cretaceous Chrysalidininae (see below) and the mainly Jurassic Paravalvulininae (see Chapter 4). This classification is followed in this and the previous chapter.

Subfamily Chrysalidininae Neagu, nom. Transl, Banner et al., 1991

This subfamily, as revised by Banner et al. (1991), is essentially triserial throughout its ontogeny (at least in the megalospheric generation), becoming biserial or quadriserial in the adult. Walls are solid but sometimes becoming canaliculate. Early Jurassic (Sinemurian) to Late Eocene.

- *Accordiella* Farinacci, 1962 (Type species: *Accordiella conica* Farinacci, 1962). Initially triserial, but later quadriserial (or multiserial) with internal pillars. Late Cretaceous (Coniacian to Santonian) (Plate 5.15, fig. 21).
- *Chrysalidina* d'Orbigny, 1839 (Type species: *Chrysalidina gradata* d'Orbigny, 1839). Triserial throughout but with internal pillars. Cretaceous (Aptian to Cenomanian) (Plate 5.15, figs 6-8, 16).
- *Dukhania* Henson, 1948 (Type species: *Dukhania conica* Henson, 1948). Initially triserial but biserial in the adult stage, with internal pillars and convex septa. Cretaceous (Hauterivian to earliest Albian) (Plate 5.15, figs 10, 19-20).
- *Praechrysalidina* Luperto Sinni, 1979 (Type species: *Praechrysalidina infracretacea* Luperto Sinni, 1979). Triserial throughout, no internal pillars. Early Cretaceous (Hauterivian to Albian) (Plate 5.15, figs 12-14).

Superfamily CYCLOLINOIDEA Loeblich and Tappan, 1964

Members of this superfamily have a test that is composed of cyclical chambers filled with numerous radial pillars. Late Jurassic (Kimmeridgian) to Late Cretaceous (Santonian).

Family Cyclolinidae Loeblich and Tappan, 1964

The test shows annular or cyclical adult chambers that may be subdivided by pillars or partitions. Cretaceous (Valanginian to Campanian).

Subfamily Cyclopsinellinae Loeblich and Tappan, 1984

The tests have cyclical chambers, with a median region with numerous radial pillars, but there are no subdivisions present in the external subepidermal area. A radial stolon system is present. Late Jurassic (Kimmeridgian) to Late Cretaceous (Santonian).

- *Cyclopsinella* Galloway, 1933 (Type species: *Cyclopsina steinmanni* Munier-Chalmas, 1887). The chamber interiors are undivided in the early stage, but later they have pillars which may bifurcate and fuse to form anastomosing pillars. Apertures form a double row of pores. Late Cretaceous (Cenomanian to Santonian).
- *Mangashtia* Henson, 1948 (Type species: *Mangashtia viennoti* Henson, 1948). The test is compressed, discoidal with numerous cyclical or annular chambers. Apertures are multiple, aligned in one row in the middle of the apertural face. The axes of the stolons are radial. Numerous subcylindrical or beam shaped pillars that are

perpendicular to the septa are present in the central zone of the chambers. The marginal zone of the chamber is not internally subdivided. Originally listed by Loeblich and Tappan in their "genera of uncertain status". They noted that the genus is "unrecognisable" because many of the essential characters were not described by Henson. Fourcade et al. (1997) revised the genus based on the study of new topotype material, as well as specimens preserved in the Henson Collection housed in the Natural History Museum (London), and placed the genus in the subfamily Cyclopsinellinae. Their description is therefore regarded as the emendation. *Mangashtia* differs from *Cyclopsinella* in the nature of the internal structure (pillars in the form of beams) and in its apertural characteristics (a single row of pores rather than a double row). Late Jurassic (Kimmeridgian) to Late Cretaceous (Turonian) (Plate 5.1, fig. 14; Plate 5.9, fig. 5).

Subfamily Ilerdorbinae Hottinger and Caus, 1982
The outer parts of the chambers are subdivided by secondary partitions. Apertures alternate in position from chamber to chamber, producing an oblique stolon system. Cretaceous (Valanginian to Campanian).

- *Dohaia* Henson, 1948 (Type species: *Dohaia planata* Henson, 1948). Simple exoskeletal partitions perpendicular to the wall subdivide the peripheral part of the chambers. Late Cretaceous (Cenomanian to Turonian) (Plate 5.16, figs 1-3).
- *Eclusia* Septfontaine, 1971 (Type species: *Eclusia moutyi* Septfontaine, 1971). Shows no alveolar hypodermis. Central pillars span flabelliform chambers with the partitions forming a series of chamberlets. Early Cretaceous (Valanginian to Barremian) (Plate 5.16, fig. 4).

ORDER ROTALIIDA Delage and Hérouard, 1896
Members of this order have tests that are multilocular with a calcareous wall, made of perforate hyaline lamellar calcite. They exhibit apertures that are either simple or have an internal tooth plate. Triassic to Holocene.

Superfamily PLANORBULINOIDEA Schwager, 1877
Tests are trochospiral in the early stages, but later stages may be uncoiled and rectilinear or biserial or with many chambers in the whorl. Apertures are intra- to extra-umbilical, but additional equatorial apertures may be present. Early Cretaceous (Berriasian) to Holocene.

Family Cymbaloporidae Cushman, 1927
Chambers occur in a single layer. Late Cretaceous (Cenomanian) to Holocene.

- *Archaecyclus* Silvestri, 1908 (Type species: *Archaecyclus cenomaniana* (Seguenza) = *Planorbulina? Cenomaniana* Seguenza, 1882). The test has a large proloculus, followed by an enrolled stage of 5 chambers per whorl, later chambers occur in an annular series. Late Cretaceous (Cenomanian to Campanian) (Plate 5.5, figs 1-7).

Superfamily ORBITOIDOIDEA, Schwager, 1876

Tests are discoidal to lenticular with prominent dimorphism, in most orbitoidal species both megalospheric and microspheric generations are found. Microspheric specimens occur with a distinctly small protoconch (usually about 20 μm), but megalospheric forms have a distinctive embryonic stage, enclosed in a thicker wall. Equatorial and lateral chambers may be differentiated or indistinguishable (Figs 5.13-5.17). Late Cretaceous (Santonian) to Oligocene.

Family Orbitoididae Schwager, 1876

Tests are large lenticular, non-canaliculate, and composed of a median layer, comprising the initial stage surrounded by concentrically arranged equatorial chambers and flanked by layers of lateral chamberlets. The embryonic chamber usually consists of two parts (a globular first chamber or protoconch and a reniform second chamber or deuteroconch) surrounded by a relatively thick wall that is in turn surrounded immediately by equatorial chambers, called peri-embryonic chambers (Fig. 5.13). The third chamber is called an auxiliary chamber, and in advanced species, there may be two auxiliary chambers. Stolons (apertures) are present and connect the equatorial chambers. Most genera of Cretaceous orbitoidal foraminifera are ornamented with rounded

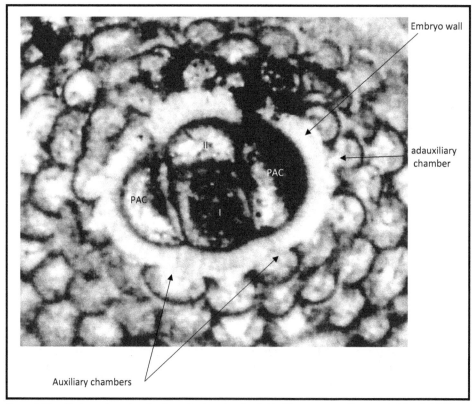

Figure 5.13. Embryonic and periembryonic chambers in *Orbitoides* spp. according to van Hinte's 1966 concept. **I:** protoconch; **II:** deuteroconch; **PAC:** principal auxiliary chamber.

pustules. Late Cretaceous (Santonian to Maastrichtian). For a comprehensive study of this group see van Gorsel (1978).

Subfamily Orbitoidinae Schwager, 1876

Lateral chambers may be present and differentiated from the equatorial layer in the post embryonal stage. Late Cretaceous (late Santonian to Maastrichtian).

- *Monolepidorbis* Astre, 1928 (Type species: *Monolepidorbis sanctae-pelagiae* Astre, 1927). *Monolepidorbis* differs from *Orbitoides* by the absence of lateral chamberlets. However, long incipient lateral chamberlets may be present. Several authors (van Gorsel, 1978; Loeblich and Tappan, 1988) have argued in favour of the inclusion of these forms in *Orbitoides*. Late Cretaceous (late Santonian to Maastrichtian) (Plate 5.17, figs 1-4; Fig. 5.14).

- *Orbitoides* d'Orbigny, 1848 (Type species: *Orbitoides media* (d'Archiac) = *Orbitolites media* d'Archiac, 1889). Lenticular to discoidal in shape. The test is made up of a single, roughly flat layer of large chambers (the median layer) with a mound of much smaller chambers on either sides (the lateral chambers). In plan view, the test is slightly asymmetric, usually with a central knob on one side and radial ridges on the test. In the central part of the median layer, there is a group of two to four chambers with thin and straight mutual walls, the whole enclosed within a thick and conspicuously porous outer wall. Epi-auxilliary chambers (chambers originating from stolons in the embryonal wall enveloping the protoconch, deuteroconch and auxiliary chambers) are present. This is followed immediately by concentric rings of simultaneously-formed smaller chambers. Stolons are big enough to be observed in vertical sections. Late Cretaceous (late Santonian to Maastrichtian) (Plate 5.17, figs 5-9; Plate 5.18, figs 1-6; Figs 5.15, 5.16).

- *Pseudomphalocyclus* Meriç, 1980 (Type species: *Pseudomphalocyclus blumenthali* Meriç, 1980). The embryo is followed by a single row of arcuate equatorial chambers, growing to two layers and three layers near the periphery. Three to four layers of lateral chamberlets are present at the centre of the test, decreasing to one or two layers near the periphery. Late Cretaceous (late Maastrichtian) (Fig. 5.14).

- *Simplorbites* de Gregorio, 1882 (Type species: *Simplorbites cupulimis* de Gregorio, 1882 = *Nummulites papyraceus* Boubeé, 1832). This form shows a median layer, arched and added to by an annular series. Lateral chamberlets occur with pillars. Equatorial chamberlets are curved and surrounded on both sides by small curved thickened chamberlets. *Simplorbites* differs from *Orbitoides* in having a larger nucleoconch, which is subdivided by numerous arcuate partitions into a multilocular form instead of a quadrilocular or a bilocular one, as in *Orbitoides*. Late Cretaceous (Maastrichtian) (Plate 5.19, figs 1-6; Fig. 5.14).

Subfamily Omphalocyclinae Vaughan, 1928

The lateral chambers are not differentiated from the equatorial layer in the post embryonal stage. Late Cretaceous (Maastrichtian).

- *Omphalocyclus* Bronn, 1853 (Type species: *Omphalocyclus macroporus* Bronn, 1853). The test is discoidal and biconcave. The central part of the test may consist of one or two layers of chambers. Additional layers of arched chambers are added in alternate stages,

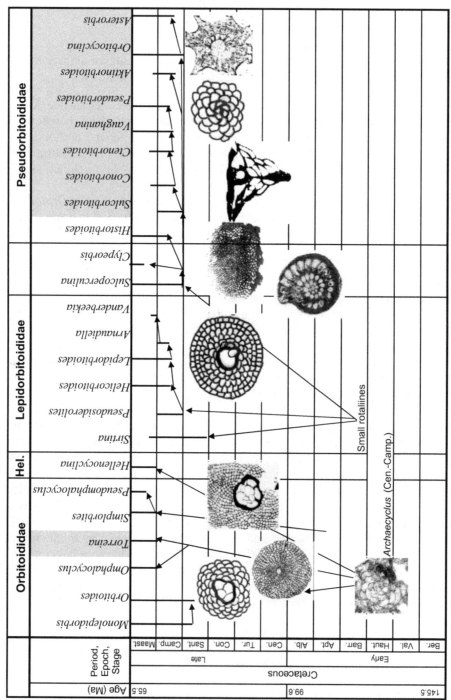

Figure 5.14. The different lineages of the orbitoids. The shaded boxes highlight forms found only in the Caribbean province.

instead of an orbitoid spiral (Fig. 5.15). The embryonic stage has two to four chambers. Equatorial chambers communicate through large marginal stolons. Late Cretaceous (Maastrichtian) (Figs 5.15A, B; Fig. 5.16C; Plate 5.20, figs 1-5; Plate 5.21, figs 2-4, 6-8).

- *Torreina* Palmer, 1934 (Type species: *Torreina torrei* Palmer, 1934). The test is globular with a centrally located embryon. The embryonic stage has four to five chambers, surrounded by a thick wall. From the embryon, lateral chambers are low and arcuate, and added uniformly in all directions. This growth pattern is characteristic of Cretaceous orbitoids, but is also found in the Neogene genus *Sphaerogypsina* (see Chapter 6). Late Cretaceous (Maastrichtian) (Fig. 5.14).

Family Hellenocyclinidae Freudenthal, 1969
Shows an orbitoid test, but without lateral chamberlets. Late Cretaceous (Maastrichtian).

- *Hellenocyclina* Reichel, 1949 (Type species: *Hellenocyclina beotica* Reichel, 1949). The test is small and slightly conical. The embryonic stage consists of five to six chambers, cruciformly arranged around the protoconch. The embryon is surrounded by arcuate equatorial chambers, arranged in a regular concentric growth pattern. The lateral walls are traversed by fine pores. It differs from *Lepidorbitoides* (below) by the absence of lateral chambers, the small size of the test and initial chambers. Late Cretaceous (Maastrichtian) (Fig. 5.14).

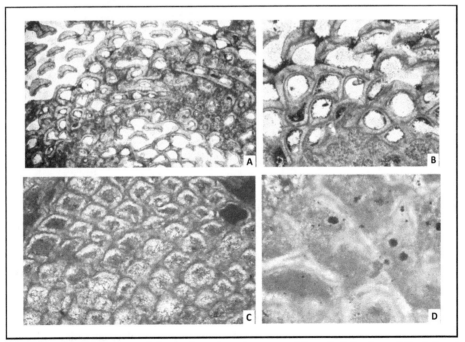

Figure 5.15. A, B) development of chamberlet cycles, enlargement of the arched chambers of *Omphalocyclus* sp. They are added in alternate stages, instead of an orbitoid spiral. The embryonic stage has two to four chambers, and equatorial chambers communicate through large marginal stolons; C-D) *Orbitoides* sp., South Africa, with concentric rings of simultaneously formed smaller chambers, and with stolons that are big enough to be observed in vertical sections.

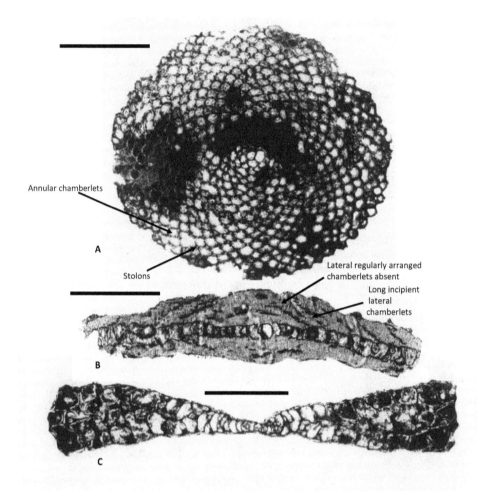

Annular chamberlets

Stolons

Lateral regularly arranged chamberlets absent

Long incipient lateral chamberlets

A

B

C

Figure 5.16.   A) Equatorial section of *Monolepidorbis*; B) Vertical section of *Monolepidorbis*; C) Vertical section of *Omphalocyclus.*

Family Lepidorbitoididae Vaughan, 1933

The Lepidorbitoididae differ from the Orbitoididae by having a distinct structure to their embryonic apparatus, and by the form of their median chambers. An embryonic stage with two chambers is followed by hexagonal or arcuate equatorial chamberlets and by differentiated lateral chambers (Fig. 5.17). Late Cretaceous (Santonian) to Middle Eocene.

- *Arnaudiella* Douvillé, 1907 (Type species: *Siderina douvillei* Abrard, 1926 = *Arnaudiella grossouvrei* H. Douvillé. 1907). The test is large, flat, lenticular with a wall much thickened into a broad flattened flange at the periphery, pierced by radial canals that open into coarse pores at the periphery, or into the chamber lumen of the succeeding whorl. Numerous pustules cover the ventral area. The embryonic apparatus is composed of a spherical protoconch and a slightly larger deuteroconch, followed

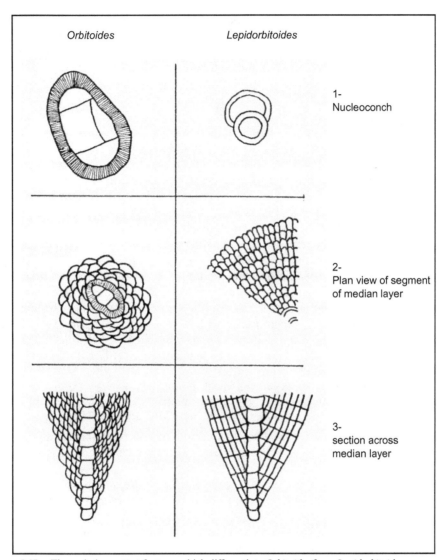

Orbitoides	Lepidorbitoides	
		1- Nucleoconch
		2- Plan view of segment of median layer
		3- section across median layer

Figure 5.17.   The main important features which differentiate *Orbitoides* from *Lepidorbitoides*.

by two to five whorls of spirally arranged primary chambers that increase rapidly from about eight chambers in the earliest whorl to up to forty five in the latest whorl. No secondary equatorial chambers are found. At the peripheral side of the whorls of the primary chambers, radial canals and rods are formed. On the lateral sides, orbitoidal chamberlets are added between the planispiral-involute spiral whorls. Late Cretaceous (late Campanian) (Fig. 5.14).

- *Clypeorbis* Douvillé, 1915 (Type species: *Orbitoides mammillatus* Schlumberger, 1902). An asymmetrical test, with a thin equatorial layer of curved chambers, with numerous hexagonal lateral chamberlets on both sides. Late Cretaceous (late Maastrichtian) (Fig. 5.14).

- *Helicorbitoides* Macgillavry, 1963 (Type species: *Pseudorbitoides longispiralis* Papp and Küpper, 1953). Species of this genus have lateral chambers and a long primary spiral, between the whorls of which secondary equatorial chambers are formed. *Pseudosiderolites*-like radial canals and plates are conspicuous in the primitive forms. Late Cretaceous (late Campanian to Maastrichtian) (Fig. 5.14).
- *Lepidorbitoides* Silvestri, 1907 (Type species: *Orbitoides socialls* Leymerie, 1851). Forms are flat, large (up to 10 mm in diameter, and rarely up to 25 mm), lenticular in shape, and made up of a single roughly flat layer of moderately large chambers (median layer), with a mound of much smaller chambers on either side (lateral chambers), ornamented with rounded pustules. In plan view the central part of the median layer usually consists of two moderately large chambers, the protoconch is the smaller of the two. It is partially embraced by the second chamber, the reniform deuteroconch. Together they make up the nucleoconch. They are followed by concentric rows of closely-spaced, simultaneously formed small spatulate to arcuate equatorial chambers arranged in alternating positions around the nucleoconch. Nepionic arrangements vary from reduced uniserial to quadriserial, with up to 15 adauxilliary chambers (small chambers originating from stolons in the deuteroconchal wall). The equatorial chambers communicate with each other via diagonal and median stolons, while the lateral chambers do so via pores. In European material it is quite easy to distinguish *Lepidorbitoides*, with its regular, porous and relatively thin lateral walls, from *Orbitoides* with it more solid walls that have a distinct black inner lining. Late Cretaceous (late Campanian to Maastrichtian) (Fig. 5.17; Plate 5.2, figs 12-14).
- *Pseudosiderolites* Smout, 1955 (Type species: *Siderolites vidali* Douvillé, 1907). The test is large and bilaterally symmetric. Walls are thick and lamellar. The umbilical region has spiral canals between the pillars. Apertures are single in the protoconch and deuteroconch, but multiple in later chambers with a single row of openings at the base of the apertural face. Late Cretaceous (Campanian) (Fig. 5.14).
- *Sirtina* Brönnimann and Wirz, 1962 (Type species: *Sirtina orbitoidiformis* Brönnimann and Wirz, 1962). Tests are lenticular with a two-chambered embryo of approximately equal size protoconch and deuteroconch, followed by a trochospiral early stage, but later they become nearly planispiral and involute. Thin vertical canals form between well developed umbilical pillars. Several layers of orbitoidal lateral chambers are present on the dorsal side in addition to the pillars. Late Cretaceous (Santonian to early Maastrichtian) (Fig. 5.14).
- *Sulcoperculina* Thalmann, 1939 (Type species: *Camerina? dickersoni* Palmer, 1914). A rotaliine without lateral chambers but with a peripheral sulcus in which small radial plates or rods can be observed. Late Cretaceous (Campanian to Maastrichtian) (Figs 5.14; 5.18).
- *Vanderbeekia* Brönnimann and Wirz, 1962 (Type species: *Vanderbeekia trochoidea* Brönnimann and Wirz, 1962). This form differs from *Sirtina* by having a thin layer of equatorial chambers between the lateral chambers of the dorsal side and by exhibiting ventral spiral chambers. Late Cretaceous (early Maastrichtian) (Fig. 5.14).

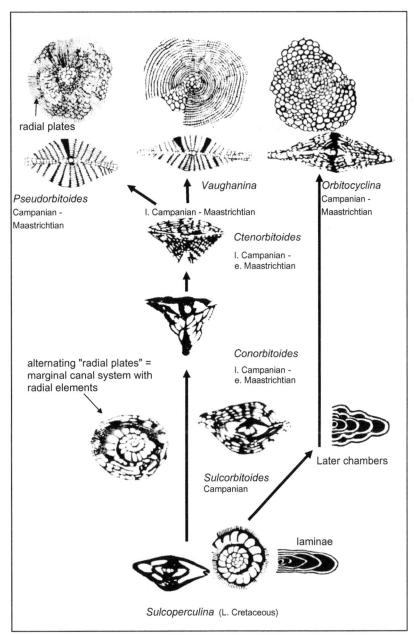

Figure 5.18. The three lineages that evolved from *Sulcoperculina* in the Late Cretaceous. The "radial plates" of Brönnimann (1956) and other US workers are in fact a marginal canal system with radial elements that is, however, not a marginal cord, thus unrelated to the Neogene nummulitids.

Superfamily ROTALIOIDEA Ehrenberg, 1839
Tests are involute to evolute, initially trochospiral or planispiral, commonly with many chambers in numerous whorls. As new chambers are added, septal flaps attach to previous apertural faces and enclose radial canals, fissures, umbilical cavities, and intraseptal and subsutural canals. The wall is perforate, hyaline calcite, and generally optically radial in structure. Primary apertures are single or multiple. Small opening into the canal system may occur along the sutures. Late Cretaceous (Coniacian) to Holocene.

Family Pseudorbitoididae Rutten, 1935
The test is lenticular, canaliculated with a two-chambered embryon followed by a spire of nepionic chambers. This group is characterized by the presence of a structure of radially arranged calcareous elements in the equatorial layer (Figs 5.14, 5.18). Late Cretaceous (Campanian to Maastrichtian).

Subfamily Pseudorbitoidinae Rutten, 1935
The equatorial (median) layer is subdivided vertically by variously arranged "radial plates or rods". The "radial plates" of Brönnimann (1956) and other US workers are in fact a marginal canal system with radial elements that are, however, not marginal cords, thus are unrelated to the Neogene nummulitids. Late Cretaceous (Campanian to Maastrichtian).

• *Conorbitoides* Brönnimann 1958 (Type species: *Conorbitoides cristalensis* Brönnimann 1958). Characterised by a conical test with a pointed apex and a neanic stage similar to *Sulcoperculina*. Late Cretaceous (late Campanian to early Maastrichtian) (Fig. 5.18).
• *Historbitoides* Brönnimann 1956 (Type species: *Conorbitoides kozaryi* Brönnimann 1958). Similar to *Pseudorbitoides*, but the single set of radial plates are irregularly interconnected laterally, and with incipient radii and interradii. Late Cretaceous (late Campanian to early Maastrichtian).
• *Pseudorbitoides* Douvillé, 1922 (Type species: *Pseudorbitoides trechmanni* Douvillé, 1922). An orbitoidal foraminifera with a circular outline and median and lateral layers, but with the median layer being double or triple towards the periphery as in *Omphalocyclus*. The radial (equatorial) layer lacks annular walls. The lateral chambers are irregular in form and relatively thin walled, and connected by numerous pores in the chamber walls. The neanic stage has a single set of radial plates, that are not interconnected radially. Late Cretaceous (late Campanian to Maastrichtian) (Fig. 5.18).
• *Sulcorbitoides* Brönnimann, 1955 (Type species: *Sulcorbitoides pardoi* Brönnimann, 1955). The equatorial chambers are trochospirally arranged in three whorls. It differs from *Pseudorbitoides* in having two sets of alternating systems of vertical radial plates, which Brönnimann referred to as "radial rods", separated by a median gap and originating from the peripheral sulcus. This form differs from *Sulcoperculina* (see above) in having lateral chambers and lengthened radial plates or rods. Late Cretaceous (Campanian) (Fig. 5.18).

Subfamily Vaughaninae MacGillavry, 1963.

Members of this subfamily have a juvenile stage similar to *Sulcoperculina* developing lateral chambers. The equatorial layers are similar to those of the pseudorbitoids. Late Cretaceous (Campanian to Maastrichtian).

- *Aktinorbitoides* Brönnimann, 1958 (Type species: *Aktinorbitoides browni* Brönnimann, 1958). The test is lenticular, with a stellate outline. The juvenarium consists of two to three whorls of trochospirally arranged chambers, followed an equatorial layer comprising seven to ten rays (radii), consisting of annular chambers traversed by two alternating systems of radial plates, separated by a medium gap as in *Vaughanina*. Late Cretaceous (Campanian to Maastrichtian) (Fig. 5.14).

- *Ctenorbitoides* Brönnimann, 1958 (Type species: *Ctenorbitoides cardwelli* Brönnimann, 1958). The test is conical with a comb-like apex and a neanic stage similar to *Vaughanina*. Late Cretaceous (late Campanian to early Maastrichtian) (Fig. 5.14).

- *Vaughanina* Palmer, 1934 (Type species: *Vaughanina cubensis* Palmer, 1934). The test is lenticular and circular with a conspicuous flange on which protruding radial plates are usually visible. *Vaughanina* differs from *Pseudorbitoides* in having its annular equatorial chambers crossed by two systems of alternating vertical radial plates, extending from the roof and floor into the equatorial layer. Late Cretaceous (late Campanian to Maastrichtian) (Plate 5.20, fig. 7; Fig. 5.18).

Subfamily Pseudorbitellinae Hanzawa, 1962

The subfamily has a pseudorbitoidal structure, but lacks plates. Late Cretaceous (Campanian to Maastrichtian).

- *Asterorbis* Vaughan and Cole, 1932 (Type species: *Asterorbis rooki* Vaughan and Cole, 1932). The test is stellate with four to eight rays. The surface has large pustules on the umbo, the bilocular embryo which is surrounded by a thick wall and two large auxiliary chambers. Equatorial chambers are diamond shaped to ogival (see Fig. 5.19) in plan, and increasing in height from the centre to the test periphery. The layers decrease in number from the umbo towards the periphery, where the lateral layer is left exposed without lateral chambers, but with well developed pillars. Late Cretaceous (late Campanian to Maastrichtian) (Fig. 5.14).

- *Orbitocyclina* Vaughan, 1929 (Type species: *Lepidorbitoides minima* Douvillé, 1927). The test is circular with a bilocular embryon, enclosed by a thick wall and followed by spiral chambers. The equatorial layer is composed of arcuate to diamond-shaped chambers, interconnected by stolons and surrounded by lateral layers of chambers on both sides. Late Cretaceous (Campanian to Maastrichtian) (Figs 5.18).

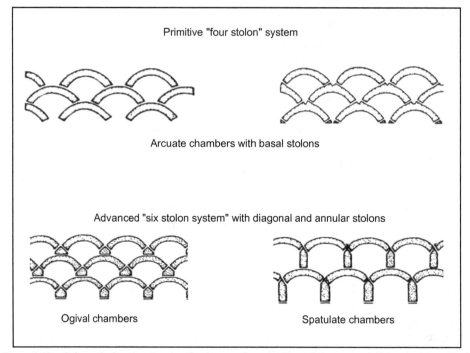

Figure 5.19. Schematic drawing showing the shape and connections of the equatorial chambers in orbitoidal foraminifera, modified after van Gorsel (1975, 1978).

Family Rotaliidae Ehrenberg, 1839

Tests are trochospiral, with umbilical plugs and radial canals throughout, or with fissures and intraseptal and subsutural canals. Apertures are umbilical, basal, single to multiple. Late Cretaceous (Maastrichtian) to Miocene.

- *Kathina* Smout, 1954 (Type species: *Kathina delseata* Smout, 1954). The test is lenticular, with a single spire of small simple chambers, without supplementary chamberlets or umbilical extensions. The umbilicus has a central plug with numerous strong vertical canals. This genus is distinguished from later Paleogene forms *Dictyoconoides* and *Dictyokathina* by having a simple spire, and from *Lockhartia* and *Sakesaria* by the lack of umbilical cavities (see Chapter 6). Late Cretaceous to Paleocene.
- *Pararotalia* Le Calvez, 1949 (Type species: *Rotalina inermis* Terquem, 1882). Biconvex, low trochospiral, smooth with a simple plug that fills the umbilicus. Loeblich and Tappan (1988), in their new description of the genus *Pararotalia,* reported the presence of a septal flap partially doubling each of the septa. They describe the aperture as interiomarginal, extending obliquely into the apertural face, and the intercameral foramen as areal, due to the attachment of an imperforate tooth plate that extends to the distal margin of the aperture. Hottinger et al. (1991) emended the description of the Pararotaliinae to include the canal system. It is also distinguished from

Neorotalia (see Chapter 6) by the lack of the closed interlocular spaces ("enveloping canal system", Hottinger et al., 1991). Late Cretaceous (Coniacian) to middle Eocene.

- *Fissoelphidium* Smout, 1955 (Type species: *Fissoelphidium operculiferum* Smout, 1955). The test is bilaterally symmetric, planispiral with numerous chambers. Deeply fissured dendritic patterns occur along the sutures and finely divided umbilical boss. Apertures are basal and multiple. Late Cretaceous (Maastrichtian) (Plate 5.2, figs 10-11).

- *Laffitteina* Marie, 1946 (Type species: *Laffitteina bibensis* Marie, 1946). The test is lenticular, asymmetric with bifurcate interseptal canals that open as two alternating rows of openings along the septal sutures on the dorsal side. Late Cretaceous (Maastrichtian) to Paleocene (Danian) (Plate 5.20, fig. 6).

- *Orbitokathina* Hottinger, 1966 (Type species: *Orbitokathina vonderschmitti* Hottinger, 1966). The test is conical, but only the microspheric generation of this genus shows orbitoidal character. Pillars are only found on the concave ventral side. The spiral stage is followed by an orbitoidal stage consisting of arcuate chambers, arranged in irregular concentric rows. Lateral chambers are absent. Late Cretaceous (Coniacian).

- *Rotalia* Lamarck, 1804 (Type species: *Rotalites trochidiformis* Lamarck, 1804). The test is trochospiral and biconvex. Secondary deposits build thick, short lamellar pillars with marked fissures that fill the umbilical area. Apertures are umbilical, basal with umbilical extensions to the umbilical canal. Late Cretaceous (Coniacian) to Eocene (Plate 5.1, fig. 13).

Family Calcarinidae Schwager, 1876

Tests are enrolled, with large inflated spines. Late Cretaceous (Maastrichtian) to Holocene.

- *Siderolites* Lamarck, 1801 (Type species: *Siderolites calcitrapoïdes* Lamarck, 1801). The test is large with a globular proloculus followed by an involute coil of about four whorls. In the final whorl two to seven large coarse spines are present. Spiral canals occur in the umbilical region on both sides of the test. Multiple foramina are present. Pillars appear in the umbilical region as solid pustules. Late Cretaceous (Maastrichtian) (Fig. 5.20; Plate 5.20, figs 8-11; Plate 5.21, figs 1, 6).

ORDER MILIOLIDA Delage and Hérouard, 1896

The miliolides have tests that are porcelaneous, imperforate and made of high Mg-calcite with fine randomly oriented crystals. They range from the Carboniferous to the Holocene.

Superfamily ALVEOLINOIDEA Ehrenberg, 1839

Tests are enrolled along an elongate axis, being initially planispiral or streptospiral, or miliolide with chambers added in varying planes. Cretaceous to Holocene.

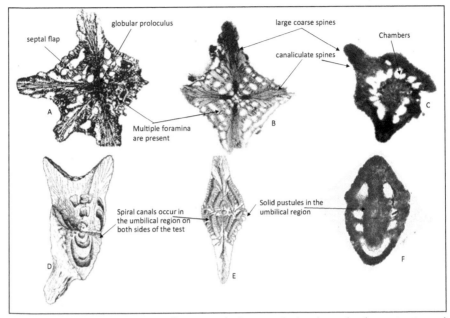

Figure 5.20. Siderolites showing canaliferous enveloping, marginal and pseudospinose structures: A-B) Horizontal section; C) Oblique horizontal thin sections; D-E) Sketch of vertical and a stereodiagram drawn by Hofker (1927); F) Axial section with broken spines.

Family Fabulariidae Ehrenberg, Munier-Chalmas, 1882, emend. Hottinger et al., 1989
The test is large, dimorphic, multi-chambered with a miliolide coiling, tending to become reduced in subsequent growth stages, either to bilocular or to monolocular chamber cycles with a trematophore as aperture. According to Hottinger (2006) "A trematophore, or a sieve constituting the face of many porcelaneous larger foraminifera, is in miliolides produced by the coalescence of teeth, covering a large pre-septal space. May be supported by residual pillars". Chambers have a thickened basal layer, subdivided by pillars or secondary partitions. Foraminifera in the adult growth stages have fixed apertural axis. For the terminology and the orientation of sections needed for detailed structural analysis see Drobne (1974; 1988) and Hottinger et al. (1989). Late Cretaceous to Early Oligocene.

• *Adrahentina* Bilotte, 1978 (Type species: *Adrahentina iberica* Bilotte, 1978). The test is large and subspherical, with a megalospheric stage with a large proloculus followed by enveloping chambers. Radial pillars arise from the chamber floor but never reach the chamber roof. Pillars are alternating on adjacent ribs. Late Cretaceous (Maastrichtian).
• *Lacazina* Munier-Chammas, 1882 (Type species: *Alveolina compressa* d'Orbigny, 1850). The test is large, discoid to ovoid. The early miliolide coiling is followed by chambers that completely overlap the early ones. Chambers are subdivided by longitudinal partitions. Apertures are cribrate at one extremity of the test. Late

Cretaceous to Eocene (Coniacian to Ypresian) of Europe, Early Oligocene of Indonesia (Plate 5.22, fig. 6; Fig. 5.21).

- *Pseudochubbina* De Castro, 1990 (Type species: *Pseudedomia globularis* Smout, 1963). The test is subspherical to flaring with parallel anastomising passages in the basal layer, and an outer layer with regular chamberlets. Late Cretaceous (Campanian).

- *Pseudolacazina* Caus, 1979 (Type species: *Pseudolacazina hottingeri* Caus, 1979). The test is globular to ovate with a quinqueloculine early stage. Microspheric chambers have pluriloculine chamber cycles to monoloculine cycles in the adult. The megalospheric stage is biloculine, with completely embracing chambers. Chambers are subdivided by longitudinal partitions or by pillars supporting the chamber roof. Late Cretaceous (Santonian to Maastrichtian) and Middle to Late Eocene.

Family Rhapydioninidae Keijzer, 1945

Tests are planispiral to streptospiral in the early stage, but become uncoiled compressed, peneropliform-flaring or cylindrical in later stages. The embryonal apparatus is simple or miliolide. The central zone shows canaliculated thickening except for a pre-septal space with buttresses (residual pillars). Apertures are multiple on the final chamber face. Late Cretaceous (Cenomanian to Maastrichtian).

- *Chubbina* Robinson, 1968 (Type species: *Chubbina jamaicensis* Robinson, 1968). Tests are large globular and streptospiral, later flaring planispiral and peneropline. Interiors of chambers are subdivided by numerous septula. Late Cretaceous (Campanian to Maastrichtian) (Plate 5.3, fig. 12; Plate 5.6, figs 10-11; Plate 5.9, figs 8-10: Fig. 5.21).

- *Murciella* Fourcade, 1966 (Type species: *Murciella cuvillieri* Fourcade, 1966). Tests are planispiral and involute, later uncoiling and becoming rectilinear. Horizontal and vertical septa subdivide the chambers into secondary chamberlets. Late Cretaceous (Campanian) (Fig. 5.21; Plate 5.4, fig. 11).

- *Pseudedomia* Henson, 1948 (Type species: *Pseudedomia multistriata* Henson, 1948). Tests are globular to lenticular with the last chambers becoming progressively longer and strongly overlapping the preceding chambers, until they become cyclical with compressed arched chambers in its final whorls. Late Cretaceous (Cenomanian to Maastrichtian) (Plate 5.1, figs 1-6; Plate 5.22, fig. 12; Fig. 5.21).

- *Raadshoovenia* van den Bold, 1946 (Type species: *Raadshoovenia guatemalensis* van den Bold, 1946). The early stage is streptospiral, then planispiral and becoming rectilinear. The centre of the test is pillared with pillars fusing laterally to form chamberlets opening into the preseptal space. Late Cretaceous (Campanian-Early Maastrichtian, Paleocene).

- *Rhapydionina* Stache, 1913 (Type species: *Peneroplis liburnica* Stache, 1889). The test is involute, lenticular, with pronounced dimorphism. Cylindrical megalospheric forms have an enrolled flexostyle followed by three to four chambers in a single planispiral coil, with a rectilinear adult stage. Planispiral to uncoiled microspheric forms have a small proloculus followed by uniserial, flabelliform and flattened adult stage with low arched chambers. The peripheral region is subdivided by vertical septula. Late Cretaceous (late Cenomanian to Maastrichtian) (Fig. 5.21).

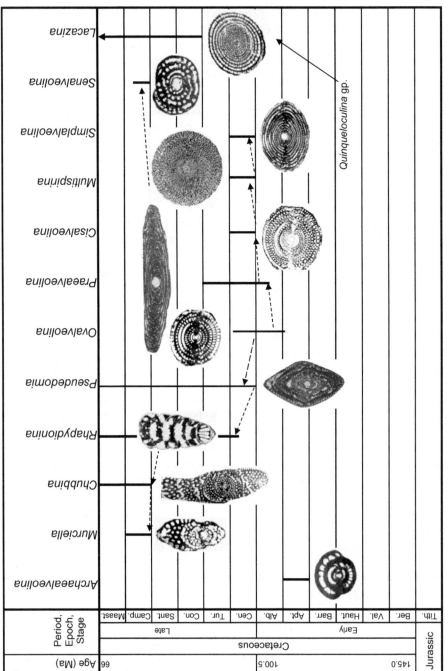

Figure 5.21. The evolution of the Alveolinoidea in the Cretaceous.

Family Alveolinidae Ehrenberg, 1839
Tests are free, large, planispiral to fusiform, cylindrical or globular, and coiled about
elongate axis. Chambers are divided into chamberlets by secondary septa (septula) perpendicular to the main septa, and connected along the chamber by passages. Apertures
are numerous, circular, and aligned in horizontal rows. Early Cretaceous (Aptian) to
Holocene.

- *Archaealveolina* Fourcade, 1980 (Type species: *Ovalveolina reicheli* De Castro, 1966).
  The test is subglobular, small, and streptospiral in early coiling, followed by a planispiral stage. Septa are oblique, with pre-septal passages present. Apertures form
  a single row of round openings at the base of the apertural face. Early Cretaceous
  (Aptian) (Plate 5.22, Fig. 2; Fig. 5.21).
- *Cisalveolina* Reichel, 1941 (Type species: *Cisalveolina fallar* Reichel, 1941). The test
  is globular and streptospiral in the early stage, but later planispiral. The megalospheric generation is planispiral throughout ontogeny. Later stages have alternating chamberlets. There are wide post-septal passages and a long, low single, slit-like
  aperture. Late Cretaceous (Cenomanian) (Fig. 5.21; Plate 5.22, figs 3-4; 13-14).
- *Helenalveolina* Hottinger, Drobne and Caus, 1989 (Type species: *Helenalveolina*
  tappanae Hottinger, et al.,1989). The test is spherical, with streptospiral coiling that
  may have a planispiral-involute adult stage. Early chambers are undivided in both
  generations, later chambers are subdivided by longitudinal partitions (septula) into
  alternating chamberlets as in *Cisalveolina*. The pre-septal passage is wide, with a
  ribbed floor, and a short post-septal passage. The distal margin of the apertural slit
  is notched. *Helenalveolina* is distinguished from the Fabulariidae by the absence
  of pillars subdividing the chambers, by its streptospiral coiling, and by the lack
  of a trematophore in its apertural face. Late Cretaceous (late Coniacian to early
  Santonian).
- *Multispirina* Reichel, 1947 (Type species: *Multispirina iranensis* Reichel, 1947). The
  test is spherical, large with numerous chambers divided into chamberlets by the septula, which is continuous from chamber to chamber. There are intercalated whorls
  and sutural apertures. Pre-septal passages are large. Apertures form a row of pores
  in the apertural face. Late Cretaceous (Cenomanian) (Fig. 5.21; Plate 5.1, fig. 9).
- *Ovalveolina* Reichel, 1936 (Type species: *Alveolina ovum* d'Orbigny, 1850). The test
  is globular or ovoid with planispiral coiling. Chambers are short with widely spaced
  septula. Pre-septal passages are large. There is one row of apertures in the apertural
  face. Cretaceous (Albian to Cenomanian) (Fig. 5.21; Plate 5.22, figs 5, 9A, 13).
- *Praealveolina* Reichel, 1933 (Type species: *Praealveolina tenuis* Reichel, 1933). Test
  ranges in shape from subglobular to fusiform. They consist of a number of chambers, planispirally coiled around an elongate axis, each chamber being broken into
  tubular chamberlets by secondary septa (septula) that run in the direction of coiling.
  It differs from *Alveolina* (see Chapter 6) in having its septula and tubular chamberlets in continuous alignment from chamber to chamber. Differences in shape
  will ordinarily distinguish *Praealveolina* from *Ovalveolina*, but where *Praealveolina*
  is represented by subglobular species, the most practical means of differentiation
  is by comparison of chamberlet proportions: *Ovalveolina* has chamberlets that are
  much higher and broader in proportion to their length than those of *Praealveolina*.

A pre-septal passage is present. Apertures form a single row of pores on the apertural face near the equator but increases pole-ward to many rows of openings. Cretaceous (late Albian to Turonian) (Plate 5.21, fig. 9, Plate 5.22, figs 9-11).

- *Senalveolina* Fleury, 1984 (Type species: *Senalveolina aubouini* Fleury, 1984). The test is globular. Early coiling is streptospiral, but later it becomes planispiral with numerous long low chambers per whorl. Narrow pre-septal passages are present. Walls and septula are relatively thick. Apertures form a row of pores in the apertural face. Late Cretaceous (early Campanian) (Fig. 5.21).
- *Simpalveolina* Reichel, 1964 (Type species: *Praealveolina simplex* Reichel, 1936). The test is small and ovoid. Numerous septula are aligned from chamber to chamber to form a single layer of chamberlets of oval section. It lacks the secondary chamberlets of *Praealveolina*. The pre-septal canal has a circular section. The aperture is a single row of pores in the apertural face. Late Cretaceous (Cenomanian) (Plate 5.21, fig. 5; Plate 5.22, fig. 8).
- *Streptalveolina* Fourcade, Tardy and Villa, 1975 (Type species: *Streptalveolina mexicana* Fourcade, et al., 1975). The test is globular and streptospirally coiled. Septa are oblique. Chambers are divided by partial septula. A pre-septal passage is present. The aperture is a row of openings in the apertural face. Late Cretaceous (early Cenomanian).
- *Subalveolina* Reichel, 1936 (Type species: *Subalveolina dordonica* Reichel, 1936). The test is spherical to fusiform. Septula have no definite arrangement. The secondary chamberlets lie below the plane of the primary chamberlets in an irregular pattern. A large pre-septal canal is present, occupying the total height of the chamber. Above the row of primary apertures in the apertural face is a row of more numerous small apertures connecting with alveoli. Late Cretaceous (late Santonian to Campanian) (Plate 5.22, fig. 7).

Superfamily SORITOIDEA Ehrenberg, 1839
Chambers are planispiral, uncoiling, flabelliform or cyclical, and may be subdivided by partitions or pillars. Late Permian to Holocene.

Family Meandropsinidae Henson, 1948
The test is planispiral to annular discoid, laterally compressed, involute with partially developed partitions. Numerous septula divide the interior of the chambers. Apertures are multiple. Late Cretaceous (Campanian to Maastrichtian).

- *Ayalaina* Seiglie, 1961 (Type species: *Meandropsina? rutteni* Palmer, 1934). The test is flaring with numerous chambers increasing rapidly in height. Late Cretaceous (Campanian to Maastrichtian).
- *Broeckina* Munier-Chalmas, 1882 (Type species: *Cyclolina dufrenoyi* d'Archiac and Haime, in d'Archiac, 1854). The test is discoidal, the early stage is planispiral, and later chambers are annular and divided by incomplete partitions. The multiple apertures form two separate rows. Late Cretaceous (Cenomanian to Maastrichtian).

- *Fallotia* Douvillé, 1902. (Type species: *Fallotia jacquoti* Douvillé, 1902). The test is involute, planispiral with chambers subdivided by numerous partitions. Late Cretaceous (Santonian) (Fig. 5.22).
- *Larrazetia* Ciry, 1964 (Type species: *Meandropsina larrazeti* Munier-Chalmas, in Schlumberger, 1898). There is an annular test, laterally flanked by a series of polar capsules, slightly biconvex with annular chambers subdivided throughout with partial partitions. Late Cretaceous (Campanian to Maastrichtian).
- *Meandropsina* Munier-Chalmas, 1898 (Type species: *Meandropsina vidali* Schlumberger, 1898). The test is large, compressed, with the initial part of the test lenticular and consisting of a planispiral coil of chambers that are completely involute and very thinly and evenly spread out over the surface of the preceding whorl. Late Cretaceous (Coniacian to Maastrichtian) (Fig. 5.22).
- *Nummofallotia* Barrier and Neumann, 1959 (Type species: *Nonionina cretacea* Schlumberger, 1900). The test is lenticular, planispiral with an umbonal plug, with early whorls evolute, later becoming involute. Late Cretaceous (Coniacian to Maastrichtian).
- *Pastrikella* Cherchi, Radoičić and Schroeder, 1976 (Type species: *Broeckina* (*Pastrikella*) *balcanica* Cherchi et al., 1976). The test is large, flat, becoming annular. The interior is subdivided by regularly arranged septula. Late Cretaceous (middle to late Cenomanian).
- *Perouvianella* Bizon, Bizon, Fourcade and Vachard, 1975 (Type species: *Orbiculina peruviana* Steinmann, 1930). The test is lenticular to discoidal, planispiral with chambers never completely annular. Interiors of chambers are subdivided by radial pillars of triangular sections. Late Cretaceous (Santonian).
- *Pseudobroeckinella* Deloffre and Hamaoui, 1969 (Type species: *Pseudobroeckinella soumoulouensis* Deloffre and Hamaoui, 1969). The test is large, planispiral with cyclic chambers in the final stage. Chambers are subdivided by three types of septula, primary and secondary septula that are vertical to the septa, and transverse septula that are parallel to the septa. Late Cretaceous (Santonian).
- *Spirapertolina* Ciry, 1964 (Type species: *Spirapertolina almelai* Ciry, 1964). The test is discoidal, planispiral, involute becoming peneropliform, then reniform. The cyclic chambers comprise most of the test, subdivided by numerous partitions, those of successive chambers alternating in position. Late Cretaceous (Santonian).

Family Soritidae, Ehrenberg, 1839
Tests are involute, planispiral to uncoiled evolute, flaring, annular discoids with partial or complete partitions. Apertures are multiple. Late Cretaceous (Cenomanian) to Holocene.

- *Cycledomia* Hamaoui, 1964 (Type species: *Edomia iranica* Henson, 1948). Exhibits a discoidal, flattened and biconcave test. The early part is involute, but later is evolute and cyclic. Chambers are divided by short septula. Late Cretaceous (late Cenomanian to early Turonian).

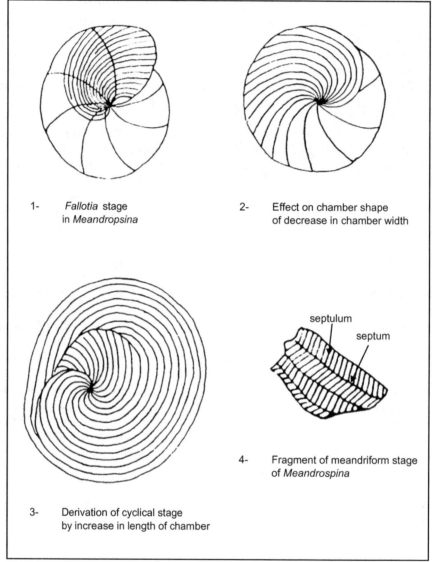

1-   *Fallotia* stage
       in *Meandropsina*

2-   Effect on chamber shape
       of decrease in chamber width

septulum

septum

4-   Fragment of meandriform stage
       of *Meandrospina*

3-   Derivation of cyclical stage
       by increase in length of chamber

Figure 5.22.   The stages of developments from the involute, planispiral form of 1-2) *Fallotia* to a complete
meandriform cyclical form of 3-4) *Meandrospina.*

- *Edomia* Henson, 1948 (Type species: *Edomia reicheli* Henson, 1948). Tests are large discoidal, planispiral and involute, later becoming evolute with cyclic chambers. Interiors have irregularly distributed pillars. Late Cretaceous (Cenomanian to Turonian) (Plate 5.16, figs 5-8).
- *Lamarmorella* Cherchi and Schroeder, 1975 (Type species: *Lamarmorella sarda* Cherchi and Schroeder, 1975). The test is discoidal, planispiral to peneropline, finally becoming annular. The interior is subdivided by partial partitions. Late Cretaceous (Coniacian to Santonian).

- *Murgella* Luperto Sinni, 1965 (Type species: *Murgella lata* Luperto Sinni, 1965). Tests are large, planispirally coiled, later uncoiling. Chambers are divided by numerous thick radially arranged septula. Late Cretaceous (Coniacian to Maastrichtian).
- *Praetaberina* Consorti, Caus, Frijia and Yazdi-Moghadam, 2015 (Type species: *Taberina bingistani* Henson, 1948). According to Consorti et al. (2015), *Praetaberina* differs from the Paleogene genus *Taberina* (see Chapter 6) in having a more complex internal structure with two orders of marginal partitions and apertures alternating with the main septula. Late Cretaceous (late Cenomanian) (Plate 5.18, figs 7-8).
- *Pseudorhapydionina* De Castro, 1971 (Type species: *Rhapydionina laurinensis* De Castro, 1965). Tests are elongate, planispiral, later uncoiling and rectilinear. Chambers have subdivided interiors with thin radial partitions. Late Cretaceous (late Cenomanian).
- *Pseudorhipidionina* De Castro, 1971 (Type species: *Rhipidionina casertana* De Castro, 1965). The test is compressed, planispiral, becoming peneropliform, then uncoiling. The interiors of chambers are subdivided by numerous short vertical septula. Late Cretaceous (Cenomanian).
- *Scandonea* De Castro, 1971 (Type species: *Scandonea samnitica* De Castro, 1971). The test is enrolled, streptospiral to planispiral with uncoiled chambers, subdivided by very short radial vertical partitions. Late Cretaceous (Campanian) to Early Paleocene.
- *Tarburina* Schlagintweit, Rashidi and Barani, 2016 (Type species: *Tarburina zagrosiana* Schlagintweit et al., 2016). The cylindro-conical test is formed by a large globular proloculus followed by few, planispirally coiled chambers. The chambers are partly subdivided by vertical (radial) partitions, often continuous from one chamber to the next. Shorter secondary partitions may be intercalated. Cretaceous (late Maastrichtian).
- *Zekritia* Henson, 1948 (Type species: *Zekritia langhami* Henson, 1948). The test has lamelliform buttresses. Late Cretaceous (Turonian) (Plate 5.22, fig. 1).

Family Keramosphaeridae Brady, 1884
Tests are globular with concentric chambers connected by stolons in the same series as well as those of successive series. Cretaceous (Berriasian to Maastrichtian), Miocene to Holocene

- *Keramosphaerina* Stache, 1913 (Type species: *Bradya tergestina* Stache, 1889). Shows a spherical test with a miliolide early stage. Microspheric tests have a proloculus and flexostyle, with post-embryonic chambers with thick walls and streptospiral coiling. Later stages are regularly concentric with radially aligned chambers. Stolons connect adjacent chambers. Late Cretaceous (Coniacian to Maastrichtian).
- *Pavlovecina* Loeblich and Tappan, 1988 (Type species: *Peneroplis karreri* Wiesner, 1923).The test is globular with a miliolide early stage. Post embryonic chambers are numerous and extremely irregular. Early Cretaceous (Berriasian).

Figure 5.23.   Morphological characteristics of key species of *Nautiloculina*

Nautiloculinidae

Planispiral, involute, biumbonate, nautiliform; chambers slowly increase in height; septa bilamellar; aperture interiomarginal, equatorial, single

Periphery rounded in axial section	Sutures little if at all depressed	*Nautiloculina oolithica*	Bajocian - Aptian
Periphery subangular in axial section	Sutures depressed, periphery lobulate in equatorial section	*Nautiloculina circularis*	Bajocian - Oxfordian
	Sutures not much depressed, periphery smooth in equatorial section; test flat, large	*Nautiloculina cretacea*	Berriasian - Early Aptian

Superfamily MILIOLOIDEA Ehrenberg, 1839

Tests are coiled commonly with two or more chambers arranged in varying planes about the longitudinal axis; later they may become involute. Advanced forms may have secondary partitions within the chambers. Late Triassic (Norian) to Holocene.

Family Nautiloculinidae Loeblich and Tappan, 1985

Tests are free, lenticular, planispiral, and involute with secondary thickening in the umbilical region. Apertures are equatorial. Middle Jurassic to Late Cretaceous.

• *Nautiloculina* Mohler, 1938 (Type species: *Nautiloculina oolithica* Mohler, 1938) (see Chapter 4). Jurassic (Late Bajocian) to Early Cretaceous (Plate 4.9, figs 6A, 7-12; Fig. 5.23).

5.3 Biostratigraphy and Phylogenetic Evolution

Cretaceous carbonate formations show sequences of foraminiferal faunas, recognizable in thin sections of marine limestones, which are of short range and so enable broad stratigraphic correlations to be made (see Charts 5.1 and 5.2 online). Three main orders flourished at different times:

• Firstly, the agglutinated larger foraminifera, which made their first appearance in the Jurassic and particularly thrived during the Early Cretaceous, where they occur with many different short ranged groups.
• Most of these forms disappeared before the Cenomanian, where they were replaced by mainly porcelaneous miliolide forms.
• Many of the miliolides disappeared in turn towards the end of the Cenomanian and were replaced gradually by a thriving population of hyaline calcareous rotaliines and orbitoids.

These three orders are discussed in detail below.

5.3.1 The Textulariides of the Cretaceous

In the Cretaceous, agglutinated foraminifera retained some of the features of their Jurassic forebears but also acquired new features, which became important in the later Cretaceous. Four main major superfamilies are of most biostratigraphic value (see Fig. 5.24):

- the Textularioidea,
- the Lituoloidea,
- the Ataxophragmioidea, and
- the Orbitolinoidea.

Some of these forms exhibit complex labyrinthic features, while others are partitioned; many were quite short ranged and so are valuable as index fossils. They have been closely studied by many workers, and detailed evolutionary lineages have been established by, amongst others, Banner (1970), Banner et al. (1990), and Simmons et al. (2000).

Cretaceous Textularioidea seem to have evolved independently from those of the Jurassic (see Chapter 4). They are represented by the Chrysalinidae (see Fig. 5.25),

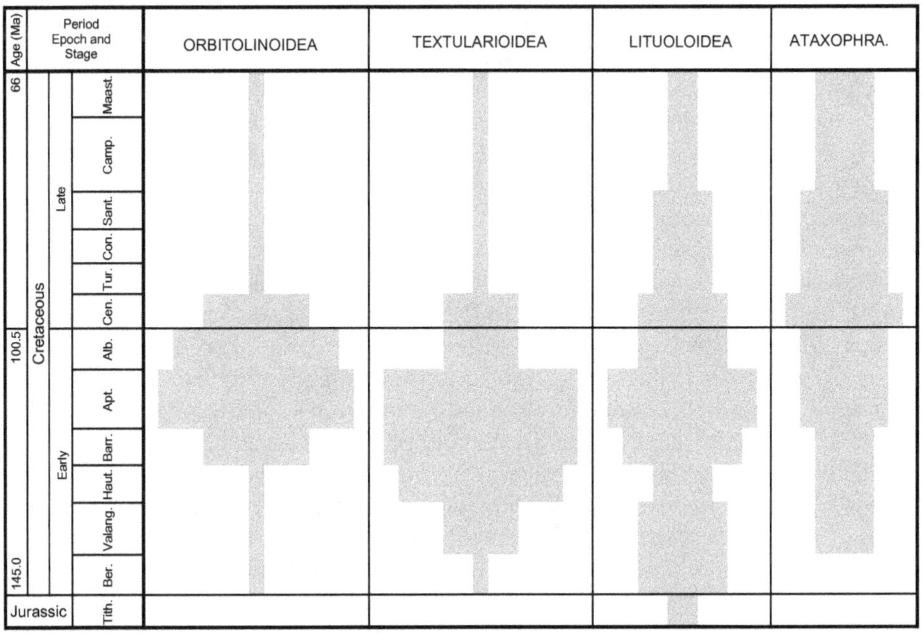

Figure 5.24. The biostratigraphic range and diversity of the main agglutinated superfamilies found in Tethys during the Cretaceous.

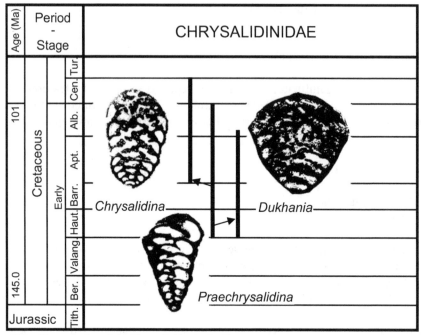

Figure 5.25.   The evolution of the Chrysalinidae in the Cretaceous.

which evolved directly from a simple *Verneulinoides* in the early Cretaceous by the appearance of the triserial, non-pillared *Praechrysalidina*. The development of pillars in the Aptian and Albian was gradual, and it was not before the Cenomanian that completely pillared forms were developed. These included the rapidly tapering, terminally biserial, *Dukhania* and the slowly tapering, terminally triserial, *Chrysalidina*. These forms had no descendants and are confined to central Tethys (Banner et al., 1991). On the other hand, their possible ancestor *Praechrysalidina* spread throughout western and central Tethys throughout the Early Cretaceous and might have been ancestral to *Accordiella* in the Coniacian to Santonian (see Fig. 5.25, and Charts 5.1 and 5.2).

Cretaceous lituolids (see Fig. 5.26) vary in shape from planispiral (e.g. *Buccicrenata, Choffatella,* Fig. 5.26 and Chapter 4) to low trochospiral (e.g. *Sornayina,* Fig. 5.26), becoming flabelliform (e.g. *Pseudochoffatella,* Plate 5.3, fig. 13), becoming spherical (e.g. *Reticulinella,* Plate 5.4, fig.3) to fusiform (e.g. *Loftusia,* Plate 5.8) in the Late Cretaceous. However, the main morphological trend in this superfamily is the development of labyrinthic structures along a number of separate lineages, and the development of central pillars in genera such as *Spirocyclina* (Plate 5.3, fig. 4). Cretaceous lituolids have their ancestors in the Jurassic. In the Sinemurian, a descendant of a species of *Ammobaculites*, with a single areal aperture, developed an alveolar wall with the septa remaining solid, to give rise to *Everticyclammina*. The early *Everticyclammina*, with a coarsely labyrinthic hypodermis (e.g. *E. praeviruliana,* Plate 4.16, fig. 3), evolved in the Early Cretaceous into individuals with a smooth, regularly structured hypodermis with much finer and more regularly formed alveolae (e.g. *E. kelleri,* Plate 4.16, fig. 4). Throughout the Cretaceous, the alveoles of *Everticyclammina* become regularly and tightly bifurcating. In the Barremian-Aptian and before its disappearance, *Everticyclammina* developed forms (*E. greigi*, Plate 4.16, fig. 5) in which the lower parts of the adult septa become oblique, then tangential to the spiral suture, and thicken and coalesce to form imperforate "basal layers" to the chambers. The reduction of the solid septa produced the Albian-Cenomanian *Hemicyclammina* (Fig. 5.26), in which the upper parts of the septa are thinned and meet the test periphery (almost at right angles), so that the chambers become almost quadrangular in equatorial section (e.g. *H. whitei*, Plate 5.6, figs 14-15). Quite independently from *Hemicyclammina* and in the Barremian-Aptian, *Everticyclammina* evolved into evolute and streptospiral *Hottingerita* (see Fig. 5.26; Plate 5.2, fig.2).

Another lituolid lineage thrived in the Cretaceous parallel to the *Everticyclammina* lineage, namely that of *Buccicrenata*. Unlike *Everticyclammina*, where the septa remained solid and angularly differentiated from the walls, *Buccicrenata* had septa with alveolar extensions of the lateral chamber wall, and seems to have directly descended from the *Lituola* group *camerata* in the Kimmeridgian, quite independently (see BouDagher-Fadel, 2000, 2001). In the Cretaceous, in *Buccicrenata* the sequence of alveolae became finer, the walls themselves became thinner but all the microspheric forms are able to become rectilinear after the first two whorls for the final three chambers. The single aperture is areal in all species, but became more compressed and slit-like as the forms became stratigraphically younger and the microspheric chambers became terminal.

Pseudocyclammina had similarly evolved from *Lituola* in the late Sinemurian by developing an areal cribrate aperture and a coarsely alveolar wall. These forms passed through the Jurassic-Cretaceous boundary and persisted to the Aptian. The Cretaceous forms developed alveolar layers more strongly, with several independent evolutionary modifications occurring; these include trends towards increased complexity in the hypodermis (see Banner, 1970), a trend which was independently followed and reached acmes in *Martiguesia*, where delicate pillars bridge from septum to septum; *Choffatella*, which developed tighter coiling and septa as complex and thick as the hypodermis; and *Torinosuella* with increased test compression and coiling rate. Parallel to these lines of evolution many forms evolved in the middle to Late Cretaceous, with flabelliform forms and even more coarsely agglutinated epidermis, such as *Pseudochoffatella*.

In the Cenomanian *Choffatella* gave rise to *Spirocyclina* (see Fig. 5.26), with a flat initial spire, becoming peneropliform, with a coarse reticulate hypodermis and a few scattered central pillars. In turn *Spirocyclina* may have given rise to *Sornayina* in the Coniacian, by increased trochospirality and the development of more regularly spaced and stronger transverse sub-hypodermal septulae (Banner, 1970). *Anchispirocyclina* (Plate 4.15, figs 8-9) may have evolved from *Pseudocyclammina* in the Oxfordian by developing a central zone filled by a complex reticulum of densely spaced pillars. *Anchispirocyclina* did not become extinct before the Campanian.

In the Late Cretaceous, lituolids became large and internally very complicated. Alveoliniform, globose spherical genera, such as *Reticulinella*, with interseptal horizontal lamellae, appeared in the Cenomanian, while *Loftusia* with interseptal pillars and the strong hypodermal development of *Martiguesia* appeared in the Maastrichtian and attained immense sizes, while acquiring an increased tightness of coiling leading

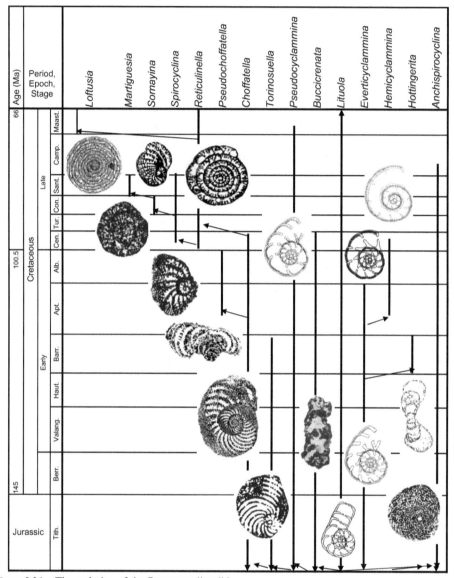

Figure 5.26.   The evolution of the Cretaceous lituolids.

to fusiform tests. However, as with so many other forms, the end of the Cretaceous saw the extinction of most of the lituolids.

The **Ataxophragmioidea** were common in the Cretaceous. They were trochospiral, tightly or loosely coiled with high spires, or conical to sub-flabelliform (see Fig. 5.27). The main morphological trend in this superfamily is the subdivision of the periphery of the chambers by radial or curved partitions and buttresses, forming occasionally small chamberlets.

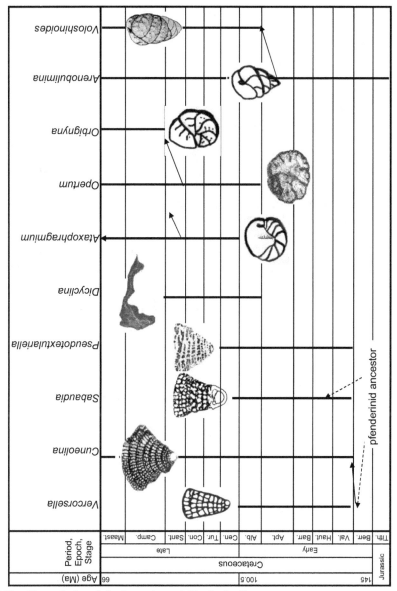

Figure 5.27. The evolution of the Ataxophragmioidea in the Cretaceous.

In the Campanian, the Cretaceous pfenderids lost their continuous composite central columella and evolved into the ataxophragmids. The tightly coiled *Ataxophragmium* acquired peripheral partitions, (e.g. *Opertum*) and buttresses (e.g. *O. orbignyna*). Quite independently in the Albian, the trochospirally enrolled simple *Arenobulimina* acquired arched partitions and buttresses that meet and form irregular chamberlets as in *Voloshinoides* (see Fig. 5.8).

Another parallel evolution from a pfenderid ancestor saw the development of the cuneolids in the Early Cretaceous (see Fig. 5.27). In the Early Cretaceous, the conical pfenderids lost their central columella and the chambers became divided by vertical partitions, or rarely by horizontal partitions (as in *Vercorsella*), or by both vertical and horizontal partitions (as in *Cuneolina*). In the Cenomanian, ataxophragmids with a trochospiral initial stage subdivided their chambers by vertical and horizontal partitions (e.g. *Pseudotextulariella*). Annular growth was achieved in the Ataxophragmioidea, and *Dicyclina* has two layers with radial partitions.

Another important agglutinated group from Early Cretaceous in the Tethyan realm are the orbitolinids (see Figs 5.10–5.12). The orbitolinids are characterized by conical tests that are usually a few millimeters in height and diameter (although they can attain diameters of 5 cm or more). They evolved from forms such as *Urgonina*, with a trochospiral early stage but simple uniserial later stage with chambers lacking partitions, to forms that are partially subdivided by radial partitions or with pillars, *Coskinolinoides*. These forms evolved into forms with peripheral tiered rectangular chamberlets, *Paleodictyoconus,* which eventually lost its initial coil and filled its centre with thick radial partitions, as in *Orbitolina* (see Fig. 5.10). The earliest formed chambers of the

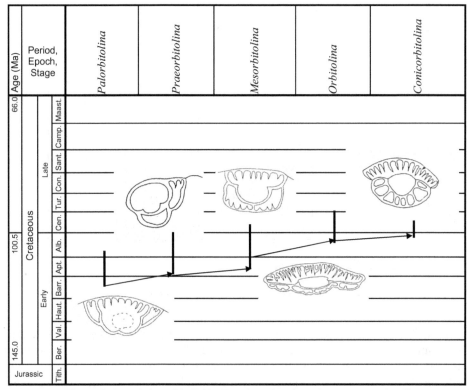

Figure 5.28. The evolution of the orbitolinids. Drawing of embryonic apparatus after Simmons *et al.* (2000).

megalospheric generation can form a complex embryonic apparatus, which can be divided into protoconch, deuteroconch, sub-embryonic zone and peri-embryonic chamberlets, which is the most important feature for their taxonomic division (see Fig. 5.28).

The orbitolinid test is defined by the shape of the embryonic apparatus, and by the size and shape of the chamber passages that can be seen in tangential sections. The chamber passages are formed in the radial part of the central zone of each chamber layer (Figs 5.11 and 5.12), where each chamber passage is subdivided by vertical main partitions, which are prolongations of the vertical main partitions of the marginal zone. In cross section, these can be triangular, rectangular or oval, or can show a gradation between shapes (Schroeder, 1975). In the radial zone of *Orbitolina,* the stolons are arranged in radial rows alternating from one chamber to the next one (see Figs 5.11 and 5.12). Their alternating position obliges the protoplasm to flow in oblique direction (Hottinger, 1978).

In axial section, the embryo is located at the apex of the cone, followed by a series of discoidal chamber layers. In transverse section, the chambers are seen divided into a marginal zone, with subepidermal partitions and a central zone with radial partitions. The radial partitions in *Orbitolina* thicken away from the periphery and anastomose in the central area, producing an irregular network. The earliest formed chambers of the megalospheric generation can form a complex embryonic apparatus that can be divided into a protoconch, deuteroconch, a sub-embryonic zone and peri-embryonic chamberlets depending on the genera involved. It evolved from a simple apparatus, consisting of a large globular fused protoconch and deuteroconch, followed by peri-embryonic chambers as in *Palorbitolina*, to an embryonic apparatus divided into a protoconch and deuteroconch but not completely divided sub-embryonic zone, as in *Praeorbitolina*. This latter evolved in turn into forms in which the deuteroconch and sub-embryonic zone are more or less of equal thickness, as in *Mesorbitolina*. In *Conicorbitolina* the marginal zone became extensively divided by vertical and horizontal partitions, while in *Orbitolina* the deuteroconch is highly subdivided and of much greater thickness than sub-embryonic zone (see Fig. 5.28; Schroeder, 1975; Hottinger, 1978; Simmons et al., 2000; BouDagher-Fadel, 2008; Schroeder et al., 2010; BouDagher-Fadel et al., 2017).

The orbitolinids are very useful biostratigraphic markers in mid-Cretaceous Tethyan carbonate platforms (Henson, 1948; Schroeder, 1975; BouDagher-Fadel et al., 2017). They have short ranges and are easily identified in thin sections (e.g. see Figs 5.28, 5.29 and 5.30). Together with the miliolides (see below), they enrich Cretaceous biota, but unlike the miliolides they show provincialism.

5.3.2 The Miliolides of the Cretaceous

For the first time in their history, with the appearance of the **Alveolinoidea,** the miliolides exhibited a fusiform morphology. In an example of convergent evolution, the external appearance the alveolinids closely resemble the fusulinides of the Permian, but they can be seen to be quite distinct when studied in axial and equatorial (median) sections. The alveolinids differ fundamentally from the fusulinides in that they have an

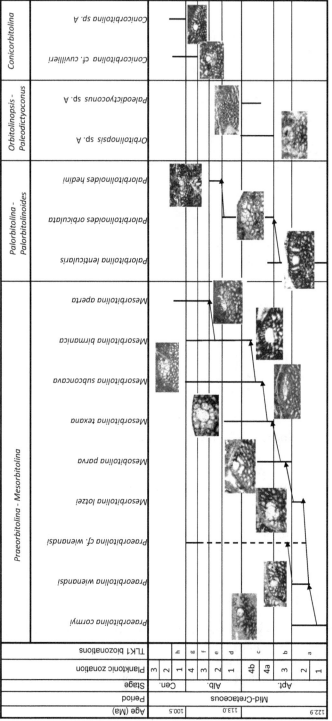

Figure 5.29. Phylogenetic evolution of Tibetan orbitolinids (after BouDagher-Fadel et al., 2017). Planktonic Zones after BouDagher-Fadel (2015).

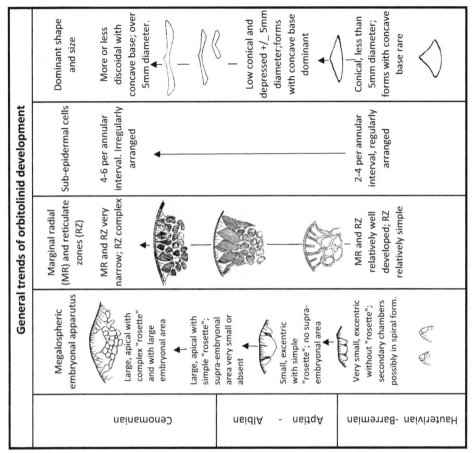

The table within the figure contains:

General trends of orbitolinid development

	Dominant shape and size	More or less discoidal with concave base; over 5mm diameter.		Low conical and depressed +/_ 5mm diameter;forms with concave base dominant		Conical, less than 5mm diameter; forms with concave base rare
Sub-epidermal cells		4-6 per annular interval. Irregularly arranged			2-4 per annular interval, regularly arranged	
Marginal radial (MR) and reticulate zones (RZ)		MR and RZ very narrow; RZ complex			MR and RZ relatively well developed; RZ relatively simple	
Megalospheric embryonal apparutus		Large, apical with complex "rosette" and with large embryonal area	Large, apical with simple "rosette"; supra-embryonal area very small or absent	Small, excentric with simple "rosette"; no supra-embryonal area	Very small, excentric without "rosette"; secondary chambers possibly in spiral form.	
	Cenomanian		Albian - Aptian		Hauterivian -Barremian	

Figure 5.30. Evolutionary trends in the orbitolinids.

imperforate, porcelaneous wall structure, which consists of an external lamina, usually light in colour and an internal darker basal layer, and spiral septula. The basic morphology of the alveolinids is shown in Fig. 5.21. Unlike the fusulinides, septal folding does not occur and the spiral septula reach the floor across the chamber. Deposition of secondary calcite (flosculinisation) occurs more evenly across the floor of the chamber, rather than being almost entirely concentrated in the axial zone as in the fusulinides. The embryonic apparatus consists of a spherical proloculus followed by a spiral tube (flexostyle), succeeded in some genera by a miliolide nepion (pre-adult stage). In the adult, there are numerous chambers planispirally coiled around an elongate axis, each chamber being broken into tubular chamberlets by secondary septa (septula) that run in the direction of coiling. Communication through the chambers is maintained through the vertical septula (canals/passages). These may be situated in front of the septum (post-septal) or behind it (pre-septal). All Cretaceous genera have pre-septal canals, a few genera, such as *Cisalveolina* (Plate 5.22, figs 3-4), have post-septal passages.

Archaealveolina, a subglobular alveolinid, with early streptospiral whorls and a single row of apertures, makes its first appearance in the Aptian. *Ovalveolina* and *Praealveolina* appear in the Albian, while most of the alveolinids appear in the Cenomanian, many with short ranges. Very few appear in the Late Cretaceous, such as *Subalveolina* in late Santonian-Campanian and *Senalveolina* in the early Campanian. *Helenalveolina* and *Subalveolina*, alveolinids with a septulate endoskeleton lacking pillars and with streptospiral early stage and planispiral-involute adult stage, may be linked phylogenetically to *Pseudonummuloculina* (a simple miliolide foraminifera with streptospiral-involute coiling) of mid-Cretaceous age (Hottinger et al., 1989). *Subalveolina* is interpreted by Hottinger et al. (1989) as an endemic offspring of American immigrants. Early chamber arrangement in alveolinids was shown to be linked with proloculus size by Pêcheux (1984, 1988) for Cenomanian streptalveolinids.

The rhapydioninids may have evolved from *Ovalveolina* via *Pseudedomia* in the Cenomanian (Hamaoui and Fourcade, 1973). *Pseudedomia* resembles *Ovalveolina* in its globular initial part, but develops compressed curved chambers in its final whorls. Hamaoui and Fourcade (1973) also suggest that *Chubbina* arose from intermediate forms of *Ovalveolina*. The trend towards uncoiling, and becoming cylindrical, continued through *Murciella* and *Rhapydionina* in the Late Cretaceous (see Fig. 5.21). The latter was described by Stache (1913) from the Late Cretaceous of the Adriatic Platform as spiroliniform shells with a planispiral-involute nepiont followed by a uniserial-cylindrical adult stage. However, he also used the name *Rhipidionina* to descibe peneropliform shells with flaring adult chambers producing a flattened fan reaching a semicircular outline. However, Stache's original specimens have identical structural patterns (Reichel, 1984) and are one dimorphic species, representing respectively a megalospheric and a microspheric generation (Hottinger, 2007).

The Fabulariidae, with round trematophore, were common in the Late Cretaceous and arose directly from the simple *Quinqueloculina* group (Haynes, 1981) or a true miliolide origin exemplified by *Idalina* (see Chapter 6). *Idalina*, with a quinqueloculine to biloculine test, a round trematophore but with a bar-like tooth, acquired incomplete partitions in *Periloculina* (see Hottinger et al., 1989) and the overlapping throughout in *Lacazina* (Plate 5.22, fig. 6).

The Soritoidea of the Cretaceous are separated by a considerable stratigraphic gap from those of their younger descendants in the Eocene and they seem to have originated from different ancestors. Haynes (1981) suggested that they may have arisen in the Senonian from a simple foraminifera *Praepeneroplis*, with a close-coiled and triangular chambers. The main morphological trend in the resulting superfamily is the uncoiling of the planispiral test to become flabelliform and cyclical.

The development of *Meandropsina* is an example of the trend of evolution followed by the Soritoidea in the Late Cretaceous. Each chamber is divided into numerous chamberlets by transverse partitions that are in continuous alignment from chamber to chamber. The resulting composite lines spiral outward from the centre of the test and maintain approximately the same spacing by periodic intercalation of new lines of septa. At this stage *Meandropsina* is exactly the same as *Fallotia* (see Fig. 5.22, stage 1). As growth continues, however, the volume of the individual chambers does not increase at a sufficiently rapid rate to maintain the same shape when stretched across the growing bulk of the test, the radial septa necessary fall closer and closer together and, as they approach parallelism, tend more and more to assume a shape like that of the outward spiraling lines of secondary septa (Fig. 5.22, stage 2). The chambers are now long and thin, and they appear at this stage to have reached a stable minimum width that is maintained throughout the remainder of the individual's development. Since the width of the chambers is now stable, any further increase in size must take the form of an increase in length. This eventually results in a cyclical form of growth in which the chambers form concentric rings that are no longer embracing to the centre of the test (Fig. 5.22, stage 3). In the final stage, the chambers no longer lie astride the periphery, but they withdraw completely to one side and wander haphazardly over the surface of the previously formed test in wavy meandriform line (Fig. 5.22, stage 4).

In conclusion, the radial partitions subdividing the long narrow chambers into chamberlets and the tendency toward uncoiling, becoming flabelliform and annular in the final stage of the test, are two important trends followed by the Soritoidea in the Cretaceous.

5.3.3 The Rotaliides of the Cretaceous

The rotaliides of the Cretaceous developed important evolutionary trends, which reappeared independently again in the Paleogene and Neogene. Large rotaliides were

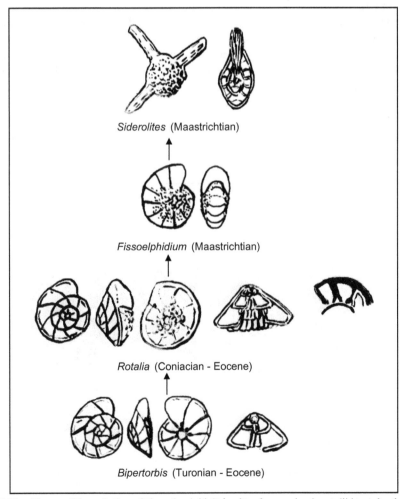

Siderolites (Maastrichtian)

Fissoelphidium (Maastrichtian)

Rotalia (Coniacian - Eocene)

Bipertorbis (Turonian - Eocene)

Figure 5.31. The possible evolution of the calcarinid *Siderolites* from a simple rotaliid test, by developing spines and multiple foramina.

derived from different families of smaller rotaliides by opening up of the umbilical area and its infilling by pillars. *Rotalia* appeared in the Late Cretaceous possibly from a discorbinellid, *Bipertorbis* (see Fig. 5.31), which appeared in the Turonian with a small plug filling the umbilical area. In *Rotalia* the plug is enlarged and composed of thick lamellar pillars with septal fissures. *Rotalia* in turn evolved rapidly given rise to the different families of the Rotalioidea. An example (Fig. 5.31) of one possible line of evolution from *Rotalia* to a calcarinid test, is the development of involute, bilaterally symmetrical forms, *Fissoelphidium* in the Maastrichtian, which may have evolved peripheral spines and multiple foramina in *Siderolites*.

Foraminifera with an orbitoidal growth first appeared in the Late Cretaceous. They have been studied extensively by many authors, and one of the most comprehensive studies was that of van Gorsel (1978). All orbitoidoids were derived from small benthic foraminifera with a simple, spiral chamber arrangement. A typical orbitoidal test (Fig. 5.32) varies from globular to lenticular, and is composed of a median layer, consisting of an initial stage surrounded by concentrically arranged equatorial chambers, flanked on either side by layers of lateral chambers.

The central part of the equatorial layer is composed of the embryonic chambers that are surrounded in turn by the peri-embryonic chambers or the neanic phase. The arrangement of the embryonic chambers is one of the most important characters needed for the identification of an individual species. The embryonic chambers usually comprise two chambers surrounded by a relatively thick wall, a protoconch and a reniform second chamber or deuteroconch (see Fig. 5.33).

In many species of *Orbitoides* there are four chambers within the embryonic wall (Fig. 5.34). The peri-embryonic chambers or nepionic chambers immediately surround

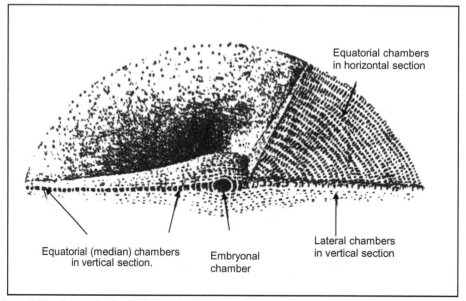

Figure 5.32. Schematic sketch of an orbitoidal foraminifera (modified after Carpenter *et al.*, 1862).

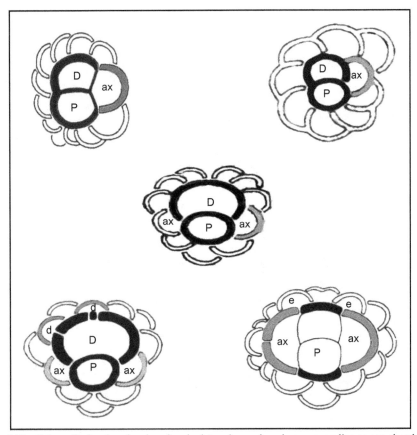

Figure 5.33. Schematic drawing showing chamberlet cycles, each cycle corresponding to one chamber and the change from spiral to concentric growth in orbitoidal foraminifera, as a consequence of the introduction of retrovert apertures, modified after van Gorsel (1978). P = protoconch, D = deuteroconch, ax = auxiliary chambers, d = adauxilliary chambers, e = epiauxilliary chambers.

the embryon, the third chamber is the auxiliary chamber. In advanced forms there may be two auxiliary chambers. In *Lepidorbitoides*, adauxiliary chambers (small chambers originating from stolons in the deuteroconchal wall) are present, while in *Orbitoides*, the chambers originating from the stolons in the thick embryonal wall are usually called epi-auxiliary. In younger species there is an increase in the number of these chambers. In the spiral growth, chambers with one aperture, "uni-apertural chambers" are one chamber-forming apertures, and their number is often designated by the parameter Y, while chambers with two apertures give rise to two chambers, one at each aperture. In the equatorial layer the shape of the chambers varies from open arcuate to ogival or spatulate (see Fig. 5.34).

The main features which differentiate *Orbitoides* from *Lepidorbitoides* can be summarized as follows (see Figs 5.17, and 5.34):

1. The nucleoconch of *Orbitoides* is larger than that of *Lepidorbitoides*, and consists of a porous, thick-walled spheroidal body subdivided internally into two or more chambers that are separated from each other by thin, straight walls. The nucleoconch of *Lepidorbitoides* is smaller than that of *Orbitoides* and consists of two chambers, of

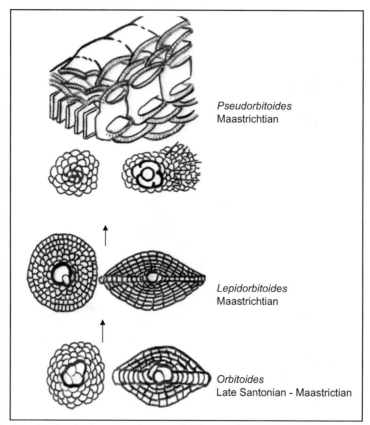

Figure 5.34. Sketches showing important features of different orbitoidal genera.

which the second (deuteroconch) is the larger and partially embraces the first chamber (protoconch).

2. Individual chambers in the rings of simultaneously-formed chambers in the median layer are widely-spaced in *Orbitoides* and crowded together in *Lepidorbitoides*. They tend to be wider than high in *Orbitoides* and higher than wide in *Lepidorbitoides*.

3. Cross-sections through the layers of lateral chambers in *Orbitoides* show them as a series of low arches, the arches of one layer overlapping the arches of the next layer. The lateral chamberlets in *Lepidorbitoides* are rectangular in section and are lined up in vertical tiers.

The origin of the orbitoids has been the subject of many studies and there are many different interpretations. Unlike previous workers, such as Küpper (1954), who regarded the initial chambers of the non-canaliculate Orbitoididae to be biserial, van Gorsel (1978) interpreted them as spiral in the microspheric generation, becoming irregularly alternate. He noted that the equatorial chambers of early *Orbitoides* species were acervuline, thus suggesting an attached mode of life. He therefore supported the idea that non-canaliculate orbitoids were derived from an unknown "*Cibicides*". Here I suggest that a planorbulinid ancestor, such as *Archaecyclus*, may have been the ancestor of the orbitoids. The evolution from *Monolepidorbis* seems to have happened gradually, by the subdivision of the solid walls of *Monolepidorbis* by lateral chamberlets (see Fig. 5.16). However, in *Torreina*, the embryon is similar to *Orbitoides* and *Omphalocyclus*, but no median layer can be distinguished; this form is only found in the Caribbean.

The *Helicorbitoides-Lepidorbitoides* lineage seems to have evolved from the helical *Pseudosiderolites*, which in turn evolved from a small rotaliide (van Gorsel, 1978) by developing incipient lateral chambers. Secondary equatorial chambers were added between the primary spires in *Helicorbitoides* before acquiring well developed lateral chamberlets in *Lepidorbitoides*. This evolutionary lineage provides one of the best examples of nepionic acceleration, where the ancestral spiral is gradually reduced and a concentric growth pattern is reached progressively. The *Lepidorbitoides* lineage shows other evolutionary trends, such as a systematic increase with time of adauxiliary chambers and the protoconch-deuteroconch diameter ratio (Caus et al., 1988).

The Caribbean forms (see Fig. 5.18) evolved from the cosmopolitan genus *Sulcoperculina* (van Gorsel, 1978) by the acquisition of lateral chambers and the further development of the system of incipient radial rods, as in *Sulcorbitoides*. This latter acquired protruding radial plates in *Vaughanina* (Plate 5.35), stellate growth pattern in *Aktinorbitoides* and arcuate secondary equatorial chambers without the radial plates in *Pseudorbitoides*. *Sulcoperculina* evolved into *Orbitocyclina* by suppressing the system

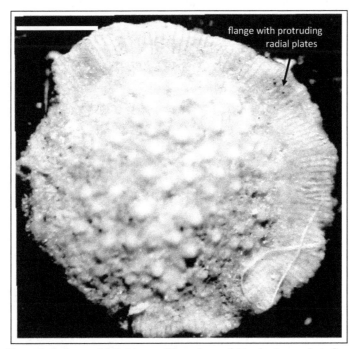

Figure 5.35. The main features exhibited by a solid specimen of *Vaughanina* sp., Late Campanian, Cuba, UCL coll. Scale bar = 0.3mm.

of plates, and by acquiring lateral and curved equatorial chambers, which in turn it evolved to a stellate growth pattern in *Asterorbis.*

Several more unrelated and rapidly evolving lineages of rotaliines thrived in the Late Cretaceous, but most of them became extinct at the end of the Maastrichtian. Due to their rapid evolution and short ranges, the orbitoids are excellent biostratigraphic markers for the Cretaceous.

5.4 Palaeoecology of the Cretaceous Larger Foraminifera

The cooling trend of the Late Jurassic continued into the Early Cretaceous, however the climate warmed up fairly rapidly after the middle Berriasian, triggering an increase in the level of oxygen in the atmosphere, which is inferred (see Ward, 2006) to have steadily increased throughout this period (see Fig. 3.10). A tropical climate existed throughout the whole subsequent Cretaceous period over the Tethyan realm (see Fig. 5.36), which was the main breeding ground of the larger benthic foraminifera and rudist reefs. While coral reefs were predominant in the Jurassic, just as they are today, during the early Cretaceous, a superheated, hypersaline ocean-climate zone favoured the proliferation of rudists over scleractinian corals (Kauffman and Johnson, 1988; Johnson et al., 1996).

The larger benthic foraminifera which survived the minor Jurassic-Cretaceous boundary crisis were mainly robust, shallow, clear-water forms. They consisted mainly of agglutinated foraminifera with large intramural alveoles, such as *Everticyclammina*, which could tolerate water rich in argillaceous suspensions. A very few forms with

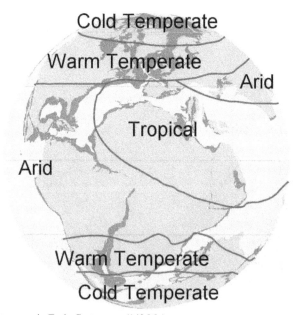

Figure 5.36.   Climate zones in Early Cretaceous (140 Ma).

narrow alveoles and a regularly labyrinthic hypodermis (e.g. *Choffatella,* see Chapter 4 and Plate 5.5, figs 8-9) were also present and occupied a deeper neritic palaeoenvironment, with reduced illumination, in which codiacean and dasyclad algae were rare or absent (Banner and Whittaker, 1991). By the Valanginian-Hauterivian, agglutinated foraminifera, largely adapted in their evolution to the increase in oxygen level and the superheated oceans, increased the sizes of their tests and the complexity of their wall structures. The Chrysalidinidae developed canaculi which had the same function as that of the alveoles in the Cyclamminidae. They facilitated ionic exchange between intrathalamous cytoplasm and external sea-water, allowing the foraminifera to survive and proliferate under different conditions.

By the Barremian, new forms and a major new lineage appeared in Tethys, the agglutinated orbitolinids. These robust forms had the ability to survive in many shallow carbonate environments (Arnaud-Vanneau, 1980), however, they were most common in the outer platform (Vilas et al., 1995). They strengthened their conical test by subdividing it into many small chamberlets, which housed within their walls symbiotic algae. By studying the size and shape of their test, Masse (1976) deduced that they had a free, epifaunal mode of life. They lived lying on the substrate, on the flat base of their conical test, by their apertural face. However, the small primitive forms of the orbitolinids may have been epiphytic (Arnaud-Vanneau, 1975). Using associated algae, Banner and Simmons (1994) noted that *Palorbitolina lenticularis* was most common in sediments thought to be deposited at depths of between 10 and 50m. Simmons et al. (1997) noted that muddy, orbitolinid-rich beds, with large, flat, orbitolinids seem to be characteristic of transgressive deposits, while more conical forms thrive in the shallowest water. The relationship between orbitolinid shape and palaeobathymetry seems to mimic that observed for Holocene larger foraminifera (e.g. *Operculina*) by Reiss and Hottinger (1984) (see Chapter 7). In association with the orbitolinids and *Choffatella*, mid-Cretaceous reefs were typically composed of rudists, a group of aberrant bivalve mollusks (Lehmann et al., 1999). Corals may also be associated with rudist reefs, but possibly only those reefs far from the equator. It has been argued at some length that equatorial surface waters during the Aptian approached 30°C, and were lethal to corals. Thus, this first generation of rudist reefs may have obtained its start because the seas were simply too hot for corals to thrive. Rudist diversity began to escalate in the Aptian-Albian time, but collapsed abruptly at the end of the Maastrichtian, and the entire taxon became extinct at the end of the Mesozoic.

The increase in the number of foraminiferal forms having large alveoles in the early to late Aptian may have been as a consequence of adaptation to the adverse conditions during the anoxic events that accompanied this interval. The dramatic increase of carbon dioxide in the atmosphere (Gradstein et al., 2004), which may have been triggered by the Ontong-Java volcanism, must have led to an increase in temperature, with possible subsequent release of marine clathrates (methane hydrates) leading a greenhouse period and water anoxia (e.g. Kerr, 2006).

Within the Cenomanian, another important group made its first appearance, namely the alveolinids. In comparing the occurrences of similar Holocene forms, we can safely deduce that these fusiform types, with large alveoles (Fig. 5.37A), were found in sediments deposited in warm, shallow water. In fact, all alveolinids are regarded as neritic (inner shelf) by Reichel (1964), restricted to littoral, tropical, protected shelf and reef

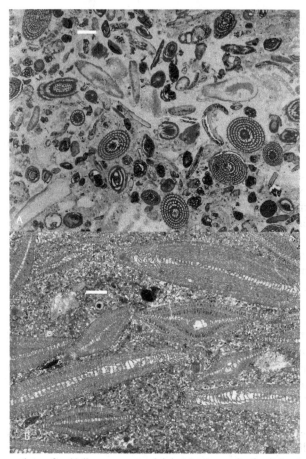

Figure 5.37.   Thin sections of a A) limestone containing *Lacazina* and a few *Monolepidorbis,* Santonian of Spain; B) sandy limestone containing *Orbitoides* and a few *Siderolites,* Maastrichtian of Spain. Scale bars = 1mm.

shoals (Dilley, 1973; Hottinger, 1973). Their appearance might have been caused by one of the most widespread oceanic anoxic events (OAE) at the Cenomanian–Turonian boundary (93 Ma), which may in turn have been triggered by the Madagascar and Caribbean provinces basaltic events (see discussion below).

In the Late Cretaceous, the global climate was warm and moist but cooler than that in Jurassic (210 to 140 Ma). The arrangement of the continents and oceans encouraged warm waters at the equator to circulate freely. During this period larger foraminifera became cosmopolitan and colonized deeper waters. They occupied a wide tropical-subtropical zone, as cool water was only found near the polar regions. Coral reefs grew 5 to 15 degrees closer to the poles than they do today. The niches favoured by the larger foraminifera, which became extinct towards the end of the Cenomanian (see discussion below), were gradually filled by new, highly evolved rotaliides, having an orbitoidal test (Fig. 38B). The introduction of this orbitoidal character, and the subsequent

development of equatorial and lateral chamberlets, may have given these foraminifera a selective advantage in adverse environments over the smaller ones. Orbitoids
now achieved large sizes, which might have reflected the high growth rates that can
be sustained by symbiotic species, with a tendency to increase the size of their protoconchs. Large-sized protoconchs are considered to be advantageous in speeding up
test growth (van Gorsel, 1975) and could be related to high light intensity in the photic
zone (Fig. 5.37B; Drooger and Raju, 1973). The tendency by the orbitoids towards
radial symmetry may be closely related to the more sedentary life possible for symbiotic
forms (Chaproniere, 1975), as it is generally accepted that orbitoids thrived in shallow
tropical and sub-tropical seas, in areas with little or no clastic influx (van Gorsel, 1975).
However, van Gorsel (1975) noted that not all of the genera had the same ecological
requirements. For most genera, a water depth of 50m is the limit of habitation, but in
the Caribbean, *Pseudorbitoides* appeared to have lived in deeper fore-reef environments,
while *Orbitoides* and *Vaughanina* (often associated in many localities with *Asterorbis*)
are found in sediments deposited in back-reef environments. In Europe, *Omphalocyclus*
probably lived in shallower waters than *Lepidorbitoides* and *Clypeorbis*, but their
depth ranges overlap (van Gorsel, 1975, 1978). Some rotaliides developed spines, as
in the calcarinids, to spread the weight of their shell, as in the case of *Siderolites* in
the Maastrichtian. The larger benthic foraminifera in the Cretaceous occupied various
facies, close to wave base, in lagoonal settings or in sediments overlying or underlying
massive reefs (Noujaim Clark and BouDagher-Fadel, 2001). Different associations of
foraminifera belong to different biofacies and different depositional environment; the
Nezzazatidae and the Cuneolinidae are found mainly in lagoonal facies together with
Miliolidae and Verneulinidae. However, their absence points to a tidal environment

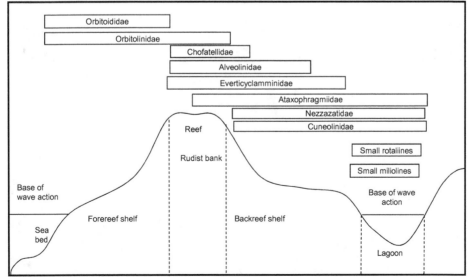

Figure 5.38.   The special distribution of the Cretaceous larger benthic foraminifera along the shelf of
Tethys.

(Amodiao, 2006). Fig. 5.38 summarizes the palaeoecological range and significance of most Cretaceous forms.

5.5 Palaeogeographic Distribution of the Cretaceous Larger Foraminifera

The breakup of the super-continent Pangaea, which began during the Jurassic, continued during the Early Cretaceous. Many of the land masses were covered by shallow continental oceans and inland seas. Europe, Asia, Africa and North America were all a series of islands. The Cretaceous saw the lengthening and widening of the proto-Atlantic Ocean, which began to spread further south, while the Alps began to form in Europe. India broke free of Gondwana and became an island continent. Africa and South America split apart, Africa moving north and closing the gap that was once the Tethys Sea. The continents began to take on their modern forms (see Fig. 5.39).

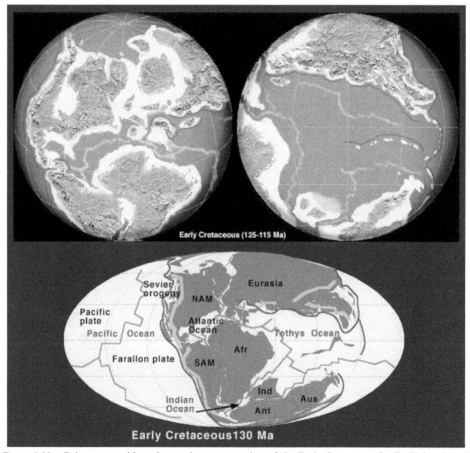

Figure 5.39. Palaeogeographic and tectonic reconstruction of the Early Cretaceous (by R. Blakey, http://jan.ucc.nau.edu/~rcb7/paleogeographic.html).

Cretaceous Speciation

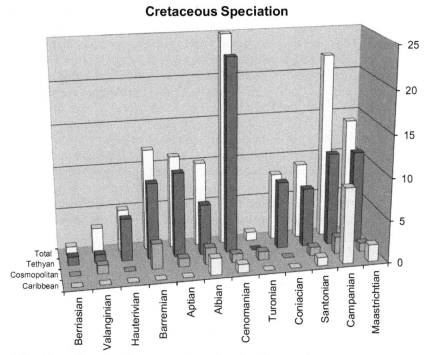

Figure 5.40. The total number of new genera found in each Cretaceous stage.

Cretaceous Genera

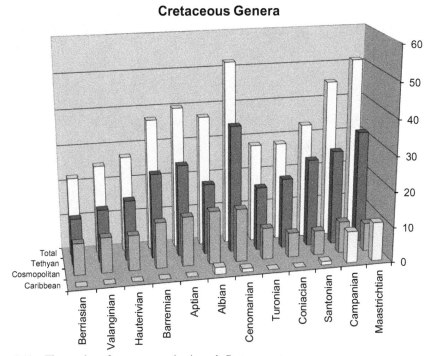

Figure 5.41. The number of genera occurring in each Cretaceous stage.

In the Berriasian, most of the characteristic larger benthic foraminifera of the Cretaceous had yet to evolve. With very few new appearances in the Berriasian (5%), nearly all of the larger benthic foraminifera were Jurassic survivors, and 55% of the Berriasian larger foraminifera were restricted to the Tethyan realm (Figs 5.40 and 5.41). They colonised, together with "aragonitic" scleractinian corals, calcareous green algae (e.g. *Actinoporella podolica*) and rudists bivalves, all early Cretaceous reefs. Most of the forms continued through to the Valanginian. Few foraminifera were to be found in the Caribbean (see Fig 5.41). The Pacific Plate was growing, but still quite small, and the entire western margin of North and South America was fringed with volcanic island arcs (Zharkov et al. 1998). Towards the late Berriasian only 5% of the larger foraminifera had disappeared from Tethys where the cooling trend of the late Jurassic was at an end.

The Valanginian was a period of transition between the relatively cold time at the end of the Jurassic to the "greenhouse" world that continued for the rest of the Cretaceous. During this period, new forms evolved (14% of the whole Valanginian component, see Figs 5.40 and 5.41), increasing the dominance of the agglutinated foraminifera in the Tethyan realm. Only a few of new forms (20%) were cosmopolitan (see Fig. 5.40 and Chart 5.1).

The Hautervian saw the continuation of forms from the Valanginian, with very few new species appearing (18%), and again mainly in Tethys. Although the end of the Valanginian and early Hauterivian, saw considerable volcanic activity, such as the development of the Paraná traps in South America (Fig. 5.42), together with their smaller severed counterpart of the Etendeka traps in Namibia and Angola (Courtillot and Renne, 2003), very few species of larger foraminifera became extinct (only ~7%), and those that did were mainly in Tethys (see Fig. 5.43). The extinct forms were mainly the strong, elongated forms with solid cores that had survived the Jurassic boundary. This is in contrast to the statement made by Courtillot and Renne (2003) that the end of the Valanginian is

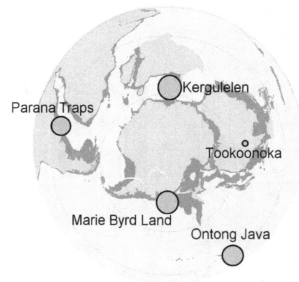

Figure 5.42. The end Barremian-early Aptian world, showing the position of the Ontong–Java LIP and other middle Cretaceous events.

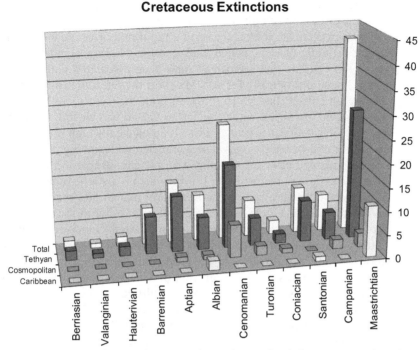

Figure 5.43. The number of genera that went extinct at the top of each Cretaceous stage boundary.

a prominent extinction level (although there may have been a major crisis for bryozoan faunas). This stage boundary may have also coincided with a cooling event, as noted by Walter (1989) that was followed by a warming event around (133–132 Ma).

At the beginning of the Barremian, foraminifera flourished and many new forms appeared (32%, see Fig. 5.40), however, they were short lived and did not have time to become established, as most of them became extinct towards the end of the Barremian. The end of the Barremian, a time of a major anoxic event and also of global sea-level rise (Courtillot and Renne, 2003), saw a modest increase in the number of extinctions (22% of the Barremian assemblages, see Fig. 5.43).

The end Barremian-early Aptian extinctions could have been the results of, or triggered by, the Ontong–Java volcanic event (see Fig. 5.42), which gave rise to the largest of all oceanic basaltic plateaus, that in places reaches a thickness of 40 km (Courtillot and Renne, 2003). Hallam and Wignall (1997) point out that most early Cretaceous biota were largely unaffected by this and other major volcanic episodes, and indeed as seen in Fig. 5.43 and Chart 5.1, there are a relatively modest number mass extinction of larger benthic foraminifera at that time. This may be because these eruptions were submarine and, as explained by Courtillot and Renne, the enormous mass of ocean water is expected to act as a strong buffer; SO_2 is not expected to reach the atmosphere but CO_2 would still lead to water anoxia and a greenhouse period (Kerr, 2006). For this reason, there are good reasons to correlate the Ontong–Java volcanism and the end-Barremian Ocean Anoxic Event (OAE), but perhaps because of their reef forming niche this led to the extinction of only a few larger foraminiferal genera.

The Aptian was an eventful, long stage, with a rapid rate of oceanic spreading in the Atlantic. The Atlantic Ocean opened wide enough to allow significant mixing of waters across the equator that seemed to have been associated with a series of OAEs, which continued periodically for ~35 Ma (into the Santonian). Black shales, signalling deep ocean anoxia, have been found in both the Central Tethys (Europe) and the western Pacific provinces, while the Tethys Ocean was closing and the Alps began to form. Ecological evidence points towards an unusually high rate of volcanic activity, especially at mid-oceanic ridges. The Kerguelen plateau (Fig. 5.42), the second largest oceanic plateau after the Ontong–Java plateau, erupted in the southern Indian Ocean in the early Aptian, around 118 Ma, which is also the time of eruption of the Rajmahal traps in eastern India. The Kerguelen plateau, Rajmahal traps, and eruptions in western Australia were located close to each other at the time of eruption, and were roughly coeval with the breakup of eastern Gondwana (Courtillot and Renne, 2003). During these eruptions the sea floors rose, forcing the sea water to rise worldwide, flooding up to 40 % of the continents. Sea levels were over 200m higher than present day levels. Such a high rate of volcanic activity released massive amounts of carbon dioxide into the ocean and ultimately the atmosphere. This quantity of carbon dioxide would have made the oceans relatively short of oxygen. The evidence for this is displayed in the geological record. The abundance of black shale and petroleum-rich formations, suggest they formed in an oxygen poor environment. Dramatic rises in temperature was recorded in the early Aptian and mid-Aptian (Jenkyns and Wilson, 1999; Jenkyns, 2003), and there is a significant turnover of larger benthic foraminifera throughout the Aptian in Tethys (see Fig. 5.43 and Chart 5.1). Eleven new agglutinated genera appeared in the Aptian, ten of them restricted to Tethys, while the other was cosmopolitan (Fig. 5.40 and Chart 5.1). Large miliolides made their first appearance in the Aptian and went on to colonise hot lagoonal environments in the Albian and Cenomanian of Tethys. Large conical, agglutinated, internally complicated orbitolinids invaded the shallow warm reefal environments of Tethys. Rudists were also very common during the Aptian replacing corals in many niches. The late Aptian saw an increase in extinctions, but some of these were replaced by new genera at the Aptian-Albian boundary. These extinctions coincided with a rapid sea-level fall at the Aptian-Albian boundary and the collapse of reef ecosystems (Walliser, 1996).

In the Albian, and for the first time in the Cretaceous, a small percentage of larger foraminifera become restricted to the Caribbean province (2%). With 50% belonging only to Tethys, the larger foraminifera set a provincialism trend that extends into the Late Cretaceous. Towards the end of the Albian, the Caribbean foraminifera became extinct and many of the Early Cretaceous Tethyan forms disappeared. This event may have been correlated to the Hess Rise volcanic event in the North Pacific Ocean (Eldholm and Coffin, 2000).

Following the late Albian event, the early Cenomanian interval is marked by a short term turnover in larger foraminifera, 25 new genera appeared in Tethys, out of which 68% were alveolinids. It is the highest faunal turnover in the Cretaceous (see Figs 5.40, 5.41 and 5.43). The main breeding grounds of the larger foraminifera were the tropical to subtropical lagoonal and shallow shelf waters of Tethys, with only one new genus appearing in the Caribbean. However, by the end of the Cenomanian many of them became extinct and indeed many of the early Cretaceous forms and Jurassic survivors failed to survive the end of the Cenomanian. These extinctions were initiated by the collapse of palaeotropical reef ecosystems, which happened according to Walliser (1996) near the middle Cenomanian. The extinction affected most agglutinated and porcelaneous foraminifera, leaving globally many empty niches. As with the epifaunal bivalves, which were virtually unaffected by these events (Harries and Little, 1999), some of the surviving taxa of larger foraminifera were also epifaunal and thus pre-adapted to low-oxygen conditions, such as *Ataxophragmium*. Bambach (2006) considered this peak of extinction to be connected with an oceanic event, while Walliser (1996) associated these extinctions with a near-peak Mesozoic eustatic sea-level high-stand (sea levels rose dramatically to over 200m higher than present levels in the Cenomanian and 300m above present stand in the early Turonian). These events might also have been affected by another major sub-marine event (the Wallaby eruption) in the Indian Ocean (Eldholm and Coffin, 2000), which again would have been associated with the high emission of CO_2 contributing to the global warming peak and greenhouse climates during that period.

In the Turonian, very few new larger foraminifera appeared. However, towards the end of the Turonian many of the foraminifera that survived the Albian-Cenomanian crisis become extinct (see Fig. 5.43). Although none of the larger foraminifera were restricted to the Caribbean in the Turonian, a major extinction among Caribbean rudist occurred at this time (Walliser, 1996), and the restriction of reef ecosystems might have contributed to the disappearance of the global foraminifera. Johnson et al. (1996) proposed that the collapse of mid-Cretaceous dominated reef ecosystems, may be attributed to the collapse of the Tethyan ocean-climate system. Courtillot and Renne (2003) stated that the volcanism, which occurred between 91 and 88 Ma in a number of short, discrete events, may be the cause of the oceanic events. The Caribbean-Colombian Cretaceous Igneous province (CCIP), the Madagascar event (see Fig. 5.44) and Phase 2 of the Ontong-Java event all occurred within this short interval, and would have certainly contributed to the extinction phase at the end of the Turonian. Kerr (2006) has proposed volcanism-related CO_2 lead to an enhanced greenhouse climate and global warming, leading eventually to the extinction of the most vulnerable reefal communities of larger foraminifera and rudists. There is an oxygen isotope anomalies at 91.3 Ma (Bornemann et al., 2008), that has been interpreted as being indicative of a glaciation event in the middle of the Turonian, one of the warmest periods of the Mesozoic. If this event occurred, it may have been triggered by short term atmospheric changes associated by flood basalt events in the Caribbean and Madagascar regions.

At the Coniacian boundary, a short-term turnover of larger benthic foraminifera occurred. About a quarter of the Coniacian foraminifera were new, and only a few genera became extinct towards the end of this stage. The foraminifera turnover in the Coniacian takes place according to Walliser (1996) during a global flooding interval and the termination of the second regional oxygen depletion event, which is recognized by many places by organic-rich dark shales. The Santonian saw the appearance of a significant number of new genera, however, by the end of the

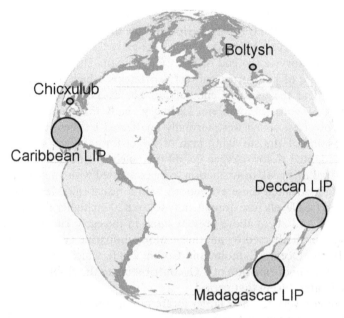

Figure 5.44. The end Maastrichtian world showing the position of the Chicxulub impact and other major Late Cretaceous events.

Santonian a high percentage of foraminifera again became extinct, mainly in the Tethyan realm. This may again have been triggered by renewal of the Kerguelen, Broken Ridge volcanic event that generated over 9 million km³ eruptive material (Eldholm and Coffin, 2000).

During the early to mid Late Cretaceous agglutinated and porcelaneous foraminifera were dominant in the shallow water of the Tethyan realm, however, in the Late Cretaceous, a major shift occurred when new groups developed and progressed to fill the empty niches left by the two successive extinction events at the end of the Cenomanian and end of the Turonian. The calcareous hyaline orbitoids increased in diversity and abundance in the Campanian and Maastrichtian, showing for the first time provincialism, until they outnumbered the agglutinated and porcelaneous groups. By this stage, larger benthic foraminifera were either cosmopolitan or restricted to two different provinces. This might be the result of the Late Cretaceous developing isolated land masses (see Fig. 5.45) because of the dramatic rise in sea level, which produced major inland seas. As a result, the Campanian saw a rapid development of new species, the second highest in the Cretaceous (see Fig. 5.40), many of which colonized deeper water than their ancestors, the orbitoids. About 50% of the Campanian genera were new. Of these new appearances, 20% belonged exclusively to the Caribbean and only 20% were cosmopolitan, while the rest thrived in Tethys. The new appearances in the Caribbean might have been the result of the filling of reefal niches previously occupied by rudists before they were destroyed in the mid-Cretaceous. Towards the end of the Campanian, 15% of the foraminifera became extinct. This might have been related by two volcanic events in the Atlantic Ocean, the Sierra Leone Rise and the Maud Rise

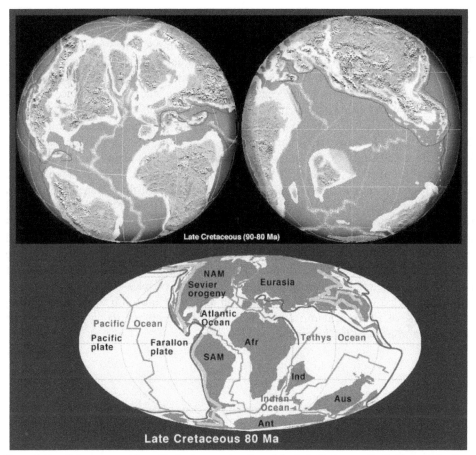

Figure 5.45. Palaeogeographic and tectonic reconstruction of the Late Cretaceous (by R. Blakey, http://jan. ucc.nau.edu/~rcb7/paleogeographic.html).

(Eldholm and Coffin, 2000). This was accompanied by a regional oxygen reduction event (Walliser, 1996) occurring during the Campanian.

The beginning of the Maastrichtian was a period of high turnover, and about a quarter of the foraminifera had their first appearance at the Campanian-Maastrichtian boundary. The larger foraminifera in the Maastrichtian were highly developed and complicated internally, in order to compensate for their large sizes. Tethys was again the main breeding ground for new forms.

The larger benthic foraminifera of the Cretaceous were brought to an end by one of the greatest mass extinctions of all time, the Cretaceous - Paleogene (K-P) event, or terminal Mesozoic extinction (Macleod et al., 1997). About 83% of the Maastrichtian foraminifera died out, including all orbitoids and alveolinids. The survivors were the smallest and toughest, 8% of the Tethyan forms and the rest being cosmopolitan. The K-P extinction has been the subject of many studies. Some have argued that it really began before the end of the Cretaceous, between 67.5 and 68 Ma, with the abrupt extinction of rudist bivalve dominated reef ecosystems (Johnson and Kauffman, 1990), with new radiometric dates putting this extinction around 68-68.5 (Walliser, 2003). Others blame the K-P extinction on a single or multiple impacts at ~66 Ma (Keller et al., 2003; Alegret et al., 2005; Stüben et al., 2005; Schulte et al., 2006; Kuroda et al., 2007), or on an associated massive volcanism, which in turn triggered significant climatic changes, inducing biotic crises and oceanic anoxia. There is certainly evidence supporting multiple impact events at the end Maastrichtian. An impact crater of 24km diameter occurred in Ukraine around 65.17 Ma, the Boltysh impact (Kelley and Gurov, 2002; see Fig 5.45). On the other hand, the largest impact, that is now widely accepted to have been the cause of the mass extinctions, occurred at Chicxulub, Yucatan in Mexico (Glikson, 2005; Macleod, 2013; see Fig 5.44). However, at about the same time as these impact events there was also a major volcanic event that formed the Deccan Traps (see Fig 5.45). Courtillot and Renne (2003) pointed to the Ir bearing layer related to the Chixculub crater impact within the traps, which indicate that Deccan volcanism began prior to this impact and straddled it in time. There was also a sea level fall ~100kyr before the Ir-defined K-P boundary. The low point occurred 10kyr before the K-P event according to Hallam and Wignall (1997), who attribute the subsequent rise and warming to Deccan eruptions and associated CO_2 release. The K-P boundary impact coincided with the sea level regression and the Chicxulub event contributing to the demise of all tropical and subtropical larger foraminifera, to say nothing of the dinosaurs and most other life forms!

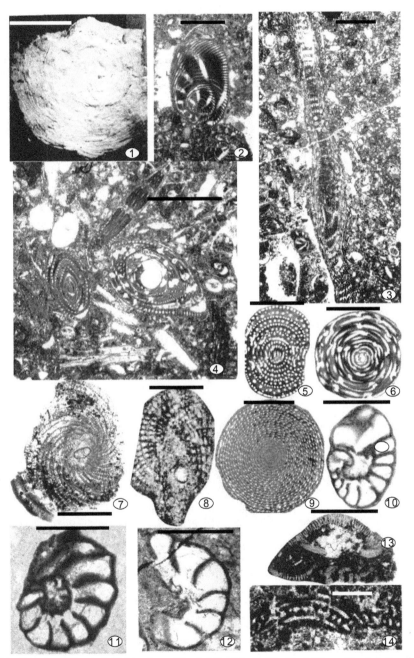

Plate 5.1 Scale bars: Figs 1, 3, 10-14 = 1mm; Figs 2, 4-8 = 0.5mm; Fig. 9 = 0.15mm. Figs 1-4. *Pseudedomia complanata* Eames and Smout, Campanian, Um Gudair, Kuwait, 1) holotype, NHM P42909; 2-4) paratypes, NHM P42916-9. Figs 5-6. *Pseudedomia globularis* Smout, Campanian, Iraq, paratypes, NHM P42646.

Fig. 7. *Meandropsina vidali* Schlumberger, Santonian, Spain, B-Form, NHM P35882. Fig. 8. *Meandrospina vidali* Schlumberger *biconcava* Henson, oblique section, Cenomanian, Wadi Meliha, Israel, paratype, NHM P35886. Fig. 9. *Multispirina* sp., Cenomanian, Qatar, NHM M2425. Figs 10-11. *Nezzazatinella picardi* (Henson), paratype, Santonian, Egypt, 10) NHM P39125; 11) NHM P39100. Fig. 12. *Nezzazata conica* (Smout), paratype, Cenomanian, Iraq, NHM M8311. Fig. 13. *Rotalia* sp., Maastrichtian, Qatar, NHM M3687. Fig. 14. *Mangashtia viennoti* Henson, syntype, Cenomanian toTuronian, Tang-i-Kurd, Iran, NHM P35881.

Plate 5.2 Scale bars: Figs 1-7 = 0.5mm; Figs 8-11 = 0.25mm; Figs 12-14 = 1mm. Fig. 1. *Montsechiana* aff. *montsechiensis* Aubert et al., Valanginian, Murban-2, Thamama V, Abu Dhabi, NHM coll. Fig. 2. *Hottingerita complanata* (Hottinger), figured by Hottinger (1967), Barremian, Switzerland. Fig. 3. *Paracoskinolina* sp., Albian, Libya, UCL coll. Fig. 4. *Neoiraqia convexa* Danilova, figured by Loeblich and Tappan (1988), Cenomanian to Turonian, Yugoslavia. Fig. 5. *Gyroconulina columellifera* Schroeder and Darmoian, Maastrichtian, Saudi Arabia, NHM M7434. Figs 6-7. *Coxites zubairensis* Smout, paratype, Cenomanian, Iraq, 6) NHM P42954; 7) NHM P42952. Figs 8-9. *Elphidiella multiscissurata* Smout, Maastrichtian, Qatar, 8) NHM P42175; 9) NHM P4217577. Figs 10-11. *Fissoelphidium operculiferum* Smout, paratypes, Maastrichtian, Qatar, 10) NHM P42167; 11) NHM P42168. Figs 12-13. *Lepidorbitoides minor* (Schlumberger), topotypes, Maastrichtian, Netherlands, MHM M3163-4. Fig. 14. *Lepidorbitoides* sp., Maastrichtian, Kuh - i Bizadan, near Jahrun, Iran, NHM P32949.

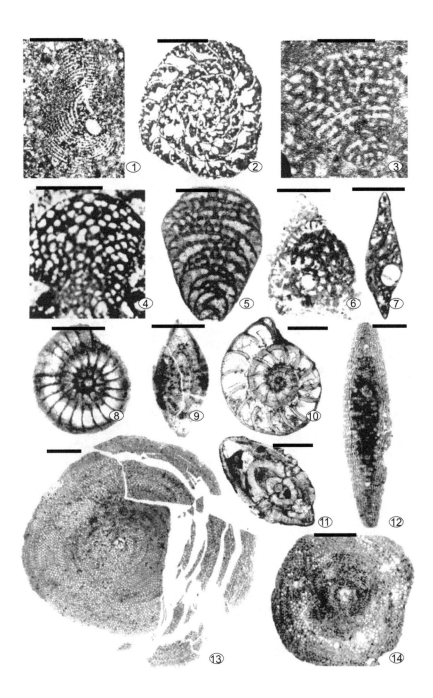

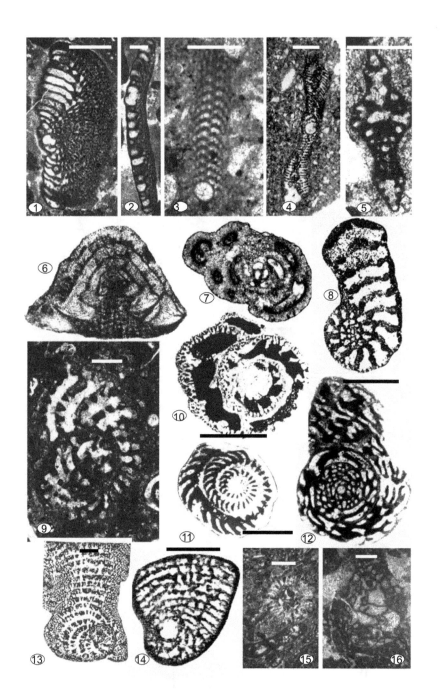

Plate 5.3 Scale bars: Figs 1-4, 8-16 = 0.5mm; Figs 5-7 = 0.16mm. Figs 1-2. *Banatia aninensis* Schlagintweit and Bucur, types figured by Schlagintweit and Bucur (2017) from the upper Barremian of Reşita, Moldova Nouă Zone, SW Romania, (Courtesy of Dr Schlagintweit). Fig. 3. *Tarburina zagrosiana* Schlagintweit et al., holotype figured by Schlagintweit et al. (2016) from the late Maastrichtian of the Tarbur Fm., Zagros Zone, SW Iran (Courtesy of Schlagintweit). Fig. 4. *Spirocyclina* sp. Middle Cretaceous, Shilaif (Khatiyah) Fm., Qatar, NHM coll. Fig. 5. *Demirina meridionalis* Özcan, Cenomanian, southeastern Turkey (courtesy of Dr Özcan). Fig. 6. *Lupertosinnia pallinii* Farinacci, type specimen from Farinacci (1996), upper Campanian limestones, southern Italy. Fig. 7. *Takkeina anatoliensis* Farinacci and Yeniay, type specimen from Farinacci and Yeniay (1994), western Taurus, Turkey. Fig. 8. *Feurtillia frequens* Maync, topotype from Switzerland figured by Hottinger (1967). The thick septa and strong apertural neck distinguish *Feurtillia* even in small wholly coiled specimens. Fig. 9. *Zagrosella rigaudii* Schlagintweit and Rashidi, type figured by Schlagintweit and Rashidi (2017) from upper Maastrichtian, Tarbur Fm. of the Naghan section, Zagros Zone, SW Iran. Fig. 10. *Everticyclammina eccentrica* (Redmond), Valanginian, Mutriba-2, core, 11,690 ft, MINAGISH D, Yamama, Arabian Peninsula, NHM coll. Fig. 11. *Daxia cenomana* Cuvillier and Szakall, figured by Maync (1972) from late Cenomanian, France. Fig. 12. *Chubbina cardenasensis* (Barker and Grimsdale), figured by Robinson (1968) as a cotype megalospheric specimen of *Chubbina jamaicensis* Robinson, Campanian, Jamaica, BMNH P48053. Fig. 13. *Pseudochoffatella cuvillieri* Deloffre, figures by Gušić (1975) from the latest Aptian-earliest Albian of North Croatia. Fig. 14. *Spiroconulus perconigi* Allemann and Schroeder, holotype figured by Allemann and Schroeder (1972) from the Bathonian of Cadiz, South Spain. Fig. 15. Textulariida, miliolids and Dasyclad sp., Aptian, Tibet, UCL coll. Fig. 16. *Bacinella irregularis* Radoičič, Aptian, West Africa, UCL coll.

Plate 5.4 Scale bars: Figs 1 – 4, 6-11 = 0.5mm; Fig. 5 = 0.25mm. Figs 1-2. *Praereticulinella cuvilleri* Deloffre and Hamaoui, figured type by Deloffre and Hamaoui (1970), Barremian, Spain, 1) paratype, axial section; 2) holotype, equatorial section. Fig. 3. *Reticulinella reicheli* (Cuvillier, et al.), metatypic topotypes, Cenomanian to Turonian, Libya, UCL coll. C2.6412. Fig. 4. *Martiguesia cyclamminiformis* Maync, topotype, Santonian, Les Martigues, equatorial section showing the progressive infilling of the chambers with extensions of the alveolar hypodermis, signalling the evolution of *Loftusia* (see 5.26), NHM coll. Figs 5-6. *Qataria dukhani* Henson, paratypes, 5) NHM P35978; 6) NHM P35984. Fig. 7. *Daxia* sp., This form has thicker test and higher chambers, than *D. cenomana* (Cenomanian) and *D. minima* (Aptian), Aptian, Zakum -1. Upper Kharaib (upper Thamama II), Arabian Peninsula, NHM coll. Fig. 8. *Mayncina* cf. *dorbignyi* (Cuvillier and Szakall), early Valanginian, Mutriba-2, core, 13,386 ft, lower MINAGISH (= Sulaiy), NHM coll. Fig. 9. *Gendrotella rugoretis* Gendrot, topotype figured by Banner (1966), Santonian, Les Martigues, NHM coll. Fig. 10. *Stomatostoecha plummerae* Maync, paratype figured by Maync (1972), Albian, Texas. Fig. 11. *Murciella cuvillieri* Fourcade, figured by Fourcade (1966), Campanian, Spain.

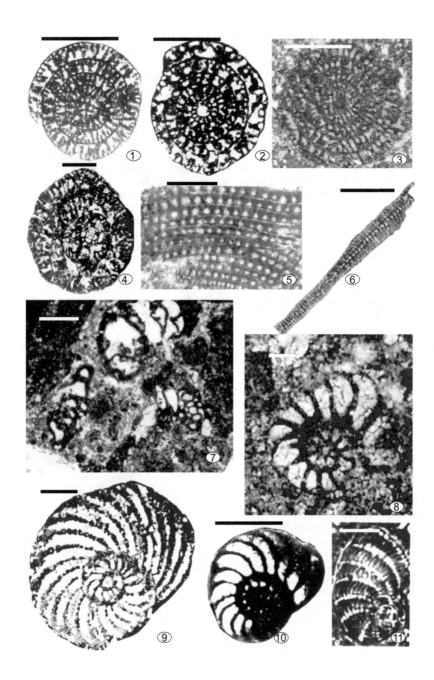

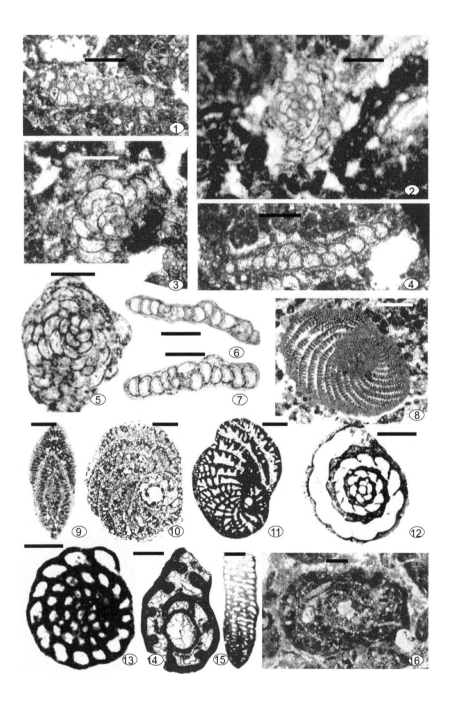

Plate 5.5   Scale bars: Figs 1-7, 12-14 = 0.25mm; Figs 8-11, 15, 16 = 0.5mm. Figs 1-7. *Archaecyclus midorientalis* Eames and Smout, Campanian, Um Gudair, Kuwait, paratypes, NHM P42924-8. Figs 8-9. *Choffatella decipiens* Schlumberger, Early Cretaceous, Dukhan no.1 well, Qatar, NHM P35780. Fig. 10. *Pseudochoffatella cuvillieri* Deloffre, A-form, figured by Cherchi and Schroeder (1982), Aptian, Spain. Fig. 11. *Sornayina foissacensis* Maries, figured by Loeblich and Tappan (1988), Coniacian, France. Fig. 12. *Charentia cuvillieri* Neumann, figured by Hofker (1965), Aptian to Albian, Spain. Fig. 13. *Debarina hahounerensis* Fourcade, paratype, figured by Fourcade *et al.* (1972), Aptian, Algeria. Fig. 14. *Lituola* sp. aff. *nautiloides* (Lamarck), Late Cretaceous, Iraq, NHM P35873. Fig. 15. *Lituonelloides compressus* Henson, paratype, Maastrichtian, Dukhan no. 1, Qatar, NHM P35876. Fig. 16. *Pseudocyclammina bukowiensis* (Cushman and Glaz.), Upper Zangura Limestone, Valanginian, Iraq, NHM P52301.

Plate 5.6   Scale bars: Figs 1-5, 7, 10-11 = 1mm; Figs 6, 8-9, 12-19 = 0.5mm. Fig. 1. *Pseudocyclammina bukowiensis* (Cushman and Glaz.), the thick-walled, coarsely agglutinated, inner hypodermal alveolae distinguish this species from *P. lituus*, Upper Zangura Limestone, Valanginian, Iraq, NHM P52301. Figs 2-4. *Pseudocyclammina lituus* (Yokoyama), Berriasian to Hauterivian, Iran, NHM P52300; 5) Valanginian, Libya, UCL coll. Fig. 6. *Pseudocyclammina cylindrica* Redmond, Berriasian to Valanginian, United Arab Emirates, NHM P52304. Fig. 7. *Pseudocyclammina rugosa* (d'Orbigny), topotype, Cenomanian, Île de Madame, Charente Inférieur, France, NHM coll. Fig. 8. *Cylammina* sp., equatorial section, Cretaceous, UCL coll. Fig. 9. *Lituola obscura* Barnard and Banner, paratype, Late Cretaceous, Labiatus Zone, Norfolk, England, 4507, UCL coll. Figs 10-11. *Chubbina jamaicensis* Robinson, lectotypes, figured by Robinson (1968), Campanian, Jamaica. Figs 12-13. *Hemicyclammina sigali* Maync, Albian, Rumaila 1, Asara Formation, Iraq, NHM M8348. Figs 14 -15. *Hemicyclammina whitei* (Henson), Aptian to Albian, Dukhan no. 3, Qatar, NHM P35797-99, 14) holotype; 15) paratype. Fig. 16. *Buccicrenata hedbergi* (Maync), paratype figured by Maync (1953), middle Albian, Venezuela. Fig. 17. *Pseudolituonella reicheli* Marie, figured by Sartorio and Venturini (1988), late Cenomanian, Iran. Fig. 18. *Ammobaculites* gr. *edgelli* Gollestaneh, late Valanginian, Fuwairat-1, core, 4150-55 ft, Yamama, Arabian Peninsula. Fig. 19. *Spirocyclina choffati* Munier-Chalmas, near-topotype, showing thick early epidermis, but rapid development of subepidermal (hypodermal) alveolae and sub-hypodermal transverse septulae in second whorl, figured by Banner (1966), from the Santonian, Carrière de Martigues, Bouches-du-Rhône, France.

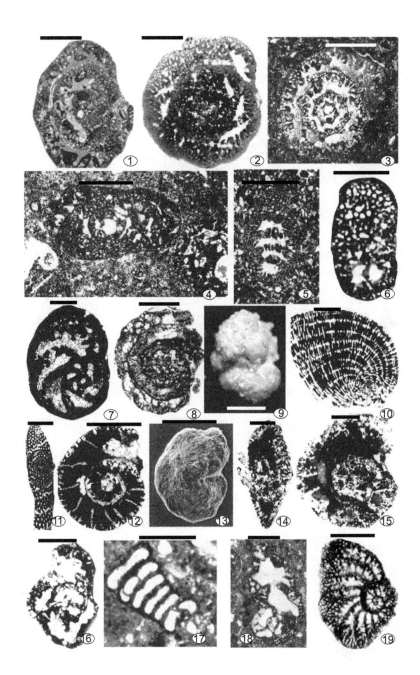

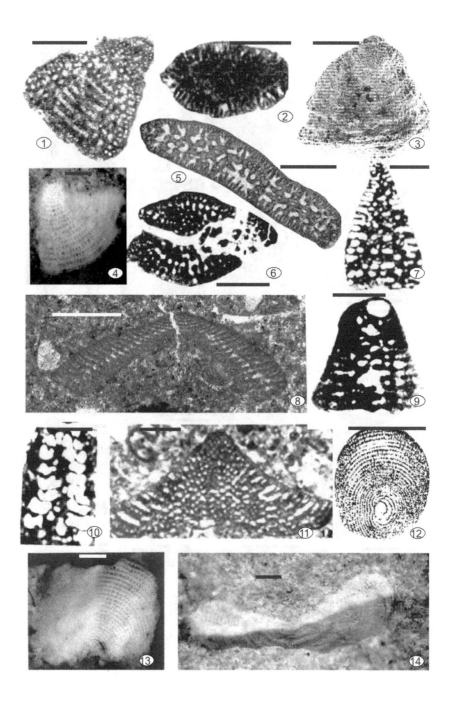

Plate 5.7 Scale bars: Figs 1-8, 11-14 = 0.5mm; Figs 9-10 = 0.15mm. Figs 1-3. *Dictyoconella minima* Henson, paratypes, Cretaceous, Dukhan no.3 well, Qatar, 1) NHM P35836; 2) NHM P35835; 3) solid specimen, NHM P35838. Figs 4-6. *Dictyoconella complanata* Henson, Maastrichtian, Dukhan no. 1 well, Qatar, 4) holotype, NHM P36832; 5-6) paratypes, 5) NHM P35831; 6) NHM P35830. Fig. 7. *Cribellopsis neoelongata* (Cherchi and Schroeder), holotype, figured by Loeblich and Tappan (1988), early Aptian, France. Figs 8, 14. *Pseudorbitolina marthae* Douvillé, Maastrichtian, Dukhan no.1 well, Qatar, 8) NHM P35964; 14) NHM P35965. Fig. 9. *Abrardia mosae* (Hofker), figured by Loeblich and Tappan (1988), Maastrichtian, France. Fig. 10. *Campanellula capuensis* De Castro, holotype figured by Loeblich and Tappan (1988), Valanginian to Barremian, Italy. Fig. 11. *Calveziconus lecalvezae* Caus and Carnella, figured by Loeblich and Tappan (1988), Campanian, Spain. Fig. 12. *Balkhania balkhanica* Mamontova, type figures from Mamontova (1966), Barremian, Great Balkan Mountain, Turkmenistan. Fig. 13. *Broeckinella arabica* Henson, holotype, Maastrichtian, Dukhan no. 1 well, Qatar, NHM P35778.

Plate 5.8 Scale bars: Figs 1-8 = 1mm. Fig. 1. *Loftusia coxi* Henson, syntype, Maastrichtian, Dukhan no.1cwell, Qatar, NHM P35378. Fig. 2. *Loftusia persica* Brady, Maastrichtian, Iran, NHM P34679. Fig. 3. *Loftusia harrison* Cox, Maastrichtian, Adiyaman, Turkey, UCL coll. Figs 4-8. *Loftusia* sp., Maastrichtian, 4, 6-8) UCL coll.; 5) Iraq, NHM 7467.

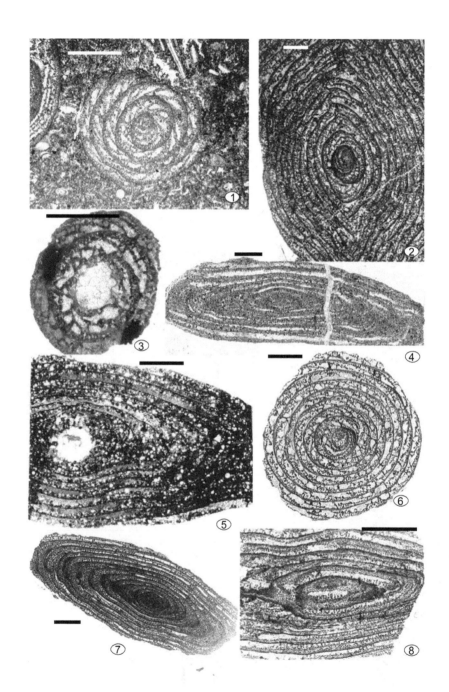

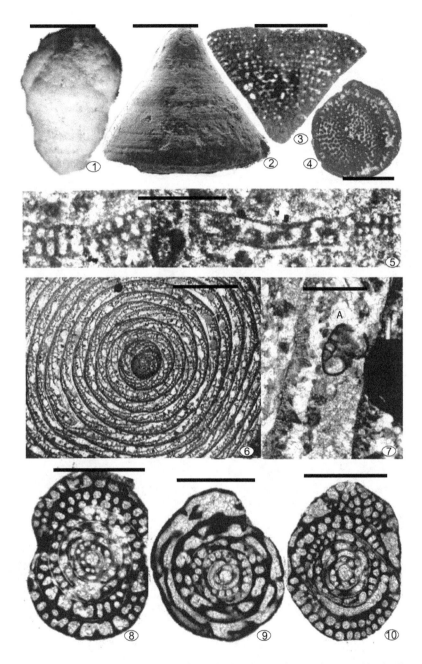

Plate 5.9 Scale bars: Figs 1-2, 4-6, 8-10 = 0.5mm; Fig. 3 = 1mm; Fig. 7 = 0.25mm. Fig. 1. *Chrysalidina* sp., Cenomanian, Iran, NHM coll. Fig. 2-4. *Montseciella arabica* (Henson), Barremian, Dukhan no.2 well, Qatar, 2) NHM P35815; 3) NHM 35753; 4) NHM 35760. Fig. 5. *Mangashtia viennoti* Henson, syntype, Cenomanian to Turonian, Tang-i-Kurd, Iran, NHM P35881. Figs 6-7. *Loftusia persica* Brady, Maastrictian/ ?Eocene, Iran, NHM coll., 6) enlargement of centre of test; 7) enlargement of one chamber to show A) *Turborotalia* sp. (an Eocene planktonic foraminifera) trapped in the test. Figs 8-10. *Chubbina cardenasensis* (Barker & Grimsdale), Late Cretaceous, Cardenas, San Luis Potosi, Mexico, NHM P33042-4.

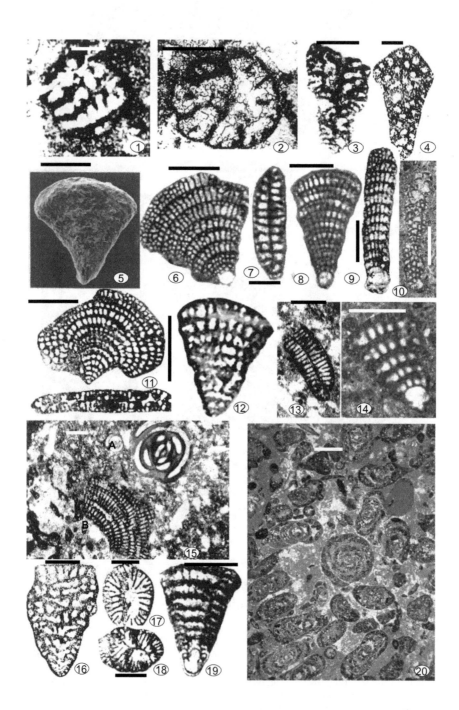

Plate 5.10   Scale bars: Figs 1, 3, 5, 6, 8, 9-11, 20 = 0.5mm; Figs 2, 4, 7, 12-19 = 0.25 coll. Figs 1-2. *Opertum* sp., Aptian, Zakum-1, 7116ft, Thamama II, Abu Dhabi, NHM coll. Fig. 3. *Voloshinoides* sp., Albian, England, NHM coll. Fig. 4. *Voloshinoides* sp., Albian, Iran, NHM coll. Figs 5-8. *Cuneolina parva* Henson, paratypes, Santonian, Sudr Heitan, Egypt, 5) solid specimen, NHM P39116; 6) vertical section, NHM P39119; 7) horizontal section, NHM P39121; 8) vertical section, NHM P39120. Fig. 9. *Cuneolina cylindrica* Henson, holotype, Maastrichtian, North Iraq, NHM P39122. Fig. 10. *Cuneolina* sp., Late Cretaceous, Spain, NHM coll. Fig. 11. *Cuneolina hensoni* Dalbiez, syntypes, vertical and horizontal sections, figured by Dabiez (1958), from the Valanginian, France. Fig. 12. *Vercorsella camposaurii* Sartoni and Crescenti, holotype figured by Sartoni and Crescenti (1962), Valanginian to Aptian, Central Apennines, Italy. Figs 13, 15. Thin section photomicrographs from Cenomanian, Audignon, near Saint- Sever, France, UCL coll; 13) horizontal section of *Cuneolina pavonia* d'Orbigny; 15) A) *Pseudonummuloculina* sp., B) fragment of *Cuneolina pavonia* d'Orbigny. Fig. 14. *Vercorsella arenata* Arnaud-Vanneau, Aptian, Tibet, UCL coll. Fig. 16. *Pseudotextulariella courtionensis* Brönnimann, holotype figured in Brönnimann (1966) from the Barremian of Switzerland. Fig. 17. *Pseudotextulariella cretosa* (Cushman), near-topotype, figured by Brönnimann (1966), Cenomanian, England. Figs 18-19. *Sabaudia capitata* Arnaud-Vanneau, figured by Chiocchini *et al.* (1984) from Aptian, Italy. Fig. 20. A packstone of *Pseudocyclammina lituus* (Yokoyama), Early Cretaceous, Makhul no.1 well Iraq, NHM P35970.

Plate 5.11   Scale bars: Figs 1-5 = 1mm; Fig. 6 = 0.25mm. Figs 1-3. *Dicyclina qatarensis* Henson, syntypes, Cenomanian, Dukhan well, Qatar, NHM P39113-5. Figs 4-6. *Orbitolina concava* (Lamarck), 4-5) Cenomanian, Spain, UCL coll; 6) late Albian, Upper Greensand, Devon, Elliot coll., NHM P43429.

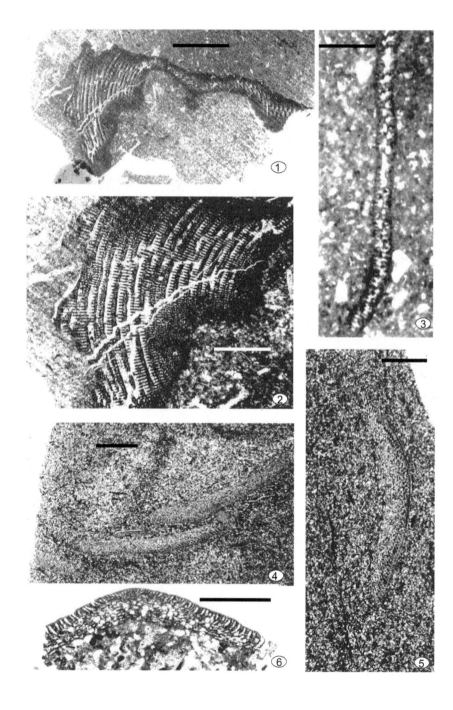

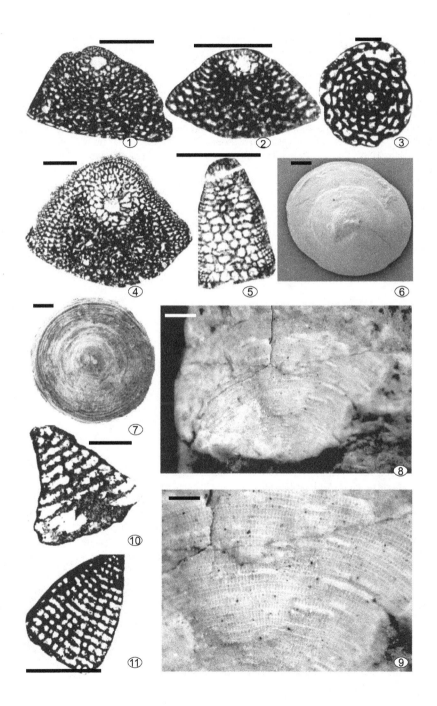

Plate 5.12   Scale bars: Figs 1-11 = 0.5mm. Fig. 1. *Orbitolina sefini* Henson, Cenomanian, NE of Erbil, Iraq, NHM P35903. Fig. 2. *Mesorbitolina texana* (Roemer), Albian, south of Husainabad, Iran, NHM M/2050. Fig. 3. *Rabanitina basraensis* Smout, Cenomanian, Zubair 1, 8600-50ft, Iraq, NHM P42961. Fig. 4. *Orbitolina* cf. *duranddelgai* Schroeder, Cenomanian, Iran, NHM M2058. Fig. 5. *Conicorbitolina cobarica* (Schroeder), Cenomanian, Iran, NHM M2074. Fig. 6. *Orbitolina* sp., SEM photograph of a solid specimen, UCL coll. Fig. 7. *Orbitolina qatarica* Henson, solid specimen, early Cenomanian, Dukhan no.1 well, Qatar, NHM P35916. Figs 8-9. *Dicyclina qatarensis* Henson, syntypes, early Cenomanian, Dukhan well, Qatar, NHM P39112, M2512. Fig. 10. *Coskinolinoides texanus* Keijzer, type species from Loeblich and Tappan (1988), Albian, USA. Fig. 11. *Paracoskinolina sunnilandensis* (Maync), paratype, from Maync (1955), Aptian-Albian, USA.

Plate 5.13   Scale bars: Figs 1-14 = 0.25mm; Figs 15-16 = 0.5mm. Fig. 1 *Simplorbitolina manasi* Ciry and Rat, type figure from Loeblich and Tappan (1988), Early Cretaceous, Spain. Figs 2-3. *Iraqia simplex* (Henson), syntype, Aptian, Iraq, NHM P35871. Fig. 4. *Orbitolinopsis kiliani* (Prever), type figure, Aptian, France. Figs 5-6. *Neorbitolinopsis conulus* (Douvillé), type species figured from Albian to early Cenomanian, Spain, 5) vertical section; 6) transverse thin section showing the second-order peripheral rectangular chamberlets. Figs 7-10. *Montseciella arabica* (Henson), Barremian, Dukhan no.2 well, Qatar, 7-8) NHM P35810; 9-10) NHM P35805. Figs 11-12. *Dictyoconus walnutensis* (Carsey), early to middle Albian, Walnut Clay, U.S.A., NHM coll. Fig. 13. *Paleodictyoconus cuvillieri* Foury, latest Aptian-earliest Albian, top Shuaiba, Saudi Arabia, NHM coll. Fig. 14. *Orbitolinella depressa* Henson, Cenomanian-Turonian, syntype, Dukhan no.2 well, Qatar, NHM coll. Figs 15-16. *Pseudocyclammina lituus* (Yokoyama), Early Cretaceous, Makhul no.1 well, Iraq, NHM P35970.

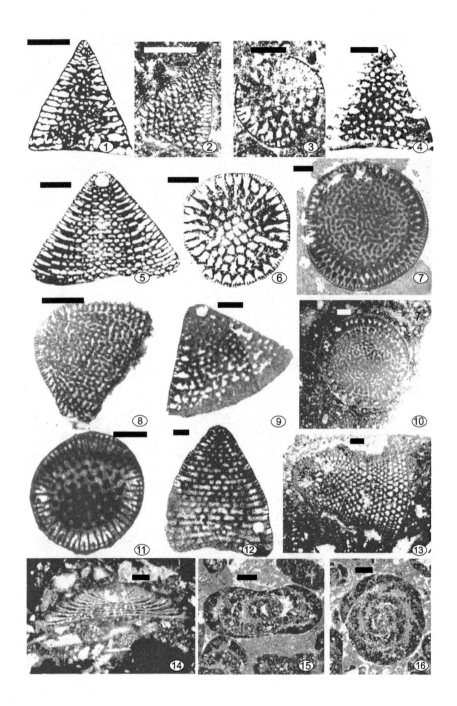

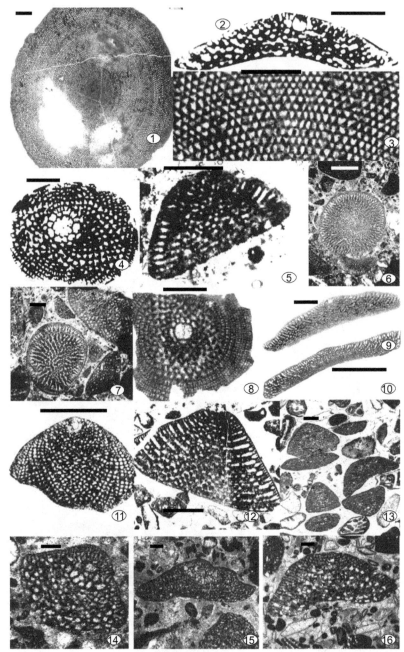

Plate 5.14 Scale bars: Fig. 1 = 1mm; Figs 2-16 = 0.5mm. Figs 1-4. *Palorbitolina lenticularis* (Blumenbach, 1805), identified as *Palorbitolina discoidea* Gras, Barremian to Aptian, Thamama Fm., Oman, NHM coll., 1) transverse section; 2) enlargement of the Fig. 1; 3) axial section; 4) oblique transverse section. Fig. 5. *Orbitolina trochus* (Fritsch), late Albian, Pilatus Kulu, Lucerne, Switzerland, NHM P38097. Figs 6-9.

Mesorbitolina delicata Henson, Hadhramaut, 6-7) ideotype, early Aptian, NHM P35918; 8) Qatar, NHM P35919; 9) Qatar, NHM coll. Figs 10-11. *Mesorbitolina libanica* Henson (= *Mesorbitolina texana* (Roemer)), 10) syntype, Aptian, Mdeireidj, Lebanon, NHM P35930; 11) Aptian, Medjel Chems, Syria, NHM M/2160. Fig. 12. *Mesorbitolina kurdica* Henson, syntype, late Aptian, Aoraman, Iraq, NHM P35935. Figs 13, 16. *Palorbitolina* cf. *Lenticularis* (Blum), Aptian, east of Beirut, Lebanon, NHM P25946; 16) Aptian, North Lattakia, Syria, NHM P35944. Figs 14-15. *Orbitolina qatarica* Henson, early Cenomanian, Dukhan no.1 well, Qatar, NHM Henson coll.

Plate 5.15   Scale bars: Figs 1 - 5, 21 = 0.25mm. Figs 6-20 = 0.5mm. Fig. 1. *Neotrocholina lenticularis* (Henson), paratype, Cenomanian, Dukhan no. 1 well, Qatar, NHM P38489. Figs 2-4. *Neotrocholina minima* (Henson), paratype, late Cenomanian, Dukhan no. 1 well, Qatar, 2) NHM M4674; 3-4) solid specimen, paratype, NHM P38550. Fig. 5. *Trocholina altispira* Henson, paratype, Cenomanian, Dukan no. 3 well, Qatar, NHM P38486. Figs 6-8, 16. *Chrysalidina gradata* d'Orbigny, 6) Cenomanian, Deh Luran, Iran, NHM P39127; 7) Cenomanian, Jebel Madamar, Natih Fm., Oman, NHM P52608; 8) Cenomanian, Auvignon, France, UCL coll; 16) Cenomanian, Île de Madame, Charente Inférieur, France, NHM P5203. Fig. 9. *Paravalvulina arabica* (Henson), Valanginian, Dukhan no. 2 well, Qatar, NHM P52645. Fig. 10. Morphologically intermediate between *Praechrysalidina infracretacea* Luperto Sinni and *Dukhania conica* Henson, figured by Banner *et al.* (1991), Albian, Northern Iraq, NHM P52591. Fig. 11. *Pseudomarssonella* cf. *plicata* Redmond, Bajocian or Bathonian, Abu Dhabi, Um Shaif, NHM P52629. Figs 12-14. *Praechrysalidina infracretacea* Luperto Sinni, 12) Aptian, Um Shaif, Upper Shuaiba Fm., United Arab Emirates, NHM P52584; 13-14) Hauterivian, Dukhan no. 2 well, Qatar, NHM P52580. Fig. 15. *Paravalvulina arabica* (Henson), Valanginian, Zakum Fm., Well-Jumayla-1, United Arab Emirates, NHM P52641. Figs 17-18. Morphologically intermediate between *Praechrysalidina infracretacea* Luperto Sinni and *Chrysalidina gradata* d'Orbigny, 17) Cenomanian, WM 99, Natih Fm. Oman, NHM coll; 18) middle Cretaceous, Iran, NHM P52595. Fig. 19-20. *Dukhania conica* Henson, paratypes, Cenomanian, Dukhan no. 3 well, Qatar, 19) NHM P52601; 20) NHM P52597. Fig. 21. *Accordiella conica* Farinacci, figured by Sartorio and Venturini (1988), Coniacian to Santonian, Italy.

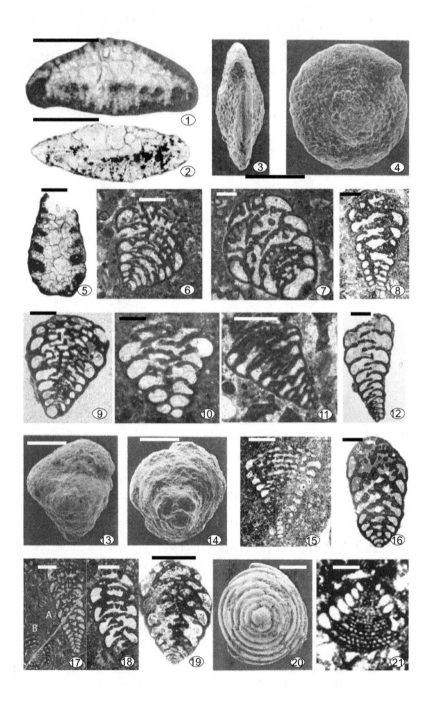

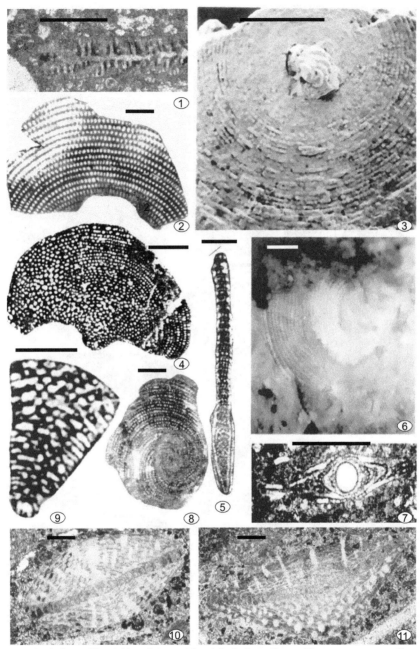

Plate 5.16 Scale bars: Figs 1-11 = 0.5mm. Figs 1-3. *Dohaia planata* Henson, paratypes, Cretaceous, Dukhan no.1, well, Qatar, NHM P35843. Fig. 4. *Eclusia moutyi* Septfontaine, figured by Azema et al. (1977), Valanginian, France. Figs 5-8. *Edomia reicheli* Henson, Cenomanian, Wadi Meliha, Israel, 5) paratype, NHM P35850; 6) holotype; 7) paratype, megalospheric form (A-form), NHM P35851; 8) paratype, NHM P35853. Fig. 9. *Naupliella insolita* Decrouez and Moullade, figured by Loeblich and Tappan (1988), late Albian, Greece. Figs 10-11. *Orbitoides* sp., Sabkha, Arab Fm., Abu Dhabi, UAE.

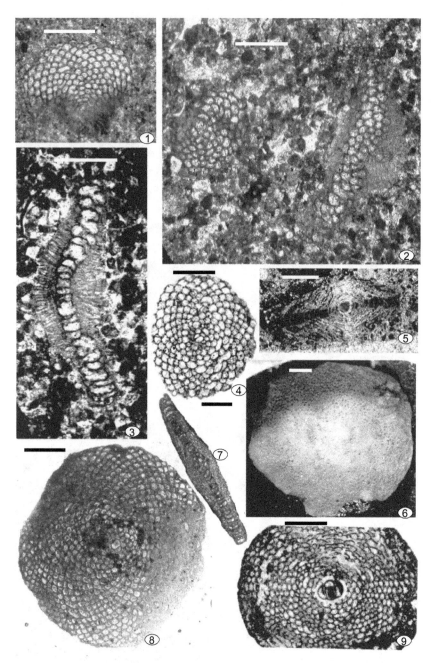

Plate 5.17   Scale bars: Figs 1-4, 6-7, 9 = 0.5mm; Figs 5, 8 = 1mm. Figs 1-4. *Monolepidorbis douvillei* (Silvestri), Late Cretaceous, 1-3) Jawan no.2, Iraq, NHM P40685-6; 4) Qatar, NHM M8239. Figs 5-6. *Orbitoides* sp., 5) Sabkha, Arab Fm. Abu Dhabi, UAE; 8) solid specimen, Gensa, France. Figs 7-8. *Orbitoides* cf. *tissoti* Schlumberger, shales below quartz sandstone, Punjab, NHM coll. Fig. 9. *Orbitoides faujasii* (Defrance), Maastrichtian, Iraq, NHM M568.

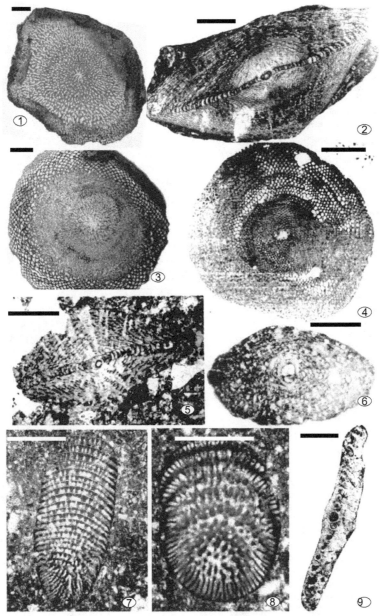

Plate. 5.18 Scale bars: Figs 1-7, 9 = 1mm; Fig. 8 = 0.5mm. Fig. 1. *Orbitoides medius* (d'Archiac), Turkey, UCL coll. Figs 2-3. *Orbitoides browni* (Ellis), Late Cretaceous, Havana Province, Cuba, NHM P33354. Fig. 4. *Orbitoides faujasii* (Defrance), Maastrichtian, NHM coll. Fig. 5. *Orbitoides faujasii* (Defrance), Maastrichtian, Iraq, NHM M7305. Fig. 6. *Orbitoides vredenburgi* Douvillé, Maastrichtian, Iraq, NHM, M7302. Figs 7-8. *Praetaberina bingistani* (Henson), syntypes, registered as *Taberina bingistani* Henson, middle Cretaceous, Kuh-i- Bingistan, Iran, NHM P35786. Fig. 9. *Sirelella safranboluensis* Özgen-Erdem, holotype figured by Özgen-Erdem (2002) from Turkey.

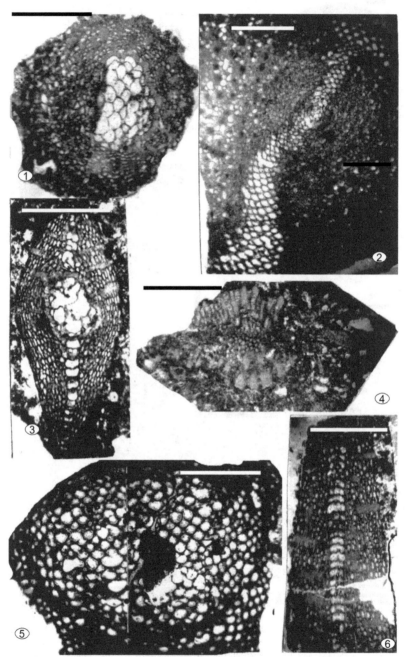

Plate 5.19 Scale bars: Figs 1-6 = 1mm. Figs 1-6. *Simplorbites gensacicus* (Leymerie), Late Cretaceous, Dukhan no. 51 well, Qatar, NHM coll.

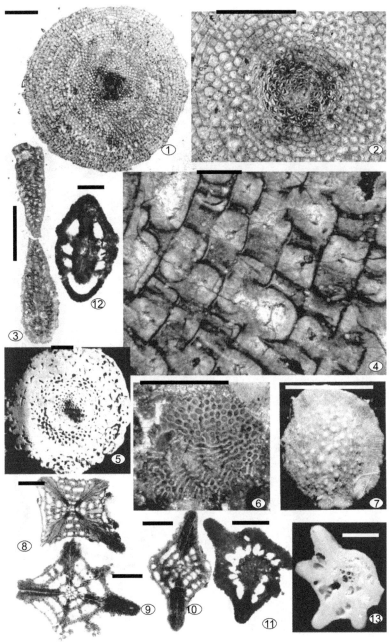

Plate 5.20 Scale bars: Figs 1, 3, 5-7 = 1mm; Fig. 2, 8-13 = 0.5mm; Fig. 4 = 0.15mm. Fig. 5. *Omphalocyclus* sp., solid specimen, Maastrichtian, France, UCL coll. Fig. 6. *Laffitteina vanbelleni* Grimsdale, Musharah no. 1 well, Iraq, NHM P40693. Fig. 7. *Vaughanina* sp., Campanian, Cuba, UCL coll. Figs 8-10. *Siderolites calcitrapoides* Lamarck, Maastrichtian, Belgium, 8) NHM P54874; 9) P41580; 10) P41575. Fig. 11-12. *Siderolites* sp., Maastrichtian, Holland, UCL coll. Fig. 13. *Siderolites spengleri* (Grimsdale), Maastrichtian, St Pietersberg Maastricht, Belgium, NHM P2780.

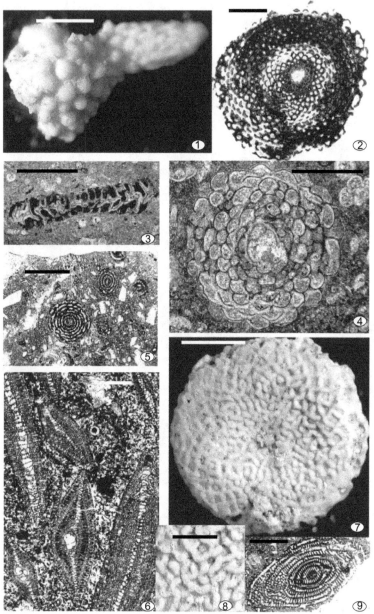

Plate 5.21 Scale bars: Figs 1- 6, 9 = 1mm; Fig. 7 = 0.5mm; Fig. 8 = 0.25 mm. Fig. 1. *Siderolites calcitrapoides* Lamarck, Maastrichtian, Maastricht Valley, Holland, UCL coll. Figs 2-4, 7-8. *Omphalocyclus* sp., Maastrichtian, Libya, UCL coll. Fig. 5. *Simplalveolina* sp., Cenomanian, France, UCL coll. Fig. 6. Thin section of *Orbitoides* sp. and *Siderolites* sp., Maastrichtian, Spain, UCL coll. Fig. 9. *Praealveolina* spp., Cenomanian, France, UCL coll.

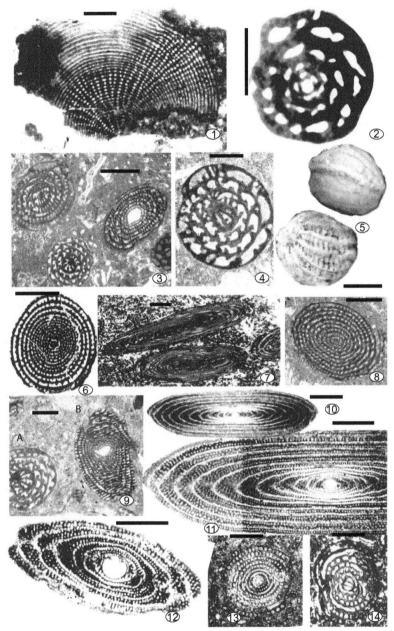

Plate 5.22 Scale bars: Figs 1-2, 6, 11 = 0.5mm; Figs 3-5, 7-10, 12-14 = 1mm. Fig. 1. *Zekritia langhami* Henson, holotype, Turonian, Dukhan no. 3 well, Qatar, NHM P36030. Fig. 2. *Archaealveolina reicheli* (de Castro), figured by Loeblich and Tappan (1988), Aptian, Italy. Figs 3-4. *Cisalveolina* sp., Cenomanian, India, UCL coll. Figs 5, 13. *Ovalveolina* sp., Cenomanian, France, 5) solid specimens; 13) thin section. Fig. 6. *Lacazina* sp., Coniacian, India, UCL coll. Fig. 7. *Subalveolina* sp., Santonian, France, UCL coll. Fig. 8. *Simplalveolina* sp., Cenomanian, France, UCL coll. Fig. 9. A) *Ovalveolina* sp., B) *Praealveolina* sp., oblique sections, early Cenomanian, Mexico, UCL coll. Figs 10-11. *Praealveolina tenuis* Reichel, Cenomanian, Qatar, NHM M4804; 11) enlargement of Fig. 12. *Pseudedomia* (= *Sellialveolina* sp.) registered at the NHM as *Praealveolina cretacea* (d'Archiac), Cenomanian, Qatar, NHM M7092. Fig. 14. *Cisalveolina lehneri* Reichel, middle Cretaceous, Shilaif (Khatiyah) Fm, Qatar, UCL coll.

Chapter 6

The Cenozoic Larger Benthic Foraminifera: The Paleogene

6.1 Introduction

As noted in the previous chapter, the Cretaceous-Paleogene crisis wiped out over 80% of the Maastrichtian larger benthic foraminifera. The Early Paleocene was a recovery period for larger foraminifera. As was the case during the recovery stage after previous extinctions, larger foraminifera were morphologically small and rare, and even the newly evolved foraminifera exhibited a morphological manifestation of post-crisis ecological stress, (i.e. the "Lilliput effect", which characterises a temporary, within-lineage size decrease after extinction events (Twitchett, 2006)). Their evolution and development occurred at different rates and followed different lines in different parts of the globe. It is now possible to recognise larger benthic foraminiferal bioprovinces in Tethys, the Americas, the Far East, in and Southern Africa (see for example BouDagher-Fadel and Price, 2010a,b,c, 2013, 2014, 2017).

In the Tethys, for example, morphologically larger miliolides and rotaliides (especially nummulitids and orthophragminids) did not appear before the Late Paleocene. Here, the miliolides included large fusiform alveolinids, which show examples of convergence with the extinct Cretaceous alveolinids, and the discoid soritids that become prominent throughout the Eocene. Some rotaliides developed a complex system of marginal cords, characteristic of *Nummulites*. These latter exhibited high rates evolutionary diversification and became very abundant in the Eocene, thriving (together with the three layered orthophragminid, *Discocyclina* and its descendants) in the forereef and shallow marine open platforms of Tethys. Parallel to these lines of evolution, the agglutinated textulariides developed into forms that imitated their Cretaceous ancestors by developing, in the case of the textulariides, internal pillars, and with the orbitolinids, complicated partitions.

In the American province, however, the recovery period was longer than in Tethys, and it was not before the Middle Eocene that the rotaliides developed a new evolutionary lineage; they evolved into the three layered *Lepidocyclina* and *Eulepidina*. These American forms migrated into Tethys in the Oligocene, however a reverse migration of alveolinids and discocylinids from Tethys to the American province never occurred. The nummulitids of the American province never reached the giant sizes of their analogues in the Tethyan realm (see BouDagher-Fadel and Price, 2010a, 2013, 2014, 2017).

Paleogene larger foraminifera from different localities around the world have been studied in detail by many workers. Among the most recent studies are those of Hottinger and Drobne (1980), Adams (1987), Racey (1995), Banerjee et al (2000), Hottinger (2001, 2013), Özgen-Erdem (2002), Meinhold, and BouDagher-Fadel (2010), Benedetti and Pignatti (2012), Politt et al. (2012), Hu et al. (2013), BouDagher-Fadel and Price (2013, 2014, 2017), Scotchman et al. (2014). In this chapter, the taxonomy of the major larger Paleogene benthic foraminifera is summarized, and then their bio-stratigraphic, paleoenvironmental and paleogeographic significance are discussed.

6.2 Morphology and Taxonomy of Paleogene Larger Benthic Foraminifera

The three orders of larger benthic foraminifera that dominated the Paleogene were the:

- Textulariida
- Miliolida
- Rotaliida

The relationship and evolution of the superfamilies of these orders is shown in Fig. 6.1.

ORDER TEXTULARIIDA Delage and Hérouard, 1896

The tests of these agglutinated foraminifera are made of foreign particles bound by organic cement. They range from Early Cambrian to Holocene.

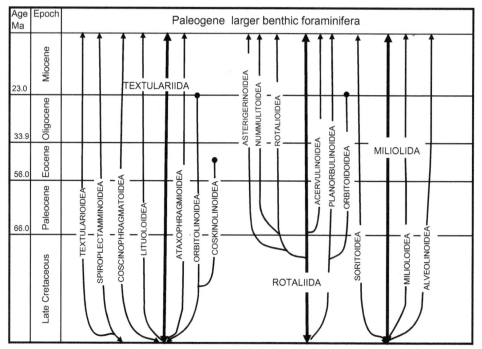

Figure 6.1. The evolution of the Paleogene orders (thick lines) and superfamilies (thin lines) of larger foraminifera.

Superfamily ATAXOPHRAGMIOIDEA Schwager, 1877
Members of this superfamily have a multilocular, trochospiral test that becomes biserial or uniserial in later stages. Middle Triassic to Holocene.

Family Globotextulariidae Cushman, 1927
Tests have a highly trochospiral stage followed by a quadriserial, triserial or biserial stage. The interior of the chambers may be subdivided by internal partitions. The wall is non-canaliculated. Late Cretaceous (Campanian) to Holocene.

- *Cubanina* Palmer, 1936 (Type species: *Cubanina alavensis* Palmer, 1936). The test is triserial to uniserial. Late Oligocene to Holocene (Plate 6.1, fig. 1).
- *Liebusella* Cushman, 1933 (*Type species*: *Lituola nautiloidea* Lamarck var. *soldanii* Jones and Parker, 1860). The test is trochospiral to uniserial. Chambers are slightly overlapping and subdivided by vertical partitions projecting inward from the outer wall and extending from chamber floor to roof. Late Eocene to Holocene (Plate 6.1, fig. 3).
- *Matanzia* Palmer, 1936 (Type species: *Matanzia bermudezi* Palmer, 1936). The test is trochospiral to biserial. Chambers are subdivided by narrow vertical partitions that radiate inward from the outer wall. Late Cretaceous to Miocene (Plate 6.1, fig. 4).

Superfamily PAVONITINOIDEA Loeblich and Tappan, 1961
Tests are coiled in the early stages, and triserial to biserial or uniserial. The interiors of the chambers are partially divided by numerous vertical partitions (beams) or septula, that project downwards from the roof, and rarely may have a few connecting horizontal partitions (rafters). Late Cretaceous and Oligocene to Pliocene.

Family Pavonitinidae Loeblich and Tappan, 1961
Tests are palmate, triangular in section, with an early stage this has triserial to biserial or uniserial coiling. Interiors of chambers are partially divided by numerous vertical partitions (beams) or septula that project downwards from the roof and rarely may have a few connecting horizontal partitions (rafters). Apertures are terminal, and single or multiple. Oligocene to Pliocene.

Subfamily Spiropsammiinae Seiglie and Baker, 1984
The test is strongly compressed, but later stages may be uncoiled. Interiors of chambers are subdivided by numerous elongate septula. Apertures are single, and terminal. Oligocene to Early Pliocene.

- *Spiropsammia* Seiglie and Baker, 1984 (Type species: *Spiropsammia uhligi* (Schubert) = *Cyclammina uhligi* Schubert, 1901). The early stage is evolute, planispirally enrolled, but later uncoiled and rectilinear. The aperture is single, and terminal. Oligocene to Early Pliocene (Plate 6.1, figs 5-7).

Subfamily Pavonitininae Loeblich and Tappan, 1961
The test is triserial, biserial or uniserial. Chambers are undivided by secondary septula. Oligocene to Miocene.

- *Pavonitina* Schubert, 1914 (Type species: *Pavonitina styriaca* Scubert, 1914). The test is broad and flattened. A large proloculus is followed by a biserial stage, composed of low and curved chambers, followed by broad and centrally curved uniserial rectilinear chambers. Oligocene to Miocene (Plate 6.1, fig, 8).
- *Pavopsammia* Seiglie and Baker, 1984 (Type species: *Pavopsammia flabellum* Seiglie and Baker, 1984). The test is palmate, flattened, with a proloculus followed by a triserial to biserial and finally uniserial stage. Oligocene (Plate 6.1, fig. 9).
- *Zotheculifida* Loeblich and Tappan, 1957 (Type species: *Textularia lirata* Cushman and Jarvis, 1929). The test is compressed, and biserial throughout. Late Oligocene to Early Miocene (Plate 6.1, fig. 10).

Superfamily COSCINOPHRAGMATOIDEA Thalmann, 1951
The test is free or attached, may be coiled in the early stages, but is later uncoiled or branched. The walls are finely agglutinated, traversed by pores, or with a coarsely perforate or canaliculate inner layer and an outer imperforate layer. Triassic to Holocene.

Family Haddoniidae Saidova, 1981
The test is attached. The aperture is terminal, simple to complex. Paleocene to Holocene.

- *Haddonia* Cushman, 1898 (Type species: *Haddonia torresiensis* Chapman, 1898). The test is large. Chambers are broad and irregular in shape. The walls are traversed by numerous large pores. The aperture is a slit. Early Paleocene to Holocene (Plate 6.1, fig. 12).

Superfamily TEXTULARIOIDEA Ehrenberg, 1838
The test is trochospiral, biserial or triserial in early stages and later may be uniserial or biserial. The walls are agglutinated, and canaliculated. Early Jurassic (Sinemurian) to Holocene.

Family Chrysalidinidae Neagu, 1968
The test is high trochospiral, with quinqueserial or quadriserial or triserial or biserial coiling modes, or with certain consecutive pairs of these. The aperture is central along the axis of coiling.

Subfamily Chrysalidininae Neagu, 1968
This subfamily, as revised by Banner et al. (1991), is essentially triserial throughout its ontogeny (at least in the megalospheric generation), becoming biserial or quadriserial in the adult. The walls are solid but sometimes becoming canaliculate. Early Jurassic (Sinemurian) to Late Eocene.

- *Pfendericonus* Hottinger and Drobne, 1980 (Type species: *Lituonella makarskae* van Soest, 1942). The early trochospiral stage is multichambered and occupies a third of the test, becoming uniserial in the final stage. Chambers are internally undivided by partitions and connected by thin vertical pillars. The apertural face is convex. Late Paleocene to Eocene (Fig. 6.2).

- *Pseudochrysalidina* Cole, 1941 (*Pseudochrysalidina floridana* Cole, 1941). The test is initially triserial, but biserial in adult with internal pillars and oblique septa. Eocene (Fig. 6.2).
- *Vacuovalvulina* Hofker, 1966 (Type species: *Marssonella keijzeri* van Bellen, 1946). It differs from *Pseudochrysalidina* in having plano-concave septa. Paleocene (Fig. 6.2).

Superfamily LITUOLOIDEA de Blainville, 1825

Members of this superfamily have a multilocular, rectilinear and uniserial test. The early stage has plani- (strepto-) or trochospiral coiling. The peripheries of the chambers have radial partitions, but centrally there are either no or scattered, separated pillars. The aperture is simple or multiple cribrate. Early Jurassic (Sinemurian) to Holocene.

Family Cyclamminidae Marie, 1941

The test is involute with alveolar walls. The aperture is near the septal face. Jurassic to Holocene.

- *Cyclammina* Brady, 1879 (Type species: *Cyclammina cancellata* Brady, 1879). The test is planispiral, and flattened. The walls are thick, with an alveolar subepidermal meshwork of a thickness exceeding that of the chamber lumen. Paleocene to Holocene (Plate 6.1, fig. 11).

Family Lituolidae de Blainville, 1827

The early stage is enrolled, but later it may be rectilinear. There are few chambers (less than 10) per whorl. Carboniferous to Holocene.

Family Spirocyclinidae Munier-Chalmas, 1887

The test is planispiral, becoming peneropliform to annular in later stages. Chambers are partially subdivided by septula. Jurassic to Eocene

- *Saudia* Henson, 1948 (Type species: *Saudia discoidea* Henson, 1948). Microspheric forms have a uniserial flabelliform last stage. The megalospheric form has a large proloculus and evolute, annular last stage. Chambers possess a superficial subepidermal cellular layer ("pigeon-hole" structure). Paleocene to Middle Eocene (Plate 6.1, figs 13-16).
- *Thomasella* Sirel, 1998 (Type species: *Saudia labyrinthica* Grimsdale, 1952). The megalospheric embryo is large and spherical, followed by a few undivided chambers, with successive chambers becoming cyclical. A very narrow exoskeletal subepidermal layer is present on both sides of the test, consisting of two or more generations of vertical partitions (beams), perpendicular to the septa, and two or more sets of horizontal partitions (rafters) parallel to the septa, producing a complex

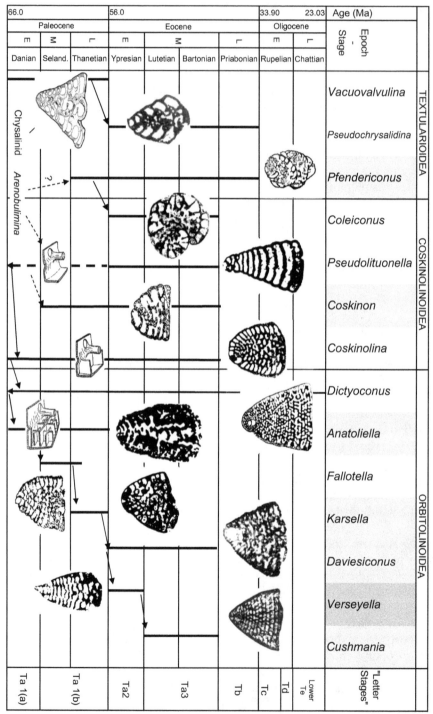

Figure 6.2. The evolution of the Paleogene textulariides. Global distribution is shown by the shading of the background to genera names: White = Cosmopolitan, Light Grey = Tethys, Dark Grey = Americas.

subepidermal network with numerous alveolar compartments near the surface. The central (labyrinthic) zone consists of numerous pillars aligned from one chamber to the next. Early Eocene (Plate 6.1, fig. 17).

Superfamily COSKINOLINOIDEA Moullade, 1965

Members of this superfamily have a conical test with a trochospiral early stage, becoming uniserial and rectilinear. Late Cretaceous to Middle Eocene (Cenomanian to Lutetian).

Family Coskinolinidae Moullade, 1965

Simple, or without, exoskeleton in the marginal chamber cavity. The rectilinear part has broad, low chambers, subdivided by pillars or irregular partitions. The aperture is a series of pores scattered over the apertural face. Late Cretaceous to Midlle Eocene (Cenomanian to Lutetian).

- *Barattolites* Vecchio and Hottinger, 2007 (Type species: *Barattolites trentinarensis* Vecchio and Hottinger, 2007). The test is highly conical with a trochospiral early part followed by a uniserial part with a slightly convex base. Chambers are subdivided by vertical partitions more or less in line from one chamber to the next. The megalospheric apparatus has simple walls and consists of two chambers divided by a very thin, straight septum, separating the protoconch from the deuteroconch. The following five to six chambers constitute a trochospiral nepiont with an exoskeleton beginning at the third chamber. Early to Middle Eocene (Plate 6.1, fig. 18).
- *Coleiconus* Hottinger and Drobne, 1980 (Type species: *Coskinolina elongata* Cole, 1942). The early stage is trochospirally enrolled, as in *Arenobulimina,* but without any partitions; later parts have few scattered pillars. There is a simple exoskeleton (beams only) in the marginal chamber cavity. Early to Middle Eocene (Figs 6.2, 6.3, 6.4; Plate 6.1, Figs 19-21).
- *Coskinolina* Stache, 1875 (Type species: *Coskinolina liburnica* Stache, 1875, = *Lituonella roberti* Schlumberger, 1905). The early stage is arenobuliminid, later uniserial with numerous chambers. The inside of the test has vertical partitions and pillars. There is no exoskeleton in the marginal chamber cavity. Paleocene to Middle Eocene (Figs 6.2, 6.3; Plate 6.1, figs 22-23; Plate 6.2, figs 1-2, 19; Plate 6.3, fig. 5A).
- *Coskinon* Hottinger and Drobne, 1980 (*Coskinolina (Coskinon) rajkae* Hottinger and Drobne, 1980). There is a very reduced arenobuliminid early growth stage, with later parts being uniserial with scattered pillars. There is no exoskeleton in the marginal chamber cavity. Middle Paleocene to Middle Eocene (Plate 6.2, fig. 3; Figs 6.2, 6.3).

Superfamily ORBITOLINOIDEA Martin, 1890

The test is conical with numerous chambers, partially subdivided by radial or transverse partitions or pillars. Middle Jurassic to Oligocene.

Family Orbitolinidae Martin, 1890

There is an initial, low trochospire, usually very much reduced, followed by a later recti-linear, broad and conical growth, made of low uniserial chambers subdivided with pillars or vertical partitions. The aperture is cribrate. Middle Jurassic to Oligocene (Fig. 6.3).

- *Anatoliella* Sirel, 1988 (Type species: *Anatoliella ozalpiensis* Sirel, 1988). The test is a low to high conical form in late ontogeny, with three chambers in each whorl. The embryonic apparatus is followed by a series of trochospiral chambers, subdivided by a network of vertical beams and horizontal partitions. The central zone is divided by numerous pillars. *Anatoliella* differs from all orbitolinid genera in having just a single series of shallow cuplike chambers in its triserial chamber arrangement. Paleocene (Plate 6.2, fig. 4; Fig. 6.2).
- *Cushmania* Silvestri, 1925 (Type species: *Conulites americana* Cushman, 1919). The early trochospiral is very reduced, with an apical protoconch. Chambers are divided by short vertical partitions, which are intersected by at least two horizontal partitions. Middle Eocene (Plate 6.2, fig. 5).

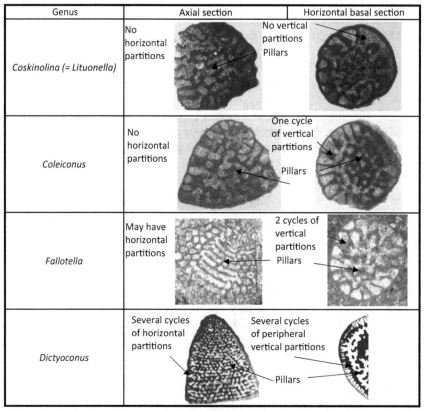

Figure 6.3.   The distribution of the vertical and horizontal partitions in the coskinolinids and orbitolinids.

- *Daviesiconus* Hottinger and Drobne, 1980 (Type species: *Coskinolina balsilliei* Davies, 1930). The initial part of the test is almost planispiral. Walls have vertical partitions in the marginal zone, which Hottinger (2007) interpreted as septula comprising a part of the endoskeleton, but lacking horizontal partitions. Early to Middle Eocene (Plate 6.2, fig. 6.6).
- *Dictyoconus* Blanckenhorn, 1900 (Type species: *Dictyoconus egyptiensis* (Chapman) = *Patellina egyptiensis* (Chapman)). Radial partitions thicken away from the periphery, and become broken up into pillars in the central zone. Peripheral tiered rectangular chamberlets are present. Aptian to Oligocene (Figs 6.3, 6.4; Plate 6.2, figs 7-9, 18; Plate 6.28, figs 5-6).

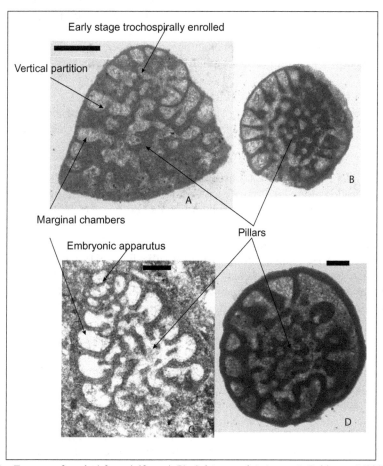

Figure 6.4. Features of conical foraminifera. A-B) *Coleiconus christianaensis* Robinson, Middle Eocene, Upper Chapelton Formation, Jamaica, type figures, NHM P52808-9; C-D) *Coskinolina* cf. *douvillei* (Davies), Middle Eocene, Upper Chapelton Formation, Jamaica, UCL coll.

- *Fallotella* Mangin, 1954 (Type species: *Fallotella alayensis* Mangin, 1954). The early trochospiral coil is very reduced. The uniserial part is pillared, and has a thick marginal wall. The exoskeleton is simple (only beams) or moderately complex (beams and rafters). Middle to early Late Paleocene (Figs 6.2, 6.3; Plate 6.2, fig. 17).
- *Karsella* Sirel, 1997 (Type species: *Karsella hottingeri* Sirel, 1997). This form differs from the Late Cretaceous *Calveziconus* Caus and Cornella, 1982 (see Chapter 5) in having a more complex internal structure, with more than two vertical (beams) and horizontal (rafters) partitions forming an irregular network under the epidermis. It also possesses a large protoconch and periembryonic chambers. It differs from *Orbitolina* d'Orbigny, 1850 (see Chapter 5) in having a more complex exoskeletal structure and from *Cushmania* Silvestri 1925 in having a shorter trochospiral early stage. Paleocene (Thanetian) (Plate 6.7, fig. 3; Fig. 6.2).
- *Verseyella* Robinson, 1977 (Type species: *Coskinolinoides jamaicensis* Cole, 1956). The early trochospiral stage is absent. The test is biserial in the early part, becoming uniserial in later stages. The interiors of chambers are subdivided by vertical partitions (beams) that are aligned from chamber to chamber, forming a ring around the central shield which is supported by pillars. Early Eocene (Plate 6.2, figs 10-11).

ORDER MILIOLIDA Delage and Hérouard, 1896
The miliolides have tests that are porcelaneous, imperforate, and made of high Mg-calcite with fine randomly oriented crystals. They range from the Carboniferous to the Holocene.

Superfamily ALVEOLINOIDEA Ehrenberg, 1839
The test is enrolled along an elongate axis, and initially planispiral or streptospiral, or milioline, with chambers added in varying planes. Cretaceous to Holocene.

Family Alveolinidae Ehrenberg, 1839
The test is free, large, planispiral to fusiform, subcylindrical or globular, and coiled about elongate axis. Early whorls may be irregular, streptospiral in monomorphic species, and may be restricted to microspheric forms in dimorphic species. Chambers are subdivided into chamberlets by longitudinal partitions (septula) perpendicular to the main septa, and connected by passages below the apertural face. The basal layer may include anastomosing canals (Hottinger et al., 1989), and may have thick deposition of secondary calcite (flosculinisation) (Figs 6.5, 6.6). The aperture is a slit parallel to the base of the apertural face, or a single row of circular openings, or numerous rows of such openings in horizontal rows, arranged in definite patterns matching the patterns of chamber divisions. Early Cretaceous (Aptian) to Holocene.

- *Alveolina* d'Orbigny, 1826 = *Fasciolites* Parkinson 1811 (Type species: *Oryzaria boscii* Defrance, in Bronn, 1825). The test is large, fusiform or cylindrical with single tier chamberlets alternating in position in successive chambers. The basal layer is thick, as deposition of secondary calcite (flosculinisation) fills most of the chamber lumen. Successive chambers communicate with post- and pre-septal passages.

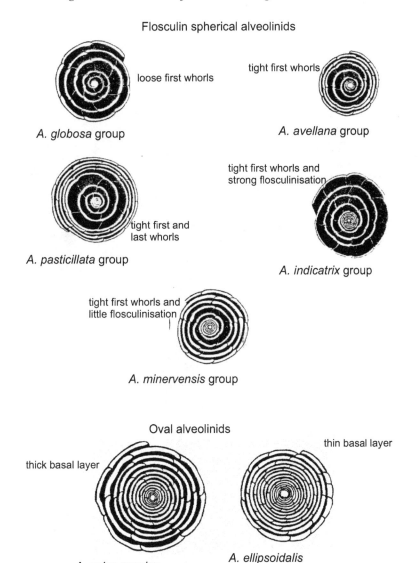

Figure 6.5.   Basal layers and flosculinisation in *Alveolina.*

The final chamber has two rows of apertures alternating in position. Late Paleocene to Eocene (Figs 6.5, 6.6; Plate 6.2, figs 15-19; Plate 6.3, figs 1-4, 6; Plate 6.4, figs 1-11; Plate 6.5, figs 1-2, 12, 14; Plate 6.6; figs 1-2; Plate 6.7, fig. 9; Plate 6.8, fig. 3A).

- *Borelis* de Montfort, 1808. (Type species: *Borelis melonoides* de Montfort, 1808). The test is small, spherical to fusiform. It differs from *Alveolina* in having only a pre-septal passage and a secondary small tier of chamberlets which alternate with the larger ones producing "Y" shaped septa in axial section. The aperture is a single row of pores. Late Eocene to Holocene (Plate 6.5, figs 5A, 6-8).

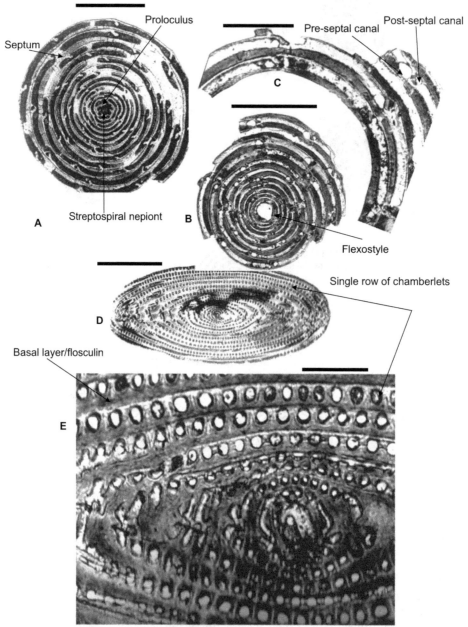

Figure 6.6. Alveolina oblonga d'Orbigny. Chaussy, Val-d'Oise, France; Figs A-B) equatorial sections; C) an enlargement of part of B; Figs D-E) axial sections. Scale bars: Figs A-B = 0.5mm; C = 0.25mm; D = 1mm; E = 160μm.

- *Bullalveolina* Reichel, 1936 (Type species: *Alveolina bulloides* d'Orbigny, 1839). The test is tiny, globular, with a streptospiral early stage which later is planispiral. Large pre-septal passages occupy about one half of the chamber floor. The final chamber has three or more rows of small apertures. Oligocene (Plate 6.5, fig. 9).

- *Globoreticulina* Rahaghi, 1978 (Type species: *Globoreticulina iranica* Rahaghi, 1978). The test is globular, planispiral, and involute with an alveolar exoskeleton and only the final whorl visible on the exterior. The outer parts of the chambers are subdivided by parallel and transverse partitions. The aperture is cribrate. Middle Eocene (Plate 6.5, fig. 10).

- *Glomalveolina* Hottinger, 1962 (Type species: *Alveolina dachelensis* Schwager,1883). The test is small, globular, consisting of a small globular proloculus followed by early streptospiral coiling, but is later planispiral. As in *Alveolina*, pre-septal and post-septal passages are present, however, unlike *Alveolina*, flosculinisation is minimal. The final chamber has a row of openings, intercalated with smaller ones. Middle Paleocene to Middle Eocene (Plate 6.2, fig. 18; Plate 6.9, fig. 1).

- *Malatyna* Sirel and Acar, 1993 (Type species: *Malatyna drobneae* Sirel and Acar, 1993). The test shows a triloculine early stage arrangement in megalospheric generations and a quinqueloculine early stage arrangement in microspheric generations. In both generations the planispiral adult stage has inflated chambers with subepidermal partitions. The aperture is cribrate. Eocene (Lutetian) (Plate 6.9, fig. 2).

- *Praebullalveolina* Sirel and Acar, 1982 (Type species: *Praebullalveolina afyonica* Sirel and Acar, 1982). The test is spherical with early streptospiral coiling. Later chambers have septula complete from floor to roof. The pre-septal passage is large. The apertural face has one row of primary apertures intercalated with smaller secondary apertures. Late Eocene to Oligocene (Plate 6.6, fig. 3).

Family Fabulariidae Ehrenberg, Munier-Chalmas, 1882, emend. Hottinger et al., 1989
The test is large, dimorphic, multichambered with miliolide coiling, tending to become reduced in subsequent growth stages, either to bilocular or to monolocular chamber cycles. They have a trematophore aperture (see Fig. 6.7), which is defined thus (Hottinger, 2006):

"A trematophore or a sieve constituting the face of many porcelaneous larger foraminifera, in miliolines produced by the coalescence of teeth, covering a large pre-septal space. May be supported by residual pillars".

Chambers show a thickened basal layer, subdivided by pillars or secondary partitions. Foraminifera in adult growth stages have a fixed apertural axis. For further terminology and detailed structural analysis see Drobne (1974; 1988).

- *Fabularia* Defrance, 1820 (Type species: *Fabularia discolites* Defrance,1825). The test is ovate, quinqueloculine in early stages and biloculine in later stages. Chambers have thick walls and are subdivided by subepidermal partitions, forming a series of chambers that are connected by pre- and post-septal passages. The aperture is cribrate (trematophore). Middle to Late Eocene (Plate 6.9, figs 4-7).

- *Lacazina* Munier-Chalmas, 1882 (Type species: *Alveolina compressa* d'Orbigny, 1850). The test is large, discoid to ovoid, and dimorphic. The early miliolide coiling is followed by chambers that partially embrace the earlier ones. Chambers are subdivided by longitudinal partitions. Longitudinal ribs are directed from one foramen to the next, so that each supports a single row of regularly spaced pillars, reaching the chamber roof and alternating in their position in adjacent rows. The aperture is cribrate (trematophore) at one extremity of the test. *Lacazina* is distinguished from *Pseudolacazina* (below) by the monocular-concentric chamber arrangement and by the alternating positions of its pillars. It is distinguished from *Fabularia* by its monolocular growth and by the presence of radial partitions. Late Cretaceous (Coniacian) to Early Oligocene (Plate 6.9, fig. 9; Plate 6.27, fig. 9).
- *Lacazinella* Crespin, 1962 (Type species: *Lacazina wichmanni* Schlumberger, 1894). The test is ovoid, with concentric chambers completely embracing and divided by low longitudinal partitions that do not reach the chamber floor and support the trematophore. Late Paleocene to Eocene (Plate 6.9, figs 10-11).
- *Neolacazopsis* Matsumaru, 1990 (Type species: *Neolacazopsis osozawai* Matsumaru, 1990). The test is ovate in outline. A large proloculus is followed by chambers in a biloculine arrangement in megalospheric forms, while microspheric forms are quinqueloculine to triloculine and finally become biloculine adults. The chambers are subdivided into arcuate to turtle-neck bottle-like form chamberlets. Walls have a finely to coarsely alveolar inner layer, and outer layers with perforations. The aperture is cribrate (trematophore). Middle to Late Eocene.
- *Pseudofabularia* Robinson, 1974 (Type species: *Borelis matleyi* Vaughan, 1929). The test is almost globular, and bilocular. The interiors of chambers are subdivided by longitudinal partitions spiralling around the test and dividing the chambers into elongate chamberlets. Middle Eocene (Plate 6.9, fig. 12).
- *Pseudolacazina* Caus, 1979 (Type species: *Pseudolacazina hottingeri* Caus, 1979). The test is globular to ovate, and dimorphic. Megalospheric forms are biloculine (with completely embracing chambers) throughout their ontogeny. Microspheric forms are quinqueloculine in the early stage, but later they show biloculine to monoloculine cycles as adult. The chambers are subdivided by longitudinal partitions or septula or by ribs with pillared extensions supporting the chamber roof. Late Cretaceous (Santonian) to Eocene (Fig. 6.7).

Superfamily MILIOLOIDEA Ehrenberg, 1839

The test is coiled in varying planes with two chambers per whorl, with the axis of coiling normal to the apertural axis and rotated, so that several angles exist between the median planes of consecutive chambers, such as 72° (quinqueloculine), 120° (triloculine) or 180° (spiroloculine or biloculine). The test may become uncoiled, cylindrical or compressed with partial partitions. The proloculus is followed by a spiral passage. The aperture is single, and may be accompanied by additional teeth that project from the opposite margins of the aperture, from the chamber roof or from the lateral wall, or with a sieve (trematophore), that in fact is present in many porcelaneous larger foraminifera.

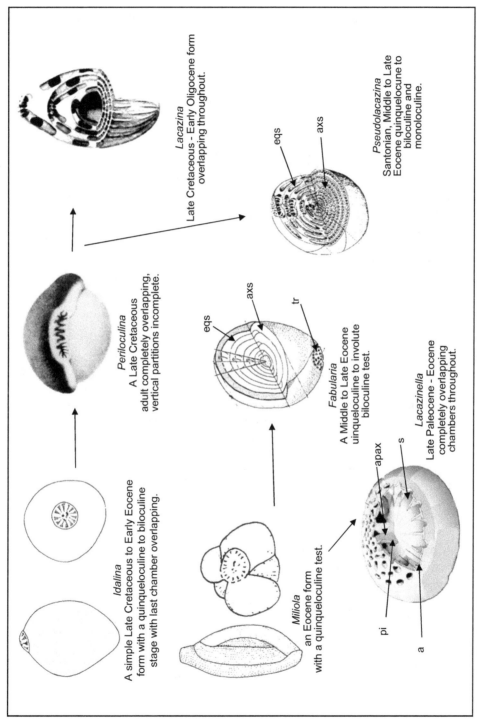

Figure 6.7. Sketches of Paleogene Fabulariidae, modified from Hottinger (2006), Drobne (1989) and Loeblich and Tappan, 1988.

Family Austrotrillinidae Loeblich and Tappan, 1986
This family has a miliolide test with coarse alveolar walls. Middle Eocene to Middle Miocene.

- *Austrotrillina* Parr, 1942 (Type species: *Trillina howchini* Schlumberger, 1893). The test is triloculine with fine to coarse blind alveoles that can bifurcate. The aperture is a simple tooth in the early chambers, branching to form smaller openings in the adult. Middle Oligocene to Middle Miocene (Langhian) (Plate 6.3, figs 7-9; Plate 6.9, figs 15-17; Plate 6.10, figs 1-2).

Family Hauerinidae Schwager, 1876
The early part of the test has a globular proloculus followed by two chambers per whorl. The chambers may be added in a quinqueloculine arrangement, but later may be uncoiled. The aperture may range from a simple to a bifid tooth, or may be a tremato-phore. Jurassic to Holocene.

- *Heterillina* Munier-Chalmas, 1905 (Type species: *Heterillina guespellensis* Schlumberger, 1905). The test is rounded, early chambers have a quinqueloculine arrangement, while later it is planispiral and evolute. The aperture is a trematophore. Middle Eocene to Oligocene (Plate 6.10, fig. 9).
- *Kayseriella* Sirel, 1999 (Type species: *Kayseriella decastroi* Sirel, 1999). The lenticular test has a milioline early stage (with a quinqueloculine embryont in the microspheric forms, and a triloculine arrangement in the megalospheric forms), followed by undivided planispiral chambers. The final stage is rectilinear. The aperture is simple, with a single basal opening with teeth and a single aperture with thick ribs in the uniserial chambers of the adult stage. This form is similar to the Late Cretaceous *Scandonea* De Castro 1971 (see Chapter 5), but the latter has thin subepidermal partitions that appear in the axial, oblique and horizontal sections of the uniserial chambers (De Castro 1971, Sirel, 1999). *Kayseriella* differs from *Heterillina* in having a simple, single aperture with teeth in the planispiral stage and uniserial chambers with a final ribbed aperture. Early Paleocene (Danian).

Family Rivieroinidae Saidova, 1981
The test is planispiral, and ovate in outline with chambers that are subdivided by oblique sutures. Middle Eocene to Holocene.

- *Pseudohauerina* Ponder, 1972 (Type species: *Hauerina occidentalis* Cushman, 1946). The test is quinqueloculine in the early stage with chambers one half-coil in length; later the chambers are planispiral with more than two chambers in each adult whorl. The interior of the test is subdivided by numerous incomplete radial subepidermal partitions that project inward from the walls for about one-third of the breadth of the chamber. The adult test has a complex trematophore with many openings. Oligocene to Holocene.

Family Spiroloculinidae Wiesner, 1920
A planispiral test consisting of a cornuspirine flexostyle, followed by a biserial part. The aperture is a simple, single basal opening with a bifid tooth.

- *Elazigella* Sirel, 1999 (Type species: *Elazigella altineri* Sirel, 1999). A lenticular test has umbonal thickening on both sides and a triangular apertural opening with a slender tooth. Late Paleocene (Thanetian) (Plate 6.10, Fig. 7).

Superfamily SORITOIDEA Ehrenberg, 1839
The chambers are planispiral, uncoiling, flabelliform or cyclical, and may be subdivided by partitions or pillars. Late Permian to Holocene.

Family Peneroplidae Schultze, 1854
The test is closely coiled in the early stage, becoming uncoiled in the later stage. Chambers have a simple interior. The aperture is single, rounded, slit-like or multiple. Late Cretaceous to Holocene.

- *Archiacina* Munier-Chalmas, 1878 (Type species: *Cyclolina armorica* d'Archiac, in Tournouër 1868). The test is a large discoid, planispiral, and semi-involute, later chambers are of peneropline shape and then cyclical. The aperture is multiple. Oligocene (Plate 6.10, fig. 4).
- *Dendritina* d'Orbigny, 1826 (Type species: *Dendritina* arbuscula d'Orbigny, 1826). The test is planispiral, and involute. The surface has numerous striae. A single areal aperture is modified by a heavily folded peristome. Middle Eocene to Holocene (Plate 6.10, fig. 5).
- *Haymanella* Sirel, 1998 (Type species: *Haymanella paleocenica* Sirel, 1999). The test is porcelaneous but with superficial, coarse agglutinated grains. The chambers are partially subdivided by irregularly disposed, radial, short partitions. The terminal stellate aperture has a protruding peristome. Paleocene (Plate 6.10, fig. 6).
- *Hottingerina* Drobne, 1975 (Type species: *Hottingerina lukasi* Drobne, 1975). A lenticular, dimorphic peneroplid, planispiral and involute test in the early stage, with later chambers arranged in an uniserial pattern in both megalospheric and microspheric generations. The interior of the chambers is subdivided by thin, short subepidermal partitions. The aperture is a simple, basal slit in the planispiral stage. Middle to Late Paleocene (Plate 6.10, fig. 10).
- *Penarchaias* Hottinger, 2007 (Type species: *Peneroplis glynnjonesi* Henson, 1950). The test is lenticular, composed of numerous planispiral-involute chambers with alar prolongations. The chambers are undivided and arranged in tight coils throughout ontogeny. In the alar prolongations of the chambers there is a single, interiomarginal row of apertures alternating with low endoskeletal ridges of the basal layer, that are perpendicular to the septal wall. The apertural face has multiple apertures. Late-Middle Eocene.
- *Peneroplis* de Montfort, 1808 (Type Species: *Nautilus planatus* Fichtel and Moll, 1798). The test is planispiral in the early part, tightly coiled and semi-involute,

lenticular and compressed, becoming uncoiled and flaring, with an umbilical depression. However, only the mineralized shell has this involute tendency, the chamber lumina are evolute and thus have no alar prolongations (Hottinger, 2007). The aperture is a single row in the median line. These apertures may be modified and/or subdivided by folded, partially fused peristomes (Hottinger et al., 1993). Eocene to Holocene (Fig. 6.8).

- *Puteolina* Hofker, 1952 (Type species: *Peneroplis proteus* d'Orbigny, 1839). The test is strongly flaring with a simple interior. It is considered by Loeblich and Tappan (1988) to be a synonym to the Miocene form *Laevipeneroplis* Šulc, 1936. Oligocene to Holocene (Fig. 6.8).

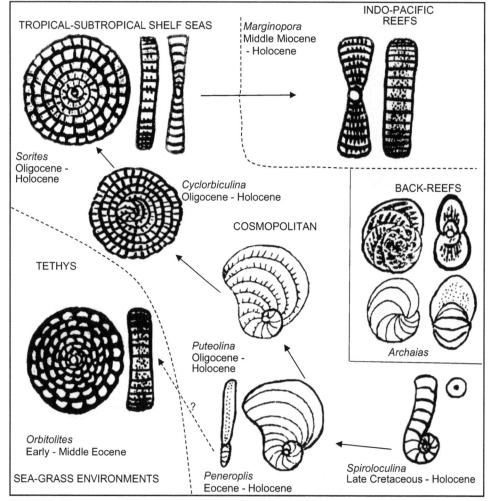

Figure 6.8. The progression from a peneropline shell to a completely annular test with chambers divided by internal partitions/crosswise-obliquely arranged endoskeletal elements in the Soritoidea (Middle Eocene to recent) with *Sorites - Amphisorus – Marginopora,* and to the archaiasines with radial endoskeletal elements. The latter have no marginal subdivision of the chambers.

- *Raoia* Matsumaru and Sarma, 2008 (Type species: *Raoia indica* Matsumaru and Sarma, 2008). Interior of chambers undivided, and marginal zone of chambers subdivided by rudimentary radial septula. Paleocene (Plate 6.11, fig. 12).
- *Spirolina* Lamarck, 1804 (Type species: *Spirolina cylindracea* Lamarck, 1804). The test is elongate, with early chambers that are planispiral, but which later are uncoiled and simple. The aperture is single, rounded. Eocene to Holocene.

Family Soritidae, Ehrenberg, 1839

The test is involute, planispiral to uncoiled evolute, flaring, annular discoid with partial or complete partitions. The aperture is multiple. Late Cretaceous (Cenomanian) to Holocene.

- *Amphisorus* Ehrenberg, 1839 (Type species: *Amphisorus hemprichii* Ehrenberg, 1839). The test is large biconcave, circular with thickened edges and two partial layers of chamberlets. The megalospheric apparatus consists of a large deuteroconch embracing a protoconch, and its wide-open flexostyle has a hemicylindrical to almost cylindrical frontal wall bearing numerous apertures. Cyclical chambers are divided by septula. Microspheric forms have a peneropline early stage with six undivided chambers, followed by ten chambers subdivided by septula that then become annular. Axial sections show an abrupt increase in the irregularity and volume of the chamberlet cavity in the last few chamberlet cycles. Crosswise-oblique stolons connect successive chambers. The aperture consists of numerous pores aligned in two alternating rows. Median apertures between the double rows of apertures may be present. Latest Oligocene to Holocene (see Chapter 7).
- *Archaias* de Montfort, 1808 (Type species: *Archaias spirans* de Montfort, 1808 = *Nautilus angulatus* Fichtel and Moll, 1798), as defined by its type species *Archaias angulatus* (Fichtel and Moll), extensively emended by Rögl and Hansen, 1984. The test is compressed, planispiral and involute, and may be partially evolute in the last whorls, with a thickened middle part and radial endoskeletal elements. Multiple apertures are flanked by irregular free and interseptal pillars. The subepidermal partitions are incomplete, and tests lack exoskeletal structures and no marginal subdivision of the chambers. Middle Eocene to Holocene (Fig. 6.8; Plate 6.10, Figs 8, 11, 12; Plate 6.12, Fig. 4).
- *Cyclorbiculina* Silvestri, 1937 (Type species: *Orbiculina compressa* d'Orbigny 1839). The test is large, and discoidal with a peneropline early stage becoming annular through most of its adult growth in both microspheric and megalospheric forms, with regular partitions. Oligocene to Holocene (Fig. 6.8; see Chapter 7).
- *Cyclorbiculinoides* Robinson, 1974 (Type species: *Cyclorbiculinoides jamaicensis* Robinson, 1974). The test is large discoid. The peneropline stage is followed by cyclical chambers with partitions aligned in successive chambers rather than alternating as in *Cyclorbiculina*. Middle to Late Eocene (Plate 6.10, Fig. 13).
- *Mardinella* Meriç and Çoruh, 1991 (Type species: *Orbitolites shirazensis* Rahaghi, 1983). This form has vertical septula dividing the chambers into small chamberlets instead of the oblique ones in *Orbitolites,* while the diagonal stolons connecting the chambers in different rows are not divided by horizontal septula. Late Paleocene (Thanetian).

- *Neorhipidionina* Hottinger, 2007 (Type species: *Rhipidionina williamsoni* Henson, 1948). A planispiral test which has an uncoiled flaring adult part. Chambers are subdivided by septula that are perpendicular to the outer chamber wall and end proximally with a slightly thickened rim. It has cribrate apertures on the apertural face with additional apertures appearing in the median plane as one or several rows. The undivided median zone of the chamber is restricted to a median annular passage. *Neorhipidionina* has a single foramen. This in contrast to *Rhabdorites* which between the septula has a radial row of foramina that almost reaches the periphery of the discoidal chambers. Middle Eocene (Plate 6.6, figs 8-9; Plate 6.10, fig. 14).

- *Neotaberina* Hottinger, 2007 (Type species: *Neotaberina neaniconica* Hottinger, 2007). The test is elongate to conical. The adult chambers are saucer-shaped and subdivided by radial partitions that are interpreted by Hottinger as "septula of an endoskeleton because they alternate regularly with radial rows of foramina", and are arranged in an uncoiled, uniserial sequence. The apertural face is strongly convex and covered by numerous areal apertures. Late Middle Eocene (Plate 6.10, fig. 14).

- *Opertorbitolites* Nuttall, 1925 (Type species: *Opertorbitolites douvillei* Nuttall, 1925). The test is lenticular, later chambers becoming cyclical and divided into numerous small chamberlets. Compact umbonal thickening, made of lateral laminae, covers the umbilical regions. Late Paleocene to Early Eocene (Plate 6.7, figs 1-5, 8).

- *Orbitolites* Lamarck, 1801 (Type species: *Discolites concentricus* de Montfort, 1808 = *Orbitolites complanatus* Lamarck, 1801). The test is a large discoid, very slightly concave, with a large proloculus and inflated nucleoconch, followed by cyclic chambers divided into small numerous chamberlets with curved thickened walls. Adjacent chambers are not interconnected with stolons, but the connections are between the obliquely adjoining chamberlets. According to Hottinger (2006) this superposition is clearly visible where the transverse section is tangential to an annular septum. Early to Middle Eocene (Plate 6.6, fig. 1; Plate 6.7, figs 6-7, 9; Plate 6.8, figs 1-3B; Plate 6.12, fig.3;).

- *Praerhapydionina* Van Vessem, 1943 (Type species: *Praerhapydionina cubana* Van Vessem, 1943). The test is elongate, sub-conical, planispiral, and later uncoiled and circular in thin section. The interior of the chambers is subdivided by radial partitions that are aligned from one chamber to the next. Additional radial partitions are intercalated between the septula, and are shorter than the primary septula, and do not reach the aperture. A single terminal aperture, on a strongly convex face, has a petaloid to stellar outline with four to six rays. Between the petals of the aperture, peristomes protrude to form toothlike features. Middle Eocene (Lutetian) to Early Miocene (Plate 6.8, figs 4-6; Plate 7.3, fig. 18).

- *Rhabdorites* Fleury, 1996 (Type species: *Rhapydionina malatyaensis* Sirel, 1976). A soritid test, with a short planispiral early part followed by a cylindrical to conical uniserial part with radial septula, and a large central pre-septal space. The primary septula of the uniserial chambers are very long, aligned in successive chambers and appear to meet in the center of the chamber, but the remainder of the central portion of the chambers remains free of partitions. The genus shows multiple apertures. Middle Eocene (Plate 6.6, figs 3-5, 7; Plate 6.8, fig. 7).

- *Somalina* Silvestri, 1939 (Type species: *Somalina stefaninii* Silvestri, 1939). The test is lenticular with cyclic chambers divided into chamberlets. The prominent lateral laminae enclose numerous cavities/chamberlets connected with stolons to the main equatorial chamber layer. Eocene (Lutetian) (Plate 6.11, figs 1-2).
- *Sorites* Ehrenberg, 1839 (Type species: *Nautilus orbiculus* Forsskal, 1775). The test is a large, discoid, with an early peneropline stage. Annular chambers are divided into numerous curved to rectangular small chamberlets, which are connected to each other and to those in adjacent chambers by stolons. The aperture is a single row of paired apertures. Oligocene to Holocene. (Plate 6.12, fig. 6).
- *Taberina* Keijzer, 1945 (Type species: *Taberina cubana* Keijzer, 1945). The test is elongate, planispiral becoming uncoiled. All chambers are subdivided by radial partitions, while pillars occupy the central areas of the chambers. Early Paleocene.
- *Twaraina* Robinson, 1993 (Type species: *Twaraina seigliei* Robinson, 1993). The test has a planispiral, compressed, early stage peneropliform, with a later part having flaring chambers crossed by irregular pillars, but lacking a cyclical stage. Eocene (Plate 6.11, figs 3-6).
- *Yaberinella* Vaughan, 1928 (Type species: *Yaberinella jamaicensis* Vaughan, 1928). The test is large, operculine to discoid, with a rapidly flaring peneropline early stage that may become cyclical in later stages. Chambers are numerous, and subdivided by oblique septula into small chamberlets that communicate through stolons. Middle to Late Eocene (Plate 6.6, fig. 11; Plate 6.10, fig.16; Plate 6.11, figs 9-11).

ORDER ROTALIIDA Delage and Hérouard, 1896

The tests are multilocular with a calcareous wall, of perforate hyaline lamellar calcite. They have apertures that are simple or have an internal tooth-plate. Triassic to Holocene.

Superfamily NUMMULITOIDEA de Blainville, 1827

The test is planispiral or cyclic, lenticular multicamerate, with septal flap and canaliculated septa. A spiral marginal cord and spiral canal system is present in early forms, but is modified in advanced forms or replaced by intraseptal canals (Figs 6.9-12). Paleocene to Holocene.

Family Pellatispiridae Hanzawa, 1937

A planispiral test, having no marginal cord, but radial and vertical canals or fissures are present. Spiral and umbilical sides are not differentiated. Planispiral-evolute chambers are connected by a single intercameral foramen. The thickened shell margin produced by a marginal canal sytem, as in *Pellatispira* (Fig. 6.10), may be overgrown by supplemental chamberlets either on the lateral flanks alone (*Biplanispira*, Fig. 6.11) or on all sides of the shell (*Vacuolispira*, Plate 6.14, fig. 7)). Paleocene to Eocene.

- *Biplanispira* Umbgrove, 1937. (Type species: *Heterospira mirabilis* Umbgrove, 1936). Biconvex, early chambers, with a planispiral involute test in the early stage, and

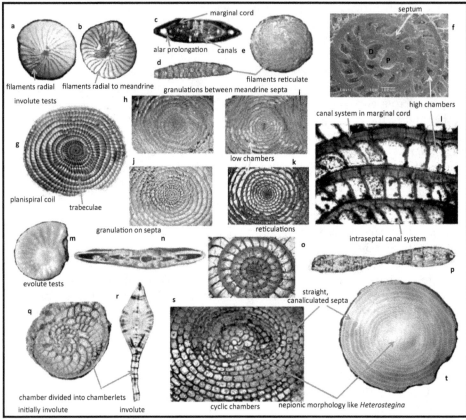

Figure 6.9. Figured by BouDagher-Fadel and Price (2014). **a**, *Nummulites deserti* De La Harpe, Egypt, P5, UCL coll., 4mm longest diameter (LD); **b**, *Nummulites irregularis* Deshayes, France, P8, UCL coll., 4mm LD; **c**, *Nummulites* sp., axial section Barton Bed, Chewton Bunny, Highcliffe, P8, UCL coll., 1.5mm LD; **d-e**, *Nummulites fichteli* Michelotti: **d**, France, P11, axial section, NHM P49522 (also figured by BouDagher-Fadel 2008), 3mm LD: **e**, solid specimen, Tang-i-Puhal area, P11, UCL coll., 5mm LD; **f**, enlargement of *Nummulites* sp. showing protoconch (P), deuteroconch (D), UCL coll., scale bar on figure; **g-j**, *Nummulites* sp., **g**, Egypt, Eocene, UCL coll., 10mm LD, **j**, Gernona, Spain, Eocene, 8mm, UCL coll. (also figured by BouDagher-Fadel 2008); **h**, *Nummulites intermedius* (d'Archiac), India, Eocene, NHM P30148, (also figured by BouDagher-Fadel 2008),12mm LD; **i**, *Nummulites gizehensis* (Forskal), Spain, Late Lutetian, UCL coll., (also figured by BouDagher-Fadel 2008),17mm LD; **k**, *Nummulites fichteli-intermedius* (d'Archiac), Lower Nari Formation, Pakistan, Oligocene, UCL coll. (also figured by BouDagher-Fadel 2008), 2.5mmLD; l, enlargement of chambers of *Nummulites* sp., France, Middle Eocene, width of field view 1.5mm; **m-n**, *Operculina aegyptiaca* Hamam, **m**, solid specimen, Egypt, Early Eocene, UCL coll., 2mm LD, **n**, axial section, megalospheric form, latest Early Eocene, Gebel Gurnah, Luxor, Egypt, paratype, NHM P49827 (also figured by BouDagher-Fadel 2008), 2.2mm LD; **o-p**, *Assilina daviesi* de Cizancourt, Lower Bhadrar Beds (Salt Range), Pakistan, Early Eocene, NHM coll. (also figured by BouDagher-Fadel 2008): **o**, equatorial setion, NHM P41529, 2.4mm LD: p, axial section, NHM P41524, 3.4mm LD; **q**, *Heterostegina (Heterostegina)* sp., Brazil, Eocene, UCL coll., 2mm LD; **r**, *Heterostegina (Vlerkina) borneensis* van der Vlerk, axial section, Borneo, Late Oligocene, UCL coll., 6mm LD; **s**, *Cycloclypeus eidae* Tan Sin Hok, Kinabatangan River, Sabah, North Borneo, Early Miocene, NHM coll., NB9067, enlargement of early part of test, width of field view 1mm; **t**, *Cycloclypeus carpenteri* Brady, off Jutanga, Holocene, UCL coll., 6mm LD.

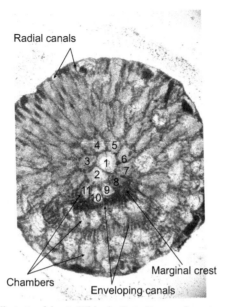

Figure 6.10. Bilamellar *Pellatispira fulgeria* Whipple, Late Eocene, Sumatra, equatorial section showing chamber arrangement, marginal crest and lateral chamberlets, early spiral chambers numbered 1-11 (1 being the protoconch), UCL coll., AS33

spiraling into two evolute spirals, with thick lateral laminae, one on each side of the equatorial plane. The thickened shell margin is produced by a marginal canal system, as in *Pellatispira* (below), that may be overgrown by supplemental chamberlets on the lateral flanks. The equatorial plane is covered by pillars with pores or canals. The later stage may become annular, with narrow marginal interlamellar cavities, formed by mostly imperforate walls suspended on radial spikes (Hottinger et al., 2001). Middle to Late Eocene (Fig. 6.11; Plate 6.13, figs 1-2).

• *Pellatispira* Boussac, 1906 (Type species: *Pellatispira douvillei* Boussac, 1906). The test is a flattened lenticular to discoidal, evolute planispiral. Double septa enclose the intraseptal canals. A supplemental skeleton forms a thick canaliculated crest, with densely grouped parallel, radial canals, extending in the equatorial direction (see Hottinger et al., 2001). Walls have coarse perforations and thick pillars perpendicular to the surface. Late Middle Eocene to Late Eocene. (Fig. 6.10; Plate 6.5, fig. 7; Plate 6.14, figs 1-6).

• *Serraia* Matsumaru, 1999 (Type species: *Serraia cataioniensis* Matsumaru, 1999). A pellatispirid test, with secondary and tertiary spiral chambers of intercalary whorls in the early growth stage. Late Middle Eocene.

• *Vacuolispira* Tan Sin Hok, 1936 (Type species: *Pellatispira inflata* Umbgrove, 1928). A thickly lenticular to globular test, with early evolute, planispiral chambers that support a heavy lateral supplemental skeleton, that is pierced by numerous radial canals. Later the spiral chambers are replaced by concentric arrangements of isolated chamberlets or interlamellar cavities, supported by pillars and covered by strong secondary lamination to form thick, perforate walls supported by pillars. The

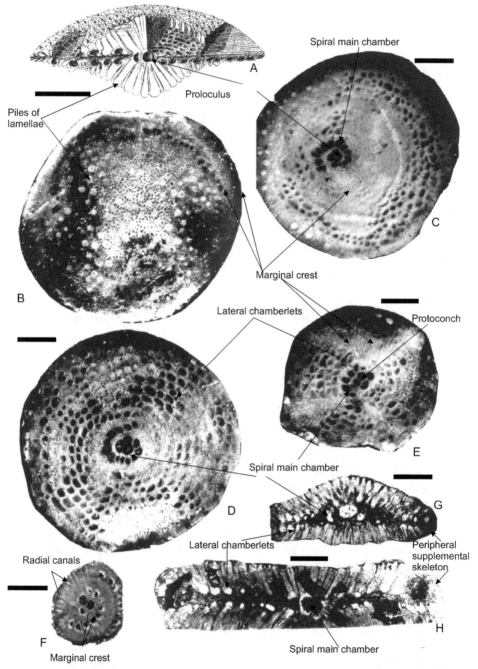

Figure 6.11. Bilamellar *Biplanipira mirabilis* (Umbgrove) growth A) Sterodiagram of *Biplanispira mirabilis* (Umbgrobe) (after Umbgrove, 1936). B-H) Type material, Utrecht University coll. The original description of *B. mirabilis* is precise except for what he called "tubules = pores, originating from the primary chambers" are in fact the tubular element of the enveloping canal system (Hottinger et al., 2001). B) External lateral view, the spiral main chambers cannot be identified from outside; C-F) Equatorial sections of megalospheric specimens showing the main spiral chambers, the lateral chamberlets/interlamellar spaces and the full extension of the marginal crest. G-H) Axial sections, G) of a lenticular specimen, H) of an almost biplanar discoidal test. Scale bars = 1mm.

thickened shell margin produced by a marginal canal system, as in *Pellatispira*, may be overgrown by supplemental chamberlets on all sides of the shell. (see Hottinger et al., 2001). Late Eocene (Plate 6.14, fig. 7).

Family Nummulitidae de Blainville, 1827
The test is planispiral involute or evolute with septal, marginal and vertical canals (Fig. 6.12). Paleocene to Holocene.

Subfamily Heterostegininae Galloway, 1933
Planispiral, with a canaliculate marginal cord, and septal canal trabeculae, but with true secondary septa, developed right across the chamber, forming chamberlets. Paleocene to Holocene.

- *Grzybowskia* Bieda, 1950 (Type species: *Grzybowskia multifida* Bieda, 1950). The test is wholly involute with favosely shaped, polygonal chamberlets. Late Eocene (Plate 6.14, fig. 12).
- *Heterostegina* d'Orbigny, 1826, sensu stricto emended Banner and Hodgkinson, 1991. (Type species: *Heterostegina depressa* d'Orbigny, 1826). The test is thick, planispiral, involute to evolute with chambers divided by secondary septa to form small chamberlets. No alar prolongations, but with raised sutures. Banner and Hodgkinson (1991) suggested three genera based on coiling:

1. *Heterostegina* (*Heterostegina*) (Type species: *Heterostegina depressa* d'Orbigny, 1826). The test is initially involute, but wholly evolute in mature whorls. Chamberlets are in equatorial view rectangular and restricted to those parts of the chambers which did not embrace earlier ones. Late Eocene to Holocene (Plate 6.14, fig. 11; Plate 6.15, figs 9-10).

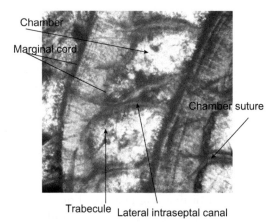

Figure 6.12. Enlargement of the lateral surface of some chambers of *Nummulites* showing part of the septal sutures between alar prolongations and their extensions, the trabeculae.

2. *Heterostegina* (*Vlerkina*) = *Heterostegina* (*Vlerkinella*) (Type species: *Heterostegina* (*Vlerkina*) *borneensis*). The test is completely involute (at least in the megalospheric forms), with rectangular lateral chamberlets, even in the chambers which embrace the preceding whorl, connected by Y-shaped intercameral stolons. Undivided sutural canals. Late Eocene to Late Miocene (Messinian) (Plate 6.14, figs 11, 14-16; Plate 6.15, fig. 2).

- *Spiroclypeus* Douvillé, 1905 (Type species: *Spiroclypeus orbitoideus* Douvillé, 1905). The test is planispiral, involute, with numerous narrow chambers increasing rapidly in height, divided into alternating chamberlets and tiers of lateral chamberlets on either side. Adjacent chamberlets of the same primary chamber, and adjacent chamberlets of successive chambers communicate with pore-like apertures. There are no alar prolongations. Late Eocene (Letter Stage "Tb") to Early Miocene (Letter Stage "Te5") (see below in section 6.3 for discussion of Letter Stages) (Plate 6.15, figs 6, 7, 11-13).
- *Tansinhokella* Banner and Hodgkinson, 1991 (Type species: *Tansinhokella tatauensis* Banner and Hodgkinson, 1991). The test is planispiral, involute, with a marginal cord, and bilamellar and canaliculated septa. It possesses groups of embracing alar prolongations, which are divided into chamberlets when seen in axial section, however, unlike *Spiroclypeus*, it lacks cubiculae (according to Banner and Hodgkinson, 1991 cubiculae are intralaminar chamberlets). Late Eocene (Letter Stage "Tb") to Early Miocene (Letter Stage "Te5") (Plate 6.15, figs 3-5; Plate 6.16, fig. 13).

Subfamily Nummulitinae de Blainville, 1827
The test is planispiral involute or evolute, with a canaliculate marginal cord, and septal canal trabeculae but without secondary septa forming chamberlets. They may become annular in more advanced forms. Late Cretaceous to Holocene.

- *Assilina* d'Orbigny, 1839 (Type species: *Assilina depressa* d'Orbigny, 1850 = *Nummulites spira de* Roissy, 1805 = *Operculina* (*Assilina*) Schaub. 1981). The test is tightly coiled, flat to biumbilicate, evolute with numerous chambers per whorl. The rate of whorl growth is slow resulting in a test without alar prolongations. The marginal cord is thick with a coarse canal system. Septa are radial, and trabeculae are absent. Late Paleocene (planktonic zone P4) to Middle Eocene (planktonic zone P14) (Plate 6.5, figs 11-12; Plate 6.14, figs 8-10; Plate 6.17, figs 2-8; Plate 6.18, figs 1-4; Plate 6.19, fig. 9; Plate 6.20, figs 6, 9-11, 14).
- *Chordoperculinoides* Arni, 1965 (Type species: *Operculina bermudezi* Palmer, 1934). The test is characterised by coarse vertical canals and a massive marginal cord. Paleocene to Eocene (Selandian to Ypresia) (Plate 6.20, figs 15-16).
- *Nummulites* Lamarck, 1801 (Type species: *Camerina laevigata* Bruguière, 1792). The test is lenticular, planispiral, involute, and tightly coiled with numerous whorls. The chambers are simple with a distinct marginal cord on the periphery and ramified canals within the sutures, which may be radial, sigmoid, meandrine or reticulate. Trabeculae are present, and there are pronounced alar prolongations. The extinction

of *Nummulites* can be correlated with the Rupelian stage and planktonic zones P18-P21a. Middle Paleocene to Early Oligocene (Plate 6.5, figs 11-12; Plate 6.12, figs 1-2; Plate 6.16, fig. 11; Plate 6.18, figs 8,10; Plate 6.19, figs 1-8; Plate 6.20, figs 1-5; Plate 6.21, figs 1-16).

- *Operculina* d'Orbigny, 1826 (Type species: *Lenticulites complanatus* Defrance, 1822). The test is planispiral, evolute lenticular to compressed and loosely coiled. The chambers are simple. Trabeculae are absent. Sutural canals are forked or branching. Late Paleocene to Holocene (Plate 6.1, fig. 1; Plate 6.12, fig. 1; Plate 6.18, figs 5-6, 9, 11-12).

- *Operculinella* Yabe, 1918 (Type species: *Amphistegina cumingii* Carpenter, 1860). The last true *Nummulites* spp. became extinct at the top of the Td "Letter Stage" with *Nummulites fichteli* Michelotti 1841 from the upper Early Oligocene of Italy. Contrary to the opinions of S. Cole (*in* Loeblich and Tappan, 1964) and Loeblich and Tappan (1988), *Nummulites* can be distinguished from *Operculinella*. Eames et al. (1962) illustrated a simple *Nummulites vascus* Joly and Leymerie (plate 1, figures A, B) to compare with *Operculinella cumingii* (Carpenter) (*Palaeonummulites nomen oblitum*). The strong dimorphism seen between microspheric and megalospheric forms of Oligocene specimens of *Nummulites* is never seen in *Operculinella* (where the microspheric and megalospheric generations are externally identical). The presence of trabeculae in *Nummulites* and their absence from *Operculinella* is noteworthy, but, most importantly, the diameter of the megalospheric protoconch of *Nummulites* (in both simple and complex forms) is much greater than the diameter of the proloculus of *Operculinella*. The megalospheric loosely coiled *Operculinella* (e.g. *Operculinella cumingii)* persists to the Holocene but the large protoconch of true *Nummulites* does not occur beyond the Early Oligocene. Oligocene to Holocene (See Chapter 7).

- *Palaeonummulites* Schubert, 1908 (Type species: *Nummulina pristina* Brady, 1874). *Palaeonummulites* are here attributed to all involute forms, but having a tight spire and lacking the developed, highly extended later chambers of *Operculinella* (see Haynes et al. 2010; BouDagher-Fadel and Price, 2014). *Operculinoides*, the American genus (type species of *Nummulites willcoxi* Heilprin, 1883; see BouDagher-Fadel and Price, 2014) is an involute, tightly coiled simple nummulitid, and is a synonym of *Palaeonummulites*. Late Paleocene to Holocene (Plate 6.19, figs 7-8; Plate 6.28, figs 18-19).

- *Planocamerinoides* Cole, 1958 (Type species: *Nummularia exponens* de Sowerby, 1870). Biumbilicate test with multilamellar thickenings over the umbonal area. Late Paleocene to Middle Eocene (Plate 6.20, fig. 14)

- *Planostegina* Banner and Hodgkinson, 1991 (Type species: *Heterostegina operculinoides* Hofker, 1927). A totally evolute, laterally more compressed form of *Heterostegina,* with chambers divided by septula into complete or incomplete subrectangular chamberlets connected by Y-shaped intercameral stolons. The test has strong ornamentation and undivided sutural canals. Late Paleocene to Holocene (Plate 6.15, fig. 1; Plate 6.18, fig. 7).

- *Ranikothalia* Caudri, 1944 (Type species: *Nummulites nuttalli* Davies, 1927). The test is lenticular with alar prolongations, initially involute, becoming evolute in the last whorls. There is a thick marginal cord with a coarse canal system that connects to simple vertical septal canals. Trabeculae are present. Late Paleocene (Plate 6.13, fig. 3; Plate 6.22, figs 1-5).

Family Cycloclypeidae Galloway, 1933 emend. BouDagher-Fadel, 2002
This family is distinguished by the development of concentric annular, wholly evolute chambers, each chamber being divided into numerous chamberlets in a median plane, and each chamberlet separated from adjacent chamberlets by straight, canaliculated walls. There is no marginal cord, except in the early stages of the microspheric generation. Eocene to Holocene.

• *Cycloclypeus* Carpenter, 1856 (Type species: *Cycloclypeus carpenteri* Brady, 1881). A nummulitid with a nepionic morphology like *Heterostegina*, but with a final growth stage with cyclic chambers. No alar prolongations occur. Early Oligocene to Holocene (Plate 6.14, fig. 4; Plate 6.17, fig. 1, see Chapter 7).

Family Orthophragminidae Vedekind, 1937
The orthophragmines have a *Cycloclypeus*-like or operculinid microspheric juvenile form. Middle Paleocene to Eocene.

Subfamily Discocyclininae Galloway, 1928
Megalospheric forms have a subspherical protoconch enclosed by a larger reniform deuteroconch. Microspheric forms have an initial spiral of small chambers, and later stages with cyclical chambers subdivided by septula into small rectangular chamberlets connected by annular and radial stolons. There is a fine equatorial layer and small lateral chamberlets. A small, intraseptal and intramural canal system is present (Fig. 6.13). Middle Paleocene to Late Eocene.

• *Actinocyclina* Gümbel, 1870 (Type species: *Orbitolites radians* d'Archiac, 1850). This form differs from *Discocyclina* (below) in having distinct rays formed by a proliferation of broad and low lateral chamberlets. Middle to Late Eocene (Plate 6.22, fig. 6).
• *Asterophragmina* Rao, 1942. (Type species: *Pseudophragmina (Astemphragmina) pagoda* Rao, 1942). Stellate in outline, with rays radiating from the centre. Radial walls are absent after five to ten annuli of cyclic chambers, which remain undivided. Late Eocene (Plate 6.22, fig. 7).
• *Athecocyclina* Vaughan and Cole, 1940 (Type species: *Pseudophragmina (Athecocyclina) cookei* Vaughan and Cole, in Cushman, 1940). *Athecocyclina* has no incipient septula. Eocene
• *Discocyclina* Gümbel, 1870 (Type species: *Pseudophragmina (Astemphragmina) pagoda* Rao, 1942). The test is discoidal, flat, with an equatorial layer composed of concentric rings of rectangular chamberlets, those of successive cycles alternating in position. Lateral chamberlets are connected with the equatorial layer by vertical stolons. Annular stolons occur at the proximal end of the radial walls and connect adjacent chamberlets. Middle Paleocene to Late Eocene (Plate 6.5, fig. 12; Plate

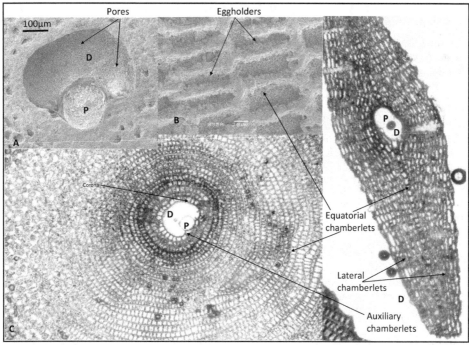

Figure 6.13. Features of *Discocyclina*; A, B) SEM photos of *Discocyclina*, the internal surface of the lateral chamber wall bears egg-holders which might have harbored symbionts; C) Equatorial section; D) Axial section of *Discocyclina* sp., France, UCL coll. P= protoconch; D = Deuteroconch

6.12, fig. 1; Plate 6.16, figs 3, 12; Plate 6.22, figs 5, 6, 8-15; Plate 6.23, figs 1-2, 4-11; Plate 6.24, figs 1, 6-8, 11; Plate 6.25, Fig. 7).

- *Hexagonocyclina* Caudri, 1944 (Type species: *Orbitoclypeus*? *cristensis* Vaughan, 1924). The test is similar to *Discocyclina* but with two symmetrical auxiliary chambers on each side of the nucleoconch and four spirals, and predominantly hexagonal equatorial chambers. Early to Middle Eocene (Plate 6.24, figs 2-5).
- *Nemkovella* Less, 1987 (Type species: *Orbitoides strophiolata* Gümbel, 1870). The test has a circular outline, with ribs constructed (as in *Actinocyclina*) exclusively from lateral layers. It differs from *Discocyclina* in lacking annular stolons. Late Paleocene (Thanetian) to Early Eocene (Ypresian) (Plate 6.16, fig. 9).
- *Proporocyclina* Vaughan and Cole, 1940 (Type species: *Discocyclina perpusilla* Vaughan, 1929). This form is characterized by the presence of well-developed, radial septula with distal annular connections. Eocene
- *Pseudophragmina* Douvillé, 1923 (Type species: *Orthophragmina floridana* Cushman, 1917). The test is circular to subquadrate in outline. An eulepidine embryo is surrounded by a single ring of large nepionic chambers, which are followed by smaller equatorial chambers. The septa are irregular. Numerous irregular layers of lateral chamberlets occur on both sides of the single equatorial layer. Although, the subgenera *P. (Proporocyclina)* Vaughan and Cole, in Cushman, 1940 and *P. (Athecocyclina)* Vaughan and Cole, in Cushman, 1940 are both considered as synonyms to

Pseudophragmina by Loeblich and Tappan, 1988, they are, however, proved to be different genera, as they in fact exhibit important, distinguishing, stratigraphically-characteristic, morphological features; thus *Athecocyclina* has no incipient septula, *Pseudophragmina* has irregular septula, while *Proporocyclina* has well-developed, radial septula with distal annular connections (see BouDagher-Fadel and Price, 2017). Eocene (Plate 6.7, figs 10-12; Plate 6.26, fig. 9).

Subfamily Orbitoclypeinae Brönnimann, 1946
Microspheric tests have an early planispiral coil, while megalospheric tests have a globular protoconch, enclosed by a larger reniform deuteroconch. Members of this subfamily may occur with or without ribs. There is a single equatorial layer of chamberlets and several layers of small lateral chamberlets, and cyclical chambers are not subdivided into chamberlets. Chambers have four stolons. Middle Paleocene to Late Eocene.

- *Asterocyclina* Gümbel, 1870 (Type species: *Calcarina? stellata* d'Archiac, 1889). The test is stellate with five to six rays which are caused by thickening and outward growth of the median layer, where normally it is thin, as in *Discocyclina*. Lateral chambers, made of axial subdivision of annuli, are present on both sides of median layer. Middle Paleocene to Eocene (Plate 6.23, figs 1-2; Plate 6.24, figs 9-11).
- *Neodiscocyclina* Caudri, 1972 (Type species: *Discocyclina anconensis* Barker, 1932). The test is flat and lenticular. Equatorial chambers are irregular, enlarging from the proloculus to the periphery, and radial walls are thin. Up to twenty layers of lateral chambers occur on both sides of the equatorial layer. Numerous pillars are visible in vertical section. Middle Eocene (Plate 6.27, fig. 1).
- *Orbitoclypeus* Silvestri, 1907 (Type species: *Orbitoclypeus himerensis* Silvestri, 1907). The test is inflated centrally. The megalospheric embryo has a deuteroconch that completely encloses the protoconch. Complete cycles of periembryonic chambers surround the embryo. Equatorial chambers are spatulate, and in concentric rings, without stolons connecting adjacent chamberlets. Ribs may be present, and constructed from lateral layers, but with a slight axial enlargement. Late Paleocene to Late Eocene (Plate 6.27, fig. 2).
- *Stenocyclina* Caudri, 1972 (Type species: *Orthophragmina advena* Cushman, 1921). The test is circular in outline. The embryo is followed by small square equatorial chamberlets. Radial walls are aligned in successive annuli. Numerous layers of low lateral chamberlets occur on both sides of a thin, but well developed, equatorial layer. Middle Eocene.

Superfamily NONIONOIDEA Schultze, 1854
Unlike in BouDagher-Fadel (2008), the miscellaneids are separated from the nummulitoids in this book and considered as belonging to the superfamily Nonionoidea because of the planispiral-involute coiling combined with an interiomarginal position of the foramina (see Hottinger, 2009).

Family Miscellaneidae Sigal *in* Piveteau, 1952
Members of this family have a planispiral, evolute test. The aperture is single or multiple, interiomarginal and symmetrical to the equatorial plane of the test (see Hottinger, 2009). Paleocene and Earliest Eocene.

Subfamily Miscellaneinae Kacharava in Rauzer-Chernoussova and Furzenko, 1959
Members of this subfamily have a single, interiomarginal, intercameral foramen that is symmetrical with respect to the equatorial plane of the shell (see Fig. 6.14).

- *Miscellanea* Pfender, 1935 (Type species: *Nummulites miscella* d'Archiac and Haime, 1853). A lenticular angular test, having few whorls, is strongly ornamented with very coarse perforation, raised rounded pustules, pillars, raised ridges and raised bars,

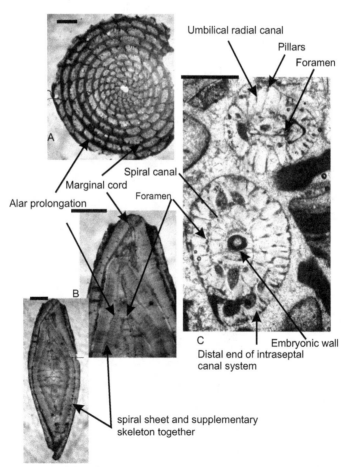

Figure 6.14. Illustrations of the differences between *Nummulites*: A) *N. masiraensis* Carter, Masira Island, Oman; B) *N.* cf. *irregularis* Deshayes, Kenya; C) *Miscellanea miscella* (d'Archiac and Haime), Paleocene, Indonesia. The straight wall between protoconch and deuteroconch indicates the biconch quality of the nepiont. UCL coll. Scale bars = 1mm.

dividing the space between longitudinal ridges. Bilamellar septa are quite distinct. The embryonic apparatus is composed of a protoconch and a slightly smaller deuteroconch, separated by a straight wall and surrounded by a common wall with canals forming a biconch. The biconch is followed by a third chamber with much smaller volume. Paleocene (Plate 6.13, figs 4-12).

Subfamily Miscellanitinae Hottinger, 2009
Members of this subfamily have a planispiral, involute test with multiple intercameral foramina. Their apertures are a single interiomarginal row in a comparatively low chamber. Late Paleocene (Thanetian).

- *Bolkarina* Sirel, 1981. (Type species: *Bolkarina aksarayi* Sirel, 1981). The test is large and discoidal. The proloculus is small, and followed by numerous rectangular peri-embryonic chambers. Septa are doubled with intraseptal canals. Late Paleocene (Thanetian) (Fig. 6.15).
- *Miscellanites* Hottinger, 2009 (Type species: *Miscellanea iranica* Rahaghi, 1983). The test is globular, planispiral and involute with low but elongate chambers reaching from pole to pole. The outer surface of the chamber wall is covered by a simple and shallow enveloping canal system with pustules. Late Paleocene (Thanetian).

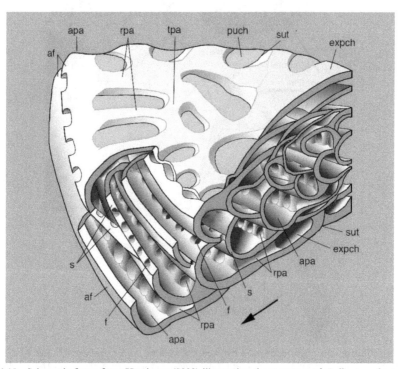

Figure. 6.15   Schematic figure from Hottinger (2009) illustrating the structure of *Bolkarina aksarayi* Sirel, 1981. **af:** apertural face; **apa:** annular passage (in pre-septal position); **expch:** expanse chamber; **f:** foramen; **puch:** penultimate (expanse) chamber; **rpa:** radial passage; **s:** septum; **sut:** (chamber) suture; **tpa:** transverse tubular passage. Arrow: direction of growth. (not to scale).

Superfamily PLANORBULINOIDEA Schwager, 1877
The test is trochospiral in the early stages, but later may be uncoiled and rectilinear, or biserial, or with many chambers in the whorl. They are found with intra- to extra-umbilical apertures, and additional equatorial apertures may be present. Early Cretaceous (Berriasian) to Holocene.

Family Eoannularidae Ferrández-Cañadell and Serra-Kiel, 1998
The test is bilamellar perforate, with a bilocular megalospheric embryo followed first by orbitoidal chamberlets, which change later to cyclical chambers. The cyclical chambers are subdivided into rectangular chamberlets, or are not subdivided. Microspheric initial chambers are arranged in a peneroplid-like spire that changes into annular chambers by a progressive increase in chamber width. Middle Eocene.

- *Eoannularia* Cole and Bermudez, 1944 (Type species: *Eoannularia eocenica* Cole and Bermúdez, 1944). The test is discoidal, flat with subdivided annular chambers and lamellar thickening only in the early stage. The megalospheric form has a protochonch that is completely enclosed by the deuteroconch, and chambers following the bilocular embryo occurring as an annular series in a single layer. Middle Eocene (Plate 6.27, fig. 3; Fig. 6.16).
- *Epiannularia* Caudri, 1974 (Type species: *Epiannularia pollonaisae* Caudri, 1974). The test is discoidal, centrally thin and thickened towards the periphery. The bilocular embryont is followed by cyclical divided chambers, later becoming undivided. Middle Eocene (Fig. 6.16).

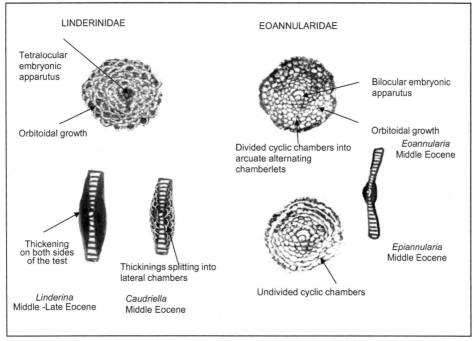

Figure 6.16. Schematic figures highlighting the differences between the Linderinidae and the Eoannularidae. Drawings not to scale.

Family Linderinidae Loeblich and Tappan, 1974, emended Ferrández-Cañadell and Serra-Kiel, 1998

The test is bilamellar with a lobate outline. The megalospheric embryo has a quadril-ocular early stage, with three initial chambers separated by flattened walls, followed by a fourth arcuate chamber with apertures at both sides. Later chambers occur in orbitoi-dal growth, with crosswise-oblique stolons. Forms may have varying amounts of calcite deposited on both sides of the central embryont, but no lateral chambers. Microspheric intial chambers are arranged in *Planorbulina*-like spire. Middle to Late Eocene.

- *Caudriella* Haman and Huddleston, 1984 (Type species: *Margaritella ospinae* Caudri, 1974). The test is lenticular with lateral chamberlets, a small proloculus followed by two larger chambers and an arcuate median chamber, with lateral, irregularly arranged chambers with thick walls and open lumina. *Caudriella* seems to differ from *Linderina* (below) by having well-developed lateral chamberlets. Caudri (1974) discussed the systematic position of her new genus. After considering and discard-ing the relationship with *Linderina* and with *Pseudolepidina* (see below), she decided to leave it "standing alone and isolated from all other groups and evolutionary lin-eages". Loeblich and Tappan (1987) included *Caudriella* into the Lepidocyclininae (together with *Astrolepidina, Eulepidina, Lepidocyclina* and *Pseudolepidina,* all of which having a bilocular megalospheric embryo). Ferrández-Cañadell and Serra-Kiel (1998) in their revision of *Linderina* decided to put the genus *Caudriella* within their emended *Linderina* on the basis of the presence of the embryonic apparatus, which is similar to that of *Linderina.* To date, only one species, *Caudriella ospinae* (Caudri, 1974), from the Middle Eocene of Margarita Island (Venezuela), is known, and more work is needed on this genus. Middle Eocene (Fig. 6.16).
- *Linderina* Schlumberger, 1893 (Type species: *Linderina brugesi* Schlumberger, 1893). The test is large, discoid but without lateral chamberlets, and with considerable thick-ening in the early stage on both sides of the test, formed by the superposition of the successive involute outer lamellae. Orbitoidal chambers, consisting of small arched chamberlets, occur in concentric series, with successive layers alternating in position. There are no annular stolons, nor stolons in the distal wall of chamberlets. The aper-tures correspond to the stolon system, with rows of apertures between chamberlets in the last chamber, from which the orbitoidal growth starts. Microspheric initial chambers are arranged in a *Planorbulina*-like spire. Middle to Late Eocene (Plate 6.27, figs 4-6; Fig. 6.16).

Family Planorbulinidae Schwager, 1877

The test is free or attached, with an early stage that is trochospiral, later becoming discoid, cylindrical or conical. The aperture is single or multiple. Eocene to Holocene.

- *Neoplanorbulinella* Matsumaru, 1976 (Type species: *Neoplanorbulinella saipanensis* Matsumaru, 1976). The test is attached, conical to concavo-convex. The microspheric proloculus is small, while the megalospheric protoconch consists of a spherical pro-loculus enclosing a reniform second chamber, followed by a third chamber with only a proximal stolon and numerous trochospirally arranged chambers. The adult form has equatorial chambers that occur in an annular series and with two or more lateral chambers on the concave side. Multiple apertures are present. Late Eocene to early Miocene.

- *Peelella* Matsumaru, 1996 (Type species: *Peelella boninensis* Matsumaru, 1996). The test is small, concavo-convex. The megalospheric apparatus has a globular protoconch and a reniform deuteroconch, surrounded by a thick wall followed by a trochospiral to planispiral symmetrical later stage. The microspheric generation has a small proloculus, followed by a trochospiral stage. Equatorial chambers form an annular series, with successive series alternating in position. Lateral chambers are irregular, curved, and well differentiated from the equatorial layer through stolons and coarse perforations on the ventral side of the test. Late Oligocene.

- *Planolinderina* Freudenthal, 1969 (Type species: *Planolinderina escornebovensis* Freudenthal, 1969). The test is attached, single-layered. Microspheric forms have a trochospiral early stage; megalospheric forms occur with a protochonch and deuteroconch. Later chambers are added in a cyclic series. The aperture is a series of basal openings. Late Oligocene to Early Miocene (Burdigalian).

- *Planorbulina* d'Orbigny 1826. (Type species: *Planorbulina mediterranensis* d'Orbigny, 1826). The test is attached, single-layered with a long initial spiral and an irregular pattern of a small number of orbitoidal chambers, with supplementary apertures in sutural positions. The protoconch and the two following chambers forming a tri-conch. The aperture is made of single slits, bordered with a narrow lip. Eocene to Holocene.

- *Planorbulinella* Cushman, 1927 (Type species: *Planorbulina vulgaris* d'Orbigny var. *larvata* Parker and Jones, 1865). The test is attached, discoidal, with alternating chambers built over the apertures of the previous ring. They may have scattered pillars dorsally and ventrally (e.g. *P. larvata*, Plate 7.10, Fig. 10) or heavy lateral thickening (e.g. *P. solida,* Plate 7.11, Figs 2-5). Chamber walls are coarsely perforate. Relapse chambers are present. The embryont has three chambers with thick walls. The aperture is made of single slits, bordered with a narrow lip. Eocene to Holocene (Plate 6.27, fig. 7).

- *Tayamaia* Hanzawa, 1967 (Type species: *Gypsina marianensis* Hanzawa, 1957). A dome-like test with the ventral hollow, filled with numerous irregular chamberlets. Only very few chambers cover the dorsal side of the dome. The median layer has a globular protoconch and arcuate chambers, interconnecting by four stolons. Upper Late Oligocene (Te) to Early Miocene.

Family Cymbaloporidae Cushman, 1927

Chambers occur in a single layer (Fig. 6.17). Late Cretaceous (Cenomanian) to Holocene.

- *Cymbalopora* von Hagenow, 1851 (Type species: *Cymbalopora radiata* von Hagenow, 1851). The test is low and conical, with an open umbilicus, thickened wall and secondary thickening, obscuring sutures and chambers on spiral side. The walls are lamellar. Late Cretaceous to Middle Paleocene (Fig. 6.17).

- *Eofabiania* Küpper, 1955 (Type species: *Eofabiania grahami* Kopper, 1955). The test is conical, with a trochospiral early stage. Later chambers have no subdivisions and may be added in annular series. Early to Middle Eocene.
- *Fabiania* Silvestri, 1924 (Type species: *Patella* (*Cymbiola*) *cassis* Oppenheim, 1896). The test is conical, with a deeply excavated centre. The early stage has two globose thick-walled and perforate chambers, later chambers are cyclical with horizontal and vertical partitions. The aperture is a single row of pores opening into the large umbilicus. Late Paleocene to Late Eocene (Plate 6.27, figs 8-11; Fig. 6.17).
- *Gunteria* Cushman and Ponton, 1933 (Type species: *Gunteria floridana* Cushman and Ponton, 1933). The test is compressed and flabelliform. The early stage has large, globular, undivided chambers, later chambers are concentric and subdivided, with radial and vertical partitions. Middle Eocene (Fig. 6.17).

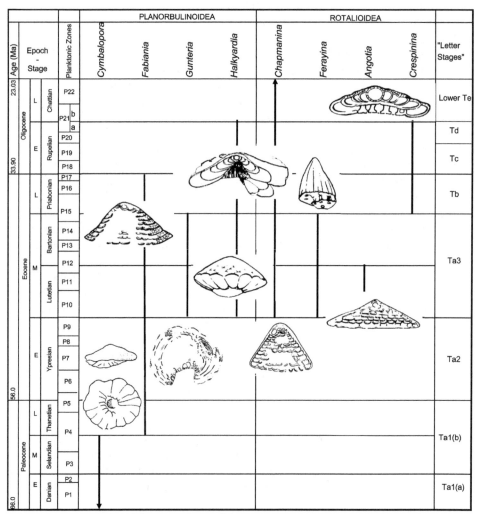

Figure 6.17 The evolution of the conical forms of the Paleogene some Planorbulinoidea (Cymbaloporidae) and Rotalioidea (Chapmaninidae). Drawings not to scale (modified after Deloffre and Hamaoui, 1973).

- *Halkyardia* Heron-Allen and Earland, 1918 (Type species: *Cymbalopora radiata* von Hagenow var. *minima* Liebus. 1911). The test is biconvex, with an embryont consisting of a large protoconch and deuteroconch, and two primary auxiliary chambers. The umbilicus is filled with horizontal bilamellae and connecting pillars. Middle Eocene (Lutetian) to Middle Eocene to Early Oligocene (Lutetian to Rupelian) (Plate 6.27, fig. 12; Fig. 6.17).

Family Victoriellidae Chapman and Crespin, 1930
The test is attached or may be free in the juvenile stage, with a trochospiral early stage, later becoming an irregular mass of chambers. Late Cretaceous (Santonian) to Holocene.

Subfamily Carpenteriinae Saidova, 1981
The test is attached, trochospiral throughout, planoconvex with a large aperture, open in the umbilicus. Paleocene to Holocene.

- *Carpenteria* Gray, 1858 (Type species: *Carpenteria balaniformis* Gray, 1858). The test has a carinate periphery, and is planoconvex with a flat spiral side and distinct rims or keels, a strongly convex, distinctly perforate umbilical side surrounded by thick pillars. Late Eocene to Holocene (See Chapter 7).
- *Neocarpenteria* Cushman and Bermúdez, 1936 (Type species: *Neocarpenteria cubana* Cushman and Bermúdez, 1936).The test is a low trochospiral and bi-evolute, with a flattened spiral side and a periphery, broadly carinate, with a keel. Late Eocene.

Subfamily Rupertininae Loeblich and Tappan, 1961
The test is attached, with a distinct flattened disk, but coiling grows out away from the site of attachment. Late Cretaceous to Holocene.

- *Biarritzina* Loeblicah and Tappan, 1964 (Type species: *Columella carpenteriaeformis* Halkyard, 1918). The test is attached by the flaring basal disk, with an early stage that is trochospirally enrolled with a distinctly loose coiling style, but later tending to become uniserial. The walls have fine perforations. The aperture is terminal with a distinctly raised lip, and may be present on one or two chambers of the final whorl. Middle Eocene to Holocene (Plate 6.26, fig. 5; see Chapter 7).

Subfamily Victoriellinae Chapman and Crespin, 1930
The juvenile stage may be free living, but later stages are attached. High spired forms develop around a hollow axis, with pillar-like thickenings in the walls. The aperture is an umbilical slit bordered by a lip. Middle Eocene to Holocene.

- *Eorupertia* Yabe and Hanzawa, 1925 (Type species: *Uhligina boninensis* Yabe and Hanzawa, 1922). The test is highly trochospiral, enrolled about an axial hollow. The walls are coarsely perforate with small pillars between the perforations. Middle to Late Eocene (Plate 6.26, figs 2-4; Plate 6.28, figs 3-4).

- *Korobkovella* Hagn and Ohmert, 1971 (Type species: *Truncatulina grosserugosa* Gümbel, 1870). The test is low trochospiral with the spiral side flattened as an attached area. Walls have coarse pits and irregular vermiform channels. Middle Eocene.
- *Maslinella* Glaessner and Wade, 1959 (Type species: *Maslinella chapmani* Glaessner and Wade, 1959). The test is low trochospiral, and semi-involute with curved sutures and fine pillars surrounding the umbilicus. Late Eocene.
- *Victoriella* Chapman and Crespin, 1930 (Type species: *Carpenteria proteiformis* Goës *var. plecte* Chapman, 1921). The test is conical, usually with a free juvenile stage and a small attachment area near the apex, consisting of a few inflated chambers with pillar-like thickenings in the wall. In the adult stage the coiling is high-spired, with three to four subspherical chambers per whorl (not enclosing), either with an umbilical depression or arranged round an axial hollow. Septa are trilamellar. Late Eocene to Early Miocene (Plate 6.26, fig. 1).
- *Wadella* Srinivasan, 1966 (Type species: *Carpenteria hamiltonensis* Glaessner and Wade, 1959). The test is subconical, with a free living juvenile stage, which later becomes fixed on the substrate. Three to five chambers occur per whorl. The walls are coarsely perforate with pores opening at small mounds. Late Eocene.

Superfamily ACERVULINOIDEA Schultze, 1854

The test is trochospiral to discoidal and encrusting, consisting of numerous irregularly formed chambers. Paleocene to Holocene.

Family Acervulinidae Schultze, 1854

The test is free or attached. A low trochospiral in the initial early stage is followed by inflated chambers spreading over the substrate in more than one layer, forming an irregular mass, that appears to possess no apertures other than the pore-like cribrate openings in their upper, distal surfaces. Paleocene to Holocene.

- *Discogypsina* Silvestri, 1937 (Type species: *Discogypsina vesicularis* Silvestri, 1937). The test is lenticular and flattened, with an equatorial layer of slightly larger thicker chambers, separating lateral layers of smaller irregularly arranged chambers with successive layers showing no alignment. The walls are coarsely perforate with no stolon system. Middle Eocene to Holocene (Plate 6.26, figs 6-7).
- *Gypsina* Carter, 1877 (Type species: *Polytrema planum* Carter, 1876). The test is attached, formed by encrusting polygonal, inflated and closely appressed chambers. Chambers of successive layers alternate in position. Late Oligocene to Holocene (Plate 6.26, fig. 10A).
- *Orbitogypsina* Matsumaru, 1996 (Type species: *Orbitogypsina vesivularis* Matsumaru, 1996). The test is concavo-convex. The megalospheric apparatus has a protoconch and deuteroconch followed by nepionic and equatorial chambers, connected by a stolon system as in the Lepidocyclinidae. Microspheric forms have a small proloculus followed by spreading and regular chambers in a globular or discoidal mass, with mural pores and a cribrate aperture in the upper part of the chambers. Successive layers of equatorial chambers are only aligned at the periphery. Late Eocene to Oligocene.

- *Protogypsina* Matsumaru and Sarma, 2008 (Type species: *Protogypsina indica* Matsumaru and Sarma, 2008). The test is spherical. The megalospheric apparutus has a spherical protoconch and a kidney-shaped deuteroconch followed by large ovoid chambers. Paleocene (Plate 6.11, fig. 8).
- *Solenomeris* Douvillé, 1924 (Type species: *Solenomeris ogormani* Douvillé, 1924). The test is large, composed of numerous branching chamberlet layers with fruticose protuberances. The fruticose branches are free from one another, or laterally coherent to varying degrees. The embryonic apparatus consists of a large proloculus surrounded by a whorl of subspherical chambers. The aperture is multiple. Eocene to Holocene.
- *Sphaerogypsina* Galloway, 1933 (Type species: *Ceriopora globulus* Reuss, 1848). The test is small, almost spherical, with chambers added in numerous layers, those of successive layers being aligned. Chamber roofs are perforate but the walls are thick and imperforate. Paleocene to Holocene (Plate 6.26, fig. 8).
- *Wilfordia* Adams, 1965 (Type species: *Wilfordia sarawakensis* Adams, 1965). This form has no true initial spire and a complex embryont with a short nepionic spiral, relatively weak 'pseudo-pillars' form in the walls of the lateral chambers, with no massive thickenings, and its chambers are clearly rectangular in section. It differs from *Sphaerogypsina* in having few spines and many pseudopillars (radial thickenings of chamberlet walls), and from *Schlumbergerella* (see Chapter 7) by having finer spines and fewer chamberlets in each tier and a less complicated embryonic apparatus. Loeblich and Tappan (1988) placed this form in the Acervulinidae because of the apparent absence of a canal system, with the communication based on wall perforation. Late Eocene (Plate 6.9, figs 12-13; Plate 6.12, fig. 7).

Family Homotrematidae Cushman, 1927

The test is attached with a trochospiral early stage, later chambers grow in a massive branching structure. Eocene to Holocene.

- *Sporadotrema* Hickson, 1911 (Type species: *Polytrema cylindricum* Carter, 1880). The test occurs with a planispiral early stage, but later chambers spiral upwards around a core of irregular tubes that open distally on the branches. Early upright spiralling chambers have a terminal aperture that remains a foramen to subsequent chambers, and develop subsequently into a complex network of stolons that in turn open to the exterior at the ends of the chambers. Multiple, fine pores at the inner surface fuse to form a coarse perforation in the outer wall. This form differs from *Victoriella* (see above) in having less inflated chambers and no pillars or pustules. Eocene to Holocene (See Chapter 7).

Superfamily ASTERIGERINOIDEA d'Orbigny, 1839

The test is trochospiral to planispiral, with a closed umbilicus. Chambers occur with internal partitions. Supplementary chamberlets develop around the umbilicus. The aperture is umbilical, and may extend up the apertural face. Late Cretaceous (Santonian) to Holocene.

Family Amphisteginidae Cushman, 1927
Chambers are numerous, with interseptal pillars. The aperture is a narrow slit. Eocene to Holocene.

- *Amphistegina* d'Orbigny, 1826 (Type species: *Amphistegina quoyii* d'Orbigny, 1826). The test is trochospiral, asymmetrically lenticular, involute with an angular, carinate periphery and lobed sutures. Chambers are strongly curved back at the periphery. Eocene to Holocene (Plate 6.18, fig. 7; Plate 6.26, fig. 10).

Family Boreloididae Reiss, 1963
Tests occur with only the later chambers divided into chamberlets. The protoconch is bilocular and is followed by an early trochospiral stage, the later stages are planispiral and involute. Late Paleocene to Middle Eocene.

- *Boreloides* Cole and Bermudez, 1947 (Type species: *Boreloides cubensis* Cole and Bermudez, 1947). The test is subspherical with a trochospiral early stage and a bilocular embryont. The later stage is annular and planispiral. The spiral wall is thick and pitted. Middle to Late Eocene.
- *Eoconuloides* Cole and Bermudez, 1944 (Type species: *Eoconuloides wellsi* Cole and Bermudez, 1944). The test is lenticular with a rectangular axial section, and involute with a bilocular embryonic stage, the final chambers are subdivided into chamberlets on the umbilical side, pillars are present over the spiral side with a thick wall. The bases of the ventral septa have multiple stolons. This genus is distinguished from *Amphistegina* (see Hanzawa, 1957, pp. 60-61, pl.6, fig. 11; Vaughan and Cole 1941, p.77, pl.45, fig.3) in having counter-septa. A counter-septum is, according to Hottinger (2006), "a kind of lower lip of an interiomarginal-basal aperture appearing in appropriate sections as a forward directed hook below the foramen and glued to the previous shell whorl". Late Paleocene to Eocene (see Chapter 6).

Family Lepidocyclinidae Scheffen, 1932
The test is discoidal, involute, and biconvex with a broad centrum, which grades into a narrow flange. Adauxiliary chambers may be present. The primary spire persists into the equatorial layer, or with annular rings of chamberlets that follow the embryont immediately. Stacks of "lateral chamberlets" (cubiculae) occur on each side of the median chamberlets. Pillars may be present between adjacent vertical stacks of cubiculae or scattered in the central region. The chamber walls are perforated by stolons, but there is no canal system. Middle Eocene to Late Miocene (Early Pliocene?).

Subfamily Helicolepidininae Tan, 1936
Members of this subfamily have tests in which the spiral arrangements completely surround the bilocular embryo, which is surrounded by a thickened wall, and is lacking adauxiliary chambers. Middle Eocene to Middle Miocene (Serravallian).

- *Eulinderina* Barker and Grimsdale, 1936 (Type species: *Planorbulina* (*Planorbulinella*) *guayabalensis* Nuttall, 1930). The test is lenticular, with an eoconuloid early stage, followed by a trochoid coil with a thick wall and many rows of arcuate median chambers that are connected by stolons. Counter-septa, as in *Eoconuloides* (see above), are present. Middle Eocene (Plate 6.26, figs 11-12; Plate 6.29, figs 2-4).

- *Helicolepidina* Tobler, 1922 (Type species: *Lepidocyclina* (*Helicolepidina*) *spiralis* Tobler, 1922). The test is lenticular, with a small eoconuloid early stage followed by a loose planispiral coil (the helicolepidine string) and a series of large imbricate chambers outside the helicolepidine string. The median layer has small arcuate chambers connected by single or double apertures. Lateral chambers are well developed. Middle to Late Eocene (Plate 6.29, fig. 5).

- *Helicostegina* Barker and Grimsdale, 1936 (Type species: *Helicostegina dimorpha* Barker and Grimsdale, 1936). The test is lenticular, with an eoconuloid early stage which constitutes the larger part of the test. The chambers in the last stage are subdivided into subsidiary chamberlets which consist of two or three rows of arcuate chamberlets growing around the last eoconuloid whorl. The counter-septa of *Eoconuloides*, which forms hooks below the foramen, is said to develop into "complete septal walls" in *Helicostegina* (see Hottinger, 2006). *Helicostegina* differs from *Eulinderina* in the increase of the size and number of whorls of the eoconuloid stage and in the decrease of the number of rows in the median chambers. Eocene (Plate 6.29, figs 6-11).

- *Helicosteginopsis* Caudri, 1975 (Type species: *Helicostegina soldadensis* Grimsdale, 1941). The test is lenticular, with an early stage eoconuloid and later numerous arcuate chamberlets, forming two or three rows just beneath the outer wall of the whorls. No lateral chamberlets are developed, nor are there any counter-septa. Late Eocene (Plate 6.29, figs 14-16).

- *Polylepidina* Vaughan, 1924 (Type species: *Lepidocyclina* (*Polylepidina*) *chiapasensis* Vaughan, 1924). The embryo consists of a protoconch and deuteroconch surrounded by a thick wall. The equatorial layer is formed by chambers arranged in two or more embryonic spires, followed by a cyclical phase. All chambers have one basal aperture from which the next chamber is formed. Later chambers have a second, or retrovert aperture. All chambers with two apertures then produce two new chambers, which eventually gives rise to cyclical growth. The equatorial chambers are arcuate with only radial stolons. Late Middle Eocene (Plate 6.23, fig. 3).

Subfamily Lepidocyclininae Scheffen, 1932

Representatives of this subfamily have a bilocular or multilocular embryonal stage, surrounded by a thickened wall and adauxiliary chambers. Microspheric tests have an early planispiral coil, while megalospheric tests have a globular protoconch, enclosed or followed by a larger reniform deuteroconch. Post-embryonic chambers evolve from cyclical, arcuate to hexagonal in shape, usually with two or more apertures. The lateral chambers are well differentiated from the equatorial layer and in the advanced forms they are arranged in tiers on either side of the equatorial layer. Surface ornaments and development of pillars seem to be of specific importance. Middle Eocene to Late Miocene (Early Pliocene?), see BouDagher-Fadel and Price (2010a).

- *Astrolepidina* Loeblich and Tappan, 1988 (Type species: *Lepidocyclina asterodisca* Nuttall, 1932). Stellate in outline, with four broad arms. The protoconch is almost equal to the deuteroconch and separated by a straight wall. Equatorial layers have ogival early chambers, that later become hexagonal. The equatorial layer increases in thickness towards periphery. Oligocene.
- *Eulepidina* Douvillé, 1911 (Type species: *Orbitoides dilatata* Michelotti 1861). The test is discoidal and biconvex, and can be extremely large with very broad lateral flanges. In the megalospheric form the small protoconch is completely enclosed by the larger deuteroconch, both are surrounded by a thick wall with many stolons connecting the embryont to the first series of auxiliary chamberlets. It has two auxiliary chambers and numerous adauxiliary chambers. In the microspheric form, the protoconch and deuteroconch are very small. The median layer is very thick and with multiple stolons. Oligocene (Rupelian, P18 in America, P19 in Tethys) to Miocene (middle Burdigalian) (Fig. 6.18; Plate 6.16, figs 7-8; Plate 6.18, fig. 7; Plate 6.25, figs 8, 10; Plate 6.26, figs 1, 10B; Plate 6.29, figs 17-19).
- *Lepidocyclina* sp. sensu lato Gümbel, 1870 emend. BouDagher-Fadel and Banner, 1997. The nomenclatural revision of *Lepidocyclina* by BouDagher-Fadel and Banner (1997) makes the genus group name *Lepidocyclina* sensu lato available for the generic naming of microspheric forms. Megalospheric forms, however, are divisible into the typical Paleogene *Lepidocyclina (Lepidocyclina)* and *L. (Nephroleidina)* which is essentially Miocene in the Far East. The test is microspheric, biconvex with a peripheral flange. In vertical section, there are numerous columns of "lateral chambers" (cubiculae) with cubicular lumina surrounding a thin median layer. In equatorial or oblique sections, the cubiculae are arcuate, hexagonal or polygonal, each with two apertures. The roof and floors of the chamberlets are perforate. It possesses a complete periembryonic ring of primary auxiliary and interauxiliary chambers. Middle Eocene (P10) to Early Miocene (N7) in America; Early/Late Oligocene to late Miocene (Early Pliocene?) in Tethys (see BouDagher-Fadel and Price, 2010a). (Plate 6.15, fig. 6).
- *Lepidocyclina (Lepidocyclina)* Gümbel, 1870 emend. BouDagher-Fadel and Banner, 1997 (Type species: *Nummulites mantelli* Morton, 1833). The test is "isolepidine", having a protoconch and deuteroconch of nearly equal size, separated by a straight wall. It has two auxiliary chambers and no adauxiliary chambers. The periembryonic chambers are arranged in four spirals. Butterlin (1981) reintroduced the subgenus *Neolepidina (Bronnimann)*, to forms with larger protoconch than deuteroconch as *Neolepidina*. However, this genus is not valid as the Late Eocene species assigned by Butterlin to *Neolepidina* do not always have a larger protoconch than deuteroconch. Middle Eocene (Lutetian, P10) to Early Miocene (Burdigalian, N7) in America, Oligocene in Tethys (P18–P22) (Plate 6.16, fig. 1; Plate 6.25, figs 1-6; Fig. 6.44).
- *Lepidocyclina (Nephrolepidina)* Douvillé, 1911 (Type species: *Nummulites marginata* Michelotti, 1841). The megalospheric test is biconvex, with an embryonic apparatus consisting of a small protoconch followed by a much larger, reniform deuteroconch. The latter forms have quadrate proloculi in the later stages of many lineages (see Chapter 7). The two chambers are separated by a thin imperforate wall with

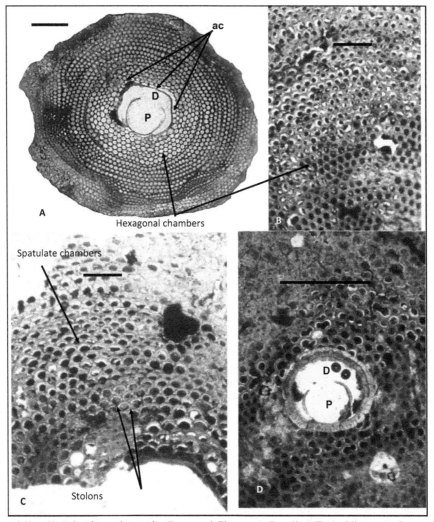

Figure 6.18. A) *Eulepidina ephippioides* (Jones and Chapman), Rupelian (Early Oligocene), Greece, SFN coll, 2001z0150/0011, showing the completely enclosed protoconch (P) within the deuteroconch (D) and the numerous small adauxiliary chambers (ac); B-D) *Eulepidina dilatata* (Michelotti), with typical embryonal development with thick-walled D almost fully embracing the thin-walled P, Chattian (Late Oligocene), France, UCL coll. Scale bars = Figs A, D = 1mm; B-C = 0.5mm.

a central foramen, and surrounded by a common thick tabulated wall. The equatorial layer of chamberlets has a basal stolon, that is arcuate in older species, but pointed or hexagonal in younger forms. Lateral chamberlets form on each side of the median layer. Middle Eocene (Lutetian, P10) to Early Miocene (Burdigalian, N7 in America), Late Oligocene in the Mediterranean (early Chattian, P21b) and Indo-Pacific (late Chattian, P22) to Late Miocene (Early Pliocene?) in the Tethyan province. (Plate 6.16, fig. 4; Plate 6.25, fig. 9; Plate 6.29, figs. 12–13; Fig. 6.45, see Chapter 7).

- *Pseudolepidina* Barker and Grimsdale, 1937 (Type species: *Pseudolepidina trimera* Barker and Grimsdale, 1937). A lenticular test having a protoconch and deuteroconch that are almost equal in size. The median layer consists of irregularly arcuate chambers, doubled beyond the protoconch and communicating with lateral chamberlets by stolons. Middle Eocene (Plate 6.16, Fig. 2).

Superfamily ORBITOIDOIDEA Schwager, 1876

The test is discoidal to lenticular with prominent dimorphism, as in most orbitoidal species both megalospheric and microspheric generations are found. Microspheric specimens have a distinctly small protoconch (usually about 20 microns), while megalospheric forms have a distinctive embryonic stage, enclosed in a thicker wall. Equatorial and lateral chambers may be differentiated or indistinguishable. Late Cretaceous (Santonian) to Oligocene.

Family Lepidorbitoididae Vaughan, 1933

The Lepidorbitoididae differ from the Orbitoididae (see Chapter 5) by the different character of the embryonic apparatus and by the form of the median chambers. An embryonic stage with two chambers is followed by hexagonal or arcuate equatorial chamberlets and by differentiated lateral chambers. Late Cretaceous (Santonian) to Oligocene.

- *Actinosiphon* Vaughan, 1929 (Type species: *Actinosiphon semmesi* Vaughan, 1929). The test is lenticular with well-developed polygonal to hexagonal equatorial and lateral chambers. The embryont consists of a large subspherical protoconch and a smaller reniform deuteroconch, and is followed by a spire of about eleven chambers. Equatorial chambers of the same cycle communicate via median stolons, while lateral chambers communicate through pores. Pillars are present forming surface papillae. Late Paleocene (Plate 6.17, fig. 10).
- *Neosivasella* Meriç and Çoruh, 1998 (Type species: *Neosivasella sungurlui* Meriç and Çoruh, 1998). The test is conical. In the megalospheric embryo the protoconch is partly enveloped by the deuteroconch. Equatorial chambers are arcuate in shape, interconnected with stolons. Late Paleocene.
- *Orbitosiphon* Rao, 1940 (Type species: *Lepidocyclina (polylepidnal) punjabensis* Davies, 1937). The arrangement of embryonic and periembryonic chambers is similar to the Cretaceous genus *Orbitoides* (see Chapter 5) in having epiauxiliary chambers, but it lacks the thick embryonic wall. Paleocene (Plate 6.17, fig. 9).
- *Sirelella* Özgen-Erdem, 2002 (Type species: *Sirelella safranboluensis* Özgen-Erdem, 2002). A trochospiral early orbitoidal test having a primitive stolon system (van Gorsel, 1978, pl.6, fig. 5a) and an umbilical plug without vertical canals. The early orbitoidal stage is similar to *Orbitokathina* (see Chapter 5) and *Neosivasella*; in *Orbitokathina*, the umbilical plug is pierced by numerous vertical canals, while *Neosivasella* possesses lateral chambers. Middle Eocene (Lutetian).

Family Orduellinidae Sirel, 1999

The test is free, dimorphic, spherical with sub-rectangular chambers arranged in multiple spirals in the early part of the test. Later chambers are arcuate and connected by stolons. They are added in concentric series in an orbitoidal manner. The aperture is a single, simple, basal slit in the early stage. Paleocene.

• *Orduella* Sirel, 1999 (Type species: *Orduella sphaerica* Sirel, 1999). The test is globular, with a large megalospheric proloculus enclosed by a thick wall, followed by sub-rectangular large chambers. The microspheric tests have a small proloculus with numerous small chambers. Late Paleocene (Thanetian).

Superfamily ROTALIOIDEA Ehrenberg, 1839

The test is involute to evolute, initially trochospiral or planispiral, commonly with many chambers in numerous whorls. As new chambers are added septal flaps attach to the previous apertural face and enclose radial canals, fissures, umbilical cavities, and intraseptal and subsutural canals. The wall is made of perforate hyaline calcite, and is generally optically radial in structure. Primary apertures occur singly or as multiples. Small opening into the canal system may occur along the sutures. Late Cretaceous (Coniacian) to Holocene.

Family Rotaliidae Ehrenberg, 1839

The test is built of radially-fibrous calcite and deposited in successive laminae. It forms a trochospiral with an evolute spiral side and an involute umbilical side. The umbilicus is filled with plugs, and throughout it has radial canals or fissures and intraseptal and subsutural canals. The aperture is umbilical, basal, and single to multiple. Late Cretaceous (Maastrichtian) to Miocene.

Subfamily Laffitteininae Hottinger, 2013

The test is trochospiral, involute to planispiral, with a enveloping canal system and vertical fissures on both sides. Umbilical and spiral sides not differentiated in structure. Umbos on both sides are pierced by numerous funnels. The aperture is a narrow slit extending obliquely over apertural face. Late Cretaceous to Miocene.

• *Cuvillierina* Debourle, 1955 (Type species: *Cuvillierina eocenica* Debourle, 1955 = *Laffteina vallensis* Ruiz de Gaona, 1948). The lenticular involute test is auriculate in outline with an angular periphery. The surface shows reticulate ornamentation and pillars. The enveloping canal system is derived from heaving feathering of the septal sutures. Vertical canals are present in the umbonal region. Bilamellar with intraseptal space widening towards the periphery. The periphery is sharp but not keeled. The aperture is an areal, comma-shaped slit. Early Eocene (Ypresian) (Fig. 6.19; Plate 6.28, figs 20-21).

• *Laffitteina* Marie, 1946 (Type species: *Laffitteina bibensis* Marie, 1946). The test is trochospiral, almost planispiral, involute, and covered on both sides with an enveloping canal system. The periphery is rounded. A single areal aperture/foramen forms a slit in the septum. Late Cretaceous to Early Paleocene (Maastrichtian to Danian) (Fig. 6.19; Plate 6.25, figs 11- 12; Plate 6.28, figs 1- 2).

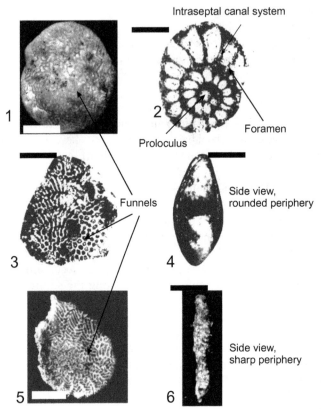

Figure 6.19. 1-4) *Laffiteina vanbellini* Grimsdale, 1952. Early Eocene, Mushoral Well 1, Iran, NHM P40691, 40692-4; 5-6) *Cuvillierina vallensis* (Ruiz). Early Eocene, Punta Iruarriaundiete, Guipuzcoa, Spain, NHM P4640. Scale bars = 0.5mm.

Subfamily Daviesininae Hottinger, 2013
This subfamily is characterized by trochospiral, bilamellar-perforate, heavily ornate tests. Middle Paleocene to Late Oligocene.

- *Daviesina* Smout, 1954 (Type species: *Daviesina khatiyahi* Smout, 1954). The test is flattened, trochospiral and large, with slightly unequal alar prolongations. The walls are thick, perforated and ornamented with thick pillars. Septa are secondarily doubled with no marginal cord. The umbilicus has a flap or plate. Prominent umbilical pillars, fissures and complex intra-septa canals are distributed unequally on both sides of the test. Middle Paleocene to Late Oligocene (Plate 6.6, fig. 10; Plate 6.27, figs 13-14; Plate 6.28, fig. 26).

Subfamily Rotaliinae Ehrenberg, 1839
The test is trochospiral with an umbilical plug formed by superposed folia fused at their tips, or by a single undivided secondary deposits or piles. Radial canals, intraseptal and subsutural canals are present. Late Cretaceous (Coniacian) to Early Miocene.

- *Medocia* Parvati, 1971 (Type species: *Medocia blayensis* Parvati, 1971). The test is lenticular and trochospiral; septa are doubled with septal passages. The spiral chambers are dorsally evolute and ventrally involute. Secondary deposits fused to a compact mass covering the umbilical area and are pierced by few tubular large funnels that are not always parallel, but connect earlier parts of the spiral side to the surface of the umbilical side. The spiral side has a thick lamellar wall. Middle Eocene (Lutetian).
- *Rotorbinella* Bandy, 1944 (Type species: *Rotorbinella colliculus* Bandy, 1944). A rotaliid with a single undivided umbilical pile with short mostly free folia (see also Revets, 2001; Hottinger, 2013). Late Cretaceous to Early Paleocene.
- *Rotalia* Lamarck, 1804 (Type species: *Rotalites trochidiformis* Lamarck, 1804). The spiral side is evolute and smooth, while the umbilicus side is filled with a columellar structure produced by the fusing of the tips of the folia. Eocene (Plate 6.30, fig. 7).

Subfamily Redmondininae Hottinger, 2013
This subfamily is characterized by coarsely perforated walls with a canal system that tends to extend onto the spiral side and a reduced umbilical filling.

- *Redmondina* Hasson, 1985 (Type species: *Redmondina henningtoni* Hasson, 1985). The spiral side is evolute and inflated. The folia are short, free or fused at the tips to form a ring of imperforate umbilical piles. Paleocene to Middle Eocene (Bartonian).

Subfamily Kathininae Hottinger, 2013
The test is lenticular or conical with chambers arranged in single or multiple spiral. The folia are small with fused folia forming a solid mass perforated by numerous funnels. Late Paleocene to Early Eocene (Thanetian to earliest Ypresian).

- *Dictyokathina* Smout, 1954 (Type species: *Dictyokathina simplex* Smout, 1954). The test has a multiple spire, as in *Dictyoconoides*. The umbilical side is covered by numerous pillars. Vertical radial canals penetrate the umbilical region and extend from the umbilical apertures of the chambers to the external pores. Paleocene to earliest Eocene (Plate 6.27, fig. 21).
- *Kathina* Smout, 1954 (Type species: *Kathina delseota* Smout, 1954). The test is lenticular in shape with a sharp un-keeled periphery. The spiral side is smooth. The ventral side is filled by a solid umbilical mass pierced by numerous slits or parallel funnels. Middle Paleocene to earliest Eocene (Selandian to early Ypresian) (Plate 6.28, figs 24-25).

Subfamily Lockhartiinae Hottinger, 2013
This subfamily is characterized by a heavily ornamented spiral side and a complex umbilical structure, where the umbilical cavities are delimited by successive foliar walls and numerous parallel umbilical piles (Late Paleocene to Middle Eocene) (Fig. 6.20).

- *Dictyoconoides* Nuttall, 1925 (Type species: *Conulites cooki* Carter, 1861). The test is conical with multiple intercalated spires of small rectangular chamberlets. The

434   *Evolution and Geological Significance of Larger Benthic Foraminifera*

umbilical side is filled by pillars separated by spaces and cavities of equal sizes. Intraseptal and subsutural canal systems are present. Middle Eocene (Plate 6.28, figs 7-8).

• *Lockhartia* Davies, 1932 (Type species: *Dictyoconoides haimei* Davies, 1927). The test is conical to lenticular with a simple spire of numerous chambers, similar to *Dictyoconoides* but lacking the intercalated spires. The dorsal side is ornamented with nodes, and the umbilicus is filled with numerous pillars with numerous cavities communicating with the chambers. Paleocene to Middle Eocene (Plate 6.28, figs 9-15).

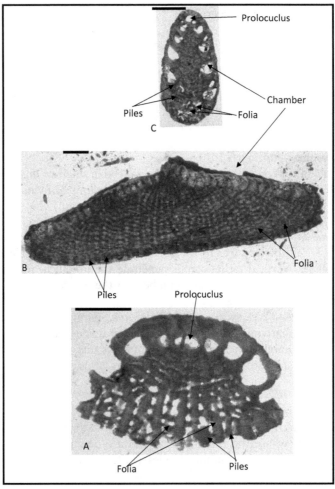

Figure 6.20. A) *Lockhartia haimei* (Davies), Paleocene, Qatar, Smout coll., NHM P40156; B) *Dictyoconoides kohaticus* Davies, Early Eocene, Kohat Shales, India, NHM P22632; C) *Sakesaria dukhani* Smout, paratype, Paleocene, Qatar, NHM P40203. Scale bars = 0.5mm.

- *Rotaliconus* Hottinger, 2007 (Type species: *Rotaliconus persicus* Hottinger, 2007). The test is trochospiral with a coarsely perforate, evolute and strongly convex dorsal side. The ventral side is involute and flattened, with a smooth or slightly pustulose, weakly perforate umbilical face. The umbilicus is covered by umbilical plates and a canal system is absent. There is a single interiomarginal aperture. Late Middle Eocene.
- *Sakesaria* Davies, 1937 (Type species: *Sakesaria cotteri* Davies, 1937). The test is elongate, with a very high trochospiral. The umbilicus is filled with pillars; cavities communicate between the chambers as in *Dictyoconoides*. Paleocene to Early Eocene (Plate 6.28, figs 1-2).

Subfamily Pararotaliinae Reiss, 1963
The test is trochospiral, with an enveloping canal system but with umbilical cavities. Late Cretaceous (Coniacian) to Holocene.

- *Camagueyia* Cole and Bermudez, 1944 (Type species: *Camagueyia perplexa* Cole and Bermudez, 1944). The test is conical with an elevated spiral side and a flattened umbilical side. Thick walls result in reduced chamber lumen. Few, but massive pillars fill the umbilical region. Middle Eocene.
- *Neorotalia* Bermüdez, 1952 (Type species: *Rotalia mexicana* Nuttall, 1928). The test is low trochospiral, with a simple umbilical boss and pillared walls, both ventrally and dorsally. The aperture is single, areal, with no complicated umbilical canal system. Early Oligocene to Early Miocene (Rupelian to Burdigalian) (Plate 6.30, fig. 17; see Chapter 7).

Family Miogypsinidae Vaughan, 1929
The test is flattened to biconvex. The microspheric form has a trochospiral or planispiral early spire, while the megalospheric form has a bilocular embryonal stage followed by a fan of median chamberlets. Middle Oligocene to Middle Miocene.
  Several related genera can be identified, thus:
  (A) Uniserial coil around the megalospheric proloculus.
  1. Initial coil around the megalospheric proloculus:
  (a) Fan of only one or two additional chambers attached to the spire in the median equatorial plane:

- *Americogypsina* BouDagher-Fadel and Price, 2010 (Type species: *Americogypsina braziliana* BouDagher-Fadel and Price, 2010b). The test is moderately small, trochospiral with a tendency to become uncoiled and biconvex in axial section with a thick granulated wall. In equatorial section, the chambers increase rapidly in size but their internal and external periphery are covered by small incomplete chamberlets. The periphery of the test is covered with pronounced fissures and granulations. Pillars cover the umbilical and spiral parts of the test. *Americogypsina* differs from *Paleomiogypsina* in having one row of small incomplete and irregular chamberlets on the periphery of the first whorl and two to three rows of mainly incomplete thick-walled chamberlets on the periphery of the last whorls. *Americogypsina* is the ancestral form of the American *Miolepidocyclina*, while *Paleomiogypsina* is the ancestor form of the typical *Miogypsina*. Middle Oligocene (see Fig. 6.21, Plate 6.30, figs 4-5, 18).

- *Paleomiogypsina* Matsumaru, 1996 (Type species: *Paleomiogypsina boninensis* Matsumaru, 1996). There is a fan of only one or two additional chambers attached to the spire in the median equatorial plane. *Paleomiogypsina* differs from *Neorotalia* (see above) in being low trochospirally coiled and in having small chamberlets scattered on the periphery of the last whorl. This the evolutionary beginning of the fan of chamberlets which was to develop in *Miogypsinella.* Early Late Oligocene (Plate 6.30, fig. 6).

(b) Fan extends, producing a broad equatorial layer of ogival chambers:

- *Miogypsinella* Hanzawa, 1940 (Type species: *Miogypsinella borodinensis* Hanzawa 1940). *Miogypsinella* differs from *Paleomiogypsina* in having a fan of equatorial chamberlets. Loeblich and Tappan (1988) considered *Miogypsinella* to be a synonym of *Miogypsinoides* (see below), but the two taxa are easily distinguished and they have very different stratigraphical ranges. *Miogypsinella* differs from *Miogypsinoides* in having a weak trochospiral initial coil, and the lateral walls of the initial spire and the succeeding fan of ogival median chamberlets are much thinner. Late Oligocene to Early Miocene (Plate 6.30, figs 3, 8-10; Plate 7.14, fig. 3).
- *Boninella* Matsumaru, 1996 (Type species: *Boninella boninensis* Matsumaru, 1996). Shows two spires of subquadrate chambers. Late Oligocene.

2. Embryont coils that are virtually planispiral, with only one whorl around the megalospheric proloculus, and a septal canal system that is weakly present:
(a) The lateral walls of the initial spire and the succeeding fan of ogival median chamberlets become very thick and solid:

- *Miogypsinoides* Yabe and Hanzawa, 1928 (Type species: *Miogypsinoides dehaarti* Van der Vlerk, 1924). The septa of all the chamberlets of *Miogypsinoides* also possess a clear intraseptal canal system. Oligocene (Rupelian, P19, to Chattian, P22, in America), Late Oligocene to Miocene (Chattian, P21b, to Burdigalian, N7, in the Mediterranean, P22 to Early Langhian, N8, in the Indo-Pacific province) (Plate 6.30, fig. 11; see Chapter 7).

(b) Lateral walls have cubiculae between the lamellae which begin to split apart:

- *Miogypsinodella* BouDagher-Fadel et al., 2000 (Type species: *Miogypsina* *(Miogypsina) primitiva* Tan Sin Hok, 1936). The embryont coil is similar to that of *Miogypsinoides* (see Chapter 6), it is virtually planispiral, but there is only one whorl around the megalospheric proloculus, and a septal canal system is present. However, the lateral walls have gaps between the lamellae, which begin to split apart and form the beginnings of lateral chamberlets. This splitting results in thick-walled irregular chamberlets, unlike the regularly formed, stacked chamberlets of *Miogypsina* (see Chapter 6). Late Oligocene (late Chattian) to Middle Miocene (Langhian) (see Chapter 7).

(B) Microspheric specimens which possess a uniserial coil around the proloculus, and a megalospheric embryont that has no coil around the proloculus, but two bidirectional coils around the proloculus, with lateral cubiculae regularly stacked on each side of the median layer, but lacking a canal system.

1. Embryont near the apex:

- *Miogypsina* Sacco, 1893 (Type species: *Nummulina globulina* Michelotti, 1841). Early species have megalospheric nepionts in which two series of chambers surround the deuteroconch unequally (e.g. *M. borneensis* Tan Sin Hok). In advanced forms (e.g. *M. indonensis* Tan Sin Hok) the series become equal, and both surround the megalospheric deuteroconch by means of equal half whorls. Microspheric specimens (e.g. the syntypic specimens of *M. borneensis* Tan Sin Hok) possess a uniserial coil around the proloculus, as in megalospheric *Miogypsinella,* the ancestral genus. Latest Early Oligocene (Rupelian, P21, in America), Late Oligocene (Chattian, P22, in Mediterranean), Early Miocene (Aquitanian, N4, in the Far East) to Early Miocene (Burdigalian, in America) and Middle Miocene (middle Serravallian, in the Indo-Pacific province) (Plate 6.30, figs 13, 15; see Chapter 7).

2. Embryont about midway between the centre of the test and the periphery:

- *Miogypsinita* Drooger, 1952 (Type species: *Miogypsina Mexicana* Nuttall, 1933). The embryont is peripheral in the microspheric test, but half way between the periphery and the centre of the megalospheric test. The embryont is divided into two unequal chambers with a straight septum and two unequal principal auxiliary chambers. Equatorial chambers have a diamond shape. Late Oligocene to Early Miocene.
- *Miolepidocyclina* Silvestri, 1907 (Type species: *Orbitoides* (*Lepidocyclina*) *burdigalensis* Gümbel, 1870). The embryonic apparatus, consisting of a large protoconch and deuteroconch, is surrounded by a thick wall. The megalospheric nepiont is similar to that of *Miogypsina,* with no coil around the proloculus but has two bidirectional coils around the proloculus. However, the nepiont is centrally placed, instead of being at the edge of the test, as in *Miogypsina.* Early Oligocene (Rupelian, P20) to Early Miocene (Burdigalian) (Plate 6.30, figs 12, 14, 16).

Family Chapmaninidae Thalman, 1938
The test is conical with a trochospiral initial part, followed by a uniserial part and a tubular apertural system. Septa are invaginated into tube pillars. Late Paleocene to Late Miocene (Tortonian).

- *Angotia* Cuvillier, 1963 (Type species: *Angotia aquitanica* Cuvillier, 1963). The embryonic apparatus is bilocular, and is followed by a series of chambers forming a high cone with a flattened base. Hollow pillars and tunnels, perpendicular to the outer margin, fill the centre of the test. Middle Eocene (late Lutetian) (Fig. 6.17).
- *Chapmanina* Silvestri, 1931 (Type species: *Chapmanina gassinensis* Silvestri, 1905). A bilocular embryonic apparatus is followed by a reduced trochospiral stage, followed

Oligocene					Early Miocene				Middle Miocene					Epoch
RUPELIAN			CHATTIAN		AQUITANIAN		BURDIG.		LANGHIAN			SERRAVALLIAN		Stage
P18	P19	P20	P21	P22	N4	N5	N6	N7	N8	N9	N10 N11	N12	N13	Planktonic Zones

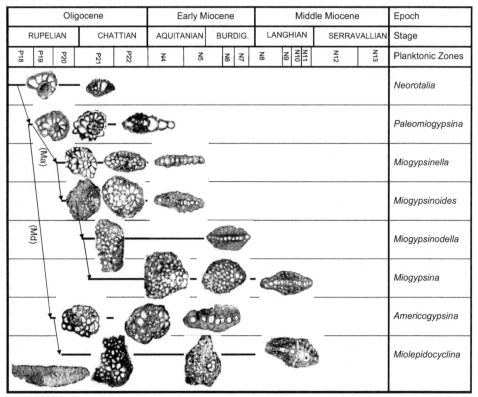

	Neorotalia
	Paleomiogypsina
	Miogypsinella
	Miogypsinoides
	Miogypsinodella
	Miogypsina
	Americogypsina
	Miolepidocyclina

Figure 6.21. Phylogenetic chart showing the evolutionary lineages of the American Miogypsinidae (from BouDagher-Fadel and Price, 2010b).

by rapidly enlarging discoidal chambers in an uniserial arrangement. Peripheral secondary septa are fissured at the umbilical surfaces. The central parts of the chambers have concentric rings of tubular pillars of identical size within a chamber. The aperture consists of multiple openings with internal tubes to the previous septum, resulting, as in *Angotia,* in the invagination of the septum. Middle Eocene to Late Miocene (Tortonian) (Fig. 6.17).

• *Crespinina* Wade, 1955 (Type species: *Crespinina kingscotensis* Wade, 1955). The microspheric test has an early planispiral coil, followed by embracing and annular chambers. The megalospheric embryo has a globular protoconch and a reniform deuteroconch, followed by a few annular undivided chambers, and later by rectilinear chambers forming a low conical test. The inside of the test is filled by unaligned hollow pillars. Late Eocene to Early Oligocene (Fig 6.17).

• *Ferayina* Frizzell, 1949 (Type species: *Ferayina coralliformis* Frizzell, 1949). The surface has saucer-like rectilinear chambers with additional ribs and costae. There is a flat imperforate apertural face, and an aperture consisting of multiple pores at the end of hollow pillars extending to the previous septum. Middle Eocene (Fig. 6.17).

- *Sherbornina* Chapman, 1922 (Type species: *Sherbornina atkinsoni* Chapman, 1922). The test is large, discoidal, with annular chambers in the adult. Lateral walls have corrugations near the sutures, which alternate from chamber to chamber. Septal and radial canals are present and terminate in coarse pores at the outer surface. Late Paleocene to Middle Miocene.

Family Calcarinidae Schwager, 1876
The test is enrolled with protruding spines. Late Cretaceous (Maastrictian) to Holocene.

- *Silvestriella* Hanzawa, 1952 (Type species: *Calcarina tetraëdra* Gümbel, 1868). The test is large, with three to four large radial spines, resulting in a tetrahedral form. Spines arise from the early whorl of the chambers and widen rapidly. Spine canals arise from the interseptal spaces of the early chambers. The spine surface has numerous large pores. Solid pillars may be present between the outermost chambers. Middle Eocene (Lutetian to Bartonian) (Plate 6.15, fig. 10).
- *Meghalayana* Matsumaru and Sarma, 2008 (Type species: *Meghalayana indica* Matsumaru and Sarma, 2008). The test is ovoidal to ellipsoidal with four spines. It resembles the Cretaceous *Siderolites* (see Chapter 5), but is different from the latter in having the raspberry-like arrangement of the megalospheric form in early stage. Late Eocene (Plate 6.11, fig. 7).

Family Elphidiidae Galloway, 1933
The test is planispiral to trochospiral and uncoiled with sutural pores, canals and supplementary apertures. The aperture is single or multiple. Paleocene to Holocene.

- *Elphidium* de Montfort, 1808 (Type species: *Nautilus macellus* var. 3 Fichtel and Moll, 1798). The test is lenticular, planispiral and involute or partially evolute with deeply incised sutures. The umbilical plug has vertical canals communicating with the spiral canals. *Cellanthus* de Montfort, 1808 is similar to *Elphidium*, but with a fully developed septal flap. Eocene to Holocene (Plate 6.29, fig. 1).
- *Pellatispirella* Hanzawa, 1937 (Type species: *Camerina matleyi* Vaughan, 1929). The test is lenticular, planispiral and involute with an umbilicus perforated by canals. Septa are folded at the base and have numerous transverse canals. Middle Eocene (Plate 6.16, figs 5-6).

6.3 Biostratigraphy and Phylogenetic Evolution

Larger benthic foraminifera are widely distributed in Paleogene carbonates (Fig. 6.22). They evolved gradually, forming a succession of biometrically large populations within important phylogenetic lineages, and have been essential in the development of Tethyan carbonate stratigraphy. Their systematics and biostratigraphy have been extensively studied and a number of zonations based on their occurrence have been erected

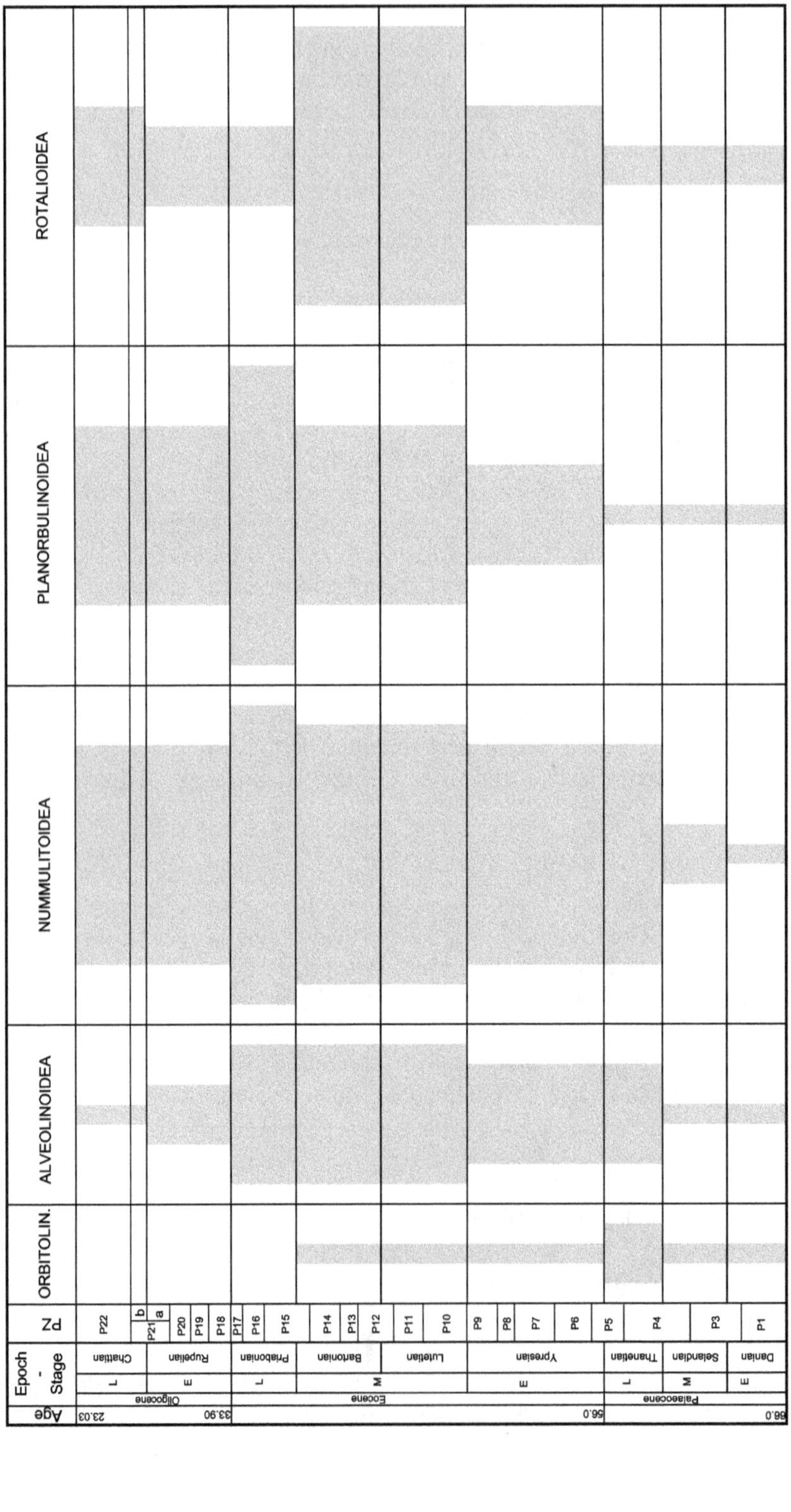

Figure 6.22. The biostratigraphic range and diversity of the main superfamilies (as shown by the horizontal scale of the spindles) found in Tethys during the Paleogene.

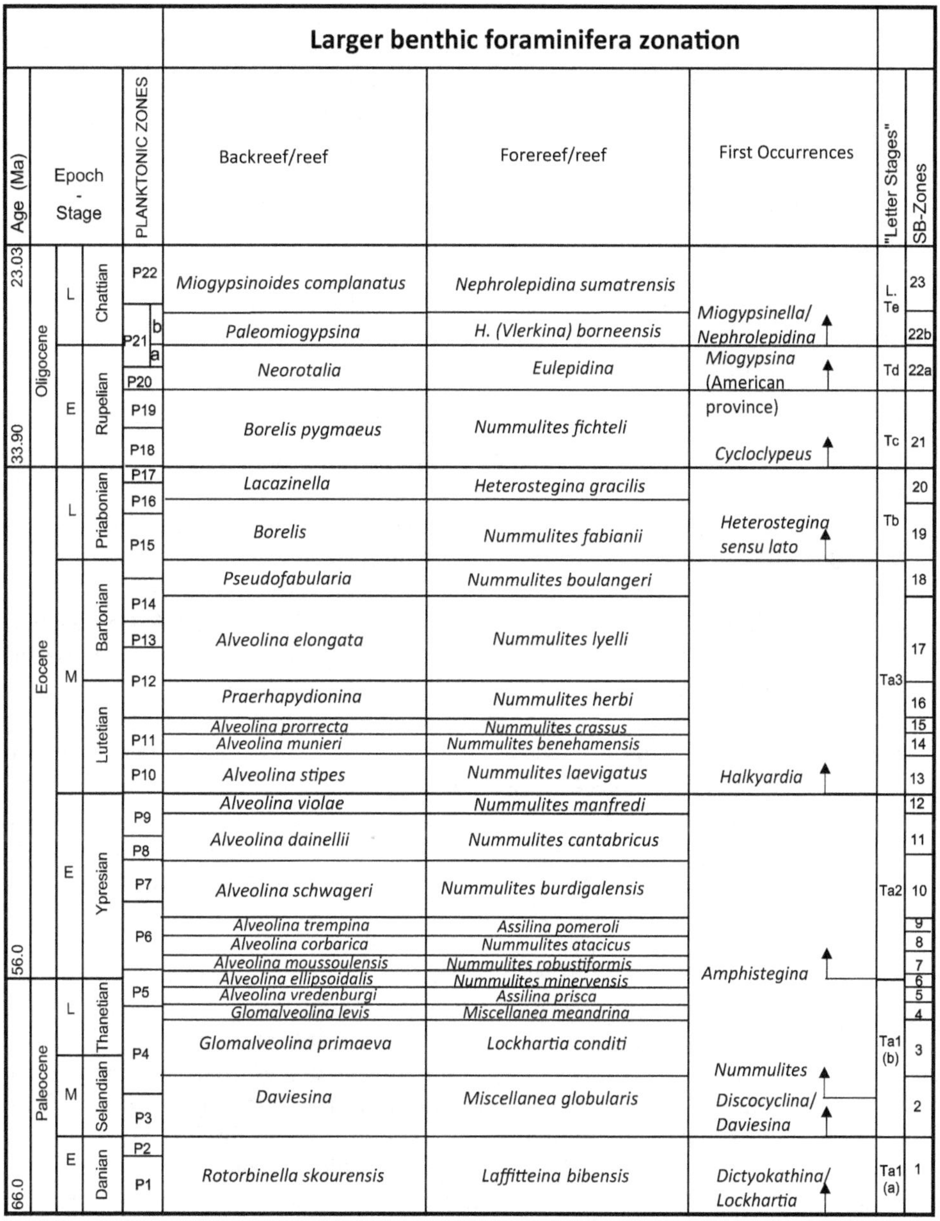

Age (Ma)	Epoch - Stage		PLANKTONIC ZONES	Larger benthic foraminifera zonation			"Letter Stages"	SB-Zones
				Backreef/reef	Forereef/reef	First Occurrences		
23.03	Oligocene	L Chattian	P22	*Miogypsinoides complanatus*	*Nephrolepidina sumatrensis*		L. Te	23
			P21 b	*Paleomiogypsina*	*H. (Vlerkina) borneensis*	*Miogypsinella/ Nephrolepidina*		22b
		E Rupelian	a P20	*Neorotalia*	*Eulepidina*	*Miogypsina (American province)*	Td	22a
33.90			P19	*Borelis pygmaeus*	*Nummulites fichteli*		Tc	21
			P18			*Cycloclypeus*		
	Eocene	L Priabonian	P17	*Lacazinella*	*Heterostegina gracilis*		Tb	20
			P16			*Heterostegina sensu lato*		
			P15	*Borelis*	*Nummulites fabianii*			19
		M Bartonian		*Pseudofabularia*	*Nummulites boulangeri*			18
			P14					
			P13	*Alveolina elongata*	*Nummulites lyelli*		Ta3	17
			P12	*Praerhapydionina*	*Nummulites herbi*			16
		Lutetian	P11	*Alveolina prorrecta*	*Nummulites crassus*			15
				Alveolina munieri	*Nummulites benehamensis*			14
			P10	*Alveolina stipes*	*Nummulites laevigatus*	*Halkyardia*		13
		E Ypresian	P9	*Alveolina violae*	*Nummulites manfredi*			12
			P8	*Alveolina dainellii*	*Nummulites cantabricus*			11
			P7	*Alveolina schwageri*	*Nummulites burdigalensis*		Ta2	10
			P6	*Alveolina trempina*	*Assilina pomeroli*			9
				Alveolina corbarica	*Nummulites atacicus*			8
				Alveolina moussoulensis	*Nummulites robustiformis*	*Amphistegina*		7
				Alveolina ellipsoidalis	*Nummulites minervensis*			6
56.0	Paleocene	L Thanetian	P5	*Alveolina vredenburgi*	*Assilina prisca*			5
				Glomalveolina levis	*Miscellanea meandrina*			4
			P4	*Glomalveolina primaeva*	*Lockhartia conditi*	*Nummulites*	Ta1 (b)	3
		M Selandian	P3	*Daviesina*	*Miscellanea globularis*	*Discocyclina/ Daviesina*		2
66.0		E Danian	P2				Ta1 (a)	
			P1	*Rotorbinella skourensis*	*Laffitteina bibensis*	*Dictyokathina/ Lockhartia*		1

Figure 6.23. Paleogene larger benthic foraminifera biozones, as refined in this study, with diagnostic first and last occurrences.

(BouDagher-Fadel and Price, 2016). A larger foraminifera zonation of the Paleocene and Eocene of the Tethyan realm was published by Serra-Kiel et al. (1998) as one of the results of the IGCP 286 *Early Paleogene Benthos* study. Cahuzac and Poignant (1997) proposed a similar larger foraminifera zonation for the Oligocene to Miocene of the west European basins. In the biostratigraphy of the Far East, the use of the "Letter Stages" is well established, although correlation with global standard bio- and chronostratigraphical scales have until recently proved difficult, but which are now integrated in Fig 6.23. These difficulties were due to the fact that the larger foraminiferal assemblages are facies controlled, and because of taxonomic confusion between the Far and Middle East assemblages, where taxonomic overlap (synonyms) is known to exist. The distribution and ranges of the Paleocene superfamilies and genera are plotted in Charts 6.1 and 6.2, and their correlation with the global Planktonic Zones (PZ) are given in Fig 6.23.

In the following sections, the biostratigraphic "Letter Stages" of Far East, and the biostratigraphic and phylogenetic evolution of the Paleogene textulariides, miliolides, and rotaliides are discussed.

6.3.1 The "Letter Stages" of SE Asia and provincial biostratigraphy

The "Letter Stages" subdivision of the Indo-Pacific Cenozoic (Leupold and van der Vlerk, 1931; Adams, 1970; Chapronière, 1984; BouDagher-Fadel and Banner, 1999; BouDagher-Fadel, 2008; Advocaat et al., 2014; Sun et al., 2015; Wang et al., 2015; An et al., 2015; BouDagher-Fadel et al., 2015; Hu et al., 2015; Li et al., 2016; BouDagher-Fadel et al., 2017) is based on larger foraminifera, some of which have also been used as range fossils in western hemisphere stratigraphy (Barker and Grimsdale, 1936; Caudri, 1996). In the Far East, where the Cenozoic sedimentary sequences are dominated by warm water, shallow marine carbonates, these deposits are largely biogenic in origin, primarily of benthic faunas and algae. The 19th and early 20th Century workers attempted to utilize new stratigraphic schemes involving benthic faunas to date and study these rocks. Martin (1880), studying the mollusc faunas of Java and surrounding areas was able to follow the work of Charles Lyell in Europe in utilizing the ratios of extant to extinct faunas to indicate relative ages. Two major contributions to Cenozoic stratigraphy came from the work of Martin. Firstly, he showed that the mollusc faunas of the East Indies developed separately from those of Europe, and secondly, as a consequence of this, the stage names of Europe could not be correlated with confidence to the Indonesian region. Martin studied a number of Javanese marine sections in detail, determined the ratios of extant to extinct molluscs, and predicted the stratigraphic order of these strata. Mapping, and studies on the other regionally significant fossil group, the larger foraminifera, supported the mollusc stratigraphy, but there was still little idea on how it fitted into the European stratigraphic column.

By the late 1920s, the larger foraminifera had become the preferred fossil group for biostratigraphy in the Indonesian area (Sharaf et al., 2014; Li et al., 2015; Gold et al., 2017a, b; Breitfeld et al., 2016; Advocaat et al., 2017; White et al., 2017). They had the advantage that they were more abundant than molluscs, and also a scheme was developed that utilised assemblage zones rather than percentages of extant forms. Using molluscs to identify and correlate sections required extensive knowledge of both living and fossil species. The larger foraminifera assemblage zones could be identified by the presence of a few key taxa, usually with a hand-lens in the field. They were proposed as alternatives to the European stages and they still are recognizable as fundamental stratigraphic units over the Indo-Pacific area.

Van der Vlerk and Umbgrove (1927) published the Letter Classification of the Indonesian "Tertiary", based on larger foraminifera. This scheme subdivided the "Tertiary" into seven parts. Six parts were labeled "a" to "f" (e.g. "Tertiary a", or "Ta" for short), and a seventh part characterized by non-orbitoidal "Tertiary" foraminifera between Tf and Tg was noted; Tg was considered Late Miocene. They equate to the Cheribonian and Sondien regional stages, which are roughly equivalent to the Early and Late Pliocene. The "Quaternary" or Pleistocene is represented in the mollusc scheme by the Bantamien regional stage, with 70 % or more of extant species. This period therefore has no equivalent in the Letter Stages (Leupold and van der Vlerk, 1931).

This scheme was immediately successful and was rapidly adopted as a workable standard biostratigraphic scheme for the SE Asian region, even if the correlation with the European epochs and stages was not adequately known. The authors pointed out that the taxonomy and detailed biostratigraphy of the genera used in this initial scheme was not adequately known, and suggested that future work would probably increase the number of recognizable subdivisions. The difference between Tf and Tg was originally based on a faunal turnover, with the common taxa *Lepidocyclina* and *Miogypsina*, and the less frequent forms *Austrotrillina* and *Flosculinella*, all disappearing from the carbonate facies, whereas no new forms appeared in "Tertiary g". The importance of this faunal turnover as a biostratigraphic event is underlined by other carbonate facies organisms showing a change at the same time as the non-*Lepidocyclina* larger foraminifera extinctions. In brief, shallow marine biohermal carbonates changed from mixed coral and coralline algal boundstones with larger foraminiferal grainstones to more dominant coral reefs with a marked increase in *Halimeda* green algae. The latter are often preserved in recognizable forms, or as an increase in micrite and fine bioclastic products from the early breakdown of aragonitic platelets. Larger foraminiferal grainstones are only rare after this event, as deeper photic *Cycloclypeus* facies or minor *Operculina/Amphistegina* or *Alveolinella* calcarenites. Most well and field sections encountering limestones of this later Miocene or Pliocene age in SE Asia do not sample rocks composed of larger foraminifera tests.

Although, the use of the "Letter Stages" is widespread, correlation with global standard bio- and chronostratigraphical scales were for a long time uncertain and problematic (Berggren, 1972; Blow, 1979; Adams, 1984). The difficulties are due to the larger foraminiferal assemblages being facies controlled, and so changes in relative sea level affecting the photic zone, wave base, clastic sediment supply, influence the larger foraminifera (and associated algal) assemblages.

However, the major planktonic foraminiferal bio-events for the Late Mesozoic and the Cenozoic were recently calibrated against the biostratigraphical time scale and the radio-isotopes by BouDagher-Fadel (2013). Those of the Cenozoic were compared with the earlier zonations of Berggren and Pearson (2005) and Wade et al. (2011). The biostratigraphic resolution of the alpha-numeric biozonations (Planktonic Zones, PZ) were determined by distinct evolutionary lineages of morpho-species, and were deduced through extensive industrial work, calibrated against Sr-isotopes and other biostratigraphic markers (see BouDagher-Fadel 2015). The resulting PZ stages, have been used to calibrate and correlate the provincial larger benthic zones, and hence link the "Letter Stages" to other bioprovinces.

In this work, the "Letter Stages" of the Paleogene are assigned as follows (see Fig. 6.23, and BouDagher-Fadel, 2015):

• The Paleocene [Ta 1] is divided into two parts:
o Ta 1(a) (corresponds to P1-P2 PZ, 66.0-61.6Ma), containing rare larger benthic foraminifera, is based on the appearance of the conical forms *Dictyokathina* and *Lockhartia* in the Western Tethys, and corresponds to Early Paleocene.
o Ta 1(b) (corresponds to P3-P5a PZ, 61.6-56.0Ma), Middle to the Late Paleocene, is based on the presence of *Fallotella* in Western Tethys and the global appearance of *Discocyclina,* and *Miscellanea globularis* and *Daviesina*. The upper part of Ta 1(b) corresponds to the appearance of *Nummulites, Alveolina* and the presence of *Ranikothalia* in Tethys. This zone is also equivalent to the appearance of the first orbitoids, *Actinosiphon* in the American province.

• Ta 2 (corresponds P5b-P9 PZ, 56.0-47.8Ma), Early Eocene, is based on the appearance of the cosmopolitan form *Amphistegina,* and the Tethyan *Orbitolites*. It is equivalent to the presence of the conical form, *Verseyella* in the American province.
• Ta 3 (corresponds to P10-P15a PZ, 47.8-35.1Ma), Middle Eocene, corresponding to the appearance of the broadly conical cosmopolitan form, *Halkyardia* and the Tethyan *Chapmanina* and *Somalina* in Tethys. The top of Ta 3 is defined on the appearance of *Pellatispira* in Tethys. This zone corresponds to the appearance of *L. (Lepidocyclina), L. (Nephrolepidina)* and *Lepidocyclina* sp. in the American province.
• Tb (corresponds to P15b-P17 PZ, 37.8-33.9Ma), Late Eocene, is based on the appearance of the cosmopolitan *Heterostegina sensu lato.*
• Tc–Td (Tc corresponds to P18-P19 PZ, 33.8-30.3Ma; Td corresponds P20-P21a PZ, 30.3-28.1 Ma): base of Tc, Early Oligocene, is based on the appearance of *Cycloclypeus* in Tethys and corresponds to the appearance of *Eulepidina* and *Astrolepidina* in the American province, and Td corresponds to the appearance of *Miolepidocyclina* and *Miogypsina* in the American province.
• Early Te (Te1–Te4) (corresponds to P21b – P22 PZ, 28.1-23.0Ma), Late Oligocene, is based on the appearance of *Miogypsinella* and *L. (Nephrolepidina)* in the Far East, *Miogypsina* in Europe and corresponds to the appearance of *Miogypsinita* in the American province. In Tethys, the boundary between Te1 and Te2 is based on the disappearance of *Paleomiogypsina* and the appearance of *Miogypsinoides complanatus. Amphisorus martini* marks the beginning of Te3, while Te4 is marked by the appearance of *Miogypsinoides formosensis* (see Chart 7.1).

Parallel to this zonation, new schemes for the Mediterranean region, based on the biogeographic zonation of shallow benthic larger foraminifera and their direct correlation with magnetostratigraphy, were defined by Serra-Kiel, et al. (1998). They presented a system of numbered units for the "Tertiary"; SBZ 1-23 for the Paleogene (see Fig. 6.23; Chart 6.2). These zonations were subsequently correlated (Less, 1998; Less et al., 2007) with the Orthophragminae (orbitoidal larger foraminifera) for the Late

Paleocene-Eocene of the Mediterranean; OZ 1a-16. These have all now been cross cor-related by the work described in BouDagher-Fadel (2015) and shown in Fig 6.23 and Chart 6.2.

Below are described the three main groups that dominated the larger foraminifera assemblages of the Paleogene, (see Charts 6.1 and 6.2), namely:

- the agglutinated textulariides (which evolved from small surviving Cretaceous forms),
- the miliolides (which became abundant in the Eocene, imitating in their evolution that seen in Cretaceous), and
- the rotaliides (mainly the orbitoids).

6.3.2 The Textulariides of the Paleogene

In the Paleogene, most of the stratigraphically important agglutinated foraminifera are included in three different superfamilies (see Figs 6.2, 6.3 and Charts 6.1 and 6.2), which can be divided into two morphological groups; the elongated forms with a trochospiral, biserial or triserial in the early stage, the Textularioidea, and the conical forms, which are homeomorphs of the Cretaceous orbitolinids, such as the Coskinolinoidea and Orbitolinoidea. The Textularioidea are mainly found in rocks from the Late Paleocene to Eocene formed in shallow water environments of Tethys, while the conical coskinolinids and orbitolinids are rare in the Middle Eocene, but absent in the Late Eocene shallow and extremely specialized facies of Tethys (Hottinger, 2007).

- The Textularioidea evolved in parallel lineages, most probably from different ances-tors. The cosmopolitan forms, *Pseudochrysalidina* and *Vacuovalvulina* followed the same trends as their ancestors, the chrysalinids of the Cretaceous, by developing interior pillars in a multiserial test. The Tethyan form, *Pfendericonus,* evolved from a simple trochospirally enrolled ataxophramid, *Arenobulimina,* by developing internal pillars (Fig. 6.2).
- The Coskinolinoidea in their turn evolved from the same ancestor, *Arenobulimina,* but followed different evolutionary trends, which involved reducing the arenobu-liminid spire with the latest part becoming uniserial in *Coleiconus* (Fig. 6.3), with chambers subdivided by scattered pillars, such as in the Tethyan *Coskinon,* or tubu-lar pillars as in the cosmopolitan *Pseudolituonella* (Fig. 6.2). In the Early to Middle Eocene, the cosmopolitan *Coskinolina* acquired, in addition to the pillars, vertical partitions.
- The Orbitolinoidea, which thrived in the Early to mid-Cretaceous of Tethys, survived the Cretaceous boundary event with a single genus, *Dictyoconus,* that give rise to many descendants. The initial trochospiral had disappeared completely in this super-family and the chambers are subdivided by vertical and horizontal partition with pillars filling the centre. The foraminiferal apertures are disposed (as in *Orbitolina,* see Chapter 5) in alternating positions from one chamber to the next one. Many of the Orbitolinoidea were short ranged and are very important biostratigraphically

in Tethys. Only *Verseyella* is unique to America. The peripheral tiered rectangular chamberlets of *Dictyoconus* seem to belong only to this genus. The rest of the Paleogene forms evolved with different kinds of partitions and different ways of scattering their pillars (see Figs 6.2, 6.3).

The larger benthic forms of the textulariids ranged throughout the Eocene, many disappearing at the Middle-Late Eocene boundary. Only *Dictyoconus* crosses the Eocene to Oligocene boundary, but then died out at the top of the Oligocene.

6.3.3 The Miliolides of the Paleogene

The stratigraphically important miliolides of the Paleogene are mainly divided into three superfamilies:

• the Alveolinoidea
• the Milioloidea
• the Soritoidea.

The Alveolinoidea of the Paleogene include two main larger benthic families, the Alveolinidae and the Fabulariidae (see Fig. 6.24).

The porcelaneous Alveolinidae evolved from a form with a simple milioline test, coiled in varying planes, with the axis of coiling normal to the apertural axis. The axis in the primitive forms is rotated so that an angle of 72° exists between the median planes of consecutive chambers in a quinqueloculine test, or 120° in a triloculine test or 180° in a spiroloculine or biloculine test. These forms evolved in turn to genera with streptospiral early coiling (e.g *Pseudonummuloculina*) (see Hottinger, Drobne and Caus 1989), but that were planispiral in the adult part and spherical to fusiform in shape (see Fig. 6.25). These alveolinids show an example of morphological convergence with those of the Cretaceous. However, a considerable gap of about 20 Ma exists in the fossil record between the Alveolinidae of the Paleocene and the mid-Cretaceous *Praealveolina* group (see Chapter 5).

The oldest Paleogene form, *Glomalveolina* (with a streptoloculine origin and with a fixed axial coiling throughout ontogeny) is considered the ancestor of *Alveolina*. *Glomalveolina* is small, globular, and non-flosculinised with a spherical proloculus in a streptospiral coil. The streptospiral coil still exists in the B forms of *Alveolina* (see Fig. 6.26). *Alveolina* is a planispiral-fusiform form with a single tier of chamberlets in each chamber and considerable thickening of the basal layer (flosculinisation) (Fig. 6.26). In the Eocene the amount of flosculinisation varies in different species of *Alveolina* and might be considered to be of specific value only. *Alveolina* evolved into larger forms through the Eocene and reached gigantic sizes before the end of the Middle Eocene. It become an essential member of Eocene carbonate fossil assemblages. However, *Alveolina* forms became smaller again before completely disappearing at the end of Eocene. (see Fig. 6.27).

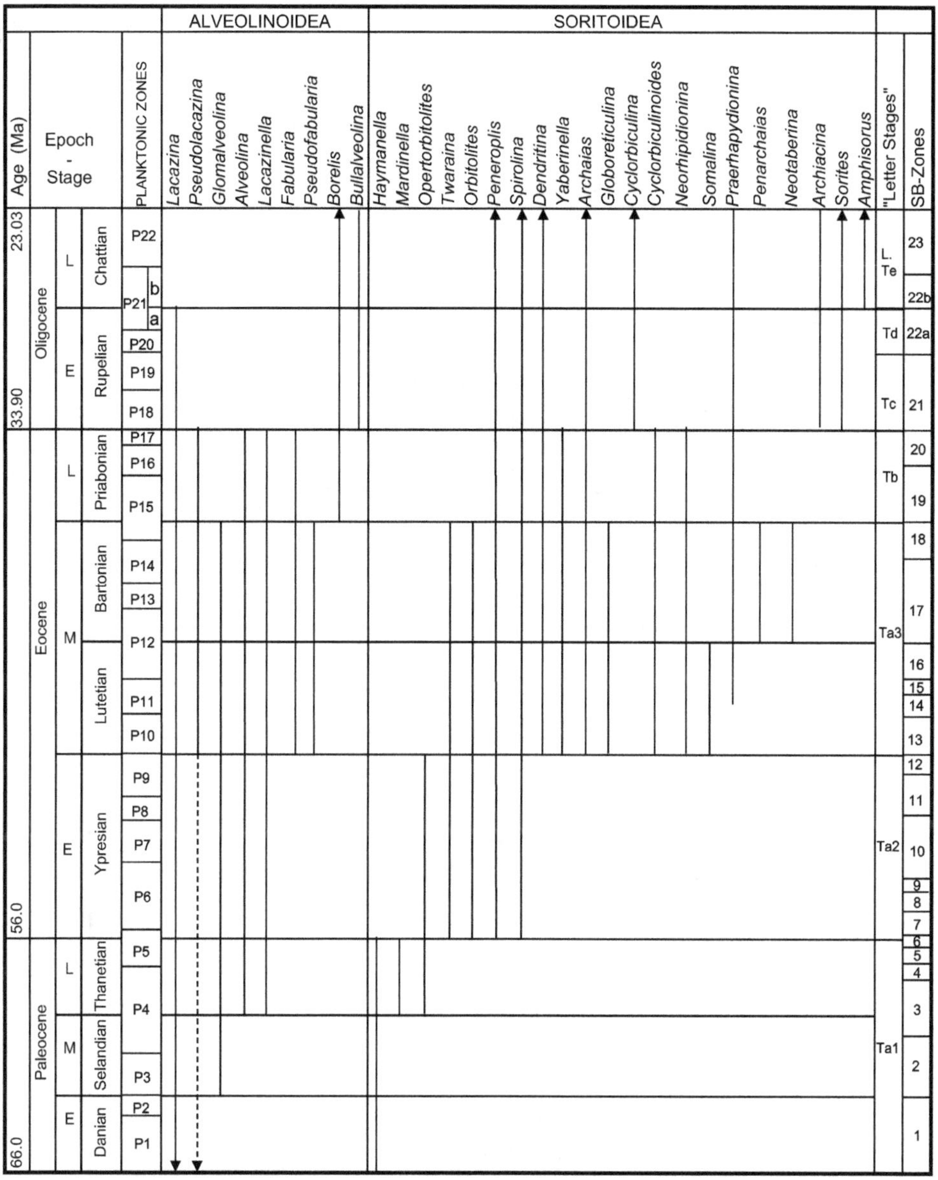

Column groups: **ALVEOLINOIDEA** | **SORITOIDEA**

Genera (left to right): *Lacazina, Pseudolacazina, Glomalveolina, Alveolina, Lacazinella, Fabularia, Pseudofabularia, Borelis, Bullalveolina, Haymanella, Mardinella, Opertorbitolites, Twaraina, Orbitolites, Peneroplis, Spirolina, Dendritina, Yaberinella, Archaias, Globoreticulina, Cyclorbiculina, Cyclorbiculinoides, Neorhipidionina, Somalina, Praerhapydionina, Penarchaias, Neotaberina, Archiacina, Sorites, Amphisorus*

Age (Ma)	Epoch-Stage		Planktonic Zones	"Letter Stages"	SB-Zones
23.03	Oligocene	L Chattian	P22	L. Te	23
			P21 b / a		22b
		E Rupelian	P20	Td	22a
			P19		
33.90			P18	Tc	21
	Eocene	L Priabonian	P17		20
			P16	Tb	
			P15		19
		M Bartonian	P14		18
			P13		17
			P12	Ta3	
		Lutetian	P11		16 / 15 / 14
			P10		13
		E Ypresian	P9		12
			P8		11
			P7	Ta2	10
			P6		9 / 8 / 7
56.0	Paleocene	L Thanetian	P5		6 / 5 / 4
			P4		3
		M Selandian	P3	Ta1	2
		E Danian	P2		1
66.0			P1		

Figure 6.24. Ranges of the main genera of the Paleogene Alveolinoidea and Soritoidea (also see Charts 6.1 and 6.2 for more details).

Borelis, an extant form and the only survivor of the terminal Middle Eocene extinction of alveolinids, is common in the Early Oligocene. A gradual stratigraphic succession of alveolinid species marked its very slow evolution in the Oligocene (Fig. 6.28). Species evolved with an incremental increase in the length of the test from pole to pole, from globular through ovoid to elongated spindle-shaped. In the Te "Letter Stage" of

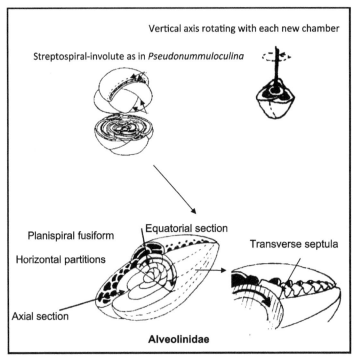

Figure 6.25. The evolution of *Alveolina* from a simple streptospiral origin (*Pseudonummuloculina*).

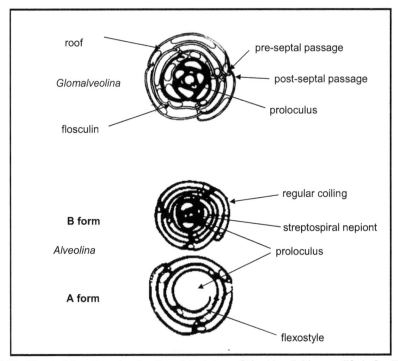

Figure 6.26. Equatorial sections showing the original stage of the Alveolinidae, modified after Hottinger (1960). "A form" is the megalospheric form and "B form" is the microspheric form.

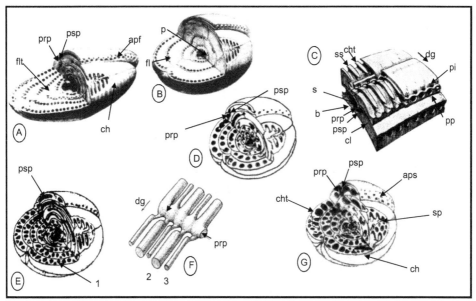

Figure 6.27. Reconstructions of different genera of the Paleogene alveolinids, modified from Reichel (1936). A: *Alveolina*; B: A flosculinized *Alveolina*; C: Stereo-diagram showing portion of two whorls; D: *Glomalveolina* with minimal flosculinisation; E: *Borelis* with only a pre-septal passage; F: Diagram of an internal cast showing the chamberlet form in *Borelis*; G: *Bullalveolina* with a large pre-septal passage (psp) and final chamber with multiple rows of small apertures. Abbreviations in figure: psp = post-septal passage, prp = pre-septal passage, flt = flexostyle, ch = chamber, cht = chamberlet, apf = apertural face, p = proloculus, fl = flosculine, ss = secondary septa, s = septum, b = basal layer, pp = main apertures, pi = supplementary apertures, dg = direction of growth, 1 = secondary small tier of chamberlets which alternates with the larger ones producing "Y" shaped septa in axial section of *Borelis*, 2, normal chamberlet, 3, displaced chamber (from Hottinger, 2006).

the Indo-Pacific, the alveolinids included *Borelis pygmaeus* (which persists to the top of the Te stage; BouDagher-Fadel and Banner, 1999), where it is suddenly replaced by a more advanced form *Flosculinella bontangensis* (see Chapter 7). In the Mediterranean, a smaller form is recorded in the Oligocene, *Bullalveolina*, which evolved from being streptospiral to having a planispiral quinqueloculine test, but unlike the extant *Borelis, Bullalveolina* disappears at the top of the Oligocene. Hottinger (2006, 2007) gathered forms that exhibit streptospiral nepionts in both generations and a planispiral-involute chamber arrangement in the adult stage of growth, such as *Subalveolina, Praebullalveolina, Bullalveolina,* and *Globoreticulina*, into the same group within his new subfamily **Malatyninae**. Sirel (2004, p. 38) classified the genus *Malatyna* into the milioline family **Rivieroinidae**, but according to Hottinger (2006, 2007) *Malatyna* can not be a milioline because it lacks the milioline pattern of growth, with two chambers per whorl and the apertural axis perpendicular to the coiling axis. *Globoreticulina*, classified by Rahaghi (1978) in the milioline subfamily **Fabulariidae**, has an architecture similar to *Malatyna* and was consequently transferred by Hottinger (2006, 2007) to the family Alveolinidae, and assigned together with *Malatyna* to the his subfamily Malatynae (which is not adopted here).

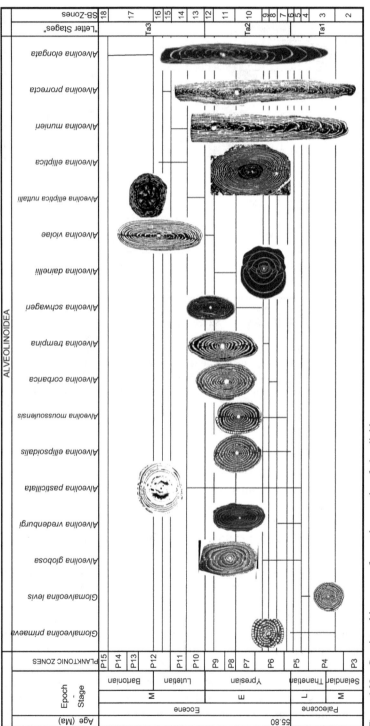

Figure 6.28.   Stratigraphic ranges of some key species of alveolinids.

The **Fabulariidae** are all linked to the Cretaceous fabulariids via *Lacazina*, which survived the across Cretaceous-Paleocene boundary (see Charts 6.1 and 6.2). *Lacazina* evolved from a miliolid-bilocular ancestor, such as the quinqueloculine to biloculine *Idalina* (see Fig. 6.7) with its overlapping chambers and fixed apertural and coiling axes (see Chapter 5). No direct phylogenetic relationship is traced between the Santonian pillared *Pseudolacazina* and the unpillared, much smaller Middle Eocene *Pseudolacazina* (Hottinger et al., 1989). All Paleogene representatives of *Pseudolacazina* are distinguished from the Cretaceous forms in having lower chambers, subdivided by continuous chamber partitions (septula) (see Drobne, 1988; Hottinger, 1989).

Fabularia is distinguished from *Lacazina* in lacking the monolocular stage and radial pillars. According to Haynes (1981), *Fabularia* evolved from the small, primitive Eocene form *Miliola* in the Middle Eocene by acquiring the involute biloculine last stage after the early quinqueloculine stage and the thick vertical partitions, which subdivide the chambers into elongated chamberlets with two tiers in outer whorls. On the other hand, *Miliola* may have given rise earlier, in the Late Paleocene, to *Lacazinella,* which has completely overlapping chambers. While many of the Alveolinoidea are still extant, most of the Fabulariidae did not survive the Eocene-Oligocene boundary, with the exception of *Lacazina,* which however died out within the Oligocene.

Middle to Late Paleogene Tethyan shallow-water foraminiferal communities are dominated by Soritoidea (see Fig. 6.24), which exhibit a completely new line of evolution for the miliolides (see Fig. 6.8). They appear to have evolved in the Eocene from forms with a simple planispiral, *Spirolina,* to a flat, evolute and flaring planispiral, *Peneroplis.* These porcelaneous, spiroliniform and peneropliform foraminifera developed uniserial cylindrical chambers with radial partitions, which varied from simple, as in *Praerhapydionina,* to combined radial partitions with pillars, as in *Neotaberina,* with apertures varying from single, stellar-shaped apertures, such as *Praerhapydionina,* to others with multiple apertures, such as *Rhabdorites.* They are very common in the Middle East, where they were originally classified by Henson (1948) as "*Rhapydionina*" and "*Rhipidionina*". However, both of these forms, previously described by Stache (1913) and later confirmed as megalospheric and microspheric forms by Reichel (1984), are different from those of the Eocene (Hottinger, 2007). In his Ph.D. thesis, Henson (1950) transferred his Eocene *Rhipidionina* species to the genus *Meandropsina.* However, the Meandopsinidae show similar morphological trends, but are confined to the Cretaceous and may have arisen from *Praepeneroplis* (see Chapter 5). Henson (1948) recognized in his Eocene material, mainly from Iraq, three taxa with similar structures:

- "*Rhapydionina*" *urensis* (Plate 6.6, figs 3-5) having a spiroliniform test,
- "*Rhipidionina*" *macfadyeni* (Plate 6.6, figs 7-8), and
- "*Rhipidionina*" *williamsoni* (Plate 6.6, fig. 9) exhibiting a peneroplid test.

All three taxa represent megalospheric forms and have the same basic patterns of chamber subdivision and a planispiral-involute nepiont (Hottinger, 2007). Hottinger emended and recognized a new genus, *Rhipidionina* designating "*Rhipidionina*" *williamsoni* as its type species. The genus *Haymanella* was originally described from the Paleocene of Turkey by Sirel (1999). However, it was shown by Hottinger (2007) that Sirel's species is the only genus of the Peneroplidae and of the Soritidae, that combines a porcelaneous wall with an agglutination of coarse grains. In this respect, *Haymanella* is similar to agglutinating miliolines such as *Agglutinella*, *Schlumbergerina* or *Siphonaperta* (see Hottinger et al., 1993). Eocene peneroplids apparently also show a progression through many lineages, leading to *Archaias*, *Sorites* and the rest of the Miocene discoid miliolines, such as *Marginopora* and *Amphisorus* (see Chapter 7).

In the Oligocene, forms with a peneropline early stage became strongly flaring, as in *Puteolina*, becoming completely annular with *Cyclorbiculina*. The peneropline nepiont still occupies more than half of the test, but the annular chambers are by now divided into small rectangular chamberlets, as in *Cyclorbiculina*. These latter forms became flat and completely discoid, as in *Sorites,* with ogival chamberlets connected by a single open space or stolon, occurring in the equatorial plane. This lineage, after adopting the circular trend, became very successful and is still living in modern seas. Another extant form evolved in the Eocene from a peneropline ancestor, *Archaias*, an involute planispiral thickened form with pillars throughout the test, but with no annular stage of growth.

A short trend occurred in Tethys during the Early and Middle Eocene, in which successive stolon layers formed a pile of discs (Hottinger and Drobne, 1980), as in *Orbitolites*. However, unlike the lineage of *Sorites*, *Orbitolites* does not have an peneropline early stage, but rather only has oblique stolons connecting the adjoining chamberlets (Hottinger, 2006) and chambers subdivided into small curved chamberlets with thick walls (see Fig. 6.8).

In the Middle Oligocene of the Far East, one widespread group of **Milioloidea** had an alveolar exoskeleton, *Austrotrillina*, in order, presumably to harbour symbiotic algae. *Austrotrillina* was revised by Adams (1968), presenting *Austrotrillina paucialveolata* Grimsdale (Plate 6.9, Figs 14-16; Plate 6.10, figs 1-2) as the most primitive species of the group, with a range starting in the Early Oligocene. However, later, it has been recorded from the Late Eocene of Iran by Rahaghi (1980), and more recently Hottinger (2007) erected a new species for Rahaghi's form, *Austrotrillina eocaenica*. According to Hottinger, this species in Rahaghi's material is "primitive", but cannot be interpreted as a direct ancestor of *A. paucialveolata,* because it is larger and has a more differentiated exoskeleton. Another doubtful occurrence of *Austrotrillina* was recorded from the Late Eocene of off-shore Tunisia (Bonnefous and Bismuth, 1982). The questionable phylogenetic relationship will not be resolved until more material is found that relates the Oligocene Far East taxa with those of the Eocene of the Middle East. The evolution of these forms in the Indo-Pacific realm will be discussed in detail in Chapter 7.

In the Caribbean, the ecological equivalents of the Tethyan Soritoidea are represented by the discoidal *Yaberinella* (Plate 6.10, fig. 16; Plate 6.11, figs 10-11), of miliolide-fabulariid origin (Hottinger, 1969), and *Cyclorbiculinoides* (Plate 6.10, fig. 13). The latter exhibits an unusual architecture with vertically superposed radial pillars (see Hottinger, 1979) of unknown origin (Hottinger, 2001).

6.3.4 The Rotaliides of the Paleogene

In the Paleogene, the rotaliides thrived in the warm shallow waters of Tethys. They became very large, reaching up to ~15cm diameter in forms such as *Cycloclypeus*. The problems associated with large size were solved by adding different structures to the foraminiferal test, such as filling the umbilical area with pillars, becoming annular by subdividing their chambers into small chamberlets, and by suspending their lamellae from a thickened marginal cord. Many rotaliides have canal systems (Fig. 6.29) within the walls of their plurilocular calcareous tests. Such constructions are called supplemental skeletons, representing infolded outer lamellae. The canals became very complicated in the Eocene with forms such as *Pellatispira* (see Hottinger et al., 2001). Their biological function has been investigated by many authors (Röttger et al., 1984; Hottinger et al., 2001). In extant nummulitids, elphidiids, calcarinids, pellatispirids, they were found filled with protoplasm with permanently differentiated microtubuli (Hottinger and Dreher, 1974; Leutenegger, 1977; Hottinger and Leutenegger, 1980; Hottinger et al., 2001). These apparent morphological changes in the rotaliines of the Paleogene provided the foundation for the stratigraphically important superfamilies (and families) discussed below:

- Planorbulinoidea
- Nummulitoidea
- Acervulinoidea
- Asterigerinoidea (especially the Lepidocyclinidae)
- Rotalioidea (especially the Miogypsinidae).

As explained in Chapter 5, nearly all groups of large rotaliine foraminifera were derived, via separate lineages, from planispiral or trochospiral small rotaliine ancestors, with a primitive canal system (Fig. 6.9), such as *Rotalia* or *Cibicides*.

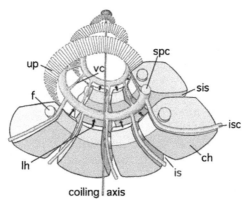

Figure 6.29. The canal system in a simple rotaliine test (after Hottinger, 2006). Abbreviations in diagram: up = umbilical plate, spc = spiral canal, sis = spiral interlocular space, isc = intraseptal canal, ch = chamber, is = interlocular, f = foramen, ih = loop-hole.

Cibicides shows a number of evolutionary trends (Fig. 6.30) leading to the development of attached forms in the Paleogene, irregularly uncoiled biserial small forms in the Eocene (*Dyocibicides*) and annular discoid small forms in the Paleocene to Miocene (*Cycloloculina*). *Cibicides* also gave rise to the **Planorbulinoidea** in the Eocene (Fig. 6.30) by developing rings of alternating chamberlets with peripheral apertures. The name of the single layered planorbulinids is derived from the genus *Planorbulina* in the family **Planorbulinidae**, which has a long trochoid, *Cibicides*-like keeled spiral and a small number of irregular later chambers in an orbitoidal pattern. It has been suggested that both the extant forms *Planorbulina* and *Planorbulinella* derived from some *Cibicides*-like ancestor (Drooger, 1993) in the Early Eocene. In both forms (*Planorbulina* and *Planorbulinella*) retrovert apertures of the later spiral chambers originated from irregular orbitoidal forms in the *Cibicides*-like ancestor. A particular feature belonging to the *Planorbulinella* morphology is the occurrence of relapse chambers, where after the occurrence of chambers with two stolons, one or more of the later chambers had reverted to having one stolon. Freudenthal (1969) introduced a morphometric analysis for *Planorbulinella*. His analysis concentrated on "Y counts" (the number of chambers with a basal opening) and measurements of the embryon expressed in diameter "d" and height "h". Y was found to vary from 8 to 2. In the Oligocene a form similar to *Planorbulinella,* but with the apertures consisting of smaller basal openings instead of single slits, *Planolinderina*, might have also evolved from a *Cibicides*-like ancestor. The relapse chambers of *Planorbulinella* are not observed in *Planolinderina,* and Drooger (1993) noted that unlike in the phylogeny of *Planorbulinella* there is no Y=2 barrier in the phylogenetic development of *Planolinderina*.

Immediately following the appearance of *Planorbulinella*, genera with thick lamellar walls and secondary deposits of calcite on both sides of the central embryont appeared in the Middle Eocene; the cosmopolitan *Linderina* and the American *Eoannularia*, with incipient orbitoidal morphology (see Figs 6.16, 6.30).

Another lineage that evolved from *Cibicides*, with conical forms having umbilical cover plates or plugs and walls thickening on the spiral side by addition of lamellae, included the Late Cretaceous-Paleocene form, *Cymbalopora* and the Eocene *Halkyardia* (Figs 6.17, 6.30). The latter has also been reported from the Early Oligocene platform and periplatform limestones of Jamaica, as well as being recorded in Cuba, Ecuador, and Mexico (Robinson, 1996).

The branching homotremids (Fig. 6.30), with planispiral and very high coiling, seem to have arisen from the Planorbulinidae in the Eocene. The extant form *Sporadotrema*, with flattened sides and cylindrical branches, first appeared in the Eocene. It has no pillar-like lamellar thickening as would be expected in *Victoriella*.

The victoriellids, with three subfamilies, the **Carpenteriinae**, the **Rupertininae** and the **Victoriellinae**, also arose from a *Cibicides*-like ancestor. The Carpenteriinae are the first, primitive forms, and evolved from *Cibicides* by acquiring a distinct keel, extending to the attachment area. The Rupertininae, with a distinctly flat spiral side, had the adult part of the test growing away from the attachment site. Members of this family are known from the Late Eocene, with *Carpenteria* still living to the present day. The Victoriellinae is a group in which the early stage was probably free living, but later became attached. These forms range from Late Eocene to Early Miocene, and are represented by high spiral forms, such as *Victoriella* and *Wadella,* and low spiral forms such as *Maslinella*.

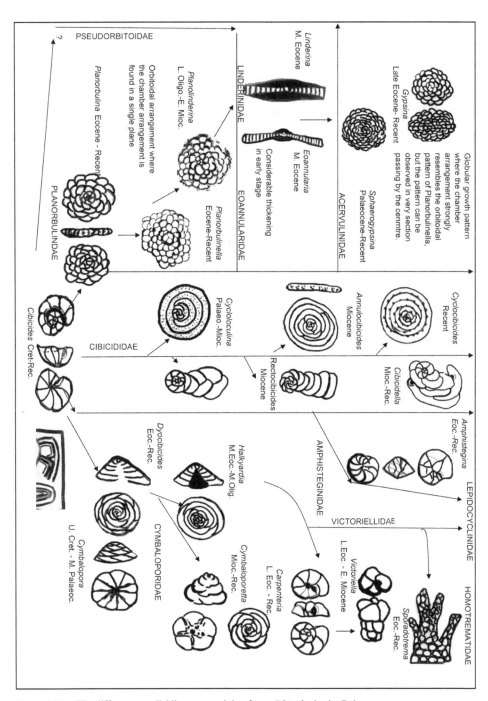

Figure 6.30. The different parallel lineages evolving from *Cibicides* in the Paleogene.

Members of the planispirally-coiled **Nummulitoidea** became very abundant in the Paleogene (see Fig. 6.31). They exhibited rapid evolutionary rates, and developed complex, large tests, which make them a very important index fossil group for shallow marine environments (Schaub, 1981; Pignatti, 1998; Serra-Kiel et al., 1998). They probably evolved from a small rotaliine, via the the Late Cretaceous *Sulcoperculina* (see Chapter 5), by developing a complex system of a marginal cord and well-developed spiral and subsutural septal canals. The development of the canaliculate marginal cord, replacing the primary aperture of the nummulitids, was essential for growth, locomotion, reproduction, excretion and protection (Röttger, 1984). This complex canal system is characteristic of all living and fossil nummulitids and gave rise to special

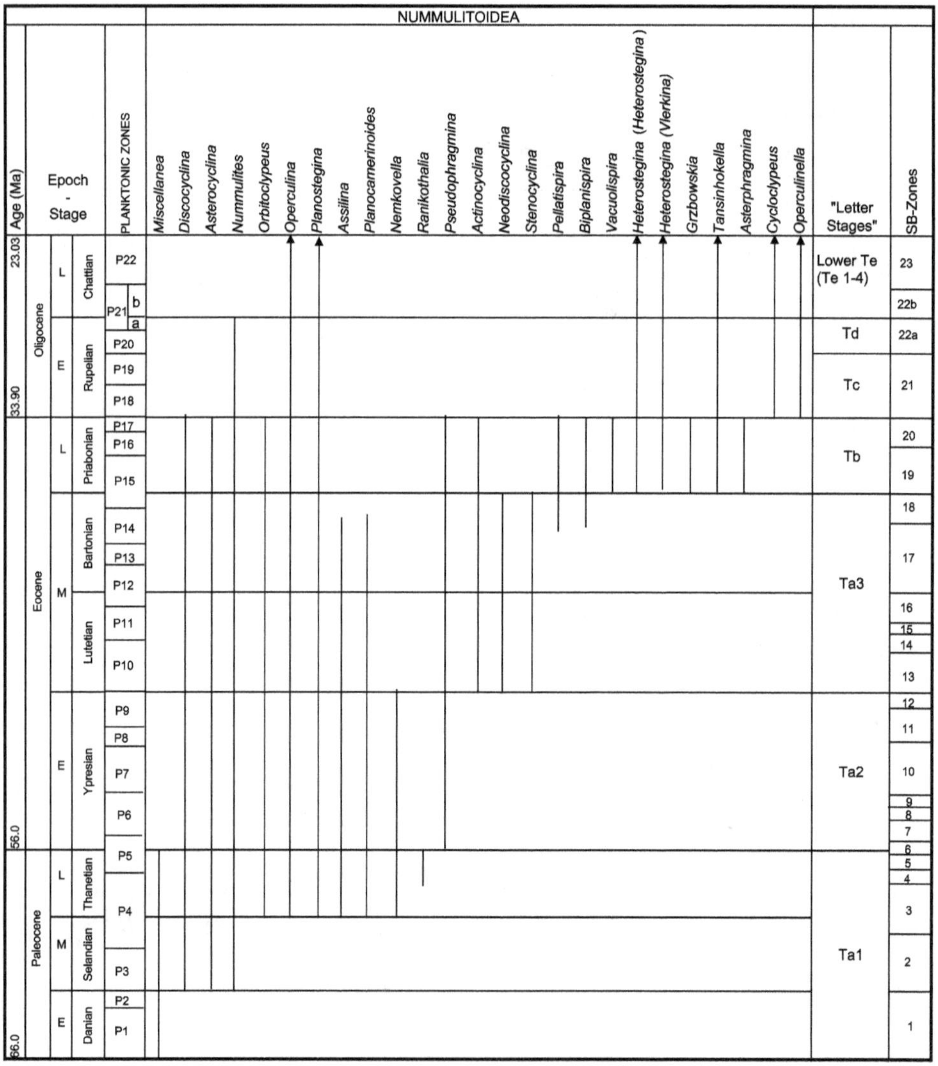

Figure 6.31. The ranges of the Nummulitoidea in the Paleogene.

lamination in the tests (Hottinger, 1977). The lamellar tests were formed during the process of chamber construction, in which each chamber wall covers the total test, including all former chambers (Hoheneger et al., 2001).

Many authors have studied the morphology of the nummulitic tests, with some detailed descriptions as early as D'Archaic (1850), Carpenter (1850), Galloway (1933), Davies (1935) and Glaessner (1945). Among the more recent studies are those of Hottinger (1977, 1978), Hottinger et al. (2001), Racey (1995) and BouDagher-Fadel and Price (2014). The morphology of *Nummulites* of the family **Nummulitidae** is illustrated in Figs 6.32 and 6.33.

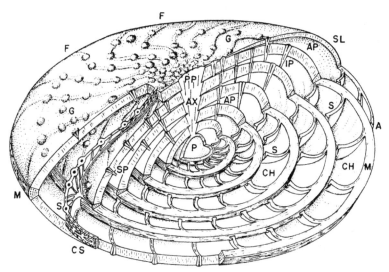

Figure 6.32. Test structure of *Nummulites* (megalospheric form, modified from Carpenter, 1850; Golev, 1961; and Barnett, 1969). Abbreviations: A = aperture; AP = alar prolongation of lumen of chambers; AX = axial plug; CH = chambers; CS = canal system in marginal cord and septa; F = filaments; G = granules on and between filaments; IP = interseptal pillars; M = marginal cord; P = proloculus; PP = coalescing granules forming polar pustule; S = septum; SP = septal pillars.

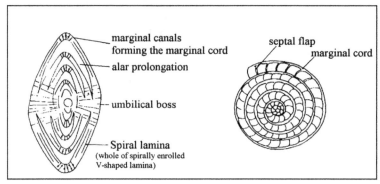

Figure 6.33. The axial and equatorial sections of *Nummulites* (after Blondeau, 1972).

Cole (1964) noted that certain features visible in axial section, such as the alar prolongations (chevron-shaped cavities on each side of the test) in involute forms, such as *Nummulites*, and the presence of thick lamellar walls on each side of the equatorial layer, as in *Cycloclypeus,* are useful in classification at the generic level. Hottinger (1977) proposed a generic classification based on the type of canal and stolon system, and the absence or presence of trabeculae ("imperforate shell material extending from an imperforate sutural zone into the perforate lateral chamber-wall and housing oblique, ramified trabecular canals opening between the pores on the surface of the lateral chamber wall" (Hottinger, 2006)). He considers the chamber formation and the type of stolon and canal system of greatest importance in the taxonomy of the nummulitic tests. Hottinger considered the stolon system and the canal structure as progressive, since they appear to become more complex with time. He distinguishes *Heterocyclina* (see Chapter 7) from *Heterostegina* on its L-shaped rather than Y-shaped stolons. However, Y-shaped stolons also occur in *Spiroclypeus* and *Cycloclypeus* (Haynes, 1981). Only the advanced *Cycloclypeus* has diagonal crossed stolons. Banner and Hodgkinson (1991, p. 105) defined the chamberlet as the division of the chamber as is seen in some species of *Nummulites* (e.g. *Heterostegina* (*Vlerkina*), etc.) and they called "cubicula" (*sensu* Banner and Hodgkinson, 1991) the separate little "chamberlets" that do not arise from the division of larger chambers and are called by others "lateral chamberlets". This applies particularly to the lateral structures of orbitoids and miogypsinids. The median layer is composed of true chamberlets (which are known to be derived from the division of primary chambers phylogenetically or ontogenetically, or both).

Within the family **Nummulitidae**, that includes the genera *Nummulites* and *Assilina*, a number of morphological evolutionary trends occurred (Racey, 1995; BouDagher-Fadel and Price, 2014), such as;

- the degree of the involution or evolution of the test;
- the degree of the development of the marginal cord;
- the extent of the opening of the spire;
- the division and subdivison of the chambers;
- the overall size of the microspheric forms (i.e. the forms produced by sexual reproduction), which increases over geological time for most lineages;
- the size of the proloculus, which increases with time in the megalospheric forms (i.e. forms produced by asexual reproduction) for most lineages;
- the development of embryonic chamber complexity tends to increase with time. Various biometric methods for studying the embryonic development of larger foraminifera have been proposed. One of them (the "Factor E method" of van der Vlerk and Bannink, 1969), involves measuring the degree of enclosure of the second chamber by the third, to give a number referred to as Factor E. It was used in studies reported by Racey (1992, 1995);
- the shape of the septa (see Adams 1988; Racey 1995), as the septal complexity increases during time with:

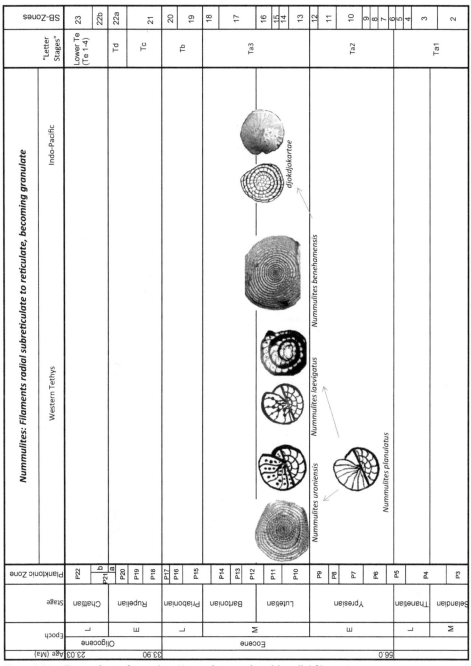

Figure 6.34. Range chart of some key *Nummulites* species with radial filaments.

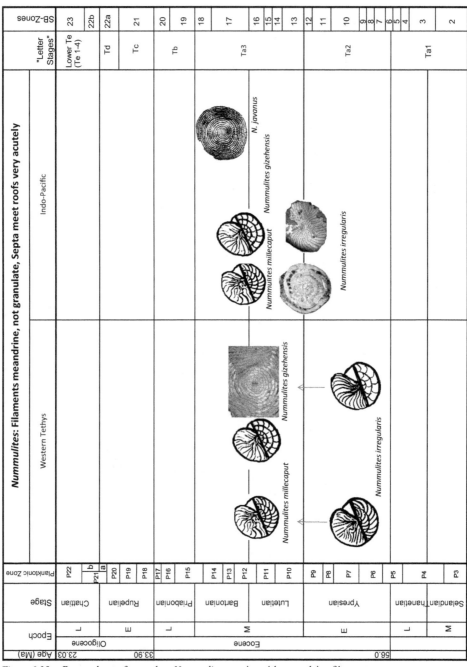

Figure 6.35.   Range chart of some key *Nummulites* species with meandrine filaments.

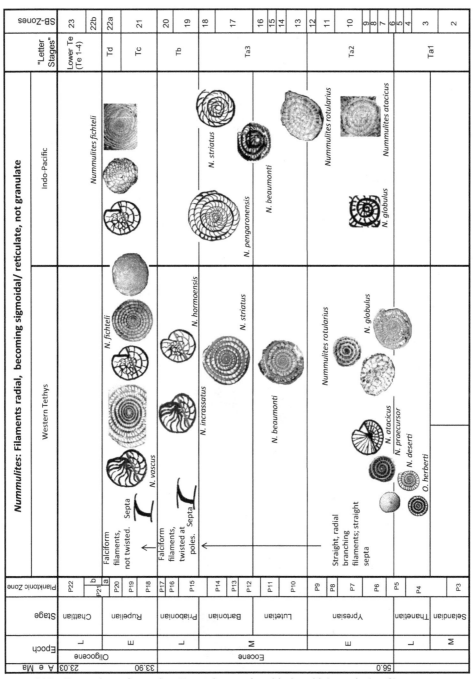

Figure 6.36. Range chart of some key *Nummulites* species with sigmoidal to reticulate filaments.

o simple radiating (striate), falciform or curved septal filaments in Late Paleocene to Early Eocene species (see Fig. 6.34);

o meandriform and complex branching in Middle Eocene forms (see Fig. 6.35);

o reticulate septal filaments in Late Eocene forms (Fig. 6.36);

o in the Early Oligocene the reticulate septa are still widespread (Fig. 6.36), but there is a tendency to the return to simple structures.

Qualitative features such as the presence/absence of granulation on the surface of the test and quantitative characteristics based on the morphometric system introduced by Drooger and Roelofsen (1982) (see Fig. 6.37) are commonly used in classifying num-mulitic species (see BouDagher-Fadel and Price, 2014, Fig. 6.37). They include:

- P: the largest diameter of the proloculus in μm, excluding the thickness of the wall,
- D: the largest diameter of the deuterconch in μm, excluding the thickness of the wall,
- X: the number of undivided, operculinid chambers before the appearance of the first partly divided planosteginid chamber, excluding the embryon (the first two chambers, protoconch and deuteroconch). This parameter indicates the degree of operculinid reduction. (see for example Fig. 6.37, in which X = 8),
- l: the maximum diameter of the first whorl along the common symmetry axis of the embryon (including the protoconch and deuteroconch),
- L: the maximum diameter of the first whorl and the subsequent half whorl as measured in μm, along the common symmetry axis of the embryon (including the protoconch and deuteroconch),
- K: the index of spiral opening, where $K = 100 \times (L-l)/(L-P)$.

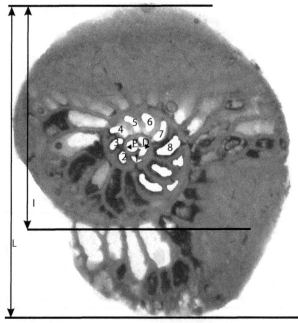

Figure 6.37. Morphometric measurements in the equatorial section of megalospheric *Planostegina africana* BouDagher-Fadel and Price (from BouDagher-Fadel and Price, 2014). P = proloculus; D = deuteroconch; Pre-planosteginid chambers (X) are 8; l = the maximum diameter of the shell in the first whorl; L= the maximum diameter of the first one and subsequent half whorl.

In the American province, *Sulcoperculina* a proposed ancestor of the Paleogene nummulitoids (see BouDagher-Fadel and Price, 2014) died out at the top of the Maastrichtian. The first Paleogene nummulitoid may have evolved in the Americas from a pre-existing rotaliid form with a trochospiral test and intraseptal passages of a canal system, typified by a genus such as *Pararotalia* (see BouDagher-Fadel and Price, 2014; Fig. 6.38). Analyses of molecular data show a close relationship of nummulitids to the rotaliids (Holzmann et al. 2003). However, the proposed process of evolving from a rotaliid such as *Pararotalia* would have been complicated and involved a series of coupled morphological changes. The trochospiral chambers of *Pararotalia* are arranged to expose the umbilical region that creates direct access to the ambient environment. The test possesses a spiral umbilical canal, formed by interconnected toothplates, with a free edge (Hottinger et al. 1991). BouDagher-Fadel and Price (2014) suggested that the Nummulitidae may have evolved from a simple rotaliid test through forms which developed a moderately thick marginal cord and a rapidly widening coil leading to, in the Late Paleocene (P4a), the long-ranging completely planispiral evolute *Operculina* (see Fig. 6.38).

The first occurrence of *Chordoperculinoides bermudezi* was recorded by BouDagher-Fadel and Price (2014) in the Middle Paleocene. This form is characterized by coarse vertical canals and a massive marginal cord, and is found throughout the Selandian (P3) and the Thanetian (P4). It gave rise to the sub-evolute *Caudrina* in the Late Thanetian (P5) of the American province (BouDagher-Fadel and Price, 2014). *Caudrina* did not survive the Paleocene-Eocene boundary, while *Chordoperculinoides* died out at the top of the Early Eocene (end Ypresian, P9). *Caudrina* has not been recorded so far from the Eastern Hemisphere, but as BouDagher-Fadel and Price (2014) infered that other forms migrated eastward from the Americas, to populate the Eastern realms (see below).

During the Thanetian (P4b), *Operculina* appears to have given rise to the evolute tightly coiled *Assilina* and the involute tightly coiled extant genus, *Palaeonummulites*. The latter had given rise to "lax" forms by the beginning of the Lutetian (P10), as seen in *Operculinella*. In a separate lineage, the Middle Eocene (P12b) also witnessed the development of an early involute test in some operculine species, which subsequently became fully involute, having chambers divided into chamberlets with secondary septa, giving rise to *Heterostegina* (see BouDagher-Fadel and Price, 2014).

In the Eocene of the Western Tethyan province, *Nummulites* and *Assilina* witnessed a major radiation and increase in test size, which persisted up to the major extinction of the last large species at the Middle-Late Eocene boundary. Few small species of *Nummulites,* and none of *Assilina,* survive into the Late Eocene, and *Nummulites* finally became extinct in the end of Early Oligocene.

The **Heterostegininae** (Fig 6.38) evolved from the **Nummulitidae** in the Late Eocene and subsequently followed a clear morphological sequence of increasing involution of the test and the enclosure, by the alar (umbilical) prolongations, of the chambers. It also showed an increase in complexity of the chamberlet development (Banner and Hodgkinson, 1991). The pattern of increasing morphological complexity was repeated in many lineages from the Paleocene to the Early Miocene, as the more advanced forms appeared later in time. The flat evolute *Operculina* gave rise to the wholly evolute

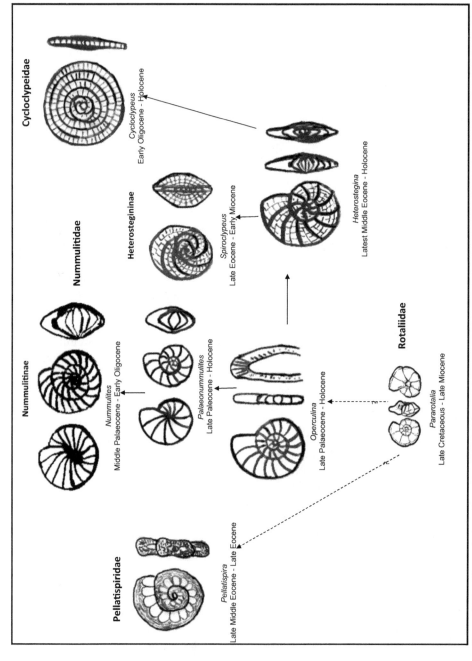

Figure 6.38.   The evolution of some nummulitid families from an rotaliid ancestor.

Planostegina in the Late Paleocene, by developing partly developed chamberlets (Fig. 6.37). *Planostegina* persisted through the Neogene to Holocene. *Heterostegina* and *Cycloclypeus* evolved separately from *Planostegina*, the first descendant becoming initially involute, while the second lost its planispirality and became annular. This inferred sequence was corroborated by Holzmann et al. (2003) from their molecular genetic analyses.

In the Late Eocene, the secondary chamberlets of *Heterostegina sensu stricto* became fully developed and the test became initially involute. This genus also persists into Holocene. In equatorial view the chambers are rectangular. However, the wholly involute *Grzybowskia* formed polygonal chamberlets in the Late Eocene. *Heterostegina (Vlerkina)* first appeared in the Late Eocene with a completely involute test and many rectangular lateral chamberlets. It disappeared within the Burdigalian (BouDagher-Fadel et al., 2000, 2001). Also during the Late Oligocene and Early Miocene, *Tansinhokella*, a wholly involute genus, with abundant lateral chamberlets and meandriform alar prolongations, became widespread. These lateral chambers proliferated and became polygonal (cubiculae, *sensu* Banner and Hodgkinson, 1991) and symmetrical on both sides of the spiral acquiring a sub-annular architecture, as in *Spiroclypeus*.

In the Early Oligocene, the evolute *Planostegina* lost its planispirality and gave rise to *Cycloclypeus,* by becoming annular discoid with thick lamellar walls on each side of the equatorial layer. *Cycloclypeus* (Fig. 6.38) is a nummulitid with a nepionic morphology like *Heterostegina*, but with a final growth stage with cyclic chambers. In the course of the ontogeny of the megalospheric individuals, the early operculine chambers are reduced and the subdivided heterosteginid chambers get longer, with an increasing number of chamberlets, until they extend backwards over the previous part of the preceding whorl (see Fig. 6.38). The evolution of *Cycloclypeus* was studied in great detail by Tan Sin Hok (1932). However, the potential for biostratigraphic application of his detailed biometrical descriptions is severely limited by the uneven distribution of suitable *Cycloclypeus* material over the geological column, and also by the complex nature of the data. The oldest species would appear to be *C. koolhoveni* (early Rupelian, Tc), possibly the direct ancestor of *C. oppenoorthi* (late Rupelian to early Chattian, Td) (see Fig. 6.39). It evolved gradually in the Oligocene to Early Miocene by reducing its nepionic whorl from two to a half, and the nepionic septa from 38 to just 2 stages in advanced forms such as *C.* cf. *guembelianus* and *carpenteri* (see Fig. 6.39). Over time, nepionic reduction in *Cycloclypeus* occured at a steady rate (Tan Sin Hok, 1932; MacGillavry, 1978; Haynes, 1981). However, it occurred within different lineages at different times. It should be noted that in general, the one *Cycloclypeus* lineage that could be continuous consists of species with very wide variations in internal morphological characters, and that the individual species, if at all separable, are very long-ranging. Holocene forms of *Cycloclypeus* have been found to possess (X-shaped) stolons (Haynes, 1981). While *Heterostegina* and *Cycloclypeus* are still living, *Spiroclypeus* became extinct in the Early Miocene.

The **Pellatispiridae** family (Figs 6.38 and 6.40) of the Nummulitoidea lived for a comparatively short period at the end of the Eocene, during the Bartonian and Priabonian stages. They all have a planispiral-evolute test connected by a single foramen, with

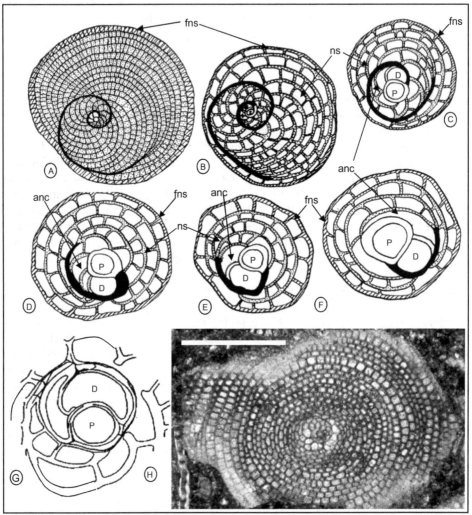

Figure 6.39. The variation in the (A) microspheric and (B–G) megalospheric embryonic and nepionic chambers of *Cycloclypeus*, modified after Tan Sin Hok (1932). Enlargement factors only refer to the nepionic stage. The last whorls shown are the first of the annular neanic stage. Here, P, stands for proloculus, and D, deuteroconch. The nepionic whorl is highlighted by a darker wall. Other abbreviation: anc, "ananepionic" chambers which are chambers of operculine shape, not subdivided into chamberlets; ns, nepionic (non-cyclic) septa; fns, first neanic (cyclic) septum. (A) An equatorial section of an early *Cycloclypeus* (Early to Middle Oligocene) with 2 anc followed by 27 heterostegine chambers, making a total of approximately 29 ns, e.g. *C. koolhoveni* Tan Sin Hok (with large central pillars) and *C. oppenoorthi* Tan Sin Hok (without central pillars). (B) *C. eidae* Tan Sin Hok with 1 anc and 18 ns (Late Oligocene to Early Miocene). (C) *C. poste-idae* Tan Sin Hok (Early Miocene) with 8 ns. (D) *Cycloclypeus indopacificus* Tan Sin Hok (Early to Middle Miocene) with 5 ns. (E) *Cycloclypeus postindopacific* Tan Sin Hok, 4 ns. (F) Advanced forms of *Cycloclypeus*, such as *C.* cf. *guembelianus* Brady with 3 ns. (G) Enlargement of the embryonic apparatus to show the development of the radial canal system into marginal cord. (H) A microspheric *Cycloclypeus* with a small protoconch, 2 operculine stages (anc) with 11 nepionic stages (ns) followed by numerous cyclic chambers, Kalimantan, Early Miocene. Scale bar = 1mm.

interlocular spaces transformed into an enveloping canal system and flying covers of perforate walls suspended on spikes or pillars (Hottinger et al., 2001). *Pellatispira* is considered by Boussac (1906) and Umbgrove (1928) to have evolved from the thick-walled assilid (planocamerinid) forms, by developing outer lamellae over the earliest chamber wall. However, the supplemental skeleton at the pellatispirid shell margin is not a marginal cord, as seen in nummulitids: it lacks the sulcus, and its canal system is radial and enveloping, not a tangential, polygonal network (Hottinger 1978, 2001). The work of Hottinger et al. (1991) clearly shows on the other hand, a close relationship between *Pellatispira* and *Biplanispira*, as emphasised by the adult of the former often forming the juvenile stage of the latter. They described the thickened shell margin produced by the marginal canal sytem in *Pellatispira* (Fig. 6.40) as being overgrown by supplemental chamberlets either on the lateral flanks alone (*Biplanispira*) or on all sides of the shell (*Vacuolispira*).

The **Orthophragminidae** are bilamellar, perforate, orbitoidal larger benthic foraminifera, and are characterized by a discoidal, lenticular test with a fine equatorial layer and small lateral chamberlets (Fig. 6.41). The growth of orthophragminids was both cyclical and involute. Each chamber had a bilamellar structure with an inner lining and an outer lamella, with the subdivision of the equatorial chamberlets provided by different inner linings (Ferràndez-Canãdell and Serra-Kiel, 1992; BouDagher-Fadel and Price, 2017). No canal system is evident (cf. the nummulitids), and connections between chambers are provided by a tridimensional stolon system (see Ferràndez-Canãdell and Serra-Kiel, 1992). Orthophragminids are classified based on the general shape of their tests, the pillar-lateral chamberlet network, the different kinds of stolons, and the size of their pillars. Moreover, their most important evolutionary parameters are associated with the shape of the embryons in the megalospheric generations, the characters of nepionic stages in the microspheric orthophragminid juvenaria "their embryons",

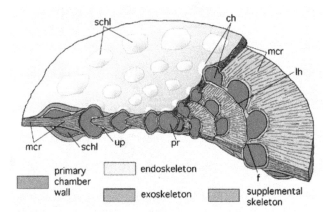

Figure 6.40. A sketch, from Hottinger (2006), showing a spiral pellatispirid shell with an enveloping canal system and a marginal crest. According to Hottinger, the primary lateral chamber walls which are "emerging" from the supplemental skeleton are covered by secondary lamellae but are perforated in continuation of the primary bilamellar wall and are not a part of the supplemental skeleton. The supplemental chamberlets are fed by canal orifices as they do not communicate directly with the spiral chambers by retral stolons. Abbreviations: ch = chamber; f = foramen; lh = loophole; mcr = marginal crest; pr = proloculus; schl = supplemental chamberlet; up = umbilical plate.

and features of the equatorial chambers (Less, 1987; Brönnimann, 1951; Ferràndez-Cañadell and Serra-Kiel, 1992; BouDagher-Fadel, 2008; BouDagher-Fadel and Price, 2017). The Orthophragminidae are divided into two subfamilies, the **Discocyclininae** and the **Orbitoclypeinae**. Loeblich and Tappan (1987) considered *Proporocyclina* and *Athecocyclina* as synonyms of *Pseudophragmina* (see BouDagher-Fadel, 2008). However, BouDagher-Fadel and Price (2017) consider them to be different genera, as they exhibit stratigraphically characteristic, distinguishing morphological features.

Dimorphism in the orthophragminids is common and is reflected in the size of the test, which is larger in sexually produced microspheric specimens than in the asexually produced megalospheric forms. The diameter of megalospheric specimens is ~3 mm or less, while microspheric specimens may reach >10 mm in diameter. As in occurrences of most larger benthic foraminifera, the megalospheric orthophragminids are more common than the microspheric ones (Ćosović and Drobne, 1995; Hottinger, 2001).

The megalospheric forms of the discocyclinines have a subspherical protoconch enclosed by a larger reniform deuteroconch (Figs 6.41, 1-6). Their microspheric forms have a *Cycloclypeus*-like microspheric juvenile (see Figs 6.41, 3-9) with an initial spiral of small chambers. Later stages exhibit cyclical chambers (annuli) subdivided by septula into small rectangular chamberlets connected by annular and radial stolons. Externally, the test surface is either smooth, with scattered pillars, or has radially developed ribs. An inflated central part or umbo (Figs 6.41, 3-4) is occasionally present.

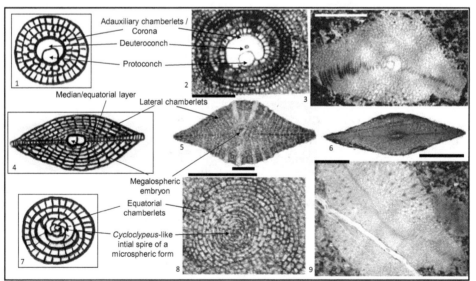

Figure 6.41. The morphological features (from BouDagher-Fadel and Price, 2017) of Discocyclininae. **1–6** megalospheric forms: **1–3** highlighting the embryonic apparatus of the equatorial section, **1** schematic figure of the embryonic apparatus, **2** *Discocyclina prattii* (Michelin), France, **3** *Discocyclina dispansa* Sowerby, Tibet. **4–6** axial sections: **4** Schematic figure showing the three layers of the test, **5** *Discocyclina sheppardi* Barker, Soldado Rock, Trinidad, **6** *Discocyclina* sp., the Hecho Group, Ainsa Basin, south central Pyrenees, Spain. **7–9** Microspheric forms: **7** schematic figure of the microspheric apparatus, **8** *Discocyclina* sp. France, **9** Oblique axial section of *Discocyclina dispansa* Sowerby, Tibet. Scale bars = 1mm, except for **1**, scale bar = 0.25mm.

Members of orbitoclypeines have an early planispiral microspheric coil, while megalospheric tests have a globular protoconch, enclosed by a larger reniform deuteroconch. Tests may occur with or without ribs. These taxa have a single equatorial layer of chamberlets and several layers of small lateral chamberlets, ranging from arcuate, spatulate to those hexagonal in shape, and cyclical chambers that are not subdivided into chamberlets.

Biometric data are frequently used in specific definitions, and were first developed for *Discocyclina* by Neumann (1958), and subsequently expanded by Brolsma (1973), Fermont (1982), and Setiawa (1983). The degree of embryonic enclosure, the dimensions of the embryonic chambers, and the number of the periembryonic chambers have all been used as characteristic orthophragminid morphometrics (Brolsma, 1973; Fermont, 1982; Setiawan, 1983). More recently the morphological features of the whole group have been revised by authors such as Less (1987, 1998), Özcan et al. (2006) and BouDagher-Fadel and Price (2017). These authors focussed on internal features found in equatorial sections, and emphasised the description of the embryo (Fig 6.19), taking into consideration a number of parameters:

- P1 and P2: the diameter of protoconch perpendicular and parallel to P-D axis,
- Dl and D2: diameter of deuteroconch perpendicular and parallel to P-D axis,
- A: the number of auxiliary chamberlets directly arising from the deuteroconch,
- the number of adauxiliary chamberlets (see Fig. 6.41), or according to Hottinger (2006) corona - the first cycle of chamberlets enveloping an embryonal apparatus completely, at least in one plane of sectioning, as in *Discocyclina*,
- the number of annuli within a 0.5mm wide stripe measured from the rim of deuteroconch along P-D axis,
- H and W: the height and width of the equatorial chamberlets in the first annulus,
- h and w: height and width of the equatorial chamberlets around the peripheral part of the equatorial layer.

The development of annular stolons in the orthophragminids is deemed to be of generic significance (Samanta, 1967). Annular stolons occur at the proximal end of the radial walls and connect adjacent chamberlets in *Discocyclina* and *Asterocyclina* (BouDagher-Fadel, 2008), and at the distal end in *Pseudophragmina* and *Asterophragmina* (Haynes, 1981).

The **Discocyclininae** have a *Cycloclypeus*-like microspheric juvenile (see Fig. 6.41) and might have evolved from an earlier nummulitid (Vaughan, 1945). However, many authors subsequently disputed the presence of true canals (Brönnimann, 1951; Samanta, 1967) but rather regarded them as the fissures of the Rotalioidea. The only reference to a primitive *Discocyclina* is that made by Caudri (1972), who describes a genus, *Hexagonocyclina* Caudri, 1944 with "two symmetric auxiliary chambers on each side of the nucleoconch and four spirals, and predominantly hexagonal equatorial chamber". However, the issue of the real ancestors of the orthophragmids is still not yet resolved as the earliest megalospheric or A-forms, from the early Thanetian, are well developed and do not yield preorbitoidal stages (Less et al. 2007; Ferrandez-Cañadell and Serra-Kiel, 1992; BouDagher-Fadel and Price, 2017).

Discocyclina was followed directly in the Late Paleocene (Thanetian) by *Nemkovella*, which differs from *Discocyclina* in lacking annular stolons. It only survived until the Early Eocene, while *Discocyclina* did not disappear before the Late Eocene. Parallel to this evolution was that of the **Orbitoclypeinae**, the first form of which to appear, in the Middle Paleocene, was the stellate *Asterocyclina*, which was directly followed by the non-ribbed *Orbitoclypeus* in the Late Paleocene. Both survived the Paleocene-Eocene boundary only to die out at the end of the Eocene.

The study of the evolutionary lineages and stratigraphic distribution of the Tethyan orthophraminids, alveolinids and nummulitids led Serra-Kiel et al. (1998) to propose 20 shallow benthic zones (SBZ) in Paleocene-Eocene time. Less (1998), and Less and Kovács (1995) documented the stratigraphic ranges of Tethyan orthophragminid foraminifera with reference to these standard zones. They separated eighteen orthophragminid zones, marked OZ, by Less et al. (in press), from OZ 1a to 16, ranging from early Thanetian to late Priabonian.

The **Acervulinidae** (Fig. 6.42) were another evolutionary lineage that occurred in the Eocene in parallel to the *Planorbulinella* lineage, that were based on globular and subglobular tests. In the Paleocene, small, spherical-concentric *Sphaerogypsina* evolved from a *Cibicides*-like ancestor by multiplying chambers in numerous layers (Fig. 6.42). The chamber arrangement strongly resembles that of *Planorbulinella*, but with the pattern being observed in every section irrespective of its direction. In the Middle Eocene,

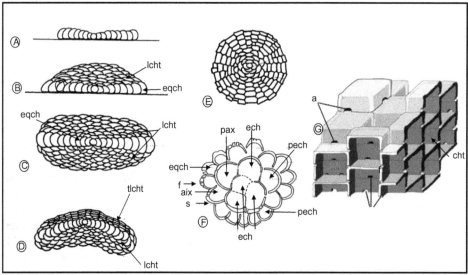

Figure 6.42. A-D sketches showing different Acervulinidae genera in axial sections, E-F in equatorial sections, modified after Bursch (1947). (A) *Planogypsina squamiformis* Bermudez showing chambers added in a single layer (see Chapter 7). (B) *Gypsina mastelensis* Borsch with arcuate lateral chamberlets form on both sides of the equatorial layer. (C) *Discogypsina discus* Goes. (D) *Discogypsina vesicularis* Silvestri with truncate lateral chambers on one side and arcuate on the other side of the equatorial layer. (E) *Sphaerogypsina globulus* Reuss, with chambers aligned in successive layers. (F) The arrangement of the first-formed chambers in *Discogypsina discus*. (G) Alternating chamberlet arrangement three dimensions, producing a chessboard pattern, as in the spherical-concentric, globular shell of *Sphaerogypsina globulus* Reuss, after Hottinger (2006).

a simple discoid, weakly symmetrical form evolved from a *Cibicides*-like ancestor into the three-layered *Discogypsina* (Fig. 6.41). In the Late Eocene, subglobular *Gypsina* developed a median layer, which can be recognized only in the central, earliest portion of the test, while a form similar to *Sphaerogypsina,* but with few spines and many pseudo-pillars, *Wilfordia,* is not yet known outside the type locality in Sarawak of Late Eocene age.

The **Asterigerinoidea** originated from an *Amphistegina*-like form that itself was derived from a *Cibicides*-like ancestor in the Eocene, and evolved into the **Lepidocyclinidae** in the Middle Eocene in the Americas, but developed later in the Oligocene-Miocene in Tethys (see Fig. 6.43; BouDagher-Fadel, 2008; BouDagher-Fadel and Price, 2010a). The **Lepidocyclinidae** comprises the subfamilies **Helicolepidininae**, in which adauxiliary chambers are absent and spirally arranged chambers completely surround the embryon, (e.g., *Eulinderina, Helicolepidina, Helicostegina, Helicosteginopsis, Polylepidina*), and the **Lepidocyclininae**, in which adauxiliary chambers are present and there is limited or no development of spirally arranged chambers (e.g., *Astrolepidina, Eulepidina, Lepidocyclina*).

The origin of the Lepidocyclininae has been the subject of many studies (e.g. Barker and Grimsdale, 1936; Tan Sin Hok, 1936; Rutten, 1941; Hanzawa, 1964; 1965; Matsumaru, 1971, 1991; Frost and Langenheim, 1974; Sirotti, 1982; Butterlin, 1987; Drooger, 1993; BouDagher-Fadel and Price, 2010a; and others). They originated in the middle Eocene in the Americas (e.g., Butterlin, 1987), and gradually migrated eastward, with their first appearances in the middle Eocene of West Africa, the early Oligocene in the Mediterranean part of the Tethys, and the latest early Oligocene in the Indo-Pacific (e.g., Adams, 1967; BouDagher-Fadel and Lord, 2000; BouDagher-Fadel and Price, 2010a). They reportedly went extinct in the middle Miocene (Serravallian)

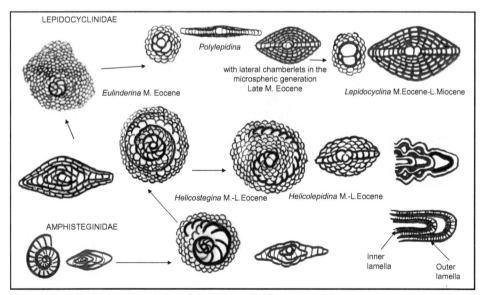

Figure 6.43. The origin and evolution of the lepidocyclinids.

in Indonesia and the northern Indo-Pacific (BouDagher-Fadel and Price, 2010a), but extended into the late Miocene (Tortonian) off Australia (Hallock and others, 2006). Reports of early Pliocene occurrences off Australia and the Pacific islands (Adams et al., 1979; Betzler, 1997) might represent reworked material and requires further biostratigraphic study to confirm their age (BouDagher-Fadel and Price, 2010a).

Lepidocyclina evolved from an amphisteginid ancestor (Grimsdale, 1959), *Eoconuloides* via *Helicostegina, Helicolepidinoides, Helicolepidina, Eulinderina* and *Polylepidina* by simultaneously gradually decreasing the size and number of whorls of the eoconuloid stage while increasing the number of rows in the median chambers. In *Helicostegina* the annular growth is derived from the eoconuloid spiral, which is reduced in *Eulinderina* and completely missing in *Polylepidina*. The persistence of the spiral character in the parallel lineage of *Helicolepidina* is accompanied by differentiation of

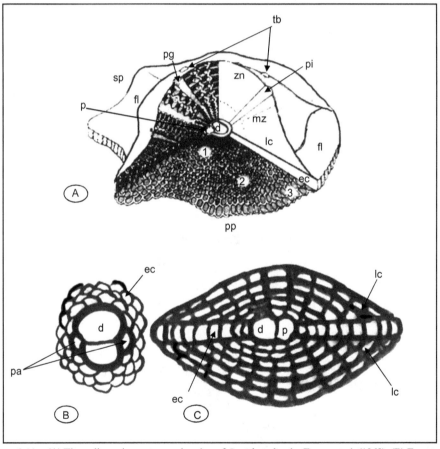

Figure 6.44.   (A) Three dimension cut away drawing of *Lepidocyclina* by Eames et al. (1968). (B) Equatorial section of a megalospheric *Lepidocyclina*. (C) Axial section of a megalospheric *Lepidocyclina*. Abbreviations: p = protoconch, d = deuteroconch, sp = stellate periphery, pp = polygonal periphery, ec = equatorial chamber, lc = lateral chamber, pi = pillar, fl = flange, tb = tubercles, zn = central zone, mz = marginal zone, pg = pillar forming an external tubercle, pa = auxiliary chambers.

lateral and equatorial zones of chambers in both stocks. During ontogeny, the equatorial chambers are mainly inside the ancestral eoconuloid wall, but later on this wall disappears and the genuine median layer is developed.

The embryonic apparatus in the megalospheric generation of all lepidocyclinids comprises the first and second chambers of the test (Fig. 6.44). These may be referred to as the proloculus and deuteroloculus when they form an integral part of the primary spire (e.g. *Helicostegina gyralis, Eulinderina* sp., see Fig. 6.43), and as the protoconch and deuteroconch when separated from the spire by a displaced third chamber (principal auxiliary) with two basal stolons. In this case, the second chamber (deuteroconch) does not form an integral part of the primary spire (e.g. *Polylepidina chiapasensis*). In all species of *Helicostegina* and *Eulinderina*, the deuteroloculus tends to be smaller than the proloculus, and the third chamber smaller than subsequent spiral chambers (Adams, 1987). As in *Discocyclina,* the protoconch of the Lepidocyclinidae is always in the centre of the test. The embryonic apparatus consists of protoconch and deuteroconch, with a strong tendency for the deuteroconch to embrace the protoconch.

Morphometric studies on the embryont of *Lepidocyclina* have been carried out by many taxonomists. Van der Vlerk (1959) proposed the parameters based on the difference between the angle of attachment of the deuteroconch, expressed as an "alpha value", and the percentage overlap, as the "Ai value". DI and DII (Fig. 6.45) are the diameter of the protoconch and deuteroconch, measured at right angles to a line joining their centre (Haynes, 1981). Nepionic chambers are considered as auxiliary if their walls rest on the embryon, and interauxiliary if they are formed from the auxiliary chambers. Accessory chambers arising over the stolons in the deuteroconch are called adauxiliary, while those arising over the protoconch are called protoconchal. The principle of nepionic acceleration seems to be the almost universal. Nepionic acceleration is a general decrease over time in the number of steps in chamber formation, or budding steps, that is needed before orbitoidal growth is reached. It is seen in all known lineages of larger foraminifera with orbitoidal growth, and can also be inferred to have analogues in other larger foraminifera, e.g. *Nummulites*. Occasionally, lineages include discontinuities or sudden apparent relapses to more primitive forms. Examples of such relapses are the relatively primitive evolutionary stage, in terms of nepionic acceleration, of the first European *Miogypsina* (see below) compared with coeval and supposedly ancestral *Miogypsinoides*, and the sharp drop in embryon size from *L. (Nephrolepidina) sumatrensis* to *N. angulosa* in the Indo-Pacific province. They can sometimes be explained by local environmental factors, or by invoking bottleneck events in which the populations are dramatically reduced and only less specialized and more primitive surviving forms provide the basis for further development. However, it may also be necessary to invoke local extinctions and re-colonization by forms which evolved elsewhere, and reached different evolutionary stages, in order to explain these relapses.

The stolon system of communication varies following the chamber shape. Arcuate chambers possess four stolons, while hexagonal chambers have up to six stolons. There have been many studies of European (Freudenthal, 1969; De Mulder, 1975) and Indo-Pacific (van Vessem, 1978) forms. However, variation of the stolon system is difficult

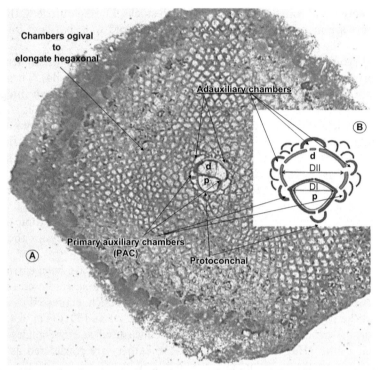

Figure 6.45. (A) *Lepidocyclina (Nephrolepidina) morgani* Lemoine and Douville, Formation bioclastique de Carry from the Nerthe area, near Marseille Petit Nid. (B) Sketch showing the embryonic apparatus of *L.(Nephrolepidina)*, modified after Haynes (1981). Abbreviations: p = protoconch, d = deuteroconch, DI = diameter of the protoconch, DII = diameter of the deuteroconch.

to see in axial section and can vary in individual specimens, which prompted some authors such as Eames et al. (1962) and van Gorsel (1975, 1978) to doubt its taxonomic usefulness.

Auxiliary chambers are primary auxiliary chambers (PAC) if they originate from stolons between the protoconch and the deuteroconch (Fig. 6.45). All subgenera of *Lepidocyclina* possess a complete ring of PAC, one on either side of the embryonic apparatus, and each PAC possess two stolons. *L. (Nephrolepidina)* develops one or more adauxiliary chambers, from a single stolon, on the deuteroconch and accessory chambers occasionally on the protoconch (protoconchal chambers).

The biometric parameters of *Lepidocyclina* have been used by many scientists since van der Vlerk (1959) first introduced the factors "A" and "B" to characterize the nucleoconch. The factor "A" measures the degree to which the protoconch is enclosed by the deuteroconch and the parameter "B" measures the extent to which the adauxiliary chambers cover the circumference of the nucleoconch. The "B" parameter was later found to be inadequate to explain the morphometric variations, and hence was rejected (van der Vlerk, 1963). Other parameters were subsequently introduced to express the phylogenetic stage of the lepidocyclinids. These includes *dc* (degree of curvature of the common wall; van der Vlerk and Gloor, 1968), INT ratio (percentages of isolepidine-nephrolepidine-trybliolepidine types in an assemblage; van der Vlerk, 1973), E (degree of curvature of the embryonic chambers; Chaproniere, 1980) and F (form number; Chaproniere, 1980).

Many of the lepidocyclinid taxa have been used by different authors in numerous ways, with different stratigraphic significance attached to each. This led to the reassessment of the genus-group names by BouDagher-Fadel and Banner (1997). These authors demonstrated that the subgenus *Isolepidina* Douvillé is a synonym of *Lepidocyclina* Gümbel, and that both are usable in the same and original sense, the latter being the senior name available for subgeneric use. They also neotypified the type species of the subgenus *L. (Nephrolepidina)* Douvillé, so that the subgenus can be used in the sense adopted by van der Vlerk and other Dutch workers. They proposed the abandonment of the name *Trybliolepidina* van der Vlerk, because a type species was subsequently designated making this genus a synonym of *Eulepidina*. (see BouDagher-Fadel and Banner, 1997). The characteristics intended for "*Trybliolepidina*" by its original author are more accurately characterized by the use of subspecific names in the nomenclature of each species with the appropriate nepionic morphology.

Many authors, such as Lunt and Allan (2004), do not agree with removing the genus *Trybliolepidina.* Others such as Renema (2005), argues that in thin sections it is hard to distinguish between *Lepidocyclina, Eulepidina* and *L. (Nephrolepidina)* and a better way of citing the species is for example *L. (Nephrolepidina)*? *rutteni*, if the A-B species (megalospheric-microspheric) couple is known or "Lepidocyclinidae gen, indet." if only the B-form (microspheric) of the species has been found. However, and as discussed by BouDagher-Fadel and Banner (1997) the microspheric forms are subgenerically indistinguishable from the Eocene to the Middle Miocene, whilst the megalospheric forms are distinguishable subgenerically in a way which is stratigraphically valuable. It is true that in the same sample both megalospheric and microspheric forms are sometimes difficult to distinguish, but as L. (*Nephrolepidina*) is stratigraphically important, it is useful to differentiate between the two forms on a subgeneric level.

The general ranges of the **Asterigerinoidea** are shown in Fig. 6.46. However, in many cases more detailed evolutionary trends have been determined. An example of the evolution of the American lepidocyclinids as developed by Barker and Grimsdale (1936) and others is given in Fig. 6.47. In the Late Paleocene *Eoconuloides lopeztrigoi* evolved gradually from relatively simple to complex forms with many lineages in the Middle Eocene. The beginning of these lineages saw the appearance of *Helicostegina gyralis*, a primitive, single long spired form, equipped with counter septa, which gradually evolved into a shorter-spire form with a complete flange of equatorial chamberlets in *H. dimorpha.* In the latter, the lateral chambers may appear in the last whorl, but a fully tiered arrangement is not attained. The lateral chambers in *H. polygyralis* were poorly developed and appeared only in later growth in *H. paucipira,* with fewer primary chambers in *H. soldadensis.* In the Middle Eocene, these forms gave rise to *Eulinderina guayabalensis,* by evolving more numerous arcuate equatorial chambers in a cyclical arrangement, with shorter primary coils and counter septa. *Helicostegina polygyralis* evolved into *Helicolepidina spiralis* by developing retrovert apertures, which in turn generated a second spire. This form evolved into *H. trinitatis,* which appears to be the most advanced member of the genus. It exhibits two subequal primary auxiliary chambers and 4 unequal periembryonic spires. According to Adams (1987) *H. trinitatis* indistinguishable from *H. nortoni* and of which it is a junior synonym.

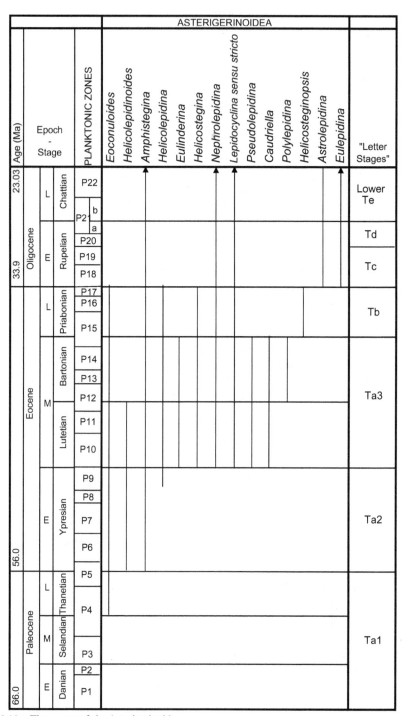

Figure 6.46. The ranges of the Asterigerinoidea.

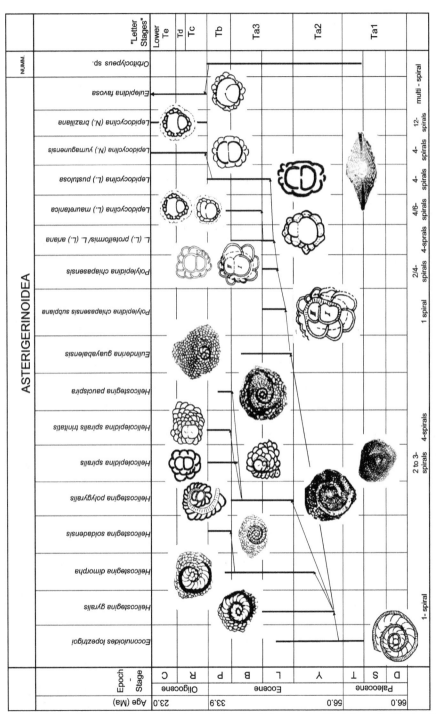

Figure 6.47. Evolution of the Asterigerinoidea during the Paleogene modified from BouDagher-Fadel (2008) and BouDagher-Fadel and Price (2010a).

Polylepidina chiapasensis evolved in the Middle Eocene from *Eulinderina guaya-balensis* by increasing the number of embryonic spires from one to two (*P. chiapasensis subplana*) or more, and by developing two principal auxiliary chambers, with basal apertures and true lateral chambers. It gradually evolved from a 2-spiraled form to being a 4 spiraled one, and gave rise to the *Lepidocyclina* (*Lepidocyclina*) *proteiformis* group, by developing 4 equal spires, the 4 spires in *Polylepidina* being unequal (see Fig. 6.47). This group evolved into *L.* (*Nephrolepidina*) *tournoueri* (see Chapter 7) and, in the rest of the European and Indo-Pacific, into the *Lepidocyclina* (*Nephrolepidina*) lineages. Quite independently, the Early Oligocene marked the first appearance of the earliest known species of *Eulepidina* (see Fig. 6.47) in America, e.g. *E. favosa* (see Chapter 7), which were the ancestors of the rest of the European and Indo-Pacific *Eulepidina* lineages (see BouDagher-Fadel and Price, 2010a; Chart 7.1). *Eulepidina* is found throughout the Oligocene and ranges to the early Miocene, but they were most common during the early to middle Oligocene. In *Eulepidina*, the adauxillary chambers are particularly numerous, relatively small and irregularly distributed (Fig. 6.18).

It is proposed that several lineages developed during the history of the evolution of *Lepidocyclina s.l.* (BouDagher-Fadel and Lord, 2000; BouDagher-Fadel and Price, 2010a), *Miogypsina* (BouDagher-Fadel et al., 2000; BouDagher-Fadel and Price, 2013), *Cycloclypeus* and *Austrotrillina,* all of which have contrasting phyloge-netic characteristics. In the early stages of evolution of *Lepidocyclina s.l.*, the mega-lospheric and microspheric forms were similar in size and shape. However, as will be discussed in Chapter 7, in the Neogene Letter Stage interval late-Te to Tf3 there was a marked development of large microspheric forms, which bear few similarities with corresponding megalospheric forms that are often smaller and less distinctive. This evolutionary history curiously parallels that of the genus *Nummulites.* In the Early and Middle Eocene, the microspheric and megalospheric generations of *Nummulites* were similar in size and shape but in the Late Eocene to Early Oligocene specimens of the microspheric generation were very large in size and dominate their (often associated) megalospheric partners. A famous example of this is the megalospheric *Nummulites fichteli* and the microspheric *Nummulites intermedius* which often occur together but are thought to be the same biological species (see van der Vlerk, 1929, p.36, figs 30a-b, 31a,b; Eames et al., pl.1).

The **Rotalioidea** (Fig. 6.48) can be traced back to *Rotorbinella* in the Late Cretaceous (see Hottinger, 2013), and possibly to a discorbid ancestor, such as *Biapertobis* (Fig. 6.49). The **Rotaliidae**, mainly with low-conical tests (with the umbilical side cov-ered by numerous pillars) appear in the Early Paleocene, and include *Dictyokathina*, *Lockhartia* and *Dictyoconoides* in the Middle Eocene. On the other hand, the high conical **Chapmaninidae** appeared in the Late Paleocene and were common in the Late Paleogene (e.g. *Angotia, Crespinina*) (see Fig 6.17).

The **Miogypsinidae** are the best known group of rotaliides foraminifera as they have been the subject of many micropaleontological investigations. Miogypsinids lived from Oligocene to Middle Miocene times, and form a largely coherent morphological and phylogenetical unit. As will be discussed in the next section, it seems that the miogy-psinid group shows distinct provincialism. Many independent lineages of *Miogypsina* have been recognised in the Mediterranean, Indo-Pacific and central American bio-provinces (e.g., Van der Vlerk, 1966; Salmeron 1972; Raju, 1974; Raju and Meijer, 1977;

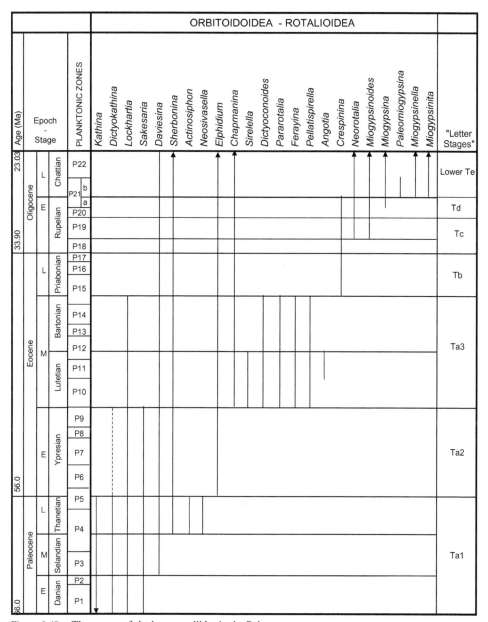

Figure 6.48. The ranges of the large rotaliides in the Paleogene.

De Bock, 1976; Chaproniere, 1983, 1984; Cahuzac, 1984; Ferrero, 1987; Wildenborg, 1991; Ferrero et al., 1994; Drooger, 1952, 1963, 1984, 1993; Drooger and Raju, 1973; Cahuzac and Poignant, 1991; BouDagher-Fadel et al., 2000; BouDagher-Fadel and Price, 2010b, 2013).

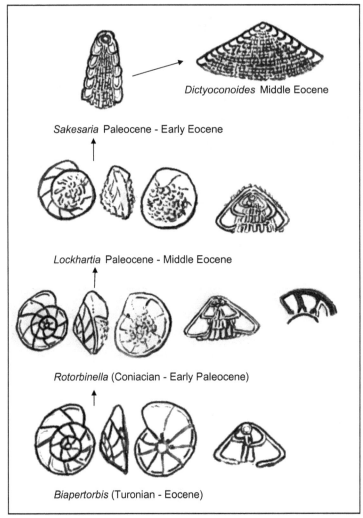

Figure 6.49. The evolution of the conical *Dictyoconoides* from a rotaliid ancestor.

It is generally accepted that the group originated from a trochoid ancestor. In the Americas, miogypsinids appear in the in latest Early Oligocene (Rupelian) (BouDagher-Fadel and Price, 2010b), while in Europe, *Miogypsina* spp. first appears in the Aquitanian (Drooger, 1963) and is not known in the underlying Oligocene. On the other hand, *Miogypsinoides* evolved independently from *Miogypsinella* in the Far and Middle East (BouDagher-Fadel and Banner, 1999) at the base of the Late Te Letter Stage. Many of the ranges presented in this and the subsequent chapter for genera occurring the Far East and in the American provinces have been revised and modified by BouDagher-Fadel and Price (2010b, 2013)

Tan Sin Hok (1936) described the evolution of the Miogypsinidae through *Miogypsinoides*, in the Early Miocene of Indonesia (then Netherlands, East Indies).

He traced the morphological development on the basis of the megalospheric nepionic chamber arrangement, but he also used a typological species concept in his classification. This typological species concept was later used by other workers such as Cole (1957). From the study of Central American collections, Barker and Grimsdale (1931) proposed that *Neorotalia* was the root stock for all the Miogypsinidae. This evolution was demonstrated by them from the presence of a septal canal in the most primitive *Miogypsinoides*. Other authors have offered different opinions about the transition from *Miogypsinoides* to *Miogypsina* (see Drooger 1993, p.75-76).

Drooger (1952) in his thesis on American Miogypsinidae introduced the population concept for classification purposes, that is to say measurements of numerous specimens were used to define taxa. Since then he has performed many biometric studies on the *Miogypsina* nepiont, as seen in equatorial section. The biometric method appeared to be a useful tool in the study of the evolution of a single lineage of *Miogypsina* and of miogypsinid genera in general (see Fig. 6.50). Indices "X" and "V" are used as the primary characteristics for subdivision, where X is the total number of spirally coiled nepionic chambers, once a second spiral appears, while the biometric factor V, which is equal to $200\alpha/\beta$, is the degree of symmetry between the two spirals (see Fig. 6.50). In *Miogypsina,* the biometric factor V is low in older species, and high in the nearly symmetrical embryon of the younger species (Drooger, 1993). In the *Miogypsinoides* sequence of assemblage-species in the Indo-Pacific (*M. complanatus - formosensis - bantamensis - dehaarti – indicus*) there is an evolutionary lineage involving the reduction of the nepionic spiral, and the increase in size of the protoconch (Drooger, 1993; see also Chart 7.1). Drooger (1993) set numerical limits between the X values to define the species, thus: *complanata - 17 - formosensis - 13 - bantamensis -10 - dehaarti - 8 - indica.*

The biometric factor "Y" is the number of spiral chambers up to the first chamber with two foramen or stolons (see Fig. 6.50), while the factor "γ" is angle of deviation of the line of symmetry of the embryont from the apical line, which is the line of symmetry of the adult test. For long-spiraled juveniles this value is considered as a negative number and positive when there is less than one juvenile whorl, and there is a second nepionic spiral. De Bock (1976) and Drooger (1993) found that *Miogypsinoides* possesses a well-developed canal system, whereas in *Miogypsina* such a system is lacking. On the other hand, *Miogypsina* was believed to develop a more complex stolon system than in *Miogypsinoides*.

Matsumaru (1996) in his studies of Japanese Paleogene foraminifera, recognised the Late Oligocene origin of *Miogypsinella* from his new genus *Paleomiogypsina* in the late Chattian, which he believed to be a descendant of *Pararotalia* (Matsumaru, 1996, p. 44). However, in common with the conclusions of Barker and Grimsdale (1931), BouDagher Fadel et al. (2000) traced the origin of the Miogypsinidae to the genus *Neorotalia*, and not to *Pararotalia* nor to *Rotalia* and its relatives (e.g. *Medocia*), which have a different umbilical structure and apertural system.

Pararotalia and *Neorotalia* have very different apertural systems from *Rotalia*, as discussed by Loeblich and Tappan (1988, p. 659), however, *Pararotalia* and *Neorotalia* are not synonymous as suggested by these authors. *Pararotalia* first occurs in the Late Cretaceous and gives rise to *Neorotalia* in the Oligocene. They both have a relatively low trochospire, but *Pararotalia* is more strongly trochospiral and smooth,

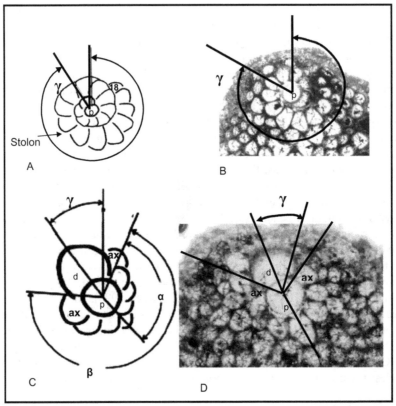

Figure 6.50.   (A). A sketch showing the earliest chamber of an older *Miogypsinoides,* such as *Miogypsinoides complanatus.* There are 18 spirally coiled nepionic chambers (biometric factor X). There are 10 spiral chambers before the first chamber with 2 stolons (biometric factor Y). The angle of rotation between the apical line of symmetry and the central line crossing the protoconch and deuteroconch is around -330°. (B) *Miogypsinoides bantamensis* Tan Sin Hok from Castlesardo section, northern Sardinia, with a biometric factor of X = 10 and a γ of approximately -290°. (C) A sketch of the early chambers of a young *Miogypsina,* such as *M. cushmani,* with 2 auxiliary chambers (ax) with a biometric factor V (= 200α/β), where γ, the positive angle of deviation of the line of symmetry of the adult test, is approximately 78°. (D) *Miogypsina cushmani* Vaughan, from Castlesardo section, northern Sardinia.

while *Neorotalia* has a much lower trochospiral form and is pillared both dorsally and ventrally. *Neorotalia,* while sharing a basic architecture with *Pararotalia,* also differs from the latter by its enveloping canal system (Hottinger et al., 1991). After *Neorotalia* acquired its dorsal, ventral and peripheral pillars, it evolved into the miogypsinid lineage by the acquisition of an additional fan of small chamberlets in a median equatorial layer (see Fig. 6.51). *Paleomiogypsina* evolved from *Neorotalia* in the Late Oligocene, by acquiring little chamberlets scattered on the periphery of the last whorl (Matsumaru, 1996, p. 54). *Miogypsinella* evolved into *Miogypsinoides* (with lateral thickening, no cubiculae, and a planispiral embryont) in the Late Te stage (BouDagher-Fadel and Banner, 1999) with the initial coil becoming planispiral, and the lateral walls of the initial spire and the succeeding fan of ogival median chamberlets becoming very thick and solid (see Fig. 6.51). In forms such as *Miogypsinoides abunensis* pillars continue on both sides of the test. In *Miogypsinoides dehaarti* the proloculus is followed by a single whorl of auxiliary chambers (Van der Vlerk, 1966) and the pillars are lost.

The evolution of the miogypsinids started with a trans-Atlantic migration of *Neorotalia* from the Americas, where miogypsinids originated. The eastward migration followed two paths: one to the south towards South Africa, where a distinct phylogenetic lineage, but similar to that found in America, developed but went extinct in the Burdigalian; the other to the north, through the Mediterranean corridor. During the Chattian and Aquitanian significant miogypsinid forms evolved in the Mediterranean from the morphologically distinct Mediterranean *Neorotalia* and migrated, within a few million years of their first appearance, into the Indo-Pacific, where they diversified further.

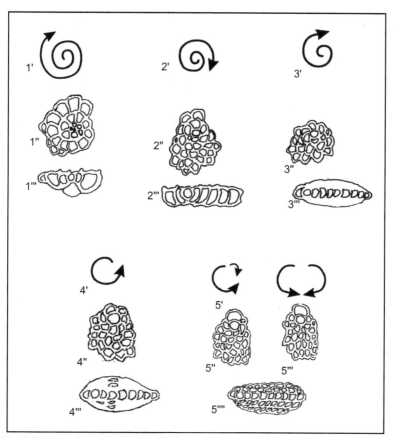

Figure 6.51. Silhouette diagrams of stages in the evolution of the Miogypsinids in the Far East, modified from BouDagher-Fadel and Lord, 2000. (1) *Paleomyogypsina.* (2) *Miogypsinella.* (3) *Miogypsinoides.* (4) *Miogypsinodella,* (5) *Miogypsina.*

5" early *Miogypsina,* 5''' advanced *Miogypsina,* e.g. *M. indonesiensis* Tan Sin Hok.
1' – 5' Direction and approximate extent of chambers coiled around the proloculus.
1' – 5", 5''' Equatorial extent of chambers coiled around the proloculus.
1''' – 4''', 5'''' Axial diagram showing form of the test.

The Mediterranean miogypsinids went extinct in the Langhian, but miogypsinids survived in the Indo-Pacific into the Serravallian. This miogypsinids exhibit an example of 'parallel speciation' as discussed by Schluter et al. (2004). As species became geographically isolated, colonizing new areas environmentally similar to each other, they thrived and evolved similar but distinct parallel lineages, taking advantages of empty niches and optimum conditions. This process probably reflects that they all shared a genetic predisposition to develop mutations of a specific, advantageous type, inherited from their last common ancestor (see BouDagher-Fadel and Price, 2010b, 2013).

6.4 Palaeoecology of the Paleogene Larger Foraminifera

The Paleogene is a transitional time during which the Earth moved from a uniformly warm Cretaceous to a cooler, more climatically heterogeneous Neogene (Berggren and Prothero, 1992). The Early Paleocene was characterized by warm, generally ice-free conditions, but slightly cooler than the preceding Cretaceous. However, temperature rose again in the Late Paleocene with an anomalously warm global climate optimum spanning some 4-5 Ma during the Early Eocene (Prothero and Berggren, 1992). In this brief period of extreme warming at the onset of the Eocene, called the Paleocene-Eocene Thermal Maximum (PETM) (Cramer and Kent, 2005), climates became hot and arid north and south of the Equator. The PETM has been attributed to a sudden release of carbon dioxide and/or methane and coincided with a major perturbation

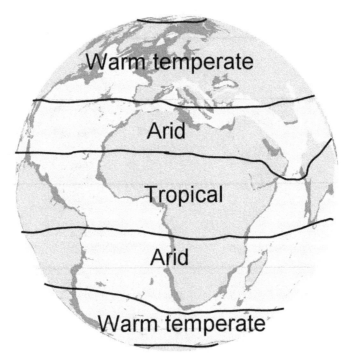

Figure 6.52.   Climate zones in the Middle Eocene.

of the carbon cycle as indicated by a sharp negative $\delta^{13}C$ excursion (CIE) (Dickens, 2001). This warm period continued through the Eocene (see Fig. 6.52), where a poleward expansion of coral reefs occurred, together with a broader latitudinal distribution of temperature-sensitive organisms, such as larger benthic foraminifera, mangroves, palms, and reptiles (Adams et al., 1990; Kiessling, 2002; Pearson et al., 2007). Coral reefs were able to grow up to 46°N, in comparison to today's coral-reef distribution that reaches to only 34°N (Kiessling, 2002).

Like most benthic fossils, the larger benthic foraminifera of the Paleogene are biofacies bound, sometimes even on a regional scale. They have biotopes closely associated with carbonate environments that represent a variety of palaeoenvironments, ranging from backreef to open marine conditions. Large scale changes in these biotopes occur in response to eustatic sea level fluctuations.

Larger foraminifera are extreme K-strategists (characterized by long individual lives and low reproductive potential, see Chapter 1) thriving in a stable, typically oligotrophic environment (Hottinger, 1983). Although all larger foraminifera are considered K-strategists, Paleogene larger foraminifera exhibit an increasing trend in K-strategy, from alveolinids to nummulitids to discocyclinids (Hottinger, 1982).

In the Cretaceous, a superheated (Fig. 6.53), hypersaline ocean-climate zone favoured the proliferation of rudists over corals (Kauffman and Johnson, 1988; Johnson et al., 1996). Following the K-P event, corals appear to have been rare and of low diversity (Johnson et al., 1996). The warm sea-surface temperatures and enhanced CO_2 levels of the Early Paleocene, would have caused symbiont loss and bleaching (Gattuso and Buddemeier, 2000), and thus prohibited the expansion of reef-building corals in the lower latitudes (Sheppard, 2003). The Early Paleocene is commonly thought to have been a time of reorganization of reefs and their communities (Newell, 1971; Talent,

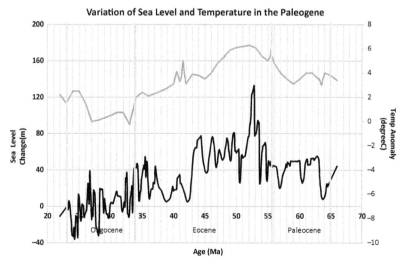

Figure 6.53. Variation in sea-level and temperature during the Paleogene based on Zachos et al. (2001) and Miller et al. (2011).

1988; Copper, 1989), and it was not until the late Thanetian that coral-algal patch-reefs became more diverse.

Following the K-P event, larger benthic foraminifera began to stage a recovery in the Early Paleocene, as unlike reef-building corals, rising summer sea-surface temperatures do not cause symbiont loss in their test (Hallock, 2000). They became the most common constituents of Late Paleocene-Early Eocene carbonate platforms, filling the empty niches left by the decline of the Cretaceous rudist-coral assemblages. They thrived on shallow, oligotrophic, circum-Tethyan carbonate platforms (Buxton and Pedley, 1989). By the Late Paleocene, nummulitids (*Nummulites*, *Assilina* and *Operculina*), orthophragminids (*Discocyclina*) and alveolinids (*Alveolina*) became important components of the carbonate Tethyan platforms. Their distributional pattern in Paleogene microfacies indicate different ecological gradients based on different faunal associations (Fig. 6.54), which in turn influenced lamellar thickness and flattening of larger foraminifera tests (Ćosović et al., 2004).

Although the Late Paleocene witnessed the beginning of the diversification of the nummulitids, it was during the Eocene that the *Nummulites* fulfilled their unique rock-forming potential. They became abundant and formed the widespread nummulitic limestones of hydrocarbon reservoirs in offshore North Africa, India, and the Middle East. Their reservoir qualities are mostly due to the preservation of the intraskeletal porosity of the *Nummulites* test. Various depositional models have been proposed, and most of them described *Nummulites* accumulations as banks, bars or low-relief banks, sometimes related to palaeo-highs. The behaviour of *Nummulites* could explain the diversity of such depositional models. Depending on local hydrodynamic conditions, autochthonous *Nummulites* deposits can be preserved as *in situ* winnowed bioaccumulations or can be accumulated offshore, onshore or alongshore, away from the original biotope (Jorry et al., 2006).

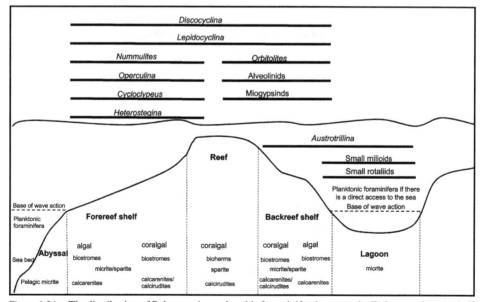

Figure 6.54.   The distribution of Paleogene larger benthic foraminiferal taxa on the Tethyan carbonate shelf.

The nummulitids included genera with flattened to stoutly lenticular and even globular species, with a periphery varying from sharp to rounded or somewhat undulose (Beavington-Penney and Racey, 2004). The globose to ellipsoidal-flat *Nummulites* thrived in warm, tropical mesophotic zone, while the large flat forms, such as *N. millecaput* are common in the oligophotic zone, either in shaded shallow water zones or deeper on the shelf (Mateu-Vicens *et al.*, 2012, Pomar *et al.*, 2017). Of the nummulitids, *Operculina* is the most "primitive" genus known and the least specialised (Cole, 1957; Chaproniere, 1975). For this reason it is thought to be less dependant on coralline algae symbionts for food, and therefore it is inferred to have had a wider environmental range than the other nummulitids (Cole, 1957; Adams, 1965). Although the living nummulitids house symbiotic microalgae, they prefer calm water conditions and avoid highly illuminated areas near the water surface, since their flat tests could easily be damaged in a turbulent hydrodynamic regime (Hohenegger et al., 2000). The living nummulitid *Operculina* appears to be restricted by oceanic salinities (Chaproniere, 1975). It has been found at depths as shallow as 14m (Hottinger, 1983). It has also been reported from the quieter parts of lagoons (McKee et al., 1959), channels on reef flats (Maxwell et al., 1961) and in off-reef shelf areas (Maxwell, 1968). *Operculina* lives on soft-bottomed substrates in the Gulf of Aqaba at depths of 30-150m, with flatter forms most common between 60-120m (Reiss and Hottinger, 1984). It dominates low-energy, muddy seabeds (Banner and Hodgkinson, 1991). In the Oligo-Miocene this genus is inferred to have inhabited environments ranging from high energy, shallow water forereef facies (BouDagher-Fadel et al., 2000) to quiet waters near the base of the photic zone (Chaproniere, 1975).

Similarly, *Cycloclypeus* has occupied a broader depth range in the past than at present. Modern species of *Cycloclypeus* are believed to live in deeper waters of 70-130m, down to the lower limit of the photic zone, crawling on firm substrates (Hottinger, 1983; Reiss and Hottinger, 1984; Hohenegger et al., 2000; Hohenegger and Yordanova, 2001; Yordanova and Hohenegger, 2002). They therefore tolerate lower light levels and temperatures than most other larger foraminifera (Cole, 1957). During the Oligo-Miocene, *Cycloclypeus* may have had a depth range of <12m to sub-euphotic depths (BouDagher-Fadel et al., 2000), and only later became restricted to the deeper range (Chaproniere, 1975). Thin, flat forms of larger benthic foraminifera are thought to be an adaptation to light attenuation with increasing habitat depth (Hallock and Schlager, 1986). In the Oligo-Miocene, *Heterostegina* inhabited high energy forereef environments (BouDagher-Fadel et al., 2000), of 20 to 30 m depth (Banner and Hodgkinson, 1991), preferring to live on hard substrates (Reiss and Hottinger, 1984). Holocene *Amphistegina* has adapted to high energy conditions, however, it is also found in mud-free sands in areas of sea grass or coralline algae and in reefal areas down to depths of 35m (McKee *et al*, 1959). Dead tests of *Amphistegina* have been found at greater depths (Chaproniere, 1975), but its main depth range is reported as 5 to 20m (Murray, 1973).

Photoautotrophic symbionts are the only food source for larger benthic foraminifera (Leutenegger, 1984; Krüger, 1994; Hohenegger, 2004) and provide the potential for calcification of large skeletons (Hallock et al., 1991). Eocene *Nummulites* has no counterpart in present-day protist groups (Jorry et al., 2006). The diameter of the largest *Operculinella venosus* microsphere is 6.4 mm, and 3.2 mm for the largest macrosphere (Hohenegger et al., 2000), whereas fossil *Nummulites* often reach several centimetres in diameter. Except for *Cycloclypeus carpenteri*, whose maximum observed diameter is 120 mm (Hohenegger et al., 2000), the largest size reported (Nemkov, 1962) is of a Mesopotamian specimen of *Nummulites millecaput* that reached 160 mm in diameter. Such gigantism is considered by Cowen (1983) as proof of an active algal symbiosis, which is also supported by the presence of microstructures similar to those observed in present-day larger benthic foraminifera, which provide shelter for symbionts and allow respiration (Bartholdy, 2002).

In deducing the palaeoecological distribution of the Orthophragminidae (such as *Discocyclina* and *Asterocyclina*), Ćosović et al. (2004) regarded these elongate, thin, flat to biconvex forms as homeomorphs to the Holocene *Cycloclypeus* and *Baculogypsinoides* (see Chapter 7). The orthophragminids at the lower limit of the photic zone are very flat, and their lateral chamberlets are particularly low in shape. The *Baculogypsinoides* individuals, with 3–4 strong spines, are similar to the genus *Asterocyclina*. They live on coral rubble and are extremely rare on sandy bottoms in Okinawa (Hohenegger, 2000; Hohenegger et al., 2000; Hohenegger and Yordanova, 2001; Yordanova and Hohenegger 2002) and the Spermonde Archipelago, Indonesia (Renema and Troelstra, 2001). Therefore, the homeomorph asterocyclinids with five to six rays might have lived on firm substrates in high energy environments, along with asterigerinids and amphisteginids, at depths of less than 30 m (Hallock, 1999; Ćosović et al., 2004). Smaller, globular nummulitid specimens are morphologically similar to the Holocene *Operculinella venosus* Fichtel and Moll, with thick lenticular tests, which inhabits sandy bottoms at moderate depths in the euphotic zone (40–80 m).

The division of the median chambers in *Cycloclypeus* has evolutionary parallels with the division of the cubiculae (or small chamberlets) of *Miogypsina*, where the illuminated chambers are also nests for diatoms (see BouDagher-Fadel et al., 2000). Each of the chambers cubiculae would act not only as a small convex lens for the focusing of sunlight, but could also act as a greenhouse for the containment and development of symbiotic diatoms. The diatoms enable these forms to acquire nutrients without food-gathering pseudopodial activity (Röttger, 1971). Although many small rotaliine foraminifera (e.g. *Rotalia, Rosalina*, etc.) gather food particles to their umbilici for extrathalamous digestion, larger foraminifera with cubiculae, but no umbilici (e.g. *Miogypsinella ubaghsi,* see Chapter 7), must have had a different method of ingestion.

In the Miogypsinidae several lineages evolved (BouDagher-Fadel et al., 2002) which followed similar evolutionary patterns, all related to the utilization of radiant light in the promotion of populations of photosymbionts. In *Miogypsinella* the flange of median chamberlets is relatively small (not much broader than the coil of the nepiont); in *Miogypsinoides* the flange grew very much broader, and would have become heavily thickened laterally. These lateral walls, becoming very thick, would have reduced the amount of sunlight which could reach the median layered chamberlets, so splitting into many layers of convex cubiculae in *Miogypsina* became competitively advantageous. Although diatom endosymbionts may initially enter the foraminifera as ingested food, they establish themselves very quickly as a permanent resident population inside the cell (Richardson, 2001). It is probable that *Miogypsinella* had active, food-gathering pseudopodia, however, it is likely that *Miogypsina* had pseudopodia which were of little effect in food gathering and instead used its cubiculae as an arable farm. The cubiculae are not linked by apertures, instead perforations occur in the walls. These perforations would be too fine to allow the passage of diatoms, which cubicular apertures would have allowed to escape. The requirement for and use of sunlight by *Miogypsina* is shown by the common occurrence of *Miogypsina* spp. only in shallow water marine limestones, where fossil algae also occur. The irregular shape of many species of *Miogypsina*, often leading in section to apparent division of their flanges (e.g. *Miogypsina bifida,* Plate 7.12, fig.14), shows they could become concavo-convex and were not growing on a flat surface. Such a shape would be developed if the species were attached on a strongly curved substrate, such as the stems or leaves of seagrass (the substrate would be bio-degradable and seagrass is not seen preserved in thin sections). On such an irregular substrate surface it would be difficult for any organism to remain fully attached and yet migrate on that surface. Therefore we believe that sedentary, attached miogypsinids grew to accommodate the shape of the vegetable substrate to which they adhered. Only in strong ambient sunlight, which would benefit both the miogypsinids and their vegetable substrate, could true *Miogypsina* flourish (BouDagher-Fadel et al., 2000).

The lineages of the lepidocyclinids evolved gradually, and newly evolved taxa persisted to become contemporaneous with their descendants, but occupied different ecological niches. They were common throughout warm-water, shallow-marine environments of the tropics. It is almost certain that their latitudinal distribution was controlled by temperature, as the 15^{0}C isotherm of the colder months limits the modern geographic distribution of Rotaliida that host algal endosymbionts (Langer and Hottinger, 2000; BouDagher-Fadel and Price, 2010a). An example of evolutionary changes influenced by ecological parameters is given by forms such as *L. (N.) ferreroi* (Plate 7.8, fig. 1; see BouDagher-Fadel et al., 2001), in which the massive pillars between lateral chamberlets may be thought to be diagnostic of living in high energy marine environments (BouDagher-Fadel et al., 2000). However, when pillars are growing on the lateral walls, as is the case in *Discocyclina*, orbitoids, lepidorbitoids, some lepidocyclinids, as well as in evolute nummulitids, they might act as true lenses, which focus light into the test for insolation of symbionts (Ferrancez-Cañadell and Serra-Kiel, 1992). On the other hand, this function is performed by thin lateral shell walls in some nummulitids (Hottinger, 1983), such as *Operculina* (Drooger, 1983) and *Cycloclypeus* (Reiss and Hottinger, 1984).

The miliolide genus *Austrotrillina* has a distinctly alveolar wall, again presumably to harbour symbiotic algae. During the Oligocene and Early Miocene, *Austrotrillina* was associated with alveolinids, such as *Borelis pygmaeus,* in the very low hydrodynamic-energy back-reefs (BouDagher-Fadel et al., 2000; see Fig. 6.54). The biological advantage of possessing an alveolar wall with increasing complexity of chamber division is not only found in Paleogene taxa, it is a well-established evolutionary trend of alveolinid foraminifera from the Cretaceous onwards. It may be analogous in function to the increased complexity of the *Austrotrillina* wall, probably increasing the efficiency of symbiosis with algae and diatoms.

Recent alveolinids occur in a wide range of habitats, from deep lagoons and to fore-reef settings, existing down to depths of about 80m. This, together with the fact that alveolinids are miliolines with a tolerance to salinity and temperature fluctuations, probably makes the group less sensitive to smaller sea level changes. The Eocene alveolinids became extinct, however, at the onset of Late Eocene rapid sea level changes, which led to the disappearance of vast carbonate platform and lagoonal areas. Oligocene and Miocene alveolinids also had a wide range of habitats, but were particularly common in deeper lagoonal settings. Their response to sea level changes might therefore be similar to that of other miliolides, were it not that extinctions during sea level falls are unlikely due to the group's extensive ecological range. Obliteration of the lagoonal habitat during sea level falls may have caused a reduction in population sizes but triggered an increase in the rate of evolution.

Extant species of soritids live in seagrass communities in tropical-subtropical, shallow water habitats, and are characterized by the possession of intracellular symbionts, such as rhodophyte, chlorophyte, or dinophyte photosymbionts (Lee and Anderson, 1991; Gast and Caron, 1996; Hallock, 1999). Species that host chlorophyte symbionts are found in relatively shallower waters (0-20m) than those that host diatom endosymbionts, which for example are found in a wider range of depth habitats (0-130m) (Leutenegger, 1984). According to Richardson (2001), the acquisition of dinophyte endosymbionts in the soritids facilitated a change in habitat from an epifaunal, free-living mode of life to one of living attached to phytal substrata.

Larger benthic foraminifera showed a diversification at a specific level, i.e. involving a rapid increase in species diversity, shell size, and adult dimorphism, during the Paleocene-Eocene transition (Hottinger, 1998). Fossil *Nummulites* are known to have developed a large range of shapes induced by reproductive strategies (small sexual A-forms and large asexual B-forms, see Chapter 1) and by environmental factors (light intensity and hydrodynamic conditions), which significantly affect the size, shape and thickness of the tests (Hallock 1979, 1981; Hallock and Glenn, 1986; Hohenegger et al., 2000, Ćosović et al., 2004). The hydrodynamic behaviour of *Nummulites*, which depends on their size, shape and density, is a fundamental parameter controlling their transport. According to Beavington-Penney and Racey (2004), foraminifera that live in shallow water produce 'robust', ovate tests with thick walls (e.g. *Orbitoclypeus*) to prevent photoinhibition of symbiotic algae within the test in bright sunlight, and/or to prevent test damage in turbulent water. However, there is a tendency towards flatter tests and thinner test walls with increasing water depth (e.g. *Discocyclina*). The change in the test shape reflects decreased light levels at greater depths. There is also a relationship between size, longevity and fecundity (Hallock 1985). It is known that larger foraminifera living in fairly turbulent waters become relatively large, with a thickness to diameter ratio of 0.6 to 0.7 (Ćosović et al., 2004). They produce up to ten times more offspring per reproduction than deeper dwelling species, with the ratio between juvenile and adult specimens also reflecting this phenomenon. Racey (2001) plotted the relationship between nummulitid accumulations and hydraulic energy; shallowing causes a progressive concentration of nummulitic tests as current energy increases, and as wave influences become dominant, imbrication gives way to chaotic stacking of the tests. As this combination of shape and foraminiferal accumulation is influenced by light intensity and hydrodynamic forces, the relationship of all these factors could be used to estimate facies type and depth in the Paleogene warm seas (Hottinger, 1997).

The Acervulinoidea (Fig. 6.55) live fixed to a substrate or most commonly to corals. The distribution of the coral-encrusting foraminiferan associations is also related to water depth gradients. Flat specimens mostly encrust the lower coral surfaces, where low light levels generally reduce competition for space with coralline algae. In contrast, globose morphotypes are successful on the coral upper surfaces, where lateral spatial competition with algae is higher (Bosellini and Papazzoni, 2003). In lamellar-perforate foraminifera, the thinning of the chamber walls results from thinning of each lamella, reflecting a decrease in the biomineralisation process with increased depth. Lamella thickness and flattening of shell shapes of larger foraminifera are products of water turbulence (Hallock 1979, 1999; Hallock and Glenn 1986; Hottinger 1977, 1997, 2000; Pêcheux 1995). It is also suggested by Ćosović et al. (2004) that the presence of fixed

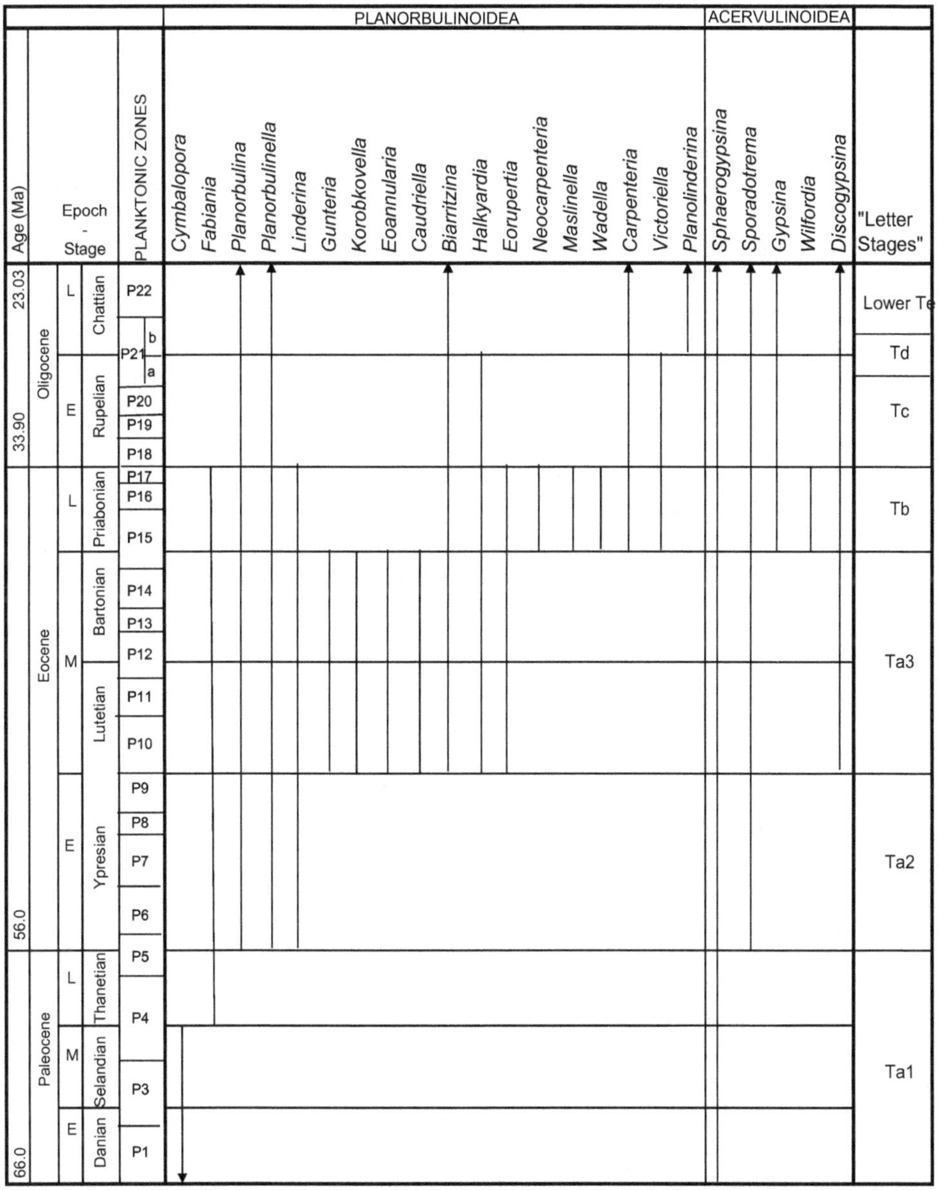

Figure 6.55. Range chart of the main species of the Planorbulinoidea and the Acervulinoidea.

(attached) foraminifera, such as *Biarritzina* and *Carpenteria*, is a good argument for clear water environments in the Eocene carbonate ramp environments, while the presence of orthophragminid dominated microfacies imply environments generally below storm wave base in middle to outer ramp areas.

The spatial distribution of Paleogene larger benthic foraminifera within the coralgal framework buildups of the Cenozoic carbonates is illustrated in Fig. 6.54. The deposition occurred entirely within the photic zone, in symbiosis with algae. The region was one of patch reefs in which the areas attacked by oceanic swell may be called "forereef". The energy distribution produced characteristic carbonate lithofacies and allowed distinctive assemblages of foraminifera to flourish:

- The reef itself, being shallowest, produced characteristic bioherms and calcirudites of corals and algae, cemented by sparitic matrix as the lime muds were flushed away by wave action.
- In the forereef shelves, proximal biostromes of corals and algae were cemented by sparite and micrite carbonates. Distally from the reefs, coral debris diminished and micrite increased, cementing the calcarenitic biogenic debris. Below the fairweather wave-base, micrite would accumulate, which contained scattered larger foraminifera swept in from the reef shelves. The proximal forereef shelf contained faunas dominated by planorbulinids, cycloclypids, lepidocyclinids, operculinids and heterosteginids.
- These shallow water carbonate facies were surrounded distally by deeper water sediment containing abundant planktonic foraminifera, which may constitute up to 35% volume. The planktonic foraminifera had flourished in the water column offshore and reflect relatively deeper water, and low energy. Here coral debris diminished and micrite increased, cementing the calcarenitic biogenic debris.
- Alveolinid, miliolid and miogypsinid larger benthic foraminifera are abundant in the backreef, sheltered from oceanic wave energies, or lagoon environment (see BouDagher-Fadel et al., 2000, p. 358; BouDagher-Fadel and Price, 2013). Here biostromes of corals and algae were cemented by sparite and micrite carbonates.

6.5 Palaeogeographic Distribution of the Paleogene Larger Foraminifera

The Paleocene ("ancient recent life") epoch marks the beginning of the Paleogene period and the Cenozoic eon. During this period, the continents continued to drift towards their present positions, with the development of the Alpine-Himalyan orogeny, the opening of the circum-Antractic seaway, but with South and North America still remaining separated (Fig. 6.56).

In the wake of the end Cretaceous crisis, when about 83% of the Maastrichtian larger benthic foraminifera (see Chapter 5) became extinct, the Early Paleocene was a period of recovery. The Paleocene exhibited a cooler climate than the Late Cretaceous. In the Early Paleocene, the absence of other reef-building, high-temperature tolerating organisms in the low latitudes, enabled larger foraminifera to occupy this vacant niche and rapidly evolve (Scheibner et al. 2005). The Danian and Selandian mark the start

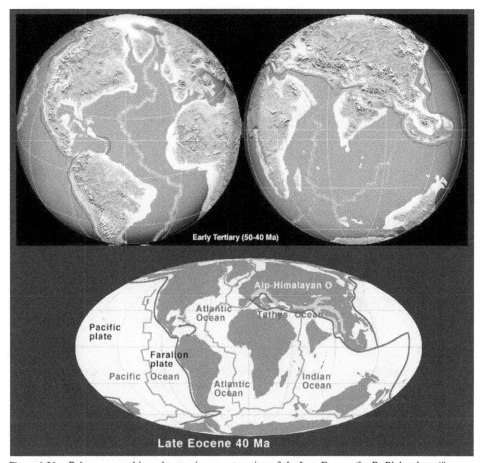

Figure 6.56. Palaeogeographic and tectonic reconstruction of the Late Eocene (by R. Blakey http://jan.ucc. nau.edu/~rcb7/paleogeographic.html).

of the recovery period for the larger foraminifera. At this stage, most of the Tethyan foraminifera (75%) were small, primitive, European-form rotaliides with low diversity (see Hottinger, 2013), and it was not until the Late Paleocene (Thanetian) that many new larger foraminifera made their first appearance. This evolutionary trend allows a closely spaced zonation of Paleogene carbonates to be defined. Alveolinids and orthophragminids, followed by nummulitic foraminifera with thick marginal cords, developed during this warm period of the Late Paleocene. They invaded the Tethyan margins during the Eocene, and become large and abundant in the forereef environments of the Early to Middle Eocene, only to disappear during the middle Oligocene. Parallel to this evolution was that of the alveolinids, which after appearing in the Late Paleocene of Tethys, became abundant, colonizing reefal and backreef environments.

The Paleocene-Eocene boundary saw the extinction of 25% of larger benthic foraminiferal species (Fig. 6.57), triggered by the thermal maximum (PETM) event, during which the sea surface temperature rose by 5°C in the tropics (Zachos et al.,

Paleogene Extinctions

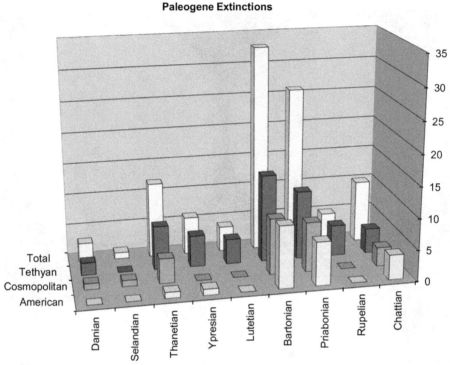

Figure 6.57. The number of larger foraminifera genera becoming extinctions at the top of each Paleogene stage boundary.

2003) and high CO_2 levels led to oceanic acidification (Zachos et al., 2005). The origins of the PETM are controversial. The Paleocene-Eocene boundary is marked by a period of intense flood basalt magmatism accompanying the opening of the North Atlantic (Eldhom and Thomas, 1993). The so-called North Atlantic Volcanic Provence (NAVP) (Fig. 6.58) is dated at ~58-55 Ma, and gave rise to volcanic fields that covered 1.3 million km[2] (Courtillot and Renne, 2003). It has been suggested that this may have generated metamorphic methane from sill intrusion into basin-filling carbon-rich sedimentary rocks (Storey et al., 2007). On the other hand, the reported presence of an iridium anomaly at the P-E boundary could be indicative of meteorite collision with Earth (Dolenec et al., 2000, Schmitz et al., 1996), which might in turn have triggered a change in the climate by releasing large amount of CO_2 from oxidized underlying marine sediments (Higgins and Schrag, 2006). Cramer and Kent (2005) argue that the very rapid onset of the PETM is best explained by such an impact mechanism, and refer to the after effects of this proposed impact as the "Bolide Summer".

However, whatever its origin, during this stressful event 30 to 50% of the deep water small benthic foraminifera suddenly became extinct. The breakdown of the stable oligotrophic environment of the larger benthic foraminifera resulted in the disappearance of most of the extreme K-strategists (Hottinger, 1983). The replacement of SBZ4 by SBZ5 faunas (see Chart 6.2), as indicated by the gradual disappearance of Paleocene taxa such *Ranikothalia* and *Miscellanea* and the rise of *Nummulites* and *Alveolina,* suggests that

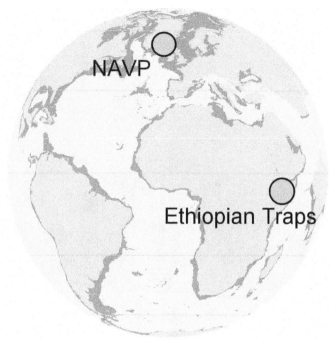

Figure 6.58. The Paleocene-Eocene world showing the position of the North Atlantic Volcanic Provence (NAVP) and the Ethiopian Traps.

such an interruption may have taken place in platform environments at low-latitude continental margins (Scheibner et al., 2006; BouDagher-Fadel et al., 2015; Li et al. 2016).

The Early Eocene witnessed the appearance of many new larger foraminifera genera (85% of which were cosmopolitan). A gradual increase of lineages followed this initial explosion of new forms, and during the Lutetian the American province (which up to the Middle Eocene had few endemic genera) experienced the highest appearance of new genera (Fig. 6.59). During the Lutetian the differences between the American province and the Western/Eastern Tethys province were considerable, and larger foraminiferal bioprovinces became pronounced. At this stage, the main province was the Tethyan (containing 35% of all genera, see Fig. 6.60), but this itself exhibited sub-provinces with some genera being restricted to for example Western Tethys or the Indo-Pacific province.

Traditionally, in the Paleogene the larger benthic foraminifera were considered to define three major, distinct palaeogeographic realms; namely, the American, the Western Tethys (which includes the modern day regions of West Africa, the Mediterranean and Tibet), and the Indo-Pacific provinces. However, a fourth distinct palaeogeographic province, the South West African realm has also been identified (see BouDagher-Fadel and Price, 2010a, 2013, 2014, 2017). Both the discocyclinids and lepidocyclinids originated in the American province (see BouDagher-Fadel and Price, 2010a, 2017), and although the discocyclinids rapidly attained a world-wide distribution, most of the earlier species became extinct at the end of the Middle Eocene in the American province, but continued elsewhere, only eventually to become extinct in the Late Eocene

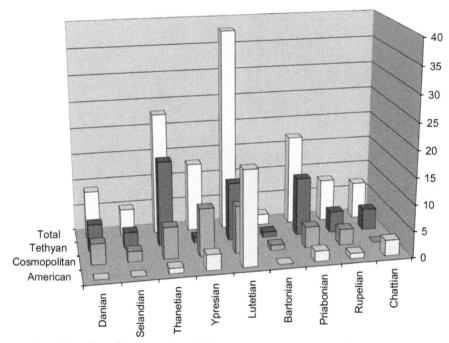

Figure 6.59.   The number of new larger foraminifera genera appearing in each of the Paleogene stages.

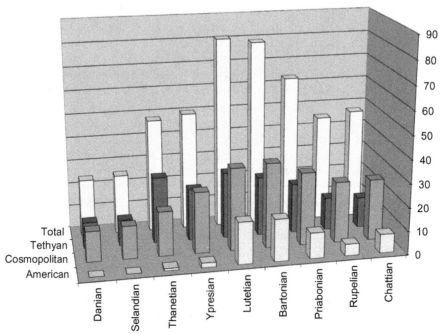

Figure 6.60.   The total number of larger foraminifera genera in each Paleogene stage.

(BouDagher-Fadel and Price, 2017). The lepidocyclinids appeared in the Lutetian and, with the exception of some forms which reached the West African shelf (Brun et al., 1982), were mainly confined to the American province in the Middle to Late Eocene (BouDagher-Fadel and Price, 2010a). Lepidocyclinids migrated eastward to the Mediterranean in the early Rupelian and reached the Indo-Pacific towards the end of the Rupelian. The earliest *L. (Nephrolepidina)* species in Tethys occurred in the late Chattian, and they were almost certainly the direct descendants of *L. (Lepidocyclina)* and *L. (Nephrolepidina)* of the Eocene and Oligocene of West Africa and America (BouDagher-Fadel and Price, 2010a). Lepidocyclinid migration from the American to the Tethyan province ended after sea-level rose following the early Oligocene sea-level minimum noted by Berggren and Prothero (1992), Miller et al. (2005), Katz et al. (2008), and Miller et al. (2011) (Fig. 6.53).

During the Eocene, the nummulitoid forms found in the Americas and South African provinces are very small (with diameters no more than 2mm), and different from those of Western Tethys and the Indo-Pacific (BouDagher-Fadel and Price, 2014). In the Western Tethyan province forms similar to the American *Pararotalia, Chordoperculinoides* and *Operculina* first appeared in West Africa in the Thanetian (P4b); later than their first appearance in the Americas. The Western Tethyan nummulitids have no apparent indigenous Tethyan ancestors, but BouDagher-Fadel and Price (2014) demonstrated that they were derived from American ancestors presumably by trans-Atlantic migration, a process also inferred (BouDagher-Fadel and Price 2010a; 2010b; 2013; 2017) to have

Figure 6.61. The inferred migration routes of orthophragminids during the Paleogene, shown by black arrows, from the Americas (1), to the Western Tethys (2), and on to the Indo-Pacific (3), and to South Africa (4) (from BouDagher-Fadel and Price, 2013).

occurred at a later geological epoch to explain the global dispersal of three other LBF groups, the discocyclinids, the lepidocyclinids and the miogypsinids; see Fig. 6.61).

During the Middle Eocene, the Tethyan (both Western Tethyan and the Indo-Pacific) province was dominated by large *Nummulites* and alveolinids. The abundance of *Nummulites* and *Assilina,* together with *Discocyclina* and *Spiroclypeus* distinguished this region (see BouDagher-Fadel et al., 2015). The nummulitoids evolved first in the Americas and then migrated eastward to Western Tethys (and then eventually on to the Indo-Pacific) and to SW Africa (BouDagher-Fadel and Price, 2014). As species became geographically isolated, they evolved parallel but distinct lineages. Eocene to Oligocene nummulitoids of Southern Africa evolved directly from American ancestors and were distinct from the Tethyan and Indo-Pacific forms, but a wave of nummulitoid migration occurred in the Miocene from the Mediterranean into the SW African province (see Chapter 7). The assilines are unknown in the Americas. The extinction of *Nummulites* can be correlated with the end of the Rupelian and planktonic zones P21.

Miogypsinids lived from Oligocene to Middle Miocene times, and form a largely coherent morphological and phylogenetical unit (BouDagher-Fadel and Price, 2013). They show a distinct provincialism: the evolutionary histories of American, European and Indo-Pacific miogypsinids, although showing roughly the same trends, are quite different (see BouDagher-Fadel and Price, 2013). They also originated in the Americas from *Neorotalia* in the Early Oligocene (Rupelian and P18; see BouDagher-Fadel and Price 2010c, 2013). During the Early Oligocene a series of sea-level regressions (Katz et al., 2008; Miller et al., 2011) reduced the effective width of the early Atlantic Ocean sufficiently to facilitate trans-oceanic migration of *Neorotalia* from the American province to the North African coast and on into the Mediterranean (BouDagher-Fadel and Price, 2013). During this time, Mediterranean shallow water niches were still occupied by the Paleogene *Nummulites*. However, towards the end of the Early Oligocene (around 31–29 Ma), environmental stresses, perhaps associated with cooling and the large flood basalt event in Ethiopia and Yemen (see Courtillot and Renne 2003), contributed to the disappearance of the last Mediterranean *Nummulites*. The Mediterranean *Neorotalia* (e.g. *N. tethyana,* see Boudagher-Fadel and Price, 2013) had by this stage become distinct from its American counterparts, and the disappearance of the *Nummulites* provided an opportunity for new phylogenetic lineages of miogypsinids miogypsinids to fill the warm reefs of the Mediterranean. As the morphologies of American and Mediterranean miogypsinids are seen to be crucially different, it follows that their evolutionary development was independent but closely parallel. After the last major regression in the early Chattian, the rising sea level and the continuing oceanic rifting effectively isolated the Mediterranean–West African shelf from the American province (around 28 Ma), ending any flow of *Neorotalia* or miogypsinids from America to the Mediterranean. It should be noted that during this time there was also a major change in oceanic circulation that resulted from the reversal of the direction of flow through the Panama Seaway (von der Heydt and Dijkstra, 2006). This reversal of flow is reported as being due to the tectonically driven widening of the Drake Passage and the narrowing of the seaway between the Mediterranean and Indo-Pacific, and that this may have also mitigated against further trans-Atlantic miogypsinid migration to the Mediterranean within the Late Oligocene (see BouDagher-Fadel and Price, 2013).

The Middle to Late Eocene boundary (37.2 Ma) witnessed other major changes in global climate and ocean circulation (Berggren and Prothero, 1992), which might have been responsible for the major extinction event of foraminifera seen at the end of the Bartonian, notably the disappearance of the large nummulitid species. The global temperature fell by an average of 2-4⁰C (see Fig. 6.53). The Eocene-Oligocene boundary temperature fall followed the onset of the Late Eocene rapid sea level changes, which led to the disappearance of vast carbonate platform and lagoonal environments and the final extinction of the Eocene alveolinids. The move into an "ice house" climate may have been triggered at this time by the opening of the Tasmanian gateway (Smith and Pickering, 2003). What ever the cause, this boundary saw a large number of extinctions, with the large *Nummulites* and *Assilina* disappearing from Tethys, and in the American province 50% of the larger foraminifera abruptly becoming extinct (see Fig. 6.57). This was accompanied a sudden crash in the taxonomic richness of the tropical phytoplankton species (Macleod, 2015).

The selectivity of Late Eocene extinctions, with the demise of mainly discoidal morphotypes of well-established calcareous larger foraminifera (40% globally), seems to constrain the possible causes of this Late Eocene event (Banerjee and Boyajian, 1997). Palaeoclimatic evidence shows a trend of cooling during late Middle Eocene and at the Eocene-Oligocene (E-O) boundary (Berggren and Prothero, 1992). Accelerated global cooling, with a sharp temperature drop of >2 °C occurred near the E-O boundary (Montanari et al., 2007). These global climate changes are attributed to the expansion of the Antarctic ice cap following its gradual isolation from other continental masses. However, multiple bolide impact events, possibly related to a comet shower lasting 2.2 Ma, may have played an important role in causing the deterioration of the global climate at the end of the Eocene epoch (Montanari et al., 2007). Indeed, the Eocene saw an unusually large number of significant impact events (see Fig. 6.62), the cumulative effect of which would have added significantly to the environmental stress during this epoch. These happened at Chesapeake Bay (C) 35.5 ± 0.3 Ma (crater diameter = 90km), Popigai (P) 35.7 ± 0.2 Ma (diameter = 100km), Mistastin (Mi) 36.4 ± 4 Ma (diameter = 28 km) ± 28 Ma, Haughton (H) 39 ± 23 Ma (diameter = 23 km), Logancha (La) 40 ± 20 Ma (diameter = 20 km), Logoisk (Lo) 42.3 ± 1.1 Ma (diameter = 15 km), Kamensk (K) 49.0 ± 0.2 Ma (diameter = 25 km), and Montagnais (Mo) 50.50 ± 0.76 Ma (diameter = 45 km). These impacts happened in parallel with major changes in global climate, beginning in the Middle Eocene and culminating in the major earliest Oligocene Oi-1 isotopic event (Montanari et al., 2007).

The largest of these impacts (Chesapeake Bay and Popigai) at ~35Ma, when associated with the later large flood basalt event in Ethiopia and Yemen (Fig 6.58) around 30 Ma (Courtillot and Renne, 2003), might have contributed to the disappearance (globally 41 %, in Tethys 17%) of many of the Eocene survivors. However, despite these impacts, and as seen in Fig. 6.57, the extinction event is not as large as that of the Bartonian-Priabonian boundary. Although some groups, such as the alveolinids, were more severely affected than others and disappeared completely at the E-O boundary, others survived this event only to become extinct within the Early Oligocene.

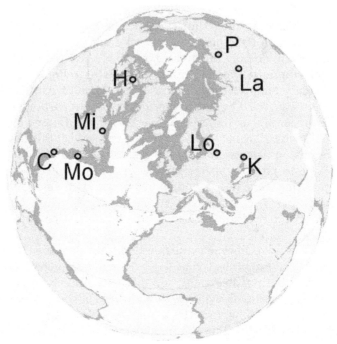

Figure 6.62.   The end-Paleogene world, showing the position of larger impacts occurring at that time.

In the Early Oligocene (from about 33.5 Ma), the Drake Passage opened and there was further significant climatic cooling and ice volume increase (Berggren and Prothero, 1992). Most larger foraminifera which had survived the E-O boundary became adapted to cooler environments, while others migrated to the warmer Tethys, such as the American lepidocyclinids and miogypsinids. At the end of the Oligocene there were very few notable extinctions. This stratigraphic boundary is probably mainly connected to plate tectonic events, such the development of the Alpine-Himalayan orogeny, which do not seem to trigger major extinctions, or to gradual changes in climate caused by the growing thermal isolation of Antartica as Australia drifted northwards (Berggren and Prothero, 1992).

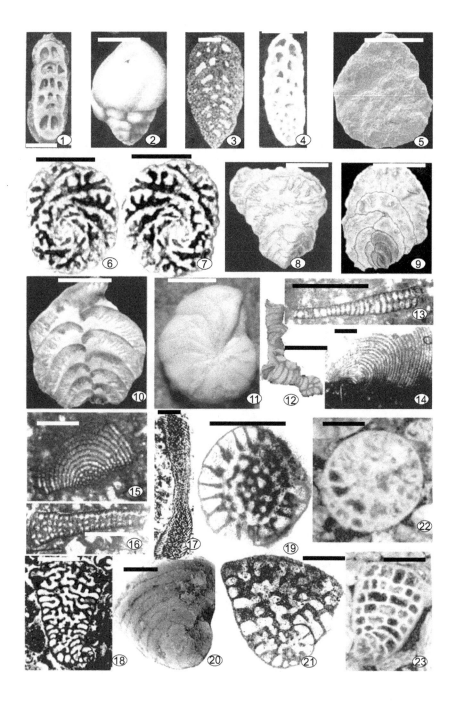

Plate 6.1 Scale bars: Figs 1, 4, 8, 11-12, 18, 20-23 = 0.5mm; Figs 2, 5-7, 9 = 0.25mm; Figs 3, 10, 13-17, 19 = 1mm. Fig. 1. *Cubanina alavensis* Palmer, figured by Loeblich and Tappan (1988), Late Oligocene, Cuba. Fig. 2. *Jarvisella karamatensis* Brönnimann, holotype, figured by Loeblich and Tappan (1988), Miocene, Trinidad. Fig. 3. *Liebusella soldanii* (Jones and Parker), figured by Loeblich and Tappan (1988), Holocene, Caribbean. Fig. 4. *Matanzia bermudezi* Palmer, paratype, figured by Loeblich and Tappan (1988), Early Oligocene, Cuba. Figs 5-7. *Spiropsammia primula* Seiglie and Baker, Congo Fan, West Africa, Kender coll., UCL. Fig. 8. *Pavonitina styriaca* Schubert, figured by Seiglie and Baker (1983), Cabinda. Fig. 9. *Pavopsammina flabellum* Seiglie and Baker, paratype, figured by Seiglie and Baker (1983), Cameroon. Fig. 10. *Zotheculifida lirata* (Cushman and Jarvis), figured by Loeblich and Tappan (1988), Late Oligocene, Trinidad. Fig. 11. *Cyclammina* sp., Eocene, Trinidad, UCL coll. Fig. 12. *Haddonia torresiensis* Chapman, lectotype, Holocene, western south Pacific, NHM 97.11.20.1. Figs 13-16. *Saudia discoidea* Henson, paratypes, Middle Eocene (Lutetian), Ansab, Iraq, 13) NHM P36020; 14) NHM P36018; 15) NHM P36019; 16) NHM 36021. Fig. 17. *Thomasella labyrinthica* Grimsdale, figured by Sirel (1988), Early Eocene, Turkey. Fig. 18. *Barattolites trentinarensis* Vecchio and Hottinger, figured by Vecchio and Hottinger (2007), Ypresian to Lower Lutetian, Trentinara Formation, Italy. Figs 19-21. *Coleiconus christianaensis* Robinson, Middle Eocene, Upper Chapelton Formation, Jamaica, 19) holotype, NHM P52805; 20-21) paratypes, 20) NHM P52808; 21) NHM P52809. Figs 22-23. *Coskinolina* sp., Late Eocene, India, UCL coll.

Plate 6.2 Scale bars: Figs 1-2, 5-6, 10-12, 14 = 0.5mm; Figs 3-4, 7-9, 13, 15, 16-19 = 1mm. Fig. 1. *Coskinolina* sp., Ypresian, Laki Formation, Pakistan, UCL coll. Fig. 2. *Coskinolina* cf. *douvillei* (Davies), Middle Eocene, Upper Chapelton Formation, Jamaica, UCL coll. Fig. 3. *Coskinon* sp., Paleocene, Meting Limestone, Pakistan, UCL coll. Fig. 4. *Anatoliella ozalpiensis* Sirel, holotype, figured by Sirel (1988), Thanetian, Turkey. Fig. 5. *Cushmania Americana* (Cushman), Middle Eocene, Oman, NHM P35802. Fig. 6. *Daviesiconus* sp., Early Eocene, lower Laki Formation, Pakistan, UCL coll. Fig. 7 *Dictyoconus* sp., Early to Middle Eocene, Cuba, NHM P40043. Figs 8-9. *Dictyoconus indicus* Davies, 8) Early Eocene, axial section, India, NHM P28105; 18) Lutetian, Sulman, S.W. Iraq, NHM 35828. Fig. 10. *Coskinolina* sp., White Limestone, Manchester, Jamaica, UCL coll. Figs 11-12. *Verseyella jamaicensis* (Cole), Early Eocene, lower Chapelton Formation, Jamaica, NHM P52823-24. Figs 13. *Coskinolina balsilliei* Davies, Lutetian, Sulman, S.W. Iraq, NHM P35781. Fig. 14. *Fallotella* sp., Middle Paleocene, Laki-Khirthar, Pakistan, UCL coll. Fig. 15, 19. *Alveolina elliptica* var. *flosculina* Silvestri, Middle Eocene, Qatar, 15) NHM P40266; 19) NHM P40256. Fig. 16. *Alveolina aramaea* Hottinger, Early Eocene, Dunghan Hill, Pakistan, NHM P52544. Fig. 17. *Alveolina katicae* White, Eocene, Oman, NHM coll. White and Racey, Wr77. Fig. 18. *Glomalveolina delicatissima* (Smout), holotype, Middle Eocene, Qatar, NHM P40265.

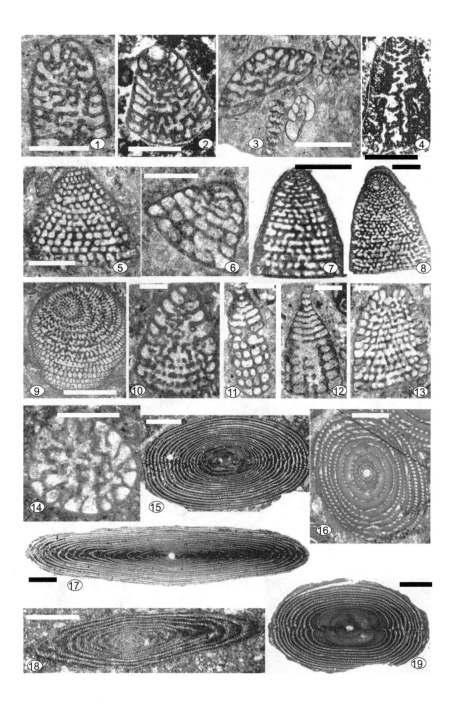

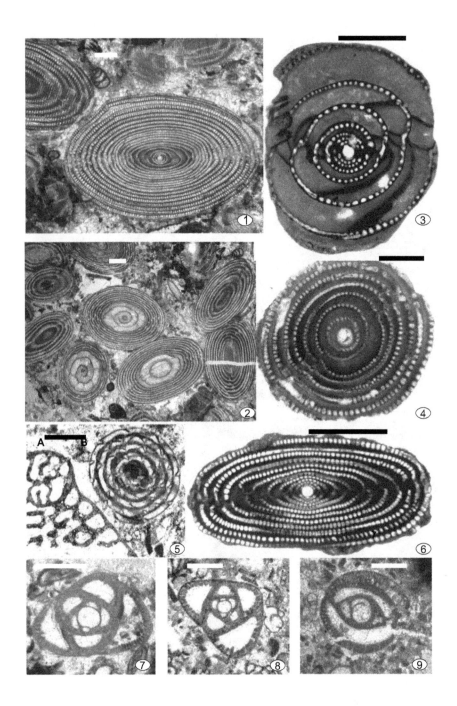

Plate 6.3 Scale bars: Figs 1-4, 6 = 1mm; Figs 5, 7-9 = 0.5mm. Figs 1-2. *Alveolina elliptica* (Sowerby) and *Alveolina elliptica nuttalli* Davies, Eocene, Afghanistan, NHM Skinner coll. P7404. Fig. 3. *Alveolina minervensis* Hottinger, Late Paleocene to Early Eocene, Montagne Noire, Aquitaine, South France, NHM Hottinger coll. Fig. 4. *Alveolina globosa* (Leymerie), Ypresian, Eocene, Laki Limestone (Laki Formation), Sakesar Peak in the Salt Range, Pakistan, NHM Davies coll. Fig. 5. A) *Coskinolina* sp., B) *Glomalveolina* sp., Ypresian, Lower Laki Formation, Pakistan, UCL coll. Fig. 6. *Alveolina vredenburgi* Davies 1937 (= *Alveolina cucumiformis* Hottinger), topotype, axial section, Late Paleocene, Aquitaine, France, UCL coll. Figs 7-9. *Austrotrillina asmariensis* Adams, Oligocene, Kirkuk Well K.93, Iraq, NHM P47581-3.

Plate 6.4 Scale bars: 1-11 = 1mm. Fig. 1. *Alveolina vredenburgi* Davies 1937 (= *Alveolina cucumiformis* Hottinger), topotype, equatorial section, Late Paleocene, Aquitaine, France, UCL coll. Figs 2, 4. *Alveolina leupoldi* Hottinger, Early Eocene, Coustouge, France, 2) solid specimens embedded in rock, 4) thin section from the same rock, UCL coll. Figs 3, 5-7. *Alveolina globosa* (Leymerie), Early Eocene, Coustouge, France, 3) thin section, 5) solid specimen embedded in the same rock, 6) Early Eocene, Khirthar, Pakistan; 7) Early Eocene, Meting Limestone, Laki group, Pakistan, UCL coll. Figs 8, *Alveolina oblonga* d'Orbigny, Dunghan, Siah Koh, NHM Davies coll. Fig. 9. *Alveolina subpyrenaica* Leymerie, Early Eocene, Limstones of Zagros, Iran, NHM coll. Fig. 10. *Alveolina subpyrenaica* Leymerie, Lutetian, India, UCL coll. Fig. 11. *Alveolina elliptica nuttalli* Davies, Early Eocene, Meting Limestone, Laki group, Pakistan, UCL coll.

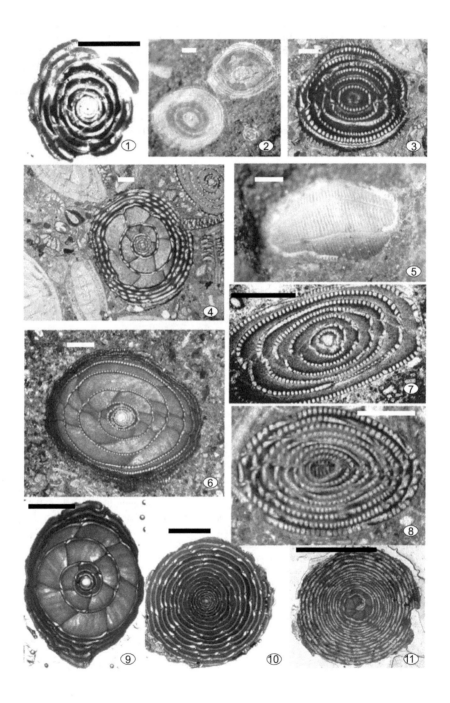

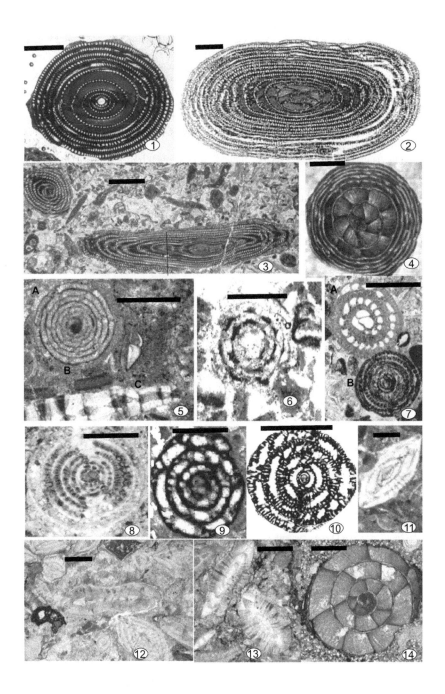

Plate 6.5   Scale bars: Figs 1-4, 6, 12-14 = 1mm; Figs 7, 9-11, = 0.5mm; Figs 5, 8 = 0.25mm. Fig. 1. *Alveolina subpyrenaica* Leymerie, Early Eocene, Limestones of Zagros, Iran, NHM coll. Fig. 2. *Alveolina elliptica nuttalli* Davies, Lutetian, India, UCL coll. Fig. 3. Thin section photomicrograph of *Alveolina elliptica* (Sowerby) and *Alveolina* cf. *stipes* Hottinger, Late Eocene, Khirthar group, Pakistan, UCL coll. Fig. 4. *Alveolina leupoldi,* Hottinger, Early Eocene, Coustouge, France, UCL coll. Fig. 5 A) *Borelis pygmaeus* Hanzawa, B) rodophytes fragments, C) *Heterostegina* (*Vlerkina*) *borneensis* van der Vlerk, Oligocene, Borneo, 65/9 Loc.205, UCL coll. Fig. 6. *Borelis* sp., Late Oligocene, Ras Chekka, Lebanon, UCL coll. Fig. 7. *Pellatispira* sp., *Borelis* sp., Late Eocene, Syria, UCL coll. Fig. 8. *Borelis haueri* (d'Orbigny), Oligocene, Iran, UCL Banner coll. Fig. 9. *Bullalveolina bulloides* Reichel, figured by Hottinger (2006), Early Oligocene, Spain. Fig. 10. *Globoreticulina iranica* Rahaghi, figured by Hottinger (2006), Middle Eocene, Shiraz, Iran. Fig. 11. *Nummulites fossulata* de Cizancourt, Middle Eocene, figured by BouDagher-Fadel et al. (2015), Shenkeza section, Tibet. Fig. 12. *Assilina leymeriei* Archiac and Haime, *Nummulites atacicus* Leymerie, *Discocyclina* sp., *Alveolina* sp., Late Eocene, Upper Khirthar, Pakistan, UCL coll. Fig. 13. *Assilina sublaminosa* Gill, Middle Eocene, figured by BouDagher-Fadel et al. (2015), Shenkeza section, Tibet. Figs 14. *Alveolina palermitana* Hottinger, Middle Eocene, Middle Khirthar, Pakistan, UCL coll.

Plate. 6.6   Scale bars: Figs 1-2, 6-8, 11 = 1mm; Figs 3-5, 9-10 = 0.5mm. Fig. 1. Thin section photomicrograph of *Alveolina elliptica nutalli* Davies, central, flosculinized parts; outer whorls missing, *Orbitolites complanatus* Lamarck, small miliolids, *Operculina* sp., Early-Middle Eocene, Laki-Khirthar, UCL coll. Fig. 2. *Alveolina moussoulensis* Hottinger, Early Eocene, figured by BouDagher-Fadel et al. (2015), Qumiba section, Tibet. Figs 3-5. *Rhabdorites urensis* (Henson), paratypes, Lutetian, Iraq, NHM P35986. Fig. 6. *Linderina burgesi* (Schlumberger), Kohat formation, Pakistan UCL coll. Fig. 7. *Rhabdorites* sp. registered as *Neorhipidionina macfadyeni* Henson, syntypes, late Lutetian, Iraq, NHM P36000. Fig. 8. *Neorhipidionina macfadyeni* Henson, syntypes, late Lutetian, Iraq, NHM P36000. Fig. 9. *Neorhipidionina williamsoni* (Henson), Lutetian, Iraq, NHM P36012. Fig. 10. *Daviesina langhami* Smout, Paleocene, figured by BouDagher-Fadel et al. (2015), Zongpubei section, Tibet. Fig. 11. *Yaberinella jamaicensis* Vaughan, Eocene, Yellow Limestone, Jamaica, Davies coll., sample J 505 M.

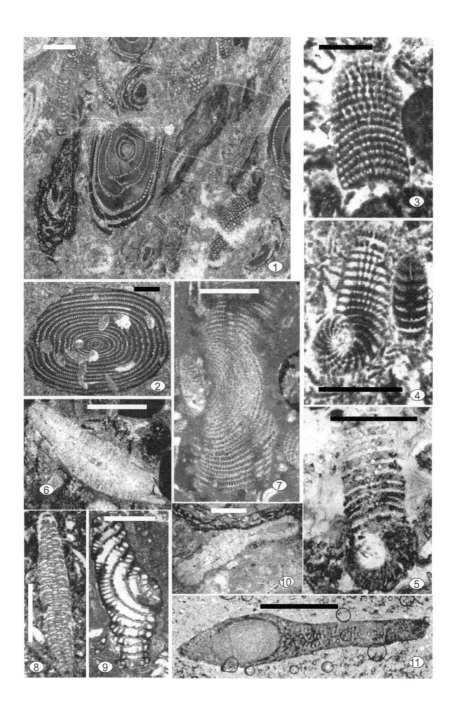

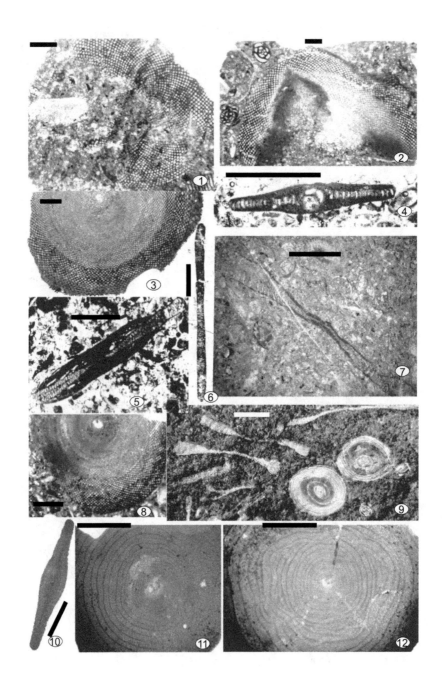

◀───────

Plate 6.7   Scale bars: Figs 1-12 = 1mm. Figs 1-3. *Opertorbitolites* sp., Early Eocene, *Alveolina Corbarica* Zone, Eastern Aquitaine, France, NHM Hottinger coll. Figs 4. *Opertorbitolites lehmanni* (Montanari), registered as "*Opertorbitolites* sp.", Eocene, Oman, White-Racey coll., NHM P52865. Fig. 5. *Opertorbitolites* cf. *douvillei* Nuttall, Eocene, Oman, White coll., NHM P52866. Figs 6-7. *Orbitolites complanatus* Lamarck, Middle Eocene, 6) Bois-Gouët, France, NHM coll.; 7) figured by BouDagher-Fadel et al. (2015), Zongpubei section,Tibet. Fig. 8. *Opertorbitolites douvillei* (Nuttall), Early Eocene, *Alveolina Corbarica* Zone, eastern Aquitaine, France, NHM coll. Fig. 9. Thin section with *Orbitolites biplanus* Lehmann and *Alveolina leupoldi*, Hottinger, Ilerdian (Lower Eocene), Coustouge (Corbières), France, UCL coll. Figs 10-12. *Pseudophragmina floridana* (Cushman), Eocene, Georgia, UCL coll.

───────▶

Plate 6.8   Scale bars: Figs 1-3 = 1mm; Figs 4-7 = 0.25mm. Figs 1-2. *Orbitolites complanatus* Lamarck, Middle Eocene, 1) Libya; 2) Coustouge, France, UCL coll. Fig. 3. Thin section photomicrograph of A) *Alveolina* sp., B) *Orbitolites omplanatus* Lamarck, Middle Eocene, Coustouge, France UCL coll. Figs 4-6. *Praerhapydionina delicata* Henson, Oligocene, Buff calcarenites, Jamaica, NHM P52829-21. Fig. 7. *Rhabdorites malatyaensis* (Sirel), figured by Hottinger (2007), Middle Eocene, Turkey.

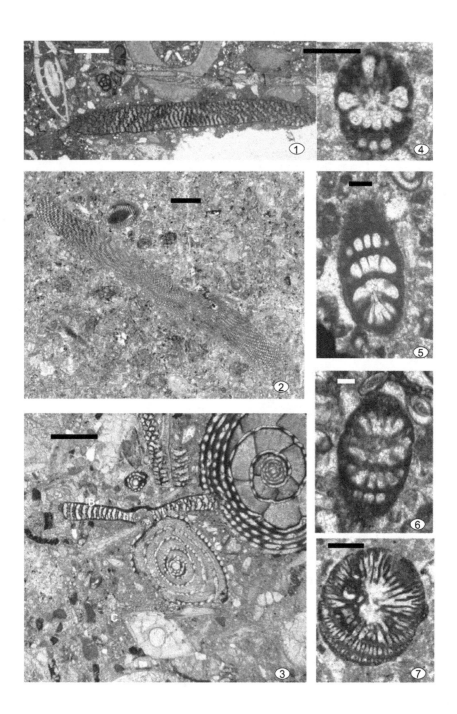

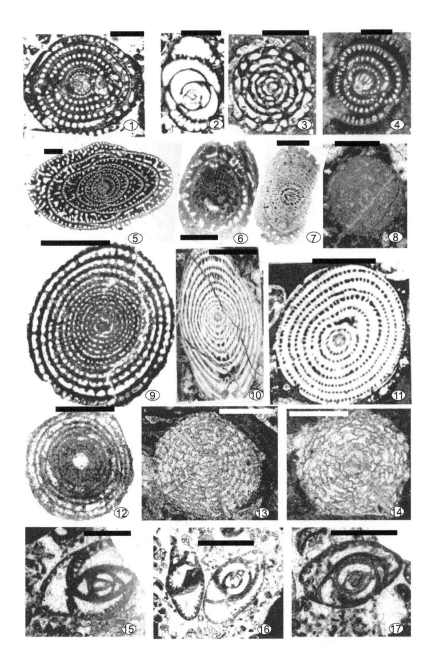

Plate 6.9 Scale bars: Figs 1-4, 12-17 = 0.5mm; Figs 5-11 = 1mm. Fig. 1. *Glomalveolina dachelensis* Schwager, figured by Loeblich and Tappan (1988), Paleocene, Egypt. Fig.2 *Malatyna drobneae* Sirel and Acar, type figured by Sirel and Acar (1993), Lutetian, Malaya. Fig. 3. *Praebullalveolina afyonica* Sirel and Acar, holotype, figured by Sirel and Acar (1982), Eocene, Turkey. Fig. 4. *Fabularia hanzawai* Robinson, cotype, Eocene, Saint Andrew Claremont Formation, Jamaica, NHM P52840. Figs 5-7. *Fabularia discolithus* Defrance, 5) Eocene, calcaire grosier de Rennes, France, Brady coll., NHM P41603; 6-7) Lutetian middle-calcaire grossier, Chaumont-en-Vexin, Paris Basin, NHM P33071. Fig. 8. *Aberisphaera gambanica* Wan, figured by BouDagher-Fadel et al. (2015) from the Paleocene, Shenkeza section, Tibet. Fig. 9. *Lacazina* sp., Eocene, Spain, UCL coll. Figs 10-11. *Lacazinella wichmanni* (Schlumberger), Late Eocene, Indonesia, UCL coll. Fig. 12. *Pseudofabularia matleyi* (Vaughan), figured by Loeblich and Tappan (1988), Middle Eocene, Chapelton Formation, Jamaica. Figs 13-14. *Wilfordia sarawakensis* Adams, paratype, Eocene, Sarawak, Malaysia, NHM P46464. Figs 15-17. *Austrotrillina paucialveolata* Grimsdale, syntypes, Oligocene, Kirkuk Well 14, Iraq, NHM P40689.

Plate 6.10 Scale bars: Figs 1-4, 6, 8, 10-11, 16 = 1mm; Figs 5, 7, 9, 12-15 = 0.5mm. Figs 1-2. *Austrotrillina paucialveolata* Grimsdale, syntypes, Oligocene, Kirkuk Well 14, Iraq, NHM P40681. Fig. 3. *Karsella hottingeri* Sirel, holotype, figured by Sirel (1997), Thanetian, Turkey. Fig. 4. *Archiacina armorica* (d'Rrchiac), figured by Loeblich and Tappan (1988), Oligocene, France. Fig. 5. *Dendritina* cf. *rangi* d'Orbigny, Early Miocene (Aquitanian), Kirkuk Well 22, Iraq, NHM P39432. Fig. 6. *Haymanella paleocenica* Sirel, figured by Sirel (1999), Thanetian, Turkey. Fig. 7. *Elazigella altineri* Sirel, type figured by Sirel (1999), Thanetian, Turkey. Fig. 8. *Archaias* sp. (pillars are present, but registered as *Heterillina hensoni* Grimsdale), syntype, Oligocene, Kirkuk Well 14, Iraq, NHM P40679. Fig. 9. *Heterillina hensoni* Grimsdale, syntype, Oligocene, Kirkuk Well 14, Iraq, NHM P40679. Fig. 10. *Hottingerina lukasi* Drobne, figured by Drobne (1975), Middle Paleocene, Yugoslavia. Fig. 11. *Archaias aduncus* (Fichtel and Moll), Oligocene, Shiranish Islam, Iraq, NHM P39651. Fig. 12. *Archaias* sp. (pillars are present, but registered as *Peneroplis* sp.), Oligocene, Buff calcarenites, Jamaica, NHM P52828. Fig. 13. *Cyclorbiculinoides jamaicensis* Robinson, holotype, figured by Robinson (1974), Eocene, Jamaica. Fig. 14. *Neorhipidionina spiralis* Hottinger, figured by Hottinger (2007), Middle Eocene, Iran. Fig. 15. *Neotaberina neaniconica* Hottinger, figured by Hottinger (2007), Middle Eocene, Iran. Fig. 16. *Yaberinella hottingeri* Robinson, Eocene, Punta Gorda-1 6270'-80', Nicaragua, NHM P52822.

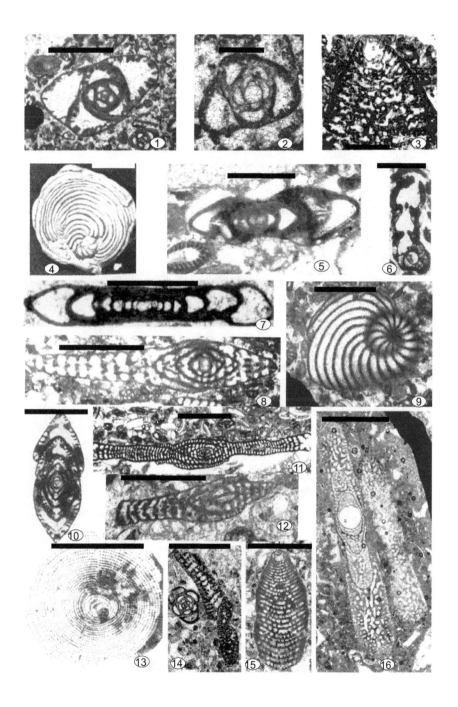

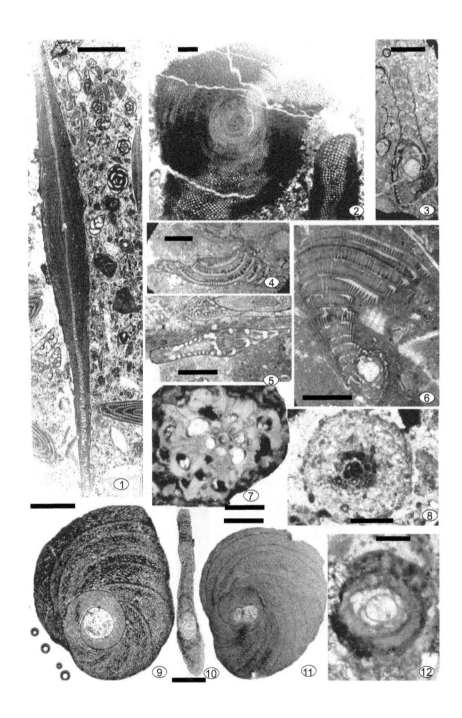

◄───

Plate 6.11 Scale bars: Figs 1-6, 9-12 = 1mm; Figs 7-8 = 0.5mm. Figs 1-2. *Somalina hottingeri* White, Eocene, Seeb Limestone Formation, Wadi Fatah, Oman, NHM P52861-2. Figs 3-6. *Twaraina seigliei* Robinson, Middle Eocene, Twara-1, Nicaragua Rise, NHM P52814-18. Fig. 7. *Meghalayana indica* Matsumaru and Sarma, holotype, Late Eocene, Meghalaya State, NE India, Matsumaru's coll., Saitama Univ, 8865. Fig. 8. *Protogypsina indica* Matsumaru and Sarma, holotype, Paleocene, Jaintia Hills, Meghalaya State, NE India, Matsumaru's coll., Saitama Univ, 8866. Figs. 9. *Yaberinella jamaicensis* Vaughan, Eocene, Yellow Limestone, Jamaica, NHM Davies coll., sample J.505 M. Figs 10–11. *Yaberinella jamaicensis* Vaughan registered in the NHM as from Lutetian, Oman, 10) axial section, NHM P52270; 11) A-form bilocular protoconch, NHM. P52266. These are undoubtedly *Yaberinella*, however, this genus is Caribbean. Their recovery in Oman is man-made: either by a confusion of labels or by the occurrence of ballast stones transported in slavery ships on their return from Jamaica. Hottinger personal communication: "I have been at the point indicated near Muscat: No trace nor even a possibility of their occurrence there considering the local geology." Fig. 12. *Raoia indica* Matsumaru and Sarma, holotype, Paleocene, Jaintia Hills, Meghalaya State, NE India, Matsumaru's coll., Saitama Univ, 8867.

───►

Plate 6.12 Scale bars: Figs 1-7 = 1mm. Fig. 1. *Nummulites intermedius* (d'Archiac), *Operculina* sp., *Discocyclina* sp., Early Eocene, Lower Nari, Pakistan, UCL coll. Fig. 2. *Nummulites atacicus* Leymerie, Early Eocene, Meting-Laki, Pakistan UCL coll. Fig. 3. *Alveolina* sp., *Orbitolites complanatus* Lamarck, Early Eocene, Meting-Laki, Pakistan, UCL coll. Fig. 4. *Archaias kirkukensis* Henson, paratype, Late Oligocene to Early Miocene (Chattian to Aquitanian), Kirkuk Well no. 57, Iraq, NHM P39645. Fig. 5. *Distichoplax biserialis* (Dietrich), Eocene, Java, UCL coll. Fig. 6. Codiacean algae, *Sorites* sp., Oligocene, France, UCL coll. Fig. 7. Thin section photomicrograph of *Wilfordia sarawakensis* Adams, Eocene, Sarawak, Malaysia, same section as of the paratypes (Plate 6.9, figs 12-14).

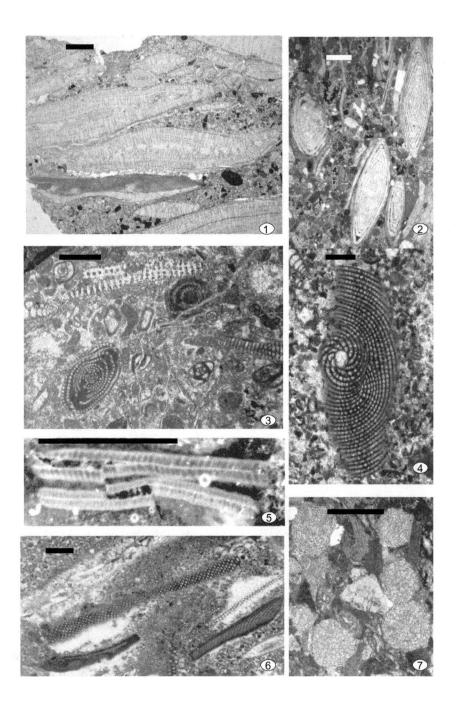

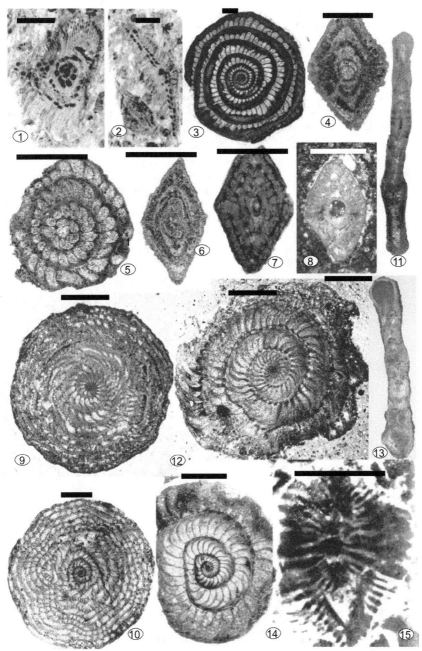

Plate 6.13 Scale bars: Figs 1-15 = 1mm. Figs 1-2. *Biplanispira mirabilis* (Umbgrove), Late Eocene, Indonesia, UCL coll. Fig. 3. *Ranikothalia nuttalli kohatica* (Davies), latest Paleocene-earliest Eocene, lower member of the Jafnayn Formation, Al Khawd, Oman, BMNHP 52442. Figs 4-7. *Miscellanea miscella* (d'Archiac and Haime, 4-6), S.W. of Hilaia, South India, UCL coll. Figs 8-10. *Miscellanea meandrina* (Carter), Paleocene, Qatar; 7) A-form, axial section, NHM P40232; 8) axial section, NHM P40234; 9) equatorial section, NHM P40230; 10) paratype, B-form, equatorial section, NHM P40239. Fig. 11. Fig. 14. *Ranikothalia margaretae* (Haynes and Nwabufo-Ene), paratype, Late Paleocene (Thanetian), El Fogaha, Libya, NHM P52253. Fig. 12. *Miscellanea miscella* var. *dukhani* Smout, paratype, Paleocene, Qatar, NHM P40251. Figs 13-14. *Miscellanea stampi* (Davies), Paleocene, Upper Ranikot Beds, the Samana Range Pakistan, NHM P41615. Fig. 15. *Miscellanea miscella* (d'Archiac and Haime), Paleocene, Indonesia, UCL coll.

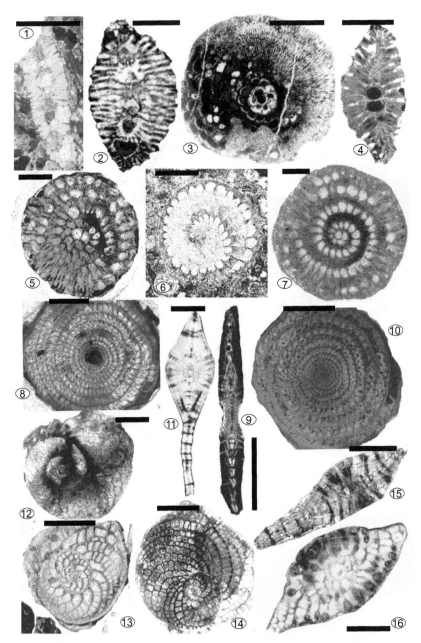

Plate 6.14   Scale bars: Fig. 1, 7, 11, 13, 15 -16 = 1mm; Figs 2-6, 8-10, 12, 14 = 0.5mm. Figs 1-5. *Pellatispira* sp., Late Eocene, 1) Pakistan; 2-5) Indonesia, UCL coll. Fig. 6. *Pellatispira fulgeria* Whipple, Late Eocene, Sumatra, UCL coll., AS33. Fig. 7. *Vacuolispira inflata* (Umbgrove), Late Eocene, Borneo, UCL coll. Figs 8-10. *Assilina cuvillieri* Schaub, B forms, Wadi Bani Khaled WK21, 8,10) equatorial sections, NHM P52513, NHM P52516, 9) axial section, NHM P52516. Figs 11, 16. *Heterostegina (Vlerkina) borneensis* van der Vlerk, Late Oligocene, Borneo, UCL coll. Fig. 12. *Grzybowskia multifida* Bieda, Eocene, Carpathians, NHM coll. Fig. 13. *Heterostegina (Heterostegina)* sp., Eocene, Brazil, UCL coll. Figs 14-15. *Heterostegina (Vlerkina) borneensis* van der Vlerk, paratypes, Te-Miocene, Borneo, NHM P45037-8.

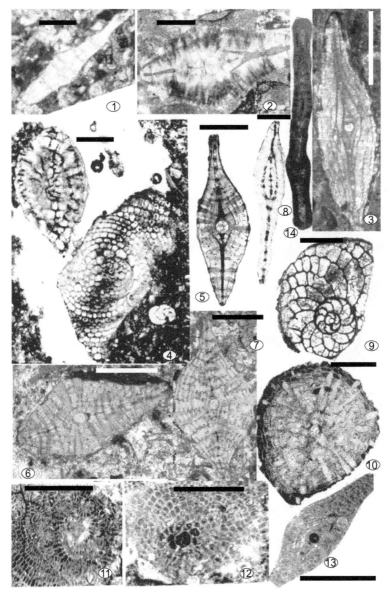

Plate 6.15 Scale bars: Figs 1-14 = 1mm. Fig. 1. *Planostegina* sp., *Operculina* sp., Oligocene, Borneo, UCL coll. Fig. 2. *Heterostegina (Vlerkina) borneensis* van der Vlerk, Late Oligocene, Borneo, UCL coll. Fig. 3. *Tansinhokella tatauensis* Banner and Hodgkinson, paratype, Eocene Limestone, Tatau Formation, Sarawak, Borneo, NHM P52296. Fig. 4. *Tansinhokella* sp., *Cycloclypeus* sp., Late Oligocene, Borneo, UCL coll. Fig. 5. *Tansinhokella yabei* (van der Vlerk), latest Oligocene, Soembal, Borneo, NHM P45042. Fig. 6. *Spiroclypeus leupoldi* van der Vlerk, *Lepidocyclina* sp., fragments of rodophyte algae, Oligocene, Java, UCL coll. Figs 7. *Spiroclypeus umbonata* Yabe and Hanzawa, Oligocene, Java, UCL coll. Figs 8-9. *Heterostegina (Hetrostegina)* sp., Oligocene, Indonesia, UCL coll. Fig. 10. *Silvestriella tetraedra* Gümbel, Eocene, Gassino, Torino, Italy, NHM P44956. Figs 11-13. *Spiroclypeus vermicularis* Tan Sin Hok, Late Eocene, East Borneo, UCL coll. Fig. 14. *Ranikothalia margaretae* (Haynes and Nwabufo-Ene), paratype, Late Paleocene (Thanetian), El Fogaha, Libya, NHM P52253.

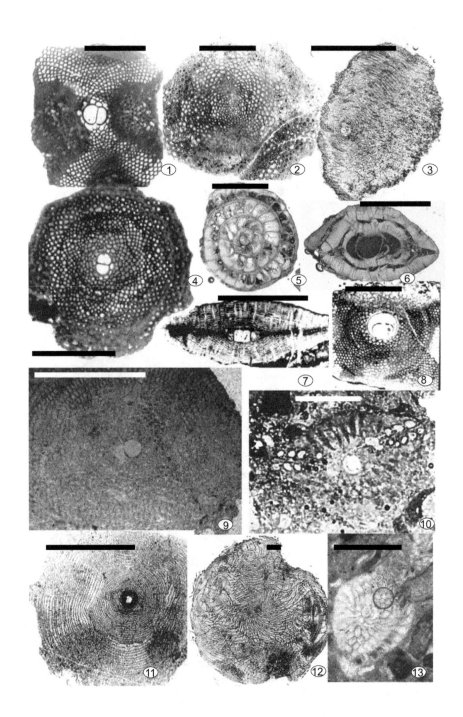

Plate 6.16   Scale bars: Figs 1, 3-4, 7-13 = 1mm; Figs 2, 5-6 = 0.5mm. Fig. 1. *Lepidocyclina (Lepidocyclina) ocalana* Cushman, Middle Eocene, Gilchrist Co., Florida, NHM P51961. Fig. 2. *Pseudolepidina trimera* Barker and Grimsdale, cotype, base of Middle Eocene, Vera Cruz, Mexico, NHM P51986. Figs 3, 12. *Discocyclina* sp., Eocene, Fontcouverte, France, UCL coll. Fig. 4. *L. (Nephrolepidina) veracruziana* (Vaughan and Cole) = *Triplalepidina veracruziana* Vaughan and Cole, topotype, Late Eocene, Arroyo Terrero, near Palma Sola Mexico, NHM P37905. Figs 5-6. *Pellatispirella antillea* Hanzawa, Middle Eocene, Soldado Rock, Trinidad NHM P33349. Figs 7-8. *Eulepidina papuaensis* (Chapman), Late Oligocene, Borneo, UCL coll. Fig. 9. *Nemkovella mcmilliana* BouDagher-Fadel and Price, holotype, Early Ypresian, figured by BouDagher-Fadel and Price (2017), South Africa, UCL MF491. Fig. 10. *Biplanispira* sp., Eocene, Tatau Formation, Sarawak, Borneo, UCL coll. Fig. 11. *Nummulites somaliensis* Nuttall and Brighton, Middle Eocene, Qatar, NHM P40244. Fig. 13. *Tansinhokella tatauensis* Banner and Hodgkinson, paratype, Eocene,Tatau Formation, Sarawak, Borneo, NHM P49525.

Plate 6.17   Scale bars: Figs 1-4, 6-10 = 1mm; Fig. 5 = 0.5mm. Fig. 1. *Cycloclypeus* sp., Late Oligocene, with a planktonic foraminifera test embedded in the broken part of the test, Java, UCL coll. Figs 2-4. *Assilina granulosa* var. *chumbiensis* Gill, Early Eocene, Lower Bhadrar Beds, Pakistan, NHM P41522; 3) enlargement of Fig. 5. *Assilina subdaviesi* Gill, Early Eocene, Lower Bhadrar Beds, Pakistan, NHM P41543. Figs 6-7. *Assilina* sp., Late Paleocene, Pakistan, Ranikot group, solid specimens, UCL coll. Fig. 8. *Assilina* sp., Eocene, axial sections, Coustouge, France, UCL coll. Fig. 9. *Orbitosiphon praepunjabensis* Adams, holotype, Late Paleocene, Khairabad Limestone, Dhak Pass, Salt Range, Pakistan, NHM P51968. Fig. 10. *Actinosiphon semmesi* Vaughan, Chicontepec Formation, El Cristo Well, Veracruz, Mexico, UCL coll.

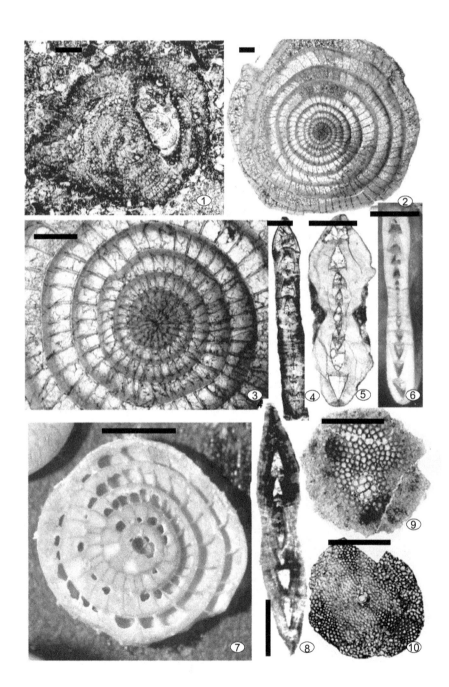

Plate 6.18   Scale bars: Figs 1-12 = 1mm. Fig. 1. *Assilina mamillata* (d'Archiac), Early Eocene, Yozgat, Turkey, UCL coll. Figs 2-4. *Assilina granulata* (d' Archiac), Middle Eocene, 2-3) Barail Formation, Bangladesh; 4) solid specimen, UCL coll. Figs 5-6. *Operculina* sp., Oligocene, Kalimantan, UCL coll. Fig. 7. Thin section photomicrographs of *Planostegina* sp., fragments of *Eulepidina* sp., *Amphistegina* sp., Loc. 130, Borneo, UCL coll. Fig. 8. *Nummulites masiraensis* Carter, Eocene, India, UCL coll. Fig. 9. *Operculina douvillei* Dorıcieux, Early Lutetian, NHM Davies coll. Fig. 10. *Nummulites fichteli* Michelotti, Oligocene, Biarritz, France, UCL coll. Fig. 11. *Operculina aegyptiaca* Hamam, paratype, megalospheric form, latest Early Eocene, Gebel Gurnah, Luxor, Egypt, NHM P49827. Fig. 12. *Operculina subgranosa* Grimsdale, Eocene, France, NHM coll.

Plate 6.19   Scale bars: Figs 1-9 =1mm. Figs 1-3. *Nummulites fichteli-intermedius* (d'Archiac) (= *Nummulites clypeus*), Lower Nari Formation, Oligocene, Pakistan, UCL coll. Fig. 4. *Nummulites fichteli* Michelotti, Oligocene, Nummulitic rock, cliffs at Biarritz south of France, NHM P49522. Figs 5-8. *Nummulites gizehensis* (Forskal), Late Lutetian, 5-7) Spain; 8) France, UCL coll. Fig. 9. *Assilina* sp., Eocene, France, solid specimen, UCL coll.

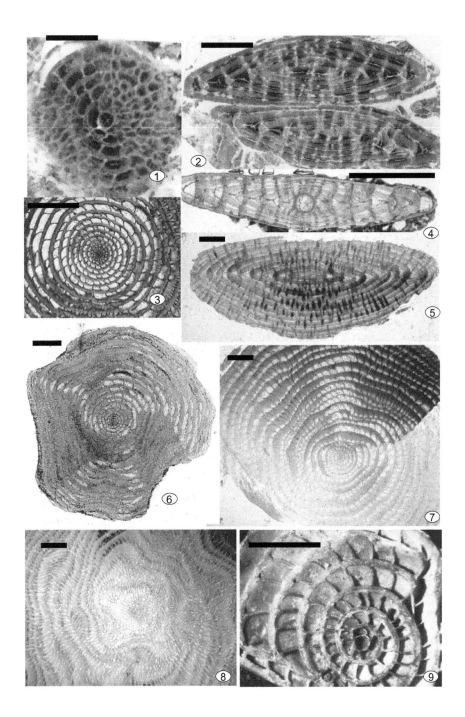

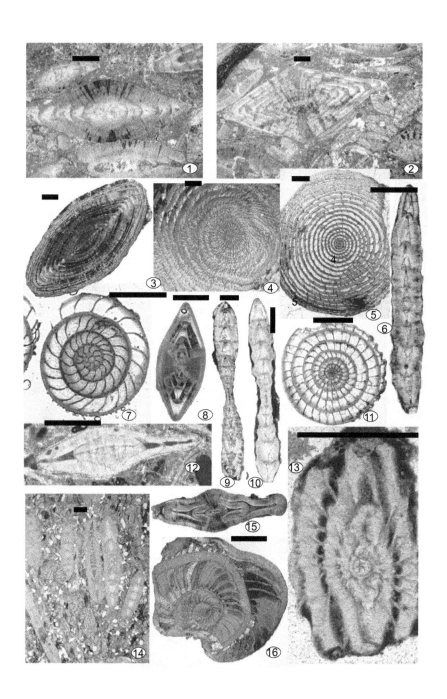

Plate 6.20 Scale bars: Figs 1-16 =1mm. Figs 1-2. *Nummulites mamilla* Fichtel and Moll, Early Eocene, Laki, Pakistan, 1) megalospheric form; 2) microspheric form, UCL coll. Fig. 3. *Nummulites perforatus* (de Montfort), microspheric form, Lutetian, San Giovanni Ilarione, Italy, NHM Davies coll. Fig. 4. *Nummulites intermedius* (d'Archiac), Eocene, India, NHM P30148. Fig. 5. *Nummulites beneharnensis* de la Harpe, B-form, middle Lutetian, Wadi Rusayl, Oman, NHM P52275. Fig. 6. *Assilina leymeriei* (d'Archiac and Haime), Early Eocene, Lower Bhadrar Beds, Pakistan, NHM P41509. Figs 7- 8. *Palaeonummulites kugleri* (Vaughan and Cole), Oligocene, Falling Waters State Park, Chipley, Florida, Suwannee Limestone, USA, UCL MF3237) equatorial section; 8) axial section. Figs 9-11. *Assilina daviesi* de Cizancourt, Early Eocene, Lower Bhadrar Beds (Salt Range), Pakistan, 9) NHM P41524; 10-11) NHM 41527-8. Figs 12-13. *Operculinoides ocalanus* (Cushman), Oligocene, Brazil, UCL coll. Fig. 14. *Assilina* sp. and *Planocamerinoides* sp., Middle Eocene, Kopili Formation, UCL coll. Figs 15-16. *Chordoperculinoides sahnii* (Davies), India, registered as *Ranikothalia sahnii* Davies, Paleocene, French West Africa, NHM P40350, 15) equatorial section showing an initial nummulitic (involute) spiral and operculine final stage; 16) axial section.

Plate 6.21 Scale bars: Figs 1-16 = 1mm. Fig. 1. *Nummulites lamarcki* d'Archiac and Haime, megalospheric form of *Nummulites laevigatus* (Bruguière), Middle Eocene, England, UCL coll. Fig. 2. *Nummulites* sp., Eocene, S.E. Coast of Arabia, UCL coll. Fig. 3. *Nummulites gizehensis* (Forskal), late Lutetian, Libya, UCL coll. Fig. 4. Nummulitc Limestone, Eocene, Libya, UCL coll. Figs 5-6. *Nummulites* sp., Eocene, Gerona, Spain; 6) enlargement of Fig. 5, UCL coll. Fig. 7. *Nummulites globulus* Leymerie, Eocene, Qatar, NHM 40258. Figs 8-9. *Nummulites vascus* Joly and Leymerie, Oligocene, Iran, UCL coll. Fig. 10. *Nummulites aturicus* Joly and Leymerie, Middle Eocene, UCL coll. Figs 11-13. *Nummulites* sp., Middle Eocene, France, 11) SEM photograph of a solid specimen showing the proloculus; 12) thin section; 13) enlargement of 12 to show the marginal cord, UCL coll. Figs 14-15. *Nummulites fichteli* Michelotti, Oligocene, from Nummulitic rock cliffs at Biaritz, France, NHM P49521. Fig. 16. *Nummulites* cf. *striatus* (Bruguière), late Middle Eocene, Priabonian, India, UCL coll.

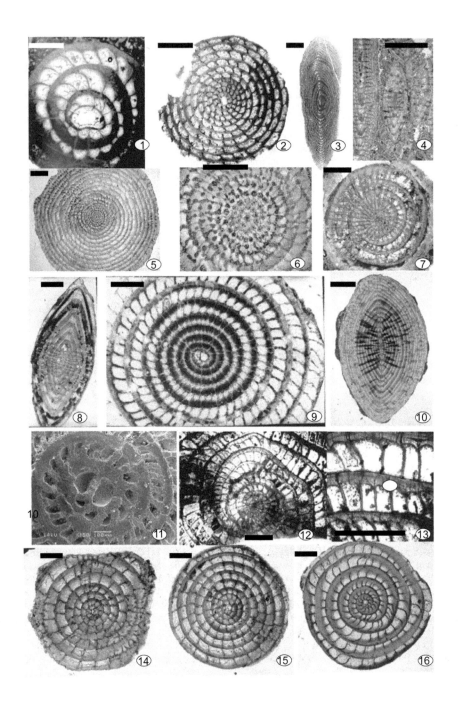

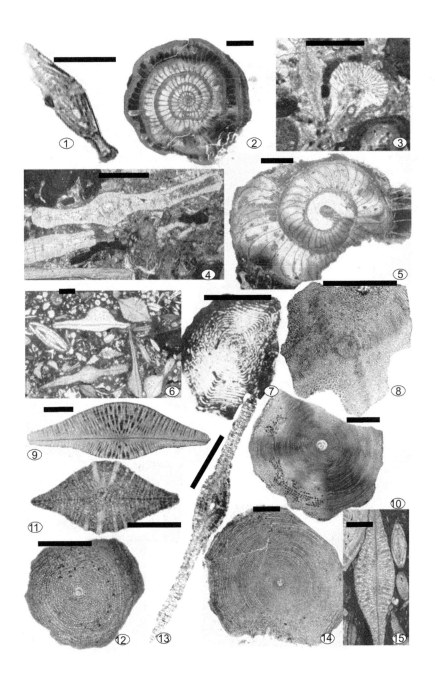

Plate 6.22 Scale bars: Figs 1-6, 9-10, 13-15 = 1mm; Figs 7-8, 11-12 = 0.5mm. Figs 1-2. *Ranikothalia nut-talli* (Davies), topotypes, Late Paleocene, Upper Ranikot Beds, Thal, Pakistan, NHM P41614. Figs 3-5. *Ranikothalia sindensis* (Davies), Late Paleocene, Upper Ranikot Beds, Pakistan, 3) a fragment showing the marginal cord, 4) a thin section with an assemblage of *Ranikothalia, Assilina,* fragments of *Discocyclina* and rodophyte algae; 5) Late Paleocene, Punjab Salt range upper part Khairabad Limestone, NHM coll. Fig. 6. Thin section photomicrograph of *Actinocyclina radians* (d'Archiac), *Discocyclina* sp., Middle Eocene, Kopili Formation, Bangladesh, UCL coll. Fig. 7. *Asterophragmina pagoda* (Rao), figured by Loeblich and Tappan (1988), Late Eocene, Burma. Fig. 8. *Discocyclina ranikotensis* Davies, Late Paleocene, Upper Ranikot Beds, Pakistan, UCL coll. Figs 9-10. *Discocyclina dispansa* Sowerby, Eocene, Goojerat, Western India, NHM P539, 9) axial section; 10) equatorial section. Figs 11-12. *Discocyclina sheppardi* Barker, Paleocene, Soldado Rock, Trinidad, NHM P33350-1, 11) axial section. Figs 13-14. *Discocyclina peruviana* (Cushman), Terebratula Bed, Peru, 13) axial section; 14) equatorial section, UCL coll. Fig. 15. Thin section photomicrograph of *Nummulites perforatus* (de Montfort), *Discocyclina dispansa* Sowerby, *Globigerina* sp., UCL coll.

Plate 6.23 Scale bars: Figs 1-2, 4-11= 1mm; Figs 3, 12-13 = 0.5mm. Figs 1-2. Thin section photomi-crograph of *Discocyclina dispansa* Sowerby, *Biplanispira* sp., *Asterocyclina* sp., Middle Eocene, Kopili Formation, Bangladesh, UCL coll. Fig. 3. *Polylepidina* sp., latest Middle Eocene, Jamaica, UCL coll. Figs 4-11. *Discocyclina* sp., microspheric equatorial section, Eocene, France,, UCL Grimsdale Coll. GS/50, 4) enlargement of fig. 5; 6-10) solid specimens; 7) enlargement of embryonic chambers; enlargement of equatorial chambers. Fig. 12. *Operculina* sp., Eocene, Philippines, UCL coll. Fig. 13. *Orbitolites* sp., Eocene, Libya, UCL coll.

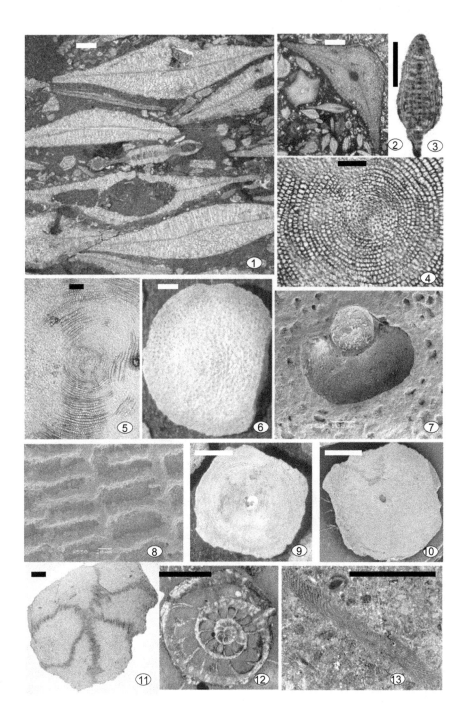

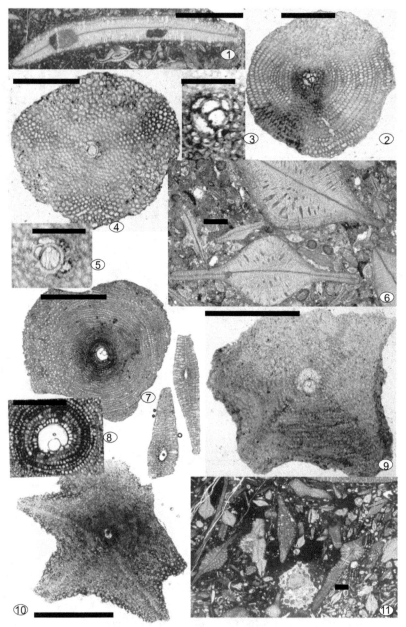

Plate 6.24 Scale bars: Figs 1, 3, 5, 8 =0.5mm; Figs 2, 4, 6-7, 9-11 = 1mm. Fig. 1. *Discocyclina sella* (d'Archiac), Early Eocene, Laki, India, UCL coll. Figs 2-5. *Hexagonocyclina cristensis* (Vaughan), 2-3) topotype, Early Eocene, Mexico, NHM P32633; 3) enlargement of the embryonic apparatus of fig. 2; 4-5) Early Eocene, well at El Cristo, Mexico, UCL coll. 5) enlargement of the embryonic apparatus of fig. 4. Fig. 6. *Discocyclina californica* (Schenk), topotype, Vaquelos Formation, California, UCL coll. Figs 7-8) *Discocyclina* sp., 7) equatorial and axial sections, Eocene, France, UCL coll; 8) enlargement of the embryonic apparatus of fig. 7. Figs 9-10) *Asterocyclina stella* (Gümbel), Late Eocene, 9) Karia, Turkey, NHM P37910; 10) Eocene, France, UCL coll. Fig. 11. Thin section photomicrograph of *Discocyclina* sp., *Asterocyclina stellata* (d'Archiac), Middle Eocene, Kopili formation, UCL coll.

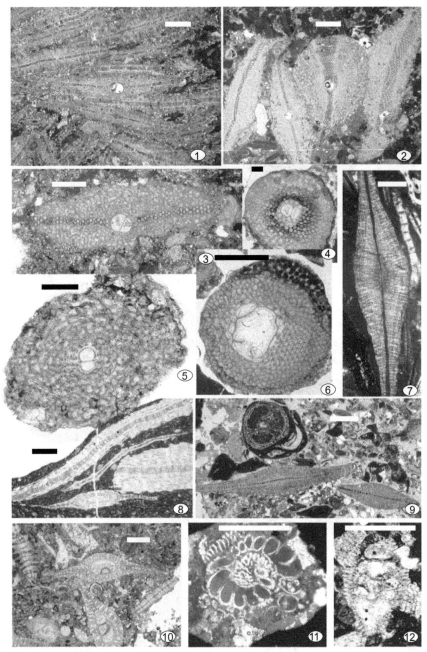

Plate 6.25 Scale bars: Figs 1-12 = 1mm. Figs. 1-3. *Lepidocyclina (Lepidocyclina) canellei* Lemoine and Douvillé, Oligocene, Brazil, UCL coll. Figs. 4-6. *Lepidocyclina (Lepidocyclina) pustulosa* (Douvillé), (4–5) Oligocene, Brazil; 6. *Pliolepidina tobleri* Douvillé, synonymous with *L. (L.) pustulosa* (Douvillé), Eocene, Masparrito Member, Venezuela, UCL coll. Fig. 7. *Discocyclina* sp., Eocene, Venezuela, UCL coll. Fig. 8. Thin section photomicrograph of *Eulepidina* sp., *Eulepidina dilatate* (Michelotti), Oligocene, Java, UCL coll. Fig. 9. *L. (Nephrolepidina) chaperi* Lemoine, microspheric form, Late Oligocene, Brazil, UCL coll. Fig. 10. *Eulepidina ephippioides* (Jones and Chapman), Oligocene, Borneo, UCL coll. Figs. 11–12. *Laffitteina vanbelleni* Grimsdale, Eocene, Kourdane, Syria, NHM P40676-7.

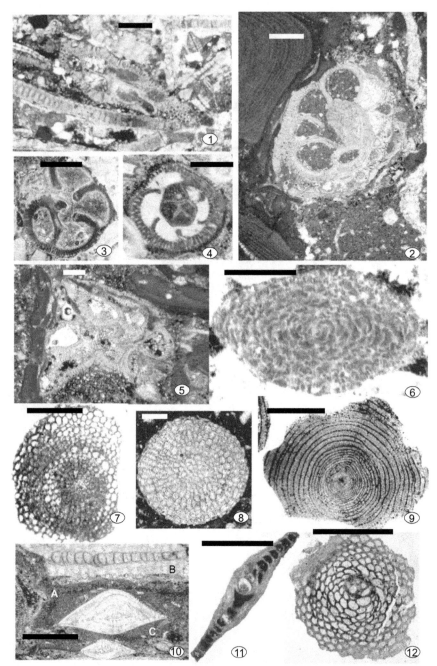

Plate 6.26 Scale bars: Fig. 1, 5, 9-12 = 1mm; Figs 2-4, 6-8= 0.5mm. Fig. 1. *Eulepidina* sp., *Victoriella* sp., Oligocene, Borneo, UCL coll. Figs 2-4. *Eorupertia* sp., Late Eocene, 2) Brazil, UCL coll. 3-4) France, UCL coll. Fig. 5. *Biarritzina* sp., Late Eocene, Brazil, UCL coll. Fig. 6-7. *Discogypsina discus* (Goës), 6) Eocene, France, UCL coll; 7) Batu, North Borneo, NHM NB 9030. Fig. 8. *Sphaerogypsina* sp., Oligocene, Kalimantan, UCL coll., 62-466B. Fig. 9. *Pseudophragmina floridana* (Cushman), Eocene, Georgia, UCL coll. Fig. 10. Thin section photomicrograph of A) *Gypsina* sp., B) fragment of *Eulepidina* sp., C) *Amphistegina* sp, UCL coll. Figs 11-12. *Eulinderina* sp., Middle Eocene, Yecuatla, Veracruz, Mexico, NHM P51959, 11) axial section; 12) equatorial section.

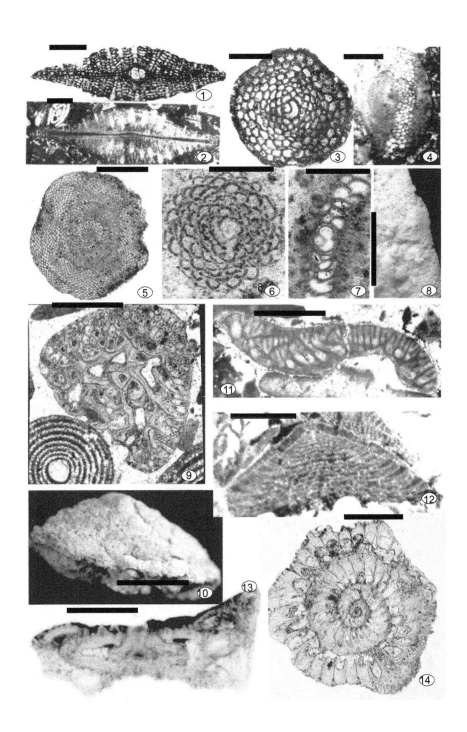

Plate 6.27 Scale bars: Figs 1-2, 5, 12-14 = 1mm; Fig. 3-4 = 0.5mm; Figs 6-11 = 0.25mm. Fig. 1. *Neodiscocyclina anconensis* (Barker), figured by Vaughan (1945), Eocene, Barbados. Fig. 2. *Orbitoclypeus nummuliticus (*Gümbel), Middle Eocene, France, UCL coll. Fig. 3. *Eoannularia eocenica* Cole and Bermúdez, Middle Eocene, figured by Cole and Bermúdez (1944), Pinar del Rio, Province. Fig. 4. *Linderina buranensis* Nuttall and Brighton, Eocene, France, UCL coll. Fig. 5 *Linderina floridensis* Cole, Late Eocene, Pakistan, NHM P48503. Fig. 6. *Linderina brugesi* Schlumberger, Middle Eocene, Qatar, NHM P40266. Fig. 7. *Planorbulinella solida* Belford, Late Oligocene, Borneo, UCL coll. Figs 8-11. *Fabiania cassis* Silvestri, Late Eocene, 8) India, solid specimen; 9-11) UCL coll.; fig. 9. with *Lacazina* sp. Fig. 12. *Halkyardia* sp., Late Eocene, France, UCL coll. Fig. 13. *Daviesina langhami* Smout, microspheric (note double periphery, registered as *Daviesina* sp.), Kohat Limestone, Pakistan, NHM coll. Fig. 14. *Daviesina langhami* Smout, Kohat Limestone, Pakistan, NHM coll., G.136, R.145.

Plate 6.28 Scale bars: Figs 1-4, 9-11, 20-21 =0.5mm; Figs 5 = 0.25mm; Figs 6-8, 12-19, 22-26=1mm. Figs 1-2. *Laffitteina vanbelleni* Grimsdale, Eocene, Kourdane, Syria, NHM P40676-7. Figs 3-4. *Eorupertia incrassata* var. *laevis* Grimsdale, syntype, Middle Eocene, Ain Zalah Well no. 1, Iraq, NHM P40695. Figs 5-6. *Dictyoconus* sp., registered as *Dictyoconoides cooki* (Carter), Eocene, Egypt, NHM P36656. Figs 7-8. *Dictyoconoides kohaticus* Davies, Early Eocene, Kohat Shales, India, 7) NHM P22632; 8) NHM P22633. Figs 9-11. *Rotalia* sp., Eocene, France, UCL coll. Fig. 12. *Lockhartia haimei* (Davies), Paleocene, Qatar, NHM P40156. Figs 13-15. *Lokhartia diversa* Smout, paratypes, Paleocene, Qatar, NHM P40192 (2) Figs 16-17. *Lockhartia conditi* (Nuttall), Paleocene, 16) Qatar, Arabia, NHM P40206; 17) Paleocene, India, NHM P38261. Figs 18-19. *Palaeonummulites pristina* (Brady), syntypes, Calcaire de Namur, Belgium. NHM P35503-4. Figs 20-21. *Cuvillierina* sp., Eocene, Bangladesh, UCL coll. Fig. 23. *Dictyokathina simplex* Smout, Eocene, Qatar Arabia, NHM P40222. Fig. 22. *Lockhartia* cf. *newboldi* (d'Ârchiac et Haime), Early Eocene, Meting-Laki, Pakistan, UCL coll. Fig. 24. *Kathina major* Smout, paratype, Paleocene, Qatar, NHM P40210. Fig. 25. *Kathina delseota* Smout, paratype, Paleocene, Qatar Arabia, NHM P40214. Fig. 26. *Daviesina khatiyahi* Smout, paratype, Paleocene, Qatar, NHM 40242.

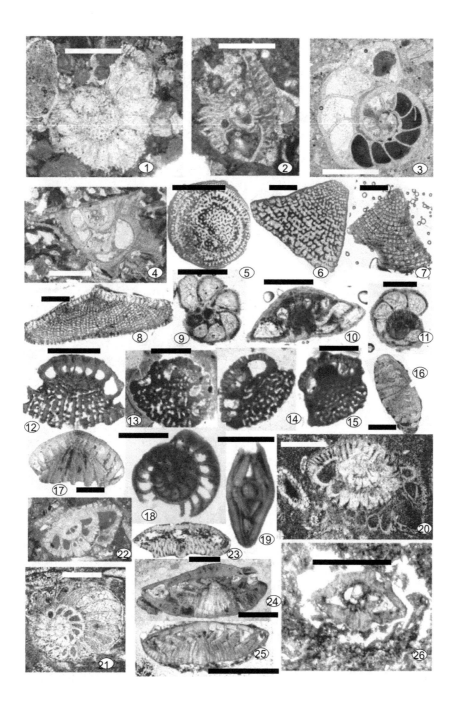

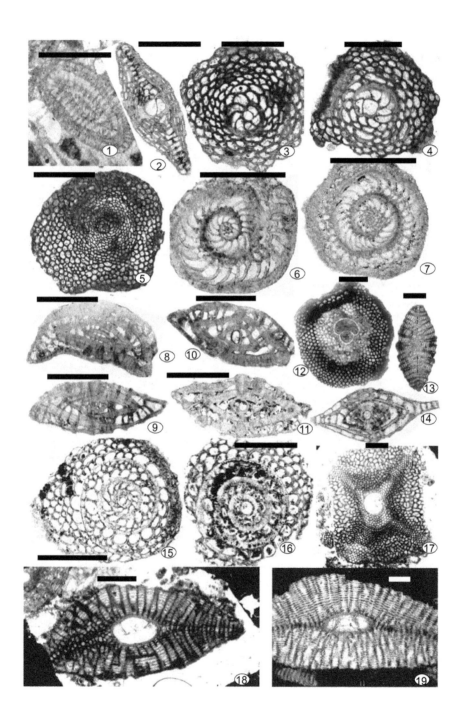

Plate 6.29   Scale bars: Figs 1-19= 1mm. Fig. 1. *Elphidium* sp., Late Oligocene, Syria, UCL coll. Figs 2-4. *Eulinderina* sp., Middle Eocene, Yecuatla, Veracruz, Mexico, 2) axial section showing thickened lateral walls, NHM P51959; 3) equatorial section with thin walls without pustules, NHM P51958; 4) equatorial section with thick walls with coarse pustules, NHM P51960. Fig. 5. *Helicolepidina spiralis* (Tobler), equatorial section, microspheric form, Eocene, El Alto, NW Peru, NHM P302478. Figs 6-11. *Helicostegina gyralis* Barker and Grimsdale, Middle Eocene, 6 -7) equatorial sections, topotypes, Yecuatla, Veracruz, Mexico, NHM 51950-6; 8-9) axial sections, topotypes, B-forms, NHM P33356; 10-11) axial sections, A-form, Sabaneta, Vercruz, Mexico, topotypes, NHM P51952. Figs 12-13. *L. (Nephrolepidina) praemarginata* (Douvillé 1908), Late Oligocene, uppermost part of the Mesolouri section, Greece, Wielandt-Schuster coll., 12) FNS 2001z0155/0010, 13) FNS 2001z0155/0006. Figs 14-16. *Helicosteginopsis soldadensis* (Grimsdale), topotype, Late Eocene, Soldado Rock Trinidad, NHM P51953-5, 14) A-form, axial section; 15-16) equatorial sections. Figs 17-18. *Eulepidina ephippioides* (Jones and Chapman), Oligocene, Kirkuk Well no.19, Iraq, 17) NHM P40667; 18) axial section, NHM P40668. Fig. 19. *Eulepidina* sp., Oligocene, Dutch New Guinea, NHM P22790.

Plate 6.30   Scale bars: Figs 1, 2-3, 10, 13, 18-19 = 1mm.; Figs 4-9, 11-12, 14-17 = 0.5mm. Fig. 1. *Sakesaria dukhani* Smout, paratype, Paleocene, Qatar, NHM P40203. Fig. 2. *Sakesaria ornata* Smout, paratype, Paleocene, Qatar, NHM P40205. Fig. 3. *Miogypsinella* cf. *borodinensis* Hanzawa, Chattian, Sukau Road Quarry, North East Borneo, UCL coll. Fig. 4. *Americogypsina braziliana* BouDagher-Fadel and Price, paratype, Early Oligocene, figured by BouDagher-Fadel and Price (2010b), offshore Brazil, UCL MF57. Fig. 5. *Americogypsina americana* BouDagher-Fadel and Price, holotype, Early Oligocene, figured by BouDagher-Fadel and Price (2010b), offshore Brazil, UCL MF63. Fig.6. *Paleomiogypsina boninensis* Matsumaru, Chattian, North East Borneo, UCL coll. Fig. 7. *Rotalia trochidiformis* Lamarck, Kairabad Limestone, Pakistan, UCL coll. Figs 8-10. *Miogypsinella* sp., Late Chattian, Java, UCL coll. Fig. 11. *Miogypsinoides formosensis* Yabe and Hanzawa, Late Oligocene, Corsica, UCL coll. Fig. 12. *Miolepidocyclina mexicana* Nuttall, Late Oligocene, from BouDagher-Fadel and Price (2010b), offshore Brazil. Fig. 13. *Miogypsina triangulata* BouDagher-Fadel and Price, holotype, Late Oligocene, figured by BouDagher-Fadel and Price (2010b), Brazil, UCL MF142. Fig.14. *Miolepidocyclina panamensis* (Cushman), Late Oligocene, from BouDagher-Fadel and Price (2010b), Brazil. Fig. 15. *Miogypsina gunteri* Cole, Late Oligocene, from BouDagher-Fadel and Price (2010b), Brazil. Fig. 16. *Miolepidocyclina braziliana* BouDagher-Fadel and Price, holotype, Late Oligocene, figured by BouDagher-Fadel and Price (2010b), Brazil, UCL MF124. Fig. 17. *Neorotalia* sp. 1, BouDagher-Fadel and Price, Early Oligocene, from BouDagher-Fadel and Price (2010b), offshore Brazil, UCL MF54. Fig. 18. *Americogypsina koutsoukosi* BouDagher-Fadel and Price, paratype, Early Oligocene, figured by BouDagher-Fadel and Price (2010b), offshore Brazil, UCL MF69.

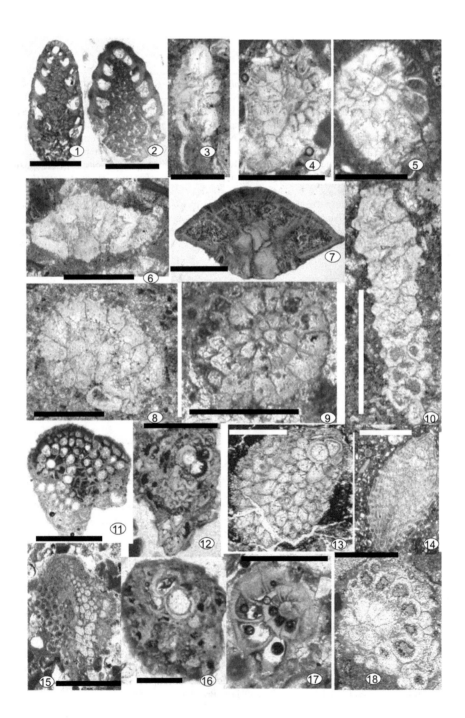

Chapter 7

The Cenozoic Larger Benthic Foraminifera: The Neogene

7.1 Introduction

As seen in Chapter 6, the Oligocene-Miocene boundary was not such a significant event for larger benthic foraminifera as the sharply defined Cretaceous-Paleocene boundary. Indeed, almost 80% of species survived the boundary into the Miocene. Although most of the Miocene superfamilies are extant, provincialism is prominent at both the generic and specific levels. Following the initial transoceanic migrations of larger benthic foraminifera between the American and the Tethyan provinces, facilitated by the series of global sea-level regressions, this migration stopped after rising sea-level in the early Oligocene separating the American from the other larger benthic provinces (see BouDagher-Fadel and Price, 2010a, b, c, 2013, 2017). As species became geographically isolated, colonizing new areas environmentally similar to each other, they thrived and evolved similar but distinct parallel lineages during the Neogene. Despite showing different evolutionary lineages at the species level, the main line of evolution follows the same patterns as in the Oligocene, and larger foraminifera become very important biostratigraphical markers globally at this time. The Tethyan seaway between the proto-Mediterranean and the proto-Indian Ocean became narrower during the Early Miocene, which restricted further migration between the two provinces and resulted in further provincialism.

During the Neogene, lepidocyclinids and miogypsinids completely disappeared from America in the late Early Miocene, and they disappeared from the proto-Mediterranean in the Serravallian. Deep water textulariides made their first appearance in America, while new genera of alveolinoids appeared in the Indo-Pacific. The development of the Indo-Pacific as a separate province continued in the Late Miocene. *Cycloclypeus* continues to range up to present, while there was a considerable proliferation of the calcarinids in the Pliocene. Most of the superfamilies continued to survive with globally spread representatives, except the American deep water Pavonitoidea, which disappeared completely in the early Pliocene.

7.2 Morphology and Taxonomy of Neogene Larger Benthic Foraminifera

In this section, the main superfamilies (see Fig. 7.1) and families of the following three Neogene orders are discussed, namely the:

- Miliolida
- Rotaliida
- Textulariida.

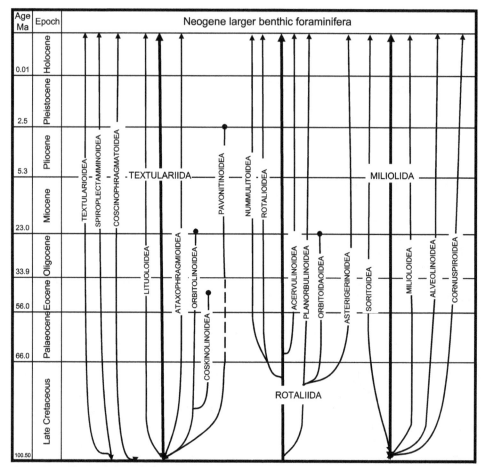

Figure 7.1. The evolution of the Neogene orders (thick lines) and superfamilies (thin lines) of larger foraminifera.

ORDER MILIOLIDA Delage and Hérouard, 1896

The miliolides have tests that are porcelaneous and imperforate made of high Mg-calcite with fine randomly oriented crystals. They range from the Carboniferous to the Holocene.

Superfamily ALVEOLINOIDEA Ehrenberg, 1839

The test is enrolled along an elongate axis, initially being planispiral or streptospiral, or milioline with chambers added in varying planes. Cretaceous to Holocene.

Family Alveolinidae Ehrenberg, 1839, emend. Hottinger at al., 1989

The test is free, fusiform, and coiled along an elongate axis (see full description Chapter 6). Early Cretaceous (Aptian) to Holocene.

- *Alveolinella* Douvillé, 1907 (Type species: *Alveolina quoyi* d'Orbigny, 1826). The test is elongate fusiform, with several rows of chamberlets in axial section. Pre-septal passages connect adjacent chamberlets, with smaller secondary pre-septal passages in later whorls. The final chamber has numerous rows of apertures. Late Middle Miocene (Serravallian) to Holocene (Fig. 7.2; Plate 7.1, fig. 20; Plate 7.2, figs 1-8).

- *Flosculinella* Schubert, 1910 (Type species: *Alveolinella bontangensis* Rutten, 1912). The early part of the test is streptospiral and similar to *Borelis* de Montfort, 1808, but with double rows of chamberlets on the floor of each chamber, one row being smaller than the other. Early Miocene (Burdigalian) to Middle Miocene (mid Serravallian) (Plate 7.2, figs 9-12).

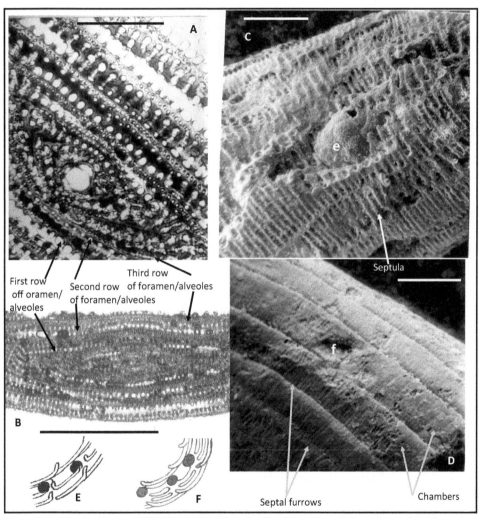

Figure 7.2 Alveolinella quoyi (d'Orbigny), Port Moresby, Coral Sea, New Guinea: A-B, axial thin sections showing elongate fusiform test, with several rows of chamberlets; C, solid specimen by Banner (1971) showing a parasite/symbiont completely enclosed by the whorls, *Planorbulinopsis parasitica* Banner (e) embedded in the test; D, a solid specimen showing a bored area (f); E-F, schematic drawing showing the single pre-septal passage in E) *Flosculinella* and F) doubled pre-septal passage of *Alveolinella*. Scale bars: A, C, D = 0.25mm, B = 1mm.

Superfamily CORNUSPIROIDEA Schultze, 1854
The test is free or attached composed of a globular proloculus, followed by a tubular enrolled chamber. The coiling is planispiral or trochospiral, evolute or involute, and may become irregular. The aperture is simple, at the end of the tube. Lower Carboniferous to Holocene.

Family Discospirinidae Wiesner, 1931
The form is discoid with a globular proloculus, followed by a planispirally enrolled test. The later chambers are annular and may be subdivided into small chamberlets. Middle Miocene to Holocene.

- *Discospirina* Munier-Chalmas, 1902 (Type species: *Orbitolites tenuissimus* Carpenter, 1870). The test is large, fragile, thin and flattened, with a peneropliform early stage, followed by annular chambers subdivided by numerous internal septa that fall short of the anterior wall of each chamber. Middle Miocene to Holocene (Plate 7.3, fig. 10).

Superfamily MILIOLOIDEA Ehrenberg, 1839
The test is coiled in varying planes or uncoiled, cylindrical or compressed with partial partitions. Upper Triassic to Holocene.

Family Rivieroinidae Saidova, 1981
The test is planispiral, ovate in outline with chambers subdivided by oblique sutures. The aperture is a single curved slit, but may be cribrate. Middle Eocene to Holocene.

- *Riveroina* Bermúdez, 1939 (Type species: *Riveroina caribaea* Bermúdez, 1939). The test has flattened sides. Chambers are one-half coil in length, subdivided by oblique septula, extending completely across the chamber lumen. The aperture is an arched slit at the end of the final chamber. Holocene.
- *Pseudohauerinella* McCulloch, 1981 (Type species: *Pseudohauerina dissidens* McCulloch, 1977). The test has a quinqueloculine early stage and adult planispiral chambers with incomplete subepidermal partitions. The aperture is terminal and cribrate. Holocene.

Superfamily SORITOIDEA Ehrenberg, 1839
Chambers are planispiral, uncoiling, flabelliform or cyclical, and may be subdivided by partitions or pillars. Late Permian to Holocene.

Family Peneroplidae Schultze, 1854
The test has a closely coiled early stage becoming uncoiled in later stages. Chamber interiors are simple. The aperture is a single rounded and slit-like, or multiple. Late Cretaceous to Holocene.

- *Laevipeneroplis* Šulc, 1936 (Type species: *Peneroplis karreri* Wiesner, 1923). The test is compressed and flaring with chambers becoming progressively broader and more curved, but increasing very little in height. The interiors of the chambers are undivided. The aperture is multiple near the base of the apertural face, becoming two rows of openings at the end of the apertural face. Miocene to Holocene.

Family Soritidae, Ehrenberg, 1839

The test is biconvex, involute-planispiral to an uncoiled evolute, flaring, annular discoid with partial or complete partitions. Apertures are multiple. Late Cretaceous (Cenomanian) to Holocene.

- *Androsina* Lévy, 1977 (Type species: *Androsina lucasi*, Lévy 1977). The test is tightly coiled, with chambers increasing rapidly in size, strongly curved, but do not become completely annular. Chambers are subdivided by flattened pillars in the median plane. Double annular passages occur in lateral positions. The apertural face has two to four rows of pores. Pleistocene to Holocene.
- *Androsinopsis* Hottinger, 2001 (Type species: *Androsinopsis radians* Hottinger, 2001). This genus is similar to *Androsina* but becomes circular at an early stage of ontogeny with heavy equatorial pillars in the annular adult stage. Late Miocene.
- *Annulosorites* Hottinger, 2001 (Type species; *Annulosorites spiralis* Hottinger, 2001). This genus is similar to *Sorites* (see Chapter 6), but with an involute spiral nepiont. The megalospheric protoconch is large, involute and planispiral with a rapidly expanding flexostyle, ending with a frontal wall bearing multiple apertures. The following deuteroconch has multiple apertures and is subdivided by two septula. The microspheric generation is not known. Whereas in *Sorites* the opposing subepidermal partitions form a continuous septulum in the same chamber, in *Annulosorites*, this connection is lacking and the two opposing subepidermal partitions form two separate septula. From one chamber to the next, the septula are alternating in a radial disposition. Apertures are large and rounded, forming a single row in the equatorial plane of the shell, in alternating position with the septula. In late, adult growth-stages, the apertures may become separated into two rows. Late Miocene.
- *Archaias* de Montfort, 1808 (Type species: *Archaias spirans* de Montfort, 1808 = *Nautilus angulatus* Fichtel and Moll, 1798), as defined by its type species *Archaias angulatus* (Fichtel and Moll), extensively emended by Rögl and Hansen, 1984. The test is compressed, planispiral and involute, and may be partially evolute in the last whorls, with a thickened middle part and radial endoskeletal elements. Multiple apertures are flanked by irregular free and interseptal pillars. The subepidermal partitions are incomplete, and tests lack exoskeletal structures and no marginal subdivision of the chambers. Middle Eocene to Holocene (Fig. 7.3; see Chapter 6).
- *Cycloputeolina* Seiglie and Grove, 1977 (Type species: *Peneroplis pertusus* (Forskal) var. *discoideus* Flint, 1899). Final chambers become circular and subdivided with vertical partitions. The aperture is one or two rows of openings, each bordered by a lip. Miocene to Holocene.

- *Fusarchaias* Reichel, 1952 (Type species*: Fusarchaias bermudezi* Reichel, 1952). The test is fusiform, planispiral with numerous chambers and endoskeleton of interseptal pillars. Oligocene to Miocene.
- *Marginopora* Quoy and Gaimard, 1830 (Type species: *Marginopora vertebralis* Quoyi and Gaimard, 1830). The test is large and biconcave. The embryonic apparatus of the megalospheric test consists of a large deuteroconch that embraces a small protoconch, including its wide-open flexostyle with an almost cylindrical frontal wall bearing numerous pores. The microspheric test has an early planispiral and peneropline stage followed by annular, concentric chambers, with thickened and folded margins. Initially with two layers of annular chamberlets, later low chambers are inserted between them. The annular chambers are subdivided by incomplete vertical septula and partitions. Oblique stolons connect the lateral chamberlets to the chambers above and below. The aperture is multiple over the peripheral wall. Miocene to Holocene (Plate 7.2, figs 16-17; Plate 7.3, figs 1-5).
- *Miarchaias* Hottinger, 2001 (Type species: *Miarchaias meander* Hottinger, 2001). The test has a pillared, radial endoskeleton and a radial exoskeleton consisting of short radial partitions (beams). The megalospheric test has a narrow flexostyle. It differs from *Archaias* in having marginal apertures and complete subepidermal partitions in alignment from one chamber to the next one. Late Miocene (Fig. 7.3).

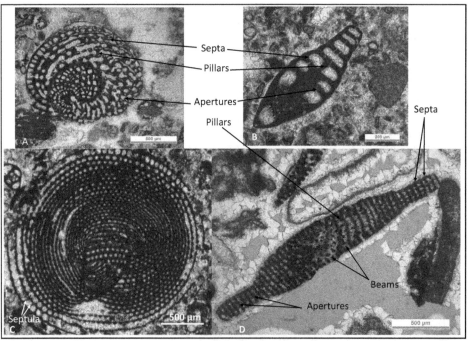

Figure 7.3.   A-B, *Archaias* sp.; C, D, *Miosorites americanus* (Cushman), C) *Miarchaias* sp. with an early peneropline evolute stage and annular chambers, divided by simple radial septula; D, *Miosorites americanus* (Cushman). All figured specimens are from Bahamas (courtesy of G.J. Fischer).

- *Miosorites* Seiglie and Rivera, 1976 (Type species: *Orbitolites americana* Cushman, 1918). The test is annular and evolute. The microspheric test has an early peneropline evolute stage. The megalospheric protoconch consists of a large, flaring, involute flexostyle and of a semilunar, subdivided, slightly involute deuteroconch (Hottinger, 2001). The following few chambers are reniform. The chambers become annular in the later stage. Annular chambers are divided by simple radial partitions, interpreted as septula, alternating in radial position from one chamber to the next. Oval apertures occur in marginal positions, with crosswise-oblique stolon axes relative to the radius of the test. The genus *Miosorites* differs from *Amphisorus* (see Chapter 6) by its narrower apertural face, confined in an equatorial depression, and by its involute embryonic apparatus, where the flexostyle envelops large, lateral surfaces of the megalosphere (see Hottinger, 2001, Fig. 8). No internal pillars are present. (Fig. 7.3) Burdigalian to?Pliocene.
- *Parasorites* Seiglie and Rivera, 1977 (Type species: *Praesorites orbitolitoides* Hofker 1930). The test is annular and evolute. The microspheric test has an early peneropline stage. The megalospheric test has a subglobular proloculus with a long flexostyle, followed by an evolute, rapidly flaring peneropline and an annular later stage. Annular chambers are subdivided by simple radial lateral partitions (interpreted by Hottinger, 2001, as exoskeletal beams), that alternate in radial positions in subsequent chambers. The aperture is one to five rows of rounded openings. Late Miocene to Holocene.
- *Pseudotaberina* Eames, 1977, emended, Banner and Highton, 1989 (Type species: *Orbitolites malabarica* Carter 1853). The test becomes cyclical in the latest growth of the microspheric form. Megalospheric forms have a large proloculus. Later chambers are not embracing and may become cyclical and evolute. Chambers show "stalagmitic" and "stalactitic" projections/pillars that discontinuously fuse across the chambers away from the lateral walls, producing chamber-subdividing structures, distinct from the separated pillars of *Archaias* (Fig. 7.3; see Chapter 6). The discontinuity of the internal structures allows a distinction to be made between *Pseudotaberina* and the Cretaceous *Larrazetia* (see Chapter 5), while *Cyclorbiculina* possesses true subepidermal partitions. Unlike the apertures of *Archaias,* which are situated in regular, parallel rows, those of *Pseudotaberina* are scattered over the apertural face and each of them is surround by a projected lip. Early Miocene (Fig. 7.4; Plate 7.4, figs 1-6, 12).
- *Sorites* Ehrenberg, 1839 (Type species: *Nautilus orbiculus* Forsskal, 1775). The test is a large discoid, with an early peneropline stage. Annular chambers are divided into numerous curved to rectangular small chamberlets, which are connected to each other and to those in adjacent chambers by stolons. The aperture is a single paired row. Oligocene to Holocene. (Fig. 7.5; Plate 7.3, figs 6-8; Plate 7.8, fig. 3; see Chapter 6).

Family Keramosphaeridae Brady, 1884
The test is globular with concentric chambers connected by stolons in the same series as well as those of successive series. Early Cretaceous (Berriasian) to Late Cretaceous (Maastrichtian), and Miocene to Holocene.

- *Kanakaia* Hanzawa, 1957 (Type species: *Kanakaia marianensis* Hanzawa, 1957). The test is large, formed of encrusting layers of chambers. Adjacent chambers are

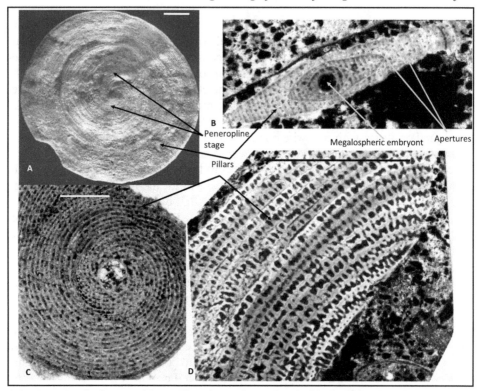

Figure 7.4. Pseudotaberina malabarica (Carter), type figures deposited in the NHM: A, solid specimen of a microspheric specimen; B, oblique axial section showing the disposition of the pillars; C, oblique equatorial section showing the alternating disposition of the foramina and the pillars; D, enlargement of the chambers showing the "stalagmitic" and "stalactitic" pillars that discontinuously fuse across the chambers. Scale bars = 1mm.

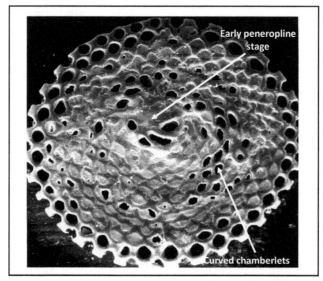

Figure 7.5. Sorites sp., a microspheric form showing a peneroplid stage followed by annular series of small chamberlets.

connected by horizontal stolons, and oblique stolons connect those of successive layers. Early Miocene (Aquitanian).

- *Keramosphaera* Brady, 1882 (Type species: *Keramosphaera murrayi* Brady, 1882). The test is globular, with irregular chamberlets, added in concentric unaligned spherical series. The adjacent chamberlets are connected by stolons. Holocene.

ORDER ROTALIIDA Delage and Hérouard, 1896

The test is multilocular, with a calcareous wall made of perforate, hyaline lamellar calcite. The aperture is simple or with an internal tooth-plate. Triassic to Holocene.

Superfamily ASTERIGERINOIDEA d'Orbigny, 1839

The test is trochospiral to planispiral, with a closed umbilicus. Chambers have internal partitions. The aperture is umbilical, and may extend up the apertural face with complex chamberlets at the centre of umbilical side. Late Cretaceous (Santonian) to Holocene.

Family Lepidocyclinidae Scheffen, 1932

The test is discoidal, involute, and biconvex with a broad centrum, which grades into a narrow flange. Adauxiliary chambers may be present. The primary spire persists into the equatorial layer, or with annular rings of chamberlets that follow the embryont immediately. Stacks of "lateral chamberlets" (cubiculae) occur on each side of the median chamberlets. Pillars may be present between adjacent vertical stacks of cubiculae or scattered in the central region. The chamber walls are perforated by stolons, but there is no canal system. Middle Eocene to Late Miocene (Early Pliocene?).

Subfamily Lepidocyclininae Scheffen, 1932

Representatives of this subfamily have a bilocular or multilocular embryonal stage, surrounded by a thickened wall and adauxiliary chambers. Microspheric tests have an early planispiral coil, while megalospheric tests have a globular protoconch, enclosed or followed by a larger reniform deuteroconch. Post-embryonic chambers evolve from cyclical, arcuate to hexagonal in shape, usually with two or more apertures. The lateral chambers are well differentiated from the equatorial layer and in the advanced forms they are arranged in tiers on either side of the equatorial layer. Surface ornaments and development of pillars are of specific importance. Middle Eocene to Late Miocene (Early Pliocene?), (see Chapter 6).

Superfamily NUMMULITOIDEA de Blainville, 1827

The test is planispiral or cyclic, lenticular, multicamerate, with a septal flap and canaliculated septa. A spiral marginal cord and spiral canal system are present in early forms, and are modified in advanced forms or replaced by intraseptal canals. Paleocene to Holocene.

Family Cycloclypeidae Galloway, 1933 emend. BouFagher-Fadel, 2002
This family is distinguished by the development of concentric annular chambers, that are wholly evolute, with each chamber divided into numerous chamberlets in the median plane, and with each chamberlet separated from adjacent chamberlets by canaliculated, straight walls. No marginal cord exists, except in the early stages of the microspheric generation. Eocene to Holocene.

• *Cycloclypeus* Carpenter, 1856 (Type species: *Cycloclypeus carpenteri* Brady, 1881). A nummulitoid with a nepionic morphology like *Heterostegina*, but with a final growth stage with cyclic chambers. No alar prolongations occur. Early Oligocene to Holocene (Fig. 7.6; Plate 7.5, figs 15-19; Plate 7.6, figs 1-11; Plate 7.7, fig. 9; see Chapter 6).
• *Katacycloclypeus* Tan Sin Hok, 1932 (Type species: *Cycloclypeus* (*Katacycloclypeus*) *martini* Tan Sin Hok, 1932). Tan Sin Hok (1932), in his description of *Katacycloclypeus*, assigned it as a new subgenus of *Cycloclypeus*. However, there is no direct evidence of intergradation, either in the modelling of the test or in the embryonic structure between *Cycloclypeus* and *Katacycloclypeus*. The latter has a trilocular embryont and a thin test with a central umbo, surrounded by several annular inflations of the solid lateral walls. The stratigraphic range is also quite different. *Katacycloclypeu*s is confined to the upper Tf1 and Tf2 Letter Stages of the Middle Miocene of the Indo-Pacific, while *Cycloclypeus* ranges from the Oligocene to the Holocene throughout all the tropics. Therefore, the two forms should be considered to be generically different. Middle Miocene (late Langhian to Serravallian) (Plate 7.6, figs 12-13; Plate 7.7, figs 1-8; Plate 7.8, fig. 7).

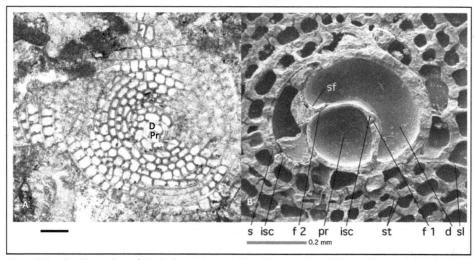

Figure 7.6. A, thin section of *Cycloclypeus indopacificus* Tan Sin Hok, Middle Miocene, Nias, Sumatra, UCL coll., scale bar = 0.25mm; B, SEM of *Cycloclypeus carpenteri* (Brady), Holocene, from Hottinger (2006) Bikini, Pacific. (Abbreviations: **d**: deuteroconch; **f 1**: foramen of protoconch; **f 2**: foramen of deuteroconch; **ics**: intraseptal canal system; **pr**: proloculus; **s**: septum; **sl**: septulum; **st**: stolon (Y-shaped)).

Family Nummulitidae de Blainville, 1827
The test is planispiral, involute or evolute, with septal, marginal and vertical canals.
Paleocene to Holocene.

- *Heterocyclina* Hottinger, 1977 (Type species: *Heterostegina luberculata* Mobius, 1880). The test is discoidal, planispiral, and evolute with whorls becoming annular in the adult. Chambers are divided by septula into rectangular chamberlets. Suture canals are unbranched and the stolon system is L-shaped. Holocene.
- *Bozorgniella* Rahaghi, 1973 (Type species: *Bozorgniella qumiensis* Rahaghi. 1973). The planispiral, involute test has two and a half whorls. Sutures are straight to slightly curved near the end, but externally obscured by pustules. Early Miocene (Aquitanian).
- *Planoperculina* Hottinger, 1977, emended Banner and Hodgkinson, 1991 (Type species: *Operculina heterostegnoides* Hofker, 1933). The test is wholly evolute, with incomplete chamber partitions. Holocene.
- *Radiocycloclypeus* Tan Sin Hok, 1932 (Type species: *Cycloclypeus neglectus* Martin var. *stellatus* Tan Sin Hok, 1932). The test is stellate with irregular rays. The embryonic apparatus is enclosed by a thick wall and consists of a proto-conch surrounded by a large deuteroconch. It is followed by about six thin-walled embryonic chambers that gradually increase in length and finally become annular in the later part of the test. Chambers are divided into rectangular chamberlets, which alternate in position. Early Miocene (Burdigalian) to Middle Miocene (Serravallian).
- *Operculinella* Yabe, 1918 (Type species: *Amphistegina cumingii* Carpenter,1860). The last true *Nummulites* spp. became extinct at the top of the Td "Letter Stage" with *Nummulites fichteli* Michelotti 1841 from the upper Early Oligocene of Italy. Contrary to the opinions of S. Cole (in Loeblich and Tappan, 1964) and Loeblich and Tappan (1988), *Nummulites* can be distinguished from *Operculinella*. Eames et al. (1962) illustrated a simple *Nummulites vascus* Joly and Leymerie (their plate 1, figures A, B) to compare with *Operculinella cumingii* (Carpenter) (*Palaeonummulites nomen oblitum*). The strong dimorphism seen between microspheric and megalospheric forms of Oligocene specimens of *Nummulites* is never seen in *Operculinella* (where the microspheric and megalospheric generations are externally identical). The presence of trabeculae in *Nummulites* and their absence from *Operculinella* is noteworthy, but, most importantly, the diameter of the megalospheric protoconch of *Nummulites* (in both simple and complex forms) is much greater than the diameter of the proloculus of *Operculinella*. The megalospheric loosely coiled *Operculinella* (e.g. *Operculinella cumingii)* persists to the Holocene but the large protoconch of true *Nummulites* does not occur beyond the Early Oligocene. Oligocene to Holocene (Fig. 7.7).

Superfamily PLANORBULINOIDEA Schwager, 1877
The test is trochospiral in early stages, but later may be uncoiled and rectilinear, or bise-rial or may have many chambers in the whorl. Intra- to extra-umbilical apertures occur,

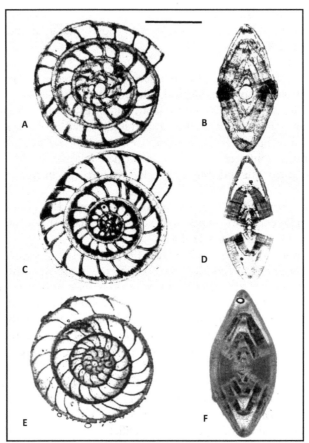

Figure 7.7. Comparison between: A-B, *Nummulites vascus* Joy and Leymerie, Oligocene, Cyrenaica, NHM P44493; C-D, *Operculinella cumingii* (Carpenter), Holocene, Port Moresby, Papua, NHM coll.; E-F, *Palaeonummulites kugleri* (Vaughan and Cole), Oligocene, Falling Waters State Park, Chipley, Florida, UCL MF323. Scale bar = 2mm.

and additional equatorial apertures may be present. Early Cretaceous (Berriasian) to Holocene.

Family Planorbulinidae Schwager, 1877

The test is free or attached. The early stage is trochospiral, but later becoming discoid, cylindrical or conical. The aperture is single or multiple. Eocene to Holocene.

• *Planorbulinopsis* Banner, 1971 (Type species: *Planorbulinopsis parasitica* Banner, 1971). The test is attached, with the early part trochospiral, the spiral side evolute, and the umbilical side involute. The umbilicus is open and deep. Later chambers alternate with those of the preceding whorl. The walls are coarsely perforate. Holocene (Fig. 7.2).

Family Victoriellidae Chapman and Crespin, 1930
The test is attached or may be free in the juvenile stage, with a trochospiral early stage, later becoming an irregular mass of chambers. Cretaceous (Santonian) to Holocene.

Subfamily Carpenteriinae Saidova, 1981
The test is attached, trochospiral throughout, and planoconvex with a large aperture, open in the umbilicus. Paleocene to Holocene.

- *Carpenteria* Gray, 1858 (Type species: *Carpenteria balaniformis* Gray, 1858). The test has a carinate periphery, and is planoconvex with a flat spiral side and distinct rims or keels, a strongly convex, distinctly perforate umbilical side surrounded by thick pillars. Late Eocene to Holocene (Plate 7.9, figs 3, 6-7; Plate 7.10, figs 1-4)).

Subfamily Rupertininae Loeblich and Tappan, 1961
The test is attached with a distinct flattened disk, but coiling grows out away from the site of attachment. Late Cretaceous to Holocene.

- *Rupertina* Loeblich and Tappan, 1961 (Type species; *Rupertia stabilis* Wallich. 1877). The test grows upright around a central column, and is trochospiral in the early stage, later becoming more loosely coiled. Miocene to Holocene (Plate 7.9, figs 8-9).

Superfamily ROTALIOIDEA Ehrenberg, 1839
The test is involute to evolute, initially trochospiral or planispiral, commonly with many chambers in numerous whorls. As new chambers are added, septal flaps attach to previous apertural face and enclose radial canals, fissures, umbilical cavities, and intraseptal and subsutural canals. The wall is made of perforate hyaline calcite, that is generally optically radial in structure. Primary apertures are single or multiple. Small openings into the canal system may occur along the sutures. Late Cretaceous (Coniacian) to Holocene.

Family Calcarinidae Schwager, 1876
The test is enrolled with protruding canaliculated spines. Free living, but they live mostly adhered to algae by a kind of plate secreted at the end of one, or more rarely two spines (Röttger and Krüger, 1990). Rows of areal foramina are found mostly near the base of the septa. Late Cretaceous (Maastrichtian) to Holocene.

- *Baculogypsina* Sacco, 1893 (Type species*: Orbitolina concava* Lamarck var. *sphaerulata* Parker and Jones, 1860). The test is biconvex, with 5 to 7 canalicular spines roughly in a single plane, radiating from the spiral juvenarium by originating as extensions of an intraseptal interlocular space (Hottinger, 2006), and continuing to enlarge with growth. Following the spiral juvenarium, elongated supplemental chamberlets overgrowing the base of the spine, are aligned in a chessboard pattern and connected to the spine canals. Pleistocene to Holocene (Plate 7.10, figs 5-7).

- *Baculogypsinoides* Yabe and Hanzawa, 1930 (Type species: *Baculogypsinoides spinosus* Yabe and Hanzawa, 1939). The test is globular with protruding canalicular spines, but unlike *Baculogypsina* they are not in a single plane, but instead they give a triangular to tetrahedral appearance to the test. Holocene.
- *Calcarina* d'Orbigny, 1826 (Type species: *Nautilus spengleri* Gmelin, 1791). The test is biconvex, trochospirally coiled throughout, with many blunt or bifurcating radial spines. There is an enveloping spiral canal system consisting of a layer formed by numerous canals that run through the spines. There are ten to twenty chambers in the final whorl, each containing passages to the canal system which is connected to the outside via many openings on the test surface. Pustules and spinules cover the whole test with the umbilicus filled by pillars. Pliocene to Holocene (Plate 7.5, figs 11-14; Plate 7.10, fig. 8).
- *Quasirotalia* Hanzawa, 1967 (Type species: *Quasirotalia guamensis* Hanzawa, 1967). The test is planoconvex with a flat spiral side. Layers of chambers are added to the periphery and the umbilical side. Pillars fill the umbilicus. Coarse pores fill the thick calcareous walls. Pliocene (Fig. 7.8).
- *Schlumbergerella* Hanzawa, 1952 (Type species: *Baculogypsina floresiana* Schlumberger, 1896). The test is globular with slightly projecting spines or tubercles. In the microspheric generation there is a planispiral coil of about two whorls, with spines. Megalospheric forms have an embryonic apparatus composed of three chambers with tetragonal spines, formed from the outer chamber walls. There are numerous dome-like lateral chamberlets, communicating with each other through stolons in the lateral walls and by large pores with chambers of the same radial wall, produce a globular test. Numerous pillars are formed between the radiating rows of chamberlets. Unlike *Baculogypsinoides*, the corners of the tetrahedral test in this genus do not support prominent spines and the canal system is much reduced, as none of the pillars of the latter have an internal canal system, but one which only occurs on the outer edge of the spines. The walls are coarsely perforated. Pleistocene to Holocene.

Family Chapmaninidae Thalman, 1938
The test is conical, with a trochospiral initial part, followed by a uniserial part and a tubular apertural system. Chambers may be annular in the adult part of the test. Septa are invaginated into tube pillars. Late Paleocene to Late Miocene (Tortonian).

- *Tenisonina* Quilty, 1980 (Type species: *Tenisonina tasmaniae* Quilty, 1980). The test is planoconvex, an early trochospire is followed by annular chambers, divided into curved chamberlets, but the final chambers are undivided. Early Miocene.

Family Miogypsinidae Vaughan, 1929
The test is flattened to biconvex. The microspheric form has a trochospiral or planispiral early spire, while the megalospheric form has a bilocular embryonal stage followed by a fan of median chamberlets. Middle Oligocene to Middle Miocene.

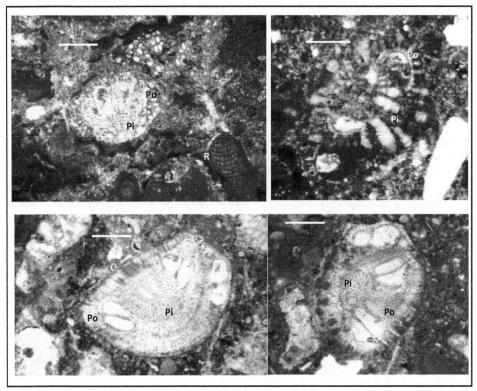

Figure 7.8. Photomicrograph of thin sections of *Quasirotalia guamensis* Hanzawa, Sulawesi, SEA coll., showing a plano-convex test with a flat spiral side, pillars (Pi) filling the umbilicus and coarse pores (Po) filling the thick calcareous walls and rodophyte algae (R). Scale bars = 0.5mm.

- *Heterosteginoides panamensis* Cushman, 1918 (Type species: *Heterosteginoides* Cushman, 1918). The test is similar to *Miolepidocyclina,* with the nepiont centrally placed, however, the nepionic spiral is much longer, overriding in its later part a variable number of equatorial chambers. Early Miocene (Aquitanian to early Burdigalian).

- *Lepidosemicyclina* Rutten, 1911 (Type species: *Orbitoides* (*Lepidosemicyclina*) *thecideaeformis* Rutten, 1911). A roughly circular test, with an embryonic apparatus made of a spherical protoconch and a reniform deuteroconch, that has a tendency to become enlarged in most advanced forms. Two sets of planispiral periembryonic chambers surround the embryo, the larger primary spiral and three unequal secondary spirals. The equatorial chamberlets are at first ogival, then rhombic and finally distinctly hexagonal. Early Miocene (Burdigalian) (Plate 7.9. fig. 15; Plate 7.11, fig. 12; Plate 7.12, fig. 11).

- *Miogypsinodella* BouDagher-Fadel et al., 2000 (Type species: *Miogypsina* (*Miogypsina*) *primitiva* Tan, 1936). The embryont coil is similar to that of *Miogypsinoides* (see Chapter 6), it is virtually planispiral, but there is only one whorl around the megalospheric proloculus, and a septal canal system is present. However, the lateral walls have gaps between the lamellae, which begin to split apart and form the beginnings of lateral chamberlets. This splitting results in thick-walled irregular chamberlets, unlike the regularly formed, stacked chamberlets of *Miogypsina* (see Chapter 6). Late Oligocene (late Chattian) to Middle Miocene (Langhian) (Plate 7.11, figs 1-4).

- *Miolepidocyclina* Silvestri, 1907 (Type species: *Orbitoides* (*Lepidocyclina*) *burdigalensis* Gümbel, 1868). The embryonic apparatus, consisting of large protoconch and deuteroconch, is surrounded by a thick wall. The megalospheric nepiont is similar to that of *Miogypsina,* with no coil around the proloculus but 2-bidirectional coils around the proloculus. However, the nepiont is centrally placed, instead of being at the edge of the test, as in *Miogypsina.* Early Oligocene (Rupelian, P20) to Early Miocene (Burdigalian) (Plate 7.11, figs 14-15; Plate 7.13, fig. 21).

- *Tania* Matsumaru, 1990 (Type species: *Tania inokoshiensis* Matsumaru, 1990). The embryonic apparatus has a globular protoconch and reniform deuteroconch in megalospheric generations, with two unequal sets of spiral nepionic chambers situated along the outer side of deuteroconch. *Tania* differs from *Miogypsinoides* in having well developed lateral chambers. It differs from *Miogypsina* by the arrangement of the embryonic chambers in the apical portion and by the development of hexagonal to spatulate chambers. It is distinguished from *Lepidosemicyclina* by the arrangement of embryonic chambers and from *Miolepidocyclina* and *Miogypsinita* (see Chapter 6) in having an apical embryonic apparatus. Early Miocene (Aquitanian).

Superfamily ACERVULINOIDEA Schultze, 1854
The test is trochospiral to discoidal, commonly with raspberry-like (or framboidal) early chambers, but with an encrusting later part consisting of numerous irregularly formed chambers. Paleocene to Holocene.

Family Acervulinidae Schultze, 1854
The early stage is followed by spreading chambers with one or more layers. Mural pores act as apertures. Paleocene to Holocene.

- *Acervulina* Schultze, 1854 (Type species: *Acervulina inhaerens* Schulze, 1854). Early chambers are coiled, later they are vermiform and irregularly arranged. The aperture is made up of coarse perforations. Miocene to Holocene.

- *Alanlordia* Banner and Samuel, 1995 (Type species: *Alanlordia niasensis* Banner and Samuel, 1995).The test is biconvex, the proloculus is followed by a single, nearly planispiral whorl of triangular chambers each successively linked by a single basal, septal aperture, but in which multiple, cribrate, pore-like apertures develop in their distal, outermost walls. The initial whorl is succeeded both dorsally and ventrally by layers of small chambers which are added in radial rows to form successive layers of small chambers. These chambers communicate with succeeding chambers in the next layers by cribrate, pore-like, small apertures. Radial pillars may develop in the median plane, both ventrally and dorsally. It is similar to *Wilfordia* (see Chapter 6) but the latter has no true initial spire. The closest homeomorph for *Alanlordia* (especially

A. niasensis) is *Vanderbeekia* (see Chapter 5), which has very similar pillaring and dorsal thickening, however, *Vanderbeekia* (like its close relatives *Sirtina, Irunites* and *Neurnannites*) appears to have had a distinct medial layer of thick-walled chambers, unlike the very short, single initial coil of *Alanlordia*. Middle Miocene (Serravallian, Tf2) to Late Pliocene (Plate 7.4, figs 7-11).

- *Borodinia* Hanzawa, 1940 (Type species: *Borodinia septetrionalis* Hanzawa, 1940). The test has more than one encrusting layer, with chambers alternating in position in successive layers. Early Miocene (Aquitanian) (Plate 7.9, figs 1-2).
- *Ladoronia* Hanzawa, 1957 (Type species: *Acervulina* (*Ladoronia*) *vermicularis* Hanzawa. 1957). Early chambers are clustered in a framboidal arrangement, followed by irregular elongate to vermiform chambers. Chambers of successive layers are connected by fine pores, but are neither aligned nor alternating in position. Miocene.
- *Planogypsina* Bermúdez, 1952 (Type species: *Gypsina vesicularis* var. *squamiformis* Chapman, 1901). The test is very thin, having globular early chambers followed by a single layer of irregular elongate to vermiform chambers. The aperture is made up of septal pores. Miocene to Holocene.

Family Homotrematidae Cushman, 1927
The test is attached, with a trochospiral early stage, later chambers growing in a massive branching structure. Eocene to Holocene.

- *Homotrema* Hickson, 1911 (Type species: *Polytrema cylindrica* Carter, 1880). A homotremid form, with a four-chambered embryo and conical projections or erect branches. Early chambers occur in clustered arrangement, but later in numerous layers have large irregular passages. The aperture is made up of larger perforations. Miocene to Holocene (Plate 7.9, figs 4-5).
- *Miniacina* Galloway, 1933 (Type species: *Millepora miniacea* Pallas, 1766). Megalospheric forms have a three-chambered juvenile stage, while microspheric forms have a trochospiral early stage. Following the free early stage, the attachment surface is narrow, and from this surface arise vertical irregularly branching structures with pillar-pore chambers (calyces) surrounding the central core. Adult forms have one to multiple rounded apertures with a bordering lip at the end of the branches. Early Miocene (Aquitanian) to Holocene (Plate 7.9, figs 10-13).

ORDER TEXTULARIINA Delage and Hérouard, 1896
The tests of these agglutinated foraminifera are made of foreign particles bound by organic cement. They range from lower Cambrian to Holocene.

Superfamily ATAXOPHRAGMIOIDEA Schwager, 1877
Members of this superfamily have a multilocular, trochospiral test to biserial or uniserial in later stages. Middle Triassic to Holocene.

Family Alveovalvulinidae Seiglie, Fleisher and Baker, 1986
The test is trochospiral to triserial and uniserial. Chambers have alveoli conjoined with vertical radial partitions or "tubiliform connections". Apertures are interiomarginal or terminal.

* *Alveovalvulina* Brönnimann, 1951 (Type species: *Alyeovalvulina suteri* Brönnimann, 1951). The test is trochospiral to triserial, with an interiomarginal aperture. Late Early Miocene to Early Pliocene.
* *Guppyella* Brönnimann, 1951 (Type species: *Goesella miocenica* Cushman, 1936). The later stage of the test is uniserial, and the aperture is terminal and circular. Late Early Miocene to Holocene.
* *Jarvisella* Brönnimann, 1953 (Type species: *Jaryisella karamatensis* Brönnimann, 1953). A trochospiral to triserial test that has chamber interiors subdivided by vertical folding of the outer wall, forming double-walled septula. Late Early Miocene to Middle Miocene.

Family Textulariellidae Grönhagen and Luterbacher, 1966
Tests have an early trochospirally enrolled stage, which later is reduced to triserial or uniserial. The chambers are overlapping and the wall is agglutinated with alveoles. Miocene.

* *Cuneolinella* Cushman and Bermúdez, 1941 (Type species: *Cuneolinella lewisi* Cushman and Bermúdez, 1941). The test is flattened, with a later stage that is biserial and compressed. Chambers increase rapidly in breadth so that the test becomes flabelliform. The aperture is a multiple row of openings. Middle Miocene.
* *Textulariella* Cushman, 1927 (Type species: *Textularia barrettii* Jones and Parker, 1876). The later stage of the test is biserial, with the interior of the chambers having numerous vertical partitions that are anastomising inward to form tiny alveoles. The aperture is a low arch. Miocene to Holocene.

Superfamily LITUOLOIDEA de Blainville, 1825
Members of this superfamily have a multilocular, rectilinear and uniserial test. The early stage has plani- (strepto-) or trochospiral coiling. The peripheries of the chambers have radial partitions, but centrally there are either no or scattered, separated pillars. The aperture is simple or multiple cribrate. Early Jurassic (Sinemurian) to Holocene.

Family Cyclamminidae Marie, 1941
The test is involute with alveolar walls. The aperture is near the septal face. Jurassic to Holocene.

* *Cyclammina* Brady, 1879 (Type species: *Cyclammina cancellata* Brady, 1879). The test is planispiral, and flattened. The walls are thick, with an alveolar subepidermal meshwork of a thickness exceeding that of the chamber lumen. The primary

aperture is a basal sutural slit, and the supplementary apertures are areal in the aperture face and septa and bordered by a lip. Paleocene to Holocene (Fig. 7.9).

- *Reticulophragmium* Maync, 1952 (Type species: *Alveolophragmium venezuelanum* Maync, 1952). The wall is thick with an alveolar hypodermis. The septa are solid and the aperture is situated in the basal suture with a lip on the upper side only. Paleocene to Holocene (Fig. 7.9).

Superfamily PAVONITINOIDEA Loeblich and Tappan, 1961
The early stage of the test is coiled, triserial to biserial or uniserial. The interiors of the chambers are partially divided by numerous vertical partitions (beams) or septula, that project downwards from the roof and rarely may have a few connecting horizontal partitions (rafters). Late Cretaceous and Oligocene to Pliocene.

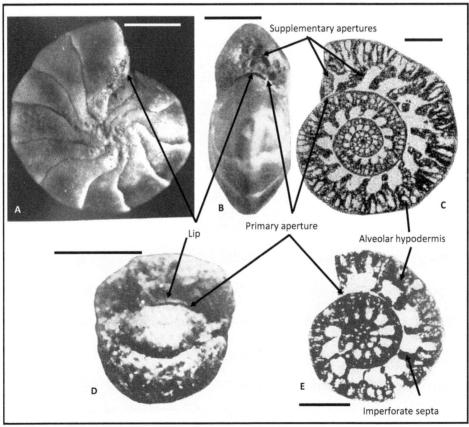

Figure 7.9. A-C, *Cyclammina cancellata* Brady, topotypes from Challenger Station 168, northwest flank of the Hikurangi Trench, NHM 1964.12.9.2-4, showing the alveolar hypodermis and the lipped aperture and areal cribrate aperture; D-E, *Reticulophragmium orbicularis* (Brady), topotypes from Challenger Station 323, south flank of the Rio Grande Rise, NHM P1964.12.9.17-18, showing basal aperture, imperforate septa and alveolar hypodermis. Scale bars = 0.5mm.

Family Pavonitinidae Loeblich and Tappan, 1961
The test is palmate and triangular in section. The aperture is terminal single or multiple. Oligocene to Pliocene.

Subfamily Pavonitininae Loeblich and Tappan, 1961
The test is triserial, biserial or uniserial. Chambers are undivided by secondary septula. Oligocene to Miocene.

• *Pseudotriplasia* Malecki, 1954 (Type species: *Pseudotriplasia elongata* Małecki, 1954). The test is triangular in thin section with concave sides, uniserial throughout. The aperture is cribrate. Miocene.

Superfamily TEXTULARIOIDEA Ehrenberg, 1838
The test is trochospiral, biserial or triserial in early stages, but later may be uniserial or biserial. Walls are agglutinated and canaliculated. Early Jurassic (Sinemurian) to Holocene.

Family Textulariidae Ehrenberg, 1839
The early stage is biserial, but may be reduced to uniserial. The aperture is single or multiple. Paleocene to Holocene.

Subfamily Tawitawiinae Loeblich and Tappan, 1961
The test is biserial throughout, compressed and palmate. Chambers are subdivided by short septula. Walls are thin and canaliculate. Apertures are multiple. Holocene.

• *Tawitawia* Loeblich, 1952 (Type species: *Textularia immensa* Cushman, 1913). The chambers are strongly overlapping in the plane of biseriality. Holocene.

7.3 Biostratigraphy and Phylogenetic Evolution

During the Neogene different faunal assemblages dominated different depositional environments, however, the most dominant fossils in all formations are warm water, shallow marine groups including larger benthic foraminifera, hermatypic corals and coralline algae. Tracing the stratigraphic distribution of the larger benthic foraminifera allows the understanding of the impact of climate, tectonic activity and volcanism on long-term (i.e. millions years) evolution of these shallow-water carbonate platforms (Courgeon et al., 2016, 2017; Gold et al., 2017a, 2017b).

Larger foraminifera and planktonic foraminifera overlap in occurrence in many localities allowing direct comparison of larger foraminifera "letter stages" biozones with oceanic planktonic scales (BouDagher-Fadel, 2002; BouDagher-Fadel, 2013, 2015; Sharaf et al., 2013).

The most important superfamilies that dominated Tethyan facies are the Nummulitoidea and the Soritoidea (see Fig. 7.10, and Charts 7.1-7.3), with the latter dominating the assemblages from Late Miocene to present day. However, in order to fully understand the Neogene, all three of the main groups need to be studied, namely:

- the porcelaneous miliolides,
- the calcareous rotaliides and
- the agglutinated textulariides.

Also, as in the Paleogene, bio-provincialism was strongly expressed in the Neogene, and the "Letter Stage" biostratigraphic sub-divisions are particularly important in the study of SE Asian assemblages (see Chat 7.1).

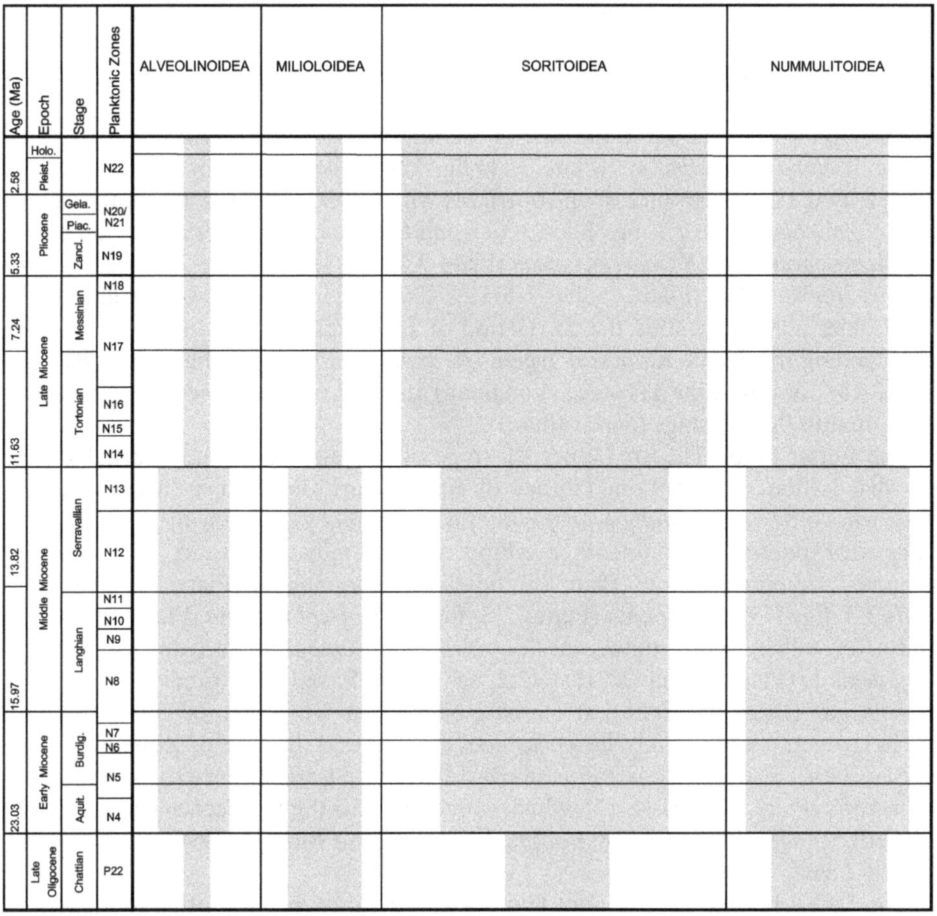

Figure 7.10. A schematic representation of the range and diversity of the main superfamilies in the Neogene (see also Charts 7.2 and 7.3 for details).

7.3.1 The Letter Stages for the Neogene of SE Asia

As introduced in Chapter 6, the "Letter Stages" subdivision of the Indo-Pacific Paleogene and Neogene (Leupold and van der Vlerk, 1931; Adams, 1970; Chapronière, 1984; BouDagher-Fadel and Banner, 1999; BouDagher-Fadel, 2008) is based on larger foraminifera, some of which have also been used as range fossils in western hemisphere stratigraphy (Barker and Grimsdale, 1936; Caudri, 1996). The details of these stages and their type fossils are given in Chart 7.1.

Many of the Neogene species are short lived and biostratigraphers have relied on them to date the Tethyan carbonates. The distinction between Oligocene and Early Miocene parts of the Te "stage" is drawn on the occurrence of *Miogypsinella borodinensis* (Plate 7.14, fig. 3) in the former, and true *Miogypsina* and *Miogypsina tani* (Plate 7.8, fig. 1) in the latter. The base of the Tf1 stage is marked by the disappearance of *Austrotrillina striata* (Plate 7.1, figs 4-7), *Eulepidina* spp. and *Heterostegina* (*Vlerkina*) *borneensis* (Plate 7.8, figs 8-9). The latter may be restricted in Melanesia to the lower part of the Te interval, but evidence from Borneo has showed that it ranges throughout the Te interval in the eastern region as a whole (BouDagher-Fadel et al., 2001). *Austrotrillina howchini* (Plate 7.1, figs 1-3, 10, 12) replaces *A. striata* and marks the beginning of Tf1 stage (see Chart 7.1).

The top of the lower Tf1 at the base of the Middle Miocene is defined by the extinction of *Lepidosemicyclina* sp. (Plate 7.11, fig. 12), while the top of the Tf1 stage is defined by the disappearance of *Austrotrillina* with the extinction of the youngest species, *A. howchini* (Plate 7.1, figs 1-3, 10, 12), and the top of the Tf2 stage is marked by the disappearance of *Katacycloclypeus* (Plate 7.6, figs 12-13; Plate 7.7, figs 1-5) and *Planorbulinella solida* (Plate 7.5, figs 2-5) (see Chart 7.1). However, BouDagher-Fadel and Lokier (2006) emended the correlation of BouDagher-Fadel and Banner (1999) by extending the age of *Katacycloclypeus* (*K. annulatus,* Plate 7.7, figs 1-5; Plate 7.16, figs 12-13) down into the Tf1 stage (Langhian) and that of *Flosculinella* (Plate 7.2, figs 9-12) up into the Tf2 stage (Serravallian).

The Letter Stage Tf3 (or Upper Tf stage of some authors) is remarkable only by what is does not contain. Faunas of this age are depleted in larger foraminifera with only rare *Lepidocyclina,* sometimes *Cycloclypeus* and, in shallower settings with quieter environments, dwelling on algal substrate, sand, dead coral or seagrass, *Alveolinella quoyi* (Plate 7.2, figs 2-4)*, Marginopora* (Plate 7.2, figs 16-17; Plate 7.3, figs 1-5)*, Operculina* (Plate 7.13, fig. 2, 6)*, Amphistegina* (Plate 7.14, fig. 1) and a few other long ranging species. *Alveolinella quoyi* arose from its ancestral *A. praequoyi* (Plate 7.1, fig. 20; Plate 7.2, figs 6-7), referred to by previous authors as *A. fennemai* (Plate 7.2, fig. 8), at the base of the Tf3, which can be correlated with the horizon at which *Katacycloclypeus* goes extinct, and the base of Zone N13 (Late Serravallian). The same horizon is marked by the first appearance of *Marginopora vertebralis* (Plate 7.2, figs 16-17), which ranges up into the Holocene. The top of the Tf Letter Stage was defined by BouDagher-Fadel and Banner (1999) on the extinction of *Lepidocyclina.*

The Late Miocene Tg does not contain any distinct larger benthic foraminiferal assemblages. The Letter Stages Tg and Th represent shallow marine biohermal carbonates that changed from mixed coral and coralline algal boundstones with larger foraminifera grainstones to more dominant coral reefs with a marked increase in *Halimeda* green algae. The latter is often preserved in recognisable forms, or as an increase in micrite and fine bioclastic products from the early breakdown of its aragonitic platelets. Larger foraminiferal grainstones are rare after this event, as deeper photic *Cycloclypeus* facies or minor *Operculina/Amphistegina* or *Alveolinella* calcarenites. Most well and field sections encountering limestones of this later Miocene or Pliocene age in Southeast Asia do not sample rocks composed of larger foraminifera tests.

7.3.2 The Miliolides of the Neogene

The miliolides of the Neogene fall into four superfamilies:

- the Alveolinoidea
- the Milioloidea
- the Soritoidea
- the Cornuspiroidea

Most of the Paleogene **Aveolinoidea** disappeared at the Eocene-Oligocene boundary and the only survivor of the terminal Middle Eocene extinction of the alveolinids was the simple yet cosmopolitan form *Borelis,* which in turn survived the Oligocene-Miocene boundary and continues to thrive to the present day. Over this period, alveolinids underwent a very slow evolution, with an incremental increase in the length of the test from pole to pole, changing from globular, ovoid to elongated spindle-shaped, and with an equally incremental increase in the number of secondary chamberlets ("mansardes") from zero in *Borelis* through one in *Flosculinella* to several rows of chamberlets in *Alveolinella.* Parallel lineages have each developed with the acquisition of an additional row of chamberlets in the *Borelis-Flosculinella* descent. This must have occurred at least twice during Te-Tf interval (Early to Middle Miocene).

An example of one of these lineages is seen with the appearance of *Flosculinella reicheli* within the Early Miocene (upper Te) and of *F. bontangensis* (Plate 7.2, figs 9-11) in the early Middle Miocene (just at the top of the Te stage). The latter gave rise to *Alveolinella praequoyi* (Plate 7.1, fig. 20) in the Serravallian, at the base of Tf2 stage. *A. praequoyi* has early whorls akin to *F. bontangensis*, but in the latter the chamberlets of each whorl are covered by at least 2 layers of smaller chamberlets. In *Alveolinella quoyi* (Plate 7.2, figs 2-4), in the Tf3 stage, all of the whorls have a multiple layer of chamberlets. *Flosculinella bontangensis* grades into *Alveolinella* forms about the same time as the disappearance of *Austrotrillina* (see Chart 7.1). This evolution parallels to that described by BouDagher-Fadel and Lord (2000) for *Lepidocyclina sensu lato.*

The genus *Austrotrillina* of the **Milioloidea** is essentially a *Quinqueloculina* (a simple small mioline), with a distinctly alveolar wall, presumably to harbour symbiotic algae. The evolution in the Indo-Pacific realm of *A. paucialveolata* (see Chapter 6) - *striata* (Plate 7.1, figs 4-7) - *asmariensis* (Plate 7.1, figs 8-9) - *howchini* (Plate 7.1, figs 10, 12) is a straightforward lineage with a gradually increasing number and complexity of alveolae. *A. asmariensis* is distinguished by its closely spaced, narrow alveoles, which were present in a single series and did not bifurcate peripherally as they would have done in

A. howchini. In many places transitional forms, in which the alveoles in the later whorls are simple and undivided (as in *A. asmariensis*) while the first whorls have thicker, bifurcating and more complex alveoli (as in *A. howchini*), co-exist with specimens typical of *A. asmariensis* with narrow alveoles. The gradual evolution from the primitive form *A. asmariensis* into more advanced form *A. howchini* occurred in the lower Tf1 stage and only the advanced form *A. howchini* persisted into the upper Tf1 stage, where it disappears completely near the top (see Chart 7.1).

The top of the lower part of lower Tf (Tf1) stage is defined by the extinction of the youngest species, *Austrotrillina howchini,* while the extinction of *Flosculinella bontangensis* occurred within the lower part of Tf2 where it occurs parallel to the appearance of *Alveolinella praequoyi*. BouDagher-Fadel and Banner (1999) summarized the ranges of individual species, and the series *Borelis pygmaeus - Flosculinella bontangensis - Alveolinella praequoyi - A. quoyi* seems to be the central evolutionary lineage. *Borelis pygmaeus* persists to the top of the Te stage (BouDagher-Fadel and Banner, 1999) where it is suddenly replaced by *Flosculinella bontangensis*, which disappears within the early part of the Tf2 stage. On the other hand, it evolves in the Tf2 into *A. praequoyi*, which is then replaced by *Alveolinella quoyi*, in the Tf3 stage.

The imperforate **Soritoidea** (Fig. 7.11) are characterized by peneropliform-planispiral, flabelliform and annular tests. They evolved by enlarging their apertural face and multiplying their apertures with increasing test size during ontogeny and phylogeny (Hottinger, 2001). This resulted in a discoidal shape with the aperture constituting the total shell margin (Hottinger, 2000), e.g *Marginopora*, or in an involute-fusiform shell in which the apertural face is enlarged at the poles by "polar torsion" (Leppig, 1992), e.g. *Archaias* (see Figs 7.3, 7.11). They are divided into two subfamilies (Loeblich and Tappan, 1987), the Soritinae and Archaiasinae; the soritids have been separated from the archaiasines by the presence of lateral, subepidermal partitions (Fig. 7.3; Henson, 1950) or intradermal plates (Seiglie et al., 1976), in contrast to a pillared endoskeleton in the archaiasines (Hottinger, 2001). However according to Levy (1977) some forms, such as *Cyclorbiculina* (Plate 7.3, figs 11-12), have both elements combined in the same shell, and it is therefore difficult to classify in such a system. For this reason, Hottinger (2001) recommended using the pattern of the stolon axes, radial or crosswise oblique, as the general, overriding character dividing the two groups. In this book, no assignations to subfamilies of the Soritidae are given at all because the morphological divisions of the soritines the archaiasines remain unclear.

Many of the cosmopolitan Soritoidea (75%) survived the Oligocene-Miocene boundary. Very few new forms made their appearance in the Miocene of Tethys. The Early Miocene (Te5 and lower Tf1) included the complex cyclical form with subdivided chambers, *Pseudotaberina* (Fig. 7.4; Plate 7.4, figs 1-6, 12), and in the latest Miocene (Tf3) its analogue *Marginopora vertebralis* made its first appearance. *M. vertebralis* evolved from *Amphisorus martini* in the Serravallian at the base of Planktonic Zone N13 (see Chart 7.1) by acquiring transverse medial partitions (Lee et al., 2004). While *A. martini* ranges from latest Oligocene to the Serravallian, *M. vertebralis* and different species of *Amphisorus* range up into the Holocene.

In the Middle Miocene of the Caribbean, the Soritoidea evolved into new forms, attaining the largest known size for member of this superfamily (Fig. 7.12). They developed a lineage resembling the Late Cretaceous meandropsinids (Ciry, 1964) by

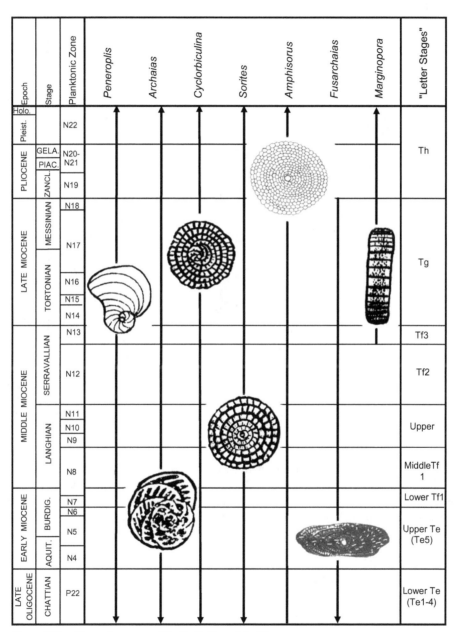

Figure 7.11. The main genera of the Soritoidea of the Neogene.

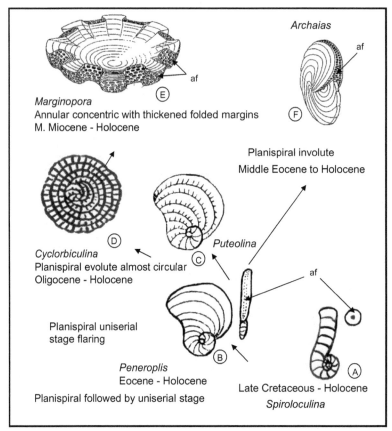

Figure 7.12.   The development of the soritids from (A) planispiral spiroliniform to (B) planispiral-evolute and flaring, (C) peneropliform, planispiral-evolute, approaching-annular stage and to annular-concentric, with concave side views and with thickened margins, as in *Marginopora vertebralis.* (af = apertural face). (E) and (F) from Hottinger (2006).

producing alar prolongations overgrowing each other on the lateral surface of the large discoidal shell (Hottinger, 2001). Most of the new forms did not survive the Late Miocene and they are not the direct ancestors of the species living today in the Caribbean. These porcelaneous soritids are more closely related to the Early Miocene forms of the Neotethys than to the Holocene Caribbean endemists (Hottinger, 2001).

7.3.3 The Rotaliides of the Neogene

The biostratigraphically important rotaliides of the Neogene form five superfamilies:

- the Nummulitoidea
- the Planorbulinoidea
- the Rotalioidea

- the Acervulinoidea
- the Asterigerinoidea.

The **Nummulitoidea** were major reef-forming organisms from the Middle Paleogene to Early Neogene. Their morphology, phylogeny and palaeogeographic evolution have been recently revised by BouDagher-Fadel and Price (2014), who described for the first time new Mediterranean-derived species of *Planostegina* in the Early Miocene (Burdigalian) of SW Africa (Fig. 7.13).

Most of the Nummulitoidea that survived of the Oligocene-Miocene boundary are extant forms, except the wholly involute *Heterostegina* (*Vlerkina*). The latter evolved in the Late Eocene from *Heterostegina sensu stricto* by losing the evolute coiling and acquiring chamberlets. It reached a maximum in abundance and diversity in the Late Oligocene (Banner and Hodgkinson, 1991), but diminished in the Lower Miocene (BouDagher-Fadel and Banner, 1999; BouDagher-Fadel *et al.*, 2000), persisting throughout the Middle and Late Miocene deposits of the Far East (BouDagher-Fadel, 2002).

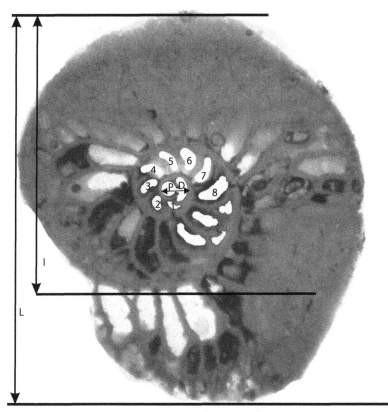

Figure 7.13. Morphometric measurements in the equatorial section of megalospheric *Planostegina africana* BouDagher-Fadel and Price, from BouDagher-Fadel and Price (2014); P, proloculus and D, deuteroconch; Pre-planosteginid chambers (X) are 8; l, the maximum diameter of the shell in the first whorl, L, the maximum diameter of the first one and subsequent half whorl.

Tansinhokella also evolved from an involute ancestor in the Eocene by developing meandriform alar prolongations, which overlapped each other, and became widespread in Tethys and the Pacific during Late Eocene, Oligocene and Early Miocene. *Spiroclypeus* and *Tansinhokella* ranged up from the Td stage to the top of the Te stage. Classical morphology-based taxonomy divides the **Nummulitidae** into two subfamilies, the **Nummulitinae** and **Heterostegininae**, according to the presence or absence of secondary septa. The evolutionary importance of this morphological feature was tested by Holzmann *et al.* (2003) by sequencing fragments of the SSU and LSU rRNA gene of many nummulitid species. According to their results, species characterized by septate chambers (*Heterostegina depressa, Planostegina operculinoides*, and *Cycloclypeus carpenteri*) either group with species lacking septate chambers (*Operculina ammonoides, Nummulites venosus*) or branch separately. They also concluded that chamber subdivisions developed several times independently in the evolutionary history of the Nummulitidae, providing another example of parallel evolution in Foraminifera.

The most primitive species of *Cycloclypeus* are very close morphologically to *Heterostegina* (see Chapter 6). Tan Sin Hok, in his monograph (1932) on Indonesian forms, documented the apparent gradual reduction of nepionic chambers in *Cycloclypeus*. *Cycloclypeus eidae* (Plate 7.5, figs 15-19) evolved in the Late Oligocene (see Chapter 6) from *C. opernoorthi,* reducing the diameter of the nepionic part of the test to about one third and the number of nepionic chambers in microspheric forms from 34 to 21 (O'Herne and van der Vlerk, 1971; O'Herne, 1972). It persisted to the Early Miocene where it was followed by a growth phase reduction from 21 to 18 in *C. posteidae* in Early Miocene times, and 18 to 9 in *C. indo-pacificus* in Early to Middle Miocene times. This group led to the younger forms which Tan Sin Hok placed within the species *C. carpenteri* (Plate 7.6, figs 1-4), with 2 to 5 nepionic chambers (see Chapter 7). In the Middle Miocene of the Far East *C. carpenteri,* with its large embryont, is common along with stellate types (e.g. *Radiocycloclypeus*). However, Tan Sin Hok (1932) recognised that this progression is not gradual and he proposed a model of minor saltations with an overall smooth transition. These saltations were not found by later workers (Drooger, 1955, 1993; MacGillavry, 1962). On the other hand, Tan Sin Hok recognised two major groups in his material on the basis of ornamentation of the test. *C. carpenteri* has pustules, while *C. guembelianus* has a smooth test. These features are hard to see in thin sections and they only have ecophenotypic importance, but they have led to an interesting line of research in recent years (Laagland, 1990). European *Cycloclypeus* has been studied in detail by many authors (Cosijn, 1938; Matteucci and Schiavinotto, 1985; Laagland, 1990; Drooger, 1993) and the successive morphometric species *C. droogeri* and *C. mediterraneus* were proposed by Matteucci and Schiavinotto (1985). More recently Laagland (1990) and Drooger (1993) remarked that *C. eidae* might have originated in the Mediterranean from *C. mediterraneus.* On the other hand, the European *C. eidae* has a thinner test and distinct pustulate ornamentation which, according to Drooger and Roelofsen (1982) might have lived at greater depth than *C. mediterraneus.* Altogether, the *Cyclolypeus* lineages appear to follow the same pattern of nepionic acceleration from Oligocene to Holocene.

Katacycloclypeus made its first appearance in the Middle Miocene (upper Tf1 and Tf2) of Tethys. In the upper Tf1 stage, a lineage of *Cycloclypeus* evolved into forms with widely separated annular inflations and a trilocular embryont (*Katacycloclypeus annulatus*), which in turn gave rise in the Tf2 stage to a form with broad closely spaced annular inflations and a trilocular embryont (*K. martini*). The co-existence of *Cycloclypeus* and *Katacycloclypeus* spp. in the Far East (BouDagher-Fadel and Lokier, 2006) demonstrates clearly parallel evolutionary sequences. Primitive specimens of *K. martini* showing the transition between both species occur together with *K. annulatus* in the same sample. These specimens link the two species and demonstrate a gradual evolution from *K. annulatus* into *K. martini*. The evolutionary sequences mean that these evolved taxa must occur one after the other in the same sequence if one is sampling the Indo-Pacific province. *Katacycloclypeus* is confined to the Tf1-Tf2 stages, upper Early Miocene to Middle Miocene of the Indo-Pacific, while *Cycloclypeus* ranges from the Eocene to Holocene throughout all the tropics.

Most of the **Planorbulinoidea**, which survived the Oligocene-Miocene boundary are extant forms, except *Neoplanorbulinella* and *Planolinderina*, both of which disappeared within the Early Miocene of Tethys. On the other hand, the forms which appeared in the Early Miocene were the short lived American form *Heterosteginoides* and the extant form *Rupertina*. Although many cosmopolitan forms of the Planorbulinoidea are still living, no new forms seem to have appeared in the Neogene, except the Holocene *Planorbulinopsis* in the Indo-Pacific.

Many of the **Rotalioidea** became extinct during the Miocene. Only the robust forms of the Rotalioidea survived to the present day, developing spines in the Pliocene to Holocene to form the calcarinids. The spiral of the calcarinid chambers is covered by a layer of canals (with the exception of the genus *Baculogypsina*), which connect the chamber lumina with the ambient sea-water and, as in the nummulitids, is necessary for pseudopodia formation, excretion, attachment to the substratum, protection and probably also growth and reproduction. (Röttger and Krüger, 1990). These canal systems which characterise the families **Calcarinidae** and Nummulitidae originate from spaces between chambers (so-called interlocular spaces), and from spaces between the many calcite lamellae deposited successively on the test surface during ontogeny (so-called secondary lamellae). The canal system of calcarinids differs morphologically from that of nummulitids, as it is an enveloping canal system consisting of a layer formed by innumerable canals (Röttger and Kriiger, 1990). It covers and often conceals the spiral of chambers (Hottinger and Leutenegger, 1980). Morphological differences in the same species, such as *Calcarina gaudichaudii* (see Chapter 1), extends to the number of spines and stages of life cycles (Röttger *et al.*, 1990). The relationship between the calcarinids and nummulitids was proposed by earlier workers such as Glaessner (1945), Pokorny (1958), Haynes (1981), and Tappan and Loeblich (1988), all of whom claim that both groups descended from a common rotaliid ancestor, and that their separation from each other and from early Rotaliidae took place in the Late Cretaceous. This was confirmed later on by Holzmann *et al.* (2003) who tested the phylogenetic relationships of five extant nummulitid genera by sequencing fragments of the SSU and LSU rRNA gene. According to their results, there is a close relationship between nummulitids, and the Rotaliidae and Calcarinidae. According to these authors, the Nummulitidae branch is a sister group to the Calcarinidae and Rotaliidae, and evolved in a parallel lineage.

In the Early Miocene, and prior to the calcarinids, the **Miogypsinidae** of the Rotalioidea made their first appearance, evolving from a rotaliid ancestor (see Chapter 6) within a very short time span in the Late Oligocene. In the Burdigalian (the uppermost part of the upper Te stage), *Miogypsinodella* evolved from *Miogypsinoides* (BouDagher-Fadel *et al.*, 2000; BouDagher-Fadel and Price, 2013) by developing splits between the lamellae in the otherwise solid lateral walls (see Chapter 6). The splits in the laminae occur throughout the thickened lateral walls in the more advanced forms (e.g. in *Miogypsinodella primitiva*: Plate 7.11, figs 2-4), but do not form stacks of cubiculae/chamberlets, as seen in *Miogypsina* spp. (with cubiculae and quadriserial embryonts). *Miogypsina* evolved at the base of the upper Te stage, Aquitanian (BouDagher-Fadel and Banner, 1999). While microspheric forms of *Miogypsina* retain the uniserial embryonic coils of their *Miogypsinella* ancestors, the megalospheric nepionts possess a deuteroconch larger than either the protoconch or first auxiliary chambers, and the latter form biserial whorls surrounding the proloculus. The cubiculae were arranged in oblique stacks and the biserial embryont evolved to achieve bilateral symmetry, (e.g. *M. indonesiensis,* Plate 7.11, fig.16). The cubiculae of *Miogypsina* are not in vertical stacks (Fig. 7.14), as they are in *Lepidocyclina*, but in obliquely developed columns. The general trend of evolution in this group is towards shorter nepionic spirals and larger embryos in successively younger species. Another descendant from *Miogypsina*, *Miolepidocyclina* has been distinguished on the more central location in the median

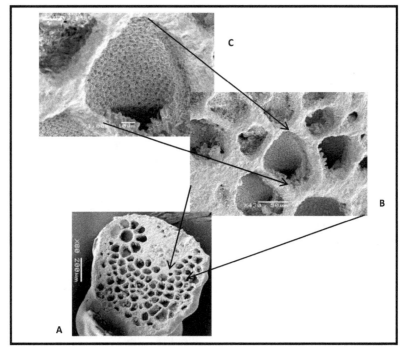

Figure 7.14.   SEM images of *Miogypsina borneensis* Tan Sin hok from BouDagher-Fadel and Price (2013) showing; (a) embryonic apparatus, and equatorial chambers; and (b, c) the internal surface of equatorial chamber wall bearing 'eggholders'.

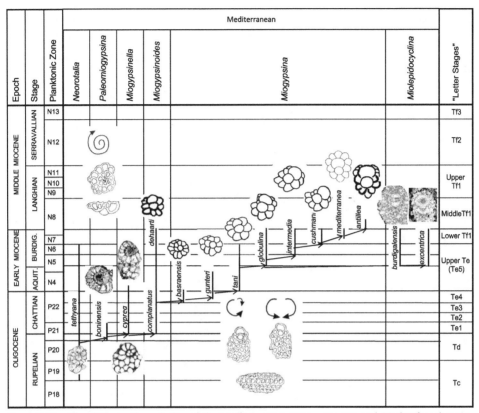

Figure 7.15. The phylogenetic evolution of the Mediterranean *Miogypsina* spp. and *Miolepidocyclina* spp., from BouDagher-Fadel and Price (2013).

plane of the nepiont. Forms with chambers with hexagonal shapes, *Lepidosemicyclina*, are found in the Burdigalian of the Far East. Miogypsinids originated in the American province (Rupelian and P18, see BouDagher-Fadel and Price, 2010b, 2013), becoming extinct in the Early Miocene (Burdigalian and the lower part of N8) in the American and South African provinces, in the earliest Middle Miocene (Langhian and the upper part of N8) in the Mediterranean, but their final global extinction occurred in the latest Middle Miocene (Serravallian and N13) in the Indo-Pacific (Figs 15-17, see BouDagher-Fadel and Price, 2013).

The biometric method (see Chapter 6) appeared to be a useful tool in the study of the evolution of a single lineage of *Miogypsina* and of miogypsinid genera in general (see Chart 7.1). The "X scale" when applied to *Miogypsina* gave rise to the subdivision in *basraensis* 12.5 – *gunteri* – 9 – *tani*. At X values between 7 and 6 and γ values close to zero, "V values" become significant and the morphometric limits of the species are as follow: *globulina* - 45 – *intermedia* – 70 – *cushmani* – 90 – *antillea* (Drooger, 1993). However, as most specimens studied by scholars from random thin sections, biometric measurements on Miogypsinidae are rarely possible. In this case, combining the broad results gained by equatorial sections of the megalospheric nepiont (as published by

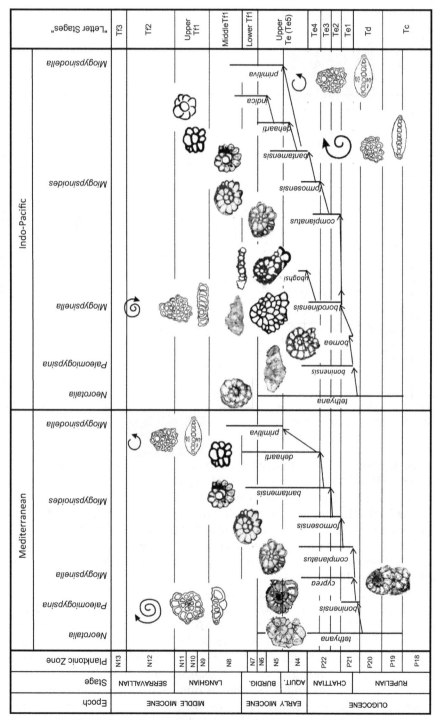

Figure 7.16. The phylogenetic evolution of the Indo-Pacific *Neorotalia*, *Paleomiogypsina*, *Miogypsinella* and *Miogypsinoides* spp., from BouDagher-Fadel and Price (2013).

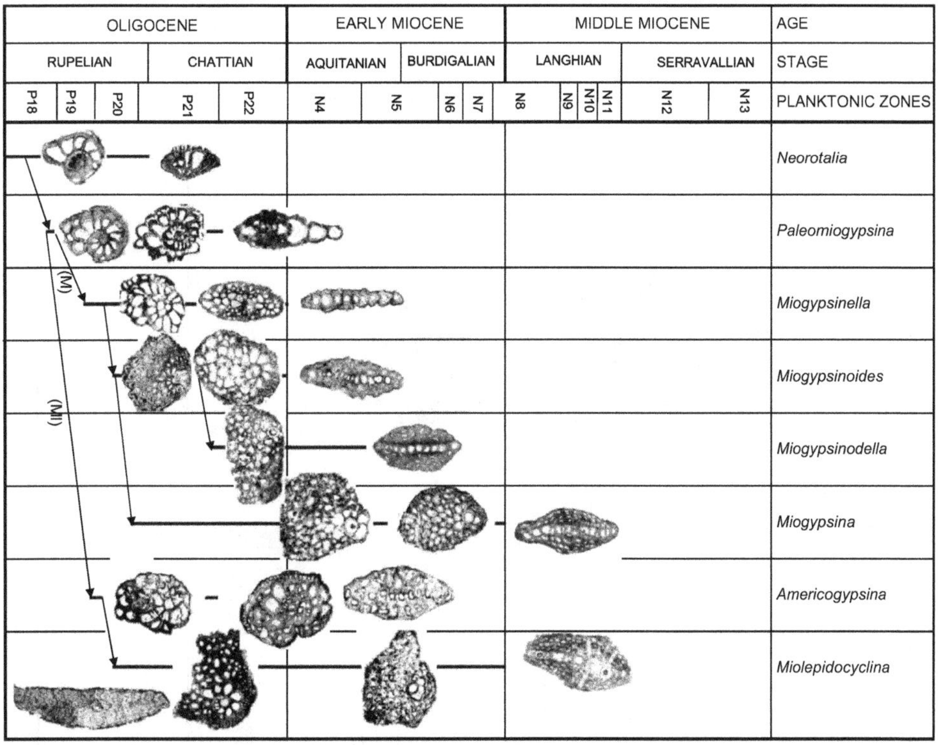

OLIGOCENE					EARLY MIOCENE				MIDDLE MIOCENE					AGE
RUPELIAN			CHATTIAN		AQUITANIAN		BURDIGALIAN		LANGHIAN			SERRAVALLIAN		STAGE
P18	P19	P20	P21	P22	N4	N5	N6	N7	N8	N9	N10 N11	N12	N13	PLANKTONIC ZONES
														Neorotalia
														Paleomiogypsina
														Miogypsinella
														Miogypsinoides
														Miogypsinodella
														Miogypsina
														Americogypsina
														Miolepidocyclina

Figure 7.17. Phylogenetic chart showing the evolutionary lineages of the American Miogypsinidae, from BouDagher-Fadel and Price (2010b).

Chaproniere 1984, and Van Vessem, 1978) with those obtained by vertical sections of the whole test (as followed by Cole 1957, 1963) is useful (see BouDagher-Fadel et al., 2000; BouDagher-Fadel and Price, 2013).

Many new cosmopolitan **Acervulinoidea** joined the Paleogene survivors of this superfamily and continued to live in the present day, for example the coloured, reefal modern *Homotrema* and *Miniacina*. These extant forms appear to have arisen from the planorbulinids in the Early Miocene and their coloration (red in the case of *Homotrema rubrum*, Plate 7.9, Figs 4-5) is due to the remnants of photosynthetic pigments of some type of green alga (Strathearn, 1986). Forms with more than one encrusting layer, such as *Borodinia*, seem to appear only in the Aquitanian of the Indo-Pacific. The acervulinid, *Alanlordia banyakensis* (Plate 7.4, figs 7-11), is first known from the Middle Miocene (Serravallian, Tf2), continues into the Pliocene where it strengthens the umbilical and spiral pillars with lamellar thickenings (*A. niasensis*, Plate 7.4, figs 9-11). *Alanlordia* resembles the Paleogene *Wilfordia,* and from which it may have descended (Banner and Samuel, 1995). It is also a gross homeomorph of the Maastrichtian *Vanderbeekia* of the Middle East (see Chapter 5).

In the **Asterigerinoidea**, *Lepidocyclina* sensu lato, an extinct three-layered Rotaliida, with a bilocular or multilocular embryonal stage, surrounded by a thickened wall and adauxiliary chambers. Microspheric tests have an early planispiral coil, while megalospheric tests have a globular protoconch, enclosed or followed by a larger reniform

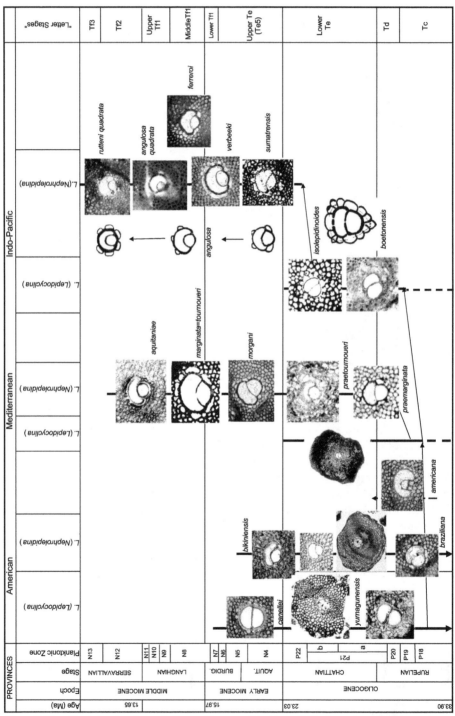

Figure 7.18. Chart showing the evolutionary lineages of lepidocyclinids throughout the bioprovinces.

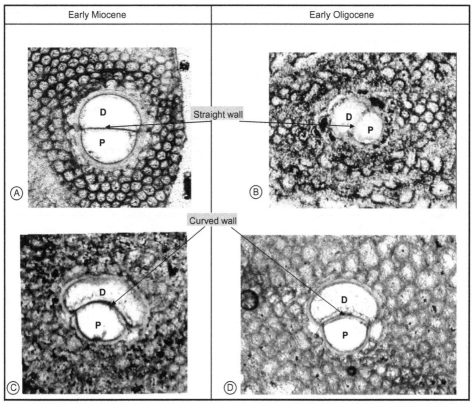

Figure 7.19. Thin sections of four specimens showing the two dominant types of lepidocyclinids offshore Brazil. A) *Lepidocyclina* (*Lepidocyclina*) *canellei* Lemoine and Douville with four equal spires, note P slightly narrower than D but slightly longer, the wall is thicker compare to B) *L.* (*L.*) *yurnagunensis* Cushman. C) *Lepidocyclina* (*Nephrolepidina*) *bikiniensis* Cole with the protoconch occupying 57% of the embryonic apparatus compared to 49% in D) *L.* (*N.*) *braziliana* BouDagher-Fadel and Price. In the latter, the arrangement of chambers is chaotic outside the periembryonic ring made of numerous short spires which are eventually replaced by the cyclic growth of hexagonal chambers. (after BouDagher-Fadel and Price, 2010a).

deuteroconch (Figs 7.18-7.22). Post-embryonic chambers evolve from cyclical, arcuate to hexagonal in shape, usually with two or more apertures (Fig. 7.19-7.20). The lateral chambers are well differentiated from the equatorial layer and in the advanced forms they are arranged in tiers on either side of the equatorial layer (Fig. 7.23) (see BouDagher-Fadel and Price, 2010). They first evolved in the American province from an amphisteginid ancestor (see Chapter 5). Following the first explosion of new forms of *L.* (*Lepidocyclina*) in the Middle to Late Eocene, the gradual evolution of *Lepidocyclina* spp. seems to have been slow with the whole stratigraphic range from the Oligocene to Miocene showing little variation. From the Oligocene to the middle Miocene, many lineages of *Lepidocyclina* can be found in Mediterranean, with a gradual stratigraphic succession of species from those with primitive to those with more advanced nepionts, and from forms with thin-walled cubiculae/chamberlets to those with more numerous cubiculae in the swollen parts of the test (Fig. 7.18). *Lepidocyclina* sensu lato first appeared in the Chattian (P21b in the Mediterranean and P22 in the

Indo-Pacific province). In the Mediterranean, the oldest forms, such as L. (*L.*) *prae-marginata*, which appeared in the Rupelian (see Fig. 7.18), are forms with an embryon intermediate in shape between L. (*Lepidocyclina*) spp. and L. (*Nephrolepidina*) spp. They are similar in shape to their ancestral form in the American province, L. (*L.*) *yurnagunensis* (Fig. 7.19), but with a slightly more curved wall separating the two embryonic chambers. On the other hand, the *Lepidocyclina (Lepidocyclina)* spp. of the Indo-Pacific province had an embryonic apparatus with a straight wall dividing the two embryonic chambers, being similar to that of their ancestors in the American province but the wall surrounding the protoconch and deuteroconch is slightly thinner and the test is smaller (see BouDagher-Fadel and Price, 2010a). While the American isolepidine embryo persisted to the early Miocene, those of Tethyan lineage disappeared completely in the late Oligocene.

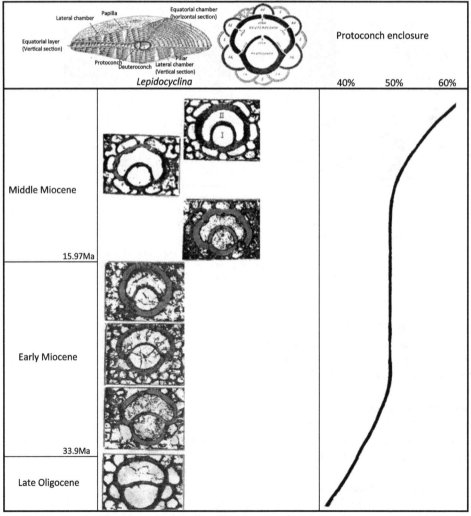

Figure 7.20.   Trend of average grade of lepidocyclinid protoconch enclosure (images not to scale).

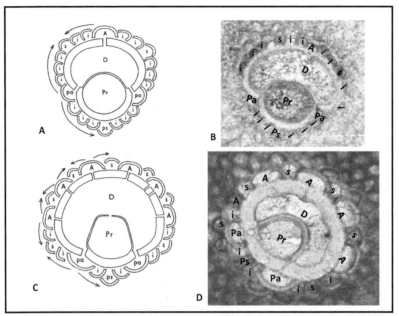

Figure 7.21. A, C) Plan diagrams to illustrate the nepionic stages of the equatorial chambers of the megalospheric primitive nepiont A) and the advanced "trybliolepidine" in *Nephrolepidina*, modified from Tan (1939), Renz and Küpper (1947) and Eames et al. (1962). B) *Lepidocyclina (Nephrolepidina) aquitaniae* Silvestri. D). *L. (N.) angulosa* Provale. Abbreviations: Protoconch (Pr); Deuteroconch (D); adauxiliary chambers (A); primary auxiliary chambers (pa), communicating directly with the nucleoconch and giving rise to the interauxiliary (i), symmetrical auxiliary (s), and protoconchal symmetrical auxiliary (pa) chambers. The arrows indicate the multispiral growth.

The lepidocyclinid lineage (Chart 7.1) also demonstrates the evolutionary development of *L. (Nephrolepidina)*, from *L. (Lepidocyclina)* at around the Late Oligocene-Miocene boundary in the Far East. Phylogenies such as this one must have occurred in other lineages of the **Lepidocyclinidae**. The *L. (Nephrolepidina)* lineage evolved many times. The earliest *L. (N.)* species in Tethys occurred in the Chattian, but they were rare (BouDagher-Fadel and Banner, 1999). Following their first appearance in the Mediterranean, there was a gradual stratigraphic evolution of species from those with the protoconch slightly enclosed by the deuteroconch (e.g., *L. (N.) praetournoueri*; see Fig. 7.19) to those with more advanced embryonic apparatus with the protoconch more enclosed within the embryonic apparatus (see Fig. 7.20). The European representatives of *L. (Nephrolepidina)* never reached the advanced phylogenetic evolutionary stages seen in the Indo-Pacific, where the protoconch became quadrate and completely enclosed by the deuteroconch (Fig. 7.21).

In the latest Early Miocene (Late Burdigalian) a lineage with irregular embryonic apparatus evolved from *L. (N.) transiens* into more extreme forms with multiloculor embryont, *Multilepidina* (included in the synonyms of *Nephrolepidina* by Loeblich and Tappan, 1988), which survived until the end of the Serravallian (top Tf3). Species of *L. (Nephrolepidina)* witnessed through the Miocene an evolutionary process involving the increase of the radial symmetry of the embryonic apparatus and the gradual enclosure of the protoconch by the deuteroconch (Figs 7.18, 7.21). Chart 7.1 shows an example of a lineage grading gradually towards forms with quadrate embryont, e.g. *L.*

(*N.*) *rutteni* (Fig. 7.22) and *L. (N.) rutteni quadrata*, which some authors call *Lepidocyclina* (*Tribliolepidina*) or just *Tribliolepidina* (see Fig. 7.21).

The diversity of forms shown in assemblages of contemporary species, e.g. as reported by Cole (1957) from the Oligocene and Miocene of Saipan, suggests that evolution must have proceeded along several different parallel lineages (see Plates 7.17-7.20). It is thought that the earliest *L.* (*Nephrolepidina*) species in Europe occurred in the latest Early Oligocene (Rupelian), reaching Tethys (from Iran to the Indo-Pacific) in the Chattian but they were rare. In Tf3, *L. gigantea* (Plate 7.8, fig. 6) became so large in Vietnam, Papua-New Guinea and East Africa that these microspheric lepidocyclinas characterize the so-called "Oyster Beds" in Vietnam. Their megalospheric partners are unknown but must have been small and undistinguished. Similar evolutionary patterns occurred in many lineages of *Lepidocyclina s.l.,* forming taxa with no confidently known megalospheric partners (see BouDagher-Fadel and Wilson, 2000; BouDagher-Fadel and Price, 2010a).

Many lineages of *Lepidocyclina* can be traced from the Eocene to the Early Miocene of the American province, and from the Oligocene to the Middle Miocene of Tethys, with a gradual stratigraphic succession of species from those with primitive to those with more advanced nepionts, and from forms with thin-walled cubiculae to those with more numerous cubiculae in the swollen parts of the test. The change from biconvexity to biclavate morphology in axial section retained the development of massive pillars, which ultimately became thinner in species such as *L. (N.) ferreroi* (Pate 7.17, figs 15-16). All Lepidocyclinidae disappear completely towards the end of the Early Miocene in the American province (BouDagher-Fadel and Price, 2010a) and in the southeastern Tethys in the Late Miocene (Hallock et al., 2006; BouDagher-Fadel and Price, 2013) or possibly in the Early Pliocene (Betzler, 1997) (see Fig. 7.18).

Nephrolepidines evolved rapidly showing many evolutionary characters with short ranges. They reached their diversification peak during the early Miocene, often dominating whole assemblages of shallow warm carbonate platforms. The European

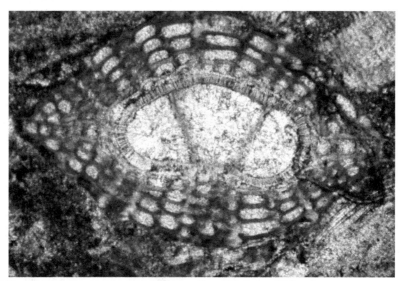

Figure 7.22.   Lepidocyclina gigantea Martin, Serravallian, Tf3, Japan, UCL coll.

representatives of *Lepidocyclina* (*Nephrolepidina*) form a well-documented lineage, with three evolutionary stages: *L.* (*N.*) *praemarginata*, *L.* (*N.*) *morgani* and *L.* (*N.*) *tournoueri* (Fig. 7.23). The *L.* (*Nephrolepidina*) development in the Indo-Pacific was distinctly different from the European lineage, and can be summarized as starting from *L.* (*L.*) *isolepidinoides*, via *L.* (*N.*) *sumatrensis*, *L.* (*N.*) *angulosa* and *L.* (*N.*) *martini* to *L.* (*N.*) *rutteni*, based on the principle of nepionic acceleration (see Chart 7.1; Fig. 7.18). This is probably an over simplification, as many more morphotypes can be distinguished, some of which may deserve species rank (BouDagher-Fadel and Lord, 2000; BouDagher-Fadel and Price, 2010a). As in *Miogypsina s.l.,* the Indo-Pacific Lepidocyclinidae show an apparent reversal of nepionic acceleration (from *L.* (*N.*) *sumatrensis* to *L.* (*N.*) *angulosa*; see Chart 7.1, Figs 7.18, 7.21) which could possibly be explained by an immigration or influx of less evolved species from elsewhere in the sub-province (BouDagher-Fadel and Lord, 2000; BouDagher-Fadel and Price, 2010b, 2013). They could sometimes be explained by local environmental factors, or by invoking bottleneck events in which only less specialized and more primitive surviving forms provide the basis for further development (BouDagher-Fadel, 2008). However, local extinctions and re-colonization by forms which evolved elsewhere and reached different evolutionary stages might also explain these relapses.

7.3.4 The Textulariides of the Neogene

The Oligocene saw the disappearance of the shallow water **Orbitolinoidea** with the extinction of *Dictyoconus*. However, by the Neogene all of the complicated textulariides had changed their habitats completely and had moved to deeper water environments (see below). They are only studied briefly in this book as they do not colonise the same shallow warm environment as other larger benthic foraminifera, however, they do draw attention to the way larger foraminifera can adapt and change environment in order to survive. The three most important superfamilies of the Neogene textulariides are:

- the Ataxophragmioidea
- the Pavonitinoidea
- the Textularioidea.

The textulariides diversified in the Neogene and many forms carried on from the Oligocene to the present day. However, a major diversification occurred in the Miocene with new forms appearing in the Early to Middle Miocene. These forms were characterised by inner structures consisting of alveoli conjoined through tubiform connections (Seiglie et al., 1986) and they include the **Alveovalvulinidae**, the **Globotextularidae** (see Chapter 6) the **Chrysalidinidae** (see Chapter 6) and the **Pavonitinidae**, the **Textulariellidae** and the **Cyclamminidae** (see Chapter 6). The latter are relict from the Cretaceous and continued to survive until the present day. The families Alveovalvulinidae and **Liebusellidae** evolved from a trochospiral agglutinated foraminifera with no inner structures. The flexible characteristic of the alveovalvulinid test contrasts, however, with the strong, thick-walled liebusellid test, suggesting that they evolved separately (Seiglie et al., 1986).

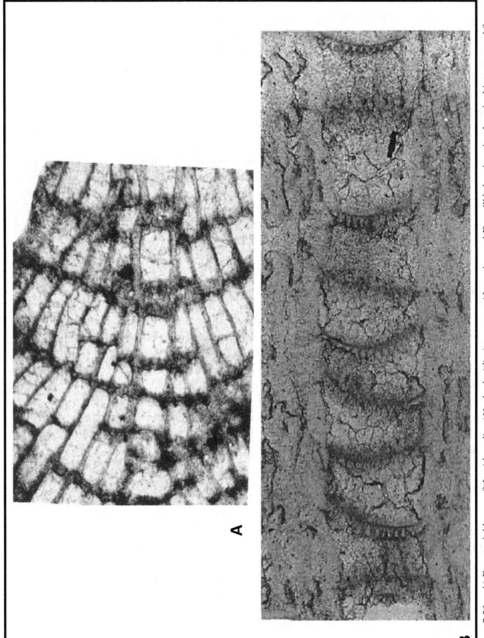

Figure 7.23. A) Equatorial layers of *Lepidocyclina* (*Nephrolepidina*) *tournoueri* Lemoine and Douvillé showing simple paired intercameral foramina and associated grooves; B) *Eulepidina* sp. showing more numerous intercameral foramina and more sharply incised grooves(modified from Eames et al., 1962).

7.4 Palaeoecology of the Neogene Larger Foraminifera

Warmer global climates than those in the preceding Oligocene, and of today, prevailed during the Miocene, which lasted for 18 million years. During this time, the modern pattern of ocean circulation was established and the mixing of warmer tropical water and cold polar water was greatly reduced. This led to well-defined climatic belts that stretch from the Poles to Equator.

In the Early Miocene, the larger benthic foraminifera were common throughout the warm-waters of the tropics, with many forms that were cosmopolitan and are still extant. Their latitudinal distribution was mainly controlled by temperature, thus the 15°C isotherm (for the colder months) limits their present day geographic distribution (Langer and Hottinger, 2000) and probably also did so in Miocene (Hughes, 2014). The distribution was also dependent on the light requirements of their endosymbionts and substrate (Murray, 2006). Their distribution in time followed parallel lineages showing gradual evolution, in which newly evolved taxa co-existed with their ancestors, but occupied different ecological niches, thus allowing palaeoenvironmental specialization into separate distinct biocenoses.

In the assemblages characteristic of the Letter Stages Te to Tf1 in the Indo-Pacific the miliolides were common and included forms such as the alveolinids *Austrotrillina, Borelis*, which avoided very shallow water and lived in sheltered habitats, such as coral rubble, in moderate energy areas (BouDagher-Fadel et al., 2000; Murray, 2006). Modern alveolinids live in a wide range of carbonate habitats, both in areas with very low hydrodynamic energy, such as deep lagoons and in forereef settings, down to depth of about 80m. This, together with the fact that alveolinids are miliolines with a tolerance to salinity and temperature fluctuations, probably makes the group less sensitive to small sea level changes. Oligocene and Miocene alveolinids also had a wide range of habitats, but were particularly common in deeper lagoonal settings. Their response to sea level changes during the Cenozoic might therefore be similar to that of other miliolines, were it not that extinctions during sea level falls are less likely due to the group's wider ecological niche. Obliteration of the lagoonal habitat during sea level falls may, however, have caused reductions in population sizes and an increase in the rate of evolution. The appearance of *Flosculinella reicheli* within the Early Miocene (late Te) and of *Flosculinella bontangensis* in the late Early Miocene (just after the top of Te) may have been the result of smaller population size coupled with an already existing tendency to increase the number of rows of chamberlets. Whatever the biological advantage of increased complexity in chamber subdivision may have been, it is a feature of alveolinid foraminifera from the Carboniferous onward, and is a well-established evolutionary trend. It may be analogous in function to the increased complexity of the *Austrotrillina* wall, probably increasing the efficiency of symbiosis with algae and diatoms. As with the fusulinines of the Permian (see Chapter 2), cyanobacterial symbionts are reported in living *Marginopora* (Lee et al., 1997).

The sequential development of new characteristics in the lepidocyclinids and miogypsinids, such as the gradual development of the lateral chamberlets, may have been driven by environmental stress. Thus, the division of the solid lateral walls into the small illuminated chamberlets of *Miogypsina* and *Lepidocyclina,* were then used in favourable warm, light-filled shelf environments as nests for diatoms (see BouDagher-Fadel 2008; BouDagher-Fadel and Price, 2010a, 2013). Each of the small chamberlet would have acted not only as a small convex lens to focus sunlight, but they could also act as a greenhouse for the containment and development of symbiotic diatoms (see Fig. 7.14). The diatoms enabled these forms to acquire nutrients without pseudopodial activity in food-gathering (Röttger, 1971). Although many small rotaliide foraminifera (e.g. *Rotalia, Rosalina*, etc.) gather food particles to their umbilici for extrathalamous digestion, larger foraminifera with chamberlets but no umbilici (e.g. *Miogypsinella* and *Miogypsina*) must have had a different method of nutrition. In the Miogypsinidae several lineages evolved (BouDagher-Fadel et al. 2000) that followed similar evolutionary patterns, all related to the utilization of radiant light to enable the promotion of populations of photosymbionts. In *Miogypsinella* the flange of median chamberlets is relatively small (not much broader than the coil of the nepiont); in *Miogypsinoides* the flange grew very much broader and became heavily thickened laterally. As in the lepidocyclinids (Fig. 7.23), the lateral walls that became very thick, would have reduced the amount of sunlight which could reach the median layered chamberlets, so they began to subdivide, as in *Miogypsinodella,* with the divisions becoming many layers of convex chamberlets in *Miogypsina*. Although diatom endosymbionts may initially enter larger foraminifera as ingested food, they establish themselves very quickly as a permanent resident population inside the cell (Richardson, 2001). This has evolutionary parallels with the division of the median chambers of *Cycloclypeus* (Fig. 7.6), and *Heterostegina* (Plate 7.15, figs 4-6) (Banner and Hodgkinson, 1991), where the illuminated chambers are also nests for diatoms. It is probable that *Miogypsinella* had active, food-gathering pseudopodia, however, it is possible that *Miogypsina* had pseudopodia which were of little effect in food gathering and instead used its chamberlets as an arable farm. The chamberlets are not linked by apertures, instead perforations are present in the walls. These perforations would be too fine to allow the passage of diatoms, while cubicular apertures would have allowed to escape. The requirement for, and use of sunlight by *Miogypsina* is shown by its occurrence only in shallow water marine limestones, associated with fossil algae.

The irregular shape of many species of *Miogypsina* suggest that their morphology is determined by the shape of the substrate on which they grew, such as the stems or leaves of seagrass (the substrate would be biodegradable and seagrass is not seen preserved in thin sections). On such an irregular substrate surface it would be difficult for any organism to remain fully attached and yet migrate on that surface. It could be concluded, that sedentary miogypsinids grew to accommodate the shape of the vegetable substrate to which they adhered. Only in strong ambient sunlight, which would benefit both the miogypsinids and their vegetable substrate, could true *Miogypsina* flourish (BouDagher-Fadel et al. 2000).

Abundant Soritoidea inhabited tropical waters of the Neogene. The extant representatives live today in the shallowest oligotrophic, tropical-subtropical, shallow-water habitats where the light intensity is high enough to produce photoinhibition in the symbionts of the foraminifera. They chose, therefore, dark substrates, most frequently on seagrass leaves (Hottinger, 1997, 2001). They are characterized by the possession of rhodophyte, chlorophyte, or dinophyte photosymbionts (Richardson, 2001).

According to Richardson, endosymbiosis has played a major role in the evolution and diversification of the clade Soritoidea. In particular, the acquisition of dinophyte endosymbionts appears to be a key innovation that facilitated a change in habitat from an epifaunal, free-living mode of life to one of living attached to phytal and non-phytal substrata. Banner (1971) described from New Guinea *Alveolinella quoii* with a small benthic foraminifera, *Planorbulinopsis parasitica,* embedded in its tests and completely enclosed by its whorls, that he concluded that it actually fed on the cytoplasm or utilised the test as a substrate.

Keramosphaera's habitat is an exception among the soritids. It lives in deep water (at about 3600m) from the South Atlantic to the Indian Ocean, in the Pacific off Japan, and in the North Sea (Loeblich and Tappan, 1988).

The test morphology of the most common encrusting foraminifera, *Miniacina* and *Homotrema*, is mainly controlled by water energy. The forms with thick globular tests dominate low-energy, more protected habitats than those of flat-encrusting tests. The latter typify exposed substrates under high-energy conditions (Brasier 1975; Ghose 1977; Reiss and Hottinger 1984; Fagerstrom 1987; Martindale 1992; Bosellini and Papazzoni, 2003). However, the coral encrusting, flat morphotypes seem to thrive on the corals' lower surfaces, where less competition with other corals occurs, allowing the foraminifera to spread laterally, while the globular forms seem to prefer the upper surface, where the competition with algae pushes them to expand their test vertically (Bosellini and Papazzoni, 2003). According to present-day data, *Miniacina* and *Homotrema* seem to prefer cryptic microenvironments, often on the underside of dead corals (Vasseur 1974; Brasier 1975; Ghose 1977; Fagerstrom 1987; Martindale 1992), although *Homotrema rubra* from Bermuda has been observed on exposed reef surfaces too (Elliott et al., 1996). *Miniacina* is an important component of the polygenic-micritic crusts that strongly bind *Porites* coral branches of the Late Miocene Mediterranean reefs (Bosellini et al., 2002; Bosellini and Papazzoni, 2003). The succession of encrusting foraminiferan assemblages in the Miocene is therefore interpreted as controlled mainly by light, competition with coralline algae, hydrodynamic energy, and coral growth fabric. Other encrusting foraminifera such as, *Alanlordia,* are confined to the carbonate-depositing, clean, middle to inner shelf, marine palaeoenvironments of the Neogene of the tropical Far East. In that area it is a potentially biostratigraphically valuable taxon.

The Nummulitidae are common throughout the Cenozoic and most of them are extant. They are in fact the largest extant calcareous foraminifera, and the morphology of some species (e.g. *C. carpenteri*) seems to be constrained by biomechanical factors that enable them to achieve relatively gigantic sizes (Yan Song et al., 1994). Living nummulitids are widely distributed in modern tropical and subtropical shallow-water seas and achieve their highest diversity in the subtropical and tropical West Pacific. Although they house photosymbionts, they prefer calm water conditions to highly illuminated areas, where their flat thin tests could be easily damaged by the hydrodynamic regime (Hohenegger et al., 2000). The relationship between their distribution and the environmental gradient "coenocline" has been investigated by some authors (e.g. Hohenegger, 2000). Test morphology depends on the depth gradient (Pêcheux, 1995) and test flattening and thinning of the walls promote photosynthesis in deeper euphotic zones (Hallock and Hansen, 1979; Hottinger, 1997).

In the upper Tf1 stage of the Indo-Pacific province a lineage of *Cycloclypeus* evolved into forms with widely separated annular inflations, *Katacycloclypeus*. The division of the median chambers of *Cycloclypeus* and *Katacycloclypeus* has evolutionary parallels with the division of the cubiculae (or "lateral chamberlets") of *Miogypsina,* where the illuminated chambers are also believed to be nests for diatoms (see BouDagher-Fadel et al., 2000). Each of the chambers cubiculae would act not only as a small convex lens for the focusing of sunlight, but could also act as a greenhouse for the containment and development of symbiotic diatoms. The diatoms enable these forms to be competitive in the supply of nutrients without pseudopodial activity in food-gathering (Röttger, 1971). Miocene shoals of the Philippines were dominated by *Miogypsina* (Hallock and Glenn, 1985, 1986).

At the top of the Tf2 stage, and within the Serravallian, *Katacycloclupeus* became extinct together with the miogypsinids. The annuli of *Katacycloclypeus* were no longer needed to focus the sparse sunlight at greater water depth into preferred areas of the equatorial layer, and therefore it seems likely that the top of Tf2 coincides with a sharp sea level drop in the Middle Miocene. It could be argued that the *Katacycloclypeus* morphotypes had ample time to migrate down slope to maintain their preferred water depth, but depth distribution was dictated by temperature rather than water depth, and the deeper water may have become too cold due to the general cooling associated with the sea level fall.

Living *Operculina* and *Heterostegina* occur over the same depth (20-130m), but the former is found to inhabit soft sediment between growths of the alga *Halophila*, while *Hetrostegina* is found to prefer living on hard substrates (Reiss and Hottinger, 1984). Hetrostegines, in developing partitioned chambers, were able to use their endosymbiotic algae more efficiently (Banner and Hodgkinson, 1991). Yordanova and Hohenegger (2004) demonstrated significant morphoclines in *Operculina* and *Planoperculina* species, and they used the different morphological characters for depth gradient estimation using regression analyses. They demonstrated that thick forms of *Operculina* with tight coiling live in shallow water (depth of 20 to 40m), while in the deeper parts of the euphotic zone (120m) thin forms with weakly coiled spirals predominate. Forms, such *Elphidium,* are suspension feeding foraminifera and they form "spiders web" between the stipes of coralline algae and utilize their pseudopodia to capture food from the water column (Langer and Lipps, 2003).

The symbiont-bearing calcarinids, such as *Calcarina* and *Baculogypsina*, are prolific contributors to reef sediment in shallow-water habitats and their distribution appears to be related to dispersal rather than habitat restriction (Lobegeier, 2002). They are common epiphytes on both macroalgae and seagrasses (Lobegeier, 2002). The thick calcareous layer of the calcarinids protects them against the light intensity and UV-radiation in sub-littoral shallow-waters (Röttger and Krüger, 1990). Their thick tests and special adhesion mechanisms (e.g. their spines) allow them to adapt to the shallowest regions, with high solar irradiation and vigorous water movement (Hohenegger, 2000).

In modern Indo-Pacific reefs, *Calcarina, Baculogypsina* and *Amphistegina lobifera* are the dominant robust larger foraminifera. The spinose *Calcarina* spp. and *Baculogypsina* populate environments with vegetation such as *Jania* mats, which are found immediately behind or within the red algal ridges (Hallock and Glenn, 1986). Algal-coated rubble leeward of the algal crest is often populated by *Amphistegina, Calcarina* and soritids, as well as abundant smaller rotaliines, peneroplids, and miliolines. Coarser, higher-energy sand flats are often devoid of living foraminifera, though the sand itself may be predominantly composed of robust foraminiferal tests. However, no living symbiont-bearing forms are found below 110m (Hallock, 1984). In the Gulf of Aqaba, these forms and *Heterocyclina tuberculata* are the deepest dwelling algal symbiont-bearing foraminifera, with depth limits of almost 130m (Reiss and Hottinger, 1984).

A relatively large number of agglutinated foraminifera with alveolar walls, such as *Lituola* (see Chapter 5), which made its first appearance in the Triassic, continued to exist in the Neogene up to present day. Attached foraminifera, such as *Haddonia* (see Chapter 6), which made its first appearance in the Early Paleogene (Danian), is common at the present day but is confined to the fringing reef assemblages (Langer and Lipps, 2003). On the other hand, the Lower to Middle Miocene saw the diversification of the agglutinated pavonitinids, which had elongate triserial to biserial tests with complicated inner structures. Unlike the rest of the larger benthic foraminiferal assemblages, these assemblages occupy chiefly bathyal to abyssal sediments of tropical waters and constitute 40% of the agglutinated taxa in Neogene tropical waters (Seiglie et al., 1986).

Neogene carbonate lithofacies represent a variety of palaeoenvironments ranging from shallow reef to relatively deep, open marine conditions. Individual outcrops often demonstrate a transition between these environments, thought to relate to changes in relative sea level, and levels of carbonate production and clastic input, which depending on location includes occasional volcaniclastic grains. Reef skeletal rich mound-shaped buildups are referred to as bioherms, while carbonate layered/tabular build-ups are instead known as biostromes. The presence of well-preserved larger benthic foraminifera within the coralgal framework buildups of the Neogene carbonates suggests that deposition occurred entirely within the photic zone. Among the many different morphologies of carbonate platform, the most widely recognised are carbonate ramps, which are gently sloping platforms, and rimmed shelves, which are flat-topped platforms bordered by rims/barriers made of reef corals. Shallow seas developed patch reefs, in which the areas attacked by oceanic swell may be called "forereef" (Fig. 7.24). The energy distribution produces characteristic carbonate lithofacies and allows distinctive assemblages of foraminifera to flourish:

- The reef itself, being shallowest, produced characteristic bioherms and calcirudites of corals and algae, cemented by sparitic matrix as the lime muds were flushed away by wave action.
- In the forereef shelves, proximal biostromes of corals and algae were cemented by sparite and micrite carbonates. Distally from the reefs, coral debris diminished and micrite increased, cementing the calcarenitic biogenic debris. Below fair weather wave-base, micrite would accumulate, which contained scattered larger foraminifera swept in from the reef shelves. The proximal forereef shelf contained faunas dominated by *Planorbulinella* spp., cycloclypids, lepidocyclinids, operculinids and heterosteginids.
- The shallow water carbonate facies are surrounded distally by deeper water sediment containing abundant planktonic foraminifera, which may constitute up to 35% volume. The planktonic foraminifera, abundant in the micritic matrix, had flourished in the water column offshore and reflect relatively deeper water, and low energy.

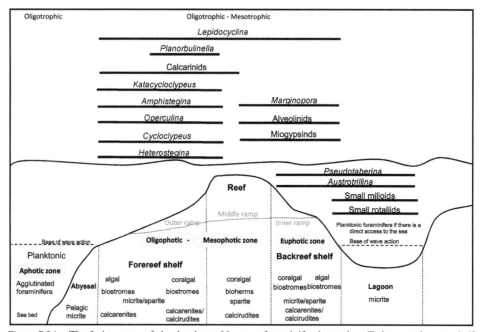

Figure 7.24. The facies range of the dominant Neogene foraminiferal taxa in a Tethyan carbonate shelf. Integrated reef/ramp model for Neogene carbonates. The ramp model is indicated by the grey dotted line. In the case of gently sloping ramp, the outer ramp lithofacies are made of mudstones and wackestones, while in the middle ramp mudstone with carbonate nodules would develop.

7.5 Palaeogeographic Distribution of the Neogene Larger Foraminifera

7.5.1 General characteristics of the distribution of Neogene larger foraminifera

The end of the Oligocene witnessed very little extinction, worldwide just 13% of the larger benthic foraminifera disappeared at that time. As explained in Chapter 6, this stratigraphic boundary is probably mainly associated with plate tectonic events, and/or gradual changes in climate (Berggren and Prothero, 1992). During the Early Miocene, modern patterns of atmospheric and ocean circulation began to be established. The proto-Mediterranean was still open, but was narrower than before (Fig. 7.25). The Early and Middle Miocene was a time of relatively warm average global temperatures (Fig. 7.26). However from the Middle Miocene, the cold phase became established as the isolation of Antarctica from Australia and South America became more pronounced, and meant the establishment of the circum-polar ocean circulation, which significantly reduced the mixing of warmer tropical water and cold polar waters, and further led to the buildup of the Antarctic ice cap. This is indicated by oxygen isotopic data obtained from deep-sea benthic foraminifera (Macleod, 2013, Fig. 7.27). Additionally, the Tibetan platform uplift (which began around the Eocene-Oligocene boundary, ~34 Ma) continued, and the Red Sea rifting accelerated (Thomas et al., 2006; Aitchison et al., 2007).

However, the tectonic changes related to the widening of the Southern Ocean passages and the closing of the Tethys seaway, influenced circulation patterns that caused a flow

Figure 7.25. Palaeogeographic map showing the proto-Mediterranean still open but narrower than during the Early Miocene (by R. Blakey, http://jan.ucc.nau.edu/~rcb7/paleogeographic.html).

reversal through the Panama Seaway between the Oligocene and Miocene (Jakobsson, et al., 2007). The global thermohaline circulation in Oligocene and Miocene models are significantly different from the present day pattern. In particular, in the Oligocene the salinity contrast between the Atlantic and Pacific oceans was reduced because of water mass exchange through the low-latitude connections between the two oceans (von der Heydt and Dijkstra, 2006). Although geochemical proxies suggest that the Drake Passage between South America and Antarctica was open earlier, in the Late Oligocene (32.8 Ma), to intermediate and deep water circulation (Lawver et al. 1992; Latimer and Filippelli, 2002), the continued opening of this gateway eventually led to the thermal isolation of Antarctica in the Early Miocene (Fig. 7.28) and the creation of the clockwise, strong Antarctic circumpolar current (Smith and Pickering, 2003).

These tectonic changes during the Early Miocene influenced the biotic distributions within the oceans, in particular that of the shallow larger benthic foraminifera. These organisms, which contribute most to the Neogene biogenic carbonate sediments, have a distinct global pattern of provincialism. The four main bioprovinces known today are those of the Americas, South Africa, Western Tethys (or the Mediterranean), and the Indo-Pacific (e.g. Adams, 1967; Rosen, 1984; BouDagher-Fadel and Price, 2010a, 2013, 2014, 2017). The distribution of the larger benthic foraminifera throughout the Neogene

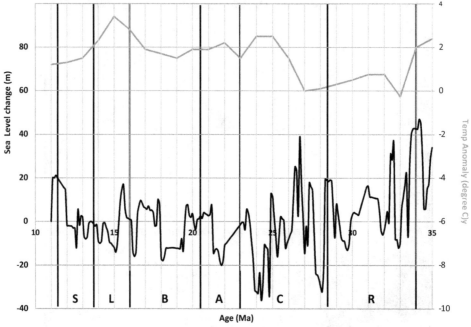

Figure 7.26. Variation in sea level and temperature during the Oligocene and Mid-Miocene based on Zachos et al. (2001), Miller et al. (2011) and BouDagher-Fadel and Price (2013).

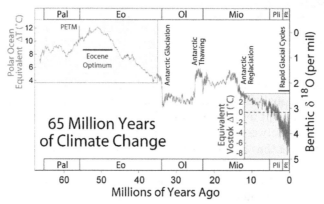

Figure 7.27. Cenozoic oxygen isotopic data in relation to global temperatures and their implications on physiographic events (from Macleod, 2013).

is associated either with the Indo-Pacific, or the Indo-Pacific and Mediterranean ("Tethyan"), or the Mediterranean and West African, or the American provinces. There are also many cosmopolitan forms (Fig. 7.29).

Although the extinctions (Fig 7.30) throughout the Neogene were not as pronounced as in other periods, each of the bioprovinces had its own patterns of evolutionary

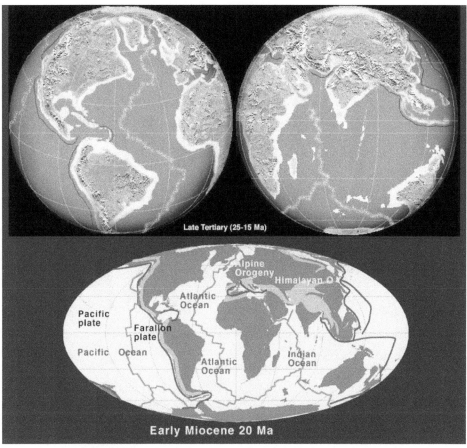

Figure 7.28. Palaeogeographic and tectonic reconstruction of the Early Miocene (by R. Blakey http://jan. ucc.nau.edu/~rcb7/paleogeographic.html).

lineages and extinctions, which at times were also punctuated by migration events from other provinces, examples of which are discussed in greater detail at the end of this section.

In the Aquitanian, with the Tethyan seaway between the proto-Mediterranean and the proto-Indian Ocean still open, 56% of the larger benthic foraminifera were cosmopolitan, of which 41% of the new Miocene forms appeared globally (Fig. 7.31). Of the new forms in the Aquitanian, 35% were encrusting foraminifera, the acervulinids, which in modern reefs are cosmopolitan, living at mid-depths in semi-cryptic (gloomy), warm-water environments, encrusting corals (Dullo et al., 1990; Martindale 1992; Bosellini and Papazzoni, 2003). One of the main changes during this period was the appearance in Tethys of what were originally American lineages, such as those of the miogypsinids and lepidocyclinids (see Chart 7.1). Such lineages became very successful and evolved many times in the Indo-Pacific (see the discussion at the end of

Neogene Genera

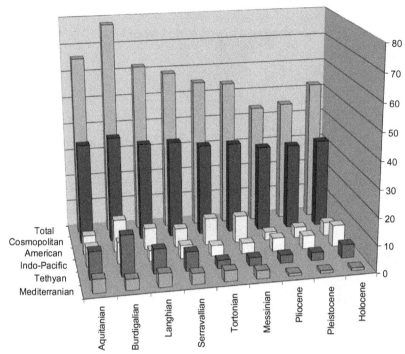

Figure 7.29. The number of Neogene larger foraminifera genera plotted by province.

Neogene Extinctions

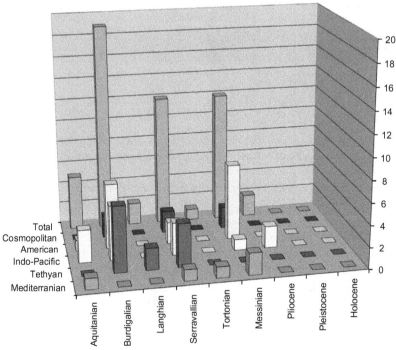

Figure 7.30 The number of extinctions of Neogene larger benthic foraminifera genera at the end of each stage.

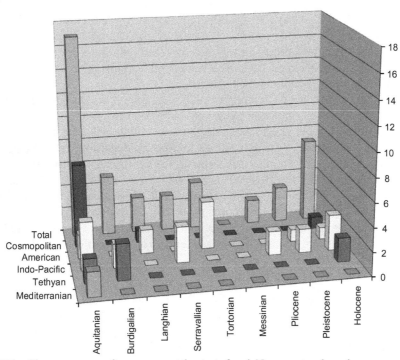

Figure 7.31. The appearances of new genera at the start of each Neogene stage boundary.

this section) and played an important part in defining the biostratigraphy of the carbonates of South East Asia, with *Miogyspsina, Lepidocyclina, Eulepidina, Spiroclypeus* appearing worldwide, along the tropical belt reaching as far south as New Zealand. Some forms, such as *Cycloclypeus*, originated in the Mediterranean, migrating raipdly towards South East Asia, but never appearing in America. Towards the end of the Aquitanian, global extinctions were low (Fig. 7.30), and only 8% of the Aquitanian genera became extinct, with the highest percentage in the Indo-Pacific (43%).

In the Burdigalian much of the diversification of Miocene larger benthic foraminifera in Tethys is at the species level. Some new rotaliid genera made their first appearance in Tethys, with only a few occurring in the Indo-Pacific province. The acervulinids and the planorbulinids were still common globally. The soritids became diverse, populating warm seagrass environments in all provinces, while the nummulitids thrived in the reef and forereef together with the lepidocyclinids. However, towards the end of the Burdigalian, 29% of the larger foraminifera became globally extinct (Fig. 7.30). The already established cosmopolitan forms, however, were not significantly affected by this event (with only 5% becoming extinct). The larger benthic foraminifera that were most affected were the provincial genera, mainly those of the American province (with 50% of them becoming extinct) and the provincial Tethyan foraminifera (with 40% extinction). Two main groups of foraminifera that became extinct in the Americas, the lepidocyclinids and miogypsinids, were living on reefs in high energy environments.

On the other hand, the Soritoidea were not affected by this extinction, most probably because they mainly lived in sheltered habitats, and they had the opportunity to change their mode of life. They changed from living attached to phytal and non-phytal substrata to an epifaunal, free-living mode of life. They also changed their endosymbionts (see above), which enabled them to swap their attached habitat to become free living in slightly deeper oligotrophic environments when reefal conditions became difficult and lack of direct sunlight hampered their growth.

The regional extinction of the larger benthic foraminifera in the Americas also coincided with the regional extinction of about half the Caribbean hermatypic corals, which died out during the latest Oligocene through Middle Miocene, about 24-16 Ma (Edinger and Risk, 1994). As was the case for the lepidocyclinids and miogypsinids, the majority of the corals that died out earlier in the Caribbean, were found in the Late Miocene of the Indo-Pacific. However, unlike the corals, which are still extant in the Indo-Pacific, the lepidocyclinids and miogypsinids died out globally at the end of the Miocene. These regional extinction, confined to the Caribbean and western Atlantic, might be related to a contemporaneous eruption of the Columbia River basalts (Fig 7.32), which was at its most intense between 17 and 15 Ma (Courtillot and Renne, 2003).

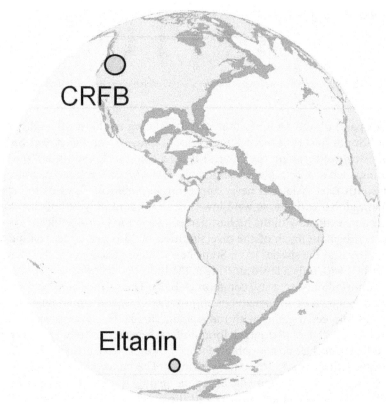

Figure 7.32.   The Neogene world, showing the location of the Columbia River Flood Basalts (CRFB) and the Eltanin crater.

The CO_2 emissions from this flood basalts could have triggered a cooling or acidification that may have affected the faunas on both patch reefs and shelf edge reefs.

Except for a small number of deep water agglutinated foraminifera appearing in the Caribbean, very few new forms appeared globally in the Langhian (Fig. 7.31). Provincialism was rare on a genus level, with 58% of the larger benthic foraminifera being cosmopolitan and 27% belonging to Tethys. However, there was diversification on a species level in Tethys, and shallow water carbonates were teaming with foraminifera. The fusiform milioline *Flosculinella bontangensis* graded into *Alveolinella* forms at about the same time as the disappearance of *Austrotrillina*, occurring alongside *Alveolinella praequoyi*, within the Serravallian in the Indo-Pacific province. Also in the Serravallian, *Cycloclypeus*, which seemed to disappear in the earliest Miocene from the Mediterranean, seems to be common in the Indo-Pacific and by the Serravallian it had developed distinct inflations along the test, as in *Katacycloclypeus*.

In the Serravallian, plate collisions sealed off the eastern proto-Mediterranean and the connections between the proto-Mediterranean and the proto-Indian Ocean closed (see Fig 7.33) (see Henderson et al., 2008, 2010; Najman et al., 2008, 2010a, b). Evaporitic deposits formed in the Red Sea, the Mesopotamian basin and along the northern margin of Africa in the Sirt basin (Gvirtzman and Buchbinder, 1978; Rögl and Steininger, 1983; Jolivet et al., 2006). Many larger foraminifera became extinct in Tethys, where 60% of the global extinctions happened. Most noticeably the immigrants from the Americas, the lepidocyclinids and miogypsinids, after thriving successfully in the shallow high energy warm waters, were completely wiped out by the end of the Serravallian. Another casualty of the closure is *Cycloclypeus,* which seems to have disappeared from the proto-Mediterranean after the closure of the sea link, but continued to thrive up to the present day in the deep photic zone of the Indo-Pacific and Australia.

In the Middle to Late Miocene the partial emergence of the Isthmus of Panama (~8.3–7.9 Ma) disrupted the Atlantic to Pacific flow, closing the intermediate depth water connection (Roth et al., 2000). During the pre-closure interval (7.6–4.2 Ma), the Caribbean was under mesotrophic conditions, with little ventilation of bottom waters and low current velocities (Jain and Collins, 2007).

This event was accompanied by the development of more provincialism, producing Caribbean endemics. Deep water textulariides, with inner complex structure, became common; they included 70% of the agglutinated larger foraminifera with inner structures found in the Caribbean. These foraminifera had structures, such as alveoles, and used to occupy shallow tropical waters of Tethys in the Early Jurassic to Cenomanian, with a brief resurgence of one of these groups *(Coskinolina, Lituonella* and *Dictyoconus)* in shallow tropical environments of the Early and Middle Eocene. In the Early Miocene, and in response to environmental stress in the Caribbean, they moved their habitats to deeper environments where they occurred in abyssal muds, poor in oxygen (Seiglie et al., 1986). In the Middle Miocene, they were mainly found in Central America (75%) with only a few occurring in West Africa and Western Mediterranean (12%), and only one genus occurring in both provinces (*Liebusella*).

In the Tortonian, more provincialism occurred in the Americas, as new soritids were limited to the Caribbean. The Soritoidea evolved into new forms and exhibited the largest sizes in their history. These new forms are morphologically related to the Early Miocene Tethyan forms, as during or prior to the Middle Miocene, and Hottinger

Figure 7.33. A map showing a permanent land barrier separating the proto-Mediterranean and proto-Indian Ocean regions developed in Middle Miocene times (by R. Blakey, http://jan.ucc.nau.edu/~rcb7/paleo-geographic.html).

(2001) has suggested that the Tethyan soritids might have migrated to the Caribbean to replace the Eocene endemics. However, given that historical migrations of LBF have always been from the Americas to Tethys (see below), this hypothesis may need further investigation. On the other hand, those in the Indo-Pacific were completely dominated by two soritids, *Amphisorus,* which appeared in the Late Oligocene and the extant newcomer, *Marginopora,* in the Middle Miocene.

By the Late Messinian (5.65 Ma) most of the larger benthic foraminifera in the Mediterranean had become extinct (Fig. 7.30), as the Mediterranean experienced a salinity crisis and became partially desiccated. It has been suggested that this was caused by a tectonic event, with no specific climatic change being responsible for the evaporation (Jolivet et al., 2006). In the Late Miocene period, the Gibraltar Straits closed, and the Mediterranean evaporated forming a deep dry basin with a bottom at some places 2 to 3 miles (3.2 to 4.9 km) below the global sea-level (Clauzon et al., 1996). The end of this crisis corresponds to the breaching of the Gibraltar Straits that resulted in an extensive and rapid flooding of the Mediterranean basin (Loget et al., 2003). During this period, the Mediterranean larger benthic foraminifera were rare and localised and by the end of the Messinian 75% of them had become extinct. The foraminifera endemic to the Indo-Pacific were not very affected, however, as only 25% of them became extinct at this time (Fig. 7.30).

The pre-closure interval of the Central American seaway (7.6–4.2 Ma) saw enhanced seasonal input of phytodetritus with even more reduced ventilation, and enhanced dissolution between 6.8 and 4.8 Ma (Jain and Collins, 2007). This lead to mesotrophic conditions and poor ventilation, abundance of the cool water and a continued drop in sea level that had begun in the Middle Miocene (Roth et al., 2000). These events might have caused the extinction of 75% of the Soritoidea, which were up to then endemic in the region. According to Hottinger (2001), these forms were more closely related to those extant in the Indo-Pacific province than those of the present day foraminifera in the Caribbean. In this case, they might not have become extinct but most probably migrated to the Indo-Pacific province to become the ancestral forms of modern Indo-Pacific soritids.

At the onset of the Pliocene, new larger foraminifera were conspicuously absent from all bioprovinces, except in the Indo-Pacific, where a new lineage, restricted to the western and central tropical Pacific, emerged. These were the calcarinids, with a preference for shallow water, high-energy environments. They were particularly abundant on coral reef flats and widespread throughout the Great Barrier Reef Province of northeastern Australia (Lobegeier, 2002). Their global distribution, according to Lobegeier (2002), seems to be limited by dispersal by oceanic currents rather than by habitat restriction. They are prevented from reaching Hawaii, and other islands lying to the east of 170°, by the westward flowing North and South equatorial currents and from the Indo-Malay region to the west or to the south by the Equatorial Countercurrent in the Indian Ocean.

In the Caribbean, the total closure of the Central American seaway at 4.2 Ma caused the formation of the Isthmus of Panama, and allowed direct land-to-land connection between North and South America. This closure blocked the Atlantic-Pacific water interchange, changing profoundly the ocean circulation, isolating the Arctic Ocean and initiating the northern polar ice (Jain and Collins, 2007). The closure of the seaway in the Caribbean produced a cascade of environmental consequences, including the reorganization of circulation in the Gulf of Mexico and much of the Caribbean, leading to reduced upwelling and palaeo-productivity (Allmon, 2001). In response, to the changes in ocean circulation and the decrease in the plankton in the Caribbean Sea, fast-growing oysters became extinct (Kirby and Jackson, 2004). Larger benthic foraminifera living in shallow-marine environments also reflected these changes by reorganizing from a suspension-feeder-dominated community to a more carbonate-rich, phototrophic-based community. The increasing divergence of milioline and textulariine groups between the Atlantic and Pacific indicates differing habitats between shallow sea grass communities and deep water benthics.

Extinctions at the end of the Pliocene were largely restricted to the Indo-Pacific. Some of these extinctions could have been triggered by the late Pliocene (2.15Myr) Eltanin asteroid impact in the Southern Ocean (see Fig 7.32). This is the only known deep ocean basin asteroid impact, and as such is suggested to have had major regional consequences (Gersonde et al., 1997). However, the late Neogene was also a time of exceptionally strong global cooling and oceanographic change (Zachos et al., 2001). Global cooling after ~3 Ma had for the first time a direct effect on tropical sea surface temperatures, resulting in high-amplitude fluctuations in global ice volume and sea levels (Fedorov et al., 2006; Johnson et al., 2007). Other than the extinctions that may have also been triggered by climate change in the Indo-Pacific, the only other effect that this cooling seemed to have had on the tropical larger benthic foraminifera was a suppression of speciation and diversification (Fig. 7.31). Long ranging species continued to thrive and adapted themselves to cooler waters, and the only bioprovince which saw a small percentage of new species (Fig. 7.31) was that of the Indo-Pacific, where empty niches from the earlier extinctions became filled with new larger foraminifera. Indeed, 40% of the endemics in the Indo-Pacific were new Pleistocene forms.

It has been suggested that at the end of the Pleistocene (at 12.9 ka), an extra-terrestrial impact event over northern North America, caused abrupt environmental changes that destabilized the Laurentide Ice Sheet, triggered cooling (the Younger Dryas), and caused large-scale mammal extinctions and the end of the human Clovis Culture (Firestone et al., 2007). However, at this time the climate was already changing rapidly, with warming and changing patterns of rainfall, which in themselves could have triggered the terrestrial extinctions. However, whatever the cause of these changes they did not seem to affect the ecology of the marine tropical belt, as extinctions of larger benthic foraminifera were rare to non-existent at this time.

In the Holocene, new forms of larger benthic foraminifera appeared through the tropical realm, with 70% of the foraminifera in the Indo-Pacific realm being new, compared with 30% in the American bioprovince. Only 6% of the cosmopolitan foraminifera are new appearances, with half of them being deeper water foraminifera (see Charts 7.2 and 7.3). Present day larger foraminifera play analogous roles in the ecosystem, where the tropical belt is divided into two parts, different assemblages colonise different environments. In particular, the Indo-Pacific larger soritids, (*Amphisorus* and *Marginopora*) are substituted in the Caribbean by *Archaias* and *Cyclorbiculina* as porcelaneous discoidal epiphytes on tropical seagrasses (Langer and Hottinger, 2000). Modern tropical western Pacific sandy shoals and beaches can be nearly pure concentrations of *Calcarina, Baculogypsina* or *Amphistegina* tests. Calcarinids can also range into the central Pacific and westward through the Indian ocean (Todd 1960; Hallock 1984; Röttger and Krüger, 1990). Analogous sediments in the Caribbean are characterized by robust peneroplids and soritids, and by thick-shelled miliolines (Brasier, 1975). Modern Indo-Pacific faunas from deeper foreslopes are characterized by *Heterostegina, Cycloclypeus, Operculina*, and the flatter species of *Amphistegina*; *Heterocyclina* replaces *Cycloclypeus* and *Calcarina* in these deeper environments of the Red Sea-East African faunal province (Reiss and Hottinger, 1984, Renema, 2007). The Caribbean lacks deep-euphotic assemblage forms. Shallower foreslopes are dominated by *Amphistegina lessonii* in the Indo-Pacific and by *A. gibbosa* in the Caribbean. Alveolinids occur on the shelf areas of both regions. *Alveolinella* in the Indo-Pacific and *Borelis* in both regions dominate quieter environments.

In the Mediterranean area, *Planorbulinella* has been reported from the Oligocene to the end of the Miocene (Messinian), when the lineage ended (Freudenthal, 1969; Drooger, 1993). It has also been reported from the Miocene of Trinidad (Cushman and Jarvis, 1930). *Planorbulinella* is widespread in the Indo-West Pacifc Ocean and at least two species occur from as far afield as the Red Sea and Hawaii (Renema, 2005).

7.5.2 Provincialism and migration of some Neogene larger benthic foraminifera

It appears that the migration of a number of larger benthic foraminifera out of the Caribbean and into the Tethyan realm, and then onward to the Indo-Pacific and South Africa is an important, repeated characteristic of the Cenozoic. The migrations appear to correlate with global eustatic sea level falls (see BouDagher-Fadel and Price, 2010a, 2013). From Middle Eocene to Early Oligocene, American larger foraminifera were mainly endemics. By the earliest Oligocene, they had crossed the ocean perhaps being carried by currents on debris, or as planktonic gametes or zygotes, across the then narrow Atlantic from America to West Africa. Migration depends on mode of life and only those forms which could settle on floatable algae or terrestrial or volcanic debris, or that had long-lived planktonic gametes or zygotes, could migrate. Migration stopped when sea level rose in the Rupelian. Repeated migration and geographical isolation drove the development of provincialism which became pronounced and characteristic of the Neogene assemblages. However, this provincialism can be highly complex, and indeed dynamic.

This Chapter ends with a discussion of specific and recently studied examples of the complexity of bio-provincialism shown by the evolutionary trends and regional occurrence of the families **Lepidocyclinidae** (BouDagher-Fadel and Price, 2010a) and **Miogypsinidae** (BouDagher-Fadel and Price, 2013), which began in the Paleogene but culminated in the Neogene.

The migration of lepidocyclinids and miogypsinids happened via three stages (Fig. 7.34). The larger foraminifera were carried first by currents on debris across the Atlantic from North America to North Africa and the Western Mediterranean in the Early Oligocene. Migration then must have occurred within the Mediterranean and along the Arabian coast by dispersal of foraminifera by algal rafting, planktonic gametes or zygotes, and long-shore drift, together with occasional migration from the Western Tethys on the Canary Current southwards towards tropical Africa, followed by the development of local stocks. They then spread rapidly in the Early Miocene, carried by ocean current from Arabia to the Indo-Pacific, or from Western Africa to South Africa.

The migration of the lepidocyclinids out of the Caribbean and into Western Tethyan is an important event in the development of Indo-Pacific faunas, and appears to have occurred immediately after (or during) the Early to Middle Oligocene eustatic sea level falls (see Fig. 7.26). The oldest Lepidocyclinidae evolved in Central America (see Chapter 6) and are differentiated into two main lineages, which can be

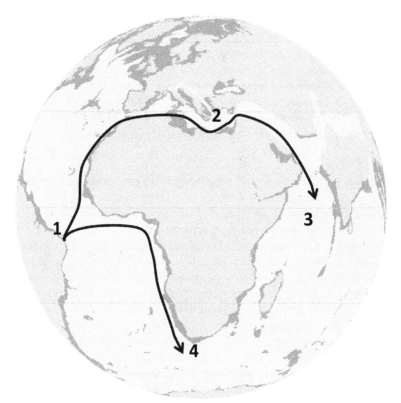

Figure 7.34. The migration of miogypsinids during Early Oligocene, shown by black arrows, from the Americas (1), to the Western Tethys (Mediterranean) (2), and on to the Indo-Pacific (3), or to South Africa (4) (see BouDagher-Fadel and Price, 2013).

called *Eulepidina* and *Lepidocyclina sensu lato*, which includes *L. (Lepidocyclina)* and *L. (Nephrolepidina)*, but from there they rapidly spread to Europe in the Oligocene. The established view is that the arrival of *Lepidocyclina* in Asia was an early to mid-Oligocene event, with some overlap of these forms with *Nummulites* (to give the Td faunas).

During the middle Eocene to early Oligocene a series of sea level falls (Katz et al., 2008) enabled the migration of some species, such as *Helicolepidina spiralis*, *Lepidocyclina (L.) pustulosa* and *L. (L.) rdouvillei*, between the American province and the West African province (Brun et al., 1982; Ly and Anglada, 1991; Mello e Sousa and et al., 2003; BouDagher-Fadel, 2008). However, after the last major regression in the early Oligocene, at 33.5 Ma, the rising sea level isolated the West African shelf from the American province, preventing further exchange between the Americas and Western African margins. Finally, they diffused in the late Rupelian, carried by ocean current from Arabia to the Indo-Pacific, where the equatorial conditions and abundant ecological niches favored rapid diversification and the development of a rich and complex range of species. Early Rupelian Tethyan lepidocyclinids evolved into a *Lepidocyclina (N.) praemarginata* and *L. (Lepidocyclina)* spp. assemblage (see

Fig. 7.35). These species, which imitated in their evolution the American Oligocene assemblages of *L.* (*L.*) *yurnagunensis* and *L.* (*N.*) *braziliana* (Plate 7.18, fig. 10), were first described by De Muller (1975) from the late Rupelian to early Chattian of Corfu. Since then, *L.* (*N.*) *praemarginata* was reported from the upper Rupelian beds in northern Spain and Chattian of Cyprus. While in the American province, the early Miocene assemblages were dominated by *L.* (*L.*) *canellei*. The American lepidocyclinids died out completely at the end of the Middle Miocene.

In Tethys, the lepidocyclinids migrated eastward from the proto-Mediterranean in the early Rupelian and reached the Indo-Pacific towards the end of the Rupelian. *L.* (*Lepidocyclina*) spp. became extinct at the end of the Oligocene in the Mediterranean province. On the other hand, *L.* (*Nephrolepidina*) continued to thrive along the Mediterranean shelf, evolving different lineages into the Early Miocene (see BouDagher-Fadel and Price, 2010a). The earliest *L.* (*Nephrolepidina*) species in Tethys occurred in the late Chattian, and they were almost certainly the direct descendants of *L.* (*Lepidocyclina*) and *L.* (*Nephrolepidina*) of the Eocene and Oligocene of West Africa and America. During the early Miocene, the Tethyan seaway between the proto-Mediterranean and the proto-Indian Ocean became narrower (BouDagher-Fadel, 2008), which restricted further migration between the two provinces. However, in the Indo-Pacific the descendants of the Mediterranean lepidocyclinids continued to thrive successfully in the shallow high energy warm waters, evolving many independent parallel lineages. The spatial and temporal separation of these lepidocyclinids from their ancestral Mediterranean forms precludes them from being the same species, but rather they are examples of parallel evolution, which occurred as the two bioprovinces developed separately.

The development of the miogypsinids is analogous to that of the lepidocyclinids, and exploited the same migratory pathway. The miogypsinids were initially provincial, being found only in the Americas and survived there into the Early Miocene. BouDagher-Fadel and Price (2010b) have demonstrated that they originated in the Americas from *Neorotalia* in the Early Oligocene (Rupelian and P18) (see Chapter 6). During the Early Oligocene a series of sea-level regressions (Katz et al., 2008; Miller et al., 2011) reduced the effective width of the early Atlantic Ocean sufficiently to facilitate transoceanic migration of *Neorotalia* from the American province to the North African coast and into the Mediterranean. During this time, Mediterranean shallow water niches were still occupied by the Paleogene *Nummulites*. However, towards the end of the Early Oligocene (around 31–29 Ma), the environmental stresses, perhaps associated with cooling and the large flood basalt event in Ethiopia and Yemen (see Chapter 6) contributed to the disappearance of the last Mediterranean *Nummulites* which provided an opportunity for new phylogenetic lineages of miogypsinids to fill the warm reefs of the Mediterranean (see BouDagher-Fadel and Price, 2013).

The oldest representatives of the genus *Miogypsina*, appeared in the Oligocene in central America, but just before the Oligocene-Miocene boundary in Europe, where its evolutionary appearance occurs at the level of *Miogysinoides formosensis* in the latest Oligocene, but in the Indo-Pacific province, the oldest *Miogypsina gunteri* is found with *Miogypsinoides bantamensis* at the start of the Miocene. *M. gunteri* shows a decrease in the size of the proloculus, which is one of several relapses (Drooger,

602   *Evolution and Geological Significance of Larger Benthic Foraminifera*

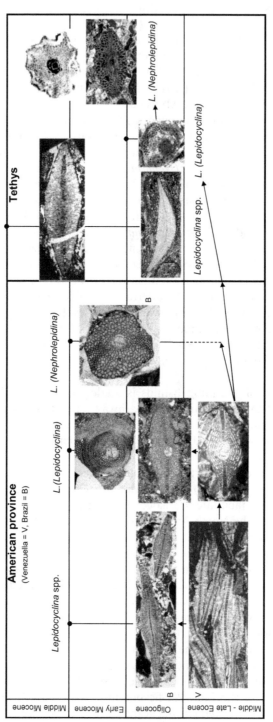

Figure 7.35.   Evolution of *Lepidocyclina sensu lato* in space and time in the Cenozoic.

1993) in the general principle of nepionic acceleration. In all three bioprovinces, there are later developments within the group of Miogypsinidae with lateral chamberlets (*Miogypsina* s.l.), towards forms in which the embryon tends to shift towards the centre of the test. These forms are distinctly different in each of the bioprovinces: the subgenera *Helicosteginoides* and *Miogypsinita* in Central America, *Miolepidocyclina* in Central America and the Western Tethys, and *Lepidosemicyclina* in the Indo-Pacific. These forms appear at different times, but are each sufficiently different, and have their origins so close to contemporaneous *Miogypsina*, that migration does not have to be invoked.

As the morphologies of American and Mediterranean miogypsinids are seen to be crucially different, it follows that their evolutionary development was independent but closely parallel, as after the last major regression in the Early Chattian, the rising sea level and the continuing oceanic rifting effectively isolated the Mediterranean–West African shelf from the American province (around 28 Ma), ending any flow of miogypsinids from America to the Mediterranean.

During the Late Oligocene and Early Miocene, successive forms of miogypsinid continued their migration eastward through the open seaway from the Mediterranean into the Indo-Pacific, where they typically arrived a million years or so after their first Mediterranean appearance (Fig. 7.34). Once in the tropical setting of the Indo-Pacific, with its diverse paleogeography, the migrants gave rise to a richer diversity of local species than seen in the Mediterranean. However, in the late Burdigalian (around 17 Ma) the eastward migration between the Mediterranean and the Indo-Pacific was interrupted as tectonic processes narrowed and closed the seaway between the Mediterranean and Indian ocean (Von Rogl 1998; see Fig. 7.28). Prior to this first closure of the Eastern Tethys seaway in the Burdigalian (17 Ma), the miogypsinids had thrived in the warm climates of the Mediterranean, reaching their peak diversity, with a maximum number of species, in the Early Burdigalian, however, with the first closure of the Tethys seaway in the late Burdigalian, the miogypsinids became isolated, with limited ecological diversity, and only very few species survived the Burdigalian–Langhian boundary (15.9 Ma) (BouDagher-Fadel and Price, 2013).

The first closure of the Tethyan seaway was short-lived, however, as a major global transgression in the Early Langhian (15.5–16.5 Ma) flooded the Mediterranean from the Indo-Pacific. This transgression led to the extinction of the remaining Mediterranean miogypsinids within the Early Langhian, as they were replaced by algal–coral patch reefs, tropical mollusc fauna and flat forms of larger benthic foraminifera, such as *Amphistegina* (BouDagher-Fadel and Clark, 2006; BouDagher-Fadel, 2008). This transgression coincided with a Middle Miocene (14–16.5 Ma) global warming (Fig. 7.26), which by contrast appeared to stimulate the development of further diversity of the miogypsinids in the Indo-Pacific province.

In the late Langhian–early Serravallian (13–14 Ma), a time of several regressions (Fig. 7.26), the short-lived marine reconnection between the Mediterranean and Indian ocean again closed (Von Rogl, 1998). This final closure coincided with the onset of global cooling (Fig. 7.26), and by the time the East Antarctic Ice Sheet was established (12 Ma), the miogypsinids had become extinct from the Indo-Pacific province (late Serravallian). This final miogypsinid extinction was also globally accompanied by the extinction of 60% of all other larger benthic foraminifera forms (BouDagher-Fadel, 2008).

In contrast to the apparent isolation of Mediterranean forms from the Americas from the Early Chattian onwards, it seems that trans-Atlantic migration to South Africa remained possible up to the late Chattian (see BouDagher-Fadel and Price, 2013), thus allowing the direct migration of the early American miogypsinids (not just the *Neorotalia* seen in the Mediterranean) to South Africa. Subsequently, the South African miogypsinids did become isolated (perhaps again owing to sea-level rises or changes in oceanic currents), and evolved lineages that were independent from their American ancestors. In the Burdigalian, South African species (Plate 7.11, figs 6-8) were still morphologically close to their American ancestors, but never evolved into the advanced forms of the megalospheric generations seen in the American, Mediterranean and Indo-Pacific *Miogypsina*. This comparatively slow rate of evolution might reflect the lack of environmental diversity and the more temperate conditions in which the South African miogypsinids found themselves. Eventually, however, increased biological stress between the Burdigalian and Langhian (Figs 7.26 and 7.27), perhaps linked to rapid sea-level changes and global tectonic events, such as the eruption of the Columbia River flood basalt (15–17 Ma), caused the extinction of the South African miogypsinids and they were replaced by other forereef larger benthic foraminifera, such as *Operculina* and *Heterostegina*. On the other hand, in the American province, the extinction of the miogypsinids, coeval with a global warming event (Fig. 7.26), the so-called mid-Miocene Climatic Maximum (Fig. 7.27), resulted in their replacement by small nummulitic forms, such as *Operculinoides* species (see BouDagher-Fadel and Price, 2010b; BouDagher-Fadel et al., 2010).

It should be noted however, that the claim of provincialism for some genera should be treated with caution, as limited observations may be due to poor preservation. In some cases, genera which were thought to be endemics to a typical province have later been found in another province. Thus, *Discospirina*, a milioline genus recorded as a fossil only from the Middle Miocene to Pleistocene of the Mediterranean, is still found living in the same area and into the North Atlantic Ocean (Adams, 1959, 1967, 1973; Adams et al., 1983). This genus has some biogeographical importance as it was considered to have been the only genus of larger foraminifera that seemed "to have originated outside the Indo-Pacific region since Early Miocene times" (Adams, 1973, p. 465). Adams et al. (1983) used the absence of records of *Discospirina* from the Indian Ocean and Indo-Pacific region to support their contention that a permanent land barrier has separated the Mediterranean and Indian Ocean regions since Middle Miocene times (Fig. 7.33). However, Chaproniere (1991) collected similar forms from the Coral Sea, separated by a considerable distance from those recorded from the Mediterranean and Atlantic Ocean. He suggested, that their presence in the Coral Sea indicates that they must have escaped from the North Atlantic or the Mediterranean, either before it was closed to the Indian Ocean in the Serravallian, or after closure, possibly during a period of high sea level. On the other hand, as *Discospirina* appears to have been able to live in deeper waters and therefore cooler temperatures, it may have been able to migrate from the southern Atlantic Ocean around the southern tip of Africa. Another alternative explanation is that this genus arose independently in the Coral Sea area. However, according to Chaproniere (1991) this appears unlikely due to the close similarity between the populations. The previous lack of records of this genus outside of the North Atlantic and Mediterranean regions may well be the result of the extreme delicacy of the very thin discoidal test, which would certainly be readily fragmented during post-mortem movement on the sea floor.

7.6 Epilogue

In modern hydrocarbon exploration, it is important to understand the environment of deposition of any carbonate sequence so as to predict the presence, quality and thickness of reservoir, seal and source rocks. As we have seen throughout this volume, larger benthic foraminiferal associations, along with light-sensitive calcareous algae, are useful as depth and age indicators. Their diversity can be used as an indicator of the locus of facies evolution (see Fig. 7.36), palaeogeography and palaeotemperature.

The study of the systematics of larger foraminifera is an essential tool for biostratigraphy. As seen in the previous chapters, in larger foraminifera growth history is recorded

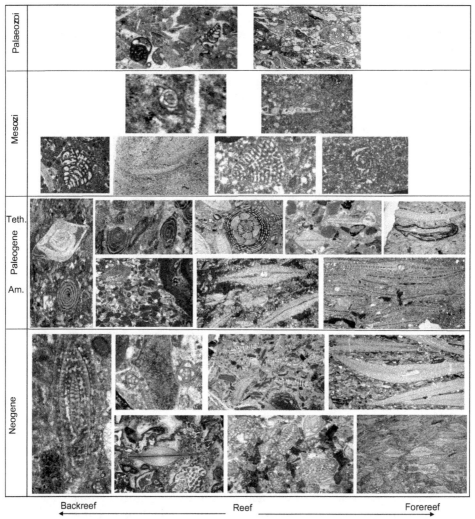

Figure 7.36. Characteristic foraminiferal assemblages in different parts of the reef ecology through time (Teth. = Tethys, Am. = Americas) for the provincialism characteristic of the Paleogene.

and preserved as a part of the test, and their evolution is recorded and defines the geological column. Therefore, well-defined lineages can be identified and traced throughout their history. They seem to follow Cope's rule of a trend towards complexity (see Chapter 1). Primitive forms persist and survive harsh conditions and major extinction events, but also give rise gradually though time and space to more complex specialist forms that flourish in stable environments. Cope's rule has been linked by many authors to being a reflection of "K-selection" and hypermorphosis (Gould, 1977). Larger foraminifera thrive in many environments, and their delayed reproductive strategy leads to a long life span, reaching a hundred years in some large *Nummulites*. This combination of a long life span, and abundance in the geological record make them a most valuable tool for biostratigraphy in shallow water environments. On the other hand, by having test features that continually change with time, larger foraminifera are useful in the study of the genetic and morphological basis of evolution. One important future development would be to combine studies of living forms with the valuable and rich fossil record in order to understand the relationship between genetic characteristics and morphology.

Work on the origin of the foraminifera and on phylogenetic analysis of verified foraminiferal DNA sequences are already underway (Wray et al., 1995; Holzmann et al., 2003). Genetic analyses are essential to understand phylogenetic relationships among larger foraminiferal genera and species. They will help to establish whether the characteristics developed by different forms are analogous (i.e. have the same appearance, but a different origin or function) or homologous (i.e. having the same origin). The tendency, noted in earlier chapters, towards evolutionary convergence is something which requires deeper analysis. Direct genetic comparison between the specimens from the European-Middle Eastern and the Far Eastern areas will also be essential to confirm the correlation recently made between the standard Planktonic Foraminiferal biozones and the geostratigraphic stages of Western Europe, that have enabled refinement of the "Letter Stages" of the Far East and recalibration of them with the European Stages (BouDagher-Fadel, 2013). This is helping to contribute to the understanding of tectonics and the timing of terrane/microplate suturing as reflected by the distribution of present day faunas. For example, the final suturing of Arabia and Eurasia is dated as Late Oligocene and shows up well in the similarity of Miocene land faunas across the suture. The Tethys was therefore finally divided into the Mediterranean and the Indian Ocean at this time. Further investigation of larger foraminifera will show if the foraminifera of the Levant demonstrate increasing divergence from those of the eastern part of Arabia and the Far East, and are shedding light on the Himalayan orogeny (e.g. Najman et al., 2010a, b; Hu et al., 2015; Najman et al., 2016 BouDagher-Fadel et al., 2017).

Culture experiments under controlled environmental conditions such as temperature, salinity, dissolved oxygen content, nutrient content, trace element concentration, and isotopic enrichment (building upon the pioneering work of Röttger) should be encouraged in order to better interpret natural ecological behaviour and environmental tolerances. The driving mechanisms behind larger foraminiferal evolution have been controlled by changes in palaeoenvironments, on a whole range of time scales from those driven by tectonics, through those associated with climate change and finally those affected by catastrophic volcanic or impact processes. Micropalaeontological correlation will clarify the role played by biogeographical restrictions. Controlled experiments on, for example, the tolerance of different forms to water depth could enable estimates of such physical variables as subsidence rates to be made. The combination of depth and time calibration would help resolve some of the great sequence stratigraphy debates as to whether particular transgressions are precisely isochronous worldwide or are due to regional eustatic sea-level rises.

Finally, it is widely held these days that human activity is destroying coral reefs worldwide. As highlighted by Hallock (2005) for example, more research on living foraminifera will give us a greater understanding of contemporary reef ecologies and will help us to understand how reef-dwelling fauna respond to changes to environmental parameters caused, for example, by climate change and/or other human generated processes.

As we have seen in this chapter, there is still much to discover about both the mode of life, evolutionary trends and the migratory habits of Neogene and existent forms. If this is the case for contemporary and geologically recent foraminifera, then it must be the case also for Early Cenozoic, Mesozoic and Palaeozoic forms. Therefore, although an attempt has been made in the preceding chapters to provide an overview of what is known about the fossil larger benthic foraminifera, it must be concluded that the study of these geologically important creatures, although mature, is far from complete, and should be pursued by future generations with vigour, as they are certainly highly versatile, biologically adaptive and flexible, and so still make a challenging subject for research.

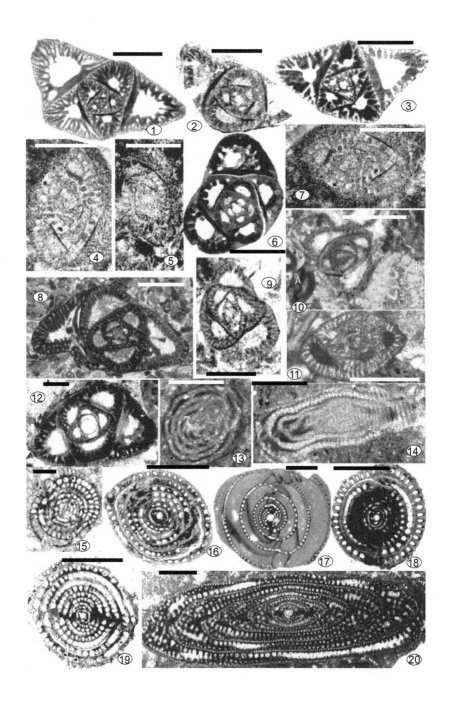

Plate 7.1 Scale bars: Figs 1-5, 7-10, 12-20 = 0.5mm; Figs 6, 11 = 0.25mm. Figs 1-3. *Austrotrillina howchini* (Sclumberger), closely spaced bifurcating alveoli, 1) Early Miocene (lower Tf), Pata Limestone, Australia, B-Form, NHM P47608; 2) Early Miocene, Lower Chake Beds, Pemba, Tanzania, NHM P22848; 3) Early Miocene, Bairnsdale, Chowilla Dam Site, Pata Limestone, Australia, NHM P47587. Figs 4-7. *Austrotrillina striata* Todd and Post, simple coarse alveoli, 4-5, 7) Late Oligocene, Borneo, UCL coll; 6) topotype, Early Miocene, Bikini Island, NHM P47596. Figs 8-9. *Austrotrillina asmariensis* Adams, numerous alveoli, Early Miocene (Aquitanian), Asmari Limestone, Luristan, Iran, NHM P47585. Fig. 10. A) *Austrotrillina howchini* (Sclumberger), B) *Miogypsinoides* sp., Early Miocene (lower Tf1), Loc. 238, Indonesia, UCL coll. Fig. 11. *Austrotrillina brunni* Marie, Early Miocene (Aquitanian), Nias, Indonesia, UCL coll. Fig. 12. *Austrotrillina howchini* (Sclumberger), Early Miocene (Tf1), Hadu Village, Coast Province, Kenya, NHM P43944. Figs 13-14. *Borelis pygmaeus* Hanzawa, Early Miocene (Aquitanian), 13) equatorial section; 14) Loc. 205, Borneo, UCL coll. Fig. 15. *Borelis melo* (Fichtel and Moll), Middle Miocene, Turkey, NHM P49087. Fig. 16. *Borelis haueri* (d'Orbigny), type, Middle Miocene, Baden-Brickyard, Baden, Australia, UCL coll. Fig. 17. *Borelis pulchra* (d'Orbigny), Holocene, Mauritius, UCL coll. Fig. 18. *Borelis melo* (Fitchel and Moll), Middle Miocene (Serravallian), Barden, Australia, NHM coll. Fig. 19. *Borelis curdica* (Reichel), Miocene, Turkey, NHM coll. Fig. 20. *Alveolinella praequoyi* Wonders and Adams, holotype, early Middle Miocene (upper Tf1-Tf2), Darai Limestone, Papua New Guinea, NHM P52658.

Plate 7.2 Scale bars: Figs 1-6, 8-12, 14-16 = 1mm; Figs 7, 13, 18 = 0.5mm; Fig. 17 = 100μm. Figs 1, 5. *Alveolinella* sp., Miocene, Port Moresby Bay, Papua New Guinea, UCL coll., 1) solid specimen; 5) equatorial section. Figs 2-4. *Alveolinella quoyi* (d'Orbigny), Holocene, 2-3) Pacific; 4) Port Moresby Bay, Papua New Guinea, solid specimen, UCL coll. Figs 6-7. *Alveolinella praequoyi* Wonders and Adams, holotype, early Middle Miocene (upper Tf1-Tf2), Darai Limestone, Papua New Guinea, NHM P52659-60. Fig. 8. *Alveolinella fennemai* Checchia-Rispoli, late Early Miocene (Burdigalian), Borneo, UCL coll. Figs 9-11. *Flosculinella bontangensis* (Rutten), 9) late Early Miocene (Burdigalian, lower Tf1), Borneo; 10-11) Middle Miocene (middle and upper Tf1), East Java, UCL coll. Fig. 12. *Flosculinella globulosa* (Rutten), Early Miocene (lowermost Burdigalian), Indonesia, UCL coll. Fig. 13. *Dendritina rangi* d'Orbigny, Early Miocene (Aquitanian), Kirkuk Well no.22, Iraq, NHM P39432. Fig. 14. *Archaias angulatus* (Fichtel and Moll), Holocene, offshore Belize, UCL coll. Fig. 15. *Cyclorbiculina* sp., Holocene, off Blackadare Bay, British Honduras, UCL coll. Figs 16-17. *Marginopora vertebralis* Blainville, SEM pictures; 17) enlargement of an equatorial chamber of the fig. 16, Papua New Guinea, UCL coll. Fig. 18. *Orbiculina* sp., Early Miocene (Burdigalian), Upper Chake Beds (Datum Stratum), Wesha Road, Pemba, Tanzania, NHM P22853.

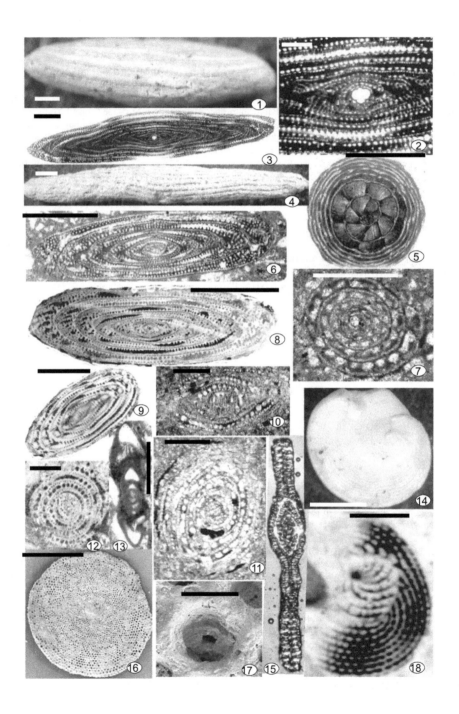

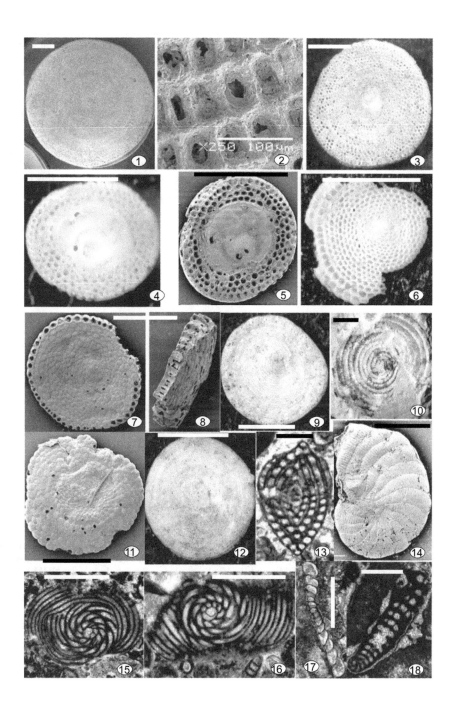

Plate 7.3   Scale bars: Figs 1-6, 9-12 = 1mm; Figs 7-8, 13, 15-18 = 0.5mm; Fig. 14 = 0.25mm. Figs 1-5. *Marginopora* sp., Holocene, Papua New Guinea, on seagrass, 1-2) SEM photograph; 2) enlargement of chambers showing rectangular chamberlets, UCL coll. Figs 6-8. *Sorites* sp., Holocene, Byblos, Lebanon, UCL coll. Fig. 9 *Archaias* sp. Holocene, Funafuti sandy beach, Tuvalu, South Pacific, UCL coll. Fig. 10. *Discospirina italica* (Costa), Late Miocene, Pakhna Formation, Anaphotia, Cyprus, UCL coll. Fig. 11. *Cyclorbiculina compressa* d'Orbigny, Holocene, off Belize, UCL coll. Fig. 12. *Cyclorbiculina* sp. Funafuti Sandy Beach, Tuvalu, South Pacific, UCL coll. Fig. 13. *Archaias kirkukensis* Henson, Kirkuk Well no. 57 Iraq, NHM P39633. Fig. 14. *Peneroplis* sp., Holocene, offshore Byblos, Lebanon, UCL coll. Figs 15-16. *Peneroplis thomasi* Henson, paratypes, Early Miocene (Aquitanian), Kirkuk Well no.57, Iraq, NHM P39634. Fig. 17. *Planorbulinella larvata* (Parker and Jones), Early Miocene, Java, UCL coll. Fig. 18. *Praerhapydionina delicata* Henson, common in both Aquitanian and Burdigalian assemblages of the Bahamas (Fischer coll., University of Geneva).

Plate 7.4   Scale bars: Figs 1-5; 7-12 = 1mm; Figs 6, 13-15 = 0.5mm. Figs 1-6, 12. *Pseudotaberina malabarica* (Carter), 1-3) Early Miocene, Malabar coast, India's southwestern coast, NHM P29875, 1) axial section, 2) equatorial section, 3) enlargement of fig. 2; 4-6) Early Miocene (Burdigalian), Darai Limestone, Papua New Guinea, UCL coll; 12) Early Miocene (Burdigalian), Java, UCL coll. Figs 7-8. *Alanlordia banyakensis* Banner and Samuel, paratypes, Middle Miocene (middle Serravallian), Banyak, Indonesia, NHM P52877-79. Figs 9-11. *Alanlordia niasensis* Banner and Samuel, topotypes, Middle Miocene (middle Serravallian), Nias, Indonesia, UCL coll., 9) with *Planostegina* sp. Figs 13-14. *Sphaerogypsina globulus* (Reuss), Middle Miocene (Serravallian), Cina Village, Indonesia, UCL coll. Fig. 15. *Sphaerogypsina* sp., Early Miocene (Burdigalian), Darai Limestone, Australia, this species differs from *S. globulus* by having thicker walls, more distinctly perforate chambers, not arranged in radial rows but alternating in annuli, and from *Discogypsina vesicularis* (Parker and Jones) but sub-spherical growth pattern.

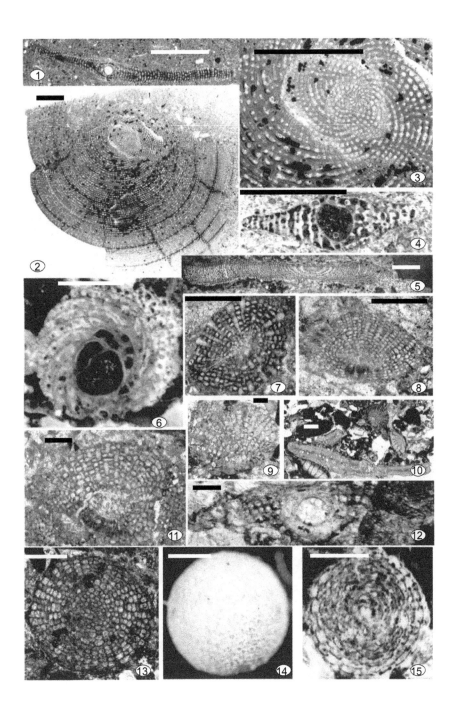

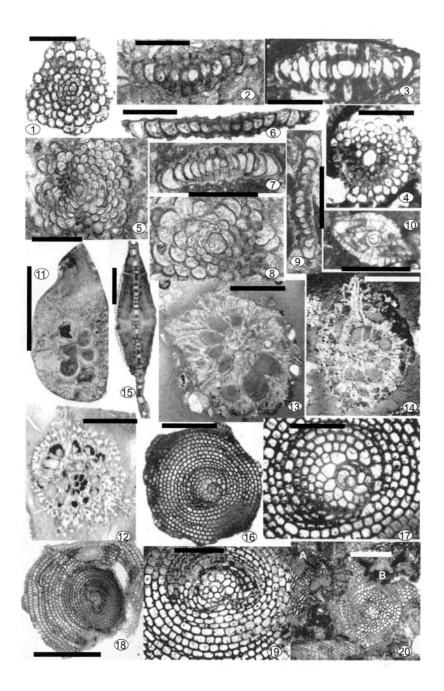

Plate 7.5 Scale bars: Figs 1-14 = 0.5mm; Figs 15-16, 18, 20 = 1mm; Figs. 17, 19 = 0.3mm. Fig. 1. *Planorbulinella kinabatangenensis* Renema, holotype, Early Miocene (Te1-4) Batu Temenggong Besar, Lower Kinabatangan River, North Borneo, NHM NB9023a. Figs 2-5. *Planorbulinella solida* Belford, Early Miocene (Tf), Nias, Indonesia, UCL coll. Fig. 6. *Planorbulinella batangenensis adamsi* Renema, Early Miocene, from loose block about 17 feet above Lower Kinabatangah River, Batu Temenggong Besar, North Borneo, NB9030a, Adams coll., NHM P67225. Figs 7-9. *Planorbulinella larvata* (Parker and Jones), Early Miocene, Java, UCL coll. Fig. 10. *Elphidium/Cellanthus craticulatus,* Jabal Terbol, Lebanon, UCL coll. Figs 11-14. *Calcarina* sp., 11) Miocene, Borneo; 12-14) Pliocene, Philippines, showing canaliferous spines, UCL coll. Figs 15-19. *Cycloclypeus eidae* Tan Sin Hok, Lower Miocene, Kinabatangan River, Sabah, North Borneo, 15) axial section, NHM coll., NB9066; 16-17) NHM P49518, 17) enlargement of early part of test of fig. 16; 18-19) NHM NB9067,19) enlargement of early part of test of fig. 18. Fig. 20. Thin section photomicrograph of A) *L. (Nephrolepidina) ferreroi* Provale, B) *Cycloclypeus carpenteri* Brady, Middle Miocene(Langhian), Nias, Sumatra, UCL coll.

Plate 7.6 Scale bars: Figs 1-4 = 1mm; Figs 5-7, 9-13 = 0.5mm; Fig. 8 = 0.3mm; Fig. 14 = 0.2mm. Figs 1-4. *Cycloclypeus carpenteri* Brady, 1-3) offshore Borneo, UCL coll; 4) Miocene, Funafuti, Tuvalu, UCL coll. Fig 5. *Cycloclypeus* sp., Dutch New Guinea, NHM P22790. Figs 6, 8, 11. *Cycloclypeus indopacificus* Tan Sin Hok, Early Miocene (Burdigalian, lower Tf1), East Borneo, UCL coll. Fig. 9. *Lepidocyclina murrayana, Cycloclypeus* cf. *orbitoitades* Schmutz, Miocene, Dutch New Guinea, NHM P22791. Figs 7, 10. *Cycloclypeus pillaria* BouDagher-Fadel, 7) Middle Miocene (Serravallian, Tf2), Bulu Mampote, Indonesia, NHM 67125; 10) Early Miocene (Burdigalian, upper Te), Loc. 620/14, Borneo, UCL coll. Figs 12-13. *Katacycloclypeus annulatus* (Martin), 12) Miocene (Serravallian, Tf2), Futuna Limestone, Vanua Mbalavu Island, Fiji, NHM P50298; 13) Middle Miocene, Java, NHM P36377. Fig. 14. *Amphistegina* sp., Holocene, offshore Lebanon, SEM photograph, enlargement of the apertural face showing pustules surrounding the aperture, UCL coll.

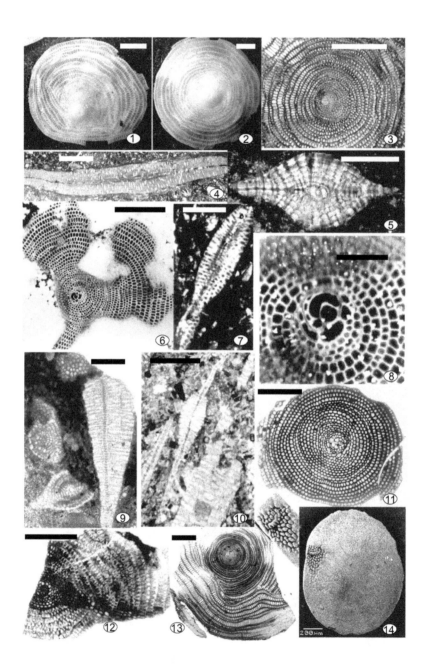

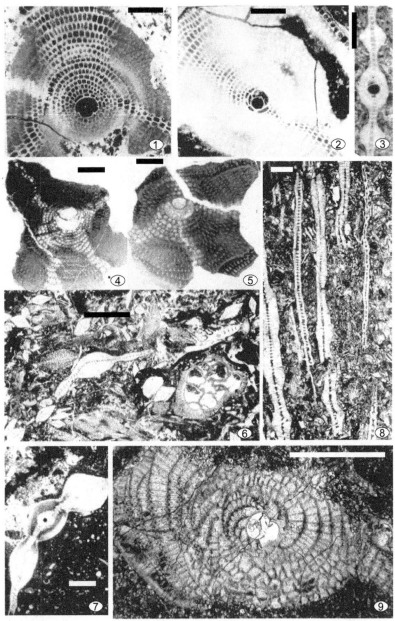

Plate 7.7 Scale bars: Figs 1-9 = 1mm. Figs 1-5. *Katacycloclypeus annulatus* (Martin), 1-3) Miocene (Serravallian, Tf2), Darai Limestone, Papua New Guinea; 4-5) late Middle Miocene Fiji, NHM P50462-3. Figs 6-7. *Katacycloclypeus martini* (van der Vlerk), 6) Miocene (Serravallian, Tf2), Sumatra; 7) Miocene (Serravallian Tf2), East Java, UCL coll. Fig. 8. *Katacycloclypeus annulatus* (Martin), *Miogypsina* sp., *Amphistegina* sp., *Victoriella* sp., Nias, Sumatra. Fig. 9. *Cycloclypeus indopacificus* Tan Sin Hok, Dutch New Guinea, NHM P22785.

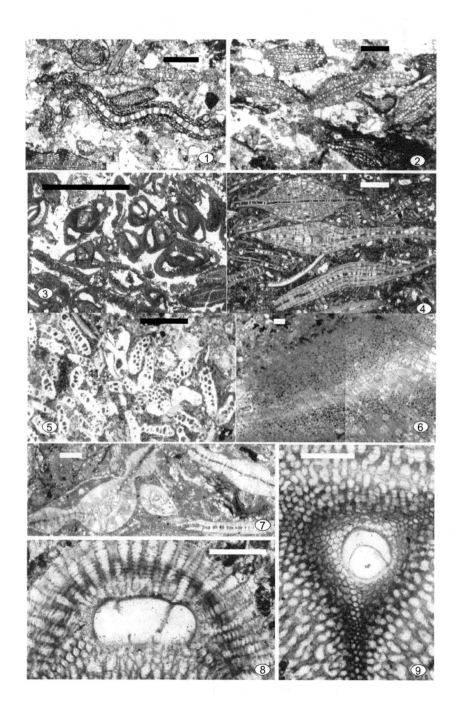

Plate 7.8 Scale bars: Fig. 1-4, 7-9 = 1mm; Figs 5-6 = 0.5mm. Fig. 1. *L.* (*Nephrolepidina*) *ferreroi* Provale, *Miogypsina tani* Drooger, Miocene (early Langhian, middle Tf1), Nias, Sumatra, UCL coll. Fig. 2. *Miogypsina* sp., Miocene (early Langhian, middle Tf1), Nias, Sumatra, UCL coll. Fig. 3. *Austrotrillina howchini* (Sclumberger), *Borelis* sp., *Sorites* sp., Miocene (Tf1), Coast Province, Kenya, NHM P43946. Fig. 4. *Spiroclypeus* sp., Oligocene, East Java, UCL coll. Fig. 5. A packstone of *Miogypsinella ubaghsi* (Tan Sin Hok), Miocene (early Aquitanian, early Te5), Darai Limestone, Papua New Guinea, UCL coll. Fig. 6. *Lepidocyclina gigantea* (Martin), Miocene (Serravallian, Tf3), Darai Limestone, Papua New Guinea, UCL coll. Fig. 7. *Katacycloclypeus annulatus* (Martin), *Lepidocyclina* sp., Miocene (Serravallian, Tf2), Borneo, UCL coll. Figs 8-9. *Eulepidina ephippioides* (Jones and Chapman), and *Heterostgina (Vlerkina) borneensis* van der Vlerk, Chattian, Mosque Quarry, Borneo, UCL coll.

Plate 7.9 Scale bars: Figs 1-4, 6, 11-15, 17 = 1mm; Figs 5, 7-10, 16 = 0.5mm. Figs 1-2. *Borodinia* sp., Early Miocene (Aquitanian), Java, UCL coll. Fig. 3. *Gypsina* sp., *Carpenteria* sp., algae, Early Miocene (Burdigalian), Loc. 251, Borneo, UCL coll. Figs 4-5. *Homotrema rubrum* (Lamarck), Holocene, Indo-Pacific, exact locality unknown, 4) vertical section from a specimen attached to *Acropora millepora* specimens, NHM coll; 5) John Murray exp.1956, NHM coll. Figs 6-7. *Carpenteria proteiformis* Goës, Holocene, off Barbados, 100 fathoms, NHM coll., 6) vertical section; 7) horizontal section of tip of right hand chamber. Figs 8-9. *Rupertina stabilis* Wallich, Holocene, cold area Faroe Channel: 8) vertical section; 9) basal horizontal section. Figs 10-13. *Miniacina miniacea* (Pallas), Holocene, 10) Tizard Bank, China Sea, Heron-Allen-Earland coll.; 11) John Murray expedition, Station 178, 91m, NHM coll; 12-13) Early Miocene, Java, UCL coll. Fig. 14. *Sporadotrema cylindricum* (Carter), Maldives Islands, Murray expedition, Station 149, NHM coll. Fig. 15. Thin section photomicrograph of A) *Lepidosemicyclina* sp., B) *Carpenteria* sp., Early Miocene (Burdigalian), Borneo, UCL coll. Fig. 16. Thin section photomicrograph of A) *L.* (*Nephrolepidina*) *sumatrensis* Brady, B) *Victoriella* sp., Early Miocene (Burdigalian), East Java, UCL coll. Fig.17. *Sporadotrema mesentericum* (Carter), Holocene, Maldives Islands, John Murray expedition, Station112, 113m, NHM coll.

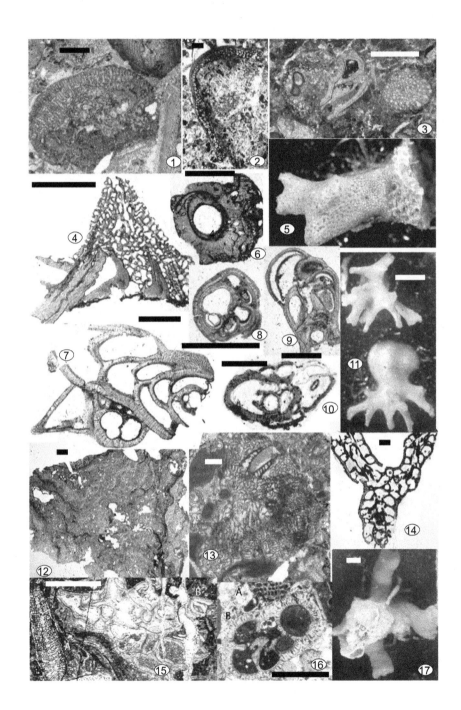

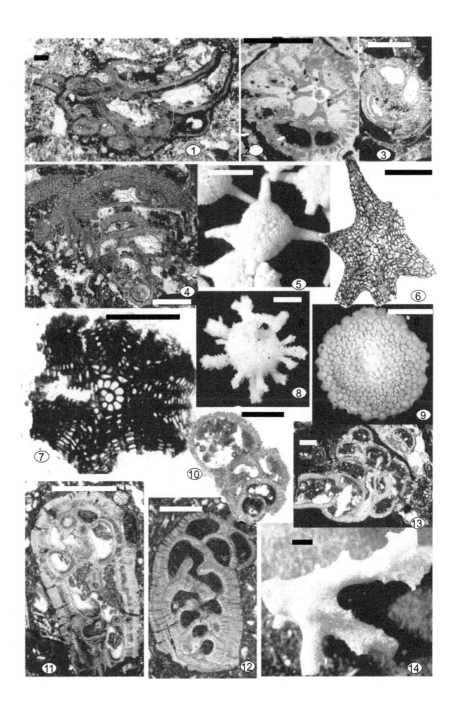

Plate 7.10 Scale bars: Figs 1, 5-8, 11, 13-14 = 1mm; Figs 2-4, 9-10, 12 = 0.5mm. Figs 1-4. *Carpenteria* sp., 1) Early Miocene, East Java; 2) Early Miocene (Burdigalian), Borneo; 3) Early Miocene (Aquitanian), Borneo; 4) Middle Miocene, UCL coll. Fig.5. *Baculogypsina* sp., Holocene, forereef, Coral Sea, Papua New Guinea, UCL coll. Fig. 6. *Baculogypsina baculatus* (Montfort), Holocene, New Zealand, NHM coll. P83. Fig. 7. *Baculogypsina sphaerulata* Parker and Jones, paralectotype, Holocene, Ex. ZF 3598, Rewa Reef, Fiji, Parker coll. Fig. 8. *Calcarina hispida* Brady var. *pulchra* Chapman, Holocene, Funafuti Atoll, Ellis Islands, UCL coll. Fig. 9. *Planorbulinella larvata* (Parker and Jones), Holocene, John Murray expedition, Station103, 101m, NHM coll. Figs 10-12. *Victoriella* sp., 10) Childs Bank Burdigalian, Ka1, South Africa, McMillan coll. 11) Early Miocene (Burdigalian), Borneo, UCL coll.; 12) Java, UCL coll. Fig. 13. *Biarritzina* sp., Early Miocene, Brazil, UCL coll. Fig. 14. *Miniacina miniacea* (Pallas), Holocene, John Murray expedition, Station 178, 91m, NHM coll.

Plate 7.11 Scale bars: Figs 1-15 = 0.5mm; Fig. 16 = 0.25mm. Fig. 1. *Miogypsinodella corsicana* Ferrandini et al., paratype, figured by Ferrandini et al. (2010b), Miocene (Burdigalian), Bonifacio, Corsica, France. Figs 2-4. *Miogypsinodella primitiva* BouDagher-Fadel and Lord, Miocene (Burdigalian), Nias, Sumatra, UCL coll. Fig. 5. *Miogypsina antillea* (Cushman), Early Miocene, Brazil, UCL coll. Fig. 6. *Miogypsina southernia* BouDagher-Fadel and Price, paratype figured by BouDagher-Fadel and Price (2013), Burdigalian, Childs Bank, Ka1, 360m, South Africa, McMillan UCL coll. MF275. Fig. 7. *Miogypsina mcmillania* BouDagher-Fadel and Price, paratype figured by BouDagher-Fadel and Price (2013), Burdigalian, Childs Bank, Ka1, 400m, South Africa, McMillan coll. UCL MF243. Fig. 8. *Miogypsina africana* BouDagher-Fadel and Price, paratype figured by BouDagher-Fadel and Price (2013), Burdigalian, Childs Bank, Ka1, 36m, South Africa, McMillan UCL coll. MF250. Fig. 9. *Miogypsina samuelia* BouDagher-Fadel and Price, holotype figured by BouDagher-Fadel and Price (2013), Nias, Indonesia. Fig. 10. *Miogypsina niasiensis* BouDagher-Fadel and Price, holotype figured by BouDagher-Fadel and Price (2013), Miocene (Early Langhian, Tf1), Nias, Indonesia, McMillan UCL coll. MF205. Fig. 11. *Miogypsina cushmani* Vaughan, Miocene (latest Burdigalian), Castelsardo section (northern Sardinia), equivalent of Cala di labra Formation, Ferrandini UCL coll. MF259–60. Fig. 12. *Lepidosemicyclina thecideaeformis* Rutten, Miocene (Burdigalian), Upper Chake Beds, Pemba, Tanzania, NHM P22849. Fig. 13. *Miogypsina sabahensis* BouDagher-Fadel, Lord and Banner, holotype, Miocne (early Burdigalian), North East Borneo, NHM P66907. Fig. 14. *Miolepidocyclina mexicana* (Nuttall), Miocene (Aquitanian), Brazil, UCL coll. Fig. 15. *Miolepidocyclina excentrica* Tan Sin Hok, Cala di Labra Formation, Cala di Ciappili section (Bonifacio area), Ferrandini coll. Fig. 16. *Miogypsina indonesiensis* Tan Sin Hok, Middle Miocene (Tf3), Yule Island Region, SW Papua, UCL coll.

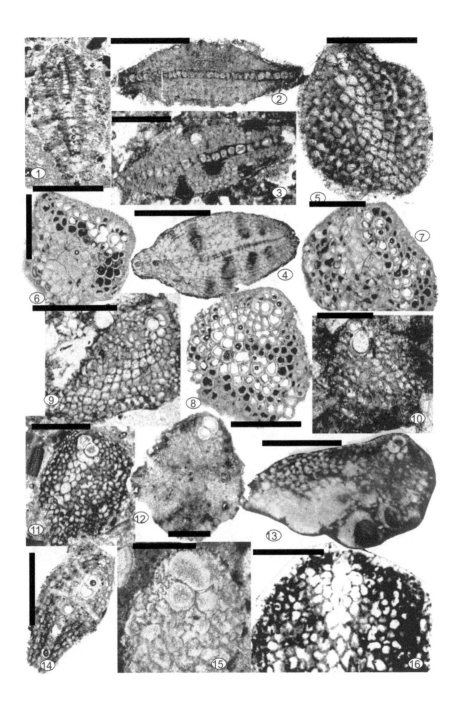

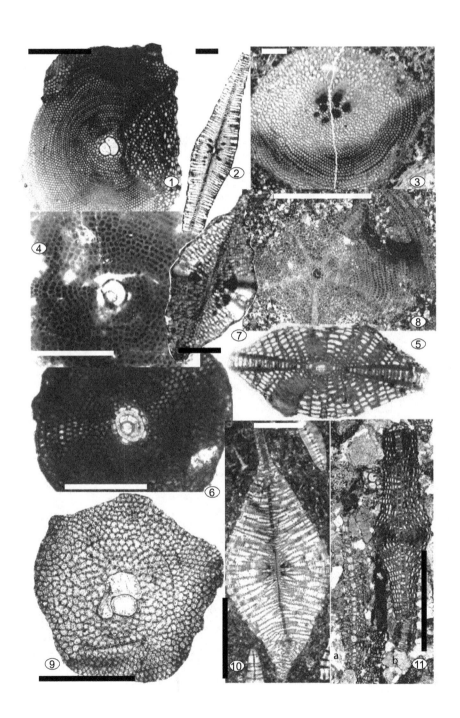

Plate 7.12 Scale bars: Figs 1-11 = 1mm. Fig. 1. *Lepidocyclina (Nephrolepidina) morgani* Lemoine and Douvillé, Miocene (Aquitanian), Nerthe, near Marseille Petit Nid, France, Ferrandini UCL coll. Figs 2-3. *Lepidocyclina delicata* Scheffen, Miocene (Serravallian, Tf2), Darai Limestone, Papua New Guinea, UCL coll. Figs 4-5. *Lepidocyclina angulosa quadrata* Provale, Miocene (Serravallian), Darai Limestone, Papua New Guinea, UCL coll. Fig. 6. *Lepidocyclina (Nephrolepidina) angulosa* Provale, NCS-3, Nam Con Son basin, Vietnam, Vova's coll. Fig. 7. *Lepidocyclina banneri* BouDagher-Fadel, Noad and Lord, holotype, Miocene (Aquitanian), Borneo, NHM P66905. Fig. 8. *Nephrolepdina radiata* (Martin), Miocene (Serravallian), Darai Limestone, Papua New Guinea, UCL coll. Fig. 9. *Lepidocyclina (Nephrolepidina) suwanneensis* Cole, Miocene (Aquitanian), Retrench member, Cipero Formation, Trinidad, UCL coll. Fig. 10. *Lepidocyclina (Nephrolepidina) sondaica* Yabe and Hanzawa, Miocene (Aquitanian), Loc. 205, Borneo, UCL coll. Fig. 11. a) *Lepidocyclina (Nephrolepidina) brouweri* Rutten, b) *Lepidosemicyclina* sp., Miocene (Burdigalian, early Tf1), Borneo, UCL coll.

Plate 7.13 Scale bars: Figs 1-21 = 0.5mm. Figs 1-3. *Miogypsinoides dehaarti* (van der Vlerk), 1-2) Miocene (Burdigalian), Waterfall Section, Christmas Island, Indian Ocean, NHM 6764 1295; 3) Early Miocene (Aquitanian), Sulawesi, UCL MF167. Figs 4-6. *Miogypsina gunteri* Cole, 4) Miocene (Aquitanian), top of Superga Mountain, Aman, about 2km S.E. of Superga, NHM coll.; 5) Miocene (Aquitanian), Superga Mountain, Aman, about 2 km SE of Superga, UCL coll. MF257; 6) *Miogypsina globulina* (Michellotti), Miocene (early Langhian), Cala di Labra Formation, Cala di Ciappili section (Bonifacio area), Corsica, France, Ferrandini UCL coll. Fig. 7. *Miogypsina orientalis* BouDagher-Fadel et al., Miocene (late Burdigalian), Kalimantan, East Borneo, UCL coll. Fig. 8. *Miogypsina kotoi* Hanzawa, Miocene (Burdigalian), Darai Limestone, Papua New Guinea, UCL coll. Figs 9-13. *Miogypsina borneensis* Tan Sin Hok, 9-10) Miocene (Aquitanian, upper Te), Darai Limestone, Papua New Guinea; SEM photographs of a microspheric specimen half dissected showing embryonic apparatus and equatorial chambers; 10) enlargement of an equatorial chamber of (9) showing modifications (called egg-holders by Hottinger, 2006) to house symbionts; 11-13) Miocene (early Burdigalian, lower Tf1), UCL coll. Fig. 14. *Miogypsina bifida* Rutten, Miocene, Sungai, Boengaloen, East Borneo, NHM. Van Vessem coll. BB 469 – 1913. Fig. 15. *Miogypsina* sp., Miocene (Burdigalian), Morocco, UCL coll. Senn27778a. Fig. 16. *Miogypsina cushmani* Vaughan, Miocene (latest Burdigalian) (note in Chart 7.1, this form appears later in time in the Far East), Castlesardo section, equivalent of Cala di Labra formation, Northern Sardinia, Ferrandini UCL coll. Fig. 18. *Miogypsina indonesiensis* Tan Sin Hok, Miocene (Serravallian), Nias, UCL MF204. *Miogypsina subiensis* BouDagher-Fadel and Price, paratype figured by BouDagher-Fadel and Price (2013), Miocene (Burdigalian), Subis Formation, Borneo. Fig. 21. *Miolepidocyclina excentrica* Tan Sin Hok, Cala di Labra Formation, Cala di Ciappili section, Bonifacio area, southern Corsica, France, Ferrandini UCL coll.

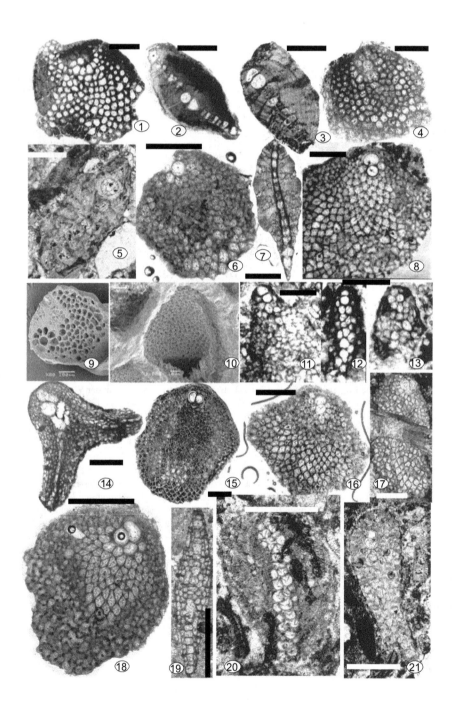

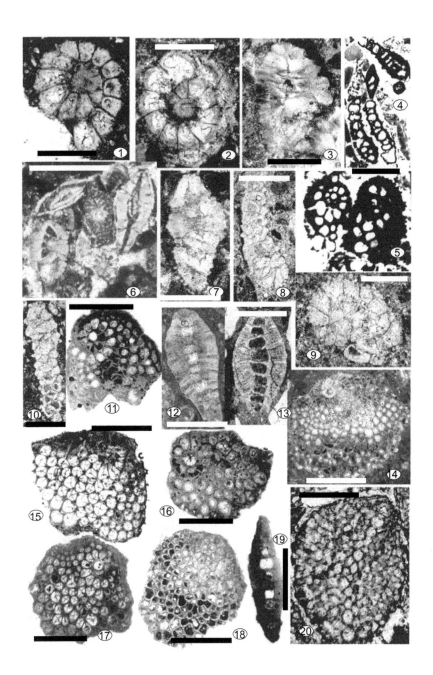

Plate 7.14 Scale bars: Figs 1-20 = 0.5mm. Figs 1-2. *Neorotalia tethyana* BouDagher-Fadel and Price, Miocene (Aquitanian), Loc. 621/3, Borneo. Fig. 3. *Miogypsinella borodinensis* Hanzawa, Late Oligocene, Loc. 204, Borneo, UCL coll. Figs 4-5. *Miogypsinella ubaghsi* (Tan Sin Hok), Miocene (early Aquitanian, Te5), Darai Limestone, Papua New Guinea, UCL coll. Fig. 6. *Boninella* sp., *Operculina* sp., *Lepidocyclina* (*L.*) *isolepidinoides* van der Vlerk, Late Oligocene, Loc. 239, Borneo, UCL coll. Fig. 7. *Paleomiogypsina* sp., Late Oligocene (Chattian), Loc. 621/3, Borneo, UCL coll. Figs 8-10. *Miogypsinella bornea* BouDagher-Fadel and Price, 8) Early Miocene (Aquitanian), Borneo, UCL coll.; 9-10) figured by BouDagher-Fadel and Price (2013) from Oligocene to Miocene (early Chattian–early Aquitanian), Java, 9) holotype; 10) paratype. Fig. 11. *Miogypsinoides formosensis* Yabe and Hanzawa, Miocene (early Aquitanian), Nerthe area, near Marseille, Petit Nid section, Formation pararecifale du Cap de Nautes, France, Ferrandini UCL coll. MF252. Figs 12-14. *Miogypsinoides dehaarti* (van der Vlerk), Miocene (Burdigalian), 12, 14) Cyprus; 13), Nias, Sumatra, 201 AS, UCL coll. Figs 15-17. *Miogypsinoides bantamensis* Tan Sin Hok, Miocene (earliest Burdigalian), 15,17) figured by Ferrandini et al. (2010) and BouDagher-Fadel and Price (2013) from Castlesardo section, equivalent of Cala di Labra formation, Northern Sardinia, Ferrandini UCL/Ferrandini coll. MF253; 16) figured by Ferrandini et al (2010), from the Nerthe area, near Marseille, Petit Nid section, Formation pararécifale du Cap de Nautes, France, UCL/Ferrandini coll. Fig. 18. *Miogypsina mcmillania* BouDagher-Fadel and Price, paratype, figured by BouDagher-Fadel and Price (2013), Miocene (Burdigalian), Childs Bank, Ka1, 400m, South Africa, UCL MKF240. Fig. 19. *Miogypsinoides mauretanica* Bronnimann, figured by Ferrandini et al. (2010) from, Castlesardo section, Northern Sardinia. Fig. 20. *Miogypsina triangulata* BouDagher-Fadel and Price, paratype, figured by BouDagher-Fadel and Price (2010b), Early Miocene (N4), offshore Brazil, UCL coll.

Plate 7.15 Scale bars: Figs 1, 4-6, 8-14 = 1mm; Figs 2-3, 7, 15 = 0.5mm. Fig. 1. *Operculinoides panamensis* (Cushman), Miocene (Aquitanian), Brazil, UCL coll. Fig. 2. *Operculina* sp., Early Miocene, Sentosa, UCL coll. Fig. 3. *Operculinoides antiguensis* Vaughan and Cole, Early Miocene, Juana Diaz Formation, La Rambla, Puerto Rico, NHM 47351-2. Figs 4-5. *Heterostegina* sp., Miocene (Te 1-4), Batu Temongong, Besar, Lower Kinabatangan River, North Borneo, 4) NHM NB9023; 5) NHM NB9020. Fig. 6. *Heterostegina* (*Vlerkina*) *borneensis* van der Vlerk, *Heterostegina* (*Vlerkina*) sp., Early Miocene, Loc. 205, Borneo, UCL coll. Fig. 7. *Tansinhokella yabei* (van der Vlerk), Miocene, S. Soembal, E. Borneo, NHM P45042. Fig. 8. *L.* (*Nephrolepidina*) *nephrolepidinoides* BouDagher-Fadel and Lord, *Heterostegina* (*Vlerkina*) *borneensis* van der Vlerk, Early Miocene, Loc. 239, Borneo, UCL coll. Figs 9-12. Thin section photomicrograph of *Spiroclypeus* sp., *L.* (*Nephrolepidina*) sp., *L.* (*Nephrolepidina*) *brouweri* Rutten, Miocene (early Burdigalian, Te5), Indonesia, UCL coll. Fig. 13. *Spiroclypeus tidoenganensis* van der Vlerk, Miocene, S. Patoeng (Antjam) E. Borneo, NHM P45039. Figs 14-15. *Discogypsina discus* (Goës), Miocene, exposure 4, Kinabatangan River, North Borneo, NHM N.B. 9050.

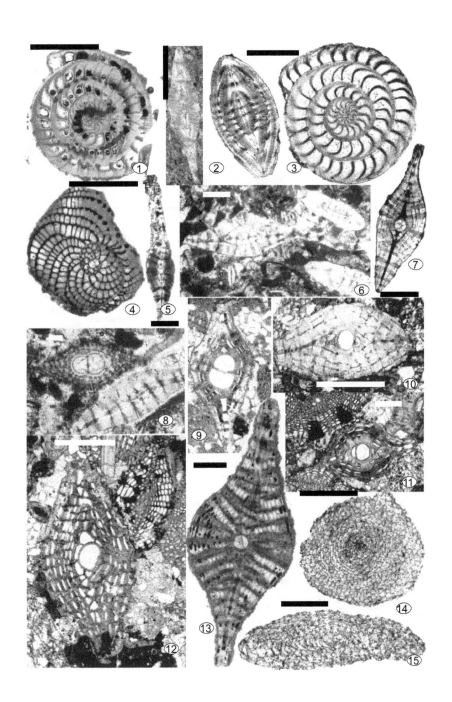

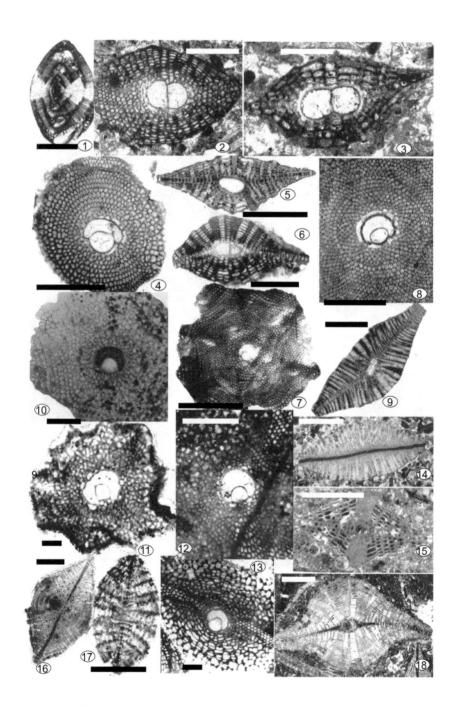

Plate 7.16   Scale bars: Figs 1-5, 15, 17 = 0.5mm; Figs 6-14, 16, 18 = 1mm. Fig. 1. *Amphistegina lessonii* d'Orbigny, Holocene, Mauritius, NHM coll. Figs 2-5. *L. (Nephrolepidina)* sp. of an amphilepidine type, Miocene (late Aquitanian or Burdigalian), Lower Chake Beds, Pemba, Tanzania 2) NHM P22846; 3) NHM P22844; 4) NHM P22483; 5) NHM 22845. Figs 6-7. *Lepidocyclina (Nephrolepidina) batesfordensis* Crespin, topotypes of tribliolepidine embryo, Miocene, Lepidocyclina Limestone, Upper Quarry, Batesford, Victoria, Australia, NHM P36060. Figs 8-9. *Lepidocyclina verbeeki* (Newton and Holland), Early Miocene, Nias, Sumatra, NHM P45073. Figs 10-11. *Lepidocyclina (Nephrolepidina) martini* (Schlumberger), Miocene, Yule Island, Papua New Guinea, UCL coll. Figs 12-13. *Lepidocyclina (Nephrolepidina) rutteni quadrata* van der Vlerk, showing a quadrate proloculus, Miocene (Serravallian, Tf2), Darai Limestone, Papua New Guinea, UCL coll. Fig. 14. *Lepidocyclina murrayana* Jones and Chapman, lectotype, Miocene, Christmas Island, NHM P22441. Fig. 15. *L. (Nephrolepidina) inflata* Provale, Miocene (Burdigalian, Tf1), Papua New Guinea, UCL coll. Fig. 16. *Lepidocyclina acuta* (Rutten), Miocene (late Burdigalian), Papua New Guinea, UCL coll. Fig. 17. *L. (Nephrolepidina)* aff. *epigona* Schuber, Miocene (Tf3), Darai Limestone, Papua New Guinea, UCL coll. Fig. 18. *L. (Nephrolepidina) soebandii* van der Vlerk, Miocene (Burdigalian, upper Te), Loc. 205, Borneo, UCL coll.

Plate 7.17   Scale bars: Figs 1-2, 8, 10, 14, 16 = 0.5mm; Fig. 3 = 0.02mm; Figs 4-7, 9, 11-13, 15 = 1mm. Fig. 1. *Lepidocyclina* sp., Miocene (Serravallian, Tf2), Darai Limestone, Papua New Guinea, UCL coll. Figs 2-3. *L. (Nephrolepidina)* sp., 3) enlargement of fig. 2, Early Miocene, Java, UCL coll. Fig. 4. *Eulepidina undosa laramblaensis* Eames et al., holotype, Early Miocene, Juana Diaz formation, La Rambla, Puerto Rico, NHM P47333. Fig. 5. *Lepidocyclina (Nephrolepidina) aquitaniae* Silvestri, Miocene (Messinian), Spain, UCL coll. Fig. 6. *Lepidocyclina canellei* Lemoine and Douville, Oligocene, base of Antigua Limestone, Antiga, NHM P28629. Fig. 7. *Lepidocyclina isolepidinoides* van der Vlerk, Late Oligocene (lower Te), Borneo, UCL coll. Figs 8-9. *L. (Nephrolepidina) bikiniensis pumilipapilla* Cole, Late Oligocene-Early Miocene, Juana Diaz Formation, La Rambla Puerto Rico, NHM P47346,9. Fig. 10. *L. (Nephrolepidina) rutteni* van der Vlerk, Miocene (Serravallian), Borneo, UCL coll. Figs 11-12. *Lepidocyclina mantelli* (Morton), Early Oligocene, Marianna Limestone, Little Stave Creek, Alabama, 11) axial section, NHM P47331; 12) equatorial section, NHM P47326. Fig. 13. *Lepidocyclina omphalus* Tan Sin Hok, Miocene (Serravallian, Tf2), Papua New Guinea, UCL coll. Fig. 14. Thin section photomicrograph of Rodophyte sp., *L. (Nephrolepidina) brouweri, Planorbulinella* sp., Miocene (Tf1), Sumatra, UCL coll. Fig. 15. *L. (Nephrolepidina) ferreroi* Provale, Miocene (Burdigalian), sample 3813 from mollusc shell Limestone at Loa Duri (W. Outcrop) 100m from River East Kalimantan, Borneo, UCL coll. Fig. 16. *L. (Nephrolepidina) ferreroi* Provale, Miocene (Langhian), Sumatra, UCL coll.

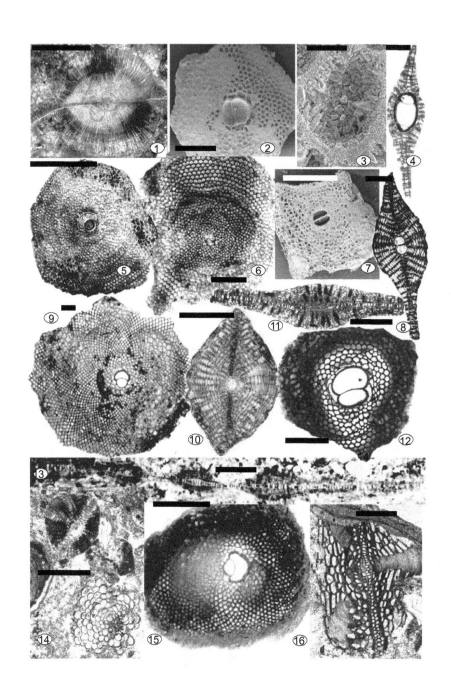

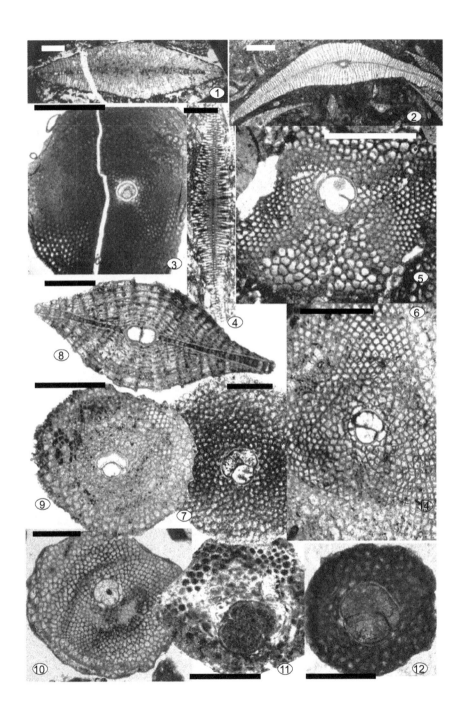

Plate 7.18 Scale bars: Figs 1-5, 9 = 1mm; Figs 6, 7-8, 10-12 = 0.5mm. Fig. 1. *Lepidocyclina ngampelensis* Gerth, Miocene (Serravallian, Tf2), Darai Limestone, Papua New Guinea, UCL coll. Fig. 2. *Lepidocyclina murrayana* Jones and Chapman, Early Miocene, Dutch New Guinea, NHM P22786. Fig. 3. *Lepidocyclina* (*Nephrolepidina*) *angulosa* Provale, Nam Con Son basin, Vietnam, Vova's coll. NCS-3. Fig. 4. *Lepidocyclina* aff. *volucris* Scheffen, Middle Miocene (Tf2), Darai Limestone, Papua New Guinea, UCL coll. Figs 5-7. *Lepidocyclina* (*Nephrolepidina*) *marginata* (Michelotti) = *Lepidocyclina* (*Nephrolepidina*) *tournoueri* Lemoine and Douvillé, 5) Oligocene, Cyprus, UCL coll.; 6) Zakas section, Greece, NHM P51963; 7) Oligocene (Rupelian, P20), Iran, Rahaghi NHM coll. Figs 8-9. *Lepidocyclina* (*Nephrolepidina*) *bikiniensis* Cole, Juana Diaz formation, La Rambla, Puerto Rico, UCL coll. Fig. 10. *Lepidocyclina* (*Nephrolepidina*) *braziliana* BouDagher-Fadel and Price, holotype figured by BouDagher-Fadel and Price, (2010a) from the early Oligocene (Rupelian, P18–P19), offshore Brazil, UCL 1MF. Fig. 11. *Lepidocyclina* (*Lepidocyclina*) *boetonensis* van der Vlerk, showing an isolepidine embryo, Middle Oligocene, Papua New Guinea, UCL coll. Fig. 12. *Lepidocyclina* (*Nephrolepidina*) *sumatrensis* Brady, Nam Con Son basin, Vietnam, Vova's coll. NCS-3.

Plate 7.19 Scale bars: Figs 1-10 =1mm. Figs 1-3. *Eulepidina badjirraensis* Crespin, Early Miocene (Burdigalian, Te5), Mosque Quarry, Loc. 205, Borneo, UCL coll. Fig. 4. *Eulepidina inaequalis* Jones and Chapman, Indian Ocean, NHM P22539. Fig. 5. *Eulepidina* sp., Oligocene, Falling Waters Sink, Florida, NHM coll. Fig. 6. *Eulepidina formosides* (Douville), Oligocene, Spain, NHM P38679. Fig. 7. *Eulepidina andrewsiana* (Nuttall), Christmas Island, Indian Ocean, NHM P22533. Fig. 8. *Eulepidina dilatata* Michelotti, Oligocene (late Rupelian), Mesohellenic Basin, Greece, SFN coll. Fig. 9. *Lepidocyclina* (*Nephrolepidina*) *brouweri* Rutten, Miocene (early Burdigalian, Te5), Borneo, Loc. 205, UCL coll.. Fig. 10. *Lepidocyclina* (*Nephrolepidina*) sp. aff. *gibbosa*, Early Miocene (early Burdigalian, Te5), Borneo, Loc. 205, UCL coll.

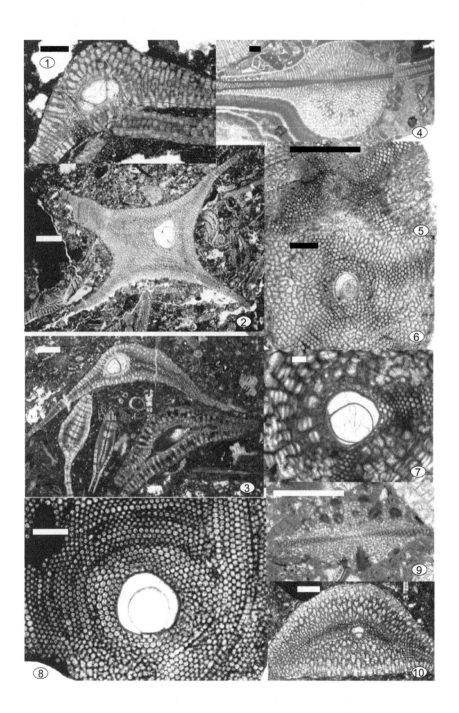

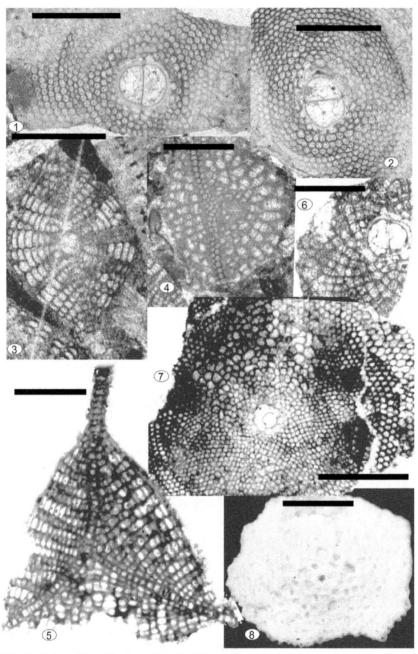

Plate 7.20 Scale bars: Figs 1-7 = 1mm; Fig. 8 = 0.5mm. Figs 1-2. *Lepidocyclina (Lepidocyclina) canellei* Lemoine and Douville´, offshore Brazil, UCL 10–12MF. Figs 3-5. *L. (Nephrolepidina) sumatrensis* Brady, 3-4) Early Miocene (Te5), Malinau River, N. of Mount Mulu, N.W. Sarawak Borneo, NHM P46529; 5) Darai Limestone, Papua New Guinea, UCL coll. Fig. 6. *Lepidocyclina (Lepidocyclina) isolepidinoides* van der Vlerk, Oligocene, Dubai, NHM P51965. Fig. 7. *L. (Nephrolepidina) pilifera* Scheffen, Miocene (Burdigalian), Darai Limestones, Papua New Guinea, UCL coll. Fig. 8. *L. (Nephrolepidina)* sp., solid specimen, Miocene (Burdigalian), Borneo, UCL coll.

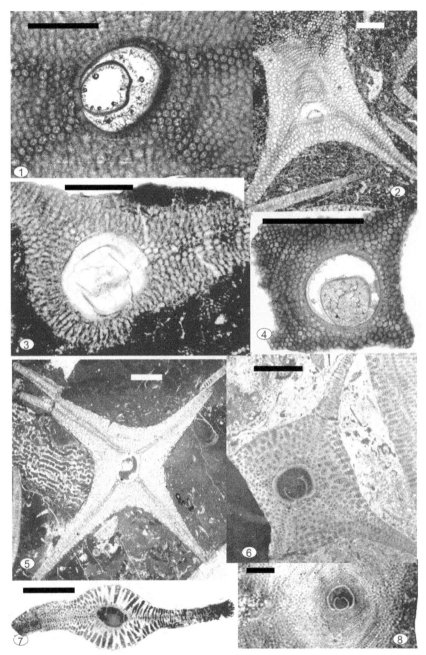

Plate 7.21   Scale bars: Figs 1-8 = 1mm. Fig. 1. *Eulepidina favosa* Vaughan, Early Miocene, Juana Diaz formation, La Rambla, Puerto Rico, NHM P47332. Fig. 2. *Eulepidina badjirraensis* Crespin, Early Miocene (Burdigalian, Te5), Mosque Quarry, Loc. 205, Borneo, UCL coll. Fig. 3. *Eulepidina parkinsonia* BouDagher-Fadel and Price, holotype figured by BouDagher-Fadel and Price (2010a), Miocene (Aquitanian), Hodges Bay, Antigua, West Indies, NHM P22753. Fig. 4. *Eulepidina undosa laramblaensis* Eames et al., holotype, early Miocene, Juana Diaz Formation, La Rambla, Puerto Rico. NHM P47353 Fig. 5-7. *Eulepidina formosa* (Schlumberger), Miocene (Aquitanian), Java, UCL coll. Fig. 8. *Eulepidina eodilatata* Douvillé, Oligocene (Rupelian), lower part of the Mesolouri section, marly facies, Greece, SFN coll. 2001z0153/0006.

References

Adams, C.G., 1959. Geological distribution of *Discospirina* (foraminifera) and occurrence of *D. italica* in the Miocene of Cyprus. Palaeontology 1, 364–368.

Adams, C.G., 1965. The foraminifera and stratigraphy of the Melinau Limestone, Sarawak, and its importance in Tertiary correlation. Quart. J. Geol. Soc. Lond. 121, 283–338.

Adams, C.G., 1967. Tertiary Foraminifera in the Tethyan, American, and Indo-Pacific Provinces. In: Adams, C.G., Ager, D. (Eds), Aspects of Tethyan Biogeography, Systematics Association Publication, special publication. London, Vol. 7, pp. 195–217.

Adams, C.G., 1968. A revision of the foraminiferal genus *Austrotrillina* Parr. British Museum (Natural History), Bulletin (Geology) 16, 73–97.

Adams, C.G., 1970. A reconsideration of the East Indian letter classification of the Tertiary. Bull. Brit. Mus. (Nat. Hist.) Geol. 19, 87–137.

Adams, C.G., 1973. Some tertiary foraminifera. In: Hallam, A. (Ed.), Atlas of Paleobiogeography. Elsevier, pp. 453–468.

Adams, C.G., 1984. Neogene larger foraminifera, evolutionary and geological events in the context of datum planes. In: Ikebe, I., Tsuchi, R. (Eds), Pacific Neogene Datum Planes. University of Tokyo Press, Tokyo, pp. 47–68.

Adams, C.G., 1987. On the classification of the Lepidocyclinidae (foraminifera) with redescriptions of the unrelated Paleocene genera Actinosiphon and Orbitosiphon. Micropaleontology 33, 289–317.

Adams, C.G., Gentry, A.W., Whybrow, P.J., 1983. Dating the terminal Tethyan event. In: Meulenkamp, J.E. (Ed.), Reconstruction of Marine Paleoenvironments. Utrecht Micropaleontological Bulletins,, 30, 273–298.

Adams, C.G., Racey, A., 1992. The occurrence and palaeobiogeographical significance of the foraminiferid *Yaberinella* from the Eocene of Oman. Paleontology 35, 237–245.

Adams, R.M., Rosenzweig, C., Peart, R.M., Ritchie, J.T., McCarl, B.A., Glyer, J.D., Curry, R.B., Jones, J.W., Boote, K.J., Allen, L.H., Jr., 1990. Global climate change and U.S. agriculture. Nature 345, 219–224.

Advocaat, E. L., Hall, R., White, L. T., Watkinson, I. M., Rudyawan, A., Boudagher-Fadel, M. K., 2017. Miocene to recent extension in NW Sulawesi, Indonesia. Journal of Asian Earth Sciences. doi:10.1016/j.jseaes.2017.07.023

Aguirre, J., Riding, R., 2005. Dasycladalean algal biodiversity compared with global variations in temperature and sea level over the past 350 Myr. Palaios 20, 581–588.

Aitchison, J.C., Ali, J.R., Davis, A.M., 2007. When and where did India and Asia collide? J. Geophys. Res.-Sol. Ea. 112(Article number), p. B05423.

Albrich, S., Frijia, G., Parente, M., Caus, M., 2014. The evolution of the earliest representatives of the genus *Orbitoides*: Implications for Upper Cretaceous biostratigraphy. Cretaceous Research 51, 22–34.

Alegret, L., Arenillas, I., Arz, J.A., Diaz, C., Grajales-Nishimura, J.M., Melendez, A., Molina, E., Rojas, R., Soria, A.R., 2005. Cretaceous–Paleogene boundary deposits at Loma Capiro, Central Cuba, evidence for the Chicxulub impact. Geology 33, 721–724.

Alexander, S.P., 1985. The cytology of certain benthonic foraminifera in relation to test structure and function. PhD thesis, University of Wales.

Al-Habeeb, K.H., 1977. Foraminifera in the carboniferous limestones in South Wales. PhD thesis, University College of Swansea, University of Wales.

Allemann, F., Schroeder, R., 1972. *Spiroconulus perconigi* n. sp. a new Middle Jurassic foraminifer of Oman and Spain. Revista Espan˜ ola de Micropaleontologia 30, 199–209.

Allmon, W.D., 2001. Nutrients, temperature, disturbance, and evolution: A model for the late Cenozoic marine record of the western Atlantic. Paleogeography, Paleoclimatology, Paleoecology, 166, 9–26.

Almazán-Vázquez, E., Buitrón-Sánchez, B. E., Vachard, D., Mendoza-Madera, C., Gómezespinosa, C., 2007. The late Atokan (Moscovian, Pennsylvanian) chaetetid accumulations of Sierra Agua Verde, Sonora (NW Mexico), composition, facies and palaeoenvironmental signals. In: Álvaro, J. J., Aretz, M., BoulvaIn: F., Munnecke, A., Vachard, D., and VennIn: E. (eds) Palaeozoic Reefs and Bioaccumulations, Climatic and Evolutionary Controls. Geological Society, London, Special Publications, 275, 189–200.

Altiner, D., Groves, J. R., O˙zkan-Altiner, S., 2005. Calcareous foraminiferal recovery from the end- Permian mass extinction, southern Turkey. Abstracts, Albertiana, International Symposium on Triassic Chronostratigraphy and Biotic Recovery, Chaohu, China, 23–25, 14–17.

Altiner D., and Savini, R., 1997. New species of *Syzrania* from the Amazonas and Solimões basins (north Brazil), remarks on the generic and suprageneric position of syzraniid foraminifers. Revue de Paléobiologie, 16, 7–20.

Altiner, D., Septfontaine, M., 1979. Micropaléontologie, stratigraphie et environnement de deposition d'une série jurassique à facies de plate-forme de la région de Punarbaşi (Taurus oriental, Turquie). Revue de Micropaléontologie, 22, 3–18.

Altiner, R.D., Rettori, R. 2003. Origin and early evolutionary radiation of the Order Lagenida (Foraminifera). Journal of Paleontology, 77, 831–843.

Amodiao, S., 2006. Foraminifera diversity changes and paleoenvironmental analysis, the Lower Cretaceous shallow-water carbonates of San Lorenzello, Campanian Apennines, southern Italy. Facies, 52, 53–67.

An W., Hu, X., Garzanti, E., BouDagher-Fadel, M., Wang, J., Sun, G., 2014. Xigaze forearc basin revisited (South Tibet): Provenance changes and origin of the Xigaze Ophiolite. Geological Society of American Bulletin, 126. 1595–1613

An, W., Hu, X., Garzanti, E., BouDagher-Fadel, M. K., Wang, J., Sun, G., 2015. Erratum to Xigaze forearc basin revisited (South Tibet): Provenance changes and origin of the Xigaze Ophiolite, [Geological Society of America, 126, 11/12, 1595–1613, 1984] 10.1130/B31020.1. Bulletin of the Geological Society of America, 127, 409. doi:10.1130/B25514.1.

Anderson, O.R., Spindler, M., Bé, A.W.H., Hemleben, Ch., 1979. Trophic activity of planktonic foraminifera. J. Mar. Biolog. Assoc. UK 59, 791–799.

Angiolini, L., Gaetani, M., Muttoni, G., Stephenson, M.H., Zanchi, A., 2006. Tethyan oceanic currents and climate gradients 300 Ma ago. Geology 35, 1071–1074.

Archibald, J.M., Longet, D., Pawlowski, J., Keeling, P., 2003. A novel polyubiquitin structure in cercozoa and foraminifera, evidence for a new eukaryotic supergroup. Molecular Biology and Evolution, 20, 62–66.

Arnaud-Vanneau, A., 1975. Reflexion sur le mode de vie de certains Orbitolinides (Foraminifères) barremo-aptiens de l'Urgonien du Vercors. Compte Rendu des Seances de la Société de Physique et d'Histoire Naturelle de Genève, 10, 126–130.

Arnaud-Vanneau, A., 1980. Micropaléontologie, paléoécologie et sédimentologie d'une plate-forme carbonatée de la marge passive de la Téthys. Géologie alpine, 11, 1–874.

Arnaud-Vanneau, A., and Peybernés, B., 1978. Les représentants Éocrétacés du genre *Nautiloculina* Mohler, 1938 (Foraminifera, Fam. Lituolidae?) dans les chaines Subalpines septentrionales (Vercors) et les Pyrénées Franco-Espagnoles. Revision de *Nautiloculina cretacea* Peybernés, 1976 et déscription de *Nautiloculina bronnimanni* n. sp.. Geobios, 11, 67–81.

Arnold, Z.M., 1978. An allogromiid ancestor of the miliolidean foraminifera. Journal of Foraminiferal Research, 8, 83–96.

Azema, J., Chabrier, G., Fourcade, E., and Jaffrezo, M., 1977. Nouvelles données micropaléontologiques, stratigraphiques et paléogéographiqies sur le Portlandien et le Néocomien de Sardaigne. Revue de Micropaléontologie, 20, 125–139.

Baldauf, S.L., 2003. The deep roots of Eukaryotes. Science 300, 1703–1706.

Bambach, R. K., 2006. Phaenerozoic Biodiversity Mass Extinctions. The Annual Review of Earth and Planetary Science, 34, 117–155.

Banerjee, A., Boyajian, E., 1997. Selectivity of foraminiferal extinction in the Late Eocene. Paleobiology 23, 347–357.

Banerjee, A., Yemane, K., Johnson, A., 2000. Foraminiferal Biostratigraphy of Late Oligocene-Miocene Reefal Carbonates in Southwestern Puerto Rico. Micropaleontology, 46 327–342.

Banner, F. T., Lord, A. R., & BouDagher-Fadel, M. K., 1999. The Terra Limestones Member (Miocene) of Western Cyprus. Greifswalder Geowissenschaftliche Beiträge, 6, 503–515.

Banner, F. T., Whittaker, J. E., Boudagher-Fadel, M. K., & Samuel, A. (1997). Socotraina, a new hauraniid genus from the upper Lias of the Middle East (Foraminifera, Textulariina). Revue de Micropaleontologie, 40, 115–123.

Banner, F.T., 1970. A synopsis of the Spirocyclinidae. Revista Espanola de Micropaleontologia 2, 243–290.

Banner, F.T., Finch, E.M., Simmons, M.D., 1990. On Lithocodium Elliott (calcareous algae); its paleobiological and stratigraphical significance. Journal of Micropaleontology 9, 21–36.

Banner, F.T., Highton, J., 1990. On *Everticyclammina* Redmond (foraminifera), especially *E. kelleri* (Henson). Journal of Micropaleontology. 9, 1–14.

Banner, F.T., Hodgkinson, R.L., 1991. A revision of the foraminiferal subfamily Heterostegininae. Revista Espanola de Micropaleontologia 23, 101–140.

Banner, F.T., Knight-Jones, E.W., and Wright, J.M., 1994. Protozoa, the marine fauna of the British Isles and North-West Europe. In: Hayward, P.J., and Ryland, J.S. (eds.), Introduction to Arthropods and Protozoans. Oxford Science Publications, 36–50.

Banner, F.T., Pereira, P.G., 1981. Some biserial and triserial agglutinated smaller foraminifera, their wall structure and its significance. Journal of Foraminiferal Research, 11, 85–117.

Banner, F.T., Samuel, A., 1995. *Alanlordia*, a new genus of acervuline foraminifera from the Neogene of Indonesia. J. Micropalaeontol. 14, 107–117.

Banner, F.T., Simmons, M.D., 1994. Calcareous algae and foraminifera as water depth indicators, an example from the early cretaceous carbonate of northeast Arabia. In: Simmons, M.D. (Ed.), Micropalaeontology and Hydrocarbon Exploration in the Middle East. Chapman and Hall, London, pp. 243–252.

Banner, F.T., Simmons, M.D., Whittaker, J.E., 1991. The Mesozoic chrysalidinidae (foraminifera, textulariacea) of the Middle East, the Redmond (Aramco) taxa and their relatives. Bull. Brit. Mus. Nat. Hist. (Geol.) 47, 101–152.

Banner, F.T., Whittaker, J.E., 1991. Redmond's "new lituolid foraminifera" from the Mesozoic of Saudi Arabia. Micropaleontology 37, 41–59.

Baranova, D.V., Kabanov, P.B., 2003. Facies distribution of fusulinoid genera in the Myachkovian (Upper Carboniferous, upper Moscovian) of southern Moscow region. Rivista Italiana di Paleontologia e Stratigrafia 109, 225–239.

Barattolo, F., Bigozzi, A., 1996. Dasycladaceans and depositional environments of the Upper Triassic–Liassic carbonate platform of the Gran Sasso. Facies 35, 163–208.

Barker, R.W., Grimsdale, T.F., 1936. A contribution to the phylogeny of the orbitoidal foraminifera, with descriptions of new forms from the Eocene of Mexico. Journal of Paleontology, 10, 231–247.

Bartholdy, J., 2002. The Architecture of Nummulites (Foraminifera) reexamined, PhD thesis, Freie Universita¨ t Berlin, 64, 1–22.

Basov, V.A., Kuznetsova, K.I., 2000. Dynamics of diversity and evolutionary trends of Jurassic foraminifers. Stratigr. Geol. Correl. 8, 593–607.

Bassoullet, J.P., Boutakiout, M., 1996. *Haurania (Platyhaurania) subcompressa* nov. subgen., nov. sp., new imperforate larger foraminifera (Hauraniidae) from lower Toarcian of septentrional Morocco (South Rif Ridges). Rev. Micropaleontol. 39, 27–40.

Bassoullet, J.P., Boutakiout, M. and Echarfaoui, H., 1999. Deux nouveaux genres, *Palaeocyclammina* et *Ijdranella*, foraminifères (Textulariina) d'un niveau Liasique à *Orbitopsella praecursor* (Gümbel) du Moyen Atlas (Maroc). Revue de Micropaléontologie, 42, 213–230.

Beavington-Penney, S.J., Racey, A., 2004. Ecology of extant nummulitids and other large benthic foraminifera, applications in palaeoenvironmental analysis. Earth Sci. Rev. 67, 219–265.

Beerling, D.J., Berner, R.A., 2002. Biogeochemical constraints on the Triassic–Jurassic boundary carbon cycle event. Global Biogeochem. Cycles 16, 1–13.

Bender, H., Hemleben, C., 1988. Calcitic elements secreted by agglutinated foraminifers grown in laboratory culture. J. Foramin. Res. 18, 42–45.

Benedetti A., Pignatti J., 2013. Conflicting evolutionary and biostratigraphical trends in Nephrolepidina praemarginata (Douvillé, 1908) (Foraminiferida). Historical Biology, 25, 363–383, ISSN: 0891-2963, doi: 10.1080/08912963.2012.713949

Benton, M.J., 2002. Cope's rule. In: Pagel, M. (Ed.), Encyclopedia of Evolution. Oxford University Press, New York, pp. 209–210.

Benton, M.J., Twitchett, R.J., 2003. How to kill (almost) all life, the end-Permian extinction event. Trends Ecol. Evol. 18, 358–365.

Berggren, W.A., 1972. Cenozoic biostratigraphy and paleobiogeography of the North Atlantic. In: Laughton, A.S., McKenzie, D.P., Sclater, J.G. (Eds), Initial Reports of the Deep Sea Drilling Project, Volume X/I, Washington, DC: U.S. Govt. Printing Office, pp. 965–1000.

Berggren, W.A., Pearson, P. N., 2005, A revised tropical to subtropical Paleogene planktonic foraminiferal zonation: Journal of Eoraminiferal Research, 35, 279–298.

Berggren, W.A., Prothero, D.R., 1992. Eocene–Oligocene climatic and biotic evolution: An overview. In: Prothero, D.R., Berggren, W.A. (Eds), Eocene–Oligocene Climatic and Biotic Evolution. Princeton University Press, Princeton, pp. 1–28.

Bergquist, H.R., 1971. Biogeographical review of Cretaceous foraminifera of the Western Hemisphere. North American Paleontological Convention, Chicago, 1969, Proceedings, pp. 1565–1609.

Berner, R.A., 2002. Examination of hypotheses for the Permo–Triassic boundary extinction by carbon cycle modelling. Proc. Natl. Acad. Sci. USA 99, 4172–4177.

Berner, R.A., Beerling, D.J., Dudley, R., Robinson, J.M., Wildman, R.A., Jr., 2003. Phanerozoic atmospheric oxygen. Earth Planet. Sci. Lett. 31, 105–134.

Betzler, C., 1997. Ecological controls on geometries of carbonate platforms: Miocene/Pliocene shallow-water microfaunas and carbonate biofacies from the Queensland Plateau (NE Australia). Facies 37, 147–166.

Blackburn, T.J., Olsen, P.E., Bowring, S.A., McLean, N.M., Kent, D.V., Puffer, J., McHone, G., Rasbury, E.T., and Et-Touhami, M., 2013, Zircon U-Pb geochronology links the end-Triassic extinction with the Central Atlantic Magmatic Province: Science, 340, 941–945, doi:10.1126/science.1234204.

Bonis, N.R., Kürschner, W.M., Krystyn, L., 2009. A detailed palynological study of the Triassic–Jurassic transition in key sections of the Eiberg basin (Northern Calcareous Alps, Austria). Review of Palaeobotany and Palynology 156, 376–400.

Bonnefous, J., Bismuth, H., 1982. Les facies carbonates de plate-forme de l'Éocene moyen et supérieur dans l'offshore tunisien nord-oriental et en mer pélagienne, implications paléogéographiques et analyse micropaléontologique. Bulletin des Centres de Recherches Exploration-Production Elf-Aquitaine 6, 337–403.

Boomer, I., Lord, A., Crasquin, S., 2008. The extinction of the Metacopina (Osatracoda). Senckenbergiana lethaea 88, in press.

Bornemann, A., Norris, R., Friedrich, O., Beckmann, B., Schouten, S., Sinninghe Damste´, J., Vogel, J., Hofmann, P., Wagner, T., 2008. Isotopic evidence for glaciation during the cretaceous supergreen- house. Science 319, 189–192.

Bosellini, A., Broglio Loriga, C., 1971. I "Calcari Grigi" di Rotzo (Giurassico inferiore, Altopiano di Asiago) e loro inquadramento nella paleogeografia e nella evoluzione tettonico-sedimentaria delle Prealpi venete. Annali del Universitario Ferrara 5, 1–61.

Bosellini, F.R., Papazzoni, C.A., 2003. Palaeoecological significance of coral-encrusting foraminiferan associations: A casestudy from the Upper Eocene of Northern Italy. Acta Palaeontol. Pol. 48, 279–292.

Bosellini, F.R., Russo, A., Vescogni, A., 2002. The Messinian Reef Complex of the Salento Peninsula (Southern Italy): Stratigraphy, facies and paleoenvironmental interpretation. Facies 47, 91–112.

Bosence, D., Procter, E., Aurell, M., Kahla, A. B., BouDagher-Fadel, M. K., & et al. (2009). A tectonic signal in highfrequency, peritidal carbonate cycles? a regional analysis of liassic platforms from western Tethys. Journal of Sedimentary Research, 79 (6), 389–415.

Boslough, M.B., Chael, E.P., Trucano, T.G., Crawford, D.A., Campbell, D.L., 1996. Axial focusing of impact energy in the Earth's interior, a possible link to flood basalts and hotspots. In: Ryder, G., Fastovsky, D., Gartner, S. (Eds), The Cretaceous–Tertiary Event and Other Catastrophes in Earth History. Geological Society of America. special paper, 307, 541–550.

BouDagher-Fadel, M.K., 2000. Benthic foraminifera of the Jurassic-Early Cretaceous of Tethys. International Workshop on North African Micropaleontology for Petroleum Exploration 1, 27–28. BouDagher-Fadel, M.K., 2001. The taxonomy and evolution of the foraminiferal genus *Buccicrenata* Loeblich and Tappan. Micropaleontology 47, 168–172.

BouDagher- Fadel, M. K., 2001. The taxonomy and evolution of the foraminiferal genus Buccicrenata Loeblich and Tappan. Micropalaeontology, 47 (2), 168- 172.

BouDagher-Fadel, M.K., 2002. The stratigraphical relationship between planktonic and larger benthic foraminifera in Middle Miocene to Lower Pliocene carbonate facies of Sulawesi, Indonesia. Micropaleontology 48, 153–176.

BouDagher-Fadel, M. K., 2008. Evolution and Geological Significance of Larger Benthic Foraminifera. Developments in Palaeontology and Stratigraphy, 21. ELSEVIER, 1–544.

BouDagher-Fadel, M.K., 2013. Diagnostic First and Last Occurrences of Mesozoic and Cenozoic Planktonic Foraminifera ((Professional Papers Series, pp. 1–4).). London, UK: UCL Office of the Vice-Provost Research.

Boudagher-Fadel, M.K., 2015. Biostratigraphic and Geological Significance of Planktonic Foraminifera (Updated 2nd Edition). London: UCL Press. doi:10.14324/111.9781910634257

BouDagher-Fadel, M. K., 2016. Genotypes of *Septatrocholina* and *Alzonorbitopsella*, two new Jurassic foraminifera: subsequent designations. Micropaleontology, 62, 87–89.

BouDagher-Fadel, M. K., Banner, F. T., 1997. The revision of some genus-group names in Tethyan Lepidocyclininae. Paleopelagos, 7, 3–16.

BouDagher-Fadel, M.K., Banner, F.T., 1999. Revision of the stratigraphic significance of the Oligocene–Miocene "Letter-Stages". Rev. Micropaleontol. 42, 93–97.

BouDagher-Fadel, M.K., Banner, F.T., Whittaker, J.E. 1997. Early Evolutionary History of Planktonic Foraminifera, British Micropalaeontological Society Publication Series, Chapman and Hall, 269p.

BouDagher-Fadel, M.K., Bosence, D., 2007. Early Jurassic benthic foraminiferal diversification and biozones in shallow-marine carbonates of western Tethys. Senckenbergiana lethaea, 87, 1–39.

Boudagher-Fadel, M. K., Hu, X., Price, G. D., Sun, G., Wang, J. -. G.,

BouDagher-Fadel, M. K., Lokier, S., 2005. Significant Miocene larger foraminifera from south central Java. Revue de Micropaléontologie, 24 (1), 291–309.

BouDagher-Fadel, M.K., Lord, A. R., 2000. The evolution of *Lepidocyclina* (*L.*) *isolepidinoides*, *L.* (*Nephrolepidina*) *nephrolepidinoides*, *L.* (*N.*) *brouweri* in the Late Oligocene-Miocene of the Far East. Journal of Foraminiferal Research, 30, 71–76.

BouDagher-Fadel, M.K., Lord, A.R., 2002. Larger foraminifera of the Jurassic Western Neotethys Ocean. Archaeology and History in Lebanon, 15, 87–94.

BouDagher-Fadel, M.K., Lord, A.R., Banner, F.T., 2000. Some Miogypsinidae (foraminiferida) in the Miocene of Borneo and nearby countries. Revue de Paléobiology 19, 137–156.

BouDagher-Fadel, M. K., Lord, A. R., 2006. Illusory stratigraphy decoded by Oligocene-Miocene autochthonous and allochthonous foraminifera in the Terra Member, Pakhna Formation (Cyprus). Stratigraphy, 3, 217–226.

BouDagher-Fadel, M.K., Noujaim Clark, G., 2006. Stratigraphy, paleoenvironment and paleogeography of Maritime Lebanon: a key to Eastern Mediterranean Cenozoic history. Stratigraphy, 3, 81–118.

BouDagher-Fadel, M.K., Price, G. D., 2009. *Loftusia persica*: an Eocene Lazarus occurrence?. Micropaleontology, 55, 75–85.

BouDagher-Fadel, M.K., Price, G. D., 2010a. Evolution and paleogeographic distribution of the lepidocyclinids. Journal of Foraminiferal Research, 40, 79–108.

BouDagher-Fadel, M.K., Price, G. D., 2010b. American miogypsinidae: An analysis of their phylogeny and biostratigraphy. Micropaleontology, 56, 567–586.

BouDagher- Fadel, M.K., Price, G. D., Koutsoukos, E. A. M., 2010c. Foraminiferal biostratigraphy and paleoenvironments of the Oligocene- Miocene carbonate succession in Campos Basin, southeastern Brazil. Stratigraphy, 7, 283- 299.

BouDagher-Fadel, M.K., Price, G. D., 2013. The phylogenetic and palaeogeographic evolution of the miogypsinid larger benthic foraminifera. Journal of the Geological Society, 170, 185–208. doi:10.1144/jgs2011-149

BouDagher-Fadel, M.K., Price, G. D., 2014. The phylogenetic and palaeogeographic evolution of the nummulitoid larger benthic foraminifera. Micropaleontology, 60, 483–508.

Boudagher-Fadel, M. K., Hu, X., Price, G. D., Sun, G., Wang, J. -. G., An, W., 2017. Foraminiferal biostratigraphy and palaeoenvironmental analysis of the mid-Cretaceous limestones in the southern Tibetan plateau. Journal of Foraminiferal Research, 47, 188-207. doi:10.2113/gsjfr.47.2.188

BouDagher-Fadel, M.K, Price, G. D., 2017. The paleogeographic evolution of the orthophragminids of the Paleogene. Journal of Foraminiferal Research, 47, 337–357.

BouDagher-Fadel, M. K., Price, G. D., Hu, X., Li, J., 2015. Late Cretaceous to early Paleogene foraminiferal biozones in the Tibetan Himalayas, and a pan-Tethyan foraminiferal correlation scheme. Stratigraphy, 12, 67–91.

Boudagher-Fadel, M., Rose, E. P. F., Bosence, D. W. J., Lord, A. R., 2001. Lower Jurassic foraminifera and calcified microflora from Gibraltar, western Mediterranean. Palaeontology, 44 (4), 601–621.

BouDagher-Fadel, M.K., Wilson, M., 2000. A revision of some larger Foraminifera of the Miocene of South-East Kalimantan. Micropaleontology, 42 (2), 153–166.

Boussac, J., 1906. Developement et Morphologie de quelques Foraminifères de Priabona. Bulletin Société Géologique de France 6, 88–97.

Bowser, S. S. and Travis, J. L., 2002. Reticulopodia: structural and behavioral basis for the suprageneric placement of granuloreticulosan protists. J. Foraminiferal Res., 32, 440–447.

Braga, J. C., García-Gomez, R., Jiménez, A. P., and Rivas, P., 1981. Correlaciones en el Lías de las Cordilleras Béticas. In: Almela Samper, A., and Meléndez Meléndez, B., (eds), Curso de conferencias sobre el Programa Internacional de Correlación Geológica, Part 2. Real Academia de Ciencias Exactas, Físicas y Naturales, Madrid, 161–181.

Brasier, M.D., 1975. An outline history of seagrass communities. Palaeontology 18, 681–702.

Breitfeld, H. T., Hall, R., Galin, T., Forster, M. A., & BouDagher-Fadel, M. K., 2016. A Triassic to Cretaceous Sundaland–Pacific subduction margin in West Sarawak, Borneo. Tectonophysics, 694, 35–56. doi:10.1016/j.tecto.2016.11.034.

Brönnimann, P., 1951. Remarks on the embryonic chambers of Upper Eocene Asterocyclinidae of Trinidad, B.W.I. Eclogae geologicae Helvetiae 44, 474–486.

Brönnimann, P., 1954a. Upper cretaceous orbitoidal foraminifera from Cuba. I. Sulcorbitoides n. gen. Contribution Cushman Foundation for Foraminiferal Research 5, 55–61.

Brönnimann, P., 1954b. Upper cretaceous orbitoidal foraminifera from Cuba II. *Vaughanina* Palmer. Contrib. Cushman Foundation for Foraminiferal Research 5, 91–105.

Brönnimann, P., 1958. New pseudorbitoidadae from the Upper Cretaceous of Cuba, with remarks on encrusting foraminifera. Micropaleontology 4, 165–186.

Brönnimann, P., 1966. *Pseudotextulariella courtionensis* n. sp., from the Valanginian of well Courtion 1, Courtion, Canton of Fribourg, Switzerland. Archives Sciences, Geneve 19, 265–278.

Brun, L., Butterlin, J., and Montell, L., 1982. Decouverte de Lépidocyclines (Foraminières) d'âge Eocène dans le Golfe de Guinée. Implications Paléobiogéographiques. Cahiers de Micropaléontologie, 2, 1–109.

Burki F, Kaplan M, Tikhonenkov D V, Zlatogursky V, Minh BQ, Radaykina L V, Smirnov A, Mylnikov P, Keeling PJ, Keeling PJ, 2016. Untangling the early diversification of eukaryotes: a phylogenomic study of the evolutionary origins of Centrohelida, Haptophyta and Cryptista, 1–10.

Bursch, J.G., 1947. Mikropalaontologische Untersuchungen des Tertiars von Gross Kei (Molukken). Schweizerische Palaeontologische Abhandlungen 65, 1–56.

Butterlin, J., 1987. Origine et e´ volution des Lepidocyclines de la re´ gion des Caraibes. Comparisons et relations avec les Lepidocyclines des autres re´ gions du monde. Rev. Micropaleontol. 29, 203–219.

Buxton, M.W.N., Pedley, H.M., 1989. A standardized model for Tethyan Tertiary cabonate ramps. J. Geol. Soc. Lond. 246, 746–748.

Cahuzac, B., 1984. Les faunes de Miogypsinidae d'Aquitaine meridionale (France), International Symposium Benthic Foraminifera (Pau, April 1983), 1984, pp. 117–129.

Cahuzac, B., Poignant, A., 1991. Morphologie des éspèces de *Pararotalia* et de *Miogypsinoides* (Foraminiferida) dans l'Oligocène d'Aquitaine me´ ridionale. Geobios, Mem. Sp. 13, 69–78.

Caplan, M.L., Bustin, R.M., 1999. Devonian–Carboniferous Hangenberg mass extinction event, widespread organic-rich mudrock and anoxia: causes and consequences. Paleogeography, Paleoclimatology. Paleoecology. 148, 187–207.

Carpenter, W.B., 1850. On the microscopic structure of *Nummulina, Orbitolites* and *Orbitoides*. Quart. J. Geol. Soc. Lond. 6, 181–236.

Carpenter, W.B., Parker, W.K., Jones, T.R., 1862. Introduction to the study of the foraminifera. Ray Soc. Lond. 319.

Castellarin, A., 1972. Evoluzione tettonica sinsedimentaria del limite tra "Piattaforma veneta" e "Bacino lombardo" a nord di Riva del Garda. Giornale di Geologia 38, 11–212.

Caudri, C.M.B., 1972. Systematics of the American Discocyclinas. Eclogae geologicae Helvetiae 65, 211–219.

Caudri, C.M.B., 1974. The larger foraminifera of Punta Mosquito, Margarita Island, Venezuela. Verhandlungen der Naturforschenden Gesellschaft in Basel 84, 293–318.

Caudri, C.M.B., 1996. The larger foraminifera of Trinidad (West Indies). Eclogae geologicae Helvetiae 89, 1137–1309.

Caus, E., 1988. Upper Cretaceous larger foraminifera, paleoecological distribution. Revue de Paleobiologie, Switzerland, volume special, 2, 417–419.

Cavalier-Smith, T., 1998. A revised six-kingdom system of life. Biol. Rev. Camb. Philos. Soc. 73, 203–266.

Cavalier-Smith, T., 2002. The phagotrophic origin of eukaryotes and phylogenetic classification of protozoa. Int. J. Syst. Evol. Microbiol. 52, 297–354.

Cavalier-Smith, T., Chao, E.E., 1997. Sarcomonad ribosomal RNA sequences, rhizopod phylogeny, and the origin of euglyphid amoebae. Archiv. Fuer. Protistenkunde. 147, 227–236.

Chapelle G., Peck, L.S., 1999. Polar gigantism dictated by oxygen availability. Nature, 399, 114–115.

Chaproniere, G.C.H., 1975. Palaeoecology of Oligo–Miocene larger foraminiferida, Australia. Alcheringa 1, 37–58.

Chaproniere, G.C.H., 1980. Biometrical studies of early Neogene larger foraminiferida from Australia and New Zealand. Alcheringa 4, 153–181.

Chaproniere, G. C. H., 1983. Tertiary larger foraminiferids from the northwestern margin of the Queensland Plateau, Australia; *in* Palaeontological papers 1983. Bureau of Mineral Resources, Australia, Bulletin 217, 31–57.

Chaproniere, G.C.H., 1984. The Neogene larger foraminiferal sequence in the Australian and New Zealand regions, and its relevance to the East Indies letter stage classification. Palaeogeog. Palaeoclimatol. Palaeoecol. 46, 25–35.

Chaproniere, G.C.H., 1991. Discospirina (miliolina, foraminiferida) from the Coral Sea off Northeastern Australia: A new record. Journal of Paleontology, 65, 332–334.

Chen, Z.Q., Benton, M.J. 2012. The timing and pattern of biotic recovery following the end-Permian mass extinction. Nature Geoscience 5: 375–383.

Chen, D., Tucker, M.E., 2003. The Frasnian–Famennian mass extinction, insights from high-resolution sequence stratigraphy and cyclostratigraphy in South China. Palaeogeog. Palaeoclimatol. Palaeoecol. 193, 87–111.

Cherchi, A., Schroeder, R., 1982. *Dictyoconus algerianus* n. sp., grand Foraminifère de l'Aptien supeérieur du plaque Africaine (marge septentrionale). Comptes Rendus de l'Académie des Sciences, Paris, Ser. II 295, 77–82.

Cherchi, A., Schroeder, R., Zhang, B.-G., 1984. *Cyclorbitopsella tibetica* n.gen., n.sp., a lituolacean foraminifer from the Lias of southern Tibet. In: Oertli, H.J. (Ed.), Benthos '83, Second International Symposium on Benthic Foraminifera. Pau, Elf Aquitaine, Esso Rep and Total CFP, Pau and Bordeaux, pp. 159–165.

Chernysheva, N.E., 1948. *Archaedisus* i blizkikh k nemu formakh iz nizhnego Karbona SSSR. Trudy Instituta Geologischeskikh Nauk, Akademiya Nauk SSSR 62, 150–158.

Chiocchini, M., Farinacci, A., Mancinelli, A., Molinari, V., Potetti, M., 1994. Biostratigrafia a foraminiferi, dasicladali e calpionelle delle successioni carbonatiche mesozoiche dell'Appennino centrale (Italia). In: Mancinelli, A., (Ed.), Biostratigrafia dell'Italia centrale. Studi Geologici Camerti, Volume Speciale 1994 (A), 9–129.

Chiocchini, M., Mancinelli, A., Romano, A., 1984. Stratigraphic distribution of benthic foraminifera in the Aptian, Albian, and Cenomanian carbonate sequences of the Aurunci and Ausoni Mountains (Southern Lazio, Italy). Benthos 83, 167–181.

Ciry, R., 1964. À propos de Meandropsina larrazeti Mun.-Chalm., generotype d'un genre nouveau Larrazetia Ciry. Rev. Micropaleontology, 6, 185–195.

Clauzon, G., Suc, J.P., Gautier, F., Berger, A., Loutre, M.F., 1996. Alternate interpretation of the Messinian salinity crisis, controversy resoved? Geology 24, 363–366.

Cloud, P., 1976. Beginnings of biospheric evolution and their biogeochemical consequences. Paleobiology 2, 351–387.

Cohen, K.M., Finney, S.C., Gibbard, P.L., Fan, J.-X., 2013, The ICS International Chronostratigraphic Chart, Episodes, 36, 199–204.

Cole, W.S., 1941. Stratigraphic and paleontologic studies of wells in Florida – no 1. Fla Geol. Surv. Bull. 19, p. 91.

Cole, W.S., 1957. Variation in American Oligocene species of epidocyclina. Bull. Am. Paleonotol. 38, 31–51.

Cole, W.S., 1963. Tertiary larger foraminifera from Guam. US Geol. Surv. Prof. Pap. 403E, p. 28.

Cole, W.S., 1964. Orbitoididae; Discocyclinidae; Lepidocyclinidae. In: Loeblich, A.R., Jr., Tappan, H. (Eds), Sarcodina, Chiefly "Thecamoebians" and Foraminiferida, Treatise on Invertebrate Paleontology (R.C. Moore, Ed.), Geological Society of America and University of Kansas Press, Part: C. Protista 2, pp. C710–C724.

Cole, W.S., Bermudez, P.J., 1944. New foraminiferal genera from the Cuban Middle Eocene. Bull. Amer. Paleont. 28, (113), 333–334.

Conway Morris, S., 2003. Life's Solution: Inevitable Humans in a Lonely Universe. Cambridge University Press, Cambridge, 464p.

Cope, E.D., 1896. The Primary Factors of Organic Evolution. Open Court Publishing Company, Chicago, 547p.

Copper, P., 1989. Enigmas in phanerozoic reef development. Memoire association Australas. Palaeontology 8, 371–385.

Copper, P., 2002. Reef development at the Frasnian/Famennian mass extinction boundary. Paleogeography, Paleoclimatology, Paleoecology, 28, 1–39.

Cosijn, A.J., 1938. Statistical Studies on the Phylogeny of Some foraminifera: Cycloclypeus and Lepidocyclina from Spain, Globorotalia from the East-Indies. Doctoral thesis, University of Leiden, pp. 1–70.

Ćosović V., Drobne, K., Moro, A., 2004. Paleoenvironmental model for Eocene foraminiferal limestones of the Adriatic carbonate platform (Istrian Peninsula). Facies 50, 61–75.

Cossey, P.J, Mundy, D.J.C., 1990. *Tetrataxis,* a loosely attached limpet-like foraminifer from the Upper Paleozoic, Lethaia, 23, 311–322.

Courgeon, S., Jorry, S. J., Camoin, G. F., BouDagher-Fadel, M. K., Jouet, G., Révillon, S., Bachèlery P, Pelleter E, Borgomano J, Poli E, Droxler, A. W., 2016. Growth and demise of Cenozoic isolated carbonate platforms: new insights from the Mozambique Channel seamounts (SW Indian Ocean). Marine Geology, 380, 90–105. doi:10.1016/j.margeo.2016.07.006.

Courgeon S, Jorry SJ, Jouet G, Camoin G, BouDagher-Fadel MK, Bachelery P, Caline B, Boichard R, Revillon S, Thomas Y, Thereau E, Guerin C., 2017. Impact of tectonic and volcanism on the Neogene evolution of isolated carbonate platforms (SW Indian Ocean). Sedimentary Geology, 355, 114–131. doi:10.1016/j.sedgeo.2017.04.008.

Courtillot, V.E., Renne, P.R., 2003. On the ages of flood basalt events. C. R. Geosci. 335, 113–140. Cowen, R., 1983. Algal symbiosis and its recognition in the fossil record. In: Teveszm, J.S., McCallp, L. (Eds), Biotic Interactions in Recent and Fossil Benthic Communities. Plenum Press, New York, pp. 431–478.

Cózar, P., Rodríguez, 2003. The palaeoecological distribution of the endothyroids (foraminifera) in the Guadiato area (SW Spain, Mississippian). Palaeogeography, Palaeoclimatology, Palaeoecology, 201, 1–19.

Cózar, P., Vachard D., 2001. Dainellinae, Subfam.nov. (early Carboniferous Foraminiferida), review and new taxa. Geobios, 34, 5: 505–526.

Cramer, B.S., Kent, D.V., 2005. Bolide summer: The Paleocene/Eocene thermal maximum as a response to an extraterrestrial trigger. Paleogeography, Paleoclimatology, Paleoecology, 224, 144–166.

Cummings, R.H., 1956. Revision of upper Paleozoic textulariid foraminifera. Micropaleontology 2, 201–242.

Cushman, J.A., Jarvis, P.W., 1931. Some new Eocene Foraminifera from Jamaica. Contrib. Cushman Found. Foramin. Res. 7, 75–78.

Cuvillier, J., Foury, G., Pignatti Morano, A.G., 1968. Foraminifères nouveaux du Jurassique supérieur du Val Cellina (Frioul Occidental, Italie). Geologica Romana 7, 141–156.

D'Archiac, A., 1850. Histoire des progrés de la Géologie de 1834 à 1849. Formation nummulitique de l'Espagne 3, p. 304.

Dain, I. G. 1958. New genera and species of foraminifera. Trudy Vsesoyuznogo Nauchno-Issled Geologorazved Institution, Gostoptekhizdat, 9, 4–81.

Davies, A. M., 1935. Tertiary faunas: A textbook for oilfield paleontologists and students of geology. The composition of Tertiary faunas. Allen and Unwin, London, Vol. 1, 1406p.

Davydov, V.I., 1984. On the problem of origin of Schwagerins. Paleontol. J. 3–16 (In Russian).

Davydov, V.I., 2011. Taxonomy, nomenclature and evolution of the early schubertellids (Fusulinida, Foraminifera). Acta Palaeontol. Pol. 56, 181–194.

Davydov, V.I., 2013. Climate fluctuations within the western Pangean tropical shelves - the Pennsylvanian/Permian record from benthic foraminifera. In: Lucas, S.G., DiMichele, W.A., Barrick, J.E., Schneider, J.W., Spielmann, J.A. (Eds.), The Carboniferous– Permian Transition. Bulletin New Mexico Museum of Natural History and Science, 73–78.

Davydov, V., 2014. Warm Water Benthic Foraminifera Document the Pennsylvanian-permian Warming and Cooling Events – The Record from the Western Pangea Tropical Shelves". Palaeogeography, Palaeoclimatology, Palaeoecology, 414, 284–295.

Davydov, V.I., Arefifard, S., 2007. Permian fusulinid fauna of Gondwanan affinity from Kalmard Region, East-Central Iran and its significance for the tectonics and paleogeography. Paleontol. Electron, 40 (http://palaeoelectronica.org/2007_2/00124/index.html).

Davydov, V.I., Wardlaw, B.R., Gradstein, F.M., 2004. The carboniferous period. In: Gradstein, F.M., Ogg, J.G., Smith, A.G. (Eds), A Geologic Time Scale. Cambridge University Press, Cambridge, pp. 222–248.

De Bock, J.F., 1976. Studies on some *Miogypsinoides – Miogypsina* s.s. associations with special reference to morphological features. Scripta geologica 36, 1–137.

De Castro, P., 1971. Osservazioni su Raadshoovenia van den Bold, e i suoi col nuovo genere Scandonea (Foraminiferida, Miliolacea). Bollettino della Societa dei Naturalisti in Napoli 80, 161–235.

De Gracianski, P.-C., Hardenbol, J., Jacquin, Th., Vail, P.R., 1998. Mesozoic–Cenozoic Sequence Stratigraphy of European Basins, Society of Economic Paleontologists and Mineralogists Special Publication, SEPM, Tulsa, OK, Vol. 60, 786pp.

De Mulder, E.F.J., 1975. Microfauna and sedimentary-tectonic history of the Oligo–Miocene of the Ionian Islands and western Epirus (Greece). Utrecht Micropaleontological Bulletins, 13, p. 147.

Deckart, K., Féraud, G., Bertrand, H., 1997. Age of Jurassic continental tholeiites of French Guyana, Surinam and Guinea: Implications for the initial opening of the central Atlantic Ocean. Earth Planet. Sci. Lett. 150, 205–220.

Decrouez, D., Moullade, M., 1974. Orbitolinides nouveaux de l'Albo-Cenomanien de Grece. Arch. Sci. Geneve 27, 75–92.

Deloffre, R., Hamaoui, M., 1969. Biostratigraphie des "Breches de Soumoulou" et description de *Pseudobroeckinella soumoulouensis* n. gen., n. sp., Foraminifère du Cretacé supérieur d'Aquitaine. Bulletin Centre Recherche Pau-SNPA 3, 5–31.

Dettmering, C., Röttger, R., Hohenegger, J., Schmaljohann, R., 1998. The trimorphic life cycle in foraminifera: Observations from cultures allow new evaluation. European Journal of Protistology, 34, 363–368.

Dickens, G. R., 2001. Carbon addition and removal during the Late Palaeocene Thermal Maximum: Basic theory with a preliminary treatment of the isotope record at ODP Site 1051, Blake Nose. In: Kroon, D., Norris, R.D., Klaus, A. (Eds), Western North Atlantic Palaeogene and Cretaceous Paleoceanography, Geological Society of London (Special Publication), London, Vol. 183, 293–305.

Dilley, F.C., 1973. Cretaceous larger foraminifera. In: Hallam, A. (Ed.), Atlas of Palaeobiogeography. Elsevier, Amsterdam, pp. 403–419.

Dolenec, T., Pavsic, J., Lojen, S., 2000. Ir anomalies and other elemental markers near the Palaeocene–Eocene boundary in a flysch sequence from the Western Tethys (Slovenia). Terra Nova 12, 199–204.

Douglass, R.C., 1977. The development of fusulinid biostratigraphy. In: Kauffman, E. G., Hazel, J. E., (Eds), Concepts and Methods of Biostratigraphy. Dowden, Hutchinson, and Ross, Inc., Stroudsbard, Pennsylvania, pp. 463–481.

Drobne, K., 1974. Les grandes Miliolides des couches paleocenes de la Yougoslavie du Nord-Ouest (*Idalina, Fabularia, Lacazina, Periloculina*). Razprave IV, razr, SAZU 17, 129–184.

Drobne, K., 1975: *Hottingerina lukasi* n. gen., n. sp. (Foraminiferida) du Paleocene moyen provennant du Nord-Ouest de la Yougoslavie. Razprave, 4.razr. SAZU 18, 242–253.

Drobne, K., 1988. Elements structuraux et repartition stratigraphique des grands miliolides de la famille des Fabulariidae. Revue de Palé obiologie, Vol. Special, Benthos '86, Vol. 2, pp. 643–661.

Drooger, C.W., 1952. Study of American Miogypsinidae. Doctoral thesis, University of Utrecht, p. 80. Drooger, C.W., 1955. Remarks on cycloclypeus. Proc. Koninklijke Nederlandse Akadamie Wetenschappen, Amsterdam, Ser. B 58, 415–433.

Drooger, C.W., 1963. Evolutionary trends in Miogypsinidae. In: *Evolutionary Trends in Foraminifera*.

Drooger, C.W., 1983. Environmental gradients and evolutionary events in some larger Foraminifera. Utrecht Micropaleontological Bulletins, 30, 255–271.

Drooger, C.W., 1984. Evolutionary partterns in lineages of orbitoidal foraminifera. Palaeontol., Proc. B 87, 103–130.

Drooger, C.W., 1993. Radial foraminifera, morphometrics and evolution. Verhandelingen der Koninklijke Nederlandse Akademie van Wetenschappen, Afd. Natuurkunde, Erste Reeks, Amsterdam, deel 41, 241p.

Drooger, C.W., Raju, D.S.N., 1973. Protoconch diameter in the Miogypsinidae. Proc. Koninklijke Nederlandse Akadamie Wetenschappen, Amsterdam, Ser. B 76, 206–216.

Drooger, C.W., Roelofsen, J.W., 1982. *Cycloclypeus* from Ghar Hassan, Malta. Proceedings of Koninklijke Nederlandse Akadamie Wetenschappen, Amsterdam, Ser. B-85, pp. 203–218.

Drooger, C.W., Rohling, E.J., 1988. Lepidocyclina migration across the Atlantic. Proceedings Koninklijke Nederlandse Akadamie Wetenschappen, Amsterdam, Ser.B, 91, pp. 39–52.

Dullo, W., Moussavian, E., Brachert, T.C., 1990. The foralgal crust facies of the deeper fore reefs in the Red Sea, a deep diving survey by submersible. Geobios 23, 261–281.

Eames, F.E., Banner, F.T., Blow, W.H., Clarke, W.J., 1962. Fundamentals of Mid-Tertiary Stratigraphical Correlation. Cambridge University Press, Cambridge, 163p.

Edinger, E.N., Risk, M.J., 1994. Oligocene–Miocene extinction and geographic restriction of Caribbean corals, roles of turbidity, temperature, and nutrients. Palaios 9, 576–598.

Eldholm, O., Coffin, M.F., 2000. Large igneous provinces and plate tectonics. In: Richards, M., Gordon, R., Van der Hilst, R. (Eds), The History and Dynamics of Global Plate Motions, American Geophysical Union Geophysical Monograph, Washington, DC, Vol. 121, pp. 309–326.

Eldholm, O., Thomas, E., 1993. Environmental impact of volcanic margin formation. Earth Planet. Sci. Lett. 117, 319–329.

Elliott, J.M., Logan, A., and Thomas, L.H. 1996. Morphotypes of the foraminifera *Homotrema rubrum* (Lamarck), distribution and relative abundance on reefs in Bermuda. Bulletin of Marine Science, 58, 261–276.

Erwin, D.H., 1993. The Great Paleozoic Crisis: Life and Death in the Permian. Columbia University Press, 327p.

Elsevier, Amsterdam, pp. 315–349.

Erwin, D.H., 1996. Understanding biotic recoveries, extinction, survival, and preservation during the end- Permian mass extinction. In: Jablonski, D., Erwin, D.H., Lipps, J.H. (Eds), Evolutionary Paleobiology. University of Chicago Press, Chicago, pp. 398–418.

Erwin, D.H., Bowring, S.A., Jin, Y.G., 2002. End-Permian mass extinctions, a review. In: Koeberl, C., MacLeod, K.G. (Eds), Catastrophic Events and Mass Extinctions, Impacts and Beyond, Geological Society of America, Special Paper, Vol. 356, pp. 363–383.

Erwin, D.H., Droser, M.E., 1993. Elvis taxa. Palaios 8, 623–624.

Fagerstrom, J.A., 1987. The Evolution of Reef Communities. Wiley, New York, 600p.

Farinacci, A. 1996. A new nezzazatid foraminifera in theSalento Upper Campanian limestones (southern Italy). Palaeopelagos, 5, 129–137.

Farinacci, A, & Yeniay, G. 1994. *Tekkeina anatoliensis* n.gen. n.sp., a new foraminifer from Susuz Dag, Western Taurus, Turkey. Palaeopelagos, 4, 47–59.

Fedorov, A.V., Dekens, P.S., McCarthy, M., Ravelo, A.C., de Menocal, P.B., Barreiro, M., Pacanowski, R.C., Philander, S.G., 2006. The Pliocene paradox (mechanisms for a permanent El Nin˜ o). Science 312, 1485–1489.

Feldmann, M., McKenzie, J.A., 1998. Stromatolite–thrombolite associations in a modern environment, Lee Stocking Island, Bahamas. Palaios 13, 201–212.

Ferrandez-Canadell, C., 2002. Multicellular-like compartmentalization of cytoplast in fossil larger foraminifera. Lethaia 35, 121–129.

Ferrandez-Canadell, C., Serra Kiel, J., 1992. Morphostructure and paleobiology of Discocyclina Guembel, 1870. J. Foramin. Res. 22, 147–165.

Ferrandini, M., BouDagher-Fadel, M. K., Ferrandini, J., Oudet, J., Andre, J. -. P., 2010. New observations about the Miogypsinidae of the Early and Middle Miocene of Provence and Corsica (France) and northern Sardinia (Italy). Annales de Paleontologie, 96, 67–94. doi:10.1016/j.annpal.2011.04.002.

Ferrero, E., 1987. Miogypsinidi della serie oligo-miocenica della Collina di Torino. Bollettino della Societa Paleontologica Italiana 26, 119–150.

Ferrero, E., Maia, F., Tonon, M., 1994. Biostratigrafia dei livelli bioclastici del Monferrato. Atti Ticinensi di Scienre della Terra, Serie Speciale 1, pp. 227–230.

Firestone, R. B., West, A., Kennett, J. P., Becker, L., Bunch, T. E., Revay, Z. S., Schultz, P. H., Belgya, T., Kennett, D. J., Erlandson, J. M., Dickenson, O. J., Goodyear, A. C., Harris, R. S., Howard, G. A., Kloosterman, J. B., Lechler, P., Mayewski, P. A., Montgomery, J., Poreda, R., Darrah, T., Hee, S. S. Que, Smitha, A. R., Stich, A., Topping, W., Wittke, J. H., and Wolbach, W. S., 2007. Evidence for an extraterrestrial impact event 12,900 years ago that contributed to megafaunal extinctions and the Younger Dryas cooling. Proceedings of the National Academy of Sciences 104, 16016–16021.

Fischer, G., Arthur, M.A., 1977. Secular variations in the pelagic realm. Soc. Econ. Paleontolog. Mineralog. 25, 19–50.

Flakowski, J., Bolivar, I., Fahrni, J., Pawlowski, J., 2005. Actin phylogeny of Foraminifera. Journal of Foraminiferal Research, 35, 93–102.

Flügel, E., 1994. Pangean shelf carbonates, controls and paleoclimate significance of of Permian and Triassic reefs. In: Klein, G. de V., (Ed.), Pangea, Paleoclimate, Tectonics and Sedimentation During Accretion, Zenith, and Breakup of a Supercontinent, GSA special paper, Boulder, CO, Vol. 288, pp. 247–266.

Fourcade, E., 1966. *Murciella cuvillieri* n. gen., n. sp., nouveau Foraminifere du Sénonien supérieur du SE de L'Éspagne. Revue de Micropaléontolgie 9, 147–155.

Fourcade, E., Mouty, M., Teherani, K.K., 1997. *Levantinella* nov.gen. et re´ vision du genre Mangashtia Henson, grands foraminife` res du Jurassique et du Creétacé du Moyen-Orient. Geobios 30, 179–192.

Fourcade, E., Neumann, M., 1966. A propos des genres *Labyrinthina* Weynschenk 1951 et *Lituosepta* Cati 1959. Rev. Micropaleontol. 8, 233–239.

Fourcade, E., Raoult, J.-F., Vila, J.-M., 1972. *Debarina hahounerensis* n. gen., n. sp., nouveau Lituolide (Foraminife` re) du Cre´ tace´ infe´ rieur constantinois (Algerie). C. R. Acad. Sci. Paris, 274, 191–193.

Fowell, S.J., Olsen, P.E., 1993. Time-calibration of Triassic/Jurassic microfloral turnover, eastern North America. Tectonophysics 222, 361–369.

Fowell, S.J., Traverse, A., 1995. Palynology and age of the upper Blomidon formation, Fundy Basin, Nova Scotia. Rev. Palaeobot. Palynol. 86, 211–233.

Freudenthal, Th., 1969. Stratigraphy of neogene deposits in the Khania Province, Crete, with special reference to foraminifera of the family Planorbulinidae and the genus *Heterostegina*. Utrecht Micropaleontological Bulletins, 1, 1–208.

Freudenthal, Th., 1972. On some larger orbitoidal foraminifera in the Tertiary of Senegal and Portuguese Guinea, Proceedings of the 4th African Micropal. Coll., (Abidjan, 1970), pp. 144–162.

Frost, S.H., Langenheim, R.L., 1974. Cenozoic Reef Biofacies. Tertiary Larger Foraminifera and Scleractinian Corals from Mexico. Northern Illinois University Press, De Kalb, pp. 1–388.

Fugagnoli, A., 1999. Cymbriaella, a new foraminiferal genus (Textulariina) from the Early Jurassic of the Venetian Prealps (Northeastern Italy). Revue de Micropaleontologie 42, 99–110.

Fugagnoli, A., 2000. First record of *Everticyclammina* Redmond 1964 (*E. praevirguliana* n. sp.; Foraminifera) from the Early Jurassic of the Venetian Prealps (Calcari Grigi, Trento Platform, Northern Italy). Journal of Foraminiferal Research, 30, 126–134.

Fugagnoli, A., 2004. Trophic regimes of benthic foraminiferal assemblages in Lower Jurassic shallow water carbonates from Northeastern Italy (Calcari Grigi, Trento Platform, Venetian Prealps). Paleogeography, Paleoclimatology, Paleoecology, 205, 111–130.

Furrer, V., and Septfontaine, M., 1977. Nouvelles données biostratigraphiques (à l'aide des foraminiferès) dans le Dogger. Faciés briançonnais des Préalpes médianes romandes (Suisse). Eclogae geol. Helv., 70, 717–737.

Gallagher S.J., 1998. Controls on the distribution of calcareous Foraminifera in the Lower Carboniferous of Ireland. Marine Micropaleontology, 34, 187–211.

Gallagher, S.J., Somerville, I.D., 2003. Lower Carboniferous (late Viséan) platform development and cyclicity in southern Ireland: foraminiferal biofacies and lithofacies evidence. Rivista Italiana di Paleontologia e Stratigrafia, 109, 159–171

Galloway, J.J., 1933. A Manual of the Foraminifera. James Furman Kemp Memorial ser. publ. 1, Principia. Press, Bloomington, Indiana, 183p.

García-Hernández, M., González-Donoso, J.M., Linares, A., Rivas, P., Vera, J. A., 1978. Características ambientales del Lías inferior y medio en la Zona Subbética y su significado en la interpretación general de la Cordillera. In: Universidad De Granada, Reunión sobre la Geodinámica de la Cordillera Bética y Mar de Alborán, Granada, Spain, May 12–14, 1976. University of Granada, Granada, 125–157.

Gargouri, S., Vachard, D., 1988. On *Hemgordiopsis* and other porcellaneous foraminifera from Jebel Tebaga (Upper Permian), Tunisia. Revue de Paléobiologie, Special vol. 2, Benthos '86, 56–688.

Gast, R.J., Caron, D.A., 1996. Molecular phylogeny of symbiotic dinoflagellates from Foraminifera and Radiolaria. Molecular Biology and Evolution, 13, 1192–1197.

Gattuso, J.-P., Buddemeier, R.W., 2000. Calcifcation and CO_2. Nature 407, 311–312.

Gaździcki, A., Trammer, J., and Zawidzka, K., 1975. Foraminifers from the Muschelkalk of Southern Poland, Acta Geologica Polonica, 25, 285–298.

Gersonde, R., Kyte, F.T., Bleil, U., Diekmann, B., Flores, J.A., Gohlk, K., Grahl, G., Hagen, R., Kuhn, G., Sierro, F.J., Völker, D., Abelmann, A., and Bostwick, J.A., 1997. Geological record and reconstruction of the late Pliocene impact of the Eltanin asteroid in the Southern Ocean. Nature, 390, 357–363.

Ghose, B.K., 1977. Paleoecology of the Cenozoic reefal foraminifers and algae - a brief review. Palaeogeography, Palaeoclimatology, Palaeoecology, 22, 231–256.

Glaessner, M.F., 1945. Principles of Micropaleontology. Melbourne University Press, 296p.

Glikson, A., 2005. Asteroid/comet impact clusters, flood basalts and mass extinctions: Significance of isotopic age overlaps. Earth Planet. Sci. Lett. 236, 933–937.

Gohrbandt, K.H.A., 1966. Some Cenomanian foraminifera from northwest Libya. Micropaleontology 12, 65–70.

Gold, D. P., Burgess, P. M., BouDagher-Fadel, M. K., 2017a. Carbonate drowning successions of the Bird's Head, Indonesia. Facies. doi:10.1007/s10347-017-0506-z.

Gold, D. P., White, L. T., Gunawan, I., BouDagher-Fadel, M. K., 2017b. Relative sea-level change in western New Guinea recorded by regional biostratigraphic data. Marine and Petroleum Geology, 86, 1133–1158. doi:10.1016/j.marpetgeo.2017.07.016.

González-Donoso, J. M., Linares, A., and Rivas, P., 1974. El Lías inferior y medio de Poloria (serie del Zegri, Zona Subbética, norte de Granada). Estudios Geológicos, 30, 639–654.

Görög, A., Arnaud Vanneau, A., 1996, Lower Cretaceous Orbitolinas from Venezuela: Micropaleontology, v. 42, p. 65–78.

Gould, S.J., 1977. Ontogeny and Phylogeny. Belknap Press of Harvard University Press, Cambridge, 357p. Gould, S.J., 2002. The Structure of Evolutionary Theory. Belknap Press of Harvard University Press, Cambridge, 1433p.

Grimsdale, T.F., 1959. Evolution in the American Lepidocyclinidae (Cainozoic Foraminifera), an interim view. Proc. Koninklijke Nederlandse Akadamie Wetenschappen, Amsterdam, Ser. B 62, 8–33.

Grønlund, H., Hansen, H.J., 1976. Scanning electron microscopy of some recent and fossil nodosariid foraminifera. Bulletin of the Geological Society of Denmark, 25, 121–134.

Groussin, M., Pawlowski, J., Yang, Z., 2011. Bayesian relaxed molecular clock analysis of divergence times: implications for the speciation timing in Foraminifera. Molecular Phylogeny and Evolution 61, 157–166.

Groves, J.R., 2000. Suborder Lagenina and other smaller foraminifers from Uppermost Pennsylvanian- Lower Permian rocks of Kansas and Oklahoma. Micropaleontology 46(4), 285–326.

Groves, J.R., 2005. Fusulinid wall structure in the *Profusulinella–Fusulinella* evolutionary transition. Bulletins of American Paleontology, 369, 199–218.

Groves, J.R., Altiner, D., 2004. Survival and recovery of calcareous foraminifera pursuant to the end- Permian mass extinction. C. R. Palevol. 4, 487–500.

Groves, R., Altiner, D., Rettori, R., 2003. Origin and early evolutionary radiation of the order Lagenida (foraminifera). Journal of Paleontology, 77, 831–843.

Groves, J.R., Altiner, D., Rettori, R., 2005. Decline and recovery of lagenide foraminifers in the Permian–Triassic boundary interval (Central Taurides, Turkey). Paleontological Society Memoir 62 [supplement to Journal of Paleontology, 79(4)], 38.

Groves, J.R., Boardman, D.R., 1999. Calcareous smaller foraminifers from the Lower Permian Council Grove Group near Hooser, Kansas. Journal of Foraminiferal Research, 29, 243–262.

Groves, J.R., Kulagina, E., Villa, E., 2007. Diachronous appearances of the Pennsylvanian fusulinid/ Profusulinella in Eurasia and North America. Journal of Paleontology, 81, 227–237.

Groves, J.R. and Lee, A. 2008. Accelerated rates of foraminiferal origination and extinction during the Late Paleozoic ice age. Journal of Foraminiferal Research, 38, 74–84.

Groves, J.R., Nemyrovska, T.I., Alekseev, A.S., 1999. Correlation of the type Bashkirian Stage (Middle Carboniferous, South Urals) to the Morrowan and Atokan series of the midcontinental and Western United States. Journal of Paleontology, 73, 529–539.

Groves, J.R., Rettori, R., Altiner, D., 2004. Wall structures in selected Paleozoic lagenide foraminifera. Journal of Paleontology, 78, 245–256.

Groves, J.R., Wahlman, G.P., 1997. Biostratigraphy and evolution of Upper Carboniferous and Lower Permian smaller foraminifers from the Barents Sea (offshore Arctic Norway). Journal of Paleontology, 71, 758–779.

Gušić, I., Velić, I., 1978. *Lituolipora polymorpha* n. gen., n. sp. (Foraminiferida Lituolacea?) from the Middle Liassic of the Outer Dinarids in Croatia and the establishment of a new family, Lituoliporidae. Geol. Vjesn. Inst. Geol. Istraz., 30, 73–93.

Gutnic, M., Moullade, M., 1967. Donnees nouvelles sur le Jurassique et le Crétacé inferieur du Barla Dag au sud de Senitkent (Taurus de Pisidie, Turquie). Ankara Maden Tetkik ve Arama Enstitusu. Bull., 69, 60–78.

Gvirtzman, G., Buchbinder, B., 1978. The late tertiary of the coastal plain of Israel and its bearing on the history of the southeastern Mediterranean coast. Initial reports DSDP 42(part 2), 1192–1222.

Hageman, S.A., Kaesler, R.L., 1998. Wall structure and growth of fusulinacean foraminifera. Journal of Paleontology, 72, 181–190.

Hallam, A., 1961. Cyclothems, transgressions and faunal changes in the Lias of northwest Europe. Trans. Edinburgh Geol. Soc., 18, 132–174.

Hallam, A., 1978. Eustatic cycles in the Jurassic. Palaeogeography, Palaeoclimatology, Palaeoecology, 23, 1–32.

Hallam, A., 1986. The Pliensbachian and Tithonian extinction events. Nature 319, 765–768.

Hallam, A., 1990. The end-Triassic mass extinctionevent. In: Sharpton, V. L., and Ward, P. D., (eds), Global catastrophes in Earth history, Geological Society of America, Special Paper, 247, 577–583.

Hallam, A., 1995. Oxygen-restricted facies of the basal Jurassic of north west Europe. Hist. Biol. 10, 247–257.

Hallam, A., Wignall, P.B., 1997. Mass Extinctions and Their Aftermath. Oxford University Press, Oxford, 307p.

Hallam, A., Wignall, P.B., 2000. Facies change across the Triassic–Jurassic boundary in Nevada, USA. J. Geol. Soc. 156, 453–456.

Hallet, D., 1966. A study of some microfossils from the Yoredale series of Yorkshire. PhD thesis, University College London.

Hallock, P., 1979. Trends in test shape in large, symbiont-bearing foraminifera. Journal of Foraminiferal Research, 9, 61–69.

Hallock, P., 1981. Production of carbonate sediments by selected large benthic foraminifera on two Pacific coral reefs. J. Sediment. Petrol. 51, 467–474.

Hallock, P., 1984. Distribution of selected species of living algal symbiont-bearing foraminifera on two Pacific coral reefs. Journal of Foraminiferal Research, 14, 250–261.

Hallock, P., 1985. Why are larger foraminifera large? Paleobiology 11, 195–208.

Hallock, P., 1999. Advantages of algal symbiosis. In: Sen Gupta, B.K. (Ed.), Modern Foraminifera. Kluwer Academic, New York, pp. 123–139.

Hallock, P., 2000. Larger foraminifera as indicators of coral-reef vitality. Environ. Micropaleontol. 15, 121–150.

Hallock, P., 2001. Coral reefs, carbonate sediment, nutrients, and global change. In: Stanley, G.D. (Ed.), Ancient Reef Ecosystems, their Evolution, Paleoecology and Importance in Earth History. Kluwer Academic/Plenum Publishers, New York, pp. 388–427.

Hallock, P., 2005. Global change and modern coral reefs: New opportunities to understand shallow-water carbonate depositional processes. Sediment. Geol. 175, 19–33.

Hallock, P., Glenn, E.C., 1985. Paleoenvironmental analysis of foraminiferal assemblages, a tool for recognizing depositional facies in lower Miocene reef complexes. Journal of Paleontology, 59, 1382–1394.

Hallock, P., Glenn, E.C., 1986. Larger foraminifera, a tool for paleoenvironmental analysis of Cenozoic carbonate facies. Palaios 1, 55–64.

Hallock, P., Hansen, H.J., 1979. Depth adaptation in *Amphistegina*, change in lamellar thickness. Bulletin of the Geological Society of Denmark, 27, 99–104.

Hallock, P., Röttger, R., Wetmore, K., 1991. Hypotheses on form and function in foraminifera. In: Lee, J.J. and Anderson, O.R., (eds), *Biology of Foraminifera*. Academic Press, New York, 41–72.

Hallock, P., Talge, H.K., 1994. A predatory foraminifer, *Floresina amphiphaga*, n. sp., from the Florida Keys. Journal of Foraminiferal Research, 24, 210-213.

Hamaoui, M., Fourcade, E., 1973. Revision de Rhapydionimae (alveolinidae, foraminiferes). Bull. Cent. Rech. Pau-SPNA 7, 361–435.

Hansen, H. J., Buchardt, B. 1977, Depth distribution of *Amphistegina* in the Gulf of Elat. *Utrecht Micropal. Bull.,* 15, 205–224.

Hanzawa, S., 1957. Cenozoic foraminifera of Micronesia. Geol. Soc. Am. Mem. 66, p. 163.

Hanzawa, S., 1964. The phylomorphogeneses of the Tertiary foraminiferal families Lepidocylinidae and Miogypsinidae. Tohoku Univ. Sci. Repts., 2nd. ser. (Geol.), 35, 295–313.

Hanzawa, S., 1965. The ontogeny and the evolution of larger Foraminifera. Tohoku Univ. Sci. Repts., ser. 2 (Geol.), 36, 239–256.

Haq, B.U., Al-Qahtani, A.M., 2005. Phanerozoic cycles of sea-level change on the Arabian platform. GeoArabia 10, 127–160.

Harney J.N., Hallock P., Talge H.K.0, 1998. Observations on a trimorphic life in *Amphistegina gibbosa* populations from the Florida Keys. Journal of Foraminiferal Research, 28, 141–147.

Harries, P.J., Little, C.T.S., 1999. The early Toarcian (Early Jurassic) and the Cenomanian–Turonian (Late Cretaceous) mass extinctions, similarities and contrasts. Paleogeography, Paleoclimatology, Paleoecology, 154, 39–66.

Haynes, J.R., 1965. Symbiosis, wall structure and habitat in foraminifera. Contributions of the Cushman Foundation of Foraminiferal Research, 16, 40–43.

Haynes, J.R., 1981. Foraminifera. MacMillan, London, 433p.

Hemleben, C.H., Be, A.W.H., Anderson, O.R., Tuntivate-Choy, S., 1977. Test morphology, organic layers and chamber formation of the planktonic foraminifer *Globorotalia menardii* (D'Orbigny). Jour. Foram. Res., 7, 1–25.

Henderson, A. L., Najman, Y., Parrish, R. R., Carter, A., BouDagher-Fadel, M. K., & et al., 2008. Constraints to the timing of India-Eurasia collision determined from the Indus Group: a reassessment. Himalayan Journal of Sciences, 5 (7), 64-?. doi:10.3126/hjs.v5i7.1266.

Henderson, A. L., Najman, Y., Parrish, R., BouDagher-Fadel, M. K., Barford, D., Garzanti, E., and Andò, S., 2010. Geology of the Cenozoic Indus Basin sedimentary rocks: Paleoenvironmental interpretation of sedimentation from the western Himalaya during the early phases of India-Eurasia collision. Tectonics, 29, TC6015. doi:10.1029/2009TC002651.

Henson, F.R.S., 1948. Larger imperforate Foraminifera of South-Western Asia. Brit. Mus. (Nat.Hist.), London, p. 127.

Henson, F.R.S., 1950. Middle Eastern Tertiary Peneroplidae (Foraminifera), with Remarks on the Phylogeny and Taxonomy of the Family. Doctoral thesis, University of Leiden, p. 70.

Hesselbo, S.P., Robinson, S.A., Surlyk, F., Piasecki, S., 2002. Terrestrial and marine extinction at the Triassic–Jurassic boundary synchronized with major carbon-cycle perturbations, a link to initiation of massive volcanism? Geology 30, 251–254.

Higgins, J.A., Schrag, D.P., 2006. Beyond methane, towards a theory for Paleocene–Eocene thermal maximum. Earth Planet. Sci. Lett. 245, 523–537.

Hofker, J. 1927. The foraminifera of the Siboga Expedition, Pt. 1, Tinoporidae, Rotaliidae, Nummulitidae, Amphisteginidae. E. J. Brill, Leiden, 74 p.

Hofker, J., 1965. Methods for study and preparation of foraminifera. In: Kummel, B., Raup, D. (Eds), Handbook of Paleontological Techniques. Freeman W.H. and Co., San Francisco, pp. 251–256.

Hohenegger, J., 2000. Coenoclines of larger foraminifera. Micropaleontology 46 (Suppl., 127–151.

Hohenegger, J., 2004. Depth coenoclines and environmental consideration of Western Pacific larger foraminifera. Journal of Foraminiferal Research, 34, 9–33.

Hohenegger, J., Piller, W., 1975, Wandstrukturen und Grossgliederung der foraminiferen, Sitzungsberichten Osterreichisch Akademie der Wissenscaften, Mathematisch-naturwissenschaftliche Klasse, Abteilung I, 184, 67–96.

Hohenegger, J., Yordanova, E., 2001. Depth-transport functions and erosion–deposition diagrams as indicators of slope inclination and time-averaged traction forces, application in tropical reef environments. Sedimentology 48, 1025–1046.

Hohenegger, J., Yordanova, E., Hatta, A., 2000. Remarks on western Pacific nummulitidae (foraminifer- ida). Journal of Foraminiferal Research, 30, 3–28.

Holzmann, M., Hohenegger, J., Pawlowski, J., 2003. Molecular data reveal parallel evolution in nummulitid foraminifera. Journal of Foraminiferal Research, 33, 277–284.

Hottinger, L., 1960. Uber paleocaene und eocaene Alveolinen. Eclogae geologicae Helvetiae 53, 265–283.

Hottinger, L., 1967. Foraminiferes imperfores du Mesozoique marocain. Notes et Mémoires Service Géologique. 209, 1–179.

Hottinger, L., 1969. The foraminiferal genus *Yaberinella* Vaughan, 1928, with remarks on its species and its systematic position. Eclogae geologicae Helvetiae 62, 743–749.

Hottinger, L., 1973. Selected Paleogene larger foraminifera. In: Hallam, A. (Ed.), Atlas of Palaeobiogeo- graphy. Elsevier, Amsterdam, pp. 443–452.

Hottinger, L., 1977. Foraminifères operculiniformes. Mem. Mus. Natl. Hist. Nat. (Paris), C 40, 1–159. Hottinger, L., 1978. Comparative anatomy of elementary shell structures in selected larger Foraminifera.

Hottinger, L., 1979. Comparative anatomy of elementary shell structures in selected larger foraminifera. In: Headley, R.H., Adams, C.G. (Eds), Foraminifera. Academic Press, London, Vol. 3, pp. 203–266.

Hottinger, L., 1982. Larger foraminifera, giant cells with a historical background. Naturwissenschaften 69, 361–371.

Hottinger, L., 1996. Sels nutritifs et biosedimentation. Mem. Soc. Geol. France n.s. 169, 99–107.

Hottinger, L., 1997. Shallow benthic foraminiferal assemblages as signals for depth of their deposition and their limitations. Bulletin de la Socie´ te´ Ge´ ologique de France 168, 491–505.

Hottinger, L., 2000. Functional morphology of benthic foraminiferal shells, envelopes of cells beyond measure. Micropaleontology 46(Suppl. 1), 57–86.

Hottinger, L., 2001. Archaiasinids and related porcelaneous larger foraminifera from the Late Miocene of the Dominican Republic. Journal of Paleontology, 75, 475–512.

Hottinger, L., 2006. Illustrated glossary of terms used in foraminiferal research. Carnets de Géologie / Notebooks on Geology, Memoir 2006/02.

Hottinger, L., 2007. Revision of the foraminiferal genus *Globoreticulina* Rahaghi, 1978, and of its associated fauna of larger foraminifera from the late Middle Eocene of Iran. Notebooks on Geology (Brest), CG2007-A06, 1–51.

Hottinger, L., 2009. The Paleocene and Earliest Eocene foraminiferal family Miscellaneidae: neither nummulitids nor rotali- ids. Carnets de Géologie (Brest), article 2009/06, 1–41.

Hottinger, L., 2013. Micropaleontology in Basel (Switzerland) during the twentieth century, rise and fall of one of the smaller fields of the life sciences. In Bowden, A.J., Gregory, F.J., and Henderson, A.S. (eds.): Landmarks in Foraminiferal Micropaleontology: History and Development. The Micropalaeontological Society Special Publications No. 6, Geological Society Publishing House, Bath, U.K., 317–335.

Hottinger, L., Caus, E., 1982. Marginoporiform structure in *Ilerdorbis decussatus* n. gen. n. sp., a Senonian, agglutinated, discoidal foraminifer. Eclogae Geologicae Helvetiae 75, 807–819.

Hottinger, L., Dreher, D., 1974. Differentiation of Protoplasm in Nummulitidae (Foraminifera) from Elat, Red Sea. Mar. Biol. 25, 41–61.

Hottinger, L., 1978. Comparative Anatomy of Elementary Shell Structures in Selected Larger Foraminifera. In: Hedley, R.H., and Adams, C.G., (eds), *Foraminifera 3*. Acad.Press, London, 203–266.

Hottinger, L., Drobne, K., 1980. Early Tertiary conical imperforate foraminifera. Rasprave IV. razr., SAZU 22, 188–276.

Hottinger, L., 1982. Larger Foraminifera, Giant Cells with a Historical Background. Naturwissenschaften, 69, 361–371.

Hottinger, L., Drobne, K., 1989. Late Cretaceous, larger complex Miliolids (foraminifera) endemic in the pyrenean faunal province. Facies 21, 99–134.

Hottinger, L., Drobne, K., Caus, E., 1989. Late Cretaceous, Larger, Complex Miliolids (Foraminifera) Endemic in the Pyrenean Faunal Province. Facies (Erlangen), 21, 99–134

Hottinger, L., Halicz, E., Reiss, Z., 1991. The foraminiferal genera *Pararotalia, Neorotalia*, and *Calcarina*, taxonomic revision. Journal of Paleontology, 65, 18–33.

Hottinger, L., Halicz, E., Reiss, Z., 1993. Recent Foraminifera from the Gulf of Aqaba, Red Sea. Opera Slovenien Acad. Sci. Arts (SAZU), 33, 1–179.

Hottinger, L., Leutenegger, S., 1980. The structure of calcarinid foraminifers. Schweiz. Palaont. Abh. Basel. 101, 115–154.

Hottinger, L., Romero, J., Caus, E., 2001. Architecture and revision of the pellatispirines, planispiral canaliferous foraminifera from the Late Eocene Tethys. Micropaleontology 47(Suppl. 2), 35–77.

Hottingerina from the Paleocene of Turkey. Micropaleontology, 45, 113–137.

Hu, X., Wang, J., Boudagher-Fadel, M. K., Garzanti, E., Wei, A., 2013. Transition from forearc basin to syn-collisional basin in Southern Tibet (Paleocene Cuojiangding Group): Implication to timing of the India-Asia initial collision and of Yarlung Zangbo ophiolite emplacement. Acta Geologica Sinica (English Edition), 87 (Supp), 27–28.

Hu, X., Wang, J., BouDagher-Fadel, M., Garzanti, E., An, W., 2015. New insights into the timing of the India-Asia collision from the Paleogene Quxia and Jialazi formations of the Xigaze forearc basin, South Tibet. Gondwana Research, 32, 76–92. doi:10.1016/j.gr.2015.02.007.

Hughes, G.W.E., 2004. Palaeoenvironments of lower Aptian agglutinated foraminifera of Saudi Arabia. In: Bubik, M., Kaminski, M. (Eds), Proceedings of the Sixth International Workshop on Agglutinated Foraminifera. Grzybowski Foundation, special publication, 8, 197–207.

Hughes, G.W., 2014. Micropalaeontology and palaeoenvironments of the Miocene Wadi Waqb carbonate of the northern Saudi Arabian Red Sea Gulf. GeoArabia, 19, 59–108.

Isozaki, Y., 2009. Integrated "plume winter" scenario for the double-phased extinction during the Paleozoic-Mesozoic transition: G-LB and P-TB events from a Panthalassan perspective. Journal of Asian Earth Sciences 36, 459–480.

Jain, S., Collins, L.S., 2007. Trends in Caribbean paleoproductivity related to the Neogene closure of the Central American Seaway. Mar. Micropaleontol. 63, 57–74.

Jakobsson, M., Backman, J., Rudels, B., Nycander, J., Frank, M., Mayer, L., Jokat, W., Sangiorgi, F., O'Regan, M., Brinkhuis, H., King, J., Moran, K., 2007. The early Miocene onset of a ventilated circulation regime in the Arctic Ocean. Nature 447, 986–990.

Jenkyns, H.C., 1988. The early Toarcian (Jurassic) anoxic event, stratigraphic, sedimentary, and geochemical evidence. Am. J. Sci. 288, 101–151.

Jenkyns, H.C., 1999. Mesozoic anoxic events and paleoclimate. Zentralblatt /ur Géologie und Palâontologie, 1997, 943–949.

Jenkyns, H.C., 2003. Evidence for rapid climate change in the Mesozoic–Palaeogene greenhouse world. Royal Society of London Philosophical Transactions Ser. A, 361, 1885–1916.

Jenkyns, H.C., Clayton, C.J., 1997. Lower Jurassic epicontinental carbonates and mudstones from England and Wales, chemostratigraphic signals and the Early Toarcian anoxic event. Sedimentology 44, 687–706.

Jenkyns, H.C., Wilson, P.A., 1999. Stratigraphy, paleoceanography and evolution of cretaceous Pacific guyots, relics from a greenhouse earth. Am. J. Sci. 299, 341–392.

Jin-zhang, S., 1990. Development of fusuline foraminifera in China. Stud. Benthic Foraminifera, Benthos 90, 11–22.

Johnson, C.C., Barron, E., Kauffman, E., Arthur, M., Fawcett, P., Yasuda, M., 1996. Middle cretaceous reef collapse linked to ocean heat transport. Geology 24, 376–380.

Johnson, C.C., Kauffman, E.G., 1990. Originations, radiations and extinctions of cretaceous rudistid bivalve species in the Caribbean province. In: Kauffman, E.G., Walliser, O.H. (Eds), Extinction Events in Earth History. Springer-Verlag, New York, pp. 305–324.

Johnson, K.G., Todd, J.A., Jackson, J.B.C., 2007. Coral reef development drives molluscan diversity increase at local and regional scales in the late Neogene and Quaternary of the southwestern Caribbean. Paleobiology 33, 24–52.

Jolivet, L., Augier, R., Robin, C., Suc, J.-P., Rouchy, J.M., 2006. Lithospheric-scale geodynamic context of the Messinian salinity crisis. Sediment. Geol. 188–189, 9–33.

Jones, A.P., Price, G.D., Price, N.J., De Carli, P.S., Clegg, R.A., 2002. Impact induced melting and the development of large igneous provinces. Earth Planet. Sci. Lett. 202, 551–561.

Jones, B., Pickering, K. T., BouDagher-Fadel, M. K., Matthews, S., 2003. Micropalaeontological characterisation of submarine fan/channel sub-environments, Ainsa system, south-central Pyrenees, Spain. The Micropalaeontological Society Special Publications, Recent Development 55–68.

Jones, E. J. W., BouDagher-Fadel, M. K., & Thirlwall, M.F., 2002. An investigation of seamount phosphorites in the Eastern Equatorial Atlantic. Marine Geology, 183 (1–4), 143–162. doi:10.1016/S0025-3227(01)00254-7.

Jorry, S., Hasler, C.-A., Davaud, E., 2006. Hydrodynamic behaviour of *Nummulites*, implications for depositional models. Facies 52, 221–235.

Kaiho, K., Kajiwara, Y., Nakano, T., Miura, Y., Kawahata, H., Tazaki, K., Ueshima, M., Chen, Z.Q., Shi, G.R., 2001. End-Permian catastrophe by a bolide impact, evidence of a gigantic release of sulphur from mantle. Geology 29, 815–818.

Kaiser, S.I., Steuber, T., Becker, R.T., Joachimski, M.M., 2006. Geochemical evidence for major environmental change at the Devonian–Carboniferous boundary in the Carnic Alps and the Rhenish Massif. Palaeogeography, Palaeoclimatology, Palaeoecology, 240, 146–160.

Kalia, P., Banerjee, A., 1995. Ocurrence of the genus *Eoconuloides* Cole and Bermudez 1944 in Rajasthan, India and its paleobiogeographic significance. Micropaleontology 41, 69–76.

Kaminski, M., 1910. The year 2010 classification of the agglutinated foraminifera. Micropaleontology, 60, 89–108.

Kauffman, E.G., Erwin, D.H., 1995. Surviving mass extinctions. Geotimes 14, 14–17.

Kauffman, E.G., Johnson, C.C., 1988. The morphological and ecological evolution of Middle and Upper Cretaceous reef-building rudistids. Palaios 3, 194–216.

Kalvoda, J., 1989. Tournaisian events in Moravia and their significance. Courier Forschungsinstitut Senckenberg 117, 353–358.

Kalvoda, J., 2002. Late Devonian–early Carboniferous foraminiferal fauna: zonations, evolutionary events, paleobiogeography and tectonic implications. Folia Facultatis Scientiarium Naturalium Universitatis Masarykianae Brunensis, Geologia 39, 1–213.

Katz, M.E., Miller, K.G., Wright, J.D., Wade, B.S., Browning, J.V., Cramer, B.S., Rosenthal, Y., 2008, Stepwise transition from greenhouse to icehouse climates: Nature Geoscience, 1, 329–334, doi: 10.1038/ngeo179.

Keller, G., Stinnesbeck, W., Adatte, T., Stueben, D., 2003. Multiple impacts across the Cretaceous–Tertiary boundary. Earth Sci. Rev. 62, 327–363.

Kelley, S.P., Gurov, E., 2002. Boltysh, another end-Cretaceous impact. Meteorit. Planet. Sci. 37, 1031–1043. Kerr, R.A., 2006. Creatures great and small are stirring the ocean. Science 313, p. 1717.

Kiessling, W., 2002. Secular variations in the Phanerozoic reef ecosystem. In: Kiessling, W., Flu¨ gel, E., Golonka, J. (Eds), Phanerozoic Reef Patterns. SEPM Special Publications, Tulsa, Vol. 72, pp. 625–690.

Kiessling, W., Flügel, E., and Golonka, J., 2003. Patterns of Phanerozoic carbonate platform sedimentation. Lethaia, 36, 195–226.

Kirby, M., Jackson, J., 2004. Extinction of a fast-growing oyster and changing ocean circulation in Pliocene tropical America. Geology 32, 1025–1028.

Knoll, A.H., Bambach, R.K., Canfield, D.E., Grotzinger, J.P., 1996. Comparative earth history and Late Permian mass extinction science 273, 452–457.

Knoll, A.H., Bambach, R.K., Payne, J.L., Pruss, S., Fischer, W.W., 2007. Paleophysiology and end-Permian mass extinction. Earth Planet. Sci. Lett. 256, 295–313.

Kochansky-Devidé, V., 1970,: Die Kalkalgen des Karbons vom Velebit-Gebirge (Moskovien und Kassimovien). Palaeont. Jugosl., 10, 32 p.

Krebs, CJ., 1972. Ecology: The experimental analysis of distribution and abundance. New York: Harper and Row.

Krüger, R., 1994. Untersuchungen zum Entwicklungsgang nrezenter Nummulitiden, *Heterostegina depressa*, *Nummulites venosus* und *Cycloclypeus carpenteri*. Hochschulschriften, 70, 1–98.

Küpper, K., 1954. Notes on Cretaceous larger foraminifera. I. Genus *Orbitoides* in America. Contributions from the Cushman Foundation for Foraminiferal Research, 5, 63–67.

Kurbatov, V.V., 1971. Foraminiferal basis for the Jurassic section of Kugitanga and contiguous regions. In: Paleontologicheskoe obosnovanie opornykh razrezov yurskoy sistemy Uzbekistana I sopredel'nukh rayonov. Sbornik, 10, 117–132.

Kuroda, J., Ogawa, N., Tanimizu, M., Coffin, M., Tokuyama, H., Kitazato, H., Ohkouchi, N., 2007. Contemporaneous massive subaerial volcanism and late Cretaceous oceanic anoxic event 2. Earth Planet. Sci. Lett. 256, 211–223.

Laagland, H., 1990. *Cycloclypeus* in the Mediterranean Oligocene. Utrecht Micropaleontological Bulletins, 39, p. 171.

Langer, M., 1995. Oxygen and carbon isotopic composition of recent larger and smaller foraminifera from the Madagan Lagoon (Papua New Guinea). Mar. Micropaleontol. 26, 215–221.

Langer, M.R., Hottinger, L., 2000. Biogeography of selected "larger" foraminifera. Micropaleontology 46(Suppl. 1), 105–127.

Langer, M.R., Lipps, J.H., 2003. Foraminiferal distribution and diversity, Madang Reef and Lagoon, Papua New Guinea. Coral Reefs 22, 143–154.

Larsen, A. R., Drooger, C. W. 1977: Relative thickness of the test in the Amphistegina species of the Gulf of Elat. Utrecht Micropaleontologica Bulletins, 15, 225–240.

Latimer, J.C., Filippelli, G.M., 2002. Eocene to Miocene terrigenous inputs and export production, geochemical evidence from ODP Leg 177, Site 1090. Paleogeography, Paleoclimatology, Paleoecology, 182, 151–164.

Lawver, L.A., Gahagan, L.M., Coffin, M.F., 1992. The development of paleoseaways around Antarctica. In: Kennett, J.P., Warnke, D.A. (Eds), The Antarctic Paleoenvironment, A Perspective on Global Change, American Geophysical Union Monograph, Vol. 56, pp. 7–30.

LeCalvez, J., 1950. Recherches sur les foraminiferes, II. Place de la meiose et sexualite´. Archis de Zoologie Experimentale et Generale 87, 211–243.

Lee, J.J. 1990. 29. Phylum Granuloreticulosa (Foraminifera). In L. Margulis et al. (eds.), Handbook of Protoctista: 524–548.

Lee, J.J., Anderson, O.R., 1991. Symbiosis in Foraminifera. In: Lee, J.J., Anderson, O.R. (Eds), Biology of Foraminifera. Academic Press, New York, pp. 157–220.

Lee, J.J., Burnham, B., Cevasco, M.E., 2004. A new modern soritid foraminifer, *Amphisorus saurensis*n. sp., from the Lizard Island group (Great Barrier Reef, Australia). Micropaleontology 50, 357–368.

Lee, J.J., Morales, J., Bacus, S., Diamont, A., Hallock, P., Pawlowski, J., Thorpe, J., 1997. Progress in characterizing the endosymbiotic dinoflagellates of soritid foraminifera and related studies on some stages in the life cycle of *Marginopora vertebralis*. Journal of Foraminiferal Research, 27, 254–263.

Lehmann, C., Osleger, D.A., Montan~ ez, I.P., Sliter, W., Arnaud-Vanneau, A., Banner, J., 1999. Evolution of Cupido and Coahuila carbonate platforms, Early Cretaceous, Northeastern Mexico. GSA Bull. 111, 1010–1029.

Leppig, U., 1992. Functional anatomy of Fusulinids (foraminifera): Significance of the polar torsion illustrated in *Triticites* and *Schwagerina* (Schwagerinidae). Palaontologische Zeitschrift 66, 39–50.

Less, G., 1987. Paleontology and stratigraphy of the European orthophragminae. Geologica Hungarica, Ser. Paleontologica Fasc. 51, p. 373.

Less, G., 1998. The zonation of the Mediterranean Upper Paleocene and Eocene by Orthopfragminae. In: Hottinger, L., Drobne, K. (Eds), Paleogene Shallow Benthos of the Tethys, Slovenian Academy of Sciences and Arts, Ljubljana, Vol. 2, pp. 21–43.

Less, G., Kovács, L.O., 1995. Age-estimates by European paleogene orthophragminae using numerical evolutionary correlation. Geobios 29, 261–285.

Less, G., Özcan, E., Báldi-Beke, M., Kollányi, K., 2007. Thanetian and early Ypresian Orthophragmines (foraminifera, Discocyclinidae and Orbitoclypeidae) from the central western Tethys (Turkey, Italy and Bulgaria) and their revised taxonomy and biostratigraphy. Rivista Italiana di Paleontogia e Stratigrafia 113, 415–448.

Leupold, W., van der Vlerk, I.M., 1931. The tertiary. Leidsche Geologische Mededelingen 5, 611–648.

Leutenegger, S., 1977. Ultrastructure de Foraminifères pérforés et impérforés ainsi que de leurs symbiotes. Cah. Micropal. 3, 1–52.

Leutenegger, S., 1984. Symbiosis in benthic foraminifera: Specifity and host adaptations. Journal of Foraminiferal Research, 14, 16–35.

Leven, E., 1993. Early Permian fusulinids from the Central Pamir. Revista Italiana di Paleontologia e Stratigrafia, 104, 3–42.

Leven, E., 1997. Permian stratigraphy and fusulinida of Afghanistan with their paleogeographic and paleotectonic implications. In: Stevens, C.H., Baars, D.L. (Eds), special paper, Geological Society of America, Special Paper, Washington, DC, Vol. 316, pp. 1–135.

Leven, E., 2003. Diversity dynamics of fusulinid genera and main stages of their evolution. Stratigraphy and Geological Correlation, 11, 220–230.

Leven, E., 2009. Origin of Higher Fusulinids of the Order Neoschwagerinida Minato et Honjo, 1966. Stratigraphy and Geological Correlation, 18, 290–297.

Leven, E., Gorgij, M.N., 2006. Upper Carboniferous–Permian stratigraphy and fusulinids from the Anarak region, central Iran. Russ. J. Earth Sci. 8, p. ES2002.

Leven, E., Korchagin, O.A., 2001. Permian–Triassic biotic crisis and foraminifers. Stratigr. Geol. Correl. 9, 364–372.

Levinton, J.S., 1970. The paleoecological significance of opportunistic species. Lethaia 3, 69–78.

Levy, A., 1977. Revision micropalontologique des Soritidae actuels Bahamiens. Un nouveau genre, Androsina. Bull. Centre Recherche Exploration-Production ELF-Aquitaine 1, 393–449.

Li, J., Hu, X., Garzanti, E., BouDagher-Fadel, M., 2016. Shallow-water carbonate responses to the Paleocene–Eocene thermal maximum in the Tethyan Himalaya (southern Tibet): Tectonic and climatic implications. Palaeogeography, Palaeoclimatology, Palaeoecology, 466, 153–165. doi:10.1016/j.palaeo.2016.11.026.

Lipina, O.A,. 1965. Sistematika Turneyellid. (Systematics of the Tournayellidae). Akad. Nauk SSSR, Geologicheskii Irzstitut, Moscou., Trudy, 130, 1–115.

Little, C., Benton, M., 1995. Early Jurassic mass extinction: A global long-term event. Geology 23, 495–498.

Lobegeier, M.K., 2002. Benthic foraminifera of the family Calcarinidae from Green Island Reef, Great Barrier Reef Province. Journal of Foraminiferal Research, 32, 201–216.

Loeblich, A.R., Jr., Tappan, H., 1949. Formainifera from the Walnut Formation (lower Cretaceous) of northern Texas and southern Oklahoma. Journal of Paleontology, 23, 245–266.

Loeblich, A.R., Jr., Tappan, H., 1961. Suprageneric classification of the Rhizopodea. Journal of Paleontology, 35, 245–330.

Loeblich, A.R., Jr., Tappan, H., 1964. Protista 2, Sarcodina, chiefly "Thecamoebians" and Foraminiferida. In: Moore, R.C. (Ed.), Treatise on Invertebrate Paleontology, University of Kansas Press, Kansas, Part C, Vols. 1 and 2.

Loeblich, A.R., Jr., and Tappan, H., 1984. Suprageneric classification of the Foraminiferida (Protozoa). Micropaleontology, 30, 1–70.

Loeblich, A.R., Jr., and Tappan, H., 1985. Some new and redefined genera and families of agglutinated foraminifera I. Journal of Foraminiferal Research, 15, 91–104.

Loeblich, A.R., Jr., Tappan, H., 1986. Some new and redefined genera and families of Textulariina, Fusulinina, Involutinina, and Miliolina (Foraminiferida). Journal of Foraminiferal Research, 16, 334–346.

Loeblich, A.R., Jr., Tappan, H., 1988. Foraminiferal Genera and their Classification. Van Nostrand Reinhold, New York, two volumes, 2047pp.

Loget, N., van Den Driessche, J., Davy, P., 2003. How did the Messinian crisis end up? Terra Nova 17, 414–419.

Lord, A. R., Harrison, R. W., BouDagher-Fadel, M., Stone, B. D., Varol, O., 2009. Miocene mass-transport sediments, Troodos Massif, Cyprus. Proceedings of the Geologists' Association, 120 (2–3), 133–138. doi:10.1016/j.pgeola.2009.08.001.

Louis-Schmid, B., Rais, P., Bernasconi, S.M., 2007. Detailed record of the mid-Oxfordian (Late Jurassic) positive carbon-isotope excursion in two hemipelagic sections (France and Switzerland), A plate tectonic trigger? Paleogeography, Paleoclimatology, Paleoecology, 248, 459–472.

Lucas, G., 1939. *Dictyoconus cayeuxi* n. sp. Foraminifère de grande taille de l'Aalénien de l'Oranie. Compte rendu sommaire des séances de la société géologiques de la France 353–355.

Lunt, P., Allan, T., 2004. Larger foraminifera in Indonesian biostratigraphy, calibrated to isotopic dating. GRDC Museum Workshop on Micropaleontology, Bandung, June, p. 109.

MacArthur, R.H., 1955. Fluctuations of animal populations, and a measure of community stability. Ecology 35, 533–536.

MacGillavry, H.J., 1962. Lineages in the genus *Cycloclypeus* carpenter (foraminifera) I and II. Proc. Koninklijke Nederlandse Akadamie Wetenschappen, Amsterdam 65 (Ser. B), 429–458.

Macgillavry, H.J., 1978. Foraminifera and parallel evolution-How or why? Geologieen Mijnbouw 57, 385–394.

Macleod, N., 2013. The great extinctions: what causes them and how they shape life. The Natural History Museum, London, 192p.

Macleod, N., Rawson, P. F., Forey, P. L., Banner, F. T., BouDagher-Fadel, M. K., Bown, P. R., . . . Young, J. R. (1997). The Cretaceous-Tertiary biotic transition. Journal of the Geological Society, 154, 265–292.

Mahoney, J.J., Duncan, R.A., Tejada, M.L.G., Sager, W.W., Bralower, T.J., 2005. Jurassic–Cretaceous boundary age and mid-ocean-ridge-type mantle source for Shatsky Rise. Geology 33, 185–188.

Mamet, B., Zhu, Z., 2005. Carboniferous and Permian algal microflora, Tarim Basin (China). Geologica Belgica 8, 3–13.

Márquez, L., 2005. Foraminiferal fauna recovered after the Late Permian extinctions in Iberia and the westernmost Tethys area. Palaeogeography, Palaeoclimatology, Palaeoecology, 229, 137–157.

Márquez, L., Trifonova, E., 2000. Tasas evolutivas de algunos subórdenes de foraminíferos triásicos del área occidental del Thetys. Revista Española de Micropaleontología, 32, 1–19.

Marriner, N., Morhange, C., Boudagher-Fadel, M., Bourcier, M., & Carbonel, P., 2005. Geoarchaeology of Tyre's ancient northern harbour, Phoenicia. Journal of Archaeological Science, 32, 1302–1327.

Marriner, N., Morhange, C., Rycx, Y., BouDagher-Fadel, M. K., Bourcier, M., Carbonel, P., Goiran J-P, Noujaim-Clark, G., 2005. Holocene coastal dynamics along the Tyrian peninsula, palaeo-geography of the northern harbour. Bulletin d'Archéologie et d'Architecture Libanaises, BAAL, Hors-Série 2, 61–89.

Martin, K., 1880. Tertiarschichten auf Java. Leiden, E.J. Brill, Lief, 3, Paleont. Theil (1879–1880), Netherland, 150–164.

Martindale, W., 1992. Calcified epibionts as palaeoecological tools, examples from the recent and Pleistocene reefs of Barbados. Coral Reefs 11, 167–177.

Martini, R., Rettori, R., Urosevic, D., Zaninetti, L., 1995. Le genre *Piallina* Rettori and Zaninetti (Foraminifère) dans des calcaires à Turriglomines du Trias (Carnien) de Serbie orientale (Domaine Carpatho-Balkanique). Revue de Pale´ obiologie 14, 411–415.

Marzoli, A., Bertrand, H., Knight, K., Cirilli, S., Buratti, N., Vérati, C., Nomade, S., Renne, P.R., Youbi, M., Martini, R., Allenbach, K., Rapaille, C., Zaninetti, L., Bellieni, G., 2004. Synchrony of the central Atlantic magmatic province and the Triassic–Jurassic boundary climatic and biotic crisis. Geology 32, 973–976.

Marzoli, A., Renne, P.R., Piccirillo, E.M., Ernesto, M., Bellieni, G., De Mind, A., 1999. Extensive 200-million-year-old continental flood basalts of the central Atlantic Magmatic Province. Science 284, 616–618.

Masse, J.-P., 1976. Les calcaires urgoniens de Provence (Valanginian-Aptien). Stratigraphie, paléontologie, les paléoenvironnements et leur évolution. Thése Doct. D'Etat, Univ. Aix-Marseille, France, Vol. I, II, p. 445.

Mateu-Vicens, G, Hallock, P, Brandano, M. 2009. Test shape variability of *Amphistegina* d'Orbigny 1826 as a paleobathymetric proxy: application to two Miocene examples. In: Demchuk T, Gary A, editors. Geologic problems solving with microfossils, 67–82. (SEPM Special Publication 93).

Mateu-Vicens, G., Pomar, L., Ferrandez-Ca~nadell, C., 2012. Nummulitic banks in the upper Lutetian "Buil level", Ainsa basin, South Central pyrenean zone: the impact of internal waves. Sedimentology 59, 527–552.

Matsumaru, K., 1991. On the evolutionary classification of the Family Lepidocyclinidae (Foraminiferida). Trans. Proc. Paleontol. Soc. Japan, new ser. 164, 883–909.

Matsumaru, K., 1996. Tertiary larger foraminifera (foraminiferida) from the Ogasawara Islands, Japan.

Matteucci, R., Schiavinotto, F., 1985. Deux nouvelles espèces de la lignée Cycloclypeus sans ornement de la région Méditerranéenne Proceedings of the Koninklke Nederlandse Akademie van Wetenschappen. Series B. Palaeontology, geology, physics and chemistry, Netherland, 88, 123–130.

Maxwell, W.G.H., 1968. Atlas of the Great Barrier Reef. Elsevier, Amsterdam, 258p.

Maxwell, W.G.H., Day, R.W., Fleming, P.J.C., 1961. Carbonate sedimentation on the Heron Island reef, Great Barrier Reef. J. Sediment. Petrol. 31, 215–230.

Maync, W., 1952. Critical taxonomic study and nomenclatural revision of the family Lituolidae based upon the prototype of the family, Lituola nautiloidea Lamarck, 1804. Contrib. Cushman Foundation for Foraminiferal Res. 3, 35–53.

Maync, W., 1953. *Hemicyclammina sigali*, n.gen. n.sp. from the Cenomanian of Algeria. Contribution. Cushman Foundation for Foraminiferal Research 4, 148–150.

Maync, W., 1959. Deux nouvelles espèces Crétacées du genre Pseudocyclammina (Foraminifères). Revue de Micropaléontologie, 1, 179–189.

Maync, W., 1966. Final remarks on C. D. Redmond's new lituolid foraminifera from Saudi Arabia. Revue de Micropaléontologie, 9, p. 56.

Maync, W., 1972. *Lituonella mesojurassica* n. sp. from the Mytilus Dogger of the Swiss Prealps. Revista Española de Micropaleontologïa 4, 251–266.

McConnaughey, T.A., 1989. ^{13}C and ^{18}O isotopic disequilibrium in biological carbonates, I. Patterns. Geochimica et Cosmochimica Acta 53, 151–162.

McConnaughey, T.A., Whelan, J.F., 1997. Calcification generates protons for nutrient and bicabonate uptake. Earth Sci. Rev. 42, 95–117.

McElwain, J.C., Beerling, D.J., Woodward, F.I., 1999. Fossil plants and global warming at the Triassic– Jurassic boundary. Science 285, 1386–1390.

McKee, E.D., Oriel, S.S., Ketner, K.B., MacLachlan, M.E., Goldsmith, J.W., MacLachlan, J.C., Mudge, M.R., 1959. Paleotectonic maps of the Triassic System. United States Geological Survey, Miscellaneous Geologic Investigations Map, Nevada, I-300, 1–33.

McKinney, M.L., McNamara, K.J., 1991. Heterochrony, The Evolution of Ontogeny. Plenum, New York, 437p.

McLaren, D.J., Goodfellow, W.D., 1990. Geological and biological consequences of giant impacts. Annu. Rev. Earth Planet. Sci. 18, 123–171.

McMenamin, M.A.S., McMenamin, D.L.S., 1990. The Emergence of Animals, The Cambrian Break- through. Columbia University Press, New York.

McMillan, I., 2000. Cainozoic planktonic and larger foraminifera distributions around Southern Africa and their implications for past changes of oceanic water temperatures. Suid-Afrikaanse Tydskrif vir Wetenskap 82, 66–69.

Meinhold, G., BouDagher-Fadel, M. K., 2010. Geochemistry and biostratigraphy of Eocene sediments from Samothraki Island, NE Greece. Neues Jahrbuch für Geologie und Paläontologie, 256, 17–38.

Meinhold, G., Kostopoulos, D., Reischmann, T., Frei, D., BouDagher-Fadel, M. K., 2009. Geochemistry, provenance and stratigraphic age of metasedimentary rocks from the eastern Vardar suture zone, northern Greece. Palaeogeography, Palaeoclimatology, Palaeoecology, 277, 199–225. doi:10.1016/j.palaeo.2009.04.005.

Mello e Sousa, S. H., Fairchild, T. R., and Tibana, P., 2003. Cenozoic biostratigraphy of larger foraminifera from the Foz Do Amazonas basin, Brazil, Micropaleontology, 49, 253–266.

Meric-, E., C-oruh, T., 1998. *Neosivasella sungurlui*, a new genus and species from the Upper Paleocene of Southeast Turkey. Micropaleontology 44, 187–194.

Mikhalevich, V. I., 1980. Sistematika i evolyuciya foraminifer v svete novyikh dannyikh po ih citologii i ultrastrukture [Systematics and evolution of the Foraminifera in view of the new data on their cytology and ultrastructure]. In: *Materialy VIII mikropaleontologicheskogo sovestchaniya "Systematika I morphologiya mikroorganismov".* 21–24 October, Baku: ELM: 77. Baku: Nauka (In Russian.).

Mikhalevich, V. I., 2000. Tip Foraminifera d'Orbigny, 1826–Foraminifery [The phylum Foraminifera d'Orbigny, 1826– Foraminifers]. *In:* Alimov, A. F. (ed.), *Protisty: Rukovodstvo po Zoologii, pt. 1,* 533–623. St. Petersburg: Nauka Publishers. (In Russian, with English summary 611–616.).

Mikhalevich, V. I., 2004. On the heterogeneity of the former Textulariina (Foraminifera). In: Bubik, M. and Kaminski, M. A., Eds., *Proceedings of the Sixth International Workshop on Agglutinated Foraminifera,* Krakow: Grzybowski Foundation. Special Publication 8, 317–349.

Mikhalevich, V. I., 2009. Taxonomic position of the superorder Fusulinoida Fursenko in the Foraminifera system. 43, 117–128.

Mikhalevich, V. I., 2013. New insight into the systematics and evolution of the foraminifera. Micropaleontology, 59, 493–527.

Miller, K.G., Mountain, G.S., Wright, J.D., Browning, J.V., 2011. A 180-million-year record of sea level and ice volume variations from continental margin and deep-sea isotopic records. Oceanography, 24, 40–53.

Mohtat Aghai, P., Vachard D., Krainer K., 2009. Transported Foraminifera in the Palaeozoic deep red Nodular Limestones; Exemplified by Latest Permian *Neoendothyra* in the Zal section (JULFA AREA, NW IRAN). *Revista Espanola de Micropaleontologia, Madrid,* 41, 197–213.

Montanari, A., Koeberl, C., Hilgen, F., Coccioni, R., 2007. Hothouse, icehouse, and impacts, the Late Eocene Earth Monte Co' nero (Ancona), Italy. GSA Today 17, 60–61.

Moore, R.C., Thompson, M.L., 1949. Main divisions of Pennsylvanian Period and System. Bull. Am. Assoc. Petrol. Geologist. 33, 275–302.

Morhange, C., Espic, K., BouDagher-Fadel, M. K., Doumet-Serhal, C. L., 2005. Les paléoenvironnements du port Nord de Sidon, tentative de synthèse. Bulletin d'Archéologie et d'Architecture Libanaises, BAAL, Hors-Série, 135–144.

Murray, J.W., 1973. Distribution and Ecology of Living Benthic Foraminiferids. Heinemann, London, 274p.

Murray, J.W., 1991. Ecology and distribution of benthic foraminifera. In: Lee, J.J., Anderson, O.R. (Eds), Biology of Foraminifera. Academic Press, New York, pp. 221–224.

Murray, J.W., 2006. Ecology and Applications of Benthic Foraminifera. Cambridge University Press, Cambridge, 426p.

Murray, J.W., 2007. Biodiversity of living benthic foraminifera, How many species are there? Mar. Micropaleontol. 64, 163–176.

Najman, Y., Appel, E., Boudagher-Fadel, M., Bown, P., Carter, A., Garzanti, E., Godin, L., Han, J., Liebke, U., Oliver, G., Parrish, R., Vezzoli, G., 2010a. Timing of India-Asia collision: Geological, biostratigraphic, and palaeomagnetic constraints. Journal of Geophysical Research: Solid Earth, 115 (12). doi:10.1029/2010JB007673.

Najman, Y., Appel, E., Boudagher-Fadel, M., Bown, P., Carter, A., Garzanti, E., Godin, L., Han, J., Liebke, U., Oliver, G., Parrish, R. Vezzoli, G., 2010b. The Age of India-Asia Collision: Biostratigraphic, Sedimentological and Paleomagnetic Constraints. U.S. Geological Survey, Open-File Report 2010–1099, 1 p.

Najman, Y., Bickle, M., BouDagher-Fadel, M. K., & et al., 2008. The Paleogene record of Himalayan erosion: Bengal Basin, Bangladesh. Earth and, Planetary Science Letters, 273, 1–14. doi:10.1016/j.epsl.2008.04.028.

Najman, Y., Jenks, D., Godin, L., Boudagher-Fadel, M., Millar I, Garzanti E, Horstwood M, Bracciali L., 2016. The Tethyan Himalayan detrital record shows that India–Asia terminal collision occurred by 54 Ma in the Western Himalaya. Earth and Planetary Science Letters, 459, 301–310. doi:10.1016/j.epsl.2016.11.036.

Nemkòv, G.I., 1962. Remarques sur la palueouecologie des Nummulites. Vopr Micropaleontol SSSR 6, 64–72.

Neumann, M., 1958. Revision des Orbitoidides du Crétacé et de l'Éocene en Aquitaine Occidentale, Mémoire de la Société géologique de France, 83. 1–174.

Newell, N.D., 1967. Revolutions in the history of life. In: Albritton, C.C. (Ed.), Uniformity and Simplicity. Geol. Soc. Am., special paper, 89, 63–91.

Newell, N.D., 1971. An outline history of tropical organic reefs. Am. Mus. Novit. 2465, 1–37.

Nikolaev, A.I., 2005. Foraminifers and zonal stratigraphy of Bashkirian Stage in the east of Timan–Pechora province. Bulletin of Paleontological and Collections of VNIGRI, 120 pp. (In Russian).

Nikolaev, S.I., Berney, C., Fahrni, J.F., Bolivar, I., Polet, S., Mylnikov, A.P., Aleshin, V.V., Petrov, N.B., Pawlowski, J., 2004. The twilight of Heliozoa and rise of Rhizaria, an emerging supergroup of amoeboid eukaryotes. PNAS 101, 8066–8071.

Noujaim Clark, G., and BouDagher-Fadel, M.K., 2001. The larger benthic foraminifera and stratigraphy of the upper jurassic/lower cretaceous of Central Lebanon. Revue de Micropaléontologie, 44, 215–232.

Noujaim Clark, G., BouDagher-Fadel, M.K., 2004. Larger Benthic Foraminifera and calcareous algae of the Upper Kesrouane Limestone Formation (Middle/Upper Jurassic) in Central Lebanon, Stratigraphy, Sedimentology and Regional Synopsis. Revue de Paléobiologie, 23, 475–504.

Noujaim Clark, G., BouDagher-Fadel, M. K. (2002). Larger foraminiferal assemblages and stratigraphy of the late Jurassic Bhannes complex, Central Lebanon | Assemblages des grands foraminifèrers et stratigraphie du Jurassique supérieur du complexe de Bhannes au Liban Central. Revue de Paleobiologie, 21 (2), 679–695.

Noujaim Clark, G., BouDagher-Fadel, M. K., 2005. Larger Benthic Foraminifera of the Upper Kesroune Limestone Formation (Middle/Upper Jurassic) in Central Lebanon, Stratigraphy, Sedimentology and Regional Synopsis. Revue de Paléobiologie, 23, 475–504.

O'Dogherty, L., Sandoval, J., Vera, J.A., 2000. Ammonite faunal turnover tracing sea level changes during the Jurassic (Betic Cordillera, southern Spain). J. Geol. Soc., Lond. 157, 723–736.

Oberhauser, R., 1956. Bericht über mikropaläontologische Untersuchungen im Herbst 1955. – Verh. Geol. B.-A., Wien, 118–119.

O'Herne, L., 1972. Secondary chamberlets in *Cycloclypeus.* — Scripta Geol., 7, 1–35.

O'Herne, L., van der Vlerk, LI.M., 1971. Geological age determinations on a biometrical basis. (Comparison of eight parameters). Bollettino della Societa Paleontologica Italiana 10, 3–18.

Olsen, P.E., Kent, D.V., Sues, H.-D., Koeberl, C., Huber, H., Montanari, A., Rainforth, E.C., Fowell, S.J., Szajna, M.J., Hartline, B.W., 2002. Ascent of dinosaurs linked to an iridium anomaly at the Triassic–Jurassic boundary. Science 296, 1305–1307.

Olsen, P.E., Koeberl, C., Montanari, A., Fowell, S., Et-Touhami, M., Kent, D.V., 2002. Continental Triassic–Jurassic boundary in central Pangea, Recent progress and discussion of an Ir anomaly. Geol. Soc. Am. special paper, 356, 505–522.

Oravecz-Scheffer, A., 1987. Triassic foraminifers of the Transdanubian central range. Geol. Hungary 50, 3–331.

Orchard, M.J., 2007, Conodont diversity and evolution through the latest Permian and Early Triassic upheavals: Palaeogeography, Palaeoclimatology, Palaeoecology, 252, 93–117, doi:10.1016/j.palaeo.2006.11.037.

Osawa Y., Fujita K., Umezawa Y., Kayanne H., Ide Y., Nagaoka T., Miyajima T., Yamano H., 2010. Human impacts on large benthic foraminifers near a densely populated area of Majuro Atoll, Marshall Islands. Mar Pollut Bull 60,1279–1287.

Ozawa, T., Watanabe, K., Kobayashi, D.F., 1992. Morphologic evolution in some schwagerinid and schubertellid lineages and the definition of the Carboniferous–Permian boundary. In: Takayanagi, Y., Saito, T. (Eds), Studies in Benthic Foraminifera, Benthos 90, Sendai, 1990. Tokai University Press, Tokyo, pp. 389–401.

Özcan, E., Less, G., Báldi-Beke, M., Kollányi, K., Kertés, B., 2006. Biometric analysis of middle and upper Eocene Discocyclinidae and Orbitoclypeidae (Foraminifera) from Turkey and updated orthophragmine zonation in the Western Tethys. Micropaleontology, 52, 485–520.

Özgen-Erdem, N., 2002. *Sirelella safranboluensis* n. gen., n. sp., a Foraminifer from the Lutetian of the Safranbolu Area (Northern Turkey). Micropaleontology, 48, 79–86.

Paleontol. Soc. Japan, special papers, 36, 239.

Palfy, J., Smith, P., 2000. Synchrony between early Jurassic extinction, oceanic anoxic event, and the Karoo–Ferrar flood basalt volcanism. Geology 28, 747–750.

Palmieri, V., 1983. Biostratigraphic appraisal of Permian foraminifera from the Denison Trough–Bowen Basin (central Queensland). Geological Society of Australia, Queensland Division, Permian Geology of Queensland 139–154.

Pawlowski, J., 2000. Introduction to the molecular systematics of foraminifera. Micropaleontology 46, 1–12.

Pawlowski, J., Bolivar, I., Fahrmi, J.F., Cavalier-Smith, T., Gouy, M., 1996. Early origin of foraminifera suggested by SSU rRNA gene sequences. Molecular Biology and Evolution, 13, 445–450.

Pawlowski, J., Bolivar, I., Fahrmi, J.F., Gouy, M., 1994. Phylogenetic position of foraminifera inferred from LSU rRNA gene sequences. Molecular Biology and Evolution, 11, 928–938.

Pawlowski J., Burki F., 2009. Untangling the phylogeny of amoeboid protists. J Eukaryot Microbiol, 56,16–25.

Pawlowski, J., Holzmann, M., Berney, C., Fahrni, J., Gooday, A.J., Cedhagen, T., Habura, A., Bowser, S.S., 2003. The evolution of early foraminifera. Proc. Nat. Acad. Sci. USA 100, 11494–11498.

Payne, J.L., Lehrmann, D.J., Wei, J., Orchard, M.J., Schrag, D.P., Knoll, A.H., 2004. Large perturbations of the carbon cycle during recovery from the end-Permian extinction. Science 305, 506–509.

Payne, J.L., Summers, M., Rego, B.L., Altiner, D., Wei, J., Yu, M., Lehrmann, D.J., 2011. Early and Middle Triassic trends in diversity, evenness, and size of foraminifers on a carbonate platform in south China: implications for tempo and mode of biotic recovery from the end-Permian mass extinction. Paleobiology, 37, 409–425.

Pearson, P.N., van Dongen, B.E., Nicholas, C.J., Pancost, R.D., Schouten, S., Singano, J.M., Wade, B.S., 2007. Stable warm tropical climate through the Eocene Epoch. Geology 35, 211–214.

Pêcheux, M.J.-F., 1984. Le Senonien superieur-Tertiaire du Chiapas (S.E. Mexique) et ses Macroforaminiferes. Theses 3ᵉ Cycle, Univ. Nice, p 154.

Pêcheux, M.J.-F., 1988. Le passage Miliolidae-Alveolinidae, mécanisme d'évolution et polymorphisme: le cas de la microfaune du Chiapas (Mexique). (Résumé). in "Benthos' 1986". Revue de Paléobiology,Volume Spécial, 2, 699–702.

Pêcheux, M.J.-F., 1995. Ecomorphology of a Recent large foraminifer, *Operculina ammonoides*. Geobios, 28, 529–566.

Pelissié B., Peybernès B., Rey J., 1984. Les grands foraminifères benthiques du Jurassique moyen/supérieur du sud-ouest de la France (Aquitaine, Causses, Pyrénées). Intérêt biostratigraphique, paléoécologique et paléobiogéographique. In: Benthos '83, 2nd int. Symp. Benthic Foraminifera (Pau, April, 1983), 479–489

Percival, L.M.E., Ruhl, M., Hesselbo, S.P., Jenkyns, H.C., Mather, T.A., Whiteside, J.H., Mercury evidence for pulsed volcanism during the end-Triassic mass extinction. Proceedings of the National Academy of Sciences of the United States of America (PNAS), 114, 30, 7929–7934 (2017) doi: 10.1073/pnas.1705378114.

Pignatti, J.S., 1998. The philosophy of larger foraminiferal biozonation – a discussion. In: Hottinger, L., Drobne, K. (Eds), Paleogene Shallow Benthos of the Tethys, Slovenian Academy of Sciences and Arts, Ljubljana, Vol. 2, pp. 15–20.

Pille, L. 2008. Foraminifères et algues calcaires du Mississippien supérieur (Viséen supérieur-Serpukhovien): rôles biostratigraphique, paléoécologique et paléogéographique aux échelles locale, régionale et mondiale. PhD thesis, Université de Lille.

Pinard, S., Mamet, B., 1998. Taxonomie des petits foraminifères du Carbonifère supérieur-Permien inférieur du bassin de Sverdrup, Arctique Canadien. Palaeontographica Canadiana, 15, 1–253.

Platon, E. and B. K. Sen Gupta. 2001. Benthic foraminiferal communities in oxygen-depleted environments of the Louisiana continental shelf., 147–164 in N. N. Rabalais and R. E. Turner, eds. Coastal Hypoxia: Consequences for Living Resources and Ecosystems. Coastal and Estuarine Studies 58.

Pokorny, J., 1958. Grundziige der zoologischen mikropaläontologie. Berlin, 582p.

Pollitt, D. A., Anthonissen, E., Saller, A. H., Boudagher-Fadel, M. K., Dickson, J. A. D., 2012. Abrupt early Eocene global climatic change as a control of carbonate facies and diagenesis: A new record of the Palaeocene-Eocene Thermal Maximum in the Umm er Radhuma Formation, Saudi Arabia and Kuwait. Terra Nova, 24, 487–498.

Pomar, L., Baceta J., Hallock, P., Mateu-Vicens, G., Basso, D., 2017. Reef building and carbonate production modes in the west-central Tethys during the Cenozoic. Marine and Petroleum Geology 83, 261–304

Pomar, L., Hallock, P., 2008. Carbonate factories: A conundrum in sedimentary geology. Earth-Sciences Reviews, 87, 134–169.

Porta Della G., Villa, E., Kenter, J.A.M., 2005. Facies distribution of fusulinida in a Bashkirian–Moscovian (Pennsylvanian) carbonate platform top (Cantabrian Mountains, NW Spain), Journal of Foraminiferal Research 35, 344–367.

Preat, A., Kasimi, R., 1995. Sédimentation de rampe mixte silico-carbonatée des couches de transition eifélienne-givétiennes franco-belges. Première partie: microfaciès et modèle sédimentaire. Bull. Centres Rech. Explor.-Prod. Elf-Aquitaine, 19/2, 329–375.

Preece, R.C., Kaminski, M.A., Dignes, T.W., 2000. *Popovia johnrolandi* n. sp., a new smaller agglutinated foraminifera from northern Venezuela, a biostratigraphic example of the second law of thermodynamics. In: Hart, M.B., Kaminski, M.A., Smart, C.W. (Eds), Proceedings of the Fifth International Workshop on Agglutinated Foraminifera. Grzybowski Foundation Special Publication, 7, 403–410.

Purton, L., Brasier, M., 1999. Giant protist *Nummulites* and its Eocene environment: Life span and habitat insights from delta ¹⁸O and delta ¹³C data from *Nummulites* and *Venericadia*, Hampshire Basin UK. Geology, 27, 711–714.

Racey, A., 1995. Lithostratigraphy and larger foraminiferal (nummulitid) biostratigraphy of the Tertiary of northern Oman. Micropaleontology 41(Suppl.), 1–123.

Racey, A., 2001. A review of Eocene nummulite accumulations, structure, formation and reservoir potential. J. Petr. Geol. 24, 79–100.

Radfords, S., 1976, Recent foraminifera from Tobago Island, West Indies. Revista Espanola de Micropaleontologia, 8, 193–218.

Rahaghi, A., 1978. Paleogene biostratigraphy of some parts of Iran. Nat. Iran. Oil Co., Geol. Lab. Publ. 7, 1–82.

Rahaghi, A., 1980. Tertiary faunal assemblage of Qum-Kashan, Sabzewar and Jahrum areas. Nat. Iran. Oil Co., Geol. Lab. Publ. 8, 1–64.

Raju, D.S.N., 1974. Study of Indian miogypsinidae. Utrecht Micropaleontological Bulletins, 9, p. 148.

Raju, D.S.N., Meijer, M., 1977. Miogypsinidae from the Cuanza basin (Angola). Akademie Van Wetenschappen. Amsterdam, Proc., Ser. B 80, 241–255.

Ramalho, M.M., 1969. Quelques observations sur les Lituolidae (Foraminifera) du Malm portugais. Bol. Soc. Geol. Port. 17, 123–127.

Ramalho, M.M., 1971. Contribution á létude micropaléontologique et stratigraphique du Jurassique supéricur el ñu Crétacé inférieur des environs de Lisbonne (Portugal). Mem Seii Geol. Portugal, NS., 19, pp 212.

Rampino, M.R., Stothers, R.B., 1988. Flood basalt volcanism during the past 250 million years. Science 241, 663–668.

Rauser-Chernoussova, D.M., 1963. Resheriya Vgorogo Kollokvyma po sistematike Endotiroidnich foraminifer, organieo-vannogo Koordinatsionnoi komissiei po mikropaleontologii a Moskva a aprele. Otdelenie Geologo-Geograficherskich Nauk, Geologicheskii Institut Akademiya Nauk SSSR, Voprosy Paleontologii 7, 223–227.

Rauser-Chernousova, D.M., Bensh, F.R., Vdovenko, M.V., Gibshman, N.B., Leven, E.Ya., Lipina, O.A., Reitlinger, E.A., Solovieva, M.N., and Chedia, I.O., 1996. *Guidebook on the Systematics of Foraminifers of Paleozoic.* Academy of Sciences of Russia. Nauka Publishing House (In Russian).

Rauser-Chernousova D.M., Fursenko, A.V., 1959. Determination of foraminifers from the oil-producing regions of the USSR. Glavnaya Redaktsya Gorno−Toplivnoy Literatury, Leningrad, Moskva, 315 pp. (in Russian)

Rauser−Chernousova, D.M., Gryslova, N.D., Kireeva, G.D., Leontovich, G.E. [Leontovič, G.E.], Safonova, T.P., and Chernova, E.I. Černova, E.I.] 1951. Srednekamennougol'nye fuzulinidy Russkoy Platformy i sopredel'nyh oblastej. Spravočnik−oprede−litel'. Geologičeskij Institut, Akademiâ Nauk SSSR, Izda−tel'stvo Akademii Nauk SSSR, Moskva., 380 pp.

Redmond, C.D., 1964. The Foraminiferal family Pfenderinidae in the Jurassic of Saudi Arabia. *Micropaleontology* 10, 251–263.

Redmond, C. D., 1965. Three New Genera of Foraminifera from the Jurassic of Saudi Arabia. Micropaleontology, 11, 133–140.

Rego, B.L., Wang, S.C., Altiner, D., Payne, J.L., 2012. Within- and among-genus components of size evolution during mass extinction, recovery, and background intervals: a case study of Late Permian through Late Triassic foraminifera. Paleobiology, 38, 627–643.

Reichel, M., 1936. Etude sur les Alveolines. Mem. Suisses Paleont. 57, 1–147.

Reichel, M., 1964. Alveolinidae. In: Moore, R.C. (Ed.), Treatise on Invertebrate Paleontology, C (Protista 2). Geological Society of America/University of Kansas Press, Boulder/Lawrence, 503–510.

Reichel, M., 1984. Le crible apertural de *Rhapydionina liburnica* Stache du Maastrichtien de Vremski-Britof, Yugoslavie. In: Oertli, H.J. (Ed.), Benthos '83, Elf-Aqitaine, Paris, France, pp. 525–532.

Reichow, M.K., Saunders, A.D., White, R.V., Pringle, M.S., Al'Mukhamedov, A.I., Medvedev, A.I., Kirda, N.P., 2002. [40]Ar/[39]Ar dates from the West Siberian Basin, Siberian flood basalt province doubled. Science 296, 1846–1849.

Reiss, Z., 1963. Reclassification of perforate foraminifera. State of Israel, Ministry of Development, Geol. Surv. Bull. 35, 1–111.

Reiss, Z., Hottinger, L., 1984. The Gulf of Aqaba: Ecological Micropaleontology. Ecological Studies, Springer-Verlag, Berlin, Vol. 50, pp. 1–354.

Renema, W., 2002. Larger foraminifera and their distribution patterns on the Spermonde shelf, South Sulawesi. Scripita Geologica 124, 1–263.

Renema, W., 2005. The genus *Planorbulinella* (Foraminiferida) in Indonesia. Scripta Geol. 129, 137–143. Renema, W., 2007. Fauna development of larger benthic foraminifera in the Cenozoic of Southeast Asia. In: W. Renema (Ed.), Biogeography, Time and Place: Distributions, Barriers and Islands, Springer, Dordrecht, 179–215.

Renema, W., Troelstra, S.R., 2001. Larger foraminifera distribution on a mesotrophic carbonate shelf in SW Sulawesi (Indonesia). Palaeogeografie, Paleoclimatologie, Paleoecologie 175, 125–147.

Rettori, R., 1995. Foraminiferi del Trias inferiore e medio della Tetide, Revisione tassonomica, stratigrafia ed interpretazi-one filogenetica. Université de Genève, Publications du Département de Géologie et Paléontologie, 18, pp 147.

Rettori, R., Zaninetti, L., Martini, R., Vachard, D., 1993. *Piallina tethydis* gen. et sp. nov. (Foraminiferida) from the Triassic (Carnian) of the Kocaeli Peninsula, Turkey. J. Micropalaeontol. 12, 170–174.

Rey, J., 1997. A Liassic isolated platform controlled by tectonics, South Iberian margin, southeast Spain. Geol. Mag. 134, 235–247.

Reytlinger, E.A., 1965. Evolution of foraminifers in the latest Permian and earliest Triassic epochs of the territory of the Transcaucasus. Akademiya Nauk SSSR, Otdelenie Nauk o Zemle, Geologicheskii Institut, Sistematika i Filogeniya Foraminifer i Ostrakod, Voprosy Mikropaleontologii 9, 45–70.

Rhumbler, L., 1909. Die Foraminiferen (Thalamophoren) der Plankton-Expedition, 1. Teil, In: Hensen, V. (Ed.), Ergebnisse der Plankton-Expedition der Humboldt-Stittung, Leipzig, Berlin, 3, 1–331.

Richardson, S.L., 2001. Endosymbiont change as a key innovation in the adaptive radiation of Soritida (foraminifera). Paléobiology 27, 262–289.

Riding, R., 1977. Skeletal stromatolites. In: Flu¨ gel, E. (Ed.), Fossil Algae. Springer, Berlin, pp. 57–60. Robinson, E., 1996. The occurrence of the foraminiferal genus *Halkyardia* in the Caribbean region. Caribb. J. Sci. 32, 72–77.

Rigaud, S., R Martini, R., Vachard, D., 2015. Early evolution and new classification of the Order Robertinida (Foraminifera). Journal of Foraminiferal Research, 45, 3–28

Rodland, D.L., Bottjer, D.J., 2001. Biotic recovery from the end-Permian mass extinction, behavior of the inarticulate brachiopod Lingula as a disaster taxon. Palaios 16, 95–101.

Rogers, A.D., 2000. The role of the oceanic oxygen minima in generating biodiversity in the deep sea. Deep- Sea Res. II 47, 119–148.

Rögl, F., and Steininger, F. F., 1983. Vom Zerfall der Tethys zu Mediterran und Paratethys. Ann. Naturhist. Mus. Wien, 85A, 135–163.

Rosen, B.R., 1988. Progress, problems and patterns in the biogeography of reef corals and other tropical marine organisms. Helgoländer Meeresuntersuchungen, 42, 269–301.

Ross, C. A., 1965. Fusulinids from the *Cyathophyllum* Limestone, Central Vestspitsbergen. Contributions from the Cushman Foundation for Foraminiferal Research, 16, 74–86.

Ross, C.A., 1967. *Eoparafusulina* from the Neal Ranch Formation (Lower Permian), West Texas. Journal of Paleontology, 41, 943–946.

Ross, C.A., 1969. Palaeoecology of Late Pennsylvanian Fusulinids (Foraminiferada), Texas. C. R., International Congress on Carboniferous Stratigraphy and Geology (Sheffield). Maastricht, E. Van Aelst, 4, 1429–1440.

Ross, C.A., 1973. Carboniferous Foraminiferida. In: Hallam, A., (ed.), *Atlas of Palaeobiogeography*. Elsevier, Amsterdam, 127–132.

Ross, C.A., 1992. Paleobiogeography of Fusulinacean Foraminifera. Studies in Benthic Foraminifera. Proceedings of the Fourth International Symposium on Benthic Foraminifera, Sendai, 1990, Tokyo, Tokai University Press, pp. 23–31.

Ross, C.A., Ross, J.R.P., 1991. Paleozoic foraminifera. BioSystems 25, 39–51.

Roth, J.M., Droxler, A.W., Kameo, K., 2000. The Caribbean carbonate crash at the Middle to Late Miocene transition, linkage to the establishment of the modern global ocean conveyor. In: Leckie, R.M., Sigurdsson, H., Acton, G.D., Draper, G. (Eds), Proceedings of the Ocean Drilling Program, Scientific Results, College Station, TX (Ocean Drilling Program), Vol. 165, pp. 249–273.

Röttger, R., 1971. Die Bedeutung der Symbiose von *Heterostegina depressa* (Foraminifera, Nummulitidae) für hohe Siedlungsdichte und Karbonatproduction. Sonderdruck aus Verhandlungsbericht der Deutschen Zoologischen Gesellschaf, 65. Jahresversammlung, Gustav Fischer Verlag, 42–47.

Röttger, R., 1983. A complicated Protozoon, *Heterostegina depressa*: Test structures and their function. German Research Reports of the DFG, 2/83, 11–13.

Röttger, R., 1984. Ökologie der Großforaminiferen. Film C 1497 des IWF, Göttingen 1983. Publikation von R. Röttger, Publ. Wissenschaftliche Film, Sekt. Biol., Ser. 16, Nr. 20/C 1497, pp 20.

Röttger, R., 1990. Biology of larger foraminifera: Present status of the hypothesis of trimorphism and ontogeny of the gamont of *Heterostegina depressa*. In: *Studies in Benthic Foraminifera*, Benthos '90, Tokai University Press, Sendai, 43–54.

Röttger, R., and Krüger, R., 1990. Observations on the biology of Calcarinidae (Foraminiferida). Marine Biology, 106, 419–425.

Röttger, R., Spindler, M., Schmal, J., Richwien, M., Fladung, M., 1984. Functions of the canal system in the rotaliid foraminifera, *Heterostegina depressa*. Nature 309, 789–791.

Ruggiero, M.A., Gordon, D.P., Orrell, T.M., Bailly, N., Bourgoin, T., Brusca, R.C., et al., 2015. A Higher Level Classification of All Living Organisms. PLoS ONE, 10, e0119248. doi:10.1371/journal.pone.0119248

Ruhl, M., Bonis, N.R., Reichart, G.J., Damste, J.S.S., and Kurschner, W.M., 2011, Atmospheric carbon injection linked to end-Triassic mass extinction: Science, 333, 430–434, doi:10.1126/science.1204255.

Runnegar, B., 1982. A molecular-clock date for the origin of the animal phyla. Lethaia 15, 199–205. Rutten, M.G., 1941. A synopsis of the Orbitoididae. Geologie en Mijnbouw 3, 34–62.

Salaj J., Borza, K., and Samuel, O., 1983. Triassic foraminifers of the west Carpathians. Geologický Ústav Dionýza Štúra, Bratislava, pp 213.

Salaj, J., Trifonova, E., Gheorghian, D., 1988. A biostratigraphic zonation based on benthic foraminifera in the Triassic deposits of the Carpatho-Balkans. Revue de Paléobiologie Spec. 2, 153–159.

Salmeron, U. P., 1972. Mutación entre los generos *Pararotalia* y *Miogypsinoides*. Rev. Inst. Mex. Pet., 4, 5 -12

Samanta, B.K., 1967. A revised classification of the family discocyclinidae galloway. Contrib. Cushman Foundation for Foraminiferal Res. 18, 164–167.

Sandberg, C.A., Morrow, J.R., Ziegler, W., 2002. Late Devonian sea-level changes, catastrophic events, and mass extinctions. In Koeberl, C.; MacLeod, K.G. Catastrophic Events and Mass Extinctions: Impacts and Beyond: Boulder, Colorado. Special Paper. 356. Geological Society of America, 473–487.

Saraswati, P.K., 2004. Ontogenetic isotopic variation in foraminifera – implications for palaeo proxy. Curr. Sci. 86, 858–860.

Saraswati, P.K., Seto, K., Nomura, R., 2003. Oxygen and carbon isotopic variation in co-existing larger foraminifera from a reef flat at Akajima, Okinawa, Japan. Marine Micropaleontology 50, 339–349.

Sartoni, S., Crescenti, U., 1962. Ricerche biostratighiche nel Mesozoica dell'Apennino meridionale. Giornale di Geologia 29, 159–302.

Sartorio, D., Venturini, S., 1988. Southern Tethys biofacies. Agip, Milan, 235p.

Schaltegger, U., Schoene, B., Bartolini, A., Guex, J., Ovtcharova, M., 2007. Precise ages for the Triassic/ Jurassic boundary and Hettangian recovery from Northern Peru. Goldschmidt Conference 2007. Geochimica et Cosmochimica Acta, Suppl., 71/158, A884.

Schaub, H., 1981. Nummulites et Assilines de la Tethys paléogene. Taxonomy, phylogenèse et biostratigraphie. Schweizerische Paläontologisches Abhandlungen, 104–106, pp 236.

Schaudinn F., 1895. Über den Dimorphismus der Foraminiferen, Sitz. Ges. Naturforsch. Freunde. Berlin, 87–97.

Scheibner, C., Speijer, R.P., Marzouk, A., 2005. Larger foraminiferal turnover during the Paleocene/ Eocene thermal maximum and paleoclimatic control on the evolution of platform ecosystems. Geology 33, 493–496.

Scheibner, C., Speijer, R.P., Marzouk, A.M., 2006. Turnover of larger foraminifera during the Paleocene–Eocene thermal maximum and paleoclimatic control on the evolution of platform ecosystems. Geology 33, 493–496.

Scherreiks, R., Bosence, D., BouDagher- Fadel, M., 2006. The Tectono- Sedimentary Evolution of the Pelagonian Carbonate- Platform- Complex. Volumina Jurassica, IV, 64–66.

Scherreiks, R., Bosence, D., BouDagher-Fadel, M. K., Melendez, G., Baumgartner, P. O., 2010. Evolution of the Pelagonian carbonate platform complex and the adjacent oceanic realm in response to plate tectonic forcing (Late Triassic to Late Jurassic), Evvoia, Greece. International Journal of Earth Sciences, 99, 1317–1344.

Scherreiks, R., Meléndez, G., BouDagher-Fadel, M., Fermeli, G., Bosence, D. 2015a. Stratigraphy and tectonics of a time-transgressive ophiolite obduction onto the eastern, margin of the Pelagonian platform from Late Bathonian until Valanginian time, exemplified in northern Evvoia, Greece. International Journal of Earth Sciences, 104, 321. doi:10.1007/s00531-014-1123-5.

Scherreiks, R., Meléndez, G., BouDagher-Fadel, M., Fermeli, G., Bosence, D., 2015b. Erratum to: Stratigraphy and tectonics of a time-transgressive ophiolite obduction onto the eastern margin of the Pelagonian platform from Late Bathonian until Valanginian time, exemplified in northern Evvoia, Greece. International Journal of Earth Sciences, 104, 321.

Scherreiks, R., Melendez, G., Boudagher-Fadel, M. K., Fermeli, G., Bosence, D.. 2016. HYPERLINK "http://iris.ucl.ac.uk/ iris/publication/1058545/1" The Callovian unconformity and the ophiolite obduction onto the Pelagonian carbonate platform of the internal Hellenides. Proceedings of the 14th International Congress, Thessaloniki, May 2016. Bulletin of the Geological Society of Greece, L, 144–152.

Schlagintweit, F., Velić, I., Sokač, B. 2013. *Robustoconus tisljari* n. gen., n. sp., a new larger benthic foraminifer from the Middle Jurassic (Early Bajocian) of the Adriatic Carbonate Platform of Croatia. Geologia Croatica, 66, 16–28.

Schlagintweit F., Bucur, I., 2017. *Banatia aninensis* n. gen., n. sp., a new complex larger benthic foraminifer from the upper Barremian of Romania. Cretaceous Research, 75, 23–30.

Schlagintweit F., Rashidi, K., 2017. *Zagrosella Rigaudii* n. gen., n sp. A new biokovinoidean foraminifer from the Maastrichtian of Iran, Acta Palaeontologica Romaniae, 3, 3–13.

Schlagintweit, F., and Wilmsen, M., 2014, Orbitolinid biostratigraphy of the top Taft Formation (Lower Cretaceous of the Yazd Block, Central Iran): Cretaceous Research, 4, 125–133.

Schlagintweit, F., Rashidi, K., and Barani, 2016. *Tarburina zagrosiana* n. gen., n. sp., a new larger benthic porcelaneous foraminifer from the late Maastrichtian of Iran. Journal of Micropalaeontology https://doi.org/10.1144/jmpaleo2016-019.

Schmitz, B., Speijer, R., Aubry, M.-P., 1996. Latest Paleocene benthic extinction event on the southern Tethyan shelf (Egypt): Foraminiferal stable isotopic ($d^{13}C$, $d^{18}O$) records. Geology 24, 347–350.

Schoene, B., Guex, J., Bartolini, A., Schaltegger, U., Blackburn, T.J., 2010. Correlating the end-Triassic mass extinction and flood basalt volcanism at the 100 ka level. Geology, 38, 387–390.

Schroeder, R., 1962, Orbitolinen des Cenomans S̈ud-westeuropas: Paliiontologische Zeitschrift, 36, 171–202.

Schroeder, R., 1964, Orbitoliniden-Biostratigraphie des Urgons nordostlich von Teruel (Spanien): Neues Jahrbuch fur Geologie und Paleontologie, Monatshefte, 462–474.

Schroeder, R., 1975, General evolutionary trends in orbitolinas: Revista espanola de Micropaleontologia, Numero especial, 117–128.

Schroeder,R.,Van Buchem, F. S. P., Cherchi, A., Baghbani, D.,Vincent, B., Immenhauser, A., Granier, B., 2010, Revised orbitolinid biostratigraphic zonation for the Barremian – Aptian of the eastern Arabian Plate and implications for regional stratigraphic correlations: GeoArabia Special Publication 4, 1, 49–96.

Schulte, P., Speijer, R.P., Mai, H., Kontny, A., 2006. The Cretaceous–Paleogene (K–P) boundary at Brazos, Texas: Sequence stratigraphy, depositional events and the Chicxulub impact. Sediment. Geol. 184, 77–109.

Scotchman, J. I., Bown, P., Pickering, K. T., BouDagher-Fadel, M., Bayliss, N. J., Robinson, S. A., 2014. A new age model for the middle Eocene deep-marine Ainsa Basin, Spanish Pyrenees. Earth-Science Reviews. doi:10.1016/ j.earscirev.2014.11.006.

Seiglie, A., Grove, K., Rivera, J.A., 1976. Revision of some Caribbean Archaiasinae, new genera, species and subspecies. Eclogae geologicae Helvetiae 70, 855–883.

Seiglie, G.A., Baker, M.B., 1983. Some West African Cenozoic agglutinated foraminifers with inner structures: Taxonomy, age and evolution. Micropaleontology 29, 391–403.

Seiglie, G.A., Fleisher, R.L., Baker, M.B., 1986. Alveovalulnidae, n. fam., and Neogene diversification of agglutinated foraminifers with inner structures. Micropaleontology 32, 169–181.

Sen Gupta, B. K., 1999. Systematics of modern Foraminifera. 7–36 in B. K. Sen Gupta, ed. Modern Foraminifera. Kluwer Academic Publishers, Dordrecht.

Sepkoski J.J., Jr., 1990. The taxonomic structure of periodic extinction. In: Sharpton, V.L., Ward, P.D. (Eds), Global Catastrophies in Earth History. Geological Society of America, special paper, 247, 33–44.

Sepkoski, J.J., Jr., 1986. Phanerozoic overview of mass extinction. In: Raup, D.M., Jablonski, D. (Eds), Patterns and Processes in the History of Life. Springer-Verlag, Berlin, pp. 277–295.

Sepkoski, J.J., Jr., 1989. Periodicity in extinction and the problem of catastrophism in the history of life. Geol. Soc. Lond. J. 146, 7–19.

Septfontaine, M., 1977. Niveaux a Foraminiferes (Pfenderininae et Valvulininae) dans le Dogger des Prealpes medianes du Chablais occidental (Haute-Savoie, France). Eclogae geologicae Helvetiae 70, 599–625.

Septfontaine, M., 1981. Les foraminiferes imperfores des milieux de plate-forme au mesozoique, determination practique, interpretation phylogenetique et utilisation biostratigraphique. Rev. Micropal. 23, 169–203.

Septfontaine, M., 1984. Biozonation (à l'aide des foraminifères imperforés) de la plate-forme interne carbonatée Liasique du Haut Atlas (Maroc). Revue de Micropaléontologie, 27, 209–229.

Septfontaine, M.,1988. Vers une classification évolutive des Lituolides (Foraminifères) Jurassiques en milieu de plate-forme carbonatée. Revue de Paléobiologie, Volume Spéciale 2, 229–256.

Serra-Kiel, J., Hottinger, L., Caus, E., Drobne, K., Ferrández, C., Jaurhi, A.K., Less, G., Pavlovec, R., Pignatti, J., Samso, J.M., Schaub, H., Sirel, E., Strougo, A., Tambareau, Y., Tosquella, J., Zakrevskaya, E., 1998. Larger foraminiferal biostratigraphy of the Tethyan Paleocene and Eocene. Bulletin de la Société Géologique de France 169, 281–299.

Severin, K.P., Lipps, J.H., 1989. The weight-volume relationship of the test of *Alveolinella quoyi*: Implications for the taphonomy of large fusiform foraminifera. Lethaia 22, 1–12.

Sharaf, E., BouDagher-Fadel, M. K., Simo (Toni), J. A., & Carroll, A. R., 2014. A revision of the biostratigraphy and strontium isotope dating of Oligocene-Miocene outcrops in East Java, Indonesia. Berita Sedimentology, 30, 44–81.

Sharaf, E., BouDagher-Fadel, M. K., Simo, T. A., & Carroll, A. R., 2006. Biostratigraphy and strontium isotope dating of Oligocene-Miocene carbonates and siliciclastics, east Java, Indonesia. Stratigraphy, 2, 239–258.

Sheppard, C.R.C., 2003. Predicted recurrences of mass coral mortality in the Indian Ocean. Nature 425, 294–297.

Simmons, M.D., BouDagher-Fadel, M.K., Banner, F.T., Whittaker, J.E., 1997. The Jurassic Favusellacea, the earliest Globigerinina. In: BouDagher-Fadel, M.K., Banner, F.T., Whittaker, J.E. (Eds), Early Evolutionary History of Planktonic Foraminifera. British Micropalaeontological Society Publica- tion Series, Chapman and Hall, pp. 17–52.

Simmons, M.D., Whittaker, J.E., Jones, R.W., 2000. Orbitolinids from Cretaceous sediments of the Middle East – a revision of the F.R.S. Henson and associated collection. In: Hart, M.B., Smart, C.W. (Eds), Proceedings of the Fifth International Workshop on Agglutinated Foraminifera. Gybrowski Foundation special publication, Vol. 7, pp. 411–437.

Sinitsyna, Z.A., Sinitsyn, I.I., 1987. Biostratigrafiâ baškirskogo ârusa v stratotipe. 72 pp. Baškirskoe otdelenie Akademii Nauk SSR, Baškirsij Intitut geologii, Ufa.

Sirel, E., 1988. *Anatoliella*, a new foraminiferal genus and a new species of Dictyokathina from the Paleocene of the Van area (East Turkey). Revue de Paléobiologie, 7, 447–493.

Sirel, E., 1993. *Kastamonina abanica* n. gen. n. sp., a complex lituolid (Foraminiferida) from the Upper Jurassic Limestone of the Kastamonu Area (North Turkey). Geologia Croatica, 46, 1–7.

Sirel, E., 1997. *Karsella*, a new complex orbitolinid (Foraminiferida) from the thanetian limestone of the Van Region (East Turkey). Micropaleontology 43, 206–210.

Sirel, E., 1999. Four new genera (*Haymanella, Kayseriella, Elazigella* and *Orduella*) and one new species of

Sirel, E., 2004. Türkije'nin Mesozoyik ve Senozoyik Yeni Bentik Foraminiferleri. Jeoloji Mühendisleri Odasi Yayinlari 84 (Chamber of Geological Engineers of Turkey, publication 84), Ankara, Emegin Bilimsel Sentezi, özel Sayi 1, pp. 1–219.

Sirel, E., Acar, S., 1982. *Praebullalveolina*, a new foraminiferal genus from the Upper Eocene from the Afyon and Çanakkale region (west of Turkey). Eclogae Geologicae Helvetiae 75, 821–839.

Sirel, E., Acar, S., 1993. *Malatyna*, a new foraminiferal genus from the Lutetian of Malatya region (East Turkey). Geologia Croatica, 46, 181–188.

Sirotti, A., 1982. Phylogenetic classification of Lepidocyclinidae: A proposal. Bollettino della Societa Paleontologica Italiana 21, 99–112.

Skipp, B., 1969. Foraminifera. In: McKee, E.D., Gutschick, R.C. (Eds), History of the Redwall Limestone of Northern Arizona. Geological Society of America Memoir, 114, 173–255.

Smith, A.G., Pickering, K.T., 2003. Oceanic gateways as a critical factor to initiate icehouse earth. Journal of the Geological Society London, 160, 337–340.

Smout, A.H., 1954. Lower tertiary foraminifera of the Qatar peninsula. Brit. Mus. (Nat. Hist.), Lond. 96 p.

Smythe, D.K., Russell, M.J., Skuce, A.G., 1995. Intracontinental rifting from major late carboniferous quartz-dolorite dyke swarm of NW Europe. Scott. Journal of Geology 31, 151–162.

Song, H., WignalL, P. B., Chen, Z. Q., Tong, J., Bond, D. P. G., Lai, X., ZhaO, X., JiaNG, H., Yan, C., Niu, Z., Chen, J., Yang, H. & Wang, Y., 2011. Recovery tempo and pattern of marine ecosystems after the end-Permian mass extinction. Geology, 39, 739-632.

Spray, J.G., Kelley, S.P., Rowley, D.B., 1998. Evidence for a late Triassic multiple impact event on earth. Nature 392, 171–173.

Stache, G., 1913. Uber Rhipidionina St. und Rhapydionina St. Jahrbuch der Geologischen Reichsanstalt 62, 659–680.

Stanley, G.D. (Ed.), 2001. Ancient Reef Ecosystems, their Evolution, Paleoecology and Importance in Earth History. Kluwer Academic/Plenum Press, New York, 299p.

Stanley, G.D., 2003. The evolution of modern corals and their early history. Earth Sci. Rev. 60, 195–225.

Stanley SM. 2009. Evidence from ammonoids and conodonts for multiple Early Triassic mass extinctions. Proc Natl Acad Sci USA. 106, 15264–15267.

Stefaniuk, L., Morhange, C., Saghieh-Beydoun, M., Frost, H., BouDagher-Fadel, M. K., Bourcier, M., Noujaim-Clark, G. (2005). Localisation et étude paléoenvironnementale des ports antiques de Byblos. Bulletin d'Archéologie et d'Architecture Libanaises, BAAL, Hors-Série 2, 19–41.

Stevens, G.R., 1971. Relationship of isotopic temperatures and faunal realms to Jurassic- Cretaceous palaeogeography, particularly of the south west Pacific: J. Roy. Soc. New Zealand, v. 1, p. 145.

Storey, M., Duncan, R.A., Swisher, C.C., 2007. Paleocene–Eocene thermal maximum and the opening of the northeast Atlantic. Science 316, 587–589.

Strathearn, G.E., 1986. *Homotrema rubrum,* symbiosis identified by chemical and isotopic analyses. Palaios 1, 48–54.

Streel, M., 1986. Miospore contribution to the Upper Famennian–Strunian event stratigraphy. Annales de la Socie´ te´ ge´ ologique de Belgique 109, 75–92.

Stüben, D., Kramar, U., Harting, M., Stinnesbeck, W., Keller, G., 2005. High-resolution geochemical record of Cretaceous–Tertiary boundary sections in Mexico: New constraints on the K/T and Chicxulub events. Geochimica et Cosmochimica Acta 69, 2559–2579.

Sun, G., Hu, X., Sinclair, HD, BouDagher-Fadel, MK, Wang, J., 2015. Late Cretaceous evolution of the Coqen Basin (Lhasa terrane) and implications for early topographic growth on the Tibetan Plateau. Geological Society of America Bulletin. DOI: 10.1130 / B31137.1.

Sweet, W.C., Yang, Z., Dickins, J.M., Yin, H., 1992. Permo-Triassic events in the eastern Tethys – An overview. In: Sweet, W.C., Zunyi, Y., Dickins, J.M., Hongfu, Y. (Eds), Permo-Triassic Events in the Eastern Tethys. Cambridge University Press, Cambridge, pp. 1–8.

Talent, J.A., 1988. Organic reef-building: Episodes of extinction and symbiosis. Senckenbergiana Lethaea 69, 315–368.

Tan Sin Hok, 1932. On the genus *Cycloclypeus*, Pt.1, and an appendix on the heterostegines of Tjimanggoe, S. Bantam, Java. Wetenschappelijke Mededeelingen van de Dienst van de Mijnbouw in Nederlandsch-Oost-Indië, 19, 1–194.

Tan Sin Hok, 1936. Zur Kenntniss der Miogypsiniden. Ingenieur Nederland. -Indië, Mijnbouw Geologie, 4, 45–61.

Tanner, L.H., Kyte, F.T., 2005. Anomalous iridium enrichment in sediments at the Triassic–Jurassic boundary, Blomidon formation, Fundy Basin, Canada. Earth Planet. Sci. Lett. 240, 634–641.

Tanner, L.H., Lucas, S.G., Chapman, M.G., 2004. Assessing the record and causes of Late Triassic extinctions. Earth Sci. Rev. 65, 103–139.

Tappan, H., 1971. Microplankton, ecological succession and evolution. North American Paleontological Convention Chicago, 1969, Proceedings H, Chicago, pp. 1058–1103.

Tappan, H., Loeblich, A., 1988. Foraminiferal evolution, diversification, and extinction. Journal of Paleontology, 62, 695–714.

Teodorovich, G.I. [Teodorovič, G.I.] 1949. The carbonate facies of upper Carboniferous–lower Permian of the Uralo−Volga area [in Russian]. Materialy k poznaniu geologičeskogo stroeniâ SSSR. Novaâ seriâ 13, 1–304.

Ter Kuile, B., 1991. Mechanisms for calcification and carbon cycling in algal symbiont-bearing foraminifera. In: Lee, J.J., Anderson, O.R. (Eds), Biology of Foraminifera. Academic Press, New York, pp. 73–89.

Thomas, E., Brinkhuis, H., Huber, M., Ro¨ hl, U., 2006. An ocean view of the early Cenozoic world. Oceanography 19, 94–103.

Thompson, M. L., 1964. Fusulinacea, *in* Loeblich, A. R., and Tappan, H., Treatise on Invertebrate Paleontology, Part C, Protista 2; Sarcodina chiefly "Thecamoebians" and Foraminiferida: Geological Society America and University Kansas Press, 1, C358–C436.

Todd, R., 1960. Some observations on the distribution of *Calcarina* and *Baculogypsina* in the Pacific Science Reports Tohoku University, Ser. 2 (Geol.), spec. 4, 100–107.

Tong, J., Shi, G.R., 2000. Evolution of the Permian and Triassic foraminifera in South China. In: Yin, H., Dickins, J.M., Shi, G.R., Tong, J., (Eds), Permian–Triassic Evolution of Tethys and Western Circum- Pacific, Developments in Palaeontology and Stratigraphy. Elsevier, Amsterdam, 18, 291–307.

Toomey D.F., Winland H.D., 1973. Rock and biotic facies associated with Middle Pennsylvanian (Desmoinesian) algal buildup, Nena Lucia Weld, Nolan County, Texas. AAPG Bull, 57, 1053–1074.

Hallock, P., Triantaphyllou, M., Dimiza, M., Koukousioura O., 2011. An invasive foraminifer in coastal ecosystems of the Eastern Mediterranean: Implications for understanding larger foraminiferal-dominated biofacies in the Cenozoic. In: GSA Annual Meeting in Minneapolis, Paper No. 231-10, Geological Society of America Abstracts with Programs, 43, 556 p.

Trifonova, E., 1978. The foraminifera zones and subzones of the Triassic in Bulgaria: Scythian and Anisian. Geol. Balcanica 8, 85–104.

Trifonova, E., 1984. Correlation of Triassic foraminifers from Bulgaria and some localities in Europe, Caucasus and Turkey. Geol. Balcanica, 13, 3–24.

Trifonova, E., 1992. Taxonomy of Bulgarian triassic foraminifera. I. Families Psammosphaeridae to Nodosinellidae. Geol. Balcanica, 22, 3–50.

Trifonova, E., 1993. Taxonomy of Bulgarian triassic foraminifera, II. Families Endothyriidae to Ophthamiidae. Geol. Balcanica, 23, 19–66.

Trifonova, E., 1994. Taxonomy of Bulgarian triassic foraminifera. III. Families Spiroloculiniidae to Oberhausereliidae. Geol. Balcannica, 24, 21–70.

Tucker, M.E., Benton, M.J., 1982. Triassic environments, climates and reptile evolution. Palaeogeography, Palaeoclimatology, Palaeoecology, 40, 361–379.

Twitchett, R.J., 2006. The palaeoclimatology, palaeoecology and palaeoenvironmental analysis of mass extinction events. Paleogeography, Paleoclimatology, Paleoecology, 232, 190–213.

Umbgrove, J.H.F., 1928. Het genus Pellatispira in het Indo-pacifische gebeid. Wetenschappelijke Mededeelingen van de Dienst van de Mijnbouw in Nederlandsch-Oost-Indië, 10, 43–71.

Urbanek, A., 1993. Biotic crisis in the history of the Upper Silurian graptolites: A palaeobiological model. Hist. Biol. 7, 29–50.

Vachard, D., 1973. Remarques sur les Foraminifères des calcaires griottes sensu lato (Frasnien inférieur-Tournaisien inférieur) du versant méridional de la Montagne Noire (Aude-Hérault). Comptes Rendus Sommaires de la Société géologique de France 1, 116–118.

Vachard, D., 1974. Contribution à l'étude stratigraphique et micropaléontologique (algues et foraminifères) du Dévonien-Carbonifère inférieur de la partie orientale du versant méridional de la Montagne Noire (Hérault, France). PhD Thesis, Université Pierre et Marie Curie (Paris VI), 1–408 (unpublished).

Vachard, D., Krainer, K., 2001. Smaller foraminifers of the Upper Carboniferous Auernig Group, Carnic Alps (Austria/Italy). Rivista Italiana di Paleontologia e Stratigrafia, 107, 147–168.

Vachard D., Massa, D., and Strank, A., 1993. Le Carbonifère du sondage A1-37 (Cyrénaique, Libye), analyse biostratigraphique, conséquences paléogéographiques. Revue de Micropaléontologie, 36,165–186

Vachard, D., Martini, R., Rettori, R., Zaninetti, L., 1994. Nouvelle classification des foraminifères endothyroïdes du Trias. Geobios, Lyon 27/5, 543–557.

Vachard, D., Munnecke, A., Servais, T., 2004. New SEM observations of keriothecal walls implications for the evolution of the fusulinida. Journal of Foraminiferal Research, 34, 232–242.

Vachard D., Pille L., Gaillot, J. 2010. Palaeozoic Foraminifera: Systematics, palaeoecology and responses to global changes. Revue de micropaléontologie 53, 209–254.

van der Vlerk, I.M., 1922. Studiën over nummulinidae en alveolinidae. Haar voorkomen op Soembawa en haar betekenis voor de geologie van oost-Azië en Australië. Wetenschappelijke Mededeelingen van de Dienst van de Mijnbouw in Nederlandsch-Oost-Indie, 6, 329–468.

van der Vlerk, I.M., 1929. Groote foraminiferen van N.O. Borneo. Wetenschappelijke Mededeelingen van de Dienst van de Mijnbouw in Nederlandsch-Oost-Indië, 9, 1–45.

van der Vlerk, I.M., 1959. Problems and principles of Tertiary and Quaternary stratigraphy. Quar. J. Geol. Soc. London, 105, 49–64. van der Vlerk, I.M., 1963. Biometric research on *Lepidocyclina*. Micropaleontology 9, 425–426.

van der Vlerk, I.M., 1966. *Miogypsinoides, Miogypsina, Lepidocyclina*, et Cycloclypeus de Larat (Moluques). Eclogae Geologicae Helvetiae 59, 421–430.

van der Vlerk, I.M., Bannink, D.D., 1969. Biometrical investigations on *Operculina*, Proceedings Koninklijke Nederlandse Akadamie Wetenschappen Amsterdam, B, Vol. 72, pp. 169–174.

van der Vlerk, I.M., Gloor, H., 1968. Evolution of an embryo. Genetica 39, 45–63.

van der Vlerk, I.M., Umbgrove, J.H., 1927. Tertiary Gidsforaminifera van Nederladisch Oost-Indie. Wetenschappelijke Mededeelingen van de Dienst van de Mijnbouw in Nederlandsch-Oost-Indië, 6, 3–35.

van Ginkel, A.C., Villa, E., 1999. Late fusulinellid–early schwagerinid Foraminifera; relationships and occurrences in the Las Llacerias Section (Moscovian/Kasimovian), Cantabrian Mountains, Spain. Journal of Foraminiferal Research, 29, 263–290.

van Gorsel, J.T., 1975. Evolutionary trends and stratigraphic significance of the Late Cretaceous helicorbitoides–lepidorbitoides lineage. Utrech Micropaleontol. Bull. 12, 1–19.

van Gorsel, J.T., 1978. Late Cretaceous orbitoidal foraminifera. In: Hedley, R. H., Adams, C.G. (Eds), Foraminifera, Academic Press, London, 3, 1–109.

van Vessem, E.J., 1978. Study of lepidocyclinidae from Southeast Asia, particularly from Java and Borneo. Utrecht Micropaleontological Bulletins, 19, 1–163.

Vasseur, P., 1974. The overhangs, tunnels and dark reef galleries of Tule´ ar (Madagascar) and their sessile invertebrate communities. Proceedings of the Second International Coral Reef Symposium, Brisbane 2, 143–159.

Vaughan, T.W., 1945. American Paleocene and Eocene larger foraminifera. Mem. Geol. Soc. Amer. 9, 1–175.

Vaughan, T.W., Cole, W.S., 1941. Preliminary report on the cretaceous and tertiary larger foraminifera of Trinidad, British West Indies. Geol. Soc. Amer., special paper, 30, 1–137.

Vdovenko, M.V., Rauzer-Chernousova, D.M., Reitlinger, E.A., Sabirov, A.A. 1993. Spravochnik po sistematike melkikh foraminifer Paleozoya (za isklyucheniem endotiroidey I permskikh mnogokamernykh lagenoidey. [A reference-book on the systematics of Paleozoic Foraminifera (excluding endothyrids and Permian multichambered lagenids)] Izdat. Nauka, Moscow, 1–125.

Vecchio, E., Hottinger, L., 2007. Agglutinated conical foraminifera from the lower Middle Eocene of the Trentinara formation (southern Italy). Facies 53, 509–533.

Vera, J. A., 1988. Evolución de los sistemas de depósito en el margen ibérico de la Cordillera Bética. Revista de la Sociedad Geológica de España, 1, 373–391.

Vermeij, G.J., 2004. Ecological avalanche and the two kinds of extinction. Evol. Ecol. Res. 6, 315–337.

Vilas, L., Masse, J.- P., Arias, C., 1995. Orbitolina episodes in carbonate platform evolution, the early Aptian model from SE Spain. Paleogeography, Paleoclimatology, Paleoecology, 119, 35–45.

von der Heydt, A., Dijkstra, H.A., 2006. Effect of ocean gateways on the global ocean circulation in the late Oligocene and early Miocene. Paleoceanography 21, p. PA1011.

Wade, B.S., Pearson, P.N., Berggren, W.A., Pälike, H., 2011. Review and revision of Cenozoic tropical planktonic foraminiferal biostratigraphy and calibration to the geomagnetic polarity and astronomical time scale. Earth-Science Reviews, 104, 111–142.

Wahlman, G.P., 2002. Upper Carboniferous-Lower Permian (Bashkirian-Kungurian) mounds and reefs. *In* W. Kiessling, E. Flugel, and J. Golonka [eds.], Phanerozoic Reef Patterns, 271–338. SEPM (Society for Sedimentary Geology) Special Publication, 72. Tulsa, Oklahoma, USA.

Wahlman, G.P., Konovalova, and M.V., 2002. Upper Carboniferous-Lower Permian Kozhim carbonate bank, Subpolar Ural Mountains, Russia: outcrop analog for some Timan Pechora Basin Reservoirs. *In* W. A. Zempolich, and H. E. Cook [eds.], Carbonate Reservoirs and Field Analogs of the CIS, 223–245. SEPM (Society for Sedimentary Geology) Special Publication 74. Tulsa, Oklahoma, USA.

Walliser, O.H., 1995. Global events and event Stratigraphy in the Phanerozoic. In: Walliser, O.H. (Ed.), Global Events and Event Stratigraphy. Springer, Berlin, 7–19.

Walliser, O.H., 2003. Sterben und Neubeginn im Spiegel der Palaeofauna. Die Bedeutung der globalen Faunenschnitte fu¨r die Stammesgeschichte. In: Hansch, W. (Ed.), Katastrophen in der Erdgeschichte. Heilbronn (Sta¨ dtische Museen), Wendezeiten des Lebens, museo, Heilbronn (Sta¨ dtische Museen), 19, 60–69.

Walter, B., 1989. Au Valanginien supe´ rieur, une crise de la faune des bryozoaires, indication d'un important refroidissement dans le Jura. Paleogeography, Paleoclimatology, Paleoecology, 74, 255–263.

Wang, J., Hu, X.., BouDagher-Fadel, M., Wu, F.Y., Sun, G.Y., 2015. Early Eocene sedimentary recycling in the Kailas area, southwestern Tibet: Implications for the initial India–Asia collision. Sedimentary Geology, 315, 1–13. doi:10.1016/j.sedgeo.2014.10.009.

Ward, P.D., 2006. Out of Thin Air. Joseph Henry Press, Washington, DC, 282p.

Ward, P.D., Brownlee, D., 2000. Rare Earth: Why Complex Life is Uncommon in the Universe. Copernicus Books, New York, 335p.

Ward, P.D., Haggart, J.W., Carter, E.S., et al., 2001. Sudden productivity collapse associated with the Triassic–Jurassic boundary mass extinction. Science 292, 1148–1151.

Wardlaw, B.R., Davydov, V., Gradstein, F.M., 2004. The Permian period. In: Gradstein, F.M., Ogg, J.G., Smith, A.G. (Eds), A Geologic Time Scale. Cambridge University Press, Cambridge, 249–270.

Wedekind, R., 1937. Einfuhrung in die Grundlagen der historischen Geologie, II. Band. Microbiostratigraphie, Die korallen- und Fora- miniferenzeit. Ferdinand Enke, Stuttgart, 136 p.

Wefer, G., Berger, W.H., 1980. Stable isotopes in benthic foraminifera: Seasonal variation in large tropical species. Science 209, 803–805.

Weidlich, O., 2007. Permian reef and shelf carbonates of the Arabian platform and Neo-Tethys as recorders of climatic and oceanographic changes. Geol. Soc. London, special publications, 275, 229–253.

White, L. T., Hall, R., Armstrong, R. A., Barber, A. J., BouDagher-Fadel, M. K., Baxter A, Wakita K, Manning C, Soesilo, J. (2017). The geological history of the Latimojong region of western Sulawesi. Journal of Asian Earth Sciences. doi:10.1016/j.jseaes.2017.02.005.

Whiteside, J.H., Olsen, P.E., Eglinton, T., Brookfield, M.E., Sambrotto, R.N., 2010. Compound-specific carbon isotopes from Earth's largest flood basalt eruptions directly linked to the end-Triassic mass extinction. Proceedings National Academy of Science USA. doi:10.1073/pnas.1001706107.

Wignall, P.B., 2001. Sedimentology of the Triassic–Jurassic boundary beds in Pinhay Bay (Devon, SW England). Proc. Geol. Assoc. 112, 349–360.

Wignall, P.B., Newton, R.J., Little, C.T.S., 2005. The timing of paleoenvironmental change and cause-and- effect relationships during the early Jurassic mass extinction in Europe. Am. J. Sci. 305, 1014–1032.

Wignall, P.B., Twitchett, R.J., 1996. Oceanic anoxia and the end Permian mass extinction. Science 272, 1155–1158.

Wildenborg, A., 1991. Evolutionary aspects of the miogypsinids in the Oligo–Miocene carbonates near Mineo (Sicily). Utrecht Micropaleontological Bulletins, 41, 1–140.

Wilson, J. L., 1975. Carbonate Facies in Geologic History. Springer Verlag, Berlin.

Woods, A.D., 2005. Paleoceanographic and paleoclimatic context of early Triassic time. Palevol. Com. Rend. Acad. Siences Paris 4, 395–404.

Wray, C.G., Martin, R., Langer, M., DeSalle, R., Lee, J.J., Lipps, J.H., 1995. Origin of the foraminifera. Proc. Natl. Acad. Sci. USA 92, 141–145.

Wyn, G., Hughes, G., 2004. Middle to Upper Jurassic Saudi Arabian carbonate petroleum reservoirs: biostratigraphy, micropalaeontology and palaeoenvironments. GeoArabia-Manama, 9, 79–114.

Yan Song, R., Black, G., Lipps, J.H., 1994. Morphological optimization in the largest living foraminifera: Implications from finite element analysis. Paleobiology 20, 14–26.

Yordanova, E.K., Hohenegger, J., 2002. Taphonomy of larger foraminifera: Relationships between living individuals and empty tests on flat reef slopes (Sesoko Island, Japan). Facies, 46, 169–204.

Zachos, J.C., Pagani, M., Sloan, L., Thomas, E., Billups, K., 2001. Trends, rhythms, and aberrations in global climate: 65 Ma to present. Science, 292, 686–693.

Zachos, J.C., Ro¨ hl, U., Schellenberg, S.A., Sluijs, A., Hodell, D.A., Kelly, D.C., Thomas, E., Nicolo, M., Raffi, I., Lourens, L.J., McCarren, H., Kroon, D., 2005. Rapid acidification of the ocean during the Paleocene–Eocene Thermal Maximum. Science, 308, 1611–1616.

Zachos, J.C., Wara, M.W., Bohaty, S., Delaney, M.L., Petrizzo, M.R., Brill, A., Bralower, T.J., Premoli- Silva, I., 2003. A transient rise in tropical sea surface temperature during the Paleocene–Eocene thermal maximum. Science, 302, 1551–1554.

Zaninetti, L. 1976. Les Foraminifères du Trias. Essai de synthèse et corrélation entre les domaines mésogéens européen et asiatique. Rivista Italiana di Paleontologia e Stratigrafia, 82, 1–258

Zaninetti, L., Rettori, R., He, Y., Martini, R., 1991. *Paratriasina* He, 1980 (Foraminiferida, Trias medio della Cina), *morfologia, tassonomia, filogenesi*. Revue de Paléobiology, 10, 301–308.

Zaninetti, L., Rettori, R., Martini, R., 1994. Paulbronnimanninae Rettori and Zaninetti, 1993 (Foraminiferida, Ammodiscidae) and other Anisian foraminifers from the Piz Da Peres section (Valdaora-Olang, Pusteria Valley, dolomites, Ne Italy). Rivista Italiana di Paleontologia e Stratigrafia 100, 339–350.

Zempolich, W.G., 1993. The drowning succession in Jurassic carbonates of the Venetian Alps, Italy: A record of supercontinent breakup, gradual eustatic rise, and eutrophication of shallow- water environments. In: Loucks, R.G., Sarg, J.F. (Eds), Carbonate Sequence Stratigraphy: Recent Developments and Applications. Memoirs-American Association of Petroleum Geologists, 57, 63–105.

Zharkov, M.A., Murdmaa, I.O., Filatova, N.I., 1998. Paleogeography of the Berriasian–Barremian of the Early Cretaceous. Stratigr. Geol. Correl. 6, 47–69.

Zhuang, S.Q., 1999. A study of stalatotheca in spirotheca of fusulinids and some new genera and species with stalatotheca. Acta Micropaleontol. Sinica 6, 357–371.

Subject Index